Southern Europe

Southern Europe
The Mediterranean and Alpine Lands

Monica and Robert Beckinsale

HODDER AND STOUGHTON
LONDON SYDNEY AUCKLAND TORONTO

Acknowledgments

The authors and publisher would like to express their thanks to the following for permission to reproduce these photographs:

Aerofilms Ltd, plates xx, xxx, xxxiv, xxxvii; Aerophoto/Aerofilms Ltd, plate x; Balair/Swiss National Tourist Office/Swiss Federal Railways, plate xxii; R. Bonnefoy/French Government Tourist Office, plate xxxiii; CETA, Madrid, plate xiii; Ente Nazionale Italiano, plates xxviii, xxxi; EPT, Vercelli/Italian State Tourist Office, plate xxv; Fotocielo, Roma, plate xxix; French Government Tourist Office, plates iv, xviii; Istituto di Coltivazioni Arboree, Università di Perugia, plate xxxii; Italian State Tourist Office, plate iii; G. Karquel/French Government Tourist Office, plate xix; Keystone Press Ltd, plate vii; Galvani di A. Marchi/Segreteria di Stato Alfari Esteri, San Marino, plate xxvii; C. Megaloconomou, Athens, plate xxxviii; Ministry of Defence (Crown copyright reserved) plates ix, xxvi; Douglas Pike, plate xiv; Portuguese Ministry of Information and Tourism, plates v, xvi, xvii; F. Catalá Roca/Spanish Ministry of Information and Tourism, plate xv; Spanish Ministry of Information and Tourism, plates ii, viii, xi, xii; Stato Maggiore Aeronautica, xxiv; Swiss National Tourist Office/Swiss Federal Railways, plates i, vi, xxi, xxiii; Yugoslav Embassy, London, plates xxxv, xxxvi.

ISBN 0 340 08001 9   Boards
ISBN 0 340 24436 4   Paperback

First published 1975
First paperbound edition 1979

Copyright © 1975 Monica and Robert Beckinsale
Maps and diagrams copyright © 1975 Hodder and Stoughton Ltd
All rights reserved. No part of this publication may be reproduced
or transmitted in any form or by any means, electronic or mechanical,
including photocopy, recording, or any information storage and retrieval system,
without permission in writing from the publisher.

Printed in Great Britain for Hodder and Stoughton Educational,
a division of Hodder and Stoughton Ltd,
Mill Road, Dunton Green, Sevenoaks, Kent,
by Hazell Watson & Viney Ltd, Aylesbury, Bucks

Preface

This study of the lands and peoples of Mediterranean and Alpine Europe is written for those who have some background of travel and geography and wish to acquire a more advanced knowledge of a general nature. We do occasionally give detailed sample regional studies but normally are more concerned with ideas and principles. The general aspects are developed sufficiently for the needs of any ordinary student who does not wish to go farther whereas the interested student can develop them at greater detail from the bibliographies of modern writings. The regional topics are treated within a political framework and these too are bolstered by a bibliography of mainly post-1950 literature.

Throughout we have insisted on statistical support but have, within reason, simplified the statistics as, from experience, we place no great reliance on the odd digit nor for that matter in large amounts on the odd thousand or two. It is to be hoped that students, if they wish, will draw their own diagrams from these statistics.

The volume has been most carefully compiled to aid 'textbook research'. The statistics form an integral part of the text, having been pruned and selected to illustrate vital themes. Similarly the general chapters complement the regional or political accounts, and the Index relates to topics and major regions. We have, wherever possible, introduced important modern aspects such as desalination, soil-moisture deficiency, pipeline transport, natural gas-fields, collectivization, land-parcel consolidation, urban spheres of influence, and so on, and have kept the regional accounts up to date by personal investigation. Lowland southern Europe has been quite transformed since 1945 and even life on the mountains is not the same. Hence the text and bibliographies do relate to modern conditions and not to those prevailing before the Second World War.

The numerical information is in the metric system but we have appended a simple table of conversions to British standards although most statistics in general use are already in metric units.

The book is the fruit of frequent combined travel in Europe for well over thirty years during which the human geography has probably changed faster than at any time since the Roman era. During this period the techniques and knowledge available to geographers have also been revolutionized. As professional geographers we have tried to keep abreast of both geographical and technical changes. Unfortunately the welcome increase in published information has been accompanied since 1945 by a serious lessening in the exchange value of the pound sterling. However, anyone can still cross the continent cheaply by car or by public bus and tourism is today a major industry.

This book, together with a companion volume on North-Western Europe, and the Danubian and COMECON countries, largely replaces Professor J. F. Unstead's *Europe*. We have made full use of his regional skill while giving the themes a much stronger political, economic and human basis. The more we tried to improve on his regional divisions the more we recognized in them the hand of a master craftsman. We have incorporated several diagrams modified from his volume.

For hospitality and assistance in various ways we would like to thank many friends, among them Paula Harris; Gillian Costelloe; Robert and Mary Beckinsale; Drs Yuri and Aida Kunaver; Dr Paul Cartledge; Dr John Wilkinson; Dr Ian Scargill; Professor Maurice Pardé; the late Professor E. W. Gilbert; Monica Foot; Dr J. M. Houston for two diagrams from *The Western Mediterranean World*; Duckworth & Co for four diagrams from R. P. Beckinsale's *Land, Air and Ocean* (1966); University of London Press Ltd for the use of a diagram from the late Professor J. F. Unstead's *Europe*; Mr K. F. Rose, the commercial director of British Aqua–Chem, for information on desalination; Mary Potter for many artistic diagrams; Elspeth Buxton, librarian of the Department of Geography, Oxford University, for never failing help; and Mr P. C. Masters for his photographic skill. In the production of this volume we would especially like to thank Mr Pendleton Campbell for his kind editorial assistance and Miss Julie Stevens for her perseverance and skill in acquiring photographs that are supplementary to the text as well as being aesthetically pleasing.

To the generous criticism and practical aid of a long succession of intrepid students we owe more than we could admit. We hope the text does at least begin to measure up to what they wanted.

Monica Beckinsale
Robert Beckinsale

Contents

Part I
General Patterns

1	Geological Structure	2
2	Relief and Landforms	8
3	The Submerged Landscape and the Fishing Industry	13
4	Climatic Characteristics	23
5	Rivers and Water Supply	37
6	Vegetation and Soils	47
7	Land Use: General Aspects and Forestry	63
8	Land Use: the Agrarian Economy	68
9	Settlers and Settlement	83
10	International and Economic Patterns	98

Part II
Western Peninsular and Alpine Lands

11	Spain: Basal Developments and Patterns	114
12	Spain: Major Geographical Regions	131
13	Portugal	147
14	Southern France: the Mountains	165
15	Southern France: the Lowlands	178
16	Switzerland and Liechtenstein	191
17	Italy: Basal Patterns	206
18	Mainland Italy: Regional Patterns	225
19	Islands of the Western Mediterranean	241

Part III
The Balkan Peninsula

20	General Balkan Patterns	260
21	Yugoslavia	267
22	Albania	286
23	Greece	291

Appendixes

I	Time Sequence of Geological Periods and Major Earth Movements	309
II	Metric Conversion Tables	310
III	Bibliography	311
	Index to Topics and Regions	322

Maps and Diagrams

1	Tentative graph of time and intensity of earth movements of the Alpine orogeny	3
2	Tectonic map of the Alps	4
3	Cross-section of the eastern Alps	5
4	Structure of southern Europe	7
5	Geological structure and morphology of the Matterhorn and the Jungfrau	11
6	Major features of submarine relief in the Mediterranean and Black Seas	13
7	Currents, temperature and salinity of sea water in Strait of Gibraltar	16
8	Airmass sources of the Mediterranean airspace	23
9	Quantitative–direction diagram of cyclonic depressions over the Mediterranean airspace	25
10	Mean actual July isotherms	29
11	Mean actual January isotherms	29
12	Season of maximum rainfall	30
13	Mean number of raindays in July	32
14	Mean number of raindays in January	32
15	The regimes of the Ebro	39
16	The regimes of the Rhône–Saône	39
17	The river regimes of Italy	40
18	Average annual maximum precipitation deficits in Europe	43
19	Distribution of chief Mediterranean oaks	48
20	Distribution of two of chief Mediterranean pines	50
21	Distribution of two characteristic maquis plants	50
22	Distribution of three characteristic garrigue plants	51
23	Development of rendzinas into brown forest soils and of deforested brown forest soils into terra rossa	57
24	Profiles of grey–brown podzolic soils and podzols	57
25	The soils of Greece	59
26	Profiles of pedocals	60
27	Forestry products in Greece	66
28	Major routes of transhumance in western Mediterranean lands	73
29	Some types of transhumant pastoral economies	73
30	Distribution of main areas of olive and maize cultivation	76
31	Olive cultivation and olive oil production in Greece	80
32	Distribution of languages in Europe	84
33	The Roman empire in the fourth century AD	90
34	The Roman empire and the Barbarian kingdoms in AD 528	93
35	The Empire of the Caliphs about AD 737	94
36	Southern Europe and Mediterranean lands about AD 1150	96
37	The Rhône–Saône improvement scheme	102
38	Distribution of electricity generation and transmission in Spain	105
39	Main railways and chief route and commercial centres of southern Europe	107
40	Crude petroleum pipelines in western Europe	108
41	Iberia: progress of Reconquista and Moorish impact on towns and place-names	115
42	Land use in Iberia	121
43	Iberia: density of stock and direction of main stock routes	124
44	Iberia: general limit of safe cultivation of date palm, olive and orange	125
45	Iberia: density of population	129
46	Geographical regions of the Iberian peninsula	131
47	Regional setting of Madrid	136
48	Irrigated tracts in the Ebro trough	139
49	Irrigated areas in the middle and lower Guadalquivir valley	141
50	Gibraltar	142
51	Vegetation and land utilization of Portugal	152
52	Farm sizes in Portugal	155
53	Portugal: provinces and provincial capitals	159

54	Characteristic structure and landforms of the French folded Jura	165
55	The tourist industry of the northern French Alps	172
56	The site, growth and industrial layout of Grenoble	173
57	Main routes and passenger traffic into Spain across the Pyrenees	175
58	Hydro-electric and irrigation developments in Mediterranean France	180
59	The urban network of Mediterranean France	181
60	Number of workers and daily commuters in the Marseille district	189
61	Territorial growth of the Swiss confederation	191
62	Geographical regions of the Alpine and neighbouring lands	193
63	Distribution of chief electricity generating stations in Switzerland	199
64	Switzerland: Cantons and cantonal capitals	202
65	Site and growth of Basle	203
66	The northeast borderlands of Italy	209
67	Some aspects of land use in Italy	216
68	Italy: some aspects of animal husbandry	218
69	Hydro-sites and industries of the west Italian Alps	219
70	Italy: natural gasfields and pipelines	220
71	Italy: general distribution of population	222
72	Geographical regions of Italy	226
73	Urban hierarchy and functions of north Italian towns	228
74	Number of workers in manufacturing and in service industries in north Italian towns	229
75	The site and setting of Venice	230
76	The canals, bridges and churches of old Venice	231
77	Geology and landforms of central peninsular Italy	233
78	Geology and landforms of southern peninsular Italy and northeast Sicily	233
79	Distribution of types of rural settlement in Italy	235
80	Nucleated settlements in the Murge	239
81	Landform types of Sicily	242
82	Land use in southern peninsular Italy and Sicily	244
83	A nuraghe	246
84	The geological structure and landform types of Sardinia	247
85	Distribution of economic activities in Corsica	252
86	Distribution of towns and nucleated villages in Corsica	253
87	Density of population in Corsica, average 1950 and 1960	255
88	The Maltese Islands	256
89	The Balkan States in 1212 and 1340	261
90	The Balkan States in 1520 and 1800	261
91	General structure of the Balkan peninsula	263
92	State farm at Bijeli Manastir, Yugoslavia	271
93	Yugoslavia: general land use	272
94	Land reclamation and improvement schemes in Yugoslavia	274
95	Yugoslavia: coal, natural gas and oil deposits	278
96	Yugoslavia: the republic and the autonomous areas, with their capitals	280
97	Yugoslavia: density and general distribution of population	281
98	Yugoslavia: distribution of towns with over 50 000 inhabitants	282
99	Albania: general land use	287
100	Albania: mineral and power resources	288
101	Albania: density of population, and chief towns	289
102	Greece: general land use	294
103	Greece: percentage cultivated area irrigated	295
104	Greece: percentage total area under cultivation	297
105	Greece: distribution of settlements with over 2000 inhabitants	301
106	Greece: variations in population density 1940–51 and 1951–61	304
107	Greece: international and national passenger traffic by sea	308

Plates

I	The Matterhorn and the Z'mutt glacier	10
II	Bermeo, a fishing port in Viscaya, Spain	20
III	The beach at Riccione on the Adriatic near Rimini, Italy	34
IV	Skiing at Chamrousse in the French Alps	35
V	Cutting cork in Alentejo, Portugal	49
VI	Transhumant herd on high alp above Zermatt, Switzerland	71
VII	Greece: a herd of goats on a mountain road	72
VIII	Olive groves at Cazorla, Jaen, southern Spain	81
IX	Roman centuriation in the Po valley near Cesina	91
X	Roman arena and theatre at Arles	92
XI	A patio in the Alhambra, Granada	95
XII	Hydro-electric dam at Picos del Sil, Orense, Spain	104
XIII	Aerial view of Madrid	137
XIV	Gibraltar from the south	142
XV	Aerial view of Barcelona	144
XVI	Port wine country near Lamêgo, Alto Douro, Portugal	150
XVII	Central Lisbon from the air	160
XVIII	Monaco	174
XIX	Aerial view of Lyon	186
XX	General view of Marseille	188
XXI	Zürich, as seen from the cathedral tower	200
XXII	Aerial view of Basle	202
XXIII	Berne, capital of the Swiss Federation	204
XXIV	Aerial view of the Po Delta	210
XXV	Transplanting rice seedlings near Turin	227
XXVI	General aerial view of Genoa	232
XXVII	San Marino	234
XXVIII	Rome: the Vatican City	236
XXIX	Rome: Ancient, Medieval and Renaissance	237
XXX	Naples as seen from the air	238
XXXI	Trulli landscape near Martina Franca, Taranto	239
XXXII	*Cultura promiscua* near Perugia, Italy	240
XXXIII	Corte, Corsica	254
XXXIV	Aerial view of Malta	257
XXXV	View of the Dinaric Alps and Yugoslavian karst	268
XXXVI	Townscape in Belgrade, Yugoslavia	282
XXXVII	Athens and the Acropolis	302
XXXVIII	Currants spread out to dry at Amaliás, Peloponnesus, Greece	303

I
General Patterns

1 Geological Structure

MAJOR MOUNTAIN-BUILDING PERIODS IN EUROPE

Four main orogenies or periods of marked mountain building are revealed in the relief of Europe. The earliest survives only as a fragment in the extreme northwest of Scotland and no doubt also in parts of the Baltic and Russian platforms. The next oldest, sometimes called the Caledonian after Scotland, affected mainly northwest Europe and has a northeast–southwest structural trend that is clearly visible in Norway, Scotland, Wales and the northern uplands of Ireland and England. The third prolonged orogeny has various names such as Variscan, Armorican (after Brittany) and Hercynian. We propose here to use the term Hercynian both for the orogeny and the structural trend of the relief. The affects of the Hercynian mountain building phase are seen in the east–west structural grain of southern Ireland, southwestern England and Brittany. In a wide zone across central Europe there are many isolated Hercynian blocks but here the structural trends are much more varied. The fourth and latest orogeny mainly affected southern Europe and created the Alpine system with its numerous upfolds and downfolds.

It will be noticed that in Europe each new major mountain belt arose mainly south of its predecessor and was composed largely of sediments eroded from the older uplands and deposited in a great geosyncline on their southern side. During an orogeny these newer sediments in the flanking geosyncline are thrust farther over and against the older blocks. The latter were worn down to their roots because in their locality the uplifting forces either ceased to be positive or became so feeble that they did not keep pace with erosion. The surviving mountain stumps consist of highly compressed strata often hardened also by metamorphic and igneous action. They are resistant to further folding and during later orogenies tend to be uplifted or depressed as a block or to be tilted and fractured into *horsts* (fracture-outlined uplands) and *graben* (fracture-lined troughs).

Old resistant massifs of this type do not occupy a large proportion of the surface in southern Europe except in Iberia (the Meseta), and the Balkan Peninsula (Rhodope upland). These large Hercynian blocks, and many others large and small not yet mentioned, greatly influenced the direction and intensity of the Alpine folding.

The Alpine orogeny

The Alpine system affords the best clues to the nature of mountain building (*orogenesis*) and continent building (*epeirogenesis*). The Alps represent the appearance of the Hercynian and Caledonian mountain belts before they were worn down to their stumps and fractured by later orogenies. The Alpine mountain building phase mainly involved sediments laid down during Mesozoic and Lower Tertiary times (that is Triassic; Jurassic; Cretaceous; and Eocene–Oligocene) in a geosyncline, often called the Tethys Sea, that stretched between pre-Alpine Europe and the African block of Gondwanaland. This depositional geosyncline was mobile and its floor rose and sank regionally rather than as a whole.

The causes of the earth movements in this mobile belt are little understood. The chief probably were slow convection currents or the lateral transference of heat and possibly also of matter in the plastic layer in the earth's upper mantle. These transferences may cause old hard blocks to shift vertically and laterally and may also cause mobile belts between old blocks to rise or sink locally. Lateral and vertical movements of the blocks lead to lateral pressure on the pliable geosynclinal sediments which are then pushed forward or slip under gravity and, if the pressure becomes severe, are compressed into folds.

There is, however, another lesser force involved in mountain uplift and basin depression. It is called isostasy and depends upon the downward pressure or weight of the surface material upon the plastic layer. The surface rocks virtually float upon this plastic layer; where they are dense they tend to be thin; where they are of lighter material, they float with deep roots. A typical young sedimentary mountain range is light and its roots protrude far into the plastic layer. When the surface of such a mountain is steadily eroded away, the whole mass tends to rise steadily under isostatic rebound. Consequently the actual height of a mountain mass is usually the balance between erosion, isostatic uplift and the negative or positive action of currents in the plastic layer of the upper mantle. In an inverse way, thick sedimentation in a geosyncline or basin will tend to depress further the basal rock floor on which the sediments are accumulating.

These various forces and movements operate very slowly and in fact much of the folding is accomplished under the sea and at low levels where and when the strata are relatively pliable. A great deal of the vertical uplift occurs after the 'mountains' are formed rather than during the compression of the sedimentary layers. The long time needed to form the present Alpine system makes most geologists certain that mountain building is a very slow process in which intermittently the forces quicken their intensity but rarely, if ever, approach convulsive violence. In the Alpine system, especially in southern Italy and the Balkans, the recent earthquakes and perhaps also the volcanic outbursts show that the orogeny is still active.

The mobile belt in which the present Alpine chains originated has in parts experienced appreciable orogenesis since at least early Cretaceous times or for some 150 million years. During this long time span, there were ten or more phases of increased orogenic activity, the chief being in the early and mid-Cretaceous, early and late Eocene, mid and late Oligocene, mid-Miocene, early Pliocene and Plio–Pleistocene or early Quaternary. A few of these affected all or almost all the Alpine chains but most of them affected different parts of the mobile belt at different times. For example, the Pyrenees, eastern Alps, Carpathians, Dinaric Alps and Apennines were severely folded

Geological Structure

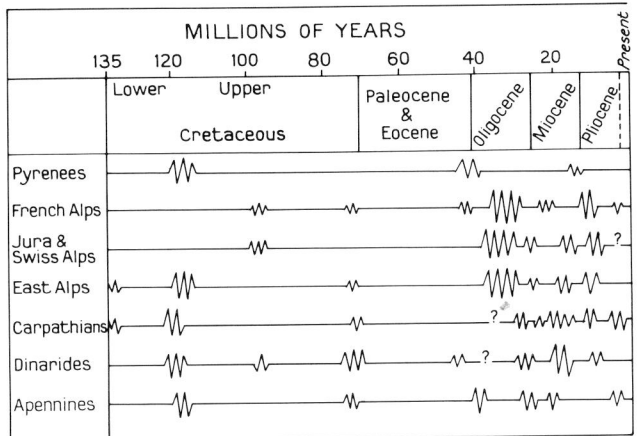

1 *Tentative graph of time and intensity of earth movements of the Alpine orogeny* (after L. J. Wills *et al.*)

as early as the mid-Cretaceous whereas the French and Swiss Alps were not excessively contorted until Oligocene times. The general sequence of these movements is given in figure 1, and for the reader who needs it we have provided a table of geological periods in Appendix I.

This long, differential time scale is important for three main reasons. First, the amount of surface erosion tends to be greatest on chains, such as the Pyrenees and Carpathians, that were the first to be elevated above sea level. Today these will not be exceptionally high unless they were strongly re-elevated later in Tertiary times.

Secondly, from the start of the Alpine orogeny large quantities of debris eroded from exposed mountains were deposited as sediments in some parts of the geosyncline and of other adjacent basins. Later orogenic movements uplifted these younger strata and often made them part of the mountain structure. The terms *flysch* and *molasse* are widely used for these newer deposits that were derived from Alpine mountains. Flysch is a dialect word used in Alpine Europe for interstratified shale and sandstone, either because it often slipped in thin slices (*fliessen*) down mountain slopes or because it was arranged in thin slabs or flags (Germ. *fliese*). Today the term is widely used for any similar Cretaceous and Tertiary deposits. The nature of the flysch may be illustrated from the Swiss Alps. Here the flysch from Eocene to early Oligocene times accumulated on dry land at the foot of the mountains which then stood much farther south than they do at present. Later the Alps were thrust forward and incorporated parts of the flysch. At the same time the site of the present Swiss plain and of the future Jura mountains subsided under water in which molasse (of mid-Oligocene to Miocene age) was deposited. The term *molasse* is a local dialect name for soft sandstone but its use has been widened to include sandstones, conglomerates, and limestones similar to those of the Swiss Plain. The molasse deposits are very thick and their coarse conglomerates are often particularly prominent in the subAlpine belt. They occur also at the foot of the Jura where the coarser conglomerates are called *nagelfluh* because where exposed they often look as if they had been studded with the round heads of giant nails. Flysch and molasse are extremely common on the outer flanks of the Alps and the Carpathians. On the inner or Mediterranean slopes of the Alpine system, Oligocene deposits are especially plentiful but here large areas of them are beneath the sea.

Thirdly, during the long intervals of standstill or relative freedom from orogenic activity when various parts of the exposed mountain system were subjected to the subaerial littoral erosion that deposited the flysch and molasse, large areas of them were peneplaned. These flat surfaces were involved in later phases of uplift and today it is quite usual for the mountain ranges to be flat-topped. The wide, level summits may later be fractured and tilted, or dissected and fragmented by erosion, as in the *rax* landforms of the Austrian Alps, but often they remain remarkably flat.

The Alpine arc

Before studying systematically the geological structure of the main northern arc of the Alpine system in Europe it is helpful to notice that severe lateral compression of beds with the formation of *nappes* is common throughout the belt although most characteristic of the Alps proper. Nappes (*decken* in German) is the French term for sheets of rock that have been thrust forward, usually by sliding and rarely by recumbent folding, over the top of relatively immobile rock. As the Alpine system has undergone several more intense phases of orogenesis, it has a complex vertical and horizontal structure which is of supreme interest to geologists and of considerable significance to geographers also.

Franco–Italian western Alps

The low east–west limestone ranges in southern Provence are a geological continuation of the Pyrenees and structurally are not a part of the Alps proper which begin at their Mediterranean end in the Maritime Alps near Monaco. From this steep coast the Franco–Italian Alps curve northwards for about 270 km to the neighbourhood of Mont Blanc. In this stretch the Alps run parallel to the faulted front of the Massif Central, a Hercynian horst from which they are separated by the narrow Rhône corridor, a typical graben.

The transverse section of the Alpine range displays three main structural belts (fig. 2).
1 A central alignment of small, separated Hercynian blocks, including those of Mercantour and Pelvoux–Belledonne. The latter continues northward to Mont Blanc. On these masses the summits exceed 3000 m and rise in Pelvoux to 4100 m. Glaciation and frost action have shaped the igneous rocks into jagged shapes, often denoted by terms such as horn, *aiguille* (needle), *dent* (tooth) and *fuorcla* (fork).

2 An eastern or inner belt, often called the Pennides after the Pennine Alps, that rises above the Po basin in Italy. This consists of the roots and nappes of strata that were intensely compressed against the central alignment of igneous cores.

3 A western or outer belt, often called the Helvetides after Switzerland (Helvetia), which contains a high proportion of limestones and consists of complicated superimposed folds that tend to ascend toward their outer (western) fronts. Most of the nappes have been separated by erosion from their roots which actually lay upon or south of the igneous blocks (belt 1). The complex sedimentary layers weather into less jagged landforms than those of the igneous massifs and small plateaux, ledges, valley-side benches and concave summits are common as is indicated by local topographical names such as *dos* (back) and *kasten* (chest). In Chapter 10 it is shown how this high triple-structured stretch of the western Alps forms a strong barrier to railways and roads.

Swiss–Italian western Alps

A vertical section across the Alps in Switzerland and northern Italy reveals additional structural belts (fig. 2). Here the lateral thrusting, as usual mainly from the Mediterranean or African side, encountered a gap between the Massif Central and the Vosges–Black Forest block. The result was a wider cordillera with at least five main structural belts in the Alps proper and two more outside the Alpine chain. The belts are as follows

1 In the south, the Dinarides, a rock series that appears to have been deposited on the African shore of the geosyncline. This belt is very narrow near Lake Maggiore and widens eastward into a broad calcareous zone.

2 A wide axial belt of Pennides or Pennine nappes of crystalline schists and gneisses and including the Pennine Alps where Monte Rosa rises to 4637 m. Some of these nappes originated in a crystalline root zone at the southern edge of the Pennides belt.

3 An alignment of crystalline Hercynians massifs, includ-

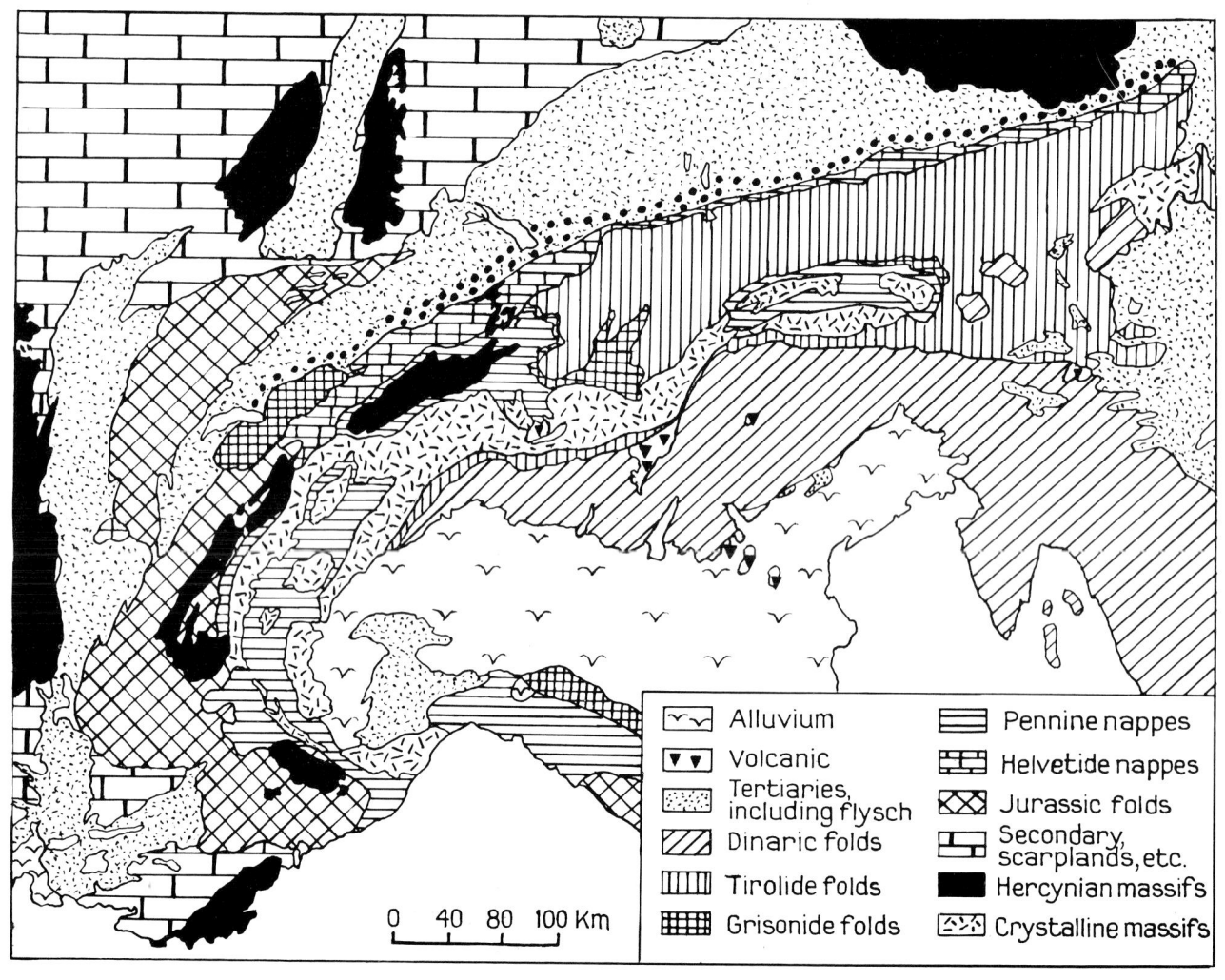

2 *Tectonic map of the Alps* (after Staub)
  Large dots denote the pre-Alpine buttress of flysch and molasse affected by Alpine movements

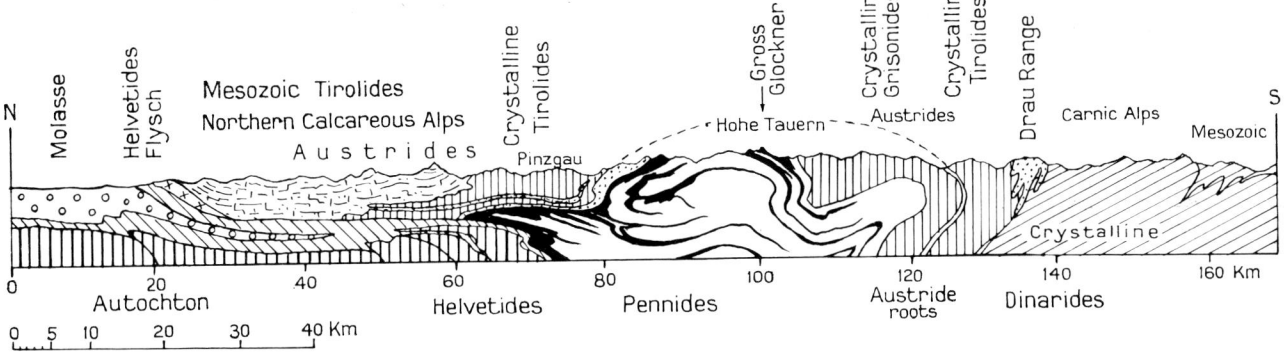

3 Tectonic cross-section of the eastern Alps
For the numerous theories on and interpretations of this arrangement, see Oxburgh, 1968

ing part of that which culminates in Mont Blanc (4807 m) and the Aar–Gotthard block.

4 The Helvetides or High Calcareous Alps which include the Bernese Oberland with its fine glaciated peaks, such as the Finsteraarhorn (4277 m), Aletschhorn (4198 m) and Jungfrau (4161 m). This very complex belt includes some autochthonous folded strata (that is, folded more or less *in situ*) and many more strata that were thrust as nappes through or over the Hercynian blocks. Most of the latter nappes had their roots in the Pennides belt but a few (called Grisonides) may have originated slightly farther south.

5 The Pre-Alps, a belt of moderate height (summits at about 2200 m) that stands up outside the main front of the Helvetides and forms a mountainous strip near the eastern end of Lake Geneva. The complexity of the structure may be judged from the fact that in parts here the molasse is covered by complicated nappes of Helvetides, Pennides and Grisonides. If, as is thought, the Grisonides originated on the African foreland they may once have covered the whole width of this stretch of the Alps and since have been removed from most of the surface by erosion.

There are two further structural belts that are associated with the Alpine orogeny but lie outside the main Alpine chain.

6 The Alpine Foreland or Swiss Plateau, a structural downfold between the Alps and the Jura in which sedimentary strata have accumulated, particularly flysch and mid-Tertiary molasse. In places the molasse rises to over 1200 m.

7 The Jura mountains, a crescent-shaped upland that protrudes into the broad gap between the Massif Central and the Vosges–Black Forest block. The deep-seated crystalline base here is covered with sedimentary beds ranging in age from Permian to recent Tertiary. Alpine folding affected only the strata of this sedentary cover that lay above the middle Trias or Muschel–chalk which acted as a plastic lubricating layer. In the south (*Jura plissé*) these upper beds were pushed into simple symmetrical folds with rather flat summit levels on the anticlines at about 1700 m; on the north (*Jura tabulaire*) the thrust was not sufficient to disturb the strata. The only area where the Jura shows signs of more intense folding is where the strata were squeezed against the southern end of the Black Forest horst.

Austro–Italian eastern Alps

The eastern Alps are usually considered to begin east of a line from Lake Como to the upper Rhine near Chur. Here, opposite the large gap between the Hercynian massifs of the Black Forest and Bohemia, the Alpine system expands to its greatest width (280 km) and consists topographically of many divergent ranges. The structure and rock composition here are very different from those of the highly compressed western Alps. In particular, the Hercynian igneous cores are absent except for a very small exposure in the Semmering district. The chief transverse structural belts are now as follows

1 The Dinarides, or southern calcareous Alps, here a broad belt of mountains as in the Bergamo Alps, the Dolomites and the Carso inland of Trieste. The strata tend to erode into block-like shapes rather than to fold into long narrow sheets as do the Pennides. These tabular minor landforms are equally well developed in the huge rectangular limestone eminences of the Dinarides on the western flank of the Balkan peninsula and in the Apennines.

The central and northern parts of the eastern Alps are composed mainly of a rock series (the Austrides) that is hardly represented in the western Alps and appears to be a geological succession of the Dinarides. The Austrides are subdivided into Grisonides (already mentioned) and slightly younger Tirolides. Each of these subdivisions originated south of the Pennides root zone and has been thrust right over the top of the Pennine nappes (figs. 2 and 3). On the basis of these rock differences and differential uplift and erosion, the main axial zone of the eastern Alps contains three distinguishable belts, here numbered 2–4.

2 Wide areas of Tirolides that consist largely of crystalline rocks. These form fine mountains as in the Silvretta group and the Otztaler but the topography is often rather uniform and not markedly ridged.

3 Where uplift and erosion have been greatest, the Tirolides have been eroded away revealing the underlying Grisonides. In the Hohe Tauern and Lower Engadine erosion has also breached the Grisonides and revealed the underlying Pennides. These 'windows' or geological *inliers* (older rocks surrounded by younger) aften have steep sides as in the precipices of the Gross Glockner (3797 m) in the Hohe Tauern.

4 The northern strip of the Tirolide belt consists of

nappes of sedimentary strata, including thick, relatively pure limestones that rise to 2996 m in the Dachstein and form the dissected summits of the Salzkammergut.

Outside or beyond this high limestone belt of the Tirolides are two additional structural belts.

5 A somewhat narrower strip of Helvetides, also consisting mainly of limestone nappes. These probably originated in a root zone not far to the south and now deeply buried under the Tirolides, as shown in figure 3. They are intricately folded Triassic and Jurassic strata that weather into sharp ridges and culminate in the Zugspitz (2963 m).

6 The northern or outer edge of the outcrop of the calcareous Helvetides is buttressed by a belt of flysch and molasse, parts of which lie in Bavaria. These Pre-Alps become quite mountainous locally. The flysch rises to 1862 m in the Hollengebirge east of Salzburg in Upper Austria and also forms, for example, the Wienerwald overlooking Vienna. The Tertiary molasse (mainly Miocene) gives rise to low ridges as in the Allgau Alps and in the Hausruck (740 m) on the Austrian foreland. The molasse eminences usually consist of nagelfluh (coarse conglomerates) and much of its outcrop has been plastered with glacial deposits.

The Carpathian arc

The Carpathian mountain system lies outside the regional framework of southern Europe but its nature and genesis throw so much light on Alpine mountain building that they will be briefly examined. The eastern Alps on approaching the great plains traversed by the middle Danube in Austria and Hungary appear to diverge into two main mountain ranges. The northern (Tirolides) belt continues eastward in an emaciated state as the Carpathians, while the southern (Dinarides) belt continues southeastward in a strengthened state as the Dinaric Alps of Yugoslavia. However, the bulk of the crystalline Tyrolide belt peters out and seems to sink beneath the middle Danube plain.

The surface of the middle Danube (Pannonian) plain has rising above its central areas a few small granitic uplands such as the Velence Hills and the Moragy block near Pecs while within its Carpathian periphery there are numerous relatively small Hercynian blocks. These old masses are thought to be the surviving fragments of a vast Hercynian plateau that formerly stretched over the area now occupied by the plain. This huge massif supplied sediments to the nearby geosyncline. The Carpathians underwent severe folding in Cretaceous times but the major massif, although lowered and fractured, persisted into Oligocene times and continued to be the source of large quantities of sediments. From late Miocene times onward the Hercynian massif slowly subsided to form a basin while at the same time the Carpathians were slowly further elevated and volcanic activity at the rim of the basin produced the most extensive volcanic belt in Europe.

The slowly subsiding basin was gradually filled up with sediments and in parts its hard floor is covered with 3000 metres of recent deposits. Borings in these young beds have revealed that subsidence was very active throughout the Pliocene and Quaternary. Yet it is also known that the Carpathians were then rising and that in the Quaternary alone they were elevated by 200 to 300 metres.

Thus the Carpathian orogeny has involved a large vanished Hercynian massif around or on the flanks of which the Carpathians to the north and Dinarides to the south were upthrust and upfolded. The foundering of this great median mass was presumably due to convexion currents or heat/matter exchanges in the plastic layer of the earth's upper mantle. These not only caused the central mass to sink and the mountains to rise but also by encouraging the relative movement of the African and European major blocks caused lateral compression in sediments on their flanks, expecially during the early stages of the orogeny. The Alpine chains of the Balkans are discussed in Chapter 20.

Probability of a major Alpine orogenic pattern

A similar structural combination of mountain ranges on either side of a median mass may perhaps be detected in other parts of the general Alpine system (fig. 4).

In the western Mediterranean Sea between Iberia and Corsica, the sea floor, in parts just over 3000 metres deep, could be a subsided Hercynian massif. If this were so, the numerous small Hercynian blocks now exposed around the edges of the basin, at Málaga, Rif, southwest Sardinia, Provence, and elsewhere, would represent the surviving marginal fragments of the vast median mass. During the orogeny the Sierra Nevada in Spain would be upthrust on the European flank and the Atlas mountains would be upthrust on the African flank of the central massif. This vanished block would have supplied enormous quantities of sediment to adjacent parts of the geosyncline and so would explain the extraordinary thickness of the Oligocene beds (up to 2000 metres) in for example the Maritime Alps, northern Apennines and other coastlands of the western Mediterranean Sea.

Similarly, the large trough of the Tyrrhenian Sea, in parts over 3500 metres deep, may be the site of another foundered Hercynian mass with surviving marginal fragments in northern Sicily and Calabria. The same structural pattern has been applied less convincingly to the Adriatic Sea and its flanking mountains. Here, however, the position of the subsided median mass, if any, is less obvious owing to the lack of possible surviving marginal Hercynian fragments, except in the Alps to the far north. The Adriatic sea today is relatively shallow but its continuation in the Po plain is known today to have sedimentary deposits up to nearly 8000 metres thick. Thus the hidden crystalline base of this depression is in fact a greater structural feature than the mountain arc around it.

The 'median mass' concept seems to apply reasonably

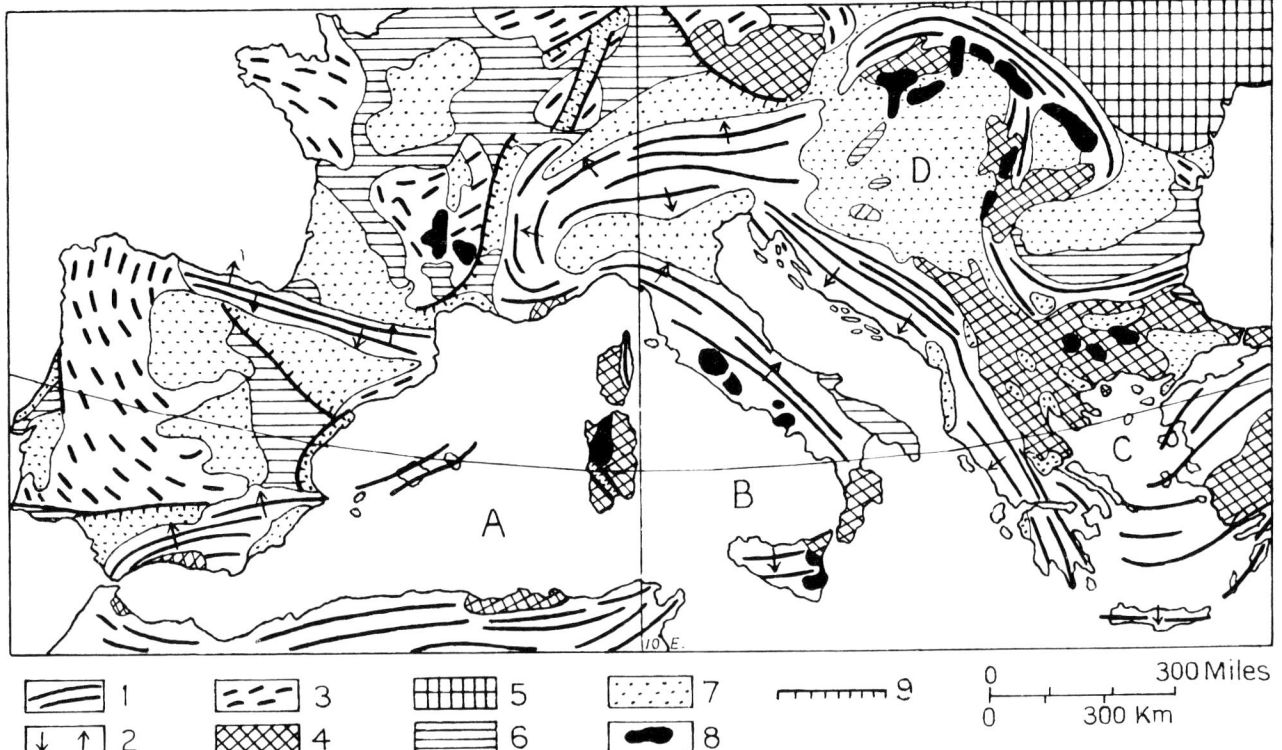

4 *Structure of southern Europe*
1 *fold mountains of Alpine type*
2 *direction of thrust or overthrust*
3 *Hercynian areas and trends*
4 *ancient blocks*
5 *east European platform*
6 *secondary strata, tilted or upraised*
7 *basins with Tertiary or later deposits*
8 *volcanic areas*
9 *major fault zones. A–D mark subsident basins*

well to the easternmost part of the Mediterranean lands, where, however, the median Hercynian massif still remains elevated in the Rhodope plateau and central Anatolia. But the Aegean Sea with its numerous protruding peninsulas and islands could satisfactorily represent an early stage in the subsidence and fracturing of this vast Hercynian massif. The geological structure of the Greek islands is described in more detail in Chapter 23.

Latest stages in the Alpine orogeny

Before leaving the structural aspects of the general Alpine system, more should be said on the most recent orogenic activity which appears to have largely caused widespread regional uplift or depression. Pliocene and Quaternary earth movements were strong but mainly vertical in the Carpathians, the middle Danube plain and the western Mediterranean. For example, the floor of the Tyrrhenian Sea subsided by about 490 metres in the Pliocene and a further 400 metres in the Quaternary. The crystalline floor of the trough beneath the Po plain sank in parts 1500 m and 100 m respectively during these two periods. Yet at the same time most of the adjacent mountains underwent appreciable uplift.

These regional elevations and subsidences were accompanied by large-scale fracturing. As a result many stretches of the Mediterranean coast are outlined by Pliocene–Quaternary faults. For example, the coastline almost all the way from the Strait of Gibraltar to the Italian riviera of Ligurian Alps is on lines of fractures. The relative absence of vulcanism around the western Mediterranean Sea compared with its abundance around the edge of the Middle Danube plain is surprising. In the west, volcanoes are only common on and near the western edge of peninsular Italy where, however, most remain active.

Geological structure and economic minerals

The recent geological composition and structure of the Alpine system as a whole has not been favourable to the accumulation and preservation of valuable minerals and metals. Coal and lignite are present only in very small pockets in, for example, the Valais and St Gallen in Switzerland. The coal seams in northern Spain are in a Carboniferous basin, downfaulted and covered by Tertiary deposits – they are not 'Alpine'. The vast expanses of limestone and of crystalline schists throughout the Alpine system are markedly lacking in large concentrations of valuable metals. On the depressed forelands, either downfolded or downfaulted, the only important strikes of petroleum have been outside the main Carpathian arc. Near the northern Mediterranean shores petroleum in small quantities has been exploited in Albania since 1918, and in Sicily, but these occurrences are exceptional. In contrast, recently natural gas in large quantities has been found in some of the older sedimentary beds now buried deeply beneath newer Tertiary sediments in several of the deeper troughs and graben. For example, in Aquitane north of the Pyrenees and in the Po basin of the Alps natural gas now greatly aids economic developments but topographically these finds belong to the plains rather than to the mountains.

2 Relief and Landforms

RELIEF

The topographical nature of the Alpine system has already been mentioned briefly in the discussion on their structural geology. However, orography or mountain topography has many aspects other than geology. The chains of the Alpine orogeny are remarkable for their great length and continuity, great height and relative narrowness as well as for their complicated cross-sections.

The length and continuity is represented by the horizontal plan or layout of the main chains which stretch in Europe, with some topographical breaks, from the Strait of Gibraltar to the Black Sea and beyond it into the Caucasus (fig. 4). The Sierra Nevada or Baetic Cordillera of southern Spain (summit, Mulhacén 3481 m) reappears northward in the Balearic Islands but southward it is no longer considered a direct geological continuation of the Atlas mountains on the African side of the Strait of Gibraltar. This cordillera is not a significant barrier except to local communications. In contrast the Pyrenees, although only 450 km long and not exceptionally high (Pic d'Aneto 3404 m), form a formidable international barrier right across the narrowest part of the European mainland. They are separated from their structural continuation in southern Provence by the downfaulted Gulf of Lions.

These two stretches of Alpine ranges differ markedly in length from the Alps proper which extend unbroken for nearly 1100 km from the riviera coast to the Danube near Vienna (fig. 4). This arc culminates in the French Alps at Mont Blanc (4807 m), the highest point of Europe outside the Caucasus where Elbrus reaches 5633 m. Across the Alps proper the lowest natural pass is the Brenner at 1375 m between Austria and Italy.

East of the Alps the northern mountain arc continues in the Carpathians for 1200 km to the Iron Gate and thence on the other side of the Danube stretches as the Balkan Mountains for a further 600 km to the shores of the Black Sea. Here, outside the regional framework of southern Europe, the summits are lower (maximum 2663 m in the Tatras) and river valleys and tectonic depressions provide relatively low passes.

Two further major Alpine arcs are important. The Apennines stretch for over 850 km along the Italian peninsula, rising to 2914 m in Monte Corno in the Gran Sasso. The Dinaric Alps and Pindus mountains extend for nearly 1400 km from Trieste to Athens and form a strong barrier on the western flank of the Balkans, with a summit height of 2751 m.

The altitudes given above are not exceptional when compared with those of Tertiary mountains in Asia but their significance is better seen by comparison with mountains in non-Alpine Europe. The Caledonian mountains culminate at 2468 m in Norway; the Hercynian blocks ascend to 1886 m in the Massif Central in France and 1894 m in the Urals. Clearly the Alpine orogeny provides the only true mountains in Europe. Moreover, the relief strength or barrier effect of the Alpine ranges is often increased by the close proximity of mountain and sea or mountain and low-lying plain. Over long stretches of the Mediterranean coast the mountains rise steeply from a deep sea floor, leaving little space for routes and houses. It is not only in Greece that 'the mountains look on Marathon, and Marathon looks on the sea'.

Fortunately for civil engineers, the Alpine chains are narrow compared with their breadth and they are deeply incised by river valleys which often give gradients that can by used by railways and roads. These gradients greatly reduce tunnelling especially as they often lead up to relatively narrow divides.

The relative narrowness of the upfolded system has another important aspect. The Alpine ranges are, as already noticed, only the exposed upper parts of the folding and thrusting. The depressed areas now buried beneath sediments at or not far above sea level are often as wide and in places wider than the nearby mountain chains. The Lombardy plain, one of the smaller basins, covers about 42 000 sq km or about the same area as Switzerland. The Pyrenees are less extensive than the Ebro trough and Aquitaine basin on their flanks. Only in peninsular Italy and the Balkans can the lofty Alpine system be said to have a strong areal dominance. Perhaps the tragedy is that in the Mediterranean area so much of the vast depressed basins is still covered by sea.

LANDFORMS

Landforms represent the interaction of orogenic movements, geological structure and external surface-shaping processes. As the first two have already been discussed, the following account will deal mainly with external (exogenetic) forces. Of these by far the chief is climate which controls weathering and erosion. It happens, however, that the climates of southern Europe have changed considerably since Tertiary times.

Climatic change

Already during the Pliocene mean air temperatures of the globe had begun to drop probably because of the continued uplift of Antarctica around the south pole and of many other continental masses in northern temperate and polar areas. Then, about 2 million years ago the air temperature fell drastically and the Quaternary ice age began. Probably the chief reason for this sudden cooling was the continued uplift of land masses in middle and polar latitudes where ice-caps could form. Once snow and ice began to cover wide areas, the intensive reflexion from the white surfaces (or in technical terms their *albedo*) caused a large loss of atmospheric heat into space. There were also small periodic decreases in the insolation received at the earth's surface and perhaps in addition a slight decrease in total solar radiation. The ultimate result was a succession of cold phases when ice sheets advanced and of warmer phases when ice sheets either melted completely or retreated back to the higher mountain tops and polar areas as today. There were four or five main cold phases when ice sheets formed on the snowy plateaux of northwestern Europe and exuded outward over adjacent seas and lowlands until they reached a zone where the climate was warm enough to melt them. These vast continental ice sheets were characteristic of northern Europe where except on the warmer Atlantic seaboard they advanced far southwards toward the arc of Alpine mountains. This Alpine arc had its own alternately expanding and diminishing ice-caps.

Climatic changes on lowlands

The climatic belts of all Europe advanced and retreated in step with the advance and retreat of the great northern ice sheets. The lowlands on the equatorward side of the ice front experienced numerous spells of cool to cold weather with frequent frosts. During glacial phases, such a climate, called *periglacial*, prevailed for example throughout southern England and in much of Northern France. In southern Europe its distribution coincided roughly with those areas that today experience rather cold winters. In the warmer Mediterranean south only the higher tracts became periglacial but in the cooler north even the lowlands, where fended from warming maritime airflow, had severe frosts, as, for example, the plains of the Po and the Danube. Here on the interior plateaux and on the lower slopes of the mountains, in non-Atlantic northern Spain and in much of the Balkans, periglacial influences were strong.

Where moist rock is exposed in periglacial climates the frequency of freeze and thaw leads to a rapid shattering of the rock face. The broken pieces soon become detached and fall as loose debris to form screes at the foot of rocky slopes. Where soil is present as is usual on plains, it is frozen downward in subzero weather. When a thaw occurs, the melting also proceeds downward and often the top soil becomes viscous or waterlogged when the subsoil remains frozen solid. Under these conditions, a shallow layer of surface material may slip downslope (the process is called *solifluxion*) into the hollows and valleys. Thus the characteristic features of landforms due to periglaciation are screes of angular debris at the foot of frost-shattered faces on rocky outcrops; and wide shallow valleys and gentle slopes encumbered with expanses of gravel and solifluctate on soil-covered lowlands. As rivers in periglacial climates flood violently in spring and early summer, the seasonal spates tend to give wide flood plains in the shallow lowland valleys.

Whereas the cooler parts of southern Europe became decidedly periglacial during ice advances, those parts that had in pre-Quaternary times and during interglacials a warm Mediterranean summer drought climate or some slight modification of it, then experienced a pluvial climate similar to that of the western Atlantic seaboard today. The ice sheets over northern Europe would tend to generate over themselves high pressure that would fend off Atlantic airmasses and divert many cyclonic depressions onto a more southerly course than they take today. This maritime airflow would be common throughout the year and would make summer drought infrequent even in the eastern Mediterranean. At the same time the present-day semi-arid and desert areas of the north African coastlands would experience a Mediterranean climate capable of supporting forests.

The effect of pluvial phases on the landforms in the parts of Europe formerly and today with a Mediterranean climate is hard to assess. At present the intense rainfall on relatively few days and the absence of frost action and of solifluxion favours the erosion of steep slopes and of deeply incised valleys. The wetter pluvial phases with their greater runoff might not appreciably lower the gradients of the slopes particularly if the rainfall remained intense.

Climatic change on mountains

During glacial phases climatic zones expand down a mountain side and in Alpine Europe the permanent snowline commonly fell by 1200 or 1300 metres. During the last main advance in the Pyrenees the glaciers descended to 400 m above sea-level on the French side and to between 800 m and 1000 m on the warmer Spanish slopes. On the drier and warmer parts of the Alps generally, snow accumulated and glaciers formed on slopes between 1500 m and 2500 m that had formerly been free of ice. On the snowier summits, especially on the Alps and the Pyrenees, the glaciers coalesced into ice caps that slipped downward far below the permanent snowline onto the adjacent foothills and plains. In the Alps the chief glacial phases were named Günz, Mindel, Riss and Würm after tributaries on the Bavarian foreland that drain to the left bank of the Danube at some distance from each other. The ice was thought to have advanced in successive phases up to these tributaries. However, today an earlier advance right to the Danube, and so named Donau, is recognized. It is obvious that most of the surviving glacial landforms will belong mainly to the Riss and Würm, the latest and two of the greatest ice advances.

Where the mountain ice caps melted on the piedmont, they deposited large quantities of moraine and drift. Meltwater assorted much of this deposition into fluvio-glacial terraces and outwash plains. In many places at the junction of mountain and plain the glaciers gouged out hollows, which, sometimes with the aid of terminal moraines, now contain ribbon-shaped lakes, such as Maggiore, Como and Garda.

When glaciers formed or expanded on the mountains they converted the pre-existing river valleys into u-shaped troughs such as the Lauterbrunnen valley, and the Val de Chamonix. These valleys owed their widening and valley-side steepening to glacial abrasion and freeze–thaw action at the glacier's side. At times when such a trough was filled to its brim with ice, any lateral tributary valleys incising the upper valley sides could not deepen their beds

1 *The Matterhorn and the Z'mutt glacier*

below the level of the top of the glacier in it. When eventually this main glacier melted – and the large thick glaciers usually persisted longest – the lateral tributary valleys were left hanging high above the trough floor. Hanging valleys where the streams end in waterfalls often 200 m or more high are common in the Alps. In places the lateral streams have cut deep gorges where they enter the main valley. Either situation is ideal for the generation of hydro-electricity.

Glaciation also emphasized the stepped nature of the long profile (*thalweg*) of many of the river valleys. The repeated uplifts of the mountains in later Tertiary times had rejuvenated the rivers, many of which had nick-points of erosion marked by a steepened gradient. When glaciers descended these valleys, they thrust downward and gouged out the valley floor at the foot of the steeper slopes into hollows which today are commonly occupied by ribbon-shaped lakes. In some Alpine valleys, such as the Swiss Engadine, valley-floor widenings with ribbon-shaped lakes alternate with gorge-like stretches.

In the present warmer climate many of the lower mountain summits and slopes fretted with cirques and other glacial features are now being slowly converted by streams into fluvial landforms. In parts the cirques contain lakes or fill with snow only in the colder months. On some peaks, as on the Jungfrau, Silvretta and Zugspitze, the glaciers only just fill the cirques. There are today about 1200 separate glaciers and *névé* (snow accumulations) in the Alps. The permanent snowline (*névé* or *firn*) lies at about 2400 to 2600 m on the snowier and northern slopes and at about 2600 to 2900 m on the drier and southern slopes. The existing glaciers in southern Europe, outside the Caucasus, cover about 3600 sq km, of which about 30 sq km are in the Pyrenees and nearly all the rest in the Alps proper. At their maximum advance the icecaps of the Alpine mountain system in southern Europe covered about 40 000 sq km of which 28 500 sq km were in the Alps. This, however, was truly insignificant compared with the 6 million sq km covered by continental icesheets in northern Europe.

Today the longest glaciers in the Alps are the Pasterze which descends 32 km from the Gross Glockner in the Austrian Tirol and the Aletsch which descends about 17 km from the Aar massif in Switzerland. These and most other Alpine glaciers have retreated markedly in the last fifty years.

The effect of geological structure and composition on landforms

Jagged and sharp-angled landforms are typical of glaciation on mountains almost irrespective of geology (fig. 5). Where fluvial processes prevail, the effect of the surface rock is easier to evaluate. We have already noticed that crystalline rock, especially schist and gneiss, often weathers and erodes into rounded forms while unmetamorphosed sedimentaries are often dissected into steep-sided tabular shapes as are common in the Dolomites and Diablerets. Limestones are especially extensive in the Alpine system of southern Europe where they probably cover as much territory as the crystalline and igneous massifs combined. In the western Alps the Jurassic limestones develop mild solution features. Where these and other limestones are intercalated with clayey beds they may weather into stepped slopes, or into parallel crests and valleys (as in the Simmental) or if the dip is steep into pyramidal peaks. But the most characteristic minor landforms are developed by solution on thick, homogeneous limestones. The Alpine system is rich in these massive limestones dating in age from the Carboniferous onwards. Many have been compacted and severely fractured during orogenesis but all are porous, relatively soluble and capable of standing in abrupt slopes. The extreme development is in the Yugoslavian karst in the

Dinaric Alps where in a labyrinth of gorges, caverns, solution basins and hollows, the drainage disappears and reappears in caves underground with an amazing intricacy. Similar karstic features occur in parts of the northern limestone belts of the Alps. Throughout these limestone outcrops caves are a noted attraction for tourists and for a growing clientele of potholers. In Yugoslavia as yet about 36 caves have been commercialized.

A minor landform, due to structure and process of another kind, is the volcano. Today vulcanism is most active in southern Italy, where the composite cones of Vesuvius (1277 m) and Etna (3269 m) vie with the excavated buried city of Pompeii as tourist attractions. There are three earth-heat power-stations at Larderello in Tuscany. Where the vulcanism has long since ceased, the ash and less-silicified lava are soon weathered and vegetated into scenery not unlike that of crystalline masses. These areas are often rich in warm mineral springs, some of which have become spas.

Landforms partly due to changes of sea level

The ice advances of Quaternary glacial phases stored up so much fresh water in a solid form that world sea level dropped markedly. This drop probably began in the Pliocene when continued uplift of Antarctica was creating a growing ice sheet there. During the Quaternary, sea level fell with each main ice advance and rose again on its retreat. On coastlines this resulted in a series of wave-cut platforms which on steep coasts may be little more than a broken tier of steps but on shallow coasts may stretch for miles inland as shallow terraces. The latter sometimes retain some of the beach material formed on them.

The coastal benches around the Mediterranean Sea range in age from at least the Pliocene onwards. Because world sea level has risen 55 m or 60 m in the last 12 000 years many former coastal benches are now drowned together with the lower courses of the river valleys that incised the lowest land areas, such as the present northern Adriatic and the shallow shelf off the Rhône delta. Such 'submarine canyons' abound off the wetter coasts. Where today coastal benches have not been drowned it is difficult to tell whether they are due to worldwide (*eustatic*) changes of sea level, such as occurred during ice advances and retreats, or whether they are the result of recent local orogenic uplift of land areas. In unstable Alpine Europe both types of change have been operative and the problem is further complicated by uncertainty as to when the Strait of Gibraltar was first opened to the Atlantic.

For decades a series of sedimentary cycles and former marine shorelines or coastal benches and cliffs has been recognized in the Mediterranean lands based on a classification by Charles Depéret. This was briefly, 90–100 m, Sicilian; 55–60 m, Milazzian; 28–30 m, Tyrrhenian; 18–20 m, Monastirian; and 6–8 m named by others Late Monastirian. Not surprisingly some of this nomenclature and sequence has proved illusory. The terms Milazzian and Monastirian have been discarded and the higher levels generally have been shown to lack accurate dating and adequate correlation over a wide area. Today the lower three stages are often called Tyrrhenian I, II and III, and increasing emphasis is put on the importance of them and of the many other terrace levels below 10 m. Undoubtedly remnants of benches do occur in places at 100 m and at 50–70 m above present sea level but the only levels at which coastal platforms occur with any regularity are at 25–50 m, 15–20 m, and 5–10 m. These are probably connected with changes of sea level during ice retreats and were exposed when world sea level fell during a glacial phase. The fact that they are still exposed (if orogenic uplift is ruled out) becomes intelligible only if we assume either that the great icesheets around the south pole and on Greenland have steadily increased in amount during the Quaternary or that our present climate is semi-glacial. Either is probable and it is certain that the lower coastal platforms of the Mediterranean would be again drowned if the vast thick icesheets of Antarctica and Greenland melted. But again we repeat that, irrespective of these global causes, it is unreasonable to expect a close and elaborate correlation of coastal bench heights in an area such as southern Europe where mountain building was active throughout the Quaternary period.

Erosional benches or terraces are not however restricted to coastal areas nor to marine planation (wave action).

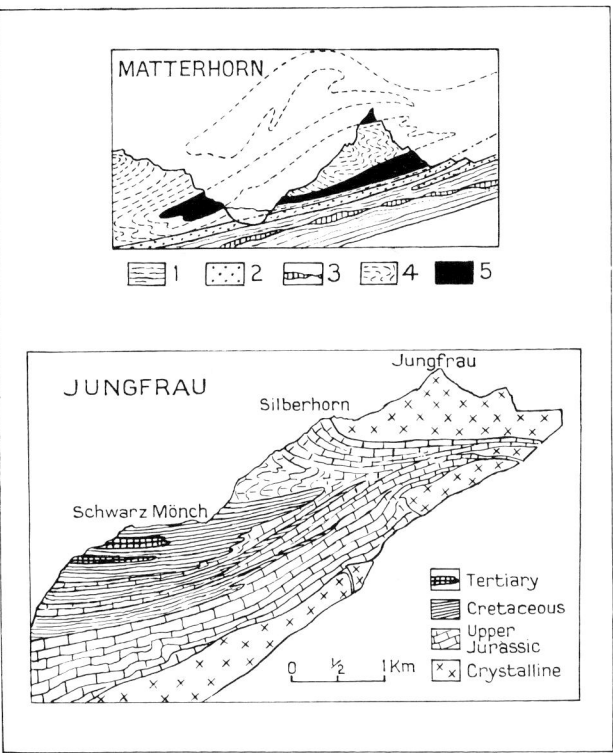

5 *Geological structure and morphology of the Matterhorn and Jungfrau* (simplified from E. Argand and L. W. Collet)

1 schistes lustrés, *crushed and in parts burnt by friction during sliding*
2 *greenstones*   3 *Trias*   4 *Arolla gneisses*
5 *Valpelline Fédos gneisses*
The sections are from northwest (left) to southeast. Northwestward the Matterhorn (4481 m) drops to the Z'mutt glacier and the Jungfrau (4161 m) to the deep U-*shaped Lauterbrunnen valley*

Whenever the level of the Mediterranean sea dropped markedly, the lower course of rivers entering it were strongly rejuvenated and began to cut new valleys within the old. Repeated rejuvenation has incised a series of terraces standing above the present floodplain. Commonly three or four terraces may be detected on the wider valley sides, the chief being at an average height of about 90m, 45 m, 23 m, and 7–9 m above the present river. On the upper, mountain courses of Mediterranean rivers, valley-side terraces also often occur but these are mainly due to rejuvenation caused by orogenic uplift and by climatic change. The result is that in southern Europe wherever the valleys widen appreciably some kind of riverine terraces afford opportunities for human settlement and routes.

The last rise of about 55 m in sea level flooded most of the pre-existing low coastal plains and many of these have been or are now being raised by sedimentation to present sea level. This rise of sea level has discouraged the rapid seaward growth of the deltas and partly explains why even the larger deltas, such as those of the Rhône, Ebro and Po have not yet protruded far beyond the general alignment of their coastlines. The wide marshes of the lower Danube and the marshlands (Las Marismas) near the lower Guadalquivir demonstrate that in some coastal areas deltaic deposition has not yet offset the recent rise of sea level.

During historical times river work has formed a low-level terrace on the lower course of most Mediterranean valleys. River deposition or valley-fill here poses many problems. During the last main glacial phase in Europe the enlarged Mediterranean streams deposited large quantities of debris on their valley floors. The streams in the extreme south became large because of less evaporation and perhaps greater rainfall; the streams in the cooler north and interiors became more powerful because of a change in their regime towards an increasingly strong flood in spring. In both areas, and especially whenever periglacial agencies prevailed, the amount of debris available for fluvial transport was greatly increased. With the return of a warmer climate the streams became smaller in volume but simultaneously were less hindered by their load because of lessened surface wash from a landscape less affected by frost and rain and better protected by a denser or more evergreen vegetation cover. This may well explain why the rivers throughout post-Neolithic and Roman times cut quite deep (but relatively narrow) channels into their debris-encumbered valley floors. But, at some stage during the Middle Ages erosion was replaced by aggradation which formed a younger valley-fill up to 10 m thick. Subsequently, renewed stream cutting has incised channels into the newer fill leaving it upstanding as a lower terrace above the present streams and floodplains.

The medieval aggradation built up much of the richer agricultural riverine lowland in Mediterranean countries and greatly extended the coastal plains and deltas. It buried to a depth of several metres famous classical sites such as Olympia on the Alfiós in the Peloponnesus. On the other hand, the modern gullying of some of the valley-fills has created badlands. The change from erosion in Roman times to medieval aggradation and back to present-day erosion is hard to explain. Medieval aggradation is usually attributed to the effects of large-scale deforestation which seriously increased surface runoff and caused extensive soil erosion. Once the existing soil or debris had been exposed to heavy rain it would be removed at a rapid rate but when the bedrock had been reached the amount of solid material available for stream transport would be greatly decreased and, because of lessened load, the rivers would again begin to incise their channels. In parts of peninsular Italy Roman structures indicate that just over one metre of surface has been removed by erosion locally in the last 1800 years while the average denudational lowering of the whole Tiber basin in that time has been computed at 31 cm. The point, however, is that erosion in lands with intense rainfalls becomes truly excessive if unprotected soils in vulnerable localities are exposed to surface wash.

But the widespread nature of the medieval aggradation and the present resumption of erosion induce some authors to consider that slight changes in climate and so of river work are also involved. It is certain that in the drier parts of the Mediterranean a slight increase in rainfall could have a large effect on the vegetation and so on the work of rivers whereas in the wetter lands to the north such slight changes, if present, probably would not have appreciable vegetal and hydrological effects.

Landforms due to wind action

In southern Europe, although winds often reach high speeds, aeolian or wind effect on landforms is largely restricted to flat floodplains and sandy littorals that dry out occasionally at the surface. The effect was greater during the Quaternary when strong winds traversed large expanses of fine silt on bare outwash plains and on the extensive floodplains of meltwater channels, especially at the southern edge of the greater northern icesheets. Some of the finer material was carried by the wind to form loess, which is of great importance in a wide belt across the whole of central Europe but in southern Europe is restricted to smaller patches, as in the Rhône valley south of Lyon. However, loess does occur in one Mediterranean state as Yugoslavia includes large areas of it near the Tisa and Danube.

Today on wide sandy beaches with strong onshore winds mobile dunes soon form unless the sand is fixed artificially with grasses and trees. Much of the sandy Atlantic coast of northern Portugal has been planted with pine forests that are worked as plantations for turpentine and resin. The marshlands (Las Marismas) of the lower Guadalquivir are bounded seaward by a belt of old sand dunes (Arenas Gordas) but as a rule in Europe mobile dunes are a thing of the past because of modern methods of fixing with marram grass and conifers.

3 The Submerged Landscape and the Fishing Industry

Shape and size

The drowned basins of southern Europe are in many ways an inverted replica of the land relief. They are, however, deserts as far as human settlement is concerned and although their water surfaces are flat they offer greater obstacles to large-scale human movements than do all but the steepest mountain chains. The geographer is especially interested in how far man has been able to overcome the marine barrier and how far he has been able to offset the great loss of potential agricultural land that it covers.

The ancient submerged geosyncline, or Tethys Sea, stretched from the present Atlantic across the site of India and the Far East to the Pacific. Many geologists consider that an extension of the warm Gulf Stream or some similar warm tropical current often flowed directly into its western parts. The warmth of the geosynclinal sea is reflected in the enormous deposition of calcium carbonates and growth of corals that form such a large proportion of the general Alpine mountain system. The present smaller geosyncline stretches through the Mediterranean to the Black Sea, Caspian Sea and beyond. We are here concerned only with the western basins which form a narrow trough between Europe and Afrasia.

The Mediterranean extends for 3700 km eastward from the Strait of Gibraltar to the coast of Syria. Its maximum width is about 1100 km between the Strait of Otranto and the coast of Libya but it often narrows to less than 650 km. Its area is about 2 976 400 sq km or ten times the size of Italy. This extensive trough is connected in the island-studded Aegean to the Black Sea by the Dardanelles, Sea of Marmara and Bosporus. The Dardanelles is a drowned river valley about 60 km long, cut in soft strata. The channel narrows to 1700 m before widening into the Sea of Marmara, a fault consisting of two parallel graben that descend in parts to over 1300 m. This sea communicates with the Black Sea by means of the Bosporus, a narrow strait, between 600 and 1600 m across, that lies along the line of a submerged river valley. The Black Sea stretches at a maximum about 1150 km east to west and 600 km north to south, and covers 424 000 sq km or about the size of Italy and Greece combined. It lies in the latitudes of the Adriatic and of the Bay of Biscay in the Atlantic and is actually situated far to the north of the easternmost parts of the Mediterranean. Perhaps, however, the stress should be laid on the southerliness of the eastern Mediterranean. This large basin of the Ionian and Levantine seas lies mainly in 37°N–31°N or latitudes for the most part equatorward of all Europe and of the western Mediterranean.

Physical and hydrological divisions

The seas of Southern Europe occupy a series of deeper basins separated by submarine sills, islands and insular arcs (fig. 6). From a physical viewpoint these basins differ markedly from each other and even in the Mediterranean proper it is necessary to emphasize its heterogeneity rather

6 *Major features of the submarine relief in the Mediterranean and Black Seas* (partly after P. Birot)
  1 *continental shelf*
  2 *main submarine precipices*
  3 *sea-floor trenches*
  4 *eminences on plains*
  5 *open circle: probable submarine volcano; black circle: volcano active in historic time*
  6 *maximum depth in metres*

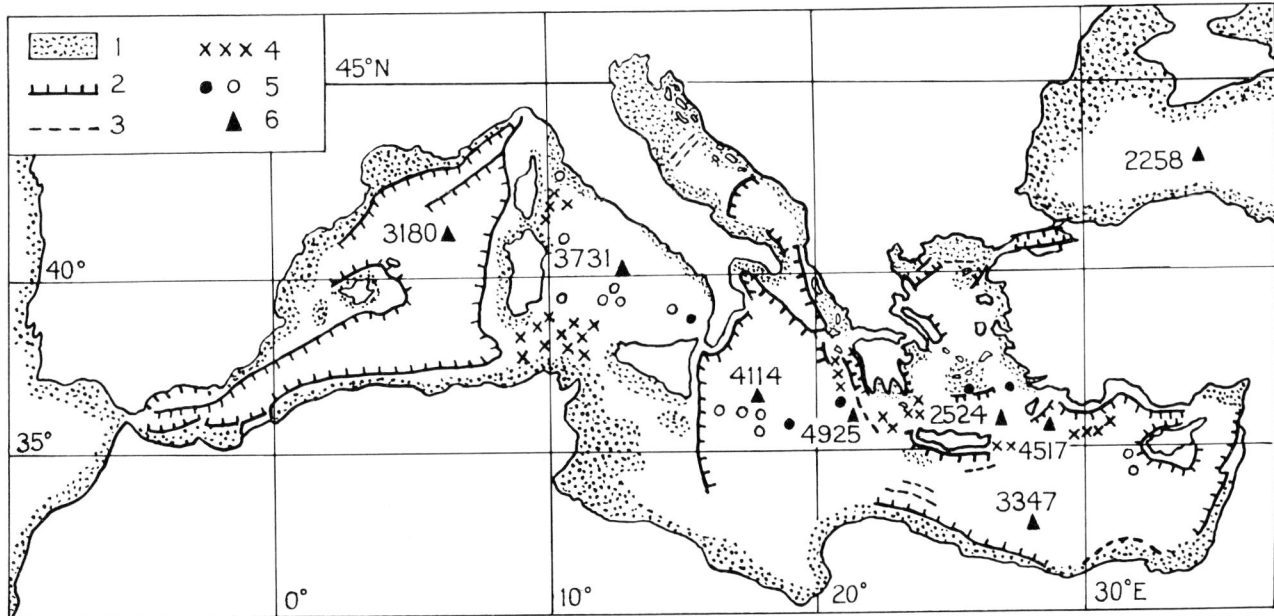

than its homogeneity. The Mediterranean consists of two major units separated by a wide submarine platform between Sicily and the African coast. The western part is relatively small and covers only one-third of the whole sea. Physically it falls naturally into:

1 The Balearic Basin, which descends sharply to 2000 m off the coasts of northern Africa, the riviera and Corsica and has a maximum depth of 3180 m off Sardinia. It is remarkable for a wide continental shelf off parts of eastern Spain, the Balearics and in the Gulf of Lions. The last-named, known as La Planasse, covers 7300 sq km between the submerged edge of the eastern Pyrenees and the Provençal Alps. Its flat surface slopes gently seaward to depths of 120–160 m to an outer edge that is fretted with numerous gorges which grow in size as they plunge down the continental slope. Such submarine canyons abound on the continental slopes off many Mediterranean coasts and presumably were cut when sea level was lower and runoff greater during the colder phases of the Quaternary period.

2 The Tyrrhenian Sea basin is relatively shallow in the north between Corsica and the Italian peninsula but deepens rapidly southward where large areas exceed 3000 m in depth, and one depression sinks to 3731 m. In the southeast of this basin, vulcanism has created many islands and submarine peaks, some of which do not reach the surface. Among the volcanic islands are the Lipari, with Vulcano and the very active cone of Stromboli.

The Tyrrhenian Sea is limited southward by Sicily and by a wide submerged platform linking Sicily with Africa. This large sill is less than 200 m deep over extensive areas and is one of the great plains of the Mediterranean floor. In its wider eastern parts the islands of Malta, Lampedusa and others rise above its shallow waters; in its western part it narrows to 138 km between Sicily and Cape Bon in Tunis, and a fall in sea level of 200 m would leave only a narrow sea strait between the two. Probably a land connexion between Africa and Europe existed here in the Quaternary when elephants lived in Italy. On one part of this extensive sill a short-lived volcano (Graham Bank) emerged in 1831 and 1863. The sill generally divides, as we have already noted, the western from the eastern Mediterranean Sea. The descent from it to depths of over 4000 m in the eastern Mediterranean is very steep, and the gradients often exceed 45° near Malta.

The Eastern Mediterranean includes the northern branches of the Adriatic and Aegean and can be subdivided into the following physical units:

3 The Ionian Basin, which stretches from the Strait of Otranto between Italy and Albania southward for 1100 km to the Libyan coast, has several troughs deeper than 4000 m and descends to 4925 m in a trench just south of Cape Matapan in the Peloponnese. It is virtually islandless.

4 The Levantine basin which forms the southeastern Mediterranean is partly separated at depth from the Ionian basin by a submarine sill between Crete and Libya. The Levantine sea descends to 4517 m in a very steep trench close to the island of Rhodes. It has a wide continental shelf off the mouth of the Nile, where the river sediments along a coastline of 900 km or more have shallowed the coastal waters. In its northeastern quarter is the large island of Cyprus.

5 Branching northward from the Ionian Basin is the Adriatic that occupies a trough in which the Po and its predecessors have steadily extended their deltas. The present sea is 770 km long and has a maximum width of only 200 km. With an average depth of less than 250 m it is by far the shallowest of the Mediterranean basins. All its northern two-thirds is under 200 m, except in a narrow fault trench off Pescara, and most of it is under 100 m. Only in the southern part near the Strait of Otranto do the depths exceed 1000 m.

6 Branching northward from the Levantine basin is the Aegean Sea, the most distinctive of the Mediterranean components. It is bounded on the south by a great mountain crescent, now splintered and submerged except in the island arc stretching from off southern Greece through Crete to Rhodes. The Aegean is an archipelago. It is dotted with islands and its coastline consists largely of long finger-like peninsulas. The fracturing and subsidence have proceeded far enough to give sea depths of between 200 and 1000 m. In the north the deepest floor is at 1311 m but in the south close to the main island arc the sea bed descends precipitously to over 2000 m in several trenches, including 2524 m near Karpathos. The islands and peninsulas continue the geological trend and structure of the adjacent mainlands. The northernmost islands and the narrow peninsulas of Chalcidice which tower in Mount Athos to 2032 m are offshoots of the Rhodope Hercynian block. Their dark colours contrast vividly with the brilliant white limestone of Lemnos, a fragment of the Dinaric folds. Similarly farther south the Cyclades are of ancient rocks whereas the Northern Sporades clearly prolong the folded limestone chains. Recent earthquakes and active vulcanism testify to the continued instability. Methana has been active in distant historic times but the main eruption occurred in 1925 when a central lava cone appeared in the sea-filled crater of the island of Santorin.

In the Aegean generally the continental shelf is restricted to patches around the coast and central islands, particularly where newer deposition has filled or is rapidly filling a bay or long narrow inlet. A famous classical locality, the pass of Thermopylae on the Gulf of Lamia, provides a small but typical example. The pass was a narrow track between the mountain wall and the sea in which small numbers of Spartans in 480 BC and of Greeks in 279 BC checked advancing armies. Today the Sperkhios river has extended its delta by 17 km into the gulf and the narrow pass has been widened into a deltaic plain several kilometres wide.

The Aegean archipelago leads northward to:

7 The Black Sea, which has no counterpart in the Mediterranean basins. It consists of two very different areas: a southern flat-floored oval basin mainly over 1500 m deep but nowhere below 2258 m and almost devoid of continental shelf; and a shallow northern shelf, including the Sea of Azov which is a recently drowned plain with extensive deltaic deposits off the mouths of the Dniepr, Dniestr and Danube. In this northern part the continental shelf is often 200 km wide.

Physical qualities of the sea water
Horizontal surface distributions

The precipitation on the Mediterranean and Black Seas is about the same as that on the adjacent lowlands. It tends to decrease markedly in the Levantine basin where the summer drought becomes excessive. The surface temperatures and salinity of the sea water increase southward and eastward. During February, the coolest month, the mean monthly surface temperatures of the western Mediterranean decrease from 12°C off Provence to about 14°C off Algeria; in the eastern Mediterranean sea water temperatures then increase from about 14°C near Sicily to over 17°C off Israel. In the same winter month the water temperatures in the northern extensions of the eastern Mediterranean decrease northwards from 13·5° to 7°C at the head of the Adriatic and to 10°C at the Dardanelles end of the Aegean, 9°C in the Sea of Marmara, 8°C in the southern Black Sea and below zero in the extreme northern Black Sea and Sea of Azov. During August, the warmest month, surface sea water temperatures are much more evenly distributed as the shallower water warms rapidly. Then, in the western Mediterranean the sea is at about 22°C off the Rhône delta and 25°C off the Algerian coast; in the eastern Mediterranean, the temperatures range from 25°C in the north to about 27° or 28°C in the east. Most of the Adriatic and Black Sea has an average monthly August temperature of 23° or 24°C.

The surface salinity shows much the same seasonal and regional variations as the temperatures, except for its marked dilution off the mouths of rivers. In a general sense the surface salinity increases eastward from 37‰ (parts per thousand) near Gibraltar to 37·5‰ near Sicily and 39‰ in the Levantine basin where evaporation is most intense. In the Adriatic the salinity decreases from 35‰ near the Strait of Otranto to 18‰ near the head of the gulf, where the Po enters and a great deal of fresh water gushes up through the sea floor off Istria. In the Black Sea the influence of the freshwater influx increases still further. Here the precipitation and the great inflow of river water exceed the evaporation and so reduce the surface salinity, that it is only 10‰ in the north and 17‰ to 19‰ in the south. In the various straits connecting the Black Sea with the Aegean the surface salinity increases from about 18‰ in the Bosporus, to 22‰ in the Sea of Marmara and 25‰ in the Dardanelles, the surface outflow of which lowers the salinity of the adjacent parts of the Aegean to 33‰.

Vertical distribution of temperature and salinity

For organic life in the sea, particularly fish, the variations of temperature and salinity with depth are of great importance. The Mediterranean is warm and saline compared with other seas as large or larger than itself. The temperature and salinity of the surface water varies appreciably daily and monthly with the amount of sunshine and precipitation or evaporation down to about 200 m in the western basins and 400 m in the Levant. Below this upper layer there can be detected in many areas an intermediate layer with a maximum of salinity and often a slight temperature rise. This in its turn often overlies a transitional layer, at between 600–1200 m in the west and 800–1300 m in the east, where the temperature drops to a minimum before rising again slightly at greater depths. In a sense this transitional layer is the upper part of the fourth water mass, the deep-water or homothermal layer. This homothermal water mass has a temperature of about 13°C in the western Mediterranean and of just over 13·5°C in the eastern Mediterranean. Its salinity near the sea floor varies from about 38·4‰ in the west to 38·7‰ in the east. The water conditions in the Black Sea are markedly different from those in the Mediterranean. In the former the temperature decreases downward from 23°C to about 7·8°C at 50 m and then rises slowly to 8·9°C at 2000 m; the salinity increases steadily downward from 18‰ at the surface to over 22‰ in the deeper layers.

Water budget

The high relative warmth and salinity of the Mediterranean are due mainly to its sunny climate, enclosed nature and moderate depth. The intense evaporation upon its surface and especially over the eastern basins far exceeds rainfall in all the warmer months and would cause a steady decrease in volume were it not for the inflow from the Atlantic. A tentative water budget for the Mediterranean, expressed in cubic metres per second is given below:

GAINS (cub m/sec)	
Inflow from Atlantic	1,750,000
Inflow from Black Sea	12,600
Precipitation	31,600
River discharge	7,300
	1,801,500

LOSSES (cub m/sec)	
Outflow to Atlantic	1,680,000
Outflow to Black Sea	6,100
Evaporation	115,400
	1,801,500

The water movement in the Suez Canal is negligible while that from the Black Sea is small but significant. Here a relatively diluted surface current flows out through the Bosporus often at a speed of 10 km per hour. Its pace lessens in the Sea of Marmara but accelerates again in the Dardanelles up to 8 km/hr in the narrowest reaches.

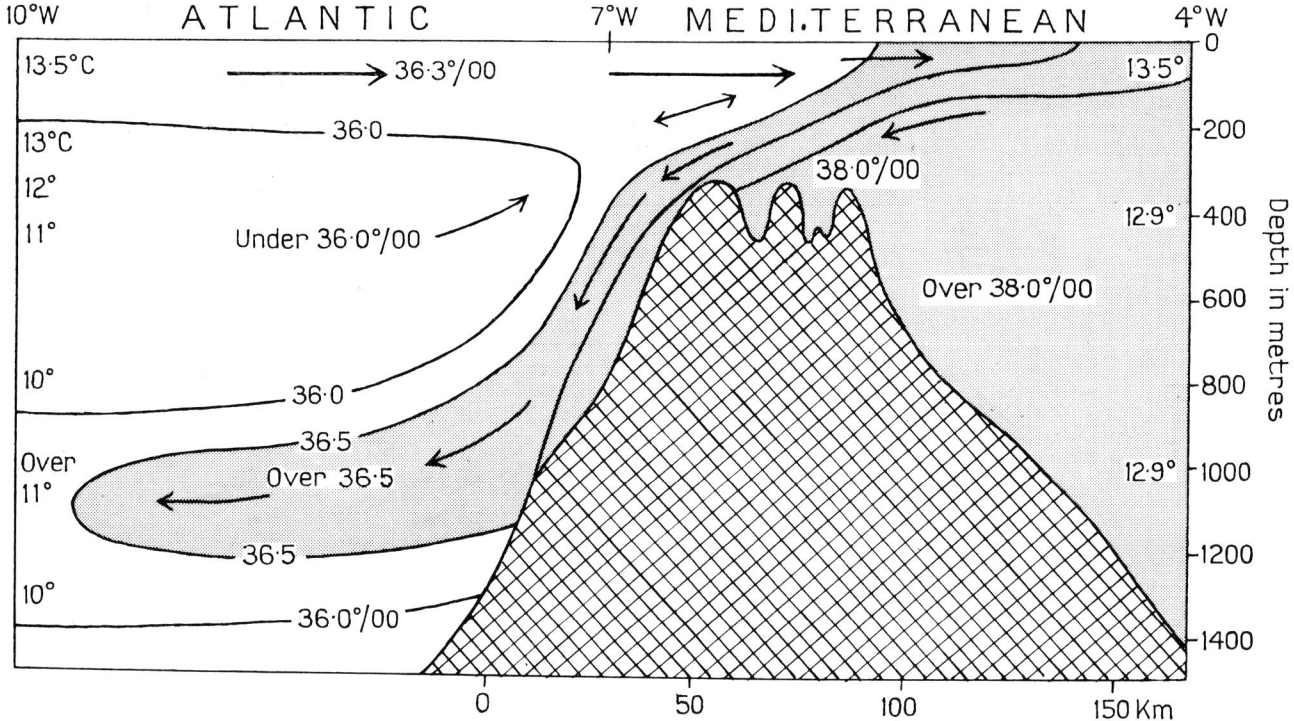

7 *Currents, temperatures and salinity of sea water in the Strait of Gibraltar* (partly after Schott)

Beneath this diluted current a denser saline current flows from the Aegean into the Black Sea. As the Bosporus and Dardanelles are shallow channels (40–90 m) there is much intermixing of the two currents and the upper flow increases in salinity from 16‰ to 30‰ while the bottom flow decreases from 38·5‰ to 30‰ during the journey. The effect of the inflowing surface current from the Black Sea is soon absorbed by the warm, saline Aegean.

The great carrier of oceanic water into the Mediterranean is the surface current which flows through the Strait of Gibraltar, partly on a slight down gradient and partly under the push of strong prevailing westerly winds and of semidiurnal tidal movements funnelled between Morocco and Spain. The Strait is 15 km across at sea level and its floor is a jagged sill with three peaks rising to within 320 m of the surface (fig. 7). This complicated relief causes pulsatory uplifts in the bottom currents which affect the base of the surface currents. The inflowing surface Atlantic water is less saline (35·3‰–36·9‰), cooler and more oxygenated than Mediterranean water. It decreases in volume and speed during easterly winds and when, for some reason, the bottom current expands. Its speed varies between 0·5 and 5 knots and its depth from 5 m up to 300 m. However, this surface current is best regarded as a great marine river, averaging about 80 m in depth and 4 knots in speed, and with a volume ten times that of the Amazon or almost double that of all the world's rivers combined. Beneath this inflow a large current of saline Mediterranean water escapes over the floor of the sill into the Atlantic. This bottom water, with its distinctive high salinity and plankton (floating organisms), enters into the deep water circulation of the Atlantic where it has been traced southward to South Africa and northward to the continental shelf west of the British Isles. It and the surface current above it act as the great rejuvenator of Mediterranean water which is completely ventilated about once every seventy-five years.

Waves and water circulation

Owing to the relatively short fetch or width of water over which the winds travel, the wave motion in the Mediterranean is moderate. Waves average only 3 or 4 m in height and even in storms seldom exceed 9 m. Yet the sea often whips up with disastrous suddenness and the sharp short waves can be distressingly uncomfortable except for large ships.

The tidal motion of the Mediterranean is also small, a range of between 15 and 30 cm being common. The tides are highest on the coast of Provence (1 m), at the head of the Adriatic, and where the water piles up on the wide continental shelf in the Gulf of Gabes (1·88 m). Often the tide is almost imperceptible and does not disturb holidaymakers. Equally it also often fails to disturb sediment in river channels and harbours. A notable exception occurs in the Strait of Messina where the lack of synchronisation of the tidal motion in the Tyrrhenian and Ionian Seas may cause a slight difference in level that induces a current of up to 5 knots to rush through the narrow strait. Hence the classical allusion to the danger of being between Scylla, a rock in southern Italy, and Charybdis, a whirlpool off the coast of Sicily. The strong currents, however, are an attraction to fish, and over 140 species have been caught in the Strait.

The water circulation of the Mediterranean depends mainly on temperature and salinity differences and on winds. Each basin tends to have its own counterclockwise surface circulation which is part of, or connected with, a marked west–east movement off the African coast. The important surface inflow through the Strait of Gibraltar often extends almost unaltered for at least 700 km along the coast of Algeria. Its speed here varies up to 3 knots and it normally has a width of 20–65 km. This 'pure' Atlantic water also flows up the Spanish coast to Valencia and to the south of Majorca. In a diffused form it can be traced in the south from Sicily and southern Italy to Genoa and Marseille. Thus Atlantic water of a pure or diffused kind almost makes a circuit of the whole western Mediterranean basin.

The eastern Mediterranean and its northern extensions in the Adriatic and Aegean also have general anticlockwise surface circulations. In the Adriatic and Aegean the more saline currents tend to be on the eastern side and the more diluted waters on the western side, where they are often pushed southward by strong winds, such as the bora. An important feature of the general Mediterranean circulation originates in the eastern basins. Here the intense surface evaporation encourages the formation of a dense intermediate layer of high salinity (38·6‰–39·1‰) and high temperatures (13·4°–14·3°C) at a depth of about 300 to 800 m. Elsewhere in the Mediterranean temperatures are rarely above 13·3°C at these levels. This intermediate layer of high salinity moves westward as a subsurface flow throughout most of the Mediterranean. It is the only permanent current in the Gulf of Genoa, near Corsica and in the southern Adriatic. It is important on the Sicily–Tunisia sill and, as we have noticed, forms the bottom flow through the Strait of Gibraltar where it occasionally reaches a speed of 4 knots, a quite remarkable velocity for such a deep current.

It is perhaps unnecessary to emphasize that the great bulk of water in the deeper, homothermal layer of the Mediterranean is not influenced by currents other than slow 'creeps'.

Effect of sea currents on coastlines

The direction of longshore currents, especially where aided by longshore drift caused by prevailing winds and waves, often influences the shape of coastlines. Active coastal currents have rounded the arcuate deltas of the Llobregat, Arno and many other rivers. In Languedoc the current moulds the river sediments into a regular coastline. At the Nile mouths the drift is eastward while at the Rhône mouths it is westward as is reflected in the direction of the longshore bars. At the heads of bays and estuaries the lack of a strong ebb tide encourages the accumulation of coastal flats which the longshore drift may convert into marshes or shallow lagoons, notorious in the past for malaria.

Effect of currents and water masses on fisheries

The fishing potential of a sea depends largely on the amount of food available for the fish. This is naturally most abundant in shallow, sunlit, well-oxygenated waters, rich in nutrient minerals such as phosphates and nitrates which favour the growth of plankton and fixed plants. The continental shelf and upper continental slope, especially where the water is well stirred by currents, are especially favourable. The amount of fish caught depends largely on the numbers, skill and needs of the fishermen who frequent the fishing grounds.

The relative value of the Mediterranean for fish production is not fully known although there are thirty-five chief marine stations and laboratories on its shores, including the world famous museum of oceanography at Monaco. A large literature now exists on its *benthos* or life on the sea floor. The high salinity (37·8‰–39·1‰) of the bulk of its waters compares most unfavourably with that of the North Sea (mostly under 35‰) and even with that of the warmest tropical regions of the Atlantic (37·2‰–37·5‰). It is generally admitted that the deep water of the Mediterranean is low in oxygen content and poor in nutrients, especially phosphates, and consequently is poor also in plankton. It is, however, far from lifeless in contrast to the deeper, almost stagnant Black Sea water where, below 1100 m, oxygen is virtually absent and hydrogen sulphide accumulates in harmful quantities. Moreover the relative warmth of the Mediterranean Sea, especially below 500 m, partly compensates for its high salinity, as species that like warm water can find it by undertaking vertical rather than horizontal migrations. Fish that must live near the surface in the Atlantic can find suitable water temperatures at hundreds of metres greater depth in the Mediterranean. Thus, when in the northern parts of the sea, water surface temperatures are cooled by northerly winds, the fish can migrate downwards into the deeper water farther offshore. Similarly the tunny migrate seaward and downward when the mouths of the Rhône discharge large quantities of cool meltwater during the early summer spate. However, depth increases the difficulty and cost of catching the fish, and the fact remains that many species common to the Atlantic and Mediterranean breed more prolifically and grow larger in the less saline Atlantic.

The lower bathyal zone of the Mediterranean (that is below about 500 or 600 m on the continental slope) is extremely poor in species and has a biomass (total of living organisms) inferior to that of the Atlantic at similar depths and similar distances from the coasts. This relative poverty makes it all the more important to reiterate that the Mediterranean is a very heterogeneous sea and has extensive areas favourable to a dense biomass. The chief favoured regions lie in the inflow of Atlantic surface water and on the continental shelf.

The surface Atlantic water entering at the Strait of Gibraltar carries a rich biomass. The 'pure' current is 2 to 5 times richer in phosphates and 5 to 12 times richer in plankton than nearby Mediterranean water masses. Not surprisingly it is also rich in sardines. Where, however, well-oxygenated Atlantic current is diffused and slowed up, it becomes little richer in plankton than the main mass of Mediterranean surface water. The advantages of this inflow to the western basins is obvious but much of the advantages lie off the African and not the European coast. No doubt the surface inflow from the Black Sea and Dardanelles has in a small way a similar effect on the fish life of the northern Aegean but information is not available.

The continental shelf, or depths down to about 200 m at which local plankton can use nitrates and phosphates, receives sunlight sufficient also for algae and fixed plants to grow. Probably the variety and extent of the Mediterranean continental shelf has been persistently underemphasized. It forms just over 22 per cent of the Mediterranean floor and although very narrow in parts, is of moderate width over large areas and of great width in several places. For richness of biomass and excellence of fisheries, four areas or locations stand out.

1 Off the river mouths, notably the Rhône, Nile and Po, where plankton abound.
2 The wider continental shelves of sedimentary deposits and recent deltaic deposition, submerged either by Pliocene downfaulting or by the post-Quaternary rise in sea-level. Among these is La Planasse, a sandy tract, 40 to 50 km wide in the Gulf of Lions which is at the moment being investigated for more intense exploitation.
3 The shallow depositional floor of the northern and central Adriatic which has distinctive planktonic associations and seasonal rhythms. Here the water is relatively cold in winter and warms up rapidly in summer. North of a line from Ravenna to Split the biomass or planktonic life is comparable with that off Western Europe.
4 The broad submarine platform between Tunisia and Sicily has a large area of continental shelf, where the tides are noticeable and currents appreciable. A surface flow eastward overlies in parts a westward intermediate undercurrent of warm, highly saline water. This warmer subsurface flow attracts tunny at breeding time and is rich in sardine and anchovy.

Although outside the Mediterranean proper, mention should also be made of the wide continental shelf, with numerous river deltas, in the northern and western parts of the Black Sea, where the rich biomass is comparable to that in Adriatic and Atlantic waters. It is the scene of prolific fisheries, especially in the Ukraine S.S.R. and the Danube delta in Romania.

Among other interesting features of the Mediterranean continental shelf are first, the inexplicable poverty of some sandy tracts on the infralittoral zone which are much poorer in biomass than are similar tracts in the North Sea; and second, the remarkable development locally of *Posidonia* meadows. *Posidonia oceanica* is an endemic (native) species of eel grass that flourishes on sandy-muddy bottoms in the infralittoral zone in many areas, for example, the central and southern Adriatic. The *Posidonia* has rhizomes which form on the sea floor a dense mat of thick roots that traps sediments. The mixture of roots and silt gradually thickens upward until it is stopped near the water surface by wave erosion. Sometimes the leaves just protrude above water level. This thick carpet advances across shallow bays and estuaries in much the same way as an offshore bar. The *Posidonia* meadows soon develop a microflora and microfauna of their own and the increased biomass makes them popular with migratory fish, especially the *Cleipides* (sardine, sprat, bluefin tunny, bonito, etc.) and the *Scombrides* (mackerel spp.).

We have yet to discuss the benthos (life) on the continental slope, or area beyond the continental shelf, at depths of 200 m to 1000 m. The continental slope forms about 21·5 per cent of the total Mediterranean floor and includes a large expanse between Tunisia and Sicily much of which is relatively flat. Here the warm, saline intermediate current moving westward along the bottom varies seasonally in strength and depth but it is a prime conveyor of shoals of sardine, anchovy, bonito and tunny. In the Mediterranean the tunny (*Thunnus*) is essentially a fish of intermediate waters and migrates southward to breed. It attracts much attention because of its size and the high quality of its flesh. It is often caught by nets in spring, by elaborate offshore traps in the breeding season and by line from boats throughout the year. However, the Mediterranean contributes less than 10 per cent of the world's tuna catch and here the seasonal migrations of the sardine (young pilchard), anchovy and mackerels (*Scombrides* spp.) are far more important.

In many parts the continental slope is steep and its lower, deeper zone has been as yet little investigated. Below about 600 m it is virtually dark but fish could live upon it. On the slope as a whole the marine life includes sponges and coral. Corallaginous banks of various colours, including red coral, occur in patches in the western Mediterranean and over considerable areas in the eastern basins. At depths of between 300 and 800 m clumps of white deep-sea corals are not uncommon. The collecting of sponges and corals, particularly the red, employs a small number of fishermen.

There remains to be mentioned the saline lagoons on the coastline and the freshwater fisheries inland. Throughout southern Europe, wherever pollution is not excessive, the coastal lagoons are stocked artificially or naturally with fish and molluscs. The saline lagoons, for example, of the Languedoc coast, the Ebro delta, the Valencian Albufera, the Mar Menor and Po delta are trapped and netted for eels, shad, seabream, golden perch, mullet and bass. The shallow lakes are ideal for aquiculture (pisciculture). Some Adriatic lagoons have produced over 150 kg of fish per hectare annually which is high except on tropical standards (500 kg/ha annually). In Languedoc the Étang de Thau has been used for aquiculture since 1936, first with mussels and later also with oysters, both grown attached to moveable rods, bars and plates.

Freshwater fishing is excessively popular in all rivers and lakes. For example, the northern Spanish rivers are rich in trout, and everywhere in reservoirs constructed for hydro-electric and irrigation purposes stocking with fish is proving worthwhile. In addition, freshwater fisheries become commercially important on the larger lakes, such as Garda, Maggiore and Como, and in the larger rivers. The advantage of inland aquiculture is that often the fish can be grown near the markets. The total recorded freshwater catch of southern Europe barely reached 60 000 tons in 1966. This excluded molluscs but included river eels which are anadromic or descend to the sea to spawn. Carp are especially important in inland lakes and lagoons in Yugoslavia and Spain. It is noticeable, however, that for fishing no Mediterranean river even approaches in importance the lower Danube and its tributaries which in 1966 yielded no less than 26 400 tons in Hungary and 22 300 tons, including some sturgeon, in Romania.

FISHING AND FISH CATCH

In 1958, the total recorded fish catch of the Mediterranean and Black Sea was about 900 000 tons compared with 11 900 000 metric tons in the North Atlantic. In 1966, the Mediterranean and Black Sea catch was 1 100 000 metric tons or 1·9 per cent of the world total and 10 per cent of the fish catch by European nations (11 500 000 metric tons). The Mediterranean and Black Sea increase of 22 per cent was far below that of almost all other major fishing regions in the world during this period.

It will be seen immediately that statistical difficulties arise when considering the recorded fish catch of southern Europe as distinct from that of the Mediterranean and Black Sea. The countries of the African and Asian coasts of the Mediterranean must be omitted from the south European total. But the latter includes Portugal that is entirely, and Spain which is largely on the Atlantic. Of France, only the southern part is involved.

The table of statistics shows the recorded live weight catch for 1966 to 1968 or the year nearest to them for which figures are available.

The live weight of fish landed by national fishermen in 1966 in the countries of southern Europe was 2 384 300 tons. Of this, 3000 tons were caught in landlocked Switzerland, about 900 000 tons (including freshwater fish)

FISH CATCH LANDED BY SOUTHERN EUROPEAN AND MEDITERRANEAN AND BLACK SEA COUNTRIES (in thousand tons)

		1966	1967	1968
Atlantic Europe	Portugal	501·6	559·8	506·4
Atlantic–Mediterranean	Spain	1357·4	1432·1	1500·1
Mediterranean Europe	Southern France	*est.* 20·0	(total French catch 804·8)	
	Italy	369·0	373·1	363·4
	Yugoslavia	45·5	47·9	45·0
	Albania	3·6 (1964)	4·0	4·0
	Greece	82·9	85·1	92·9
	Malta	1·3	1·4	1·6
Inland Alpine Europe	Switzerland	3·0	3·0	3·5
African Mediterranean	Morocco	*est.* 30·0 (total catch 295·6)		
	Algeria	20·3	21·0	18·2
	Tunisia	25·8	33·1	28·0
	Libya	0·4 (2·0 av. 1951–62)		5·5
	U.A.R.	99·0	85·0	
Asiatic Mediterranean	Cyprus	1·0	1·0	1·4
	Israel	24·5	21·9	26·1
	Syria	1·5 (1952)	0·8	1·0
	Jordan	0·2	0·1	0·1
	Lebanon	2·5	1·8	2·5
Mediterranean–Black Sea	Turkey	122·7	188·7	135·4
Black Sea	Bulgaria	26·8		
	Romania	36·4		
	R.S.S. Moldavia	1·6		
	R.S.S. Ukraine	387·4		
	R.S.S. Georgia	45·2		

were landed in European Mediterranean countries; and 1 507 000 tons were landed in Atlantic Iberia. The importance of the Atlantic fisheries could hardly be better emphasized, especially when it is realised that a considerable proportion of fish landings in Italy and Greece consists of cod and hake caught and frozen on the Atlantic.

The leading south European fishing nations are Spain, Portugal, Italy and Greece. France has a large catch but it is concentrated almost entirely into Atlantic ports outside southern Europe. Of the Spanish catch (1 357 400 tons in 1966), over three-quarters are caught and landed in Atlantic ports, and in 1968 the proportion was 85 per cent even when the landings in the Canaries were excluded. Nearly 40 per cent of all landings are cod, hake and haddock, much of which is either frozen or salted; about 17 per cent is pilchard (sardine) and anchovy; 9 per cent redfish and bass; and 5 per cent tunas, bonito and similar species. The haul of molluscs (14 per cent total or 190 000 tons) includes large quantities of common mussel and octopus The Portuguese catch consists mainly of Atlantic cod (nearly 40 per cent total), pilchard (26 per cent), horse mackerel (11 per cent) and hake (3 per cent). Tunas and molluscs, including oysters, have in recent years been relatively unimportant.

The fishing industry of Italy illustrates the transition between Mediterranean and Atlantic fisheries. Its total annual catch of 364 000 tons is far less than Spain's catch of Atlantic cod and hake alone. The Italian fish landings are dominated by anchovy and pilchard (21 per cent total); molluscs (18 per cent total); cod, hake, haddock and various unidentified frozen fish, almost certainly from the Atlantic (13 per cent total); and redfish and bass (9 per cent). Mullets, species of mackerel and of freshwater fish are of some importance and bluefin, tunny and bonito form only 1 per cent of the total landings.

A peculiar Mediterranean fishing economy prevails in Yugoslavia where the island-studded Dalmatian coast of the Adriatic is flanked by the formidable Dinaric Alps.

SPAIN: LANDINGS OF FISH BY REGIONS, 1968

Region	Weight (thousand tons)	Value (million pesetas)
Cantabria	207	4 767
Galicia	434	7 547
South Atlantic	213	4 374
South Mediterranean	49	894
Levante	45	1 356
(NE) Tramontana	53	1 523
Baleares	5	165
Total	1 001	20 628

11 *Bermeo, a fishing port in Viscaya, northern Spain*

Here nearly 40 per cent of the total national catch is obtained from freshwater sources, often lakes conveniently near the more populated regions beyond the coastal mountain range. The rest of the catch is dominated by pilchard (20 per cent total), anchovy (10 per cent total) and sprat (7 per cent total). Again the tunny forms about 1 per cent of the fish landings. A pure Mediterranean pattern of fishing must be sought outside southern Europe. Thus, the pilchard provides nearly 65 per cent of the total catch in Algeria and 38 per cent in Tunisia while the anchovy forms 53 per cent of the catch landed in Turkish ports.

On all counts the chief fish of the Mediterranean are the pilchard or sardine and the anchovy. These species haunt the warmer waters of the Eurafrican coastline but they are also truly at home in the warm, saline Mediterranean and, of course, are named after Sardinia. The sardine is especially sensible to thermal conditions, its optimal water temperatures being 14°C to 16°C. The total world catch of *Sardinia pilchardus* in 1966 was 611 000 tons. Of this 41 per cent was landed in Morocco, 20 per cent in Portugal, nearly 20 per cent in Spain, 6 per cent in France, just over 5 per cent in Italy, 3 per cent in Greece, 2 per cent in Algeria and the remainder in other Mediterranean countries. In the same year the total world catch of the North Atlantic anchovy (*Engraulis*) was 358 000 tons and of this 26 per cent was accredited to Spain, 20 per cent to Turkey, 11 per cent to Italy and nearly 4 per cent to Greece. Here again, however, it must be emphasized that the total world catch was far less than Spain's landings of cod and hake alone; and that in Spain over 75 per cent of the catch of sardine and of anchovy is caught in the Atlantic.

The problem always is to decide how much of the relatively small Mediterranean harvest is due to the poor nature of the sea itself or to the inadequate techniques of its fishermen. There seems no doubt that the benthos is mainly to blame. The Mediterranean has no true equivalents of the cold water cod and of the Atlantic flat fish. The sardine and anchovy are small substitutes for the herring. In addition the warm or hot climates around the Mediterranean cause the catch to deteriorate rapidly unless dealt with expeditiously.

Impact of fishing on industry and life

The impact of fishing on south European societies may be judged from the number of persons engaged and on the part fish plays in the national trading economies and in the national diets. Much of the fishing is part-time and reliable employment statistics are not readily available, nor can the size of the personnel be judged adequately from the number of fishing boats. The Italian fishing fleet consists of 19 000 motor vessels (205 000 gross tons) and 28 000 sailing boats; Portugal has 11 500 good-sized fishing vessels; in Spain in 1969 the registered fishing vessels numbered 13,500 with a total of 605,000 tons. The employment is highest in Spain with 106,000 professional fishermen, and in Portugal where 43 000 persons are regularly engaged in fishing.

Of the total world catch of fish, nearly one-third is sold fresh for human consumption, about one-third processed in some way (dried, cured, smoked, frozen, canned, etc.) for human food, and just over one-third processed for industrial purposes, such as fish meal. The use of the south European fish catch may be viewed against this global pattern. Here about 60 per cent of the catch is marketed fresh, about 35 per cent cured or canned, and 5 per cent treated for industrial purposes. However, if Portugal and Spain are omitted, the percentage sold fresh becomes enormous. In Italy 77 per cent of the catch is marketed fresh and in several countries there is no other method of disposal.

Spain is the only country in southern Europe that processes fish (128,500 tons) for industrial uses. For frozen fish, Spain (135 800 tons; 180 000 tons in 1968), Italy (48 600 tons) and Greece (27 000 tons) represent the nations with long-distance Atlantic trawlers equipped with freezing apparatus. For fish curing, especially cod, Portugal, with nearly 200 000 tons (230 000 tons in 1968) and Spain with 175 000 tons are quite outstanding but Italy (16 000 tons) and Greece (5000 tons) also cure smaller quantities. For canning and preserving in airtight containers, Spain (136 500 tons), Portugal (61 400 tons) and Italy (42 500 tons) have long been the chief south European producers but today Yugoslavia (18 200 tons) and Greece also partake in this industry.

The trade, as distinct from the landings, in fish reveals the value placed on fish as a food, as well as the culinary tastes of the societies. The total fish imports of southern Europe are about 930,000 tons while the exports are 335 000 tons. However, the exports are usually in a cured or canned form of relatively high value. Thus in Yugoslavia the value of the 13 000 tons of exports, mainly canned, almost offsets the cost of importing nearly four times that tonnage of fish. The chief importers of fish are Italy (294 000 tons), Spain (158 000 tons) and Switzerland (65 000 tons). By far the chief exporters are Spain (193 000 tons) and Portugal (106 000 tons); in both the value of fish exports was at least one-third greater than that of fish imports in 1968.

The part played by fish in the diets of the people of southern Europe varies greatly. Between 1960 and 1962, the average quantity of fish eaten per person annually was 48·4 kg in Portugal and 26·4 kg in Spain, both Roman Catholic countries where fish remains a popular and important item of the diet, although, of course, no longer favoured by papal decrees concerning Friday's diet. In Greece, the annual per capita fish consumption was 19 kg although this may be an underestimate in an archipelago where some of the catch may not be recorded. The corresponding statistics for Italy and France are 10·4 kg, and for Switzerland 8·7 kg. Yugoslavia has one of the lowest consumption rates in Europe (1·5 to 1·8 kg per head per annum) and the inhabitants of the Mohammedan states of northern Africa also eat very little fish.

The progress made in the fishing industry of southern Europe has been considerable but it has declined in world importance. Thus the Mediterranean and Black Sea catch rose from 700 000 tons in 1938 to 1 100 000 tons in 1966 but its proportion of the world catch declined from 3·3 to 1·9 per cent. In 1960 the total south European landings comprised 4 per cent of the world catch; in 1966 the percentage was 3·9 in spite of considerable national expansions. Thus since 1951 the Spanish landings have more than doubled, those of Portugal almost doubled and those of Italy increased by over 50 per cent.

During the last few years much initiative has been shown by many nations as well as by the Spanish and Portuguese who are by tradition great Atlantic fishermen. The Italians have also gone to the North Atlantic and West Africa with freezing factory vessels. The Yugoslavs have been encouraged in fish curing and in tunny catching by Japanese equipment and instruction. The Greeks have also been assisted by Japan and have been operating large-scale factory vessels in the Atlantic since 1960.

The distribution of the numerous fishing ports is described in the regional chapters. Hundreds of small fishing ports and a few larger ones retain old customs and a picturesque flavour of the sea. The culinary delights of many tourist resorts on islands, coasts and lake shores add greatly to the charm of small boats and fishable waters.

4 Climatic Characteristics

General pressure and surface airflow
Cool season: October–May

During the cooler months, especially in winter, the subtropical Azorean high pressure covers a large expanse of the Atlantic in and south of the latitudes of Iberia. North of this vast anticyclone stretches a wide belt of general low pressure associated with the westerly wind system. Here at the Polar Front and Arctic Front where airmasses of different temperatures meet, numerous cyclonic depressions develop over the Atlantic and move northeastward to the Arctic and eastward across Europe. When conditions in the upper troposphere are favourable these depressions intensify over the broad ocean into marked low pressure patterns with a warm front on their advancing side and a cold front at their rear. At the warm front the warmer southern or Tropical (mT) air is rising slowly above the colder northern (Polar or Arctic) air; at the cold front the colder Polar (mP) or Arctic (mA) air is pushing strongly into the warmer Tropical air. The warm front brings a progression of increasing cloud and gentle rain whereas at the cold front the weather is showery, variable and squally, with intense rainfall, and normally is followed by a change of wind direction to northerly, a drop of temperature and a relatively clear sky. The frontal pattern of a single cyclonic depression commonly extends for 3000 km or more. The succession of 'lows' carried by the Atlantic westerly belt is broken temporarily by the expansion northward of the Azorean anticyclone or of offshoots of it and by invasions of cold air (also of high pressure) from polar regions. Then for a short while drier, calmer, sunnier weather interrupts the successive spells of cloud, rain and changeable winds brought by the passage of warm and cold fronts.

Northwestern and central Europe lie well within this westerly conveyor belt throughout the whole year whereas the Mediterranean lands come fully under its influence only in the cool season and, for the most part, are free of it in the summer months. Yet even in the cool season the Mediterranean airspace tends to be different and distinct from that of the Atlantic. This is partly because the Mediterranean is weather-fended from the west by the great upland block of Iberia. At times depressions from off the Atlantic enter the Mediterranean Sea and move eastward along it but these are few in number. Much more commonly depressions either pass across northwestern and central Europe or originate over the Mediterranean. These two circumstances are closely interconnected. The cold fronts of large Atlantic depressions passing over northwestern and central Europe often send tongues of cold air into the relatively warm Mediterranean. Throughout the cool season these cold air invasions rush southward to the warm Mediterranean airspace. Usually they are concentrated into topographic gaps such as the Saône–Rhône corridor but, as will be shown later, the cold airmass sometimes glides right over the mountain rim on the north side of the sea.

These irruptions of cold air partly explain why so many depressions form in the cool season over the Mediterranean itself. The meeting of cold northerly airflow with warm Mediterranean air causes instability and a strong tendency to frontogenesis and cyclogenesis. But other cyclogenic forces also exist locally. In the Gulfs of Lions and Genoa the lee side of the great convex arc of the Alps tends, when cold air moves towards or against the mountains from the north, to experience low pressure and tropospheric con-

8 *Airmass sources of the Mediterranean airspace*

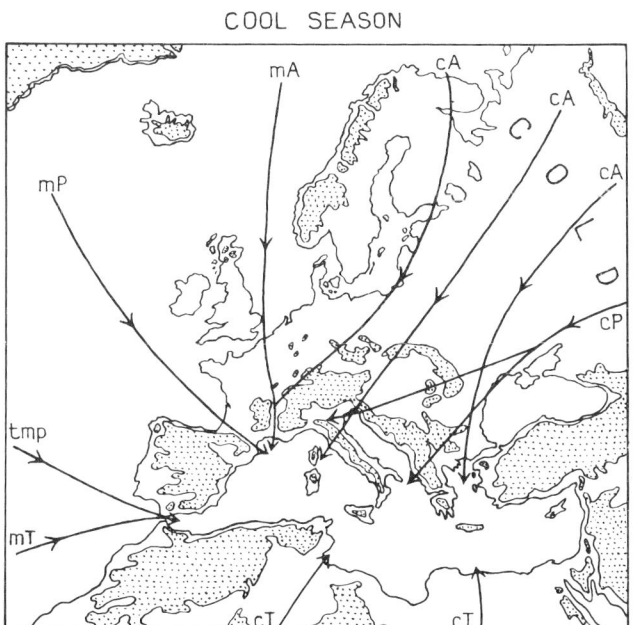
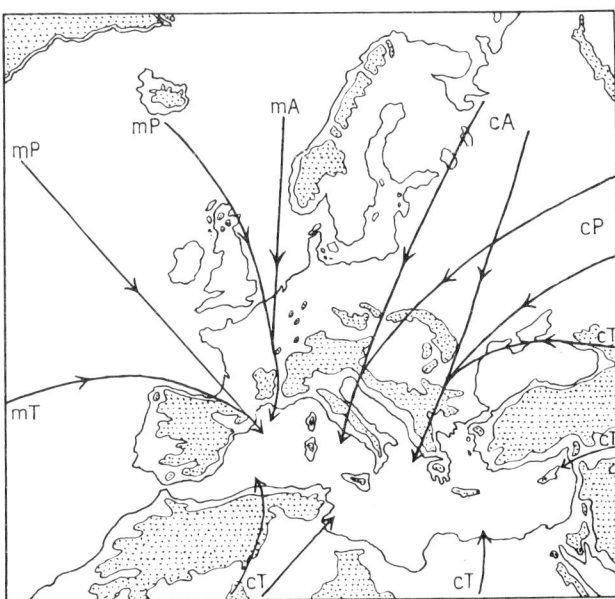

ditions favouring cyclogenesis. Normally, as mentioned above, the mountain mass diverts the lower atmospheric layers around its flanks. Whether the numerous depressions that form over the Gulfs of Lions and Genoa actually originated there *de novo* or whether many are the revitalized parts of fronts belonging to distant large depressions does not really concern us. It is, of course, common for secondary depressions to form upon fronts towards the perimeters of large primary 'lows'. Whether secondary or primary, the Mediterranean depressions soon develop into important weather systems that in the cool season dominate the airspace of the Mediterranean lands.

So far we have emphasized that the cool season weather of southern Europe comes mainly from the Atlantic and the Mediterranean. There is, however, some influence from Africa particularly when depressions form south of the Atlas Mountains and move eastward over the Ionian Sea and the Levant. Often also depressions moving eastward in the Mediterranean draw in hot African air (cT) on their advancing side as sirocco-type winds (fig. 8). In addition, there is a considerable influence from interior Eurasia. In winter a vast anticyclone tends to form over the parts of the landmass that are snow-clad for long periods. This high pressure, caused by the subsidence of air chilled from below, often expands over the Danubian plains. Then cold air masses (cP) may intrude also into the Mediterranean airspace, bringing bitter weather, occasionally with snow.

Warm season: May/June–September

In the warm season most of southern Europe is still dominantly under the influence of the Atlantic airspace but the distribution of general pressure has changed greatly. During these months, and especially from June to August, the Azorean high pressure moves northward and eastward and normally expands over Iberia and much of the western Mediterranean. Simultaneously over the Atlantic the Polar Front also migrates northward until it almost merges with the Arctic Front in the Iceland–Greenland area. Now, however, cyclogenesis is less intense and, although numerous 'lows' move eastward, they often lack the size and vigour of the winter depressions.

The pressure change over Eurasia is far more brutal. In the high sun season the heated Eurasian interior develops low pressure which encourages the Atlantic depressions to move inland across central Europe. Consequently the summers of northwestern and central Europe have an appreciable summer rainfall particularly as the heated land surface encourages the formation of thermal 'lows' or convective thunderstorm cells. But two important complications must be noted with regard to the airspace of central and southern Asia. First, much of the area is desert or semi-desert and when in summer their dry airmasses (cT) invade the Danube lands they bring hot, dry conditions. Secondly, in the furnace-like expanses of southwestern Asia and the Sahara there develops a marked low pressure, particularly over the Persian Gulf. Probably the Inter-Tropical Front or Convergence Zone is now situated in this area. Toward this great low pressure belt air flows from the lower Danubian plains and all the eastern Mediterranean. Then throughout most of the Adriatic, Ionian and Levant seas there exists a kind of Trade wind movement, with the etesians blowing strongly and steadily from a northerly quarter above a parched landscape. Thus it happens that in the warm season the airflow of the western Mediterranean is largely controlled by Atlantic Azorean high pressure while the eastern Mediterranean belongs indisputably to the tropical Asian and African airspace. In each the airflow is dominantly from the north. In each the airmasses are stable and subsident, partly because the sea is cooler than the European landmass and largely because in most areas the upper air is also subsident, being part of the normal atmospheric conditions where upper tropospheric subsidence in the subtropical anticyclones encourages at the surface an equatorward airflow.

The above account deals mainly with surface atmospheric patterns whereas in recent years meteorologists have been equally interested in conditions in the upper troposphere. Here jet streams or winds of over 30 m/sec. 110 km p.h. or 67 m.p.h.) are fairly common in polar and especially in subtropical latitudes. These normally blow eastward and are most persistent above latitudes 25°–35° but the position of the subtropical jet stream (as with that of the polar jet stream) migrates long distances largely in accordance with the overhead sun. In summer the subtropical jet stream often blows in the upper troposphere along approximately latitudes 42° or 43°N in the Mediterranean (Pyrenees–Tuscany). Almost all the Mediterranean lands are then under its right flank where the upper air has a tendency to be convergent or subsident and discouraging to cyclogenesis at lower levels. In winter the main branch of the jet stream commonly forms over latitudes 25°–30°N and the whole of southern Europe is then upon its left flank where the upper air tends to be ascending or divergent and favourable to cyclogenesis. The main jet stream and its various branches meander over all southern Europe at this season and their horizontal and vertical movements affect the formation and development of cyclonic depressions. The correlations, however, are by no means understood, particularly where high mountains influence upper airflow.

South European airmasses and weather systems

In the landlocked or almost landlocked regions of southern Europe, especially in Yugoslavia and the Po basin, the surface affects local weather but lacks sufficient uniformity and vastness to form its own airmass. The only depressions likely to develop over them are convective thermal lows in summer, which although small, often give heavy local thunderstorms. Consequently the general sequence of weather depends on airmass invasions from the Atlantic, the Mediterranean, the Arctic, and continental Eurasia.

In contrast the Mediterranean is wide enough (average 800 km) to form its own airmass but because of its landlocked nature its weather is also much affected by airmass invasions from outside. Indeed, the main role of the Mediterranean is to encourage instability and cyclogenesis in the cooler months and stability in the hottest months. The numerous cyclonic depressions formed especially in the Gulfs of Lions and Genoa in the cool season deepen rapidly and give strong frontal weather. However, typical Atlantic warm fronts with a relatively long period of increasing cloud, drizzle and steady rain are unusual even in winter. The warm fronts that develop over the Mediterranean often have massive cumulus clouds and, when associated with warm, moist southerly inflow, may bring torrential rain. Most of the fronts are of the cold type, with dense clouds, heavy precipitation, violent wind changes and a relatively quick return to clear conditions.

The actual tracks of the cyclonic depressions in the Mediterranean are hard to analyse as few maintain their original course and compactness for any length of time, and a small proportion fill up locally. The depressions are often classified as (1) Atlantic lows entering from the West; (2) thermal lows, which form mainly from May to October over the hotter parts of Spain, the Po plain and peninsular Italy; (3) lee depressions which develop on the lee side of the Alps, Dinaric Alps and other chains when a deep northerly airmass (particularly Polar and Arctic) moves towards and sends tongues of cold air through gaps in the mountain rim; and (4) wave depressions which form when a cold air mass traverses the mountain rim.

On a long-term average, about 76 separate cyclonic depressions occur annually in the Mediterranean area (fig. 9). Of these 7 (9 per cent total) enter from the Atlantic, 4 coming via the Strait of Gibraltar and 3 via the Toulouse gap. The other 69 (91 per cent) originate mainly as lee or lee-wave depressions within the Mediterranean area. About 14 (19 per cent total) originate south of the Atlas Mountains and of these Saharan depressions 8 move northward into the western and central Mediterranean and most of the others migrate eastward toward the Levant. No less than 52 (69 per cent total) originate over the Gulfs of Genoa and Lions and the northern Adriatic.

From this main area of cyclogenesis about 60 depressions move out annually. Of these about 11 move northeastward toward Hungary; another 4 or 5 migrate some distance down the Adriatic before turning northeastward across the Balkans to Romania. The remaining 45 move southeast to the central Mediterranean or Ionian Sea. Here about 51 depressions are recorded in an average year. Of these 30 take a track northeast toward the Black Sea and 20 move east to Cyprus and the Levant where several fill up. These tracks are shown in figure 9.

The average annual number of depressions gives a reliable picture of the distribution of unsettled weather in the Mediterranean. The proportion 60 for western Mediterranean and northern Adriatic, 50 for the central

9 *Quantitative-direction diagram of cyclonic depressions over the Mediterranean airspace* (modified from *Weather in the Mediterranean*, HMSO, 1962)

W *denotes winter occurrence. The movement of depressions from the northern parts of the western Mediterranean basin could be further refined into about 18½ per year southeastward from the Gulf of Lions and 26 per year southeastward from the Gulf of Genoa and northern Adriatic*

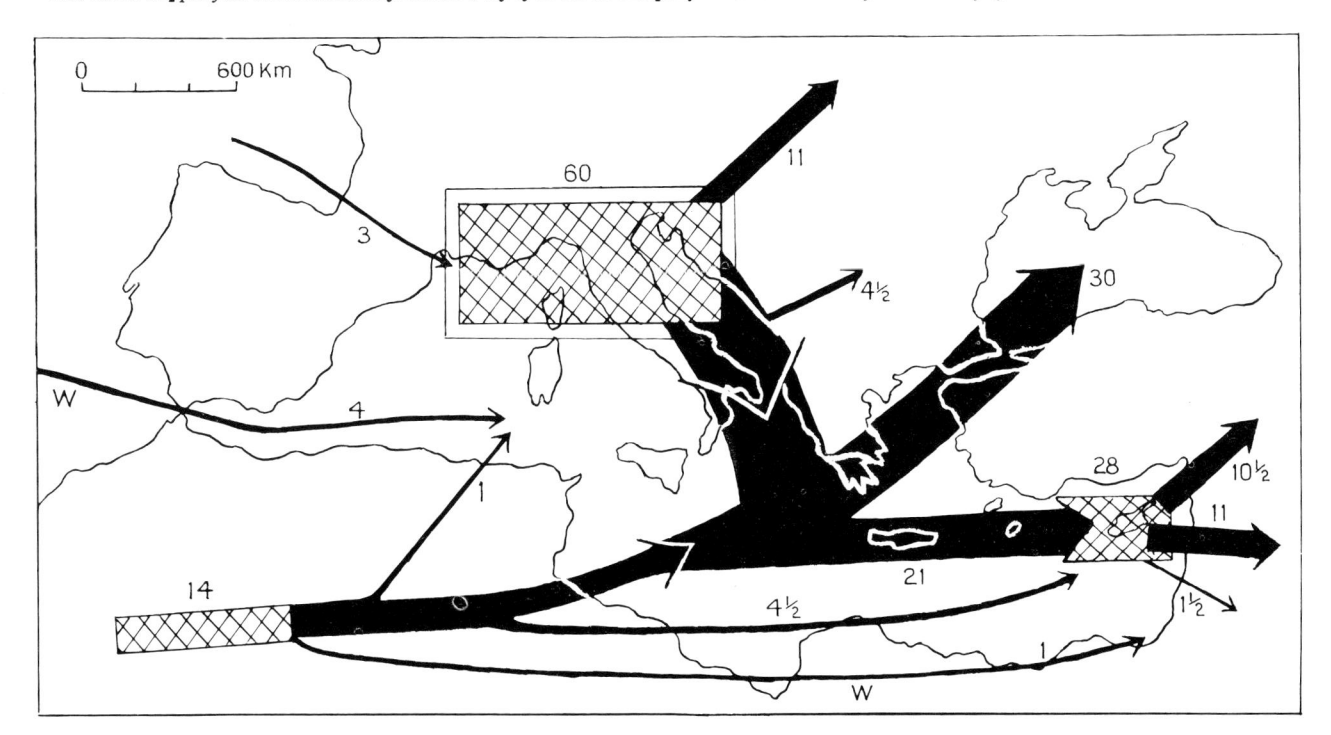

Mediterranean and 28 for the eastern Mediterranean reflects closely the decline southeastward of cloud, rainfall, raindays and length of rainy season or inversely the increase southeastward of sunshine, aridity and length of drought. In stressing unsettled weather, the frequent short and numerous long spells of settled weather must not be forgotten. These stable high-pressure conditions associated with clear skies and steady airflow are best described under surface winds, maxima temperatures and sunshine.

Airflow and orography

The interactions of airflow and relief are greatly complicated in southern Europe by the height of the mountain barriers and of the three Mediterranean peninsulas. As a rule the high western sides of these peninsulas extract much moisture from airmasses advancing in a general eastward direction. At these times the eastern sides are often in the rainshadow.

The main mountain systems are unfavourable to the maintenance of cyclonic depressions which begin to die out rapidly over mountainous relief. On the other hand, the formation of lee depressions over the Gulfs of Lions and Genoa is a major cyclogenetic factor in the Mediterranean, as also are the cold air invasions through the topographic breaks in the mountain rim.

When airflow crosses a mountain range, the leeside winds are warmed adiabatically by descent. Much is known today of this transmontane flow and its föhn effects. Where precipitation occurs on the windward side the latent heat set free partly offsets the cooling due to ascent: on the lee side the unsaturated air is warmed steadily at the adiabatic rate. The air arrives at the leeside foothills appreciably warmer than at similar altitudes on the windward side. This simple scheme applies to airmasses that traverse a mountain barrier with a more or less steady flow. Most airflow, however, is distinctly turbulential and creates large waves and eddies on the leeside of a barrier. These eddies may carry the airflow far higher than the mountain crests and then force it down to ground level where it arrives as an exceptionally dry, warm air. Exceptional föhn effects in many Swiss and Italian Alpine valleys must be attributed to the influence of lee-wave eddies.

The föhn effect is absent when cold continental air moves through low topographic gaps and when cold air resting on a plateau or snow-clad mountain surface gravitates downward to the piedmont. In these circumstances the descending airflow warms at the normal abiabatic rate but it has not been warmed by the addition of latent heat from precipitation on the windward side nor by descent from high levels on the lee side. It normally remains bitterly cold when it reaches the plains. Sometimes, as in the Rhône corridor, cold air gravitating downhill from the adjacent mountains accompanies a cold northerly airmass deflected by the Alpine barrier down the topographic gap.

The commonest routes for cold northern airmass invasions into the Mediterranean are the Saône–Rhône corridor and the break in the Dinaric Alps inland of Trieste. But it is important to notice that the latter gap is also followed by depressions from the Gulfs of Genoa and northern Adriatic passing northeastward to Yugoslavia and central Europe. In spring and autumn, depressions on this track may bring heavy rains to eastern Austria, Czechoslovakia and the middle Danube plains.

The general effect of orography on airflow is of great importance and, as will now be shown, is augmented by its widespread influence on surface winds.

Surface winds

The surface winds that most affect the inhabitants of southern Europe are on two scales, regional and local. Regional winds are associated either with a large-scale, seasonally persistent barometric gradient as in the eastern Mediterranean in summer or with the intermittent invasions of airmasses connected with weather systems. Local winds are caused mainly by the differential heating of local ground surfaces of a different nature or altitude.

Regional winds

The main large-scale pressure wind of the Mediterranean is the *etesian* of the Greeks and *meltemi* of the Turks, which blows from a northern quarter over Greece, the Aegean, the Ionian Sea, the southern Adriatic and the Levant, throughout the warmer months. The source is continental Eurasia, including the Caspian area, and in its southward progress the airmass remains dry and very stable. The winds bring a clear sky and perfect visibility and not until they have travelled for some distance over the sea do they begin to develop small cumulus clouds similar to those in the Trade wind belt. The airflow is as a rule steady, of moderate force, only occasionally whipping up to gale strength. When coming from seaward the etesians pleasantly moderate the summer heat but when coming over the land, as near Salonika and Athens, they can be hot and dusty. In the southern Adriatic 55 to 65 per cent of all winds in July come from some northern quarter. At Athens the etesians blow uninterruptedly on about five days in each of the months May, June and October, and on about twelve days in each of the hottest months July to September. In the southern Aegean they may prevail on 80 to 90 per cent of all days from mid-July to mid-August.

Most of the other regional winds of the Mediterranean are especially associated with the passage of cyclonic depressions along the Sea, whereby cold northern air is encouraged to move in at the rear of a depression and warm southern air to move toward its advancing edge.

The warm or sirocco-type airmass is derived from the Saharan and Arabian deserts. The most frequented tracks or directions are from Cyrenaica northwestward to Spain and from Tripolitania northward or northeastward toward Italy and the Balkans. In the western Mediterranean about

50 siroccos are experienced annually. The humidity of the airflow on its arrival in Europe depends largely on the speed and distance it has travelled over the Mediterranean Sea. If fast moving, it may bring relatively dry, hot, dusty air, with a yellow haze, to parts of southern Europe. The mountains of central Yugoslavia experience 'red snow' in most years and even in southern England dust particles and very rarely locusts are deposited. More commonly, however, the southerly airmass moves fairly slowly and picks up much moisture from the sea. Then it occasionally forms over the Mediterranean a warm front with low cloud, drizzle and rain typical of Atlantic depressions. Commonly, however, the front is moist and active, and on steep coasts brings heavy rain and thunderstorms. The general effect is always unpleasant.

In southeastern Spain this wind is called the *leveche*. Here it comes mainly in late spring and early summer and is normally dry and dusty owing to its short sea crossing. It seldom penetrates far inland but its desiccating effects on orchards and vineyards are dreaded. In the Gulf of Lions, the longer sea-trajectory causes the sirocco-type airmass (here called the *marin*) to be moist, with low cloud, rain and poor visibility. In the Gulf of Genoa, the same conditions are to be expected with the sirocco, except that the rainfall tends to be heavier and at times in winter the south wind brings gales and high waves. In Sicily the sirocco varies from very moist air with overcast skies and drizzle to very dry air with dust haze. The heat conditions are normally most oppressive on the south coast but occasionally the sirocco traverses the mountain backbone and descends on the north coast like a blast from a furnace. On 29 August 1885 during these föhn conditions Palermo had a relative humidity of 10 per cent and a temperature of 49°C. In the Adriatic the sirocco is nearly always moist, and fog and heavy rain are common in autumn and early winter.

The reverse of the sirocco is the cold northern airflow of mistral type. The mistral proper affects the coastlands of Mediterranean France and to a lesser extent other littorals near gaps in the mountains behind the coast between the Ebro delta and Genoa. In the eastern Pyrenees and the Balearic Islands it is called the *tramontana*. Around the Gulf of Lions these cold air invasions are mainly of maritime Polar and Arctic origin (mP; mA), and have lost much of their moisture in crossing Europe (fig. 8). The cold airmass often encourages a cyclonic depression to develop over the western Mediterranean and the resultant low pressure serves to strengthen the mistral inflow. Thus the mistral often begins as a showery, cold front and then settles in as a moderate, gusty wind bringing cool, dry, clear weather. However, in topographic gaps in highlands, funnelling may greatly increase the wind's speed as happens in the Rhône corridor and between Tortosa and Tarragona at the seaward end of the Ebro valley. In addition, where cold mountain air can gravitate down the same gap, this local *katabatic flow* may strengthen the general flow into gale force. This often happens between Montélimar and Valence in the Rhône corridor where the canalised wind rushes at high speeds through the defiles and accelerates further, with fierce local eddies, in the small basins alongside the river. In Mediterranean France mistral-type airflow is experienced on about 100 to 150 days annually. Strong mistral (over 21 knots) occurs normally on about 100 days a year, being commonest from December to April (average 10 days each) and least frequent in September and October (5 days each). Its spells may last for a few hours only but usually they persist for three or four days and rarely for several weeks. The wind reaches gale force at Marseille on about 10 to 15 times a year. The mistral landscape is characterized by tall windbreaks, usually of thickset cypress, and by windowless walls on the windward side of buildings.

The bora is a similar wind that enters the Adriatic from the northeast through a topographic break inland of Trieste and, to a lesser extent, over the Dinaric Alps father south. This invasion of continental Arctic and Polar air (cA; cP) is induced by relatively high pressure over central and eastern Europe and low pressure over the Mediterranean. As with the mistral, the cold air invasions often trigger off cyclogenesis over the warm sea and the resultant depressions strengthen the bora inflow. On its first arrival as a cold front, the air may cause heavy rain or occasionally in winter hail and snow but, once the bora has settled in, the sky over the lowland soon clears and will remain clear normally for several days and rarely for several weeks. Meanwhile the cold dry wind blows strongly and steadily apart from the occasional irruption of minor cold fronts which bring sudden squalls and gales with scarcely any cloud or rain. The bora frequently affects the western shores of the Adriatic between Venice and Ancona but here its velocity has diminished appreciably. Its prime domain is the eastern Adriatic shore from Trieste almost to Durazzo, where it is particularly violent in the north and wherever high mountains rise close to the coast. Bora spells commonly last for three or four days in winter and one day in summer. At Trieste in an average year, this wind occurs on about 40 days, of which half are experienced in December and February and only 3 from May to August. Here gusts have exceeded 110 knots and the corners of exposed pavements are provided with railings as handholds for buffeted pedestrians. In exposed localities in the eastern Adriatic, where a well developed bora will normally reach 130 km p.h., railways and agricultural fields are protected by high stone walls. On many islands off the Dalmatian coast the most exposed slopes are treeless.

In the northern Aegean, strong bora-type winds enter from a northerly quarter through the Dardanelles and down the Vardar valley especially in winter. The latter, known locally as the *vardarac*, occasionally brings cold, dusty weather to Salonika. Where this airflow passes transversely across mountainous peninsulas and islands, such as Mount Athos and the higher Cyclades, it sometimes forms on the lee side descending air currents (eddies) of a destructive and dangerous nature.

In the Tyrrhenian and Ionian Seas the bora-type wind from across the Balkans extends fairly frequently as far as Sardinia and Malta. However, here they are called *tramontana* if from the north and *greco*, *grecale* or *gregale* (the wind from Greece) if from the northeast.

In the western Mediterranean this northeasterly airflow continues often as far as the eastern coastlands of Spain, where together with other northeast winds due to other causes, it is called the *levante*. This wind in winter may bring cloud and heavy rain, with occasional snow, to eastern Spain. In spring and autumn it reaches gale force at times but in summer is usually mild and gentle with fine weather.

The regional winds described above originate outside the Mediterranean whereas there are some winds which blow for such long distances over the Mediterranean Sea that they may be considered to originate upon it. These commonly have an east or west trajectory as would be expected from the elongated shape of that sea.

Between Spain and Morocco in and near the Strait of Gibraltar the winds are commonly east or west. The east wind or *levanter* is a modification of the levante of southeastern Spain. It brings heavy rain in winter and even in summer may give low cloud, fog and sometimes drizzle. The west wind (*vendavale*) is an Atlantic wind that blows mainly in winter and not infrequently causes heavy rain.

On the western coasts of Italy and of the islands around the Tyrrhenian Sea west to southwest winds known as *libeccio* are common in all months. They are strongest in winter when they can be boisterous and wet. In summer they sometimes bring very hot weather to the lee side of islands, such as Corsica, when they have been superheated on descent to the coast.

The general temperatures of the regional winds or airflow described above may be judged from the table (left, below).

Local winds

Purely local winds are of great importance throughout southern Europe. The most widespread in occurrence is the katabatic flow of cold air downhill in mountains on calm clear nights. This *mountain breeze* then drains into topographic hollows and basins, including polja and doline, and displaces the warmer air on their floors with cold air so causing an inversion of temperature. Frost pockets of this kind abound in, for example, the Dinaric Alps. The reverse is the upslope or anabatic airflow (valley breeze) which during calm sunny daylight hours can cause convective clouds and thunderstorms on mountain summits particularly in the hotter months of the year. The mistral and bora are often strengthened by local katabatic winds. The summer maximum of rainfall on most central European mountains is partly associated with local convective updraughts and many of the lofty Mediterranean islands develop mountain and valley breezes. The fact that the downdraughts are more effective than updraughts at least in the eastern Mediterranean is due to the great stability of the summer etesian winds and the general upper air anticyclonic conditions at that season.

On the flatter parts of the Mediterranean seaboards the most characteristic local winds are land and sea breezes which may strengthen or weaken the prevailing regional airflow. The sea breezes are especially important as they blow during the hottest hours of the day and then make the climate of hot littorals bearable. They may extend inland for 15 or 35 km where the coastlands are low and flat but rarely develop strongly on steep coasts unless conjoined with valley breezes. At Malta land and sea breezes blow on about 60 days annually. At Athens, in spite of the prevailing northeasterly etesians, the sea breeze blows from the southwest uninterruptedly on about 69 days each year. In summer it may persist almost from dawn to dusk whereas in winter, when it is much less frequent, it blows only between noon and sunset. Without it at times of high sun Athens and the Acropolis would be infernos.

THE CLIMATIC ELEMENTS
Distribution of temperature
Summer

The long periods of daylight and the high angle of the sun's rays at noon (over 70° in June at 40°N) combine to make southern Europe definitely warm. The only areas with a mean temperature of less than 20°C for the hottest

TEMPERATURE CHARACTERISTICS OF MEDITERRANEAN AIRMASSES

Associated Wind	Airmass	Level 1000 mb Temperature °C	850 mb Temperature °C
JANUARY			
Bora: mistral	cA	1	−8
Mistral	mA	4	−6
Bora: crivat	cP	7	−2
	mP	11	
Vendavale	mPw	12	2
Vendavale	mT	–	10
Sirocco	cT		19
Libeccio. Levanter	Mediterranean	14	3
JULY			
Mistral	mP	–	18
Etesian	cP	26	13
Sirocco	cT	–	26
Libeccio	Mediterranean	29	19

month are parts of the maritime strip of northern Spain, the forelands of Switzerland and the uplands above 500 m in the north and 1000 m in the south (fig. 10). The summer warmth increases southeastward in the Mediterranean. Mean June temperatures of about 21° to 23°C prevail on the north Italian plain and Mediterranean France. But the southern peninsulas are much warmer, with average temperatures of 24°C or more for the hottest month, increasing to 26° or 27°C in the sunniest parts of the Aegean. The highest temperatures experienced locally do not exceed 37°C in coastal districts except in southeastern Spain, the Aegean and in lowlands liable to be affected by föhn or hot etesian winds. Everywhere in southern Europe on interior lowlands and low plateaux sheltered from maritime influences the absolute maxima exceed 38°C and in many areas rise to well over 40°C. In southern localities the highest known temperatures become truly impressive. Iraklion in Crete has experienced 46°C, Elvas in Portugal 47°C, Palermo 49°C and Seville 50·5°C.

Yet this summer heat is recorded in the shade and solid objects in the sun are considerably hotter. For humans this sunbaking is hard to endure and has encouraged shade-seeking customs, notably the habit in rural areas of rising early and finishing a job before the sun is high. More widespread are the midday siesta, and the late afternoon and evening perambulation when in the delightful coolness most shops and premises keep open, and restaurants are thronged far into the night.

Winter

During winter and spring the relative warmth of the Mediterranean Sea takes control of temperatures in the south and there are marked differences between the mild littorals and bleak interiors. The warmest coastal plains have January means of over 10°C, but averages of 7° to 9°C are more widespread, decreasing to about 5°C on coastlands exposed to the mistral and bora (fig. 11). Inland the temperatures drop rapidly yet in the lowlands the January mean remains at or just above freezing point even in the Po basin, by far the coldest Mediterranean plain in winter.

Beyond the mountains flanking the Mediterranean on the north, winters become appreciably colder. Average January temperatures on the Alpine forelands are just below freezing point ($-3°C$ to $-1°C$). On the interior mountains and plateaux the mean monthly temperatures remain below zero for four or five months at 1500 m and for eight or nine months above 3000 m. Throughout these colder interior lands the length of the growing season is controlled by frost and not by drought. The winter chill and spells of snow cover are a normal part of life. The houses are carefully heated in winter and the stock often kept indoors as the cold can be severe. There is, however, much exaggeration in the talk of the 'great cold' of the upland interior of the Spanish Meseta and the Balkans. The absolute minima, or lowest temperatures ever recorded locally, show this clearly. At Madrid it is $-12°C$, at Saragossa $-13°C$, at Burgos $-20°$ (at 887 m altitude) while just outside southern Europe in Hungary it is $-34°C$ at Baja on the Danubian plain and in Romania $-38·5C$ at Bod in the Brasov basin. Though it must be admitted that at Christmas and in early spring the night time chill at Madrid can be very piercing.

The Mediterranean coastlands appear mild by comparison. On the coldest northern coastlands of the Mediterranean the bora and mistral may rarely lower temperatures to $-7°C$ and very rarely to $-12°C$ near Marseille and about $-13°C$ at the head of the Adriatic. Everywhere on the coastal plains, the incidence and severity of frost depend on exposure to cold air invasions. At the head of the Gulf of Lions and of the Adriatic frosts occur in 25 to 30 days annually between early October and

10 *Mean actual July isotherms*

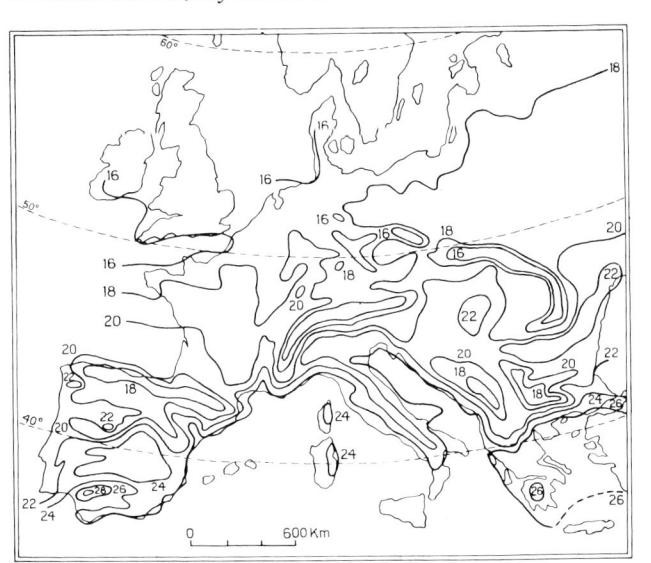

11 *Mean actual January isotherms* (after Köppen)

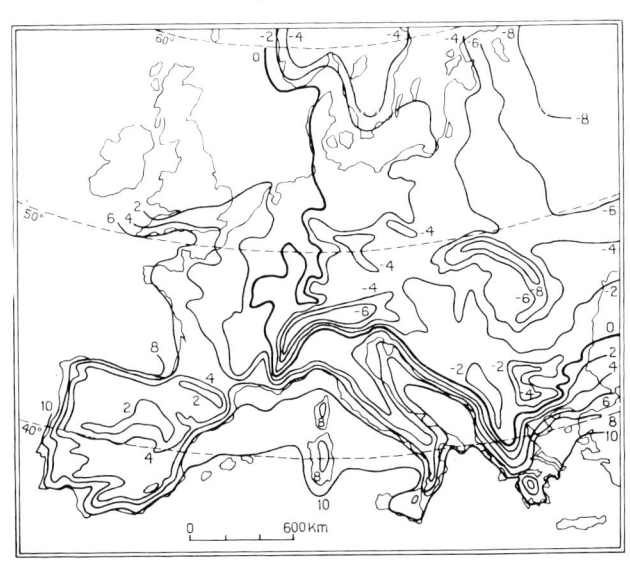

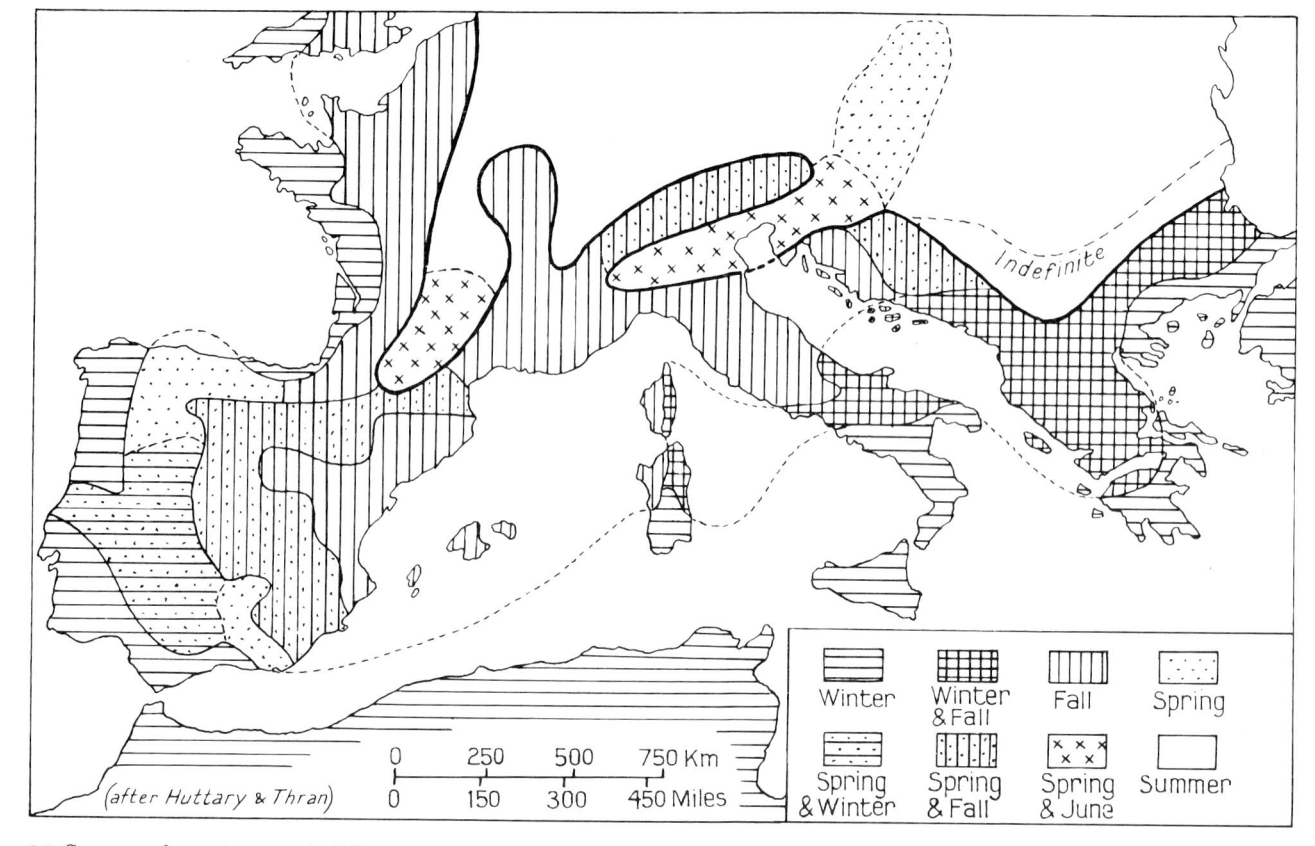

12 *Season of maximum rainfall*
The area marked as indefinite often lacks any marked seasonal precipitation maximum

March or April. On the riviera in the shelter of an unbroken mountain barrier, and on coastlands farther south, as near Barcelona, Rome and Athens, the frost period extends from November to March and the 'severity' rarely exceeds −6°C. Yet even in southern Iberia and the islands of the south Aegean the occasional frost still deserves the name for the nights feel uncomfortably cold to the inhabitants and ice may form on the pools beneath the fruiting orange trees in the Alhambra. The fact that in Mediterranean Europe only the more southerly islands, such as Malta, and coastal plains well protected from the north, never have frost is a tribute to the strength of the cold air invasions from over the Eurasian continent in winter.

Although infrequent, frosts play an important negative role in Mediterranean life. Where fairly common they prohibit the cultivation of citrus fruits, and strong cold winds occasionally damage existing citrus orchards. The olive is more tolerant but will not yield commercial fruits where cold spells are common. The vine has nothing to fear and succeeds well in the lower areas. On the other hand, many of the early market-gardening crops (*primeurs*) are disastrously vulnerable to unseasonable frost.

Distribution and nature of precipitation

The average annual precipitation in southern Europe ranges from under 250 mm in parts of southeast Spain to over 4500 mm on high exposed slopes of the Dinaric Alps in Yugoslavia. However, the extreme raininess of some coastal mountain ranges must not be allowed to hide the fact that most of the lowlands have annual rainfalls of less than 750 mm and that large areas in central and southern Iberia, and the southern Aegean normally receive less than 500 mm a year. The mean annual precipitation exceeds 1000 mm only in northwestern Iberia and throughout the more exposed parts of the Pyrenees and of the various mountain ranges north of the Mediterranean. A few localities elsewhere, such as Genoa, may normally receive over 1000 mm of precipitation annually but they are quite exceptional and experience exceptionally frequent frontal weather together with a local convergence of airflow.

When related to the high evaporation rates in the hotter months, the annual precipitation of most south European lowlands appears quite inadequate to maintain sufficient soil moisture for plant growth all the year. There is a total annual soil moisture deficiency of about 170 mm at Oporto; 200 mm at Coimbra, Nice and Barcelona; 300 mm or more in southern Portugal and Sicily; and well over 400 mm in the drier parts of Spain and Greece (fig. 18). If the water budget is expressed crudely as the ratio between the mean annual evaporation from a water surface and the mean annual precipitation, the general picture is equally discouraging. Whereas in pluviose Spain at Santiago the ratio is 890/1650 or about one-half, denoting a very satisfactory surplus for runoff and plant growth, throughout arid Spain the ratios (1520/432 at Madrid and 1880/360 at Murcia) show that the potential evaporation is at least four or five times the annual rainfall. However, this relationship is far too crude as the inadequacy or adequacy of precipitation for crops will depend also on the regime (monthly distribution) and nature of the rainfall.

Rainfall regimes

The true Mediterranean climate is unique among world climates in having its driest season when the sun is highest. Except on the plain of Lombardy and the coasts around the head of the Adriatic, the Mediterranean coastlands normally have less than 20 per cent of their annual precipitation in summer. Towards the south of the peninsulas the proportion dwindles to less than 15 per cent and equatorward of about 41°N latitude to less than 5 per cent. In the more southerly areas the drought lasts much longer than summer.

If 30 mm or less of rain in a warm month is considered a drought, the dry season lasts for one or two months in Mediterranean France, for two or three months in Campania and southwest Yugoslavia, for three or four months in lowland Sicily and Sardinia, for five months in southeastern Spain and six months in southeastern Greece and the southern Aegean. Throughout the extreme south the summers may be almost rainless.

This Mediterranean high-sun aridity contrasts vividly with conditions on the mountains and plains to the north where summer is usually the wettest season. In July the Danube plains normally have over 50 mm of rainfall and the Alps and Carpathians have 100 to 200 mm or more. In Italy the change from summer drought to appreciable summer rainfall occurs in the south of the Po basin where Milan, for example, has 180 mm of rain in summer. Here about twelve depressions pass over in a normal summer and although weak they aid local thermal convection to give rainstorms, often with thunder. The change, however, quickens north of the Alpine system where summer usually becomes the wettest season of the year. The summer proportion of the total annual rainfall increases rapidly inland from the Mediterranean seaboard. At Rijeka (Fiume) it is 20 per cent; at Zagreb 30 per cent; and at many places in Austria over 40 per cent. Moreover, on the mountains and interior plains from Basle eastward to Vienna, June or July is usually the wettest month of the year. Thus the rainfall regimes emphasize that in summer southern Europe has two different airspaces, a southern dominated by equatorward surface flow toward the Sahara and deserts of the Near East, and a northern still under the influence of the westerly belt that is conveying depressions toward interior Eurasia. The changeover between the two begins somwhere in latitudes 41° to 43°N.

The rainfall regimes in the seasons other than summer are difficult to summarize. Winter rainfall amounts seem in many ways inverse to those of summer (fig. 12). On all the interior mountains and plains of central Europe the coldest months are definitely the driest. In contrast, throughout the coastlands of all the southern parts of the Mediterranean peninsulas and of the Aegean, winter is the wettest season. Thus two clear-cut rainfall regimes can be distinguished: a truly Mediterranean with summer drought and winter maximum; and an interior or central European with a summer maximum and winter minimum. Between the two lies a broad transitional zone where the heaviest rainfall normally comes in autumn or spring.

This zone with equinoctial maximum rainfall includes some regions such as New Castile, where spring is the rainiest season but in most areas autumn is appreciably the wettest period. Throughout eastern Spain, Mediterranean France, central and northern Italy and western Yugoslavia the maximum rainfall is in autumn. Then the rainfall of exposed coastal districts in these extensive lands increases to 120 mm or more for at least one month and normally fails to reach 80 mm monthly in only the driest localities. The arrival of the autumn rains is a great feature of the climate of Mediterranean lands. During autumn an abrupt change occurs in the upper air circulation over the Mediterranean and the westerly belt shifts rapidly southward. At this season the Mediterranean Sea is usually warmer than airmasses from off Eurasia, and airflow from the north becomes warmed and moistened from below over its water. The first autumn rains often come with a cold front undercutting a moist sirocco-type airmass. The convective instability usually gives a short-lived downpour that in the following weeks is repeated with increasing frequency and duration. The change, however, is essentially of rainfall and not of temperature, for the autumns are genial and long. It should be mentioned that there is also a marked break in the precipitation regime in Alpine lands but this occurs at the spring and early summer rains and is often called 'the onset of the European monsoon'.

These differing rainfall regimes are of great human significance. Drought and sunshine in summer and early autumn are ideal for harvesting and holidaymaking. Precipitation maxima in winter, spring and early summer are favourable to crop growth and high yields. Rainfall in autumn creates almost as many problems as it solves. It replenishes springs and watercourses but it is often too late for beneficial crop growth, except the orange and olive, and too early for late harvests of grape and cereal; if it falls on bare stubble or fallow it may cause serious soil erosion.

In this respect the nature of the precipitation – its rate of fall (intensity) and number of raindays – is important. The precipitation of southern Europe is mainly associated with cyclonic depressions. Extensive warm fronts with gentle rain are common on the Atlantic seaboards and in central Europe but in the Mediterranean lands relatively few warm fronts bring this steady progression of cloud, drizzle and light rain. Here most of the frontal precipitation is of a heavy, showery nature and where the frontal uplift is increased by steep relief the resultant rains can be of tropical intensity. In the Po basin and throughout the Alpine countries during hot spells, thermal convection may occur which with, and sometimes without, the aid of weak passing lows, may develop into heavy thunderstorms. Consequently these areas tend to have much intense rainfall in the warmer season, and everywhere in southern Europe, except in northwest Spain, the number of raindays

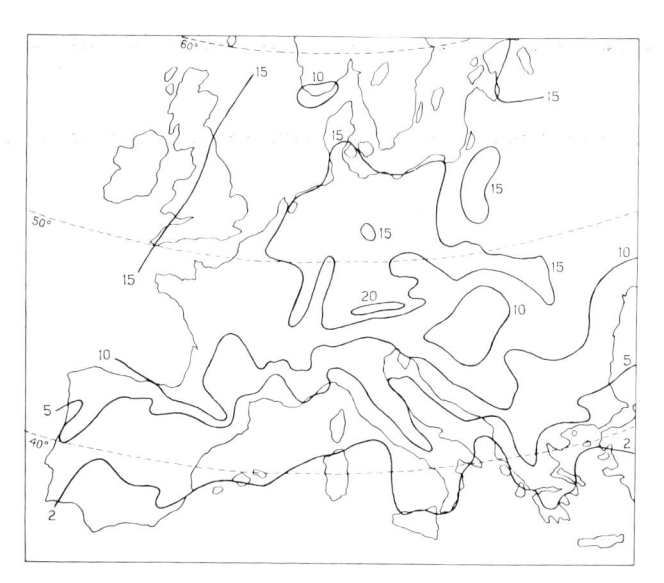

13 *Mean number of raindays in July*

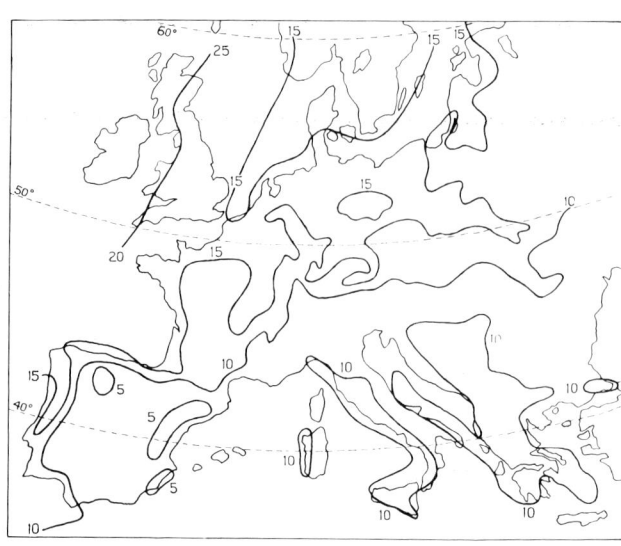

14 *Mean number of raindays in January* (after Köppen et al.)

is few compared with those at similar localities in the Atlantic northwest (figs. 13 and 14).

Coastal southeast Spain and low islands in the southern Aegean have less than 50 raindays in a normal year. The annual frequency rises to about 80 days in the southern peninsulas and to between 90 and 100 on most other coastlands, except those around the Tyrrhenian sea and northern Adriatic where the number increases to between 110 and 120. Apart from local rainshadows, the Pyrenees, Alps and Dinaric Alps have over 140 raindays annually with a maximum of about 180 on the higher, most exposed slopes.

If the annual precipitation is divided by the average annual number of raindays, the result gives a general idea of the intensity of rainfall. On the European coastlands of the Mediterranean the fall averages 6 to 10 mm per rainday and 4 to 6 mm in the drier eastern basin. These annual averages are, of course, greatly exceeded in the warm season in the Alpine countries and during the cool seasons in the Mediterranean lands. The normal intensity on Mediterranean coasts in the rainiest month may be judged from the following statistics. Cartagena 52 mm and 5 raindays; Málaga 102 mm and 6 raindays; Nice 147 mm and 10 raindays; Gibraltar 163 mm and 11 raindays; and Rome 138 mm and 12 raindays. Crkvice in the rainiest part of the Dinaric Alps has an average of 664 mm and 13 raindays in November. Similar rainfalls in northwestern Europe would be associated with at least twice this number of raindays.

The maximum rainfall in one day usually occurs in autumn or winter in the Mediterranean. Here probably all places have experienced more than 90 mm in one day; most localities have recorded over 115 mm and many up to 150 mm in twenty-four hours. Marseille once had 221 mm, Malta 294 mm, and Crkvice 441 mm in a single day.

In southern Europe the relatively intense rainfall is coupled with the irregular occurrence of heavy rain spells. Granada has had 630 mm and Jaén 500 mm in one month. Equally serious for human activities is the tendency of the rainfall to vary considerably from year to year and season to season. The mean annual variability of the rainfall is about 10 up to 15 per cent in central Europe and 15 to 28 per cent in Mediterranean lands. The annual rainfall at Granada has varied from 1220 to 305 mm and at Seville from 865 to 152 mm.

Snow

Snow is unknown in the warmest parts of the Mediterranean and does not occur every year on the southern coastlands of Italy and Greece. Its annual frequency increases to two or three days on the coast of the Gulfs of Lions and Genoa and to about six days at the head of the Adriatic and of the Aegean. Inland and upward, snowfall and snow cover become more common. Occasionally it may be deep enough to block roads on the Apennines and on the higher mountains in the various islands around the Tyrrhenian Sea. Above 1000 m snow lies for two or three weeks or more on the Abruzzi and on the Greek mountains north of the Peloponnesus. Some summits such as the Gran Sasso may keep patches of snow almost throughout the year but the Sierra Nevada is the only Mediterranean range high enough to maintain a permanent snowcap.

None of the Mediterranean lowlands has much snow, the snowiest being the plain of Lombardy where it falls on about nine days annually. However, on the interior uplands of northern Spain and the central Balkans snow is important in winter. On the northern Meseta in Old Castile most places have 10 to 20 snowy days annually and road passes across the central sierras may be blocked occasionally. On the Pyrenees the snowline descends to below 1000 m in midwinter and retreats to 2600 or 2900 m in late summer. The effect of snow on traffic along the main Pyrenean passes is indicated in the following table.

Road	Pass	Height in metres	Period liable to road-blocking
Barcelona–Puigcerda	Tosas	1800	Dec.–March
Andorra–Ax les Thermes	Envalira	2407	Nov.–June
Balaguer–Saint-Béat	Bonaigua	2072	Nov.–June
Biescas–Laruns	Sallent*	1791	Nov.–June
Jaca–Oloron	Somport	1632	Nov.–May
Pamplona–Saint-Jean	Roncesvalles	1057	Nov.–Feb.
Pamplona–Bayonne	Velate	868	Dec.–Feb.†

* Pourtalet   † occasionally

Snow cover becomes of more general importance on the interior mountains and plains of southern Europe. In the western Alps the summer or permanent snowline is at 2750 to 3200 m according to orientation and precipitation. The snow cover descends to about 2000 m by late September and to about 600 m in early December, except in the mild climate of Provence. Throughout the Italian Alps heavy snowfalls may occur above 1500 m from October to April. Here the snowfall tends to be heaviest at 2000 to 2500 m and in the east where, on the Rolle pass (1970 m) in the Dolomites its maximum depth normally reaches 4 m. During the snowiest weeks of the cold season snow lies more or less everywhere in the Alpine lands. Basle on the Rhine normally has 30 days with snow. Similarly on the Danube plains in Yugoslavia snow falls on about 15 to 30 days and lies on the ground for about 25 to 50 days between November and March. Yet here occasionally the colder months may be almost snowless and then the autumn-sown crops lack protection from frosts.

Today the deeper more persistent snow on mountains has acquired a great economic importance for skiing. This happens, for example, on the Gran Sasso and Sierra Nevada as well as on the more noted resorts of the Pyrenees, Alps and Carpathians. The tendency for precipitation to be least in winter and of snow to create a subsident airmass with clear sunny skies enhances the attractiveness of snowclad slopes. What were up to the early twentieth century deserted landscapes are now the meccas of an ever-growing body of tourists. Nor do these profitable sporting uses detract from the snow's prime advantage for man. The storage of cool season precipitation as snow and ice provides meltwater for irrigation purposes in spring and summer.

Solar radiation and sunshine

The potential sunshine is a factor of latitude (altitude of sun), but the amount actually received at ground level depends largely on the thickness of the cloud cover.

INSOLATION IN WESTERN EUROPE AND MEDITERRANEAN

	Mean hours sunshine		
	July or June	Nov. or Dec. or Jan.	Annual total
London [Kew] (England)	203	38	1466
Berlin (Germany)	247	34	1614
Prague (Czechoslovakia)	262	42	1773
Berne	256	58	1781
Vienna (Austria)	265	46	1782
Kalocsa (Hungary)	281	57	1986
Coimbra	312	133	2510
Lisbon	345	132	2741
Madrid	387	157	2909
San Fernando	335	144	2697
Perpignan	271	113	2201
Turin	220	56	1618
Rome	348	107	2362
Messina	336	102	2358
Lecce	351	109	2468
Gorizia	258	63	1843
Trieste	315	94	2205
Split	354	115	2658
Ulcinj	360	68	2693
Athens	362	125	2737
Zakinthos	420	145	3107
Sarajevo	243	55	1686
Sofia (Bulgaria)	313	52	2049
Salonika	365	112	2555
Andros	417	94	2748
Thira	431	102	2943
Iraklion	419	113	2804
Alexandria (Egypt)	373	161	3119

Cloudy climates as in northwest Europe and the equatorial selvas receive relatively small amounts of solar radiation or sunshine. The maximum annual solar radiation (200 to 220 kilolangleys*) at the earth's surface is in

---

* A langley or 1 ly/min, is equal to 1 cal/sq cm/min; a calorie or gram-calorie being the heat required to raise the temperature of 1 gram of water by 1°C (from 14·5° to 15·5°C).

Recently, the langley and calorie have been replaced by standard international units expressed as energy (Joules) and mechanical power (Watts). 1 W is equal to 1 J of work done per second. 1 calorie is equal to 4·186 J and 1 cal/sq cm/min is equal to 0·06976 W/sq cm.

cloudless deserts near the Tropics. For horizontal surfaces the corresponding amounts are 70 to 100 kilolangleys in northwestern Europe, 120 to 160 in the equatorial selvas and 130 to 175 in southern Europe. Much of southern Spain, Sicily and southern Greece receive annually considerably more solar radiation than do the hottest parts of the Congo and Malaya.

These thermal conditions are well expressed also by the duration of sunshine. The following table shows that in a general sense sunshine amounts increase eastward and southward in Europe. The statistics relate to the sunniest and the cloudiest month.

The statistics more or less speak for themselves, except perhaps the relative lowness of the number of hours of sunshine in the interior parts of the plain of Lombardy (Turin) and on the highlands in interior Yugoslavia (Sarajevo). These quite exceptional areas, situated close to high mountain ranges, are liable to experience thick rolls of cloud formed when airmasses cross the mountains, and have rainfall maximum in June or May which are elsewhere very sunny months.

The Mediterranean lands except in the Po basin, are slightly sunnier than interior Europe in summer and appreciably sunnier in winter. Cicero said that the sun shone at least once every day in Sicily. Today, as then, the frequent short sunny spells in the cool season, which even in December raise noonday temperatures to between 12° and 16°C on the coasts, attract numerous visitors. The sunny summers are, however, the main climatic attraction in the Mediterranean. In August few Mediterranean towns have more than 3 cloudy days and throughout the southern peninsulas at least 21 to 26 days can be confidently expected to be cloudless. Over much of the etesian-swept Aegean and Ionian Seas the sky remains blue for long periods. Nor does the western Mediterranean lag far behind in high-sun aridity and sunshine. Normally during the whole summer there are only 2 days with rain at Malta, 4 at Cartagena, 5 at Syracuse, 8 at Palermo and 13 on the west coast of peninsular Italy.

This arid, sunbaking season has many interesting aspects apart from its attraction for tourists and the deficiency of soil moisture for crops. The burden of irrigating certain crops is at least partly offset by the high quality of the fruits and the perfect harvest weather. The sunshine also benefits the inhabitants, as direct solar radiation is, in moderation, beneficial to health being a means whereby the body acquires vitamin D.

The intensity of solar radiation has an important relation

III *The beach at Riccione on the Adriatic near Rimini, Italy. The light canvas erections act as sunshades*

IV *Skiing at Chamrousse in the French Alps*

to surface slopes. Hillsides oriented toward the sun gain in insolation and those facing away from the sun lose insolation in proportion to their gradient. In latitude 40°, as the maximum angle of the noonday sun is 73° (in June), southward slopes with a gradient of about 17° receive perpendicular rays from the sun. Such slopes in the southern Mediterranean peninsulas and islands during clear weather have the maximum insolation possible for the latitude and because of the length of daylight, become excessively heated. Throughout Provence and the northern Apennines the maximum insolation would be on southward slopes of 25°. Slopes of this steepness could not be cultivated without elaborate terracing but on cliffed coasts they help to explain the intense warmth of coastal inlets and platforms backed by mountains. The fact that gentle equatorward slopes gain considerably in insolation on sunny days is, however, probably of more importance.

The use of direct solar radiation for energy purposes has made great strides in recent decades. At Mont Louis and at Odeillo high up in the eastern Pyrenees the French have developed large solar furnaces. Here by means of mirrors arranged in a parabolic shape (like a shallow funnel) the solar radiation is reflected and concentrated on to one spot. The resultant beam can give temperatures up to 3500°C, capable of melting steel. As yet, small solar stoves of a crude, simple, portable type, which are used in many tropical countries, have not become popular in southern Europe where people prefer to have their main meal in the cool of the evening. However, solar energy is being used here increasingly for the evaporation of sea water for drinking supplies (see Chapter 5) and the traditional collection of salt from coastal saltpans continues to flourish as well as the outdoor drying of certain fruits.

Climatic regions

The climate of southern Europe is mesothermal except on the higher mountains. The climatic regions and their variations are as follows:

1. Mesothermal with all-the-year rainfall

This climate is characterized by a frequent succession of cyclonic depressions, with numerous fronts, appreciable cloud cover and rainfall all the year. Mean monthly temperatures rarely exceed 22°C. The three main variations in southern Europe are:

1*a. The Atlantic seaboard* of pluviose Spain where rainfall is usually least in summer and most in late autumn and winter. Winters are mild, summers cool to warm and raindays relatively frequent.

1*b. The interior plains and forelands of central Europe* where summers are warm with a rainfall maximum and winters cool to moderately cold with a precipitation minimum.

1*c. The Plain of Lombardy* where normally most rainfall

comes in autumn and spring and least in winter. Summers are warm and moderately wet and cloud cover relatively abundant all the year.

2. Mesothermal summer-dry climates

These are the coastlands and the peninsulas and islands of the Mediterranean with a marked summer drought, relatively abundant sunshine and a low number of raindays. The chief subdivisions are:

2a. Southernmost areas with a summer drought and marked winter maximum of rainfall.

2b. Interior and transitional areas with summer drought and a marked maximum rainfall in autumn or spring or both.

3. Highland climates

Everywhere in southern Europe altitude brings an average decrease in temperature of about 5° or 6°C per 1000 m and an appreciable increase in precipitation. This causes the true Mediterranean climate with a marked summer drought (one month or more in hot season with 30 mm or less or precipitation) to be restricted mainly to areas below 1000 m. Thus owing to the increase in rainfall to over 30 mm in all months the higher parts of the central and northern Apennines are often grouped climatically with the Po basin, whereas in fact the central Apennines are merely a wetter version of a Mediterranean summer-dry climate, and the higher northern Apennines are transitional between it and the climate of the plain of Lombardy. Similar difficulties of classification in mountains arise in Alpine Europe. Here, however, increase in altitude soon lowers the mean temperature of the coldest month to below −3°C and lengthens the duration of snow cover to several months, so that most of the mountains can be safely classified as microthermal.

However, at about 1500 m or 1600 m mountains begin to make their own climates. Above these altitudes the thinness of the atmosphere usually allows intense heating by day and rapid cooling by night. At night the temperatures drop rapidly and frosts may be common in summer and normal in winter. By day during clear weather objects in the sun get hot and those in the shade remain cold. Up to about 2000 m or 2400 m the mean temperature of the warmest month (recorded in the shade) may exceed 10°C and trees grow well in spite of cold nights. However, higher up the mountain side, tree growth gives way to stunted shrubs and Alpine meadows. Here for a few months the mean monthly temperatures just exceed freezing point although frosts may occur every night. At about 2600 m or 2900 m, according to latitude and orientation, is the permanent snowline where on clear summer days the sun burns the skin fiercely but shade temperatures barely rise above freezing point and surface temperatures on the snow and ice remain at zero and drop far below zero after nightfall. The meadow zone has long been used for pastures in summer by transhumant flocks and herds but the sterile summits have only become popular since the early nineteenth century. Today mountains and mountaineering are as important to highland economies as snow and skiing.

5 Rivers and Water Supply

Rivers and drainage basins

The rivers of southern Europe, as demarcated in this volume, drain mainly to the Mediterranean. In addition the major part of Iberia drains to the Atlantic, much of Switzerland to the North Sea via the Rhine, and much of Yugoslavia to the Black Sea via the Danube.

The European Mediterranean is remarkable for the smallness of its drainage basins which are generally short and steep. Only the Ebro, Rhône–Saône and Po rise any distance inland and these rivers are in a sense dissimilar exceptions. The Ebro (927 km; 86 000 sq km basin) crosses a sizeable mountain range near its mouth; the Rhône–Saône (820 km long) follows a narrow tectonic trough; the river Po (675 km; 70 000 sq km) flows mostly over a flat plain built up recently from its own alluvium. The drainage away from the Mediterranean rim is far more integrated. In Iberia, the Tagus (1008 km; 81 000 sq km basin), the Douro (937 km; 97 000 sq km), the Guadiana and the Guadalquivir have a reasonable length but there is not a single river in Mediterranean Europe with a drainage basin exceeding 100 000 sq km. Conditions are very different in central Europe where the Danube is 2776 km long and drains 817 000 sq km. This fine river collects up the drainage of nearly all the Austrian Alps, the Carpathians, the Dinaric Alps and the great alluvial basins in Hungary, Yugoslavia, Bulgaria and Romania. After the Volga it is easily the second largest river in Europe for length and basin size.

It seems obvious that the runoff from hundreds, if not thousands, of small drainage basins on the Mediterranean seaboard will be expensive to control as engineering costs are likely to exceed the economic value of the benefits received. On the other hand the larger basins everywhere are likely to be more attractive and profitable for would-be controllers. However, the general difficulties facing hydraulic engineers in southern Europe are considerable as may be judged from the following account of the hydrological characteristics of some of the rivers.

The Iberian peninsula has 1800 streams of which about 90 exceed 100 km in length and only one, the Tagus, exceeds 1000 km. Most of the rivers, except in the north and northwest have a relatively small and irregular flow. Their valleys are deeply incised and steep-sided, being either old incisions in a semi-arid lofty plateau or new gashes cut into recently uplifted coastal mountains. In arid Spain most of the smaller drainage basins dry up in summer but in the cool season their watercourses may occasionally fill with startling rapidity. Then for a few hours or days or weeks a turbid torrent rushes to the sea or nearby plain.

The general drainage slopes of Iberia differ widely in character. The Cantabrian slope carries short swift streams, with relatively abundant volume all the year and a reputation for trout and salmon. In Galicia the Minho (310 km) also has a fairly constant volume and a large runoff (18 litres/sec/sq km). At its mouth it has double the flow of the Guadalquivir which has a drainage basin four times as big. On the main Atlantic slope of Iberia, the Douro benefits from its headstreams in pluviose Spain. Its flow at Oporto is about 650 cub m/sec which is equivalent to about 6·5 litres/sec for each square kilometre of its basin, a quantity greater than that of the Seine in France. The other three large rivers of the Atlantic slope mainly traverse arid Spain and in spite of high headstreams carry less water than the Douro. The Tagus at Lisbon has an average annual flow of 500 cub m/sec, or the equivalent of 5·7 litres/sec/sq km. The annual runoff of the Guadiana is only 1·7 litres/sec/sq km and that of the Guadalquivir, in spite of snowmelt from the Sierra Nevada, averages only 3·5 litres/sec/sq km. In some ways the Guadalquivir is the most useful river in Spain as its lower course lies over a coastal plain with wide fertile terraces and a gentle gradient that allows the tide, and shipping, to reach Seville 85 km upstream. The Tagus and Guadiana typify the rivers of the southern Meseta. Their banks are steep and their flow, previous to the erection of huge regulating dams, varied from a raging flood in winter to a series of gentle reaches, not hard to ford in summer. Headstreams such as the Manzanares at Madrid often dwindle to a wide sandy *rambla* (dried-up river bed) in time of drought but may become a raging torrent after heavy rains. Hence the Spanish joke: 'You have to run to see the Manzanares'.

The irregularities of these Meseta streams are, however, far eclipsed by those of the short rivers draining to the Mediterranean. In Andalusia some of the streams drop 2400 m in 65 km. The steep upper courses thread narrow canyons while the lower courses embouch on flat deltas. Beds that have been dry for long periods may carry sudden floods heralded by a wall of water several metres high. By far the chief river entering this coast is the Ebro which because of its numerous central Pyrenean tributaries has a fairly regular volume and a flow at its mouth of 615 cub m/sec or the equivalent of 7·4 litres/sec/sq km for its drainage basin. The value of heavy rain and snowfall is demonstrated still more clearly in the Alps where the Rhône has the high average annual runoff of 18 litres/sec/sq km and the Po of 24 litres/sec/sq km including water taken for irrigation.

The rivers of peninsular Italy exemplify clearly the nature of the drainage of a newly uplifted narrow range and of newly upraised marine platforms. From the uplands numerous relatively small streams rush independently to the coast and wherever the Pliocene platform is exposed quite gentle short streams also drain independently particularly to the Adriatic. On the higher backbone only the Tiber and Arno have developed sizeable basins with a more complex river pattern. But the great floods at Florence in 1966 revealed the liability of small steep basins to rapid runoff.

Perhaps the nature of rivers in a land newly affected by orogenesis can be seen even better in the Balkan Peninsula. Here a coastal strip of the lofty Dinaric Alps and Pindus

Mountains drains to the Adriatic. Much of the terrain consists of massive limestones, riddled with faults and joints, where most of the drainage is underground. The few surface rivers are short and swift and in the whole length of the Yugoslav coast only the Neretva (230 km) breaks through from the interior ranges to reach the Mediterranean. In parts copious freshwater springs rise in the sea floor.

The Balkan drainage to the Aegean bears witness to the effect of recent faulting. The Vardar valley, for example, is related to a wide fracture zone along the western edge of the Rhodope block. Its upper basin is a series of broad flat plains (fault-basins formerly occupied by lakes) that alternate with deep winding gorges. Similarly in Greece the valleys of most rivers flowing to the Aegean consist of an alternation of wide plains with sluggish rivers and narrow ravines with fast torrents. In some places rivers, such as the Pinios, have maintained their courses across a newly uplifted coastal range. There often seems to be a mixture of antecedence and fault disruption. Indeed the characteristic Greek river has a flat, wide upper course on a peneplained highland, a stepped middle course where flat-floored fault-basins alternate with deep gorges, and a flat lower course that meanders wildly upon a small coastal plain. In some areas the fault-basins still have interior drainage but in most the water has accumulated until it overflowed and eventually cut gorges that link up with seaward rivers.

The relatively new formation of the majority of rivers and river plains in southern Europe is of great human significance. The flat alluvial plain of the Po and of alluviated graben (fault-basins) elsewhere need artificial drainage and elaborate protection from floods. On the other hand, the gorges often are too steep and too narrow to provide easy routeways and in many uplands the most formidable elements of the landscape are the canyon walls and fault scarps. Except in glaciated tracts, it is not uncommon for routes to seek the adjacent summits, which often are flat plateaus, rather than thread a tortuous path along the precipitous sides of the gorges. However these gorges with steeply graded floors provide excellent sites for storage dams and hydro-electricity stations, particularly in areas of impervious rock.

River regimes

The usefulness of a river to the people living nearby depends on its regime as well as on its volume. The regime, or variations in volume during the year, largely depends on the rainfall, size, steepness, surface cover and geology of the drainage basin. It is, in other words, a factor of runoff at the surface and from springs. In southern Europe the average annual runoff is equal to over 800 mm of precipitation in the Alps and to less than 200 mm in the extreme south of the Mediterranean peninsulas.

The types of river regime fall into four main groups:

1 pluvial oceanic with a slight minimum in summer
2 Mediterranean with a marked minimum in summer
3 highland glacial and nival, with a marked maximum in spring and summer and marked minimum in winter
4 nivo–pluvial and pluvio–nival with two maxima due to snowmelt in spring and rainfall in the cool season

1 The pluvial oceanic regime is characteristic of the coastlands of pluviose Spain. Here a summer minimum, when evaporation is greatest, is followed by maximum flow in autumn or winter when rainfall is heaviest. However, even in summer, except in abnormally dry years, the rivers and streams maintain a vigorous flow.

2 The Mediterranean regime is characterised by marked low water in the hotter months when watercourses in many drainage basins of up to about 3000 or 4000 sq km may become dry and in large areas even the main rivers may dwindle to small streams. The summer low water is strongest and longest in the drier southern parts of the peninsulas where there is a winter rainfall maximum. The Mediterranean regime does not have a characteristic season of maximum runoff. In areas with a double maximum of rainfall in autumn and spring the rivers usually have most volume in spring particularly if some snowmelt occurs. Thus even in eastern Spain, southern France, central and northern Italy and western Yugoslavia where normally most rain falls in autumn, the soil then imbibes so much moisture and the water table needs so much replenishment that the higher river flow comes in spring. Only where the autumn rains are exceptionally heavy, as in the Cévennes, southeastern Spain, the Ligurian coast and parts of coastal Yugoslavia does the autumn spate exceed that of spring. But it must be repeated that irrespective of whether the higher water comes in spring or in autumn, the distinctive criterion is the marked summer minimum (fig. 17). The only appreciable natural modification of this disastrous deficiency occurs where thick limestones store water underground and give it off gradually as on the Nera and Tiber near Rome and on the upper Guadiana.

3 Glacial and nival regimes that are dominated by snow and ice are largely restricted to the highest mountains above the permanent snowline. However, their extent is increased by the fact that glaciers descend far below the permanent snowline before melting. The characteristic glacial stream has a strong maximum of flow in the hottest months and a very strong minimum in the coldest months, when the only surface movement may be as ice in the glaciers. A further characteristic of a highland glacial regime is a diurnal increase in flow in the early afternoon and a decrease after sundown during warm days in summer. This regime occurs on many small headstreams of the upper Rhône and Rhine in the Alps and, for example, in a few higher basins in the central Pyrenees.

The highland nival regime, which depends on snow

only, is very similar to the glacial in geographical distribution and type except that the spring and early summer high water drops more rapidly owing to the absence of continued icemelt. The streams have a large volume from May to June and a marked low water during the snowy months.

4 Nivo–pluvial and pluvio–nival regimes are much more widespread than highland regimes. The influence of snowmelt is weak in the warmer parts of the Mediterranean and is best seen on the Genil, a tributary of the Guadalquivir from the Sierra Nevada. However, the spring meltwater becomes important in the colder parts of northern Spain, and in the Po basin and the interior Balkans. In fact most of the highland rivers in Europe are controlled partly by snowmelt which proceeds up the mountain slopes in spring. In Alpine Europe a complication occurs because the mountains and plains tend to have a marked warm season maximum of rainfall. Consequently the coincidence or succession of snowmelt and rainy months tends to offset the increase of evaporation which no longer dominates river regimes. Here the prime features of river flow are the warm weather high water and the cold weather low water, a direct contrast to the Mediterranean regime. Inevitably there are many complications especially where streams with a warm-season spate flow into Mediterranean climates.

Nivo–pluvial regimes occur on the middle mountain slopes and on the plains and plateaux of colder areas where snowmelt in spring (usually May) causes a maximum flow that exceeds a secondary maximum in autumn or winter which is due to increase of rainfall or decrease of evaporation.

Pluvio–nival regimes originate on the lower slopes and cool plains where snowmelt in March or April causes a sharp flood but the flow rises to a slightly higher level in autumn or winter due to heavy rain. The winter flow also is distinctly less impoverished than in the nival and nivo–pluvial regimes. A retrogression of month of maximum flow will be noticed in all these regimes that are influenced by snow, from June-July in glacial, May-June in nival, April-May in nivo–pluvial and March-April in pluvio–nival.

15 *The regimes of the Ebro* (after Pardé and Guilcher)

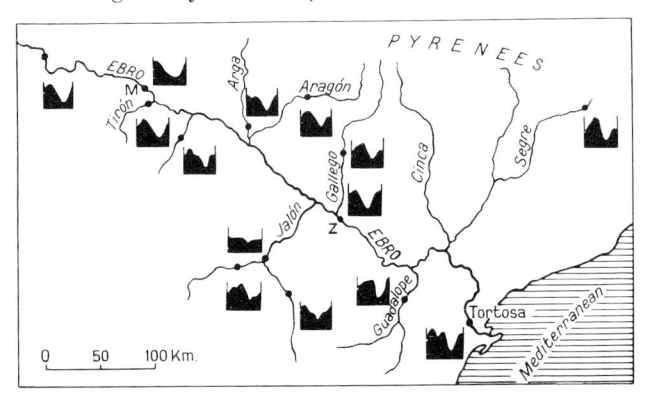

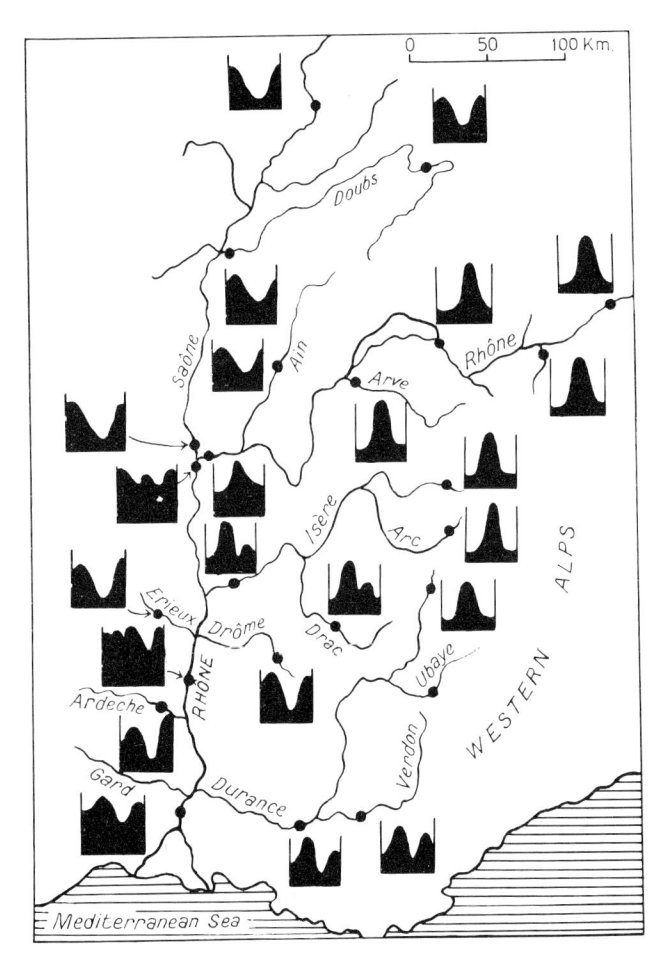

16 *The regimes of the Rhône and Saône* (after Pardé and Guilcher)

The glacial and nival areas are of little agricultural use but are of great tourist and hydrological importance. The nivo–pluvial and pluvio–nival regimes affect the major part of southern Europe and bring to would-be irrigators a maximum volume at the time of maximum crop needs. The benefits are obviously most striking in Mediterranean summer drought lands but as figure 18 shows the crops throughout lowland southern Europe usually suffer a summer deficiency of moisture for optimum growth, and would benefit from supplementary watering.

The effect of snowmelt is seen on all the main rivers of southern Europe except in southern Spain. On the Ebro the central Pyrenean tributaries such as the Segre, Cinca, Gállego and Aragón, are nivo–pluvial (fig. 15). The Rhône has a glacial and nivo–glacial (nival) regime as far as Lyon, although Lake Geneva moderates the difference between the summer spate and winter low water (fig. 16). At Lyon, the Saône (410 cub m/sec) enters with a pluvial oceanic regime and immediately downstream of the confluence (total flow 1020 cub m/sec) the Rhône has two maxima, one in midsummer and another from November to February. Farther downstream the Isère (350 cub m/sec)

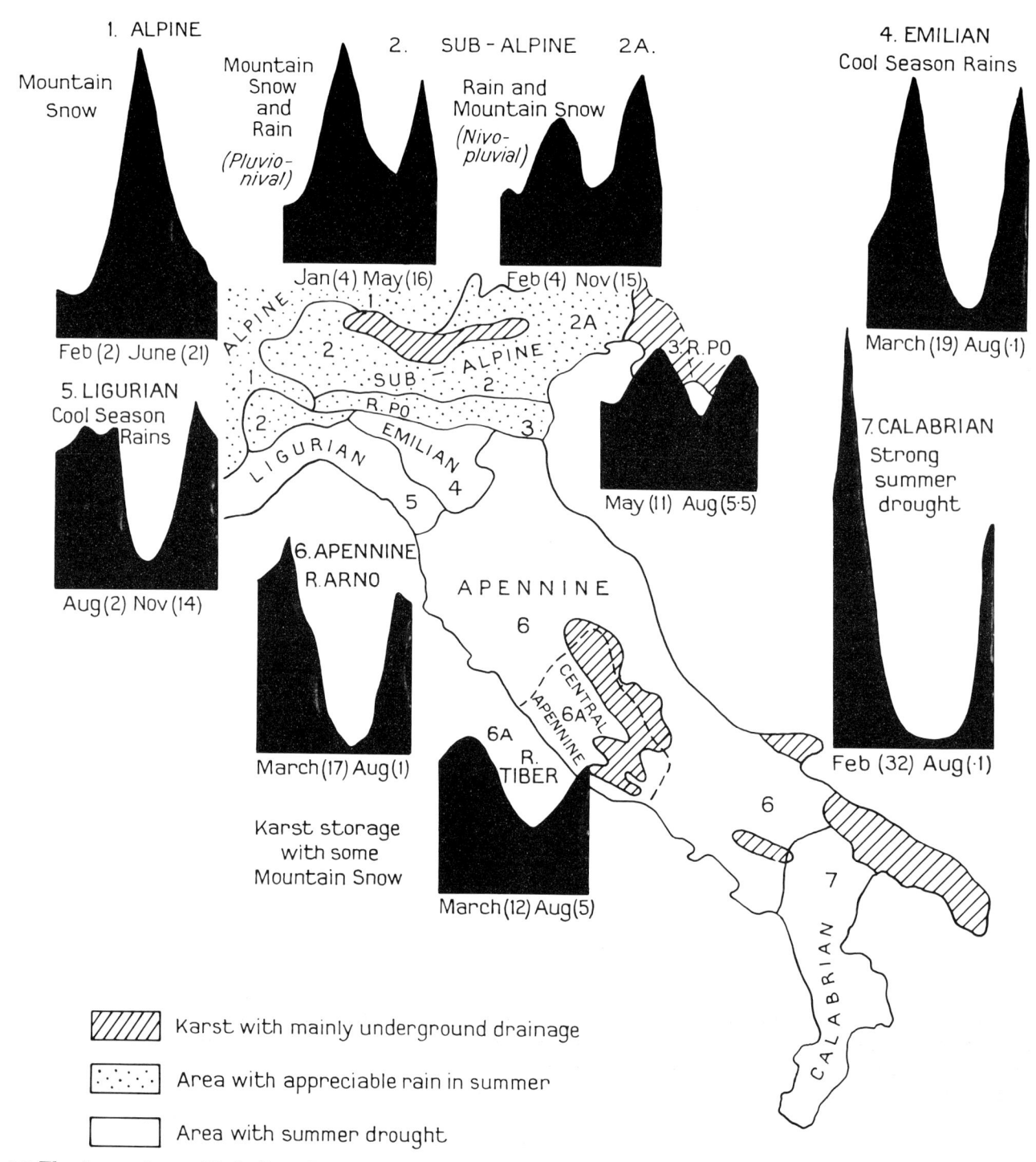

17 *The river regimes of Italy* (from R. P. Beckinsale, *Land, Air and Ocean*, 4th ed., 1966)

comes in from the French Alps and its markedly nival regime gives the Rhône at Valence a single maximum in June. Lower down the tributaries entering from the west, such as the Erieux, are distinctly pluvio–nival because the runoff from their heavy autumn rains exceeds that from spring snowmelt. The ultimate result is that near Montélimar (1500 cub m/sec) the Rhône has a remarkably even regime with no less than four maxima, in November, January, March and June. Near Avignon the Durance (220 cub m/sec) enters from the Alps with a complex nivo–pluvial regime giving a maximum in May and June. The effect of this addition is reflected in the variations of the Rhône at Arles (1850 cub m/sec) where the regime is nivo–pluvial but the so-called low water in September remains a time of ample flow. Then the extensive exposure of gravel and pebbles on the wide valley floor tell of the great floods that can occur almost at any season.

Few rivers have so complex a regime as the Rhône. The Po is more like the Ebro as its northern tributaries come from high, wet mountains and its southern from Mediterranean climates (fig. 17). The Alpine tributaries of the Po begin according to their altitudes as glacial or nival streams. The Po itself at Turin is clearly nivo–pluvial and downstream is joined by large Alpine rivers with nival regimes. However, many of these rivers are moderated naturally by great piedmont lakes which delay the high water by several weeks until late June or July and lower it considerably. They also prevent the winter minimum from being excessively low. On the other hand, the tributaries entering the Po from the Ligurian Alps and northern Apennines have a marked Mediterranean summer minimum and a maximum usually in autumn. This cool season spate from the Apennines coincides with the Alpine low water and gives the Po below Turin two maxima, one in May–June and the other in late autumn. Thus the main artery of the Po has a remarkably even regime. Near its mouth the river has an average annual flow of about 1460 cub m/sec in spite of the abstraction of 246 cub m/sec for irrigation purposes.

Human uses of rivers

The chief human uses of rivers are for transport, irrigation, power, water supply, fishing, effluent disposal, recreation and religious purposes. Of these, fishing is discussed in Chapter 3 and water power in Chapter 10; the details of effluent disposal are better left to engineers, while river worship seems defunct in Europe, although in popular rhymes the Romans are said to have prayed to Father Tiber.

River transport

Flotation or the simple floating of timber is not important in Southern Europe. Navigation, however, dates from prehistoric times. Today in western Europe the chief commercial artery is the Rhine (233 million tons of cargo in 1965) which concerns us as, after canalization in its middle course, it can take 2000-ton boats as far as Basle, the chief port of Switzerland. In 1965 Basle handled 8·6 million tons of waterborne cargo, a record amount which consisted mainly of imports. Thus Switzerland with only 21 km of national navigable waterway has by far the greatest river port in Southern Europe. At present the Rhine is being improved upstream of Basle.

The chief European rivals of the Rhine for traffic are the Volga (U.S.S.R.) and Danube. The latter is under the control of the Danube Commission which has its headquarters at Budapest and consists of members of all the riparian states. The total annual freight traffic on the river was 43½ million tons in 1962 and 45 million tons in 1966. The only Mediterranean country with a frontage on the Danube is Yugoslavia where barge transport is especially used for building sand, coal and fertilizers. The total Yugoslavian goods traffic on the main river and its tributary the Sava amounted to 5·36 million tons in 1963, exclusive of 3·85 million tons in transit.

The great possibilities of waterway transport on the Danube have no counterparts in Mediterranean Europe. In Portugal navigation for small boats ceases at the rapids near the international boundary. In Spain, the rivers have steep gradients and lack coastal plains with the notable exception of the Guadalquivir which has been made navigable for vessels drawing 7·5 m as far as Seville. In Italy less than 2000 km of navigable waterway, including 850 km of canals, are in use. Small shallow draft vessels can use the Tiber almost to Rome and the Arno, except in summer, to Florence. However, the lower Arno can be avoided by using the Pisa–Leghorn canal which is 17 km long, 3 m deep and navigable by 600-ton barges. Apart from these short stretches inland navigation is restricted to the Po basin where the main stream to Casale Monferrato (500 km) as well as the lower courses of some Alpine tributaries, the larger lakes, and deeper artificial canals can be used by shallow craft. From Mantua on the Mincio to the Po and down it to the Adriatic the channel has a depth of over 2 m and is usable by 600-ton barges.

The main navigable waterway of the Mediterranean is the Rhône, a river long notorious for the rapid current and dangerous shallows in its defiles and for shifting sandbanks at its delta. In contrast the slow-moving Saône was more easily improved and was linked by canal to the Seine drainage in 1776, to the Rhine in 1784 and later also to the Marne and Moselle. In 1926 a new river port was opened on the Saône at Lyon and proved so successful that within a few years the Compagnie Nationale du Rhône built another port downstream on the left bank of the Rhône just below its confluence with the Saône. In 1936, Lyon handled 1·25 million tons of waterway cargo but less than 8 per cent of this used the lower Rhône where the mean depth was 1·6 m only. The C.N.R. decided to improve the Rhône navigation by constructing a series of dams across the chief defiles with lateral canals around the most difficult stretches (fig. 37). The Donzère scheme was the first completed on the lower Rhône and included a lateral canal 29 km long and wider than the Suez Canal. At the power station on this canal the fall is 26 m and the navigable lock is one of the world's highest shift locks. In 1967, the five main projects of this series of lateral canals from near Donzère upstream were finished and linked to the Saône by the completion of a barrage with a lock and a 35-km lateral canal at Pierre-Bénite just south of Lyon. It is hoped that the total waterway cargo trade at Lyon (1 877 000 tons in 1964) will now increase rapidly. However, the Rhône waterway and its connexions south of Avignon still need improving. Here the various branch canals to Sète (completed in 1934) and Port de Bouc are under 2 m deep. The chief hope is a busy connexion with Marseille via the Étang de Berre where the new industrial growth is already linked to France's greatest Mediterranean seaport by a canal for 1000-ton barges. There is also hope of widening and deepening the southern stretch of the Canal de l'Est which connects the Moselle to the Saône. Improvements on the Moselle have already been made and when applied to the existing waterways as far as the middle Saône will allow boats of 1000 tons to travel from the North Sea to the Mediterranean. It may well be that with modern high-powered diesel engines and lightweight boat-building techniques this inland interocean link will attract a large tourist traffic. Tourism is today a major form of profit and waterway pleasure cruising is increasing rapidly in popularity.

Irrigation and land drainage

Irrigation was important in the European parts of the Mediterranean throughout Roman times and was later extended and perfected by the Moors in the drier tracts. The simple traditional methods of seasonal inundation canals and of raising water by means of levers (*cigonal* in Spain) and of rotating wheels with cups (*noria* in Spain) are still practised. But it was the spread of hydro-electricity generation and transmission (after 1880) and of modern concrete-strengthened dam construction and of national and international financial loans that revolutionised irrigation practices in the twentieth century. At the same time, the internal combustion engine by facilitating earth-moving and water-pumping, greatly intensified irrigation and revolutionized land drainage.

On floodplains and lowlands, irrigation can seldom be divorced from land drainage and reclamation except in arid areas where crops will only grow when watered artificially. However, as will be shown, southern Europe has relatively small truly arid tracts and is for the most part humid enough to support some natural vegetation. Today, irrigation in Europe is undertaken for three main purposes. To grow crops in areas too dry all the year or in summer for successful agriculture; to supplement the natural rainfall in the more evaporative months in order to increase crop yields and sometimes to produce an extra crop; and to grow special crops, such as rice, which have excessive water needs.

The general volume of water required by plants may be expressed as the potential evapotranspiration or the total amount of water evaporated and transpired from their exposed surfaces when their roots are never short of soil moisture. Except in arid climates, during the cool season precipitation normally exceeds evapotranspiration and some surplus water flows off from the soil as springs and streams. In the drier, hotter months evapotranspiration

ANNUAL MOISTURE BUDGET FOR SOUTHERN AND WESTERN EUROPE

Country	Station	Annual (in millimetres) Precipitation	Potential evaporation	Moisture deficit	Moisture surplus
Britain	Greenwich	609	618	212	203
	Cardiff	1043	548	55	550
Germany	Berlin	587	644	246	189
Austria	Vienna	670	678	202	194
Switzerland	Berne	977	605	8	380
	Zurich	1044	628	–	416
France	Paris	576	683	252	145
&	Lyon	796	780	186	202
Corsica	Perpignan	554	1043	574	85
	Ajaccio	746	1106	572	212
Portugal	Oporto	1157	953	378	582
	Lisbon	629	1155	648	142
Spain	Santander	840	899	277	218
	Zaragoza	297	1136	842	3
	Madrid	425	1035	680	70
	Murcia	359	1371	1012	–
Italy	Milan	1071	902	222	391
	Rome	903	1087	517	333
	Palermo	748	1271	698	175
Albania	Shkoder	1455	1018	412	849
Yugoslavia	Mostar	1416	1013	364	767
	Zagreb	902	824	192	270
	Belgrade	623	843	354	134
Greece	Salonika	486	1152	728	63
	Athens	384	1285	946	44

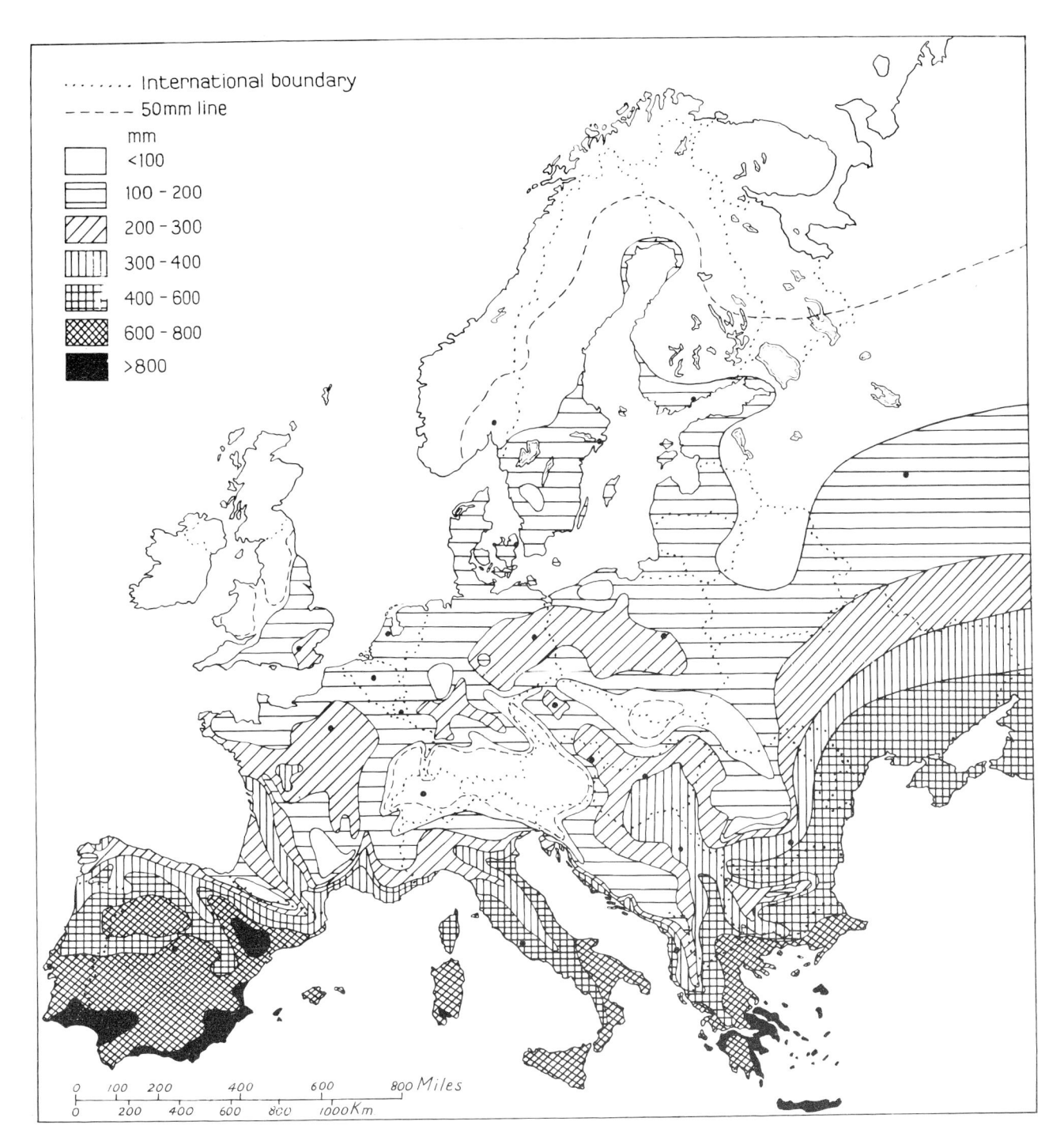

18 *Average annual maximum precipitation deficits in Europe* (after L. J. Mohrmann and I. J. Kessler, International Institute for Land Reclamation, Wageningen, Netherlands)

usually exceeds precipitation and then for various periods the vegetation must occasionally depend for growth on moisture stored in the soil. The amount of soil moisture available to roots depends largely on the depth and pore space of the soil and on the depth of the root system. Deep-rooted plants such as the olive and vine survive drought far longer than shallow-rooted vegetables.

Most plants can make use of 50–100 mm of soil moisture before they begin to wilt and deep-rooted crops can obtain up to 100–200 mm. However, once this available soil moisture is used up the plant will wilt and eventually will die unless watered. In arid climates irrigation is needed constantly to renew the soil moisture whereas in humid climates it is needed only during dry spells. The average annual maximum moisture deficits or the greatest differences between evapotranspiration and precipitation during the year in several European towns are shown in figure 18 and the average yearly water budget is given in the table for Mediterranean and some non-Mediterranean towns (opposite). The statistics are, however, averages and some years will be appreciably wetter or drier than others. In localities such as Murcia the crops will need irrigating every year whereas in humid tracts they may not require additional water more than five or seven years in ten. Mountainous and upland districts except in the southern halves of the Mediterranean peninsulas will not normally demand irrigation as their average deficiencies are less than 100 mm, or the amount stored in the soil. Thus, for example, at Berne and Zurich crops will flourish without irrigation except in very dry summers. In contrast, areas with a maximum annual moisture deficit of 200–400 mm will normally suffer severe moisture shortages occasionally even on good soils. Thus, in northwestern Spain, the Po plain and the northern Apennines, irrigation may occasionally save the crop and will normally greatly increase yields. In areas with normal annual moisture deficiencies of about 400–600 mm, such as most of 'arid' Iberia and the lowlands of southern France, peninsular Italy and Greece, irrigation assumes great importance. Where the annual deficiencies exceed 600 mm irrigation is essential and the amount of water needed for crops begins to be excessive for small drainage basins.

Irrigation works are expensive, and countries with an adequate food supply can hardly be expected to undertake elaborate river-control schemes for the sake of crops alone. It happens that hydro-electricity is usually profitable and often the barrages necessary for it can also be used for water storage, for irrigation and flood control. Today most large dams are multi-purpose, and the large reservoir has become a characteristic feature of the landscape of southern Europe.

The extent of irrigated land is hard to assess as so much irrigation is supplementary rather than essential. The estimates given in the table are for all types of irrigation and include paddy, which however is also shown separately as its area is carefully recorded. In the following account it will be obvious that the use of irrigation in any national territory depends on more than climatic influences. Portugal has vast tropical colonial territories and at home has needed power more than extra food. By 1953, only 21,054 hectares were fully irrigated but subsequently, owing mainly to hydro-electric needs, irrigation projects have been speeded up. However, the statistics given in the table relate mainly to supplemental irrigation of arable and horticultural crops. Reservoirs primarily for irrigation have recently been completed on the Sorraia and the paddy fields on the lower Tagus and lower Sado have been considerably extended. The aim was polyculture and small proprietorship, but rice, which needs units of 150 to 200 ha, became dominant with large producers and migrant workers at harvest time.

In arid Spain, long famous for its *huertas* with their perennial canal irrigation, the last thirty years have brought more advances than the previous three thousand. The bare statistics of dam building tell their own story. In the early twentieth century, the 13 chief reservoirs had a total capacity of 98 million cub m; by 1936, the 74 reservoirs stored 4000 million cub m of water. Thereafter the barrages and reservoirs increased rapidly in size and capacity; many old schemes were given a better water supply and many new schemes were begun. By 1956, some 138 dams stored 14 000 million cub m and 30 other dams were under construction. By the early 1960s the storage capacity had been raised to 18 000 million cub m and 50 dams were under construction, 8 of which were over 100 m high. By 1968 the completed works had increased the storage capacity to 30 000 million cub m. The irrigated area (850 000 ha in 1858) had been raised to 1 750 000 ha by 1958 and to 2 350 000 ha by 1968. Thus during the twentieth century the irrigated area has been doubled and now comprises nearly 12 per cent of the cultivated land. Probably another 2 million ha could be irrigated but the returns would be less rerunerative. Spain is already a land of great reservoirs near which extensive irrigated farms are dotted with new villages. The large schemes in the Ebro basin, where wide flat river terraces encourage water distribution, are described in the regional chapters. Here more than 100 new villages and towns have been created (fig. 48). Elsewhere the social repercussions have been equally beneficial. The middle Guadiana scheme, based on the large Cijara dam (storage capacity 1760 million cub m) and on several subsidiary barrages, supplies water for the perennial irrigation of 300 000 ha, besides providing electricity for industrial development in towns between Mérida and Badajoz. It has allowed the resettlement of 60 000 families into 40 new villages and towns. The upper Guadalquivir scheme involved nearly 100 000 ha of irrigable land and gives a living to more than 4000 families, grouped into 12 new villages.

In southern France, in spite of a considerable national

MEDITERRANEAN AND DANUBIAN COUNTRIES:
ESTIMATED IRRIGATED AREA (1971) AND PADDY
AREA (1968)

(in thousand hectares and thousand metric tons)

Country	Irrigation area	Rice area	production
Albania	160	4	10
France (Southern)	2 600	24	85
Greece	630	21	102
Italy	3 300	156	639
Portugal	636	33	149
Spain	2 400	60	362
Switzerland	23		
Yugoslavia	215	5	18
Malta	0·7		
Southern Europe	9 965	280	1 369
Bulgaria	955	14	39
Hungary	230	21	41
Romania	531	25	60
Total non-Soviet Europe	11 681	363	1 505
U.S.S.R.	11 000	312	1 063

surplus of some agricultural products, irrigation has expanded greatly since 1942. Nearly half the value of the irrigated crops here consists of rice, vegetables, fruits, forage and flowers. In 1942 the government guaranteed the price of rice and within ten years the paddy area in the Camargue rose to 21 000 ha and the production to 80 000 tons. Several of the administrative departments such as Bouches du Rhône, Vaucluse and Pyrénées-Orientales have over 50 000 ha under irrigation (fig. 58) and the impetus has increased with an influx of resettled patriots from Algeria when the colony became independent. These settlers were accustomed to irrigation techniques. The recent schemes include a considerable extension of pasture and forage in the 'stony desert' of the Crau, the Pleistocene delta of the Durance. Here all told 18 500 ha are mostly under hay which can be cut three times in a season. The vegetables usually reach Paris markets overnight by road whereas many of the fruits, such as peaches and apricots, go to newly built packing stations.

In Italy, which is on the whole the best watered peninsula of the Mediterranean, the rivers and springs of the Po plain are ideal for summer inundation and supplementary irrigation. Here large-scale schemes began in the 1860s when political unification allowed the construction of the Cavour Canal with 150 000 ha of new perennial irrigation. The late nineteenth century schemes included those drawing water from the Adige, which irrigated a further 366 000 ha. By 1905, about 1 365 000 ha were irrigated in all Italy. During the next forty years considerable areas of the Po delta and of the eastern Po plains were also irrigated and by 1948 the total area supplied with irrigation water in Italy had increased to 2 185 000 ha. Subsequently plans were put forward to add large areas in the north and in the peninsula and islands. These have progressed well and by 1962 the total irrigated area was 3·1 million ha.

The nothern plain is the chief area for irrigation and the Po system is bled of 250 cub m/sec for irrigation annually or nearly 70 per cent of the water used for these purposes in Italy. Here too the wide belt of water seepage (*fontanili*) adds greatly to the irrigation potential. Lombardy remains the ricebowl of Europe and, in spite of the remarkable progress made recently in Spain, Italy will probably always be the chief state for irrigation in Mediterranean Europe. The following statistics for 1962 show the spatial predominance of northern Italy which climatically and hydrologically is not Mediterranean.

	Area irrigated (1000 ha)	Per cent of total agriculture land
Northern Italy	2240	30·0
Central Italy	274	7·2
Southern Italy	387	6·7
Sicily	164	7·1
Sardinia	35	1·7
ITALY	3100	14·4

Of the national total, 69 per cent is irrigated by water from rivers, 28 per cent from wells and fontanili and the remaining 3 per cent by water from reservoirs. For the north alone the actual areas are 1 731 000 ha by river water, 485 000 ha by fontanili and well water, and 24 000 ha by reservoirs.

Throughout the eastern Mediterranean, good farmland is relatively scarce and the relief is against large-scale irrigation schemes. Yet, in spite of proneness to earthquakes dam building has been active in Yugoslavia and Greece. In Greece the newly completed schemes include a large dam on the Acheloos river. In all the Balkan countries the irrigation of cotton, rice, tobacco, maize, fruits and vegetables proves profitable and in Greece some success has been achieved with rice production on alkaline soils hitherto thought unsuitable. Hence it seems that the F.A.O. plan to double the area under irrigation in the Mediterranean between 1960 and 1975 may well be achieved as far as Europe is concerned.

The countries on the middle and lower Danube have also made irrigation and land drainage important parts of their various four- or five-year plans since 1948. In Yugoslavia, apart from the huge conjoint scheme with Romania at the Iron Gate, there has been a large multi-purpose development under way since 1956 in the Backa and Banat lowlands where the loess and sandy tracts have very little surface water and the few large rivers may flood disastrously in spring. The scheme is aimed at flood control, land drainage, irrigation, industrial water supply and navigation (fig. 94). A navigable canal from Bezdan on the Danube 130 km to Becej on the Tisa and thence eastward 135 km to the Danube again at Banakska is under construction and, with nearly 400 km of auxiliary canals, will improve the drainage of 1 million ha and allow the irrigation of about 360 000 ha. Here the water deficiencies in the growing season are often 250 mm and the supplementary water will prevent occasional poor harvests and will improve crop yields generally. It will also extend double cropping in the year, expecially of maize and lucerne (alfalfa), and so increase the numbers of stock.

Land drainage and reclamation

The chief disadvantage of a riverine location, the danger of flood, is aggravated throughout southern Europe by the height of the spates and the flat nature of many of the richer alluvial plains. On such flat terrains the sluggish rivers, even after straightening and deepening are liable to silt up their beds above the level of the adjacent floodplain.

On the middle and lower Danube and its tributaries a remarkable system of embankments has been built artificially and naturally. The total length of these levees exceeds 35 000 km. In Yugoslavia the fertile plain province of Vojvodina has 1300 km of protective embankments and on the Sava strong dykes protect the reclaimed riverine lands all the way from Zagreb to Belgrade.

The Mediterranean schemes appear small against those on the Danube. In Spain the most ambitious development has been to drain 136 000 ha of the salt marshes (Las Marismas) at the mouth of the Guadalquivir (fig. 49). In Italy the richness of the alluvial plains where malaria was rife was always a challenge. The Romans attempted to drain the Pontine marshes, the Maremma and parts of the Po plain. Land reclamation has been popular since the sixteenth century but malaria remained a problem. Between 1861 and 1915 about 330 000 ha were drained in north Italy and 2000 ha in the south. In the 1930s a fine effort was made to redress the balance. The large southern reclamation schemes included the Pontine marshes (75 000 ha), numerous lowlands in the Roman Campagna (200 000 ha), the Sele plain (41 000 ha), in Campania and the Tirso plain in Sardinia (126 000 ha). Most of these were not completed when war broke out in 1939. Today there are about 2·8 million ha of reclaimed drained land (bonifica) in Italy and the lower Po has 2000 km of protective dykes. Most of the post-1930 reclamation has been provided with substantial isolated dwellings set pleasantly beside clear watercourses. The traditional daily trek from the lowland to the hill-top village at dusk is no longer necessary for health reasons, and service villages provide local shopping and educational facilities.

In Greece public investment in agriculture takes about one-quarter of the national funds available for investment and most of this is spent on irrigation, drainage and land reclamation. Until recently Greek irrigation projects were mainly private schemes based on small streams, wells and springs but from about 1955 to 1965 half of the expansion (totalling well over 150 000 ha) was financed by public money. The last five-year Draft Economic Development Plan (1966–70) expected public works to make available nearly 150 000 ha of new irrigation lands. On these new projects about 65 per cent of the water supplied was to be used for field crops, such as cotton, maize, rice and sugar-beet, 16 per cent for vegetables, particularly potatoes and tomatoes, and 14 per cent for citrus and deciduous fruits. At the same time some processing and packing facilities have also been provided, where necessary.

Water supply

A good supply of drinking water has always been hard to maintain in Mediterranean cities and the Romans built magnificent aqueducts and storage tanks. Modern urban growth and the influx of health-conscious tourists have increased the problem in cities. Small islands and limestone tracts also have great difficulty in summer. On some of the drier islands where excessive pumping of deep wells may lead to an influx of salt water, drinking supplies are still imported by boat in the hottest months. Freshwater upwellings in the sea bed serve some Adriatic islands.

Today scientists are providing a means of overcoming these shortages. To the heightening of existing dams, the lengthening of water pipelines, and diesel pumping of deep bores, they have added desalination of sea water. In 1965 throughout the world about 635 land-based desalination plants were at work and most of them produced over 100 000 litres of freshwater daily. Numerous different processes can be used to separate the salt and water but in all, the freshwater obtained proves relatively costly unless cheap fuel is available. The usual fuel is petroleum in some form, and plants of this kind are now in use at, for example, Gibraltar and Malta. However, any heat which could evaporate the water would be suitable and the cheapest form in sunny climates would be solar energy. At present the use of solar heat is impracticable on a large scale as it requires an extensive water surface for a relatively small output of freshwater. However, on a small scale it has already been very successfully applied on the Greek Aegean islands of Patmos, Aegina, Salamis and Simi, the finances being provided by American philanthropists. The general idea can be illustrated from the solar still at Simi, a small Greek island in the Dodecanese, which has long imported drinking water from Rhodes. An American philanthropic society, wishing to improve the living conditions of the 3000 islanders, persuaded the Simians to give up their village square of half a hectare for the site of a solar desalination still. Long shallow troughs were built and were lined with sand, and covered with plastic film. Sea water is pumped into the troughs where the sun's heat causes evaporation which recondenses on the underside of the plastic cover and runs off into side gutters. The distilled water is eventually pumped into the village reservoir. Each night the troughs are washed free of salt and refilled with new seawater. In winter and at night when the air temperature is too low to cause much evaporation, the distillation is continued by means of a diesel electric power plant. In winter when rain is likely the solar still can be used to collect rain water. These schemes on Simi, Aegina and elsewhere have been so successful that there are plans for small installations for individual hotels and houses. However, the space needed to produce a small supply is considerable and desalination plants driven by petroleum fuel, some with a freshwater output of $4\frac{1}{2}$ million litres (1 million gallons) a day, are becoming common throughout the drier Mediterranean coastlands. At present the chief installations in the European Mediterranean are at Gibraltar, Malta (with projected daily total of 45 million litres) and Taranto, where plants supply the great steel works with over 5 million litres of freshwater daily and the Shell oil refinery with over 2 million litres. Plants also exist or are under construction at Ermoupolis on the island of Syros, at Porto Tórres in Sardinia, Gela in Sicily, and in Cyprus.

6 Vegetation and Soils

VEGETATION

The four or five major ice advances during the last two million years, or Quaternary period, virtually obliterated the pre-existing Tertiary vegetation of much of northwestern Europe whereas, except on the higher mountains and piedmonts, southern Europe lay outside the icesheets. During an ice advance the climate of regions just outside the ice front was periglacial or cold to cool, with liability to frequent frost even in summer. The vegetation here resembled that of the tundra, with stunted birch and fir in the warmer Atlantic west. This periglacial zone graded southward into a boreal belt of pine and fir, beyond which stretched mixed deciduous–coniferous forest. Farther south was evergreen forest, no doubt restricted in a glacial phase to the southernmost tips of the Mediterranean peninsulas. These warmer vegetation belts expanded or migrated northward during each interglacial period and after the last ice retreat they formed the vegetation zones encountered by early man. Probably in Neolithic times, when the climate was slightly wetter than today, 80 per cent or more of southern Europe was forested. During the last five thousand or six thousand years the action of man has been more important than the influence of any slight contemporary fluctuations in climate.

In Europe during this time, man has rapidly altered the natural vegetation of large areas by burning, felling, cultivating, by developing grasses and trees that he has found to be useful to him, and by importing plants from long distances. The major part of southern Europe is now cultivated, most of the pastures are made and maintained by man, and many of the forests are planted by him. In deforestation, man has been greatly aided by hordes of servile or domestic beasts, most of whom eat leaves, saplings and bark. With increases of population, the destruction of the woodland continued no matter whether the chief tribes were primarily pastoral or cereal farmers. Typically the great early civilisations of Mediterranean Europe, especially the Phoenician, Greek and Roman, overexploited the tall treegrowth and seriously damaged their natural environment. This irrevocable damage was all the more unpardonable as even then some authors saw that trees prevented soil erosion. Only in recent centuries and especially since 1920, has the scientific conservation of forests and protection of soils on watersheds and hillslopes become common national policies.

Origin of present-day vegetation

The budget of the species that compose the flora of a region has two main components, natural survivors and imported colonists. The number of natural survivors depends mainly on the severity of the climate during the last ice advance, on the ease with which the vegetation could return in the postglacial warmth and the extent to which it has escaped destruction by man. In Iberia which was farthest from the northwest ice sheet and was relatively maritime and warm, the surviving flora is richer in species than in Italy which was largely shut in northward by an icecap on the Alps. Similarly in the Balkans, which is slightly richer in species than is Iberia, plants could advance relatively easily from the north and east. In Spain about 5600 different species have been recorded and of these about 40 per cent are central European and Alpine in origin, 25 per cent are endemic (local), 20 per cent Mediterranean and 8 per cent African.

The Balkans may be taken as the other floristic extreme of southern Europe. In this peninsula, including Bulgaria, about 6530 species of flowering plants have been classified, making it, for its size, the richest floristic area in Europe. Here about 37 per cent of the species are also found in central, eastern and Alpine Europe; a further 11 per cent occur also in other lands near the Caucasus, Caspian and in the Near East; and no less than 27 per cent are endemic, some being Tertiary survivals. The proportion of North African species is negligible.

Many of the natural survivors were not useful to man and indeed some were a nuisance to him, particularly if spiny and inedible. On the other hand, some proved most useful and these as a rule he protected and extended and in many cases improved, as for example, the wild olive. Each new migratory wave of peoples tended also to bring new plants. Today many of these are a natural feature of the landscape and an integral part of the agricultural economy.

In the Mediterranean lands early invaders from southeast Asia, and particularly from Persia, introduced the vine, mulberry and opium poppy. Later from this same quadrant of Asia the Arabs brought rice, sugar-cane, cotton and orange. From Latin America the Spaniards introduced from the late fifteenth century onwards useful plants such as maize, tobacco, potato, groundnut, red pepper and prickly pear cactus. Probably the Portuguese introduced the agave or aloe (sisal) and other fleshy plants from Africa. In the last hundred years an appreciation of the value of Australian plants has led to the extensive introduction, for example, of certain acacias and eucalypts. The eucalypt is an excellent aid to draining marshes and acts also as a windbreak in Campidano and Andalusia. The list could, of course, be greatly extended if more exotic species introduced solely for decorative purposes were included.

Major vegetation types

The two main climatic types of southern Europe give rise to two major vegetation belts. The severe summer drought of the Mediterranean climate favours xerophilous evergreens, while the winter chill of the 'West European' climate favours leaf fall in autumn. The most common complication in this dual scheme is caused by relief which generates boreal and arctic climates with their characteristic plant associations. There is also in the east a close approach to continental steppe climate which, with the aid of porous soils and of man, gives some areas of grassland.

Thus southern Europe is usually divided into the following vegetation provinces: Mediterranean evergreen forest and shrub; mixed deciduous–coniferous forest; steppe; and mountain associations of boreal forest and of Alpine tundra.

Mediterranean evergreen vegetation

The warm season drought in Mediterranean lands is partly offset by anatomical adaptations such as leathery leaves, waxy coatings, thick bark, and deep root systems. The thick cuticles and waxy coatings have little value in reducing transpiration as long as the stomata (pores) are open but once the plant begins to wilt the stomata close and loss of water takes place only by cuticular (skin) evaporation. Then the anatomical modifications so reduce water losses that the wilted plant will remain alive for a long time. In the Mediterranean area 250 genera of plants have been recorded and no less than half of these are endemic. The oak with 30 species and pine, with over 20 species, are particularly well represented. Apart from changes imposed by relief, the present Mediterranean plant associations are distinguished, mainly on dominant species, and tallness and density of growth, into scrub woodland; macchia; garrigue; and sub-arid steppe. We shall see, however, that these are gradations rather than divisions and that the transition to mixed deciduous–coniferous forest is especially important in Europe.

Scrub woodland

The typical Mediterranean woodland today is a rather open association of low-crowned trees rising as individuals or in scattered clumps to heights of about 10 to 12 m and rarely, in favourable spots, to 15 m. In a few localities, as on the Gargano peninsula in Italy and some high interfluves in Catalonia, patches of more mature forest associaciations survive but it is not certain that these are natural climaxes (with complete evolution and adaptation to the natural conditions). Here evergreen oaks of up to 15 m or 16 m rise above a lower tier of tree shrubs, mainly buckthorn (*Phillyrea media*), strawberry tree (*Arbutus unedo*) and junipers, with an interlacing of lianes, particularly the spiny smilax and ivy. Beneath are a tier of low-growing evergreen shrubs such as the mastic or lentisk (*Pistacia lentiscus*) laurustinus (*Viburnum tinus*) and privet (*Rhamnus alaternus*), and a basal layer of shrubs barely 1 m high mainly of butcher's broom (*Ruscus*). More usual is a scattering of dominant trees with a ground flora all too obviously influenced by man, particularly in the extensive plantations of fruit trees and of quick-growing timber species, mainly conifers. The olive, evergreen oaks (*Quercus* spp.) and Aleppo pine occur everywhere. Other pines and cypress, sweet chestnut, lentisk and strawberry tree are common if less widespread.

The olive still grows wild in parts but has been largely replaced by the cultivated varieties which cover so large an area that in recent years they have yielded 5·5 million tons of fruit, producing over 1 million tons of oil annually in Mediterranean Europe. The evergreen oaks vary more than the olive in distribution (fig. 19). The most widespread is the holm oak (*Quercus ilex*) which bears a thick crop of small shiny leaves and a heavy yield of acorns. It will tolerate several months' drought as well as winter frosts

19 *Distribution of chief Mediterranean oaks, and annual yield of raw cork*

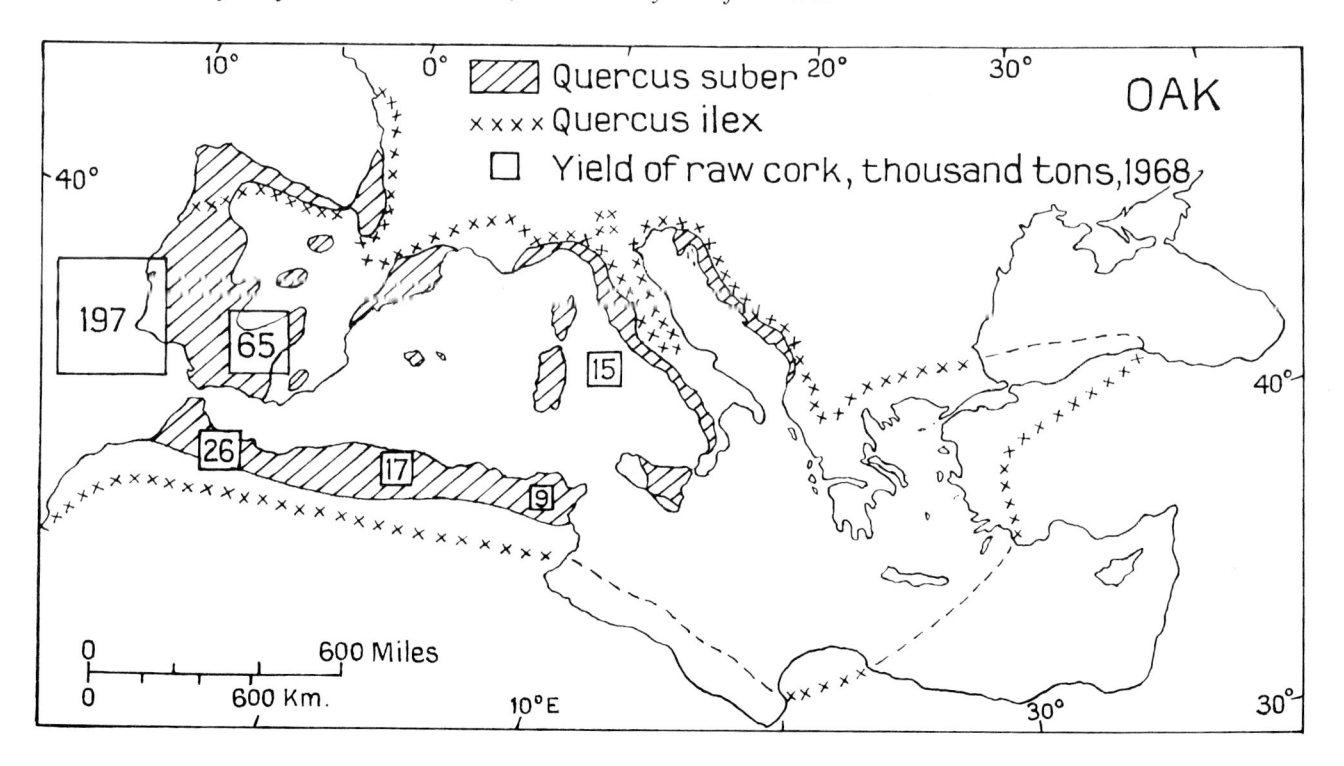

v *Cutting cork in Alentejo, Portugal*

and, in fact, occurs also outside the Mediterranean climate as it survives in the interior Balkans and in the southern British Isles. This species covers over 3 million hectares in Spain alone. The cork oak (*Q. suber*) is the dominant tall tree on the coastlands of southern Iberia and of most of the Tyrrhenian Sea and the eastern Adriatic. In Iberia and Sardinia its thick bark is stripped at about ten-year or thirty-year intervals according to the thickness of layers needed. This economic use, and its large crop of acorns, has encouraged the spread of the tree in conjunction with agricultural practices. The undergrowth is usually removed to provide pasture for sheep and pannage for swine. In considerable areas of southern Iberia and Sardinia the cork oak is retained at convenient intervals amid the arable fields, particularly those under wheat. The tree is estimated to cover nearly 700 000 ha in Portugal, over 500 000 ha in Spain and 100 000 ha in Italy. The many other species of evergreen oak include the prickly leaved kermes (*Q. coccifera*) which is common everywhere except in Italy. It is rarely a tree 15 m high but on poor soils can readily assume a bush or cushion habit. The part played by the cork oak in the western Mediterranean is replaced on a small scale in the eastern Mediterranean by the Valona oak (*Q. aegilops*) which forms extensive forests on the lower hill slopes, below about 400 m, in the southern Balkans and on some of the Aegean islands. These oak woods are often used as grazing grounds and also supply acorn cups for use in tanning and acorns for human consumption.

The evergreen oaks because of their annual or regular yield of useful products have survived better than the Mediterranean conifers which were in great demand since early times for shipbuilding, and especially for masts. However, in recent centuries large areas of conifers have been planted for various commercial uses as well as to stabilize sandy soils, to act as windbreaks and to provide shade in coastal resorts. The three chief species are the umbrella or stone or domestic pine (*Pinus pinea*), the maritime pine (*P. pinaster*) and Aleppo pine (*P. halepensis*). The umbrella pine has spread along the coastal strip of all the European Mediterranean and even grows around the northern Adriatic where the climate is central European. It has a strong resinous odour and is a familiar sight on sandy littorals and in the warmer southern parts of the peninsulas on mountain slopes up to 600 or 1000 m above sea level. The maritime pine, which thrives up to 500 m, belongs essentially to the western Mediterranean (fig. 20). The Aleppo pine is widespread and is a favourite pioneer tree for dry windy tracts and for reafforestation schemes as

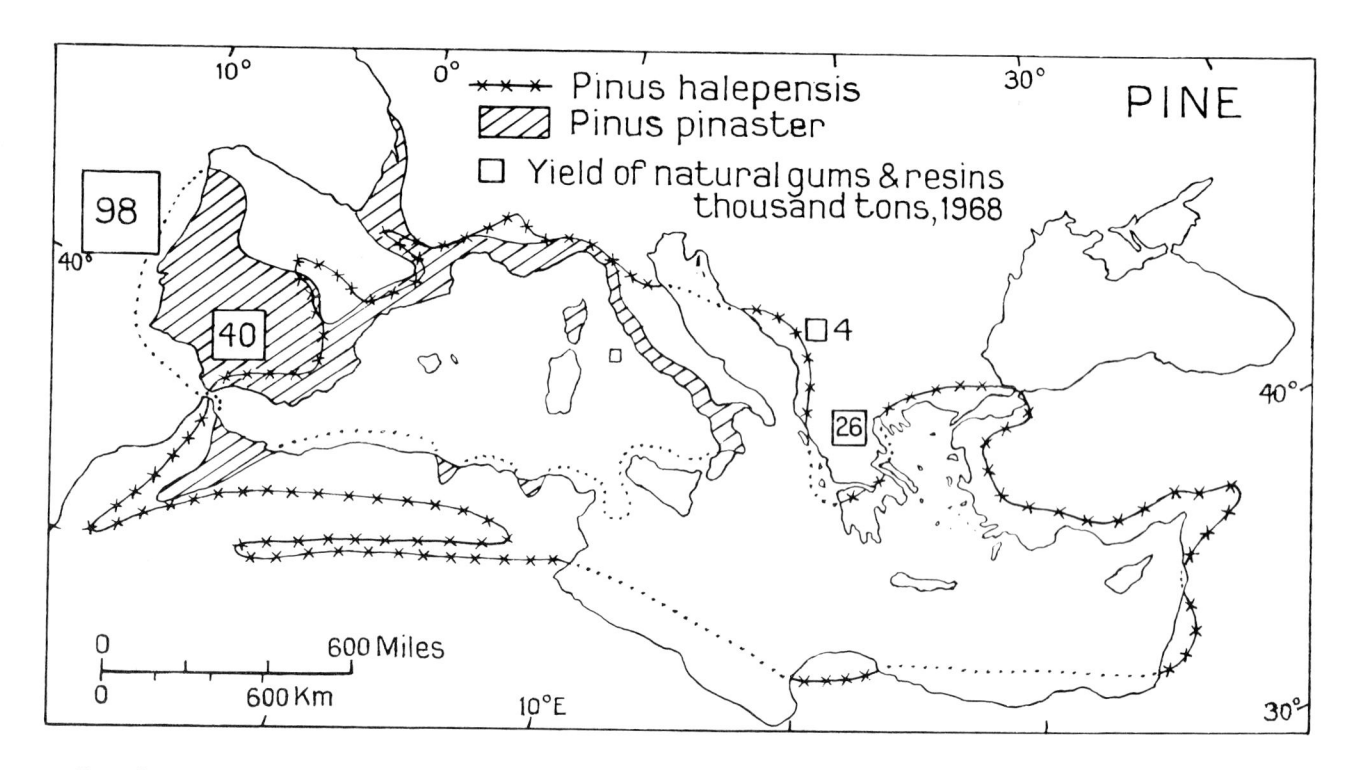

20 *Distribution of two of chief Mediterranean pines, and total annual yield of natural gums and resins from all types of conifer*

its seedlings quickly develop deep roots. In the southern Balkans it often dominates an undergrowth of broom, thyme, lentisk and other shrubs and may almost be a plant climax. It is especially common in Greece where its deformed and stunted shape is often due solely to the crude cuts made in it for collecting the resin which is used to preserve wine (*retsina*). To these well-known pines should be added a fourth conifer, the cypress, which grows fast and is often used for decorative purposes and for windbreaks, as in Tuscany and the Midi.

Macchia

The felling of the taller trees, the constant burning of low scrub growth and the pasturing of flocks and herds have destroyed most of the Mediterranean high forest. Today, except in reafforested tracts, the characteristic

21 *Distribution of two characteristic maquis plants*

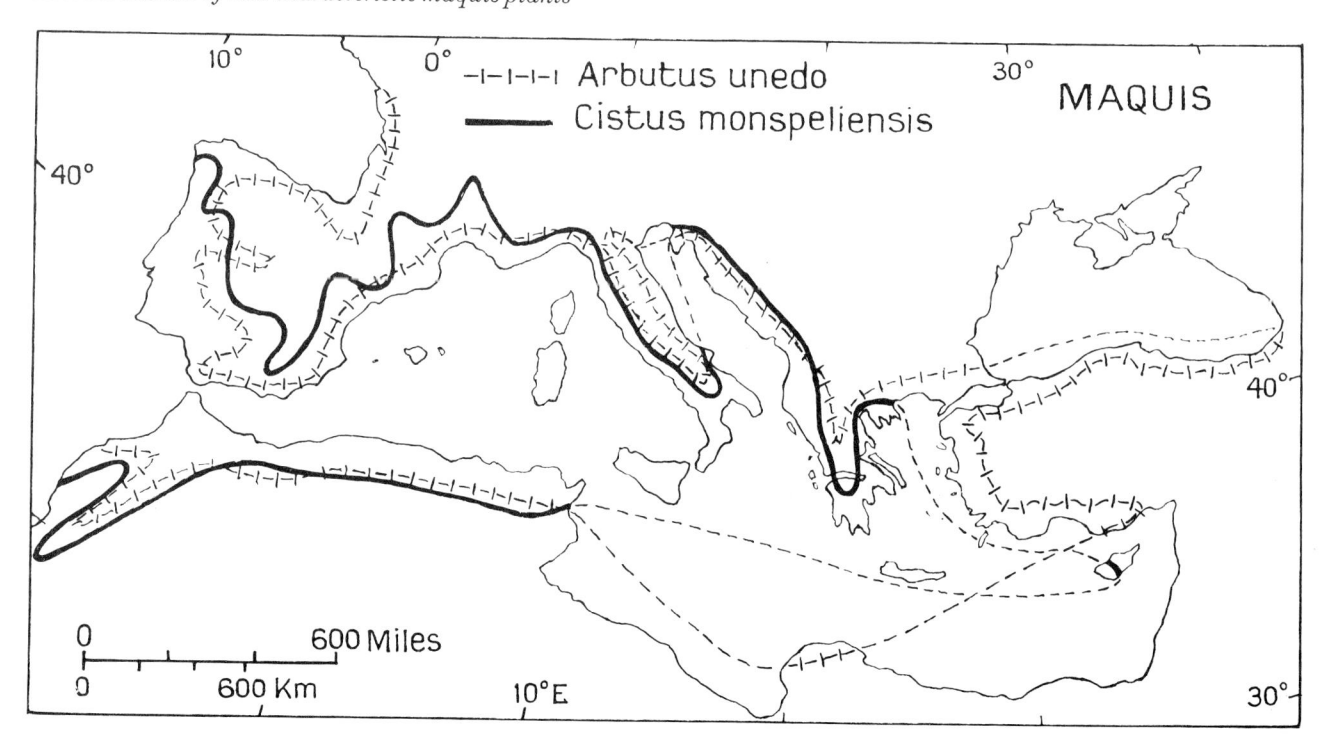

wild plant association is a stunted woodland, 3 to 4 m high, and dense thickets, 2 to 3 m tall, of evergreen shrubs and lianes, known as *macchia* in Italy, *maquis* in France and *matorral* in Spain. These scrub thickets vary widely in species and density, but characteristically are evergreen, leathery-leaved, spiny, impenetrable and highly flammable. The wild olive, heather (*Erica* spp.) lentisk, rock rose (*Cistus* spp.) strawberry tree, gorse (*Ulex* spp.), broom (*Ruscus*), carob, oleander (*Nerium*) and sweet bay (*Laurus*) are widespread. In some of the hotter, windy littorals of southern Italy, Sicily and Sardinia, where the potential evaporation and moisture deficiency are greatest, the macchia is a low, stunted 'forest' dominated by wild olive, lentisk and carob, all of which are true thermophyles. Elsewhere isolated specimens of maritime, umbrella and Aleppo pine in small soil pockets rise above what might once have been their undergrowth. The plants in the macchia association usually show the expected edaphic preferences. The oleander prefers the water courses, the laurel the damper hollows, the cistus, butcher's broom and gorse the more siliceous soils. On rocky tracts, as in Calabria, yellow euphorbia may be dominant. Species that man wishes to protect for fruits tend to occur everywhere. Thus the carob or locust (*Ceratonia siliqua*) is retained and also often cultivated for its beans which make excellent animal fodder. The lentisk yields a sticky resin (gum mastic) which is a valued ingredient of varnish for oil painting, as well, in Greece, as a flavour for a certain liqueur and sweetmeat. On the other hand, species that animals dislike as fodder, particularly the cistus and lentisk, are more likely to survive than species such as myrtle which they devour greedily. A great deal also depends on the natural ability to regenerate after burning. In this respect the rock rose or cistus seems most adept because it yields multitudes of small seeds which spring up in the protection of any surviving spiny branches (fig. 21).

These dense scrub growths would probably lead eventually to a slow regeneration of scrub woodland or even of forest were they not part of a socio–agricultural system. In some districts they are burnt and the ground tilled for a year or two until soil exhaustion makes the crop unprofitable and the macchia returns. In others they are burnt at regular intervals to promote new growth for grazing. In others, after successive burning, soil erosion may occur or a few species or even a single species may dominate and the macchia begins to lose its thicket nature and to degenerate into garrigue. In Italy nearly 1 million hectares or about 16 per cent of the wooded area is classed as macchia but another 2·5 million hectares are classified as 'other Mediterranean scrub'. In Spain, one-fifth of the total area is Mediterranean scrub and here matorral plays a key role in the pastoral economy and over 7 million hectares are used for grazing. A large proportion of this is little better than garrigue but obviously it is impossible to distinguish between the two.

Garrigue

Where the thickets and dense plant associations of the macchia degenerate into a thin or incomplete cover of low

22 *Distribution of three characteristic garrigue plants* (after Rikli)

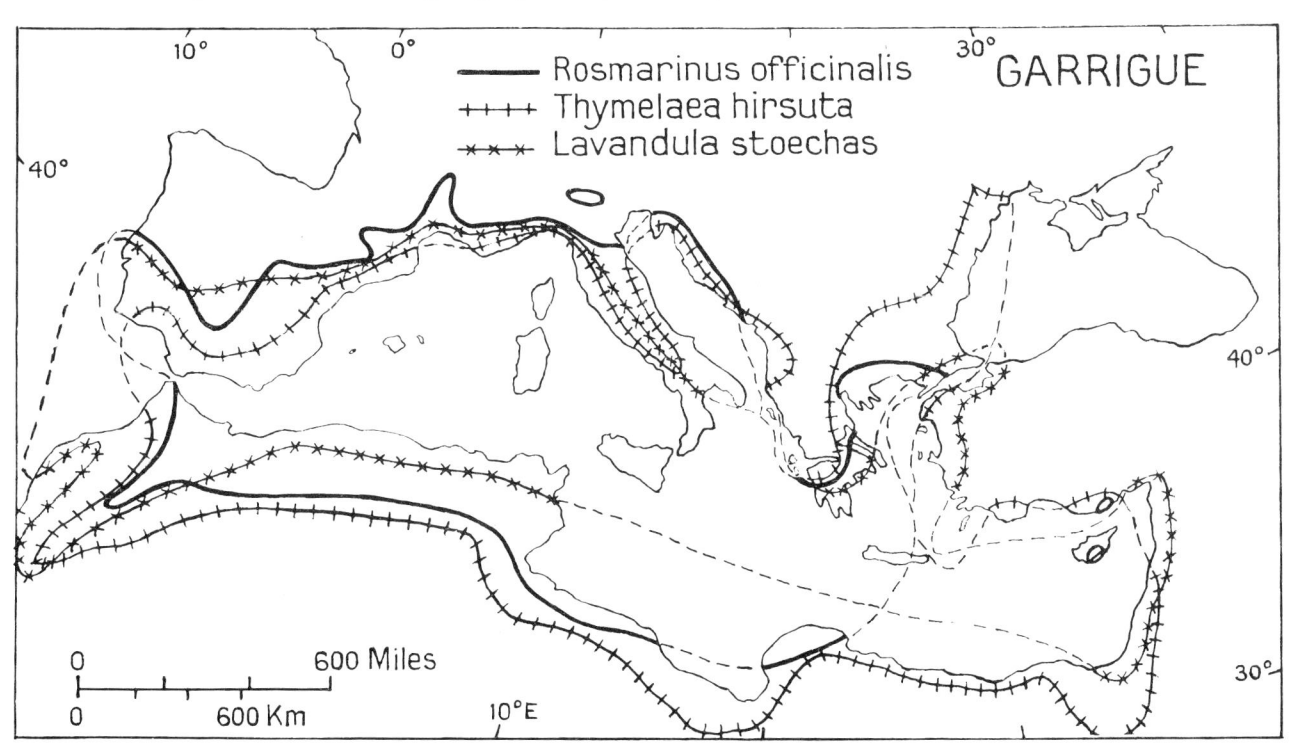

bushes and tufty plants the name *garrigue* is applied to it in Provence and *phrygana* in Greece. This open plant growth covers large areas in Spain, Italy and Greece especially where excessive burning has been followed by severe soil erosion that has exposed porous limestones, or infertile granite or very siliceous soils. The characteristic species are the more xerophytic types such as cistus, rosemary, lavender, gorse, broom (*Helichrysum* spp.), palmetto or dwarf palm (*Chamaerops humilis*) and *Euphorbia spinosa*. Garrigue which consists largely of one species is often more useful for beekeeping or for scents and liqueurs than for grazing as most animals dislike thyme, lavender, rosemary and other *labiatae* (fig. 22). These monochromatic expanses are often given special names such as *tomillares* (thyme), *jarales* (*cistus* spp.) and *goscojales* (kermes oak) in Spain.

In Greece the macchia changes its floristic composition appreciably towards the north where the myrtle, arbutus, lentisk and wild olive are replaced by the Macedonian oak, box, terebinth (*Pistacia terebinthus*), juniper, jasmine and cherry laurel (*Prunus laurocerasus*). However, the hardier species of the macchia, such as the kermes, are present, together with the usual lianes such as clematis, ivy and the very spiny *Smilax aspera*. Throughout Greece and in parts of Dalmatia these denser plant associations degenerate into *phrygana* particularly where excessive soil erosion has occurred. This type of low, open growth of thorny and woody plants occupies large areas on the mainland and on the islands, where almost bare limestone or other rock is dotted with low clumps of grey–green spiny plants. The chief species are the spiny burnet (*Poterium spinosum*), Cretan thyme (*Coridothymus*), stunted Kermes oak, cistus spp., genista and marjoram, all of which grow little more than knee-high unless goats are absent.

Mediterranean steppe

Where the rainfall is very low and the soils permeable, infertile or saline, bush growth may eventually be replaced by bunch grasses typical of the sub-arid steppe. The only sizeable areas in the European Mediterranean are in the driest parts of the Ebro basin, eastern Andalusia, New Castile and southeast coastal Spain. Here expanses of bare soil, often eroded by rainwash, are dotted with a sparse covering of dry tufted grasses and of widely spaced low isolated bushes, such as the dwarf palm in Andalusia. The chief bunch grasses are a coarse esparto (*Lygeum spartum*) and alfa (*Stipa tenacissima*) which grow in tufts less than 1 m high. Alfa is the main source of esparto products in Spain and covers about 500 000 hectares.

The above description illustrates the degradation of Mediterranean vegetation with increase of aridity and of soil-moisture deficiency. However, the summer-dry climate also undergoes a transition on its cooler wetter fringes both on the mountain slopes and in more northerly latitudes where the summer drought is less severe and the winters may bring appreciable chill. Thus the Mediterranean vegetation province includes large areas of mixed evergreen and deciduous plants. The most extensive of these transitional tracts are in northern Portugal, Old Castile, Tuscany and the north Aegean coastlands. But a belt of varying width forms the submontane vegetation of the Iberian mountains and of the lower mountain slopes facing the Mediterranean almost throughout southern Europe. The intermingling of evergreen and of deciduous species can be readily appreciated when it is realized that the olive and cork oak survives up to about 500 m altitude in Old Castile and Tuscany and to over 600 m in southern Italy. The ilex is still hardier and survives at well over 1000 m throughout the Apennines. These altitudes are also fully within the realm of the sweet or Spanish chestnut (*Castanea sativa*) and numerous species of deciduous oak, some of which drop their leaves at the beginning of spring. The chief deciduous oaks in the western Mediterranean transitional belt are the hairy oak (*Q. pubescens*), and the Turkey oak (*Q. cerris*), Portuguese oak (*Q. lusitanica*), sessile oak and pedunculate oak. In the Balkans the valona oak, macedonian oak, and other deciduous species, some with downy leaves, are well developed.

Large areas of this mixed evergreen–deciduous formation have been either brought under cultivation or converted by cutting into brushwood interspersed with patches of arable and tree crops. Where severe erosion has occurred brushwood takes over. This coppice growth usually becomes more deciduous upwards and forms a characteristic of the middle and upper slopes of all the less lofty uplands in peninsular and insular Italy and the Balkans. It belongs essentially to central Europe climatically and floristically, and will be described later.

Mixed deciduous–coniferous forest

The transitional mixed evergreen–deciduous forest belt of the Mediterranean lands gives way upward and northward to either deciduous forest or to mixed stands of deciduous and coniferous species. The designation 'deciduous–coniferous' causes difficulties as over extensive tracts of richer soils the trees are purely deciduous while the conifers occupy only the higher areas or poor sandy districts. However, there is a certain uniformity of species and towards the upper limits of deciduous tree growth on the mountains a true mixture of deciduous and coniferous trees usually occurs.

There is an important difference between this humid mixed forest and the true Mediterranean plant associations. In the former, most woodland glades and clearings are clothed in mat-forming grasses and herbs. Unless regularly grazed or tilled they would soon revert to scrub and forest. Here, owing to rapid warm-season growth of the ground carpet and its cohesive nature, soil erosion is far less a problem than in the summer-dry Mediterranean climates.

The vast mixed deciduous–coniferous belt stretches

from Atlantic Spain to the Black Sea and sends long extensions southward into the Mediterranean down the Apennine backbone and the interior plateaux of the Balkans. Throughout its domain, the deciduous trees drop their leaves mainly at the first severe frosts of autumn or winter and the evergreens are chiefly needle-leaved conifers of boreal origin. However, one of these conifers, the larch, is deciduous. The commonest broadleaved, deciduous species are pedunculate and sessile oaks and beech (*Fagus silvatica*), but also widespread and sometimes occurring in almost pure stands are hornbeam (*Carpinus*), ash (*Fraxinus excelsior*), elm (*Ulmus* spp.), lime (*Tilia* spp.), Norway or mountain maple (*Acer platanoides*), the so-called sycamore (*Acer pseudo-platanus*), alder (*Alnus* spp.) and silver birch (*Betula alba*). The most common conifers are the Scots pine (*P. silvestris*), silver fir (*Abies excelsa*) and Norway spruce (*Picea excelsa*).

Throughout this vegetation belt the dominant trees assume a general altitudinal zoning. Deciduous oaks and Spanish chestnut occupy the lowlands and lower slopes; beech associations favour the middle altitudes and commonly intermingle with boreal conifers towards their upper limit. Above on the highest slopes boreal species dominate until, on the coldest summits, alpine or arctic species take over. These purely boreal and alpine associations will be described separately.

The altitudinal gradation of the normal mixed tree growth causes marked complications in so large an area because in Europe the winters get rapidly colder eastward and warmer southward. Thus the general level of the beech drops eastward so that in Hungary and Romania it descends to the lower slopes whereas in the Apennines and the Balkans it flourishes on the upper slopes above the Mediterranean vegetation belt. Thus in the hotter south the deciduous plant associations grow near summits and watersheds in districts notorious for heavy rains and climatically rather unfavourable to natural grass swards. Here trees and bushes form most of the forage whereas in pluviose northern Spain and the humid tracts of central Europe animals find fodder in grasses and herbs at least throughout the warm growing season.

It is usual on floristic grounds to subdivide the vast deciduous–coniferous belt in Europe into Atlantic and Central European types but for the reasons given above a Mediterranean subtype might well be added.

Atlantic type of mixed deciduous–coniferous woodland

In most of pluviose Spain and maritime Europe west of the Rhine and Rhône, associations of deciduous oak (*Q. robur*; *Q. sessiliflora*) dominate on the lowlands and Spanish chestnut and deciduous beech on the uplands. These species, however, often occur together in mixed stands, and with maple, birch and ash. Some Mediterranean evergreens grow but the moist summers and lack of adequate sunshine prevent the olive from yielding commercial fruits. Conifers, in spite of their common use in plantations, are far less abundant than deciduous trees.

Central European type of deciduous–coniferous woodland

On the Alpine piedmont and in central Europe generally between the Rhine and the Black Sea, deciduous oaks, with birch, hornbeam, lime, elm, and ash, dominate on the lowlands and the beech, often in almost pure stands, on the foothills. In the west on the lower slopes of the Cévennes and southern Alps, Spanish chestnut forms fine woods but on the Swiss plateau and lower slopes of the northern Alps beech used to dominate.

On the higher mountains throughout southern Europe the beech at its upper levels usually grades into a mixed beech–conifer tier. This transition or ecotone is narrow in the Alps and widens eastward.

Mediterranean subtype of mixed deciduous–coniferous woodland

Peninsular Italy and the Balkan peninsula are on the whole less afflicted with aridity than is Iberia where Mediterranean matorral covers extensive tracts. In northern Italy the complex tiers of vegetation on the Alps grade southward into a zone of Spanish chestnut which unfortunately since 1948 has suffered from a blight disease. The Po plain, originally under oak (*Q. robur*), chestnut, elm and ash, has been largely cleared. In its present woods the poplar dominates north of the river and the chestnut in the drier south.

Throughout the peninsula the evergreen–deciduous ecotone grades upward into a belt of deciduous hardwoods between 750 and 1400 m. In this broad altitudinal range, the Spanish chestnut dominates on the lower slopes and beech on the upper. In central Italy the chestnut abounds from 600 to 1000 m and then gives way to beech which dominates up to about 1700 m although it may share the higher slopes with white fir. Over all the less dissected terrain the deciduous woods have been felled and replaced by arable fields and brushwood. Deciduous coppice, known as *bosco*, covers one-eighth of Italy or half as much again as all the high forests. It is the prime landscape feature of the Apennines where it clothes the erodable interfluves and all the uncultivable declivities at middle altitudes. Its distribution even extends to the higher slopes where beech growth is stunted naturally and mingles with low-growing alder, juniper and rhododendron. On the numerous small farms the privately owned deciduous coppice yields a rapid return of firewood, charcoal, forage and fruits (especially chestnuts for flour) and at the same time prevents excessive soil erosion. In recent years coppiced woodland has covered 5·6 million hectares in Italy and mixed coppice nearly another 2 million hectares. Their combined extent – double that of Mediterranean macchia and scrub – demonstrates the importance of central European plant associations in Italy.

In the Balkan peninsula the tier of deciduous oak–beech forests lies between 750 and 1600 m in the south and 450 and 1600 m in the interior and north. Deciduous oaks, ash (*Fraxinus excelsior*), chestnut, hornbeam, plane (*Platanus orientalis*), walnut and poplar occupy the lower levels and in the hotter parts dominate up to the coniferous forest belt. But throughout the more humid and cooler parts of the Balkans the beech takes over. It dominates the interior plateau above about 1000 m in Albania and Thrace and forms a tier on the Yugoslav mountains at 700 to 1600 m. Everywhere except in inaccessible spots this deciduous oak–beech forest has been maltreated in much the same way as the Italian *bosco*. In Yugoslavia and Albania the trees are usually stunted or coppiced because of the harvesting of young trees and of branches for fodder and fuel. The stumps and mutilated branches often sprout into a dense brushwood. Throughout the Balkans these deciduous brushwoods of a central European type are given the Serbian term *shiblyak*. Some *shiblyak* is a natural secondary growth after deforestation and is dominated by the very prickly Christ's thorn (*Paliurus*) with an admixture of other shrubs such as oriental hornbeam, lilac (*Syringa*), the hairy oak (*Q. lanuginosa* or *pubescens*) and sumac (*Rhus*). Upon large areas of karst the porous rocky soils support only an open brushwood dominated by small specimens of ash-elm (*Fraxinus ornus*) and hairy oak. On the colder summits the beech forms a natural brushwood with dwarfed conifers.

The importance of brushwood is well seen also in Greece where it covers more than half of the national territory and is the key to the pastoral economy. Here after deforestation natural regeneration of tall trees is difficult because of the frequency of summer drought, the rapidity of severe soil erosion on unprotected soils, the common practice of cutting bushes for fuel, and the constant devouring of woody plants and bark by sheep and goats. In Greece, as in Yugoslavia and Albania, the brushwood includes evergreen macchia, the evergreen–deciduous pseudo-macchia, and the degeneration of each into *phrygana*. These cover a large proportion of Mediterranean southern and coastal Greece. Here *shiblyak* or deciduous brushwood is restricted to small areas at high altitudes. However in Macedonia and Thrace *shiblyak* is widespread on the uplands and in parts descends even to the lowlands. The ubiquity of this brushwood and the insignificance of mat-forming grasses demonstrate how indispensable it is as fodder.

Boreal and Alpine mountain associations

On high mountains where forests and snowfall prevail for long periods the vegetation closely resembles in physical structure and species that of the boreal coniferous forest and polar tundra. The tall-tree coniferous zone probably has a warmest month or months with a mean temperature exceeding 10°C whereas in the Alpine [arctic: tundra] zone the mean of the warmest month is between 0° and 10°C. The frequent recurrence of frost at night and high wind velocity appear to be the most potent climatic factors as on clear sunny days the insolation is far more intense than at sea level.

The *boreal mountain zone* contains species such as Scots pine and Norway spruce, identical with those in the boreal forest of the northern European lowlands but many of its species, although closely related to their northern counterparts, have become distinct, as for example the European larch (*Larix decidua*) and silver fir. The conifers advanced far southward during the glacial phases and today the Scots pine dominates at 1500 to 2000 m in the Sierra Nevada and the white fir (*Abies alba*) at 1000 to 1800 m in the southern Apennines. Similarly the black or Austrian or Serbian pine (*Pinus nigra*) is found over much of central and Mediterranean Europe.

The boreal belt on mountains usually persists up to the timber line if the limit of tall treegrowth is reached locally. As already noticed, the conifers appear in a deciduous–coniferous ecotone in which with increasing altitude they gradually replace beech and other deciduous species. As the conifers, unless repeatedly felled by man or eaten by goats, regenerate naturally in most areas, they survive as original forests on, for example, parts of the Alpine flanks. Much spruce and white fir remains at 1600 m to 2100 m in the Pyrenees. Throughout the interior central Alps spruce, sometimes with larch, is common at 1000 to 1850 m and above it larch and cembran pine form closed forests up to 2150 m.

In the northern and central Balkans conifers cover extensive areas on the higher mountains up to the timber line at 1700 to 1900 m. Sometimes the stands are pure but more often two or more species intermix. In the north on steep limestone slopes the spruce (*Picea abies*) grows tall and slender like the cypress. Changes, however, are evident in the coniferous species. The white-barked pine (*Pinnus leucodermis*) is common and on lofty tracts the five-needled or peuke pine (*P. peuce*) flourishes. The latter is endemic to the Balkans.

Albania marks more clearly the change to the conifers characteristic of the southern Mediterranean peninsulas. The larch is not native and the spruce is unimportant being restricted, as in Greece, to a few summits in the north. But the Serbian or Austrian pine was common before it was felled for shipbuilding, and the white-barked pine and five-needled peuke pine occur up to the treeline. In the south, many of the woods are dominated by the Greek fir.

The southern parts of the Mediterranean peninsulas experience a tendency to drought in occasional years even on the lofty mountains. Here endemic species which seem to show a long evolution on the spot are commonly found. In the Sierra Nevada the firs are represented by the Spanish fir (*Abies pinsapo*). In Corsica, Sicily and La Sila, the finest tree is the Corsican pine (*P. laricio*), a close relative of the Austrian or Serbian pine. In Greece, the Peloponnese highlands are largely covered with Greek fir and

Austrian pine. The former (*Abies cephalonica*) is named from the Greek island, today called Kefallinia, where it forms magnificent forests. It tolerates limestones whereas the Austrian pine keeps to siliceous soils. Some authors think that these coniferous stands deserve to be classified as a 'supra-Mediterranean type'.

Alpine mountain associations are restricted to the highest ranges. Here wind speeds are higher than in the Arctic but solar radiation is very intense during sunny spells and permafrost (permanently frozen sub-soil) is absent. Differences in exposure to sun and wind, in gradient and in drainage cause the upper tree limit to vary greatly in altitude. Forests generally peter out at about 2000 to 2400 m in the less snowy, less exposed central massifs of the Alps proper but the treeline drops to 1800 m in the Dinaric Alps and to between 1400 and 1600 m in the northern parts of the Alpine–Carpathian system. Today in large areas of the Alps, and formerly on most high ranges, the normal gradation of plant associations consisted of open coniferous forest, particularly of spruce, cembran and other pines, degrading upward into a *sub-Alpine tier* of scattered clumps of mountain pine and juniper interspersed with swards. In this 'elfin woodland' the branches of the shrubs spread close to the ground. Typically the dwarf mountain pine sends out long zigzag stems that often root at contact with the earth and rarely exceed a metre in height. The juniper is equally dwarfed and the five-needled pine (in the Dinaric Alps) will contort itself into a shrunken girdle of brushwood. In Yugoslavia there are considerable areas of true mountain brushwood consisting mainly of dwarf juniper, mountain pine and very squat scrubs such as whortleberry, crowberry, mountain blueberry and bearberry, all of which prefer noncalcareous soils. Throughout southern Europe this scrub growth on the mountain summits has been partly or largely replaced by grasses.

The *upper Alpine tier* or *Alpine meadows* have a predominance of grass species and arctic plants. In Yugoslavia and Albania, the Alpine meadows are characterized by *Poa*, *Festuca*, and mat grass (*Nardus*). Here and, elsewhere, the meadows are being extended by transhumant flocks. Almost inevitably there are differences between the mountain pastures of Alpine and Mediterranean Europe. The herbaceous and scrub flora is less rich in the peninsulas. Of the Alpine Arctic species, about 67 per cent have migrated into the Balkans, 54 per cent into Iberia, 40 per cent into peninsular Italy and very few into Sicily. The resultant swards are far less nourishing partly also because in the south the snow melt is often followed by summer drought whereas in non-Mediterranean Europe thunderstorms frequently moisten the meadows during the warm season. Thus in the south the Alpine pastures are better suited to grazing than to hay. However, the areas of mountain pastures in the Mediterranean lands are relatively small except in the north. In Iberia the main summer grazing lands are in or near pluviose Spain. In peninsular Italy they are on the Abruzzi where the species (*Poa*, *Festuca*, etc.) are decidedly Pontic (Asian) in origin. In the Balkans the abundance of mountain pastures in northern Yugoslavia contrasts vividly with their paucity in southern Greece, where the relatively small tracts above the treeline are largely bare rock with local patches of dwarf juniper and daphne (*D. oleoides*).

SOILS

In southern Europe generally the complex mosaic of local soil types has been greatly altered by the activities of its inhabitants, particularly by deforestation and cultivation during the last five or six millenia. Only where the same type of vegetation has persisted throughout late Quaternary times will the natural soil cover persist in its original condition. Owing to the widespread exploitation by man and to the appreciable climatic change from cold glacial to the present warmer interglacial phase the present soils are not always a product of the present climate. The postglacial vegetation in much of the Mediterranean and central Europe was rapidly readjusting its distribution to the warmer climatic conditions when man began to use fire and tree-felling techniques on a sizeable scale. In some areas, as in the boreal mountain zones, the wet climate and virtual impossibility of using the land for other than tree growth, favored the natural regeneration of forests. But in most areas, once cleared, the chance of natural reafforestation was less as the surface was now exposed to erosion and to the more intense evaporation of a warmer climate, while the increasing population needed more wood for domestic and constructional purposes. No doubt forest would eventually return but the time taken for natural regeneration is so long that the chances of survival are small. Thus over the major part of southern Europe, the soils below the boreal zone reflect human or biotic influence as well as the more natural influences of climate, geology, vegetation, relief and the relative duration of present and past soil-forming processes. In practice the natural factors if left alone work together to form a soil profile with horizons that tend to assume distinctive characteristics.

The climatic influence is best seen in the appreciable differences in the vegetation and soils of the cool climates with rain all the year and of the warm summer-dry climates. In the former the frequent downward percolation eluviates soil particles and soluble constituents from the upper (A) soil horizon into the lower (B) horizon, and so eventually right out of the soil. The lower horizons become enriched with compounds of *Al*uminium and of iron (*Fe*) and the soils are called *pedalfers*. Thus in these continuously wet cool climates the acidic organic compounds combine mainly with *Al* and *Fe* which, as said, are washed downward leaving above them a layer with a high proportion of whitish–grey quartz silica. Such a soil profile is termed a podzol, from the Russian *zola* (ashes) and *pod* (under).

In warm humid climates with a marked dry season the

podzol layer does not form, partly because silica dissolves more readily in warm solutions and partly because in the hot season the soil may at times dry out and may also then experience some upward capillary movement of soil moisture. The effect of the drought and capillary action on certain soil minerals is to spread aluminium (especially kaolinite), iron oxide crystals and silica throughout the soil profile. The crystallized iron oxides give a vivid reddish colour to the soils.

In drier climates where for long periods evapotranspiration prevails over rainfall and downward percolation often does not pass right through the soil profile, an accumulation of calcium carbonate or of some other alkali occurs somewhere in the soil. Such soils, called *pedocals*, are important, for example, outside Mediterranean Europe in the chernozems of the steppes of the lower Danube plains. The relative importance of downward percolation and capillary upward movement is not certain but probably the effect of the latter is exaggerated as highly soluble salts occur at greater depths than moderately soluble salts. Frequently in the Danubian steppes the calcium carbonate enrichments form scattered nodules only. In the coastlands of Mediterranean North Africa a thick layer of lime or calcrete is often present but in the peninsulas of Mediterranean Europe, where aridity is much less severe, calcrete bands are not common. The distribution of calcium carbonate in the soil profile is greatly influenced by topography and the porosity of the parent material. On well-drained loess, which is highly calcareous, the lime is yusuall washed down to the normal water table. In flat and basin-shaped depressions the percolation may soon be arrested by a high water table. Here during dry spells upward capillary movement, coupled with lateral seepage and hydraulic pressure from the underground aquifer, may bring salts to or near the surface. However, these surface accumulations are given special names (see p. 61).

The influence of geology or parent material on soils weakens with time if the natural vegetation cover is not interfered with. In the early stages of soil formation the parent rock, no matter whether residual rock or superficial deposit, affects the chemical elements available and the rate and nature of soil-forming processes. It happens that over much of Europe man has destroyed, permanently or temporarily, the natural vegetation and at least some of the soil cover has been removed by erosion. Hence it is common in upland areas for soil profiles to be truncated or to have lost some of their upper horizons. Over considerable mountainous areas the soil has virtually gone and the parent rock acts as skeletal soil. Equally common is the covering of lowland riverine and coastal tracts with new deposits, which to a certain extent are also skeletal soils and indeed are of great importance on the deltas and river terraces of the Mediterranean lands. Thus the action of man has lessened the influence of climate and increased the influence of geology on the present soil cover.

The general influence of parent material will be briefly exemplified here. Sandy, siliceous and gravelly soils are as a rule permeable, naturally acidic and favourable to podzolization. This is seen, for example, on the fluvio–glacial and glacial deposits of the Swiss foreland and of the sub-Alpine *piedmonte* north of the Po. However, it will be noticed that, compared with glaciated northern Europe, southern Europe is peculiarly devoid of coarse glacial debris except around the chief mountain ranges.

On granitic and gneissic rocks under warm humid conditions weathering will form a thick mantle of debris above a rock surface pitted with fissures and hollows that holds up large quantities of ground water. The fine forests of cork oak and chestnut of La Sila in southern Italy are on granitic soils. On the other hand, some granitic uplands are capped by bare rock, sometimes sculptured into bizarre shapes. These probably were fashioned by groundwater solutions when the granite was covered by soil, and serve as a reminder of the erodability of such soils when the vegetation is destroyed.

Volcanic rocks vary greatly in compactness and content. Where the tuffs consist mainly of fine ash and dust they are porous and will weather into fertile soils.

Shales and schistose rocks are, as a rule, so rich in clay minerals that they are very impervious. On slopes the lubricated clay particles when wet are liable to induce landslips and when dry to flake off at the surface. In parts of southern Italy the purer clays may produce almost sterile badlands and throughout the thick shales of the peninsula and Sicily, landslips (*frane*) and gullying are common.

Southern Europe is exceptionally rich in limestones. Where these are soft and loamy, leaching may remove the loam or clay fraction almost as fast as weathering frees it, and the soil remains dark-coloured, highly calcareous, and studded with rock fragments at its base. These are *rendzinas*, many of which, if kept free of biotic interference, would develop into brown forest soils (fig. 23). The innumerable patches of dark grey to reddish clayey soils (*terra rossa*) that occupy hollows and depressions in the various karsts overlie relatively pure, compacted limestones. They may well represent a degradation of the brown forest soils formed during the last glacial phase, and subsequently exposed by deforestation and overgrazing. Then surface erosion removed the top soil horizons leaving pockets of terra rossa in protected hollows. However, the actual age of the terra rossa is problematical and will be discussed again later.

Soil types

In southern Europe most of the recognizable major zonal soil groups are podzolic as would be expected of a humid area that during the last glacial phase was fairly close to a continental or a mountain icesheet. Intrazonal soil types, or those which have not developed in a way normal to their zone because of the excessive influence of

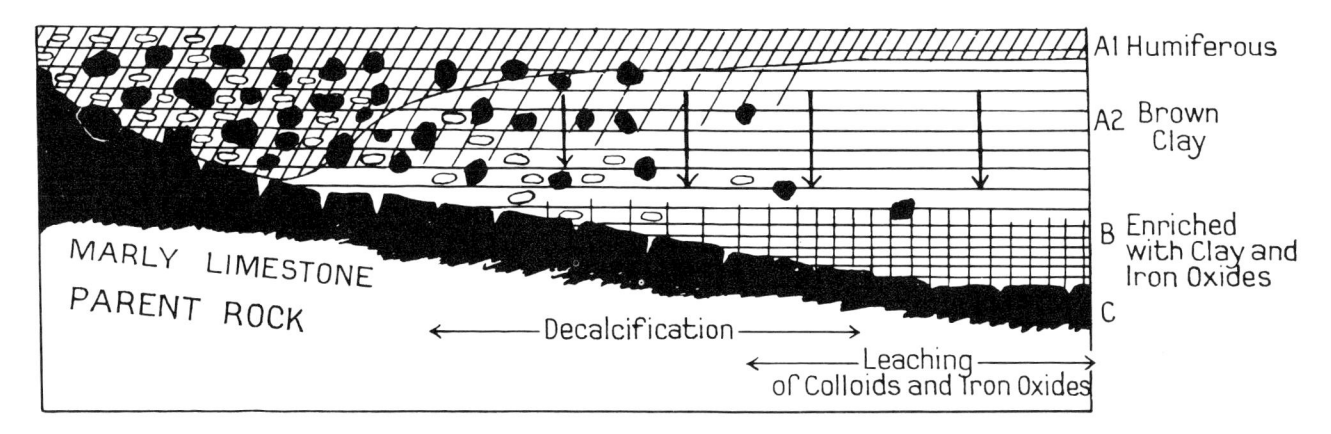

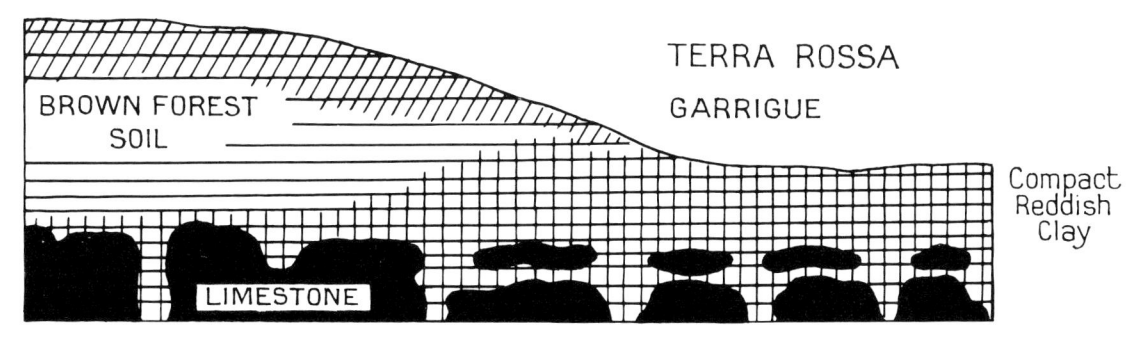

23 *Development of rendzinas into brown forest soils and of deforested brown forest soils into terra rossa* (after Duchaufour)

Oblique shading denotes humiferous horizon, horizontal shading denotes clays, and vertical shading iron oxides. White particles in and near rendzina are lime mottles. Vertical arrows show main direction of leaching process

some local factor, also cover large areas particularly where there is an excess of either calcium carbonate or of groundwater. In addition, azonal or undeveloped soil types such as rocky lithosols are widespread, particularly on deforested mountains near the Mediterranean. Some soil maps of Spain go so far as to show the major part of the Meseta and Betic Cordillera as skeletal or 'little developed' soils.

24 *Profiles of grey-brown podzolic soils and podzols*

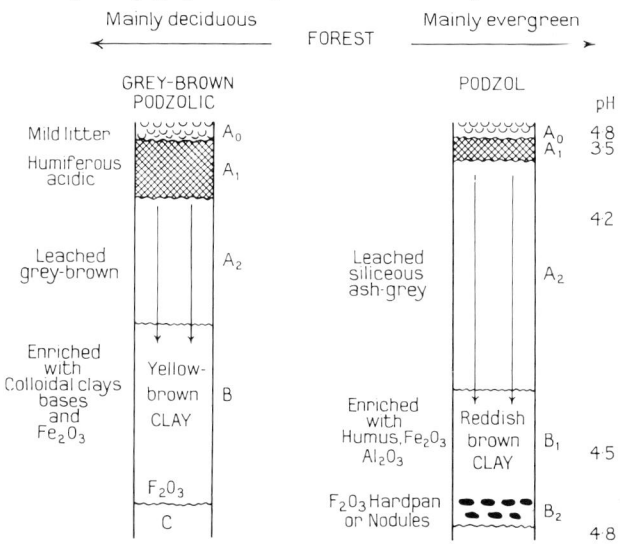

Podzolic soils (fig. 24)

In forested areas with all-the-year rainfall the frequent and dominant downward percolation creates podzols. However, soil differences arise with differences in temperature as the solvency of silica and the decay in humus increase with warmth. Podzols may form under any type of forest, even tropical selva, but they are best developed in cool to cold climates under conifers and under mixed forests on siliceous soils. In Europe the influence of a rise in temperature is reflected in the gradation from podzols, to brown forest soils > grey–brown > grey > reddish-brown > and red podzolic soils.

Podzols

Podzols proper are rare in southern Europe except on wet mountains where they are widespread beneath boreal coniferous and mixed forest. The characteristic podzol develops under dense forest and is heavily leached because of abundant rainfall, spring meltwater and high acidity. A typical soil profile is

A_0 A litter layer, 5 to 8 cm thick, often mainly conifer needles.

A_1 Grey–reddish brown acidic layer, with some humus, 2 to 5 cm thick.

A_2 A well-marked, leached, white to ash-grey horizon,

from a few cm up to 65 cm thick; loose, structureless or on clays laminated, and relatively rich in quartz silica. This is the *podzol horizon* where eluviation occurs.

B The illuvial horizon, packed with organic colloids, bases, and to a lesser extent with iron and clay minerals (mainly illite) washed down from above. It is often reddish brown in colour, nutty in structure and enriched locally with oxides of iron and aluminium that sometimes form nodules in the lower (B_2) layer and sometimes cement it into ochreish brown hardpans (*ortstein*). The whole illuvial horizon is often 40 to 80 cm thick.

Under natural conditions, the organic decay is slow, the physical structure ill-developed, the acidity excessive, drainage poor and, not surprisingly, the fertility low. But, when cleared and liberally supplied with lime and other fertilizers, the podzol yields moderately. The sowing of grass and legumes weakens the ironpan or hardpan where present and careful ploughing to a gradually increasing depth, with the addition of organic matter, will intermix the soil horizons and improve the drainage. Podzols then are suitable for grass, small grains, and vegetables, especially potatoes.

Brown and grey–brown forest soils (fig. 24)

Under slightly warmer and drier climates the dominantly coniferous forest gives way to dominantly deciduous forest and eventually to almost pure stands of beech, oak and other hardwoods which yield an annual leaf-fall. In southern Europe the brown podzolic soils are commonest in the central and western areas where the winter is milder and annual rainfall higher while the grey–brown podzolics become more prevalent towards the east or continental interior where the winters are colder and spring snowmelt is important. The brown podzolic soils cover large areas of the plains and lower and middle mountain slopes of the Alpine foreland, the non-Mediterranean Saône–Rhône valley and northwestern Iberia. They are common also on marls and clays beneath deciduous forest in the northern and central parts of Italy and the Balkans.

Compared with the boreal podzol, the brown or grey–brown podzolic soils are less leached, less acidic (almost neutral, *pH* values of 6 to 7 being common) and more retentive in their upper horizons of iron hydroxides which, with the organic matter, give the grey to brown colour. As the rate of leaching and of the dissociation of clay minerals is relatively slow, the grey–brown soils do not show podzolization. Clay eluviation goes on but the lower *B* horizon is not normally cemented or compacted. The characteristic profile is arranged as follows:

A_0 A thin cover of mainly broadleaved litter.

A_1 About 7 to 17 cm of dark grey to brown silty loam, rich in humus.

A_2 About 8 to 20 cm of brownish-grey, leached, platey-textured silty loam.

B 20 to 55 cm of yellow–brown to brown silty clay loam, acidic, enriched by illuviated clays and bases, and often arranged in coarse, angular, blocky aggregates which increase in size downwards.

Although only moderately fertile in a natural state, these grey–brown podzolic soils respond rapidly to good man-management. Today they are mainly under a wide variety of cereals, fodder grasses, roots, fruits, legumes and vegetables. They are considered well above average soils and some have been part of agricultural systems for over a thousand years but they need occasional liming and regular heavy manuring for high yields.

The brown forest soils are more truly typical of central Europe but they pose certain problems in Mediterranean lands where they are less continuous, for example on the mountains between 1700 and 2100 m. The characteristic podzolic profile occurs commonly in the humid zones of north Portugal, the Asturias, much of peninsular Italy, especially on the Adriatic slopes, and in the central and north Balkans. But on the wetter parts of these areas, as on the southwest slopes of the Sierra Nevada and parts of pluviose Spain, the brown forest soils may be degraded by heavy leaching, until most of the upper horizon has gone. But it is a third form of brown earth in the Mediterranean countries that poses the main problem. This is often called Southern Brown Earth or Brown dry forest soil. It is paler and far less stable than the brown podzolics and is thought to have been formed under evergreen oak forest. When these forests were felled, as was usual throughout the Mediterranean, the upper soil horizons were quickly removed. The typical Mediterranean brown earth is deficient in humus and retarded in chemical weathering. It is formed especially on hard silicate rocks such as granites, other igneous rocks, gneiss, mica schists and clay schists, usually today under a sparse dry vegetation of dwarf evergreen oak (ilex), juniper, umbrella pine and thyme. The absence of stable binding substances make the soil mass easily erodable and, as the plant cover is open, soil wash and gullying are common on it. Upon the southeastern slopes of the Sierra de Guadarrama, these soils have lost almost all their upper horizons and in places are almost totally removed. Their erodability contrasts with the distinct stability of brown forest soils in central Europe and is all the more serious as Southern Brown earths occur widely on silicate rocks in the hilly regions of Greece (fig. 25), southern Italy and 'arid' Iberia.

Mediterranean terra rossa

In hot humid climates bacterial decay quickens and silica dissolves more rapidly. Podzolization gives way to desilicification or the washing out of silicates from the soil. Simultaneously the soil profile reddens and becomes relatively rich in iron and clay minerals (kaolinite). These red soils are often called laterites in the tropics.

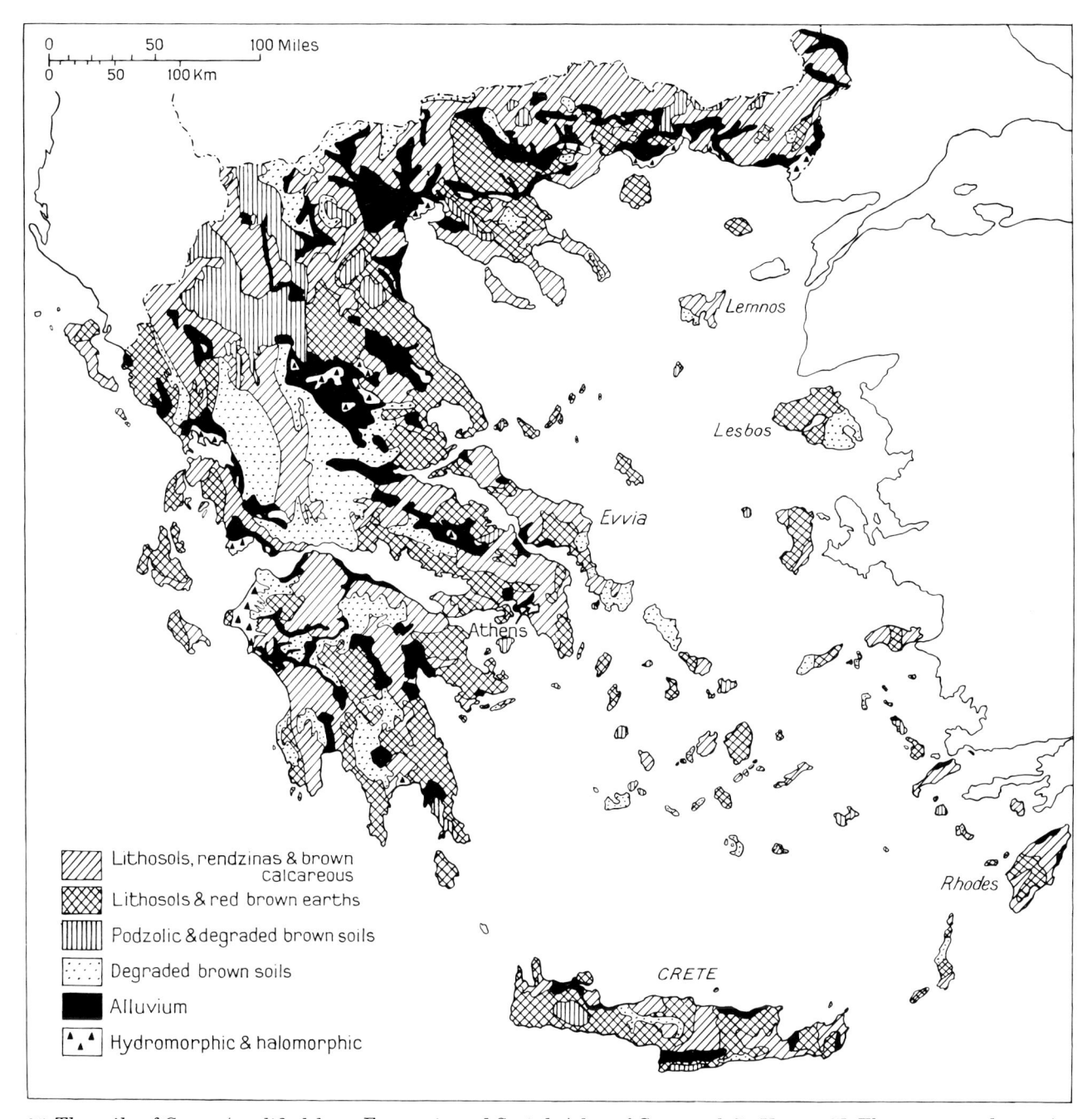

25 *The soils of Greece* (modified from *Economic and Social Atlas of Greece*, ed. B. Kayser, K. Thompson *et al.*, 1964)

However, the Mediterranean today is not tropical and does not develop tropical laterites. Yet in summer it is hot – much hotter usually than at the equator – and then normally has a distinct period of drought. Consequently it poses problems of both a transitional climate and of a climate that has changed greatly during recent geological times. In the late Tertiary the Mediterranean area was much warmer than it is today, whereas in the Quaternary it had several long periods with a cold to cool, humid climate.

The characteristic Mediterranean soil is the terra rossa or red earth (Italian, *ferreto*) which lacks a soil profile in the normal sense. This absence of distinctive horizons in profile is puzzling and has led to numerous theories. The chief are that terra rossa is:

1 the degradation of brown forest soils and of brown earths

2 the residue of lateritic soils dating back to the hotter tropical climate of later Tertiary times, and
3 the characteristic or normal product of the present Mediterranean climate and vegetation

It will be seen that the theories overlap. Brown forest soils and brown earths could have deveoped on existing 'red earths', or fossil soils. Terra rossa in pockets in the limestone karsts might well date back to later Tertiary times as it is the residue of the solution of almost pure limestone and estimates suggest that in parts 400 m of limestone would have to be dissolved to produce 1 m of it. Yet soils formed in late Tertiary climates would be tropical, with a clay mineral content mainly of kaolinite whereas terra rossa usually has a heterogenous collection of clay minerals, including much illite that is characteristic of brown forest soils. Many pedologists disagree with the idea that terra rossa is in true equilibrium with the present

Rainfall decreasing →

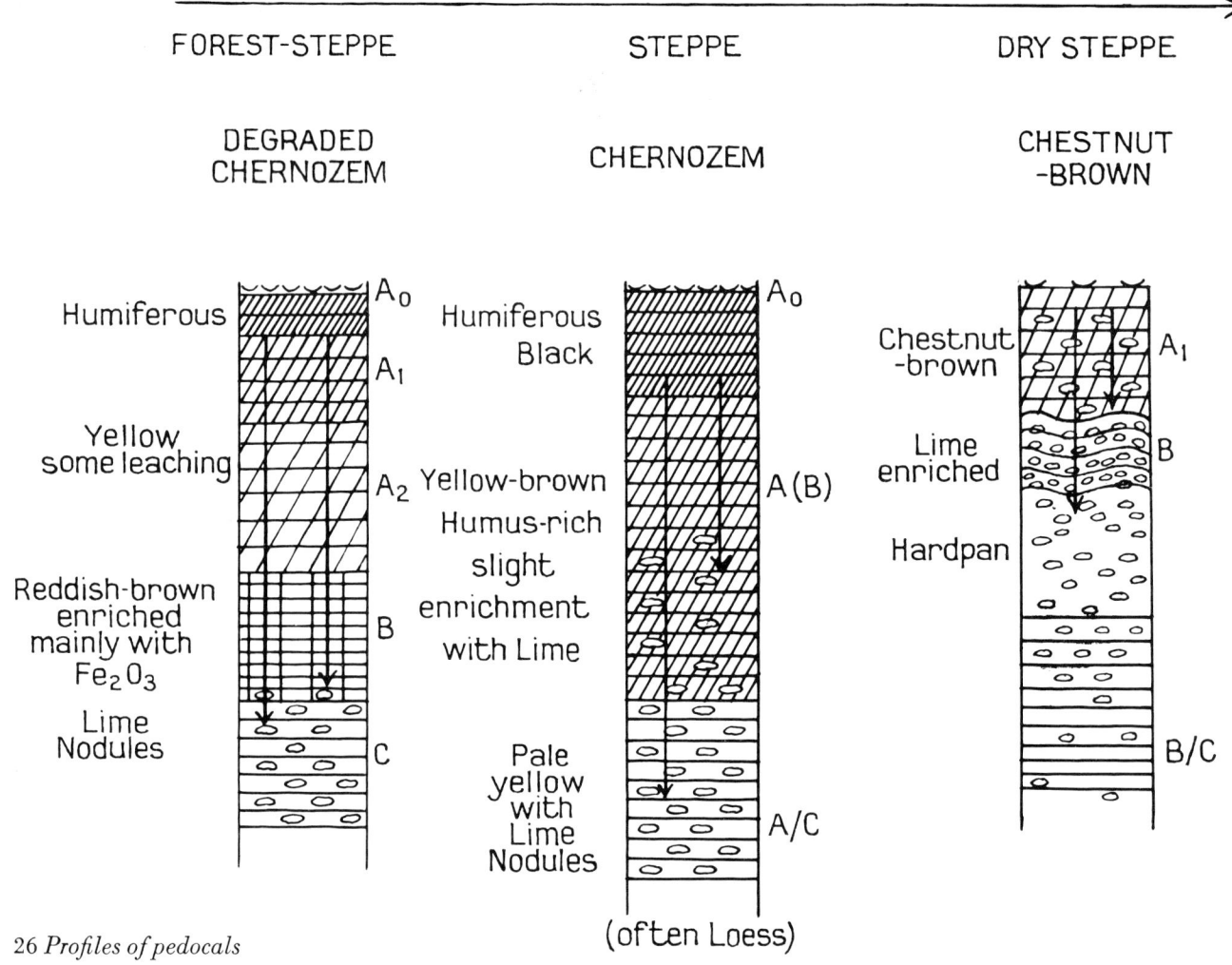

FOREST-STEPPE	STEPPE	DRY STEPPE
DEGRADED CHERNOZEM	CHERNOZEM	CHESTNUT-BROWN

26 Profiles of pedocals
For key see figure 23. White blotches denote lime mottles and nodules (figs. 23, 24 and 26 are from R. P. Beckinsale, *Land, Air and Ocean*, 1966)

Mediterranean climate. Yet it is generally admitted that reddening has occurred on alluvial terraces and mountain screes both in the warmer phases of the Quaternary period and in the present postglacial warmer phase.

It seems obvious that during the cool phases of the Ice Age, brown forest soils would develop in the Mediterranean lands at the expense of the terra rossa. With each return of a warmer climate the more open Mediterranean vegetation would tend to reassert itself and the soils would be more exposed to insolation, evaporation and to reddening. Unfortunately, during the last warmer phase, man deforested vast areas at an incredible rate. Thus terra rossa may in fact be either a fossil or palaeo-soil, or the lower layers of a truncated soil profile or a late Quaternary soil development.

Terra rossa varies considerably in calcium and silica content as well as in texture and porosity. Typically it develops on purer limestones and on schists. It occurs occasionally on alluvial terraces but rarely forms on marls and probably never on granites. On a few rocks, such as red Triassic clay and sandstone, Miocene sandstones (often bright red) and some red volcanic earths, the soil cover is naturally red but these outcrops are small in southern Europe. The redness of most of the Mediterranean red earths is attributable to climatic influence. The red colour is due to the abundance of iron, especially as minute crystals of haematite (sesquioxide of iron; Fe_2O_3) which usually in association with a silica solution coats the other soil particles. The haematite–silica gels (solutions) are dispersed in the soil partly by upward capillarity movement during the drought, the heat of which encourages the crystallization of the iron sesquioxides. During the cool-season rains the haematite–silica complex is little liable to dissolution as the groundwater then is cooler and much of the crystallization process is irreversible. The relative absence of humus is also favourable to the retention of the haematite–silica varnish and deforestation encourages soil reddening.

The typical red earth is decalcified, humus-deficient, and rich in ferric hydroxide granules and concretions scattered throughout the profile which is a dazzling red or browny red. The soil is slightly acid to neutral and very stable. It is widespread in fissures, hollows and depressions in the karst landscapes of Mediterranean Europe and rarely in central Europe. Another form termed 'siallitic' or red loam (*bollo* in Apulia) is rich in silica and highly adhesive or plastic. This too is common in Greece and arid Spain. Generally, however, terra rossa is friable or non-plastic and very liable to erosion. It forms medium to heavy clay soils which where shallow are tricky to manage but where deep and adequately manured will produce excellent crops of wheat, vines, other fruits and tobacco. On steep slopes terracing is essential and in many Mediterranean districts the small stonewalled enclosures are as much to keep soil in as to keep intruders out. Manuring is often easier in small units and much skill and effort goes into the crop production.

Chernozems

In central and eastern Europe the grey–brown podzolic forest soils grade into chernozems. The transition coincides with the change of vegetation from open oak or beech forest to wooded steppe and treeless steppe. The wooded or silvo–steppe soil is generally regarded as a degraded chernozem (fig. 26) formed when the grasslands of a cold dry climate near the ice front were invaded by forest during the warmer, wetter climate that followed the retreat of the last main European icesheet. However, the various kinds of chernozem hardly concern us except in the Danube plains of Yugoslavia. The Mediterranean lands are sadly lacking in these rich soils.

On the drier steppes, as in large areas of the lower, more arid parts of the Ebro basin and of Old and New Castile in Spain, the scanty rainfall often penetrates to only a short depth and the sparse vegetation does not form a sward. The soil horizons become less leached of lime, less rich in humus and much lighter in colour. These chestnut-brown pedocals often have an accumulation of lime at a depth of 30 or 60 cm yet, especially on loess, they retain a prismatic structure and are excellent for agriculture provided adequate irrigation is available.

In very dry areas in Spain these chestnut-brown soils may be replaced by grey earths with a shallow humus horizon, 5 to 20 cm thick, overlying a calcium-cemented crust. The typical vegetation is sparse and xerophytic, *artemisia* and bunch grasses being common. These shallow soils are, however, hard to distinguish from the salt encrustations known as solonetz, described later in this chapter.

Intrazonal soils

In any zonal or major soil region there are extensive scattered areas where some special local factor predominates over the general climatic influence. In southern Europe the chief are river floodplains; hollows with a concentration of local drainage; and porous limestone surfaces.

Hydromorphic soils

Alluvial floodplains, liable to additions of new sediments, are of considerable importance on Mediterranean lowlands, especially the Po basin. Where seldom flooded or drainable, they usually are under either rich pasture or cultivation. Where waterlogging is frequent they usually have warp or meadow soils under pasture or swamp woodland. If the water table remains near the surface for long periods, the soils become blotched with ferrous oxide accumulations and contain a *glei* horizon where greenish–grey ferrous compounds accumulate. These hydromorphic or glei soils are usually marsh and gallery forests mainly of willow, osier, alder and poplar. In most European countries they form an important part of the lowland landscape.

Conditions are very different in the drier parts of the Mediterranean lands, where the rivers dwindle in summer, often exposing wide expanses of river alluvium. This alluvium is commonly derived from the erosion of terra rossa and other red and reddish–brown earths. The deposits in exposed river beds become further reddened and on terraces are often ochreous yellow to bright red in colour. The very thin upper horizon is low in humus content while the thick lower horizon is dense and quite decalcified. It is unfavourable to the deep spread of roots and is troublesome to farmers, especially horticulturists. It is also very erodable.

The alluvium from terra rossa is redder than that from brown earths. When derived from earthy terra rossa the alluvium may be less rich in silicic acid and slightly richer in humus and, what is more significant, have a crumb structure not unfavourable to penetration by roots. But when the alluvium comes from red loams its silicic acid content is higher and the soil very plastic These various kinds of red 'vega' soils are widespread, particularly in Spain, southern Portugal and Greece.

Halomorphic soils

In localities where downslope drainage concentrates in hollows and where summer drought further encourages the upward movement of saline solutions, the upper soil horizons may become saturated with salts. The most saline soils (solonchaks), rich in sodium chloride, are restricted mainly to localities in depressions in New Castile. More widespread in occurrence are the patches of solonetz, particularly in the Spanish steppes. Here the rainfall transfers the salts (mainly sodium carbonate and calcium carbonate) to the lower horizon, and a thin platey A horizon (pH of about 7) overlies a B horizon with a prismatic columnar structure and a pH of 9 or more. It is unusual for strong crusts of salts or lime to form. Indeed, as already noticed, except in semi-arid southeastern Spain, strong lime crusts at or near the surface are uncommon in southern Europe. Yet they abound in the arid non-European Mediterranean countries where they may be strengthened by silicic acid and may form a rock-hard calcrete band up to 2 m thick.

Calcimorphic soils

Upon extensive outcrops of fairly pure limestone throughout southern Europe, clay particles are removed by rainwash almost as rapidly as they are set free from the limestone and the soil is constantly recalcinated by contact with the parent rock. The resultant soil, known as *rendzina,* develops slowly partly because the high lime content impedes chemical weathering. It has only one horizon, which is blackish to dark grey in colour, rich in microbial humus and markedly calcareous (often pH 8 or more). It may thicken to 50 cm but normally is much shallower and often quite stony. As it rests directly on the raw limestone or dolomitic grit, it tends to dry out rapidly in hot dry weather. Rendzinas on dolomites may constitute all the soil cover whereas on some other limestones they may alternate with brown forest soils. In dry karst districts in the Mediterranean peninsulas a whitish ash–grey rendzina, with a very high lime content, may form.

It will be noticed that the term rendzina is not used for brown loams (*terra fusca*), red loam (*terra gialla*) and red earth (*terra rossa*) that are also weathered from limestones. The brown loams occur mainly on limestones and dolomites with a high rainfall especially in the northern Alps, and the Balkans. They have a thin A horizon and a dense reddish B horizon, often considered to be a relict or fossil soil. It may, however, be a degraded calcimorphic form of brown forest soil.

Black earths or *tirs* are usually grouped with calcimorphic soils as they are rich in calcium and magnesium bases. These peculiar soils contain little organic matter and over 50 per cent clay, the blackness of which is due to the abundance of manganese. The main clay mineral is montmorillonite which can absorb much moisture and swells when wet. The prime occurrence of tirs in Europe is on the upper terraces of the lower Andalusian plain downstream of Cordoba, an area which may well have been waterlogged in early postglacial times. Today these soils produce excellent crops. Their distribution and relation to the irrigated floodplain are shown in figure 49.

7 Land Use: General Aspects and Forestry

The importance of agriculture and forestry to a state may be judged approximately on the proportion of the national territory they occupy, the percentage of the total workers they employ and the proportion of the gross national product they provide.

GENERAL LAND USE

The area under agriculture in any national territory reflects closely the relief and climate as well as the pressure of population. The table of land use includes for purposes of comparison statistics for non-Soviet Europe and the Danubian countries. It reveals that the proportion under farmland in southern European states is lowest in the most mountainous countries, Switzerland and Albania, and highest in Italy and Spain. The high proportion of arable land and permanent crops in Italy is due partly to the large Po plain and partly to population pressure and to the warmth of the south that allows cultivation to persist up to fairly high altitudes on the mountain sides. On the other hand, the relatively large extent of agricultural crops in Portugal is attributable mainly to low relief, a relatively warm, humid climate, and the common practice of growing cereals beneath tree crops. There is not a great pressure of population. In Albania and Yugoslavia the areas under arable crops could be extended appreciably without excessive cost whereas in Spain and Italy, for example, extensions usually call for heavy expenditure on irrigation or drainage. The fact that Spain with two-thirds of its area 'arid' has the same proportion under arable crops as in rich humid lands such as Czechoslovakia and Romania demonstrates the influence of population pressure.

The proportion of each State under permanent meadow and pasture, as well as that under forest is discussed in detail later. The proportion not under some form of agriculture and treegrowth includes lakes, other water bodies, roads, built-up areas and so on. This land lost to farming is most extensive in the Alps where many of the rock summits are above the vegetation line. Yet only in Switzerland does the non-vegetated part equal or exceed the average of Europe outside the U.S.S.R. The reasons for this low proportion of bare mountain are a southerly location, a long tradition of pastoralism and agriculture, including terracing, and the use of scrub-dotted mountain slopes as pasture for sheep and goats. In Spain and Greece mountain crests with a scattering of bushgrowth are classified as rough grazing.

Percentage population employed in agriculture

In large areas of southern Europe agriculture is still the chief source of a livelihood. The table of economically active population shows that in the Balkans well over half of the labour force is employed in agricultural pursuits. Yugoslavia alone has 4 690 000 agricultural workers. In Iberia the land still supplies one-third of all jobs and only in the mountainous Alpine states and in other highly industrialized countries such as Italy does the proportion

LAND USE BY COUNTRIES, 1966 (in thousand hectares)

	Total Area	Agricultural Arable and permanent crops	Agricultural Permanent meadow and pastures	Forested	Other unused or built on	Percentage total Agric	Pasture	Forest	Other
Portugal	8 886	4 370	530	2 500	1 486	49	6	28	17
Spain	50 475	20 594	14 175[1]	11 616[2]	4 090	41	28	23	8
Southern France	13 662	5 000	1 500	5 000	2 000[3]	37	11	37	15
Italy	30 122	15 258	5 147	6 099	3 618	51	17	20	12
Yugoslavia	25 580	8 266	6 450	8 812	2 052	32	25	34	9
Albania	2 875	501	729	1 256	389	18	25	44	13
Greece	13 194	3 854	4 824	2 668	1 848	29	37	20	14
Switzerland	4 129	404	1 774	981	970	10	43	24	23
Austria	8 385	1 686	2 249	3 203	1 247	20	27	38	15
Czechoslovakia	12 787	5 373	1 771	4 450	1 193	42	14	35	9
Hungary	9 303	5 642	1 285	1 442	934	61	14	15	10
Romania	23 750	10 502	4 333	6 371	2 544	44	18	27	11
Bulgaria	11 093	4 564	1 238	3 617	1 674	41	11	33	15
Europe outside U.S.S.R.	493 000	151 000	91 000	139 000	112 000	31	18	28	23

[1] includes 12 792 000 ha of rough grazing
[2] includes 7 million ha used for grazing
[3] all statistics approximate

ECONOMICALLY ACTIVE POPULATION, 1965

(in thousands)

	Total	In agriculture forestry and fishing	Per cent total
Portugal	3 530	1 410	40
Spain	12 184	4 143	34
France	20 010	3 600	18
Italy	19 920	5 005	25
Yugoslavia	8 780	4 690	53
Albania	840	490	58
Greece	3 685	1 935	53
Malta	85	8	9
Switzerland	2 775	275	10
Austria	3 410	690	20
Czechoslovakia	6 477	1 366	21
Hungary	4 930	1 540	31
Romania	10 855	6 450	59
Bulgaria	4 435	2 615	59
United Kingdom	25 301	961	4
U.S.S.R.	119 170	38 910	33

drop to one-quarter or less. A comparison of the United Kingdom and Italy demonstrates the difference between agricultural employment in a highly industrialized state in northern and Mediterranean Europe respectively. In the United Kingdom, continual mechanization of farming techniques and a commercial policy of buying foodstuffs from abroad in return for manufactured goods, has reduced the agricultural labour force to 961 000. Here there is a striking difference between the insignificant part agriculture plays in the national employment and the large part it plays in the landscape. In Italy, on the other hand, from a much smaller national labour force, over 5 million people find work on the land and one family in four is directly dependent on it for a living.

Since 1950 the relative number of persons employed in farming has fallen sharply in almost every European state and would, no doubt, have fallen more but for numerous irrigation and land reclamation schemes. The decrease reflects the rapid growth of manufactures and of distributive and other services as well as the effect of increasing mechanization of farming. In most countries the proportion of the national labour force engaged in farming fell by at least 10 per cent between 1955 and 1965. In Italy the decline was over 15 per cent. However, it is important to notice that modern techniques and better manuring actually increase the output even with a lesser labour force. The mechanization has been on a massive scale and there are now about 1 million tractors in use. The saving in man and animal power and the speeding up of farming operations can be judged from the use of over 500 000 tractors in Italy and 200 000 in Spain. But to these must be added the smaller type of garden tractor as well as the combined harvester–threshers. The table shows how popular the garden tractor is becoming in Mediterranean and Alpine Europe where apparently the combined harvester–thresher has made relatively little impact except on the upland basins in Spain and the flat Danubian floodplains in Yugoslavia. This machinery needs fuel, often imported at high cost, but it sets free labour and fodder for other uses and stimulates manufactures.

COMBINED HARVESTER–THRESHERS (CH), GARDEN TRACTORS (GT) AND TRACTORS (T) IN USE IN NON-SOVIET EUROPE IN 1967

(in thousands)

	CH	GT	T
Albania	–	–	20
Austria	25	23	218
Belgium	7	7	77
Bulgaria	8	–	76
Czechoslovakia	14	–	195
Denmark	37	6	171
Finland	19	8	136
France	117	245	1 107
E. Germany	18	–	139
F.R. Germany	147	100	1 257
Greece	4	27	52
Hungary	10	–	68
Ireland	6	–	66
Italy	15	127	509
Netherlands	–	20*	135
Norway	10	19	79
Poland	8	2	151
Portugal	1	–	18
Romania	42	–	93
Spain	21	38	181
Sweden	39	–	251
Switzerland	3	105*	66
U.K.	60	73	352
Yugoslavia	13	–	47
Malta	–	2	–
Total			5 512
U.S.S.R.	553	–	1 739
U.S.A.	870	–	4 822
World total			14 416

* estimate

Agricultural contribution to gross national product

The contribution made by agriculture, forestry and fishing to the total national product, reckoned at factor cost, often seems rather disappointing. In 1950, this sector supplied about one-third of the value at source of the total production of most southern European states. Only in Switzerland was the proportion less than one-quarter. In 1960, after great progress had been made in land reclamation and farming techniques, the advancement of manufactures had been so rapid that the contribution of agriculture and forestry to the gross national product decreased to about one-quarter in Spain, Portugal, Greece and Yugoslavia. In Italy and Switzerland, where manufacturing industries had been pushed forward at a greater speed, agriculture now supplied less than 16 per cent of the gross national product. By 1967, great progress had again been made in the output both of agriculture and of manufactures. The result was that in Yugoslavia the relative national importance of farming actually increased slightly. Elsewhere in southern Europe it decreased. Today in the Balkan countries about 23 per cent of the value of the gross national product comes from agriculture and forestry. Elsewhere the proportion is just under 20 per cent in Iberia, 13 per cent in Italy and 10 per cent in Switzerland.

Since 1960 a Mediterranean Regional Project set up under the auspices of the Organization for Economic Cooperation and Development (OECD) for the benefit of European Mediterranean states and Turkey, has encouraged overall growth in the national economies, including education and all other services. It is estimated that under this project the value of agricultural products in each state will increase on an average by 2 or 3 per cent annually (5 per cent in Yugoslavia) up to 1975. However, most other sectors of the national economy, such as manufacture, construction, electricity, gas and water and services, will grow twice or three times as fast. Consequently the direct contribution of agriculture to the gross national product will fall to 20 per cent in Greece, 17 per cent in Spain, 16 per cent in Yugoslavia and 11 per cent in Italy. Yet even then the farmers in the countries of southern Europe will be playing a role in their national economies far greater than that of their counterparts in several states in northern Europe. (In 1967: in United Kingdom, 3 per cent; in Belgium, 6 per cent.)

FORESTRY

The large extent of woodland in southern Europe is surprising in view of the widespread exploitation of timber in recent millenia by conquering invaders, shipbuilders, cultivators and pastoralists. The evils of treefelling on erodable slopes were recognized by the Greeks and Romans and terracing of hillslopes was introduced quite early. Yet by the mid-nineteenth century tall forest had been restricted mainly to the more inaccessible tracts unsuitable for agriculture. Subsequently reafforestation of hillslopes and watersheds became common. Everywhere the tendency was to plant useful fruit-yielding trees, such as oak and sweet chestnut, or quick-growing species such as conifer, poplar and eucalypt, excellent for sawlogs or woodpulp. Today in most areas the general distribution of treegrowth follows a fixed topographical pattern. The flatter districts have been cleared, on the lowlands for crops and on the highlands for pasture; the steeper slopes are left wooded except in warmer locations where terracing may be worthwhile. However, the construction of terraces on sunny slopes is more a question of finance than of gradient and with modern machinery could be greatly extended.

Often the woodland, particularly in Mediterranean countries, is mainly in private hands and, as described in the chapter on Vegetation, plays an important part in the farming economy. However, the bigger reafforestation schemes are usually undertaken by the national government or with the aid of official funds. Normally state laws protect forests, and state education may teach the value of their conservation. In these respects the Swiss authorities are exemplary.

The extent of forest in each state depends largely on competition with agriculture and on the economic and social attitudes to land use. Thus, in Albania, the most forested country in southern Europe, the population pressure on land for agriculture is least and undoubtedly the proportion of forest will decrease steadily with population growth. On the other hand, the high proportion of forest in Yugoslavia represents more clearly the influence of highland topography with an appreciable rainfall. The exceptional mountainous countries are Switzerland and Greece. The Swiss forests are quite extensive and carefully managed but their possible extent has been encroached upon by pastoralists. Often on the steeper terrains they are restricted mainly to the northern, less sunny slopes. In Greece forests have had to be truly inaccessible to survive. Felling has been excessive and reafforestation, natural or artificial, is often prevented by severe soil erosion on steep limestone slopes, by summer drought and the rapacity of athletic goats.

As already noticed, it is difficult to distinguish between forestry and agriculture in many parts of the Mediterranean lands. In Spain, if the forested lands used for grazing (7 million hectares) are omitted, tall forest covers only 9 per cent of the state. In Portugal a peculiar coastal factor, the vast accumulation of littoral sands, has encouraged extensive lowland coniferous plantations, the counterparts of which occur in the Landes of Gascony in France.

The economic uses of the forests vary in importance regionally and nationally. The condensed table of forest products merely gives certain national totals, and includes for purposes of comparison statistics for Sweden, the chief seller, and the United Kingdom, the chief buyer, in Europe. Throughout southern Europe fuelwood remains important and in Italy, Spain, Greece (fig. 27) and Albania its annual cut exceeds that of industrial timber. Yugoslavia is by far the chief exporter and Italy the main importer of fuelwood. The spread in some rural areas of the use of natural gas (purchased in containers) and of electricity has not greatly diminished the general popularity of wood fuel which is very cheap and often is at least partly obtained gratis from private copses or communal woodlots. In the Mediterranean lands the supply of industrial timber, especially of sawn logs, has long been inadequate and most building is mainly of stone and concrete. Italy and Spain, for example, import considerable quantities of sawlogs, particularly of hardwoods.

When the total trade in forest products is studied it becomes clear that Portugal and Yugoslavia are the only countries in Mediterranean Europe to make a considerable profit from forestry. The extensiveness of forests in Yugoslavia explains its surplus trade. In Portugal, a country one-third the size of Yugoslavia, the surplus revenue comes mainly from coniferous timber and cork. Here the forests in 1968 included 1·4 million hectares of

27 *Forestry products in Greece* (simplified from *Economic and Social Atlas of Greece*, ed. B. Kayser, K. Thompson et al., 1964)

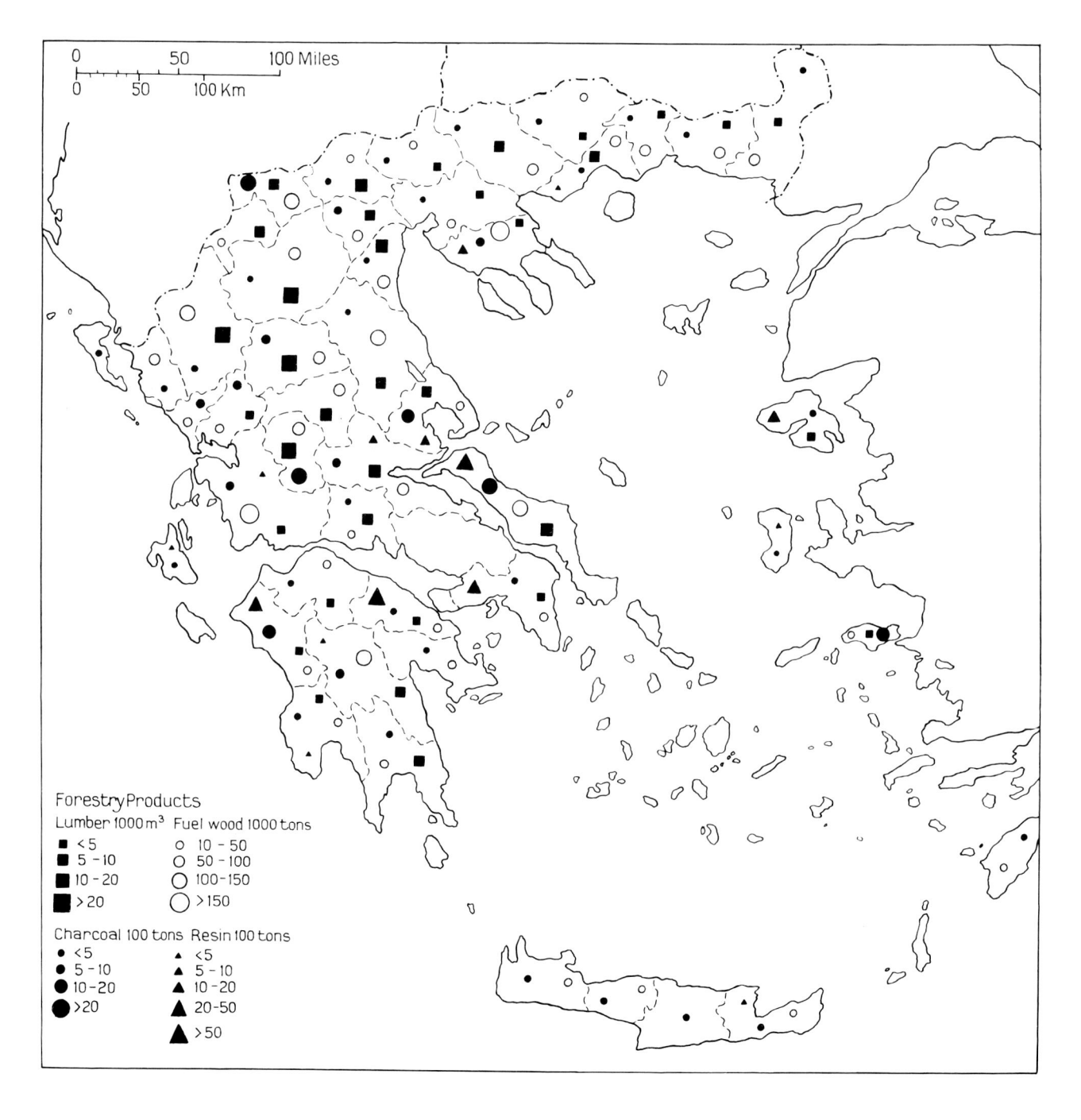

FOREST PRODUCTS, 1966

	Amount Cut (thousand cub m)			Trade in (u.s. $1000)	
	Industrial	Fuel	Total	Exports	Imports
Albania	490	1 050	1 540	?	?
France	25 113	20 000	45 113	275 640	554 076
Greece	360	2 300	2 660	1 442	77 787
Italy	5 975	9 866	15 841	128 047	498 500
Portugal	4 100	2 000	6 100	98 272	20 742
Spain	5 227	8 132	13 359	38 418	131 890
Switzerland	2 630	1 290	3 920	71 694	190 364
Yugoslavia	10 176	7 930	18 106	86 258	19 256
All Europe, outside u.s.s.r.	235 618	85 135	320 753	4 534 540	5 683 627
Austria	10 067	1 554	11 621	241 505	57 549
u.k.	3 078	413	3 491	171 060	1 476 617
Sweden	45 900	3 700	49 600	1 155 781	83 620

conifer, mainly maritime pine, which yield valuable sawlogs, pitprops and resin. Oak forests, mainly cork oak and ilex, cover another 1·5 million hectares and sweet chestnuts nearly 100 000 ha. Western Alentejo alone supports one-third of the world's cork-oak forests and produces over 40 per cent of the world supply of cork (fig. 19).

The woodpulp and papermaking industries in Mediterranean countries are small compared with those of northern Europe and depend mainly on modern plantations and imports. Eucalypts and poplars have been widely planted in Iberia and Italy for woodpulp as well as for drainage and shelter. In new forests in northern Portugal the eucalyptus is often intermixed with the maritime pine. In Spain this quick-growing tree abounds in Huelva province while the poplar is more common in the Ebro basin and along the riverine tracts in the Douro basin. In Italy the poplar dominates the Po watercourses but the eucalyptus is more favoured in coastal reclamation schemes.

Although Italy is the chief Mediterranean buyer of forest products, the extent of its tall treegrowth (over 6 million hectares) says much for its modern reafforestation schemes. Hardwood species predominate, the chief being the ilex. A notable feature is the widespread introduction of the sweet chestnut which now covers 530 000 hectares. The conifers (well over 1 million hectares) are mainly in the Alps and in new plantations in Tuscany. Unlike ownership in Spain and Portugal, private forests predominate (64 per cent total) over communal and state-owned woodland. Since 1950 the State has been attempting, with a considerable measure of success, to reafforest 425 000 hectares and to improve the quality of 700 000 hectares of existing woodland.

Throughout the Alpine lands the production of sawlog timber predominates over other forms of forest products. House building is commonly mainly or entirely of wood, even to shingle roofs, but wood fuel consumption is relatively small. Here Switzerland is in fact exceptional as it has a large deficit in forest products although its exports of them are considerable. Its close neighbour, Austria, which is not included in our survey, makes large profits from forestry.

8 Land Use: the Agrarian Economy

Grassland and pasture

As already noticed, in the humid climates of central and Atlantic Europe, pastures consist mainly of grasses and other herbaceous plants while in the summer-dry Mediterranean climate, boskage or brushgrowth often predominates. Pastures in the north will supply some feed almost all the year except where the winters are cold, whereas in the south the summer drought prohibits growth in the hot season unless groundwater or irrigation water is available. Growth in the cool season probably ceases when the mean daily temperature falls below 6°C. Consequently in central Europe the main pasturage problem is to retain or store enough food for the winter months. This may be done by storing hay or other fodder, including arable fodder crops. In the Mediterranean the bushgrowth is perennial but is much more palatable in spring and after rains. In the arid summer, stock that prefers grass may be sent to upland pastures.

The feeding habits of the main domestic animals allow them to be less competitive with each other than appears at first sight. Cows prefer long grass and sheep short grass, while the goat, although almost omnivorous and indifferent to obstacles, has a distinct liking for twigs and leaves. Almost all herbivorous animals enjoy tender tree bark and shoots and the habit of pollarding trees, to ensure numerous branches to throw to stock, is still common in parts of the eastern Mediterranean. The pig is rather exceptional as although truly omnivorous it prefers its food at ground level and cannot emulate the high-leaping antics of a hungry goat. Unlike other farm animals, the swine soon grub up pastures in search of roots, and where the ground is softened frequently by rain will rapidly ruin a good sward. The British reader trying to understand Mediterranean pastoralism could do worse than recall agricultural conditions in England at Domesday Survey. Then many woods were assessed according to their pannage for swine and the great dairy farms were on the sheep pastures as, for example, in eastern Essex.

The table of land use by countries in Chapter 7 shows the percentage under pasture and grassland of each state in southern Europe. The lowest proportion, 6 per cent, is in Portugal. This may be explained largely by the dominance of maize cultivation and pigkeeping on smallholdings in the humid north; by the extensiveness of cork-oak forests and ilex–cereal cultivations (all favourable to swine) on large holdings in the south and by a relative lack of interest in dairy cattle. Pastures with a high watertable all the year are rare throughout 'arid' Iberia and in the warmer areas such tracts where not marshy are more profitably devoted to vegetable crops. Yet, good quality meadows abound in the pluviose north and northwest where two-thirds of all the cattle are kept. Here, except on the central Pyrenean uplands, the mild Atlantic winters allow cattle to be grazed in the fields all the year and encourage dairying rather than stock raising, rearing and fattening. The main disadvantages are too much bracken and in parts too much gorse, although prized for litter. This permanent grassland economy has, as yet, been little converted to temporary leys, and contrasts markedly with the sylvo–pastoral economy of the drier parts of Spain. Here two-thirds of the total livestock are said to be supported by matorral for nearly two-thirds of the year, which is not surprising, as of the 14 million hectares classified as pastures in Spain nearly 13 million are rough grazing, mainly bushgrowth and semi-arid steppe. The high proportion of permanent meadow and pasture in Greece (37 per cent total area) is also due largely to the inclusion of bushgrowth or boskage which provides free-range grazing for the finest collection of goats in Europe.

The pastoral industry of Italy resembles that of Spain in so far as the north has extensive meadows but the Po basin has no equal in Mediterranean lands. It grows two-thirds of the Italian output of cultivated forage, supports 70 per cent of the state's cattle and produces over 90 per cent of its milk and most of its butter. Dairying predominates in southern Lombardy where, upon the wide floodplains and lower terraces irrigated by river and fontanili water, the *marcite* pastures yield six or seven cuttings of hay annually. The management in large units and on a cooperative system is highly efficient and some of the products, for example Gorgonzola and Parmesan cheese, have an international reputation. In Lombardy away from the Po and fontanili, and in western Emilia, the role of irrigated watermeadows is partly taken over by leguminous crops (lucerne, clover, etc.) and maize. Thus the pastoral side of the Italian economy falls into a central European and a Mediterranean type. The Po basin, with its warm-season rains and floods, contains most of the 3 per cent of Italy classified as 'meadow' while the peninsula and islands contain most of the additional 14 per cent (over 4 million ha) classified as 'other pastures'. The national proportion of true meadow is slightly higher than in Spain but the main difference lies in the economic approach. The north Italian pastoralism is highly commercial and capitalistic. In northern Spain, except near one or two larger towns, smallholding and individualistic marketing prevails and the ancient forms of transhumance are not entirely extinct.

In the Alpine lands and other countries, such as Albania and Yugoslavia, with extensive high-level mountain pastures the national proportion under pasture is well above the average for Europe. Switzerland (43 per cent total area) is exceptionally rich in meadow, a reflection of its ample precipitation, mountainous relief and traditional skill in dairying. These pastures occur on the foreland as permanent grassland, and on the mountains as warm-season alps. Even at moderate altitudes the intensity of insolation through the rarefied atmosphere is obvious in the grass economy. The growth is cut green several times in the summer and is cured naturally into a nutritious fodder. A little irrigation adds greatly to the yields.

Throughout the more humid parts of central Europe, except on the mountains, the agriculturalists have the choice of growing both cereals and fodder crops on the same land in a year. But this facility is not available to Mediterranean 'dry' farmers, who usually introduce into their crop rotations a long period of fallow which is deliberately aimed at *not* having fodder crops or any other crop because these would use the subsoil moisture. The economic advantage of having a fodder crop instead of a fallow is obvious. Thus, in central European climates the presence of meadows can be taken for granted and green forage crops only become distinctive in summer where set as verdant rectangles and riverine ribbons amid expanses of golden cereals.

It is important to emphasize also the striking contrast in Europe between the vegetation and land use of river floodplains with a summer spate and those with a summer low water. Throughout lands with a Mediterranean climate in the Old World there is, except in a few coastal or estuarine marshes, a notable lack of floodplains with a high watertable in summer. During medieval times the valley floors here experienced aggradation up to 10 m thick. Subsequently the rivers have begun to erode into this new sediment, leaving it upstanding as a lower terrace above the present incised rivers and floodplains. The low terraces if irrigated may be excellent for agriculture while the present narrower floodplain, where exposed in the summer low water, may be quite stony. In a few lowlands the present floodplain may consist mainly of fine alluvium and can be utilized as irrigated *huerta* or *regadio* but commonly it is too coarse to be fertile. In, for example, much of upland Greece, the edges of the winter flood channel commonly consist of pebbles and boulders that support only shrubs.

Livestock

The inhabitants of any country include the human beings and all other animals. This sum total might be called the national biomass. In discussing it we are mainly concerned with the units that are of most economic importance, primarily the people and the other domesticated animals, which together comprise the consumers, producers and workers for practical purposes. Today in many countries and in industrialized areas the biomass falls into two different types: an urban, of small extent, where most of the people and few of the other animals live, and a rural, of large extent, where most of the other animals and relatively few people live. In a few special areas both people and animals occur together in some profusion. To shorten the summary we have omitted small animals, such as rabbits and poultry. The latter play an appreciable part in the national diets by supplying eggs and meat but they are so widespread that their presence locally can be assumed. The following account deals largely with equine, bovine, caprine, ovine and porcine beasts in that order.

The horse, mule and ass are important beasts of burden and draught throughout southern Europe. The table of livestock or farm animals shows statistics for the whole of France and of Europe and, for purposes of comparison, of some of the Alpine and Danube states nearest to the countries described in this volume. Rough estimates of one-fifth to one-quarter have to be made for proportions of French livestock in the Midi and Rhône–Saône and French Alpine areas that are included in our survey, except for mules, donkeys and goats which are essentially southern beasts. The table indicates that well over one-third of the

NUMBER OF LIVESTOCK IN 1966 (in thousands)

	Horses	Mules	Donkeys	All cattle	Cows and heifers mainly for milk	Pigs	Sheep	Goats	Buffaloes
Albania	44	20	60	428	158	150	1 700	1 200	5
Greece	294	213	441	1 085	694	558	7 848	3 898	38
Yugoslavia	1 131	30	140	5 584	1 900	5 118	9 868	185	59
Italy	388	280	400	9 600	3 450	5 150	7 900	1 225	43
Portugal	75	138	255	912	–	1 600	5 793	588	–
Spain	305	694	442	3 694	–	4 681	18 785	2 309	–
France (all)	1 162	44	46	20 640	9 716	9 239	9 056	1 014	–
Switzerland	67	1	–	1 795	1 060	1 513	267	74	–
Malta	2	1	2	8	–	23	11	25	–
Non-Soviet European *Total*	9 073	1 466	2 180	121 445	–	116 429	133 650	13 390	353
Austria	85	1	1	2 441	1 103	2 752	142	98	–
Romania	689	7	10	4 935	2 008	5 365	13 125	800	80
Bulgaria	240	29	287	1 450	683	2 408	10 312	436	127

horses, over 70 per cent of the donkeys and nearly all the mules in Europe live in its Mediterranean countries. The mule, a cross between a horse and donkey, is common only in Iberia (57 per cent total), Italy and Greece. It is hardier than the horse and thrives on a coarser diet. The donkey or ass abounds in all the Mediterranean lands and, as the table shows, is common also in Bulgaria, an almost 'Mediterranean' state. The donkey is a useful, economical beast of burden and is in many ways the poor man's jeep. In contrast, the horse in southern Europe is most common in Yugoslavia which embraces extensive plains near the Danube, a steppe-like domain where much of the economy used to depend on the horse. The large numbers shown for France live mainly in the cooler, wetter northern and central regions of the country. Yet throughout Mediterranean Europe the horse is widely used both as a draught animal and as a traditional and most popular form of personal transport, particularly on farms and in mountains. It must be remembered, however, that whereas on the Danubian plains in Yugoslavia, as in the Russian and Asiatic steppes, the horse was the dominant animal, in Mediterranean lands it was always a luxury. It was introduced into Greece as a war machine – a one-man tank! – and, together with bulls, explains why the ancient Greeks prized so highly the rich summer pastures of their few coastal plains. Today everywhere as agriculture becomes more mechanized the number of horses declines, but in many rural districts the horse remains a status symbol equivalent to the automobile in urban societies. Perhaps the closest connexion between equine beasts and humans is in Greece where on an average there is 1 for every 9 persons. In the remainder of the Balkans the ratio of 1 to 15 is still amazingly high and reflects *inter alia* the lack of good roads in mountainous terrains.

However, there is also a spatial side to the question. In many districts, for example in the Balkans, where the people live in nucleated villages on the lower mountain slopes or less frequently on the lowlands, they may have to go long distances to their fields. In much of Yugoslavia this journey is commonly done in carts drawn by two light horses or a pair of oxen. In Greece most families move to their fields as a procession of beasts, not uncommonly the father on a horse, the wife and children on donkeys, with a donkey or mule as the pack animal. The time has almost gone when the man rode and the weaker sex followed on foot! Where tractors are used the whole family sits upon the vehicle or a whole group of workers fill the trailer. But even today the most striking feature of the rural Balkans at sundown in summer is the procession homeward of horses, donkeys and mules, some prodded rhythmically by the heels of their sidesaddle riders and others piled high with firewood and fodder. In Danubian Yugoslavia motorists trying to forestall darkness must constantly beware of horsemen and of drays and oxcarts moving laden to their barns in the distant village.

Cattle, sheep and goats are together far more numerous than the equine tribe. They provide meat and hides, and are today the chief suppliers of milk and hair (wool and mohair). The part they, and the pig, play in the human diet and farming economy is not exaggerated by their large numbers. They are the inhabitants of the fields and humbler dwellings and in districts with cold winters inhabit at that season the lower stage or extension of the domestic buildings.

In most southern European countries cattle or oxen play a notable part in draught and tillage. A difference between Mediterranean and north European livestock attitudes (before the latter became highly mechanized) is reflected in the English phrase 'to put the cart before the horse' which in, for example, French is *mettre la charrue devant les bœufs* – to put the plough before the oxen.

Generally in districts with a Mediterranean climate, working oxen far outnumber dairy cattle. The exception is Greece where nearly 70 per cent of all cattle are kept for milking purposes. This high proportion, however, is due to the relative absence of working oxen which here are largely replaced by donkeys and mules. The dairy cow probably consumes more water than a camel and to maintain high milk yields must be regularly supplied with abundant water or juicy forage. In countries with a central European climate, water supply is less difficult and dairy cows predominate, although in many rural areas oxen are still widely used for draught. In these territories with rain-all-the-year, the supply of forage is also easier and the numbers of store cattle (for beef and veal) also increase. The milk yield per cow is greatest in Switzerland (3410 kg/year) and lowest in Albania (560 kg/year) and Greece (1100 kg/year).* The output of cheese will be discussed after the account of goats and sheep.

Buffaloes (Asiatic variety) thrive on the lusher meadows in southern Europe but have been introduced in considerable numbers only in the Balkans, lower Danube lands and central Italy. These fine beasts are excellent for draught and tillage and yield nourishing milk which in, for example, central Italy makes an appetizing cheese (*mozzarella*).

Wherever the summer drought is intense, the goat becomes the prime producer of milk for fresh consumption. Over 90 per cent of Europe's goats live in the south and Greece alone has 30 per cent of the continental total. The frequent herds in Iberia, southern France, southern Italy and Albania are typically associated with free range grazing largely on brushwood partly created or maintained by the goats themselves. The milk is said to be free of tuberculosis (but see page 257) and is quite indispensable, particularly in Malta and Greece. This hardy animal also yields mohair, good meat and, on death, an excellent hide. Its importance declines in countries where dairy cows flourish. On the other hand, in Albania and Greece there are on an average 3 goats and 2 goats per family respec-

---

* In northern Europe many countries exceed 3500 kg per year, e.g. U.K. 3779; Denmark 3913; Netherlands 4188 kg.

VI *Transhumant herd on high alp above Zermatt, with view of the Findelen glacier*

tively. Yugoslavia had well over 1 million goats in the late 1940s. Here, in 1954, the keeping of goats was forbidden in the hope that their absence would encourage the rafforestation of karst scrubland. By 1968 the numbers had dwindled from 600 000 to 175 000.

Sheep are in many ways the rivals of goats but being less athletic usually fit better into arable and horticultural economies. Both can be moved long distances on foot or by other means and will feed en route but in their migrations the sheep cause much less destruction to tree growth. The ewe has less prolific lactation periods but her milk is often highly prized for cheesemaking. The individual yield of fleece, white or black, is far heavier than that of the goat, and dry climates and dryish pastures seem highly conducive to a fine fleece, as the Spanish merino exemplifies.

Southern Europe has over 40 per cent of the continent's sheep, and Iberia alone has just over 14 per cent. The Spaniards are renowned sheep farmers, being the founders of the Mesta and breeders of the merino. But in relation to the national population, sheep are just as important in Portugal and far more important in Greece and Albania. Whereas in Danubian Europe, as in Bulgaria (where sheep outnumber people) and in Romania, the large flocks reflect the abundance of stubble and of pastures, in the Mediterranean peninsulas they reflect mainly the presence of upland grazing for use in summer. The relative unpopularity of sheep in Switzerland demonstrates the difficulty of fitting sheep into a dairy cow economy. The rich pastures here are best left to cattle and in the lowlands the complement of an arable crop–cattle economy is pig-keeping rather than sheepfolding.

The general relation between cows, sheep and goats may be judged crudely on the cheese production in each country. In 1966, Switzerland, the home of Emmentaler, produced 80 000 tons entirely from cows' milk. In Spain, out of 81 000 tons, 41 per cent came from ewes and 12 per cent from goats. In Portugal, out of 21 000 tons 76 per cent came from ewes and 5 per cent from goats. In Greece, out of 116 000 tons nearly 90 per cent came from sheep and goats. The importance of cheese in local diets may be judged from the total production in Italy (437 000 tons). The major part of the cheese output comes from the north, the home of Gorgonzola, Parmesan and Bel Paese, all made of cows' milk whereas the south is famous for *pecorini* cheeses made of ewes' milk. For world output Italy is exceeded only by the United States and France. In 1966 all France produced 651 000 tons but what proportion of this was made in southern France is hard to assess. Roquefort, the most famous type of cheese made from ewes' milk, is named from the town in the southern Causses near the junction of Bas Languedoc and the Massif Central. Here milk is processed from a wide area including the Pyrenees and Corsica. The lowlands of Languedoc and the extensive garrigue-covered limestone slopes of the Cévennes afford little pasture in summer and then many flocks migrate to the moister crystalline parts of the Massif Central.

Pigs. The pig population exceeds that of sheep in Switzerland. Elsewhere in southern Europe pigs are also common except in Albania and Greece where their relative fewness may partly reflect the lingering influence of a Muslim hostility to porcine beasts. The tremendous herds of pigs on the Danube plains are associated with a large output of maize. A very different system of rearing exists

VII *Greece: a herd of goats on a mountain road near the village of Dimochori*

in Iberia where pigs forage beneath the cork and holm (ilex) oaks for the heavy falls of acorns. Ham and pork play a large part in the local diet. It happens that the pig is also the great consumer of organic garbage from households and this is partly responsible for the establishment around most great cities of a ring of piggeries intended to supply the local market. When sty-fed, the pig provides an excellent natural manure quite unexcelled for farming uses. Its skin also is valued for leather.

Movements of flocks and herds

In many areas the flocks and herds take part in seasonal migrations (figs. 28 and 29). Throughout mountains with snow cover only in the cold season, the chief movements are upward following the retreat of the snow cover in spring and downward in front of the advancing snow cover in autumn. This vertical transhumance makes the best use of the Alpine meadows and allows the lower pastures to be cropped for hay. The length of stay on the upper meadows depends on their freedom from snow. In the cold and snowy season the beasts are usually stall-fed in stables in the valley or lowland. Transhumance of this kind persists throughout the higher mountains of southern Europe and normally involves local owners and dairy cattle. It is a characteristic, for example, of Alpine economies in Switzerland and northern Italy.

In the Mediterranean lands the animals involved are mainly sheep and the horizontal distances travelled are often greater, being in parts of Spain over 500 km. A reason for this change lies partly in the mediocre quality of mountain pastures experiencing summer drought. Thus the grazing on the Gran Sasso in central Italy will only support 3 sheep per hectare. But the long distances involved must be also imputed partly to the traditional power of the pastoralists, and to the existence of a northern zone with rainfall all the year in Iberia, a large national territory where vertical relief differences are relatively small. In Spain during the early sixteenth century the Mesta controlled the seasonal movements of over 4 million sheep. Today the normal forms of long-distance sheep transhumance tend to decline, especially where the lower mountain slopes are increasingly being brought under cultivation. Many of the movements today are undertaken partly by truck or by train to dispersal and collection stations, and the governments tend to encourage this less leisurely means of travel.

In the Mediterranean summer-dry areas an inverse form of transhumance, moving downward in winter, is today probably more common than the summer ascensional type. The former arises basically from the winter chill which when much below a daily mean of 6°C no longer stimulates the growth of grass. Often the chill is not 'cold' in a continental sense: it merely does not promote an outdoor bite of grass or green fodder. The result may be seen on the Meseta where flocks of Old Castile move southward to winter pastures in the Guadiana basin in Extremadura. Winter transhumance is also relatively important in the Pyrenees, especially from Navarre to the Aragonese steppes and from the French slopes to the Garonne lowlands, where the pasturage includes harvested vineyards. In Sardinia flocks regularly move in winter down to the coastlands. Similar movements from the Abruzzi down to Apulia have been going on for several millenia. In some parts of the Balkans, particularly Macedonia, the permanent villages are in the uplands and the animals are driven down to the valleys in winter.

In many districts of Mediterranean Europe the upland and lowland farmers have standing agreements whereby the former rents land from the latter who also benefits from the manure and perhaps also from the ewes' milk. On the Meseta and in Sardinia the cultivator may give the pastoralist right of pasturage on fallow and stubble. In parts, as in Serbia, the expansion of arable crops on the lowlands is lessening the availability of winter pastures. In Bas Languedoc the specialization of the lowland cultivators on viticulture seriously upset their traditional co-operation with the pastoralists of the garrigues of the Cévennes and Causses. But this winter transhumance seems likely elsewhere to endure for a long time yet.

A third form of periodic livestock movement, called double transhumance, is often distinguished by agronomists. In this the permanent homestead lies somewhere between the upward limit of the summer migration and downward limit of the winter migration. Presumably the system in Old Castile whereby flocks go north to the Cantabrians in summer and south to Extremadura in winter may fall into this category. The pastoral economy of the Tremp area in Lerida on the piedmont between the upland Pyrenean pastures and the lowland steppes of the Ebro basin, and the pastoralists dwelling in the sweet

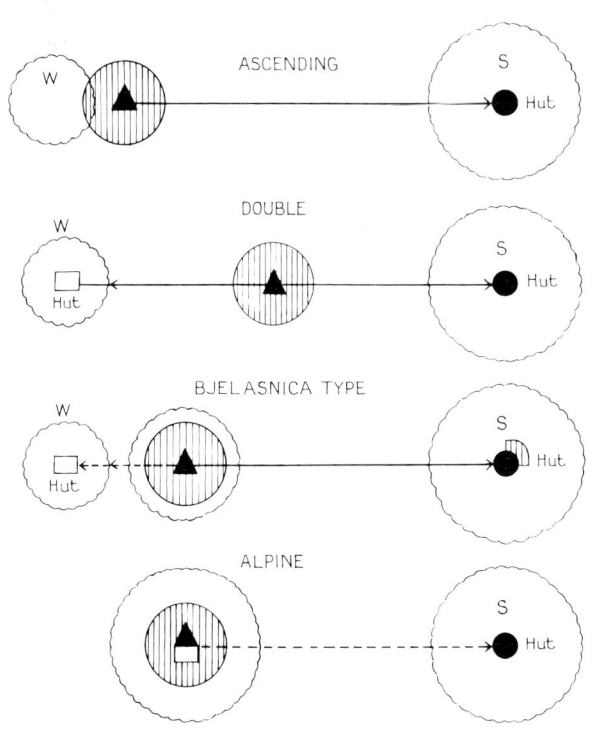

29 *Some types of transhumant pastoral economies*
Vertical shading denotes agriculture; white circles denote pastoral only; solid arrows show movement of whole society, and pecked arrows of shepherds and herders only

28 *Major routes of transhumance in western Mediterranean lands* (from J. M. Houston, *The Western Mediterranean World*, 1967)

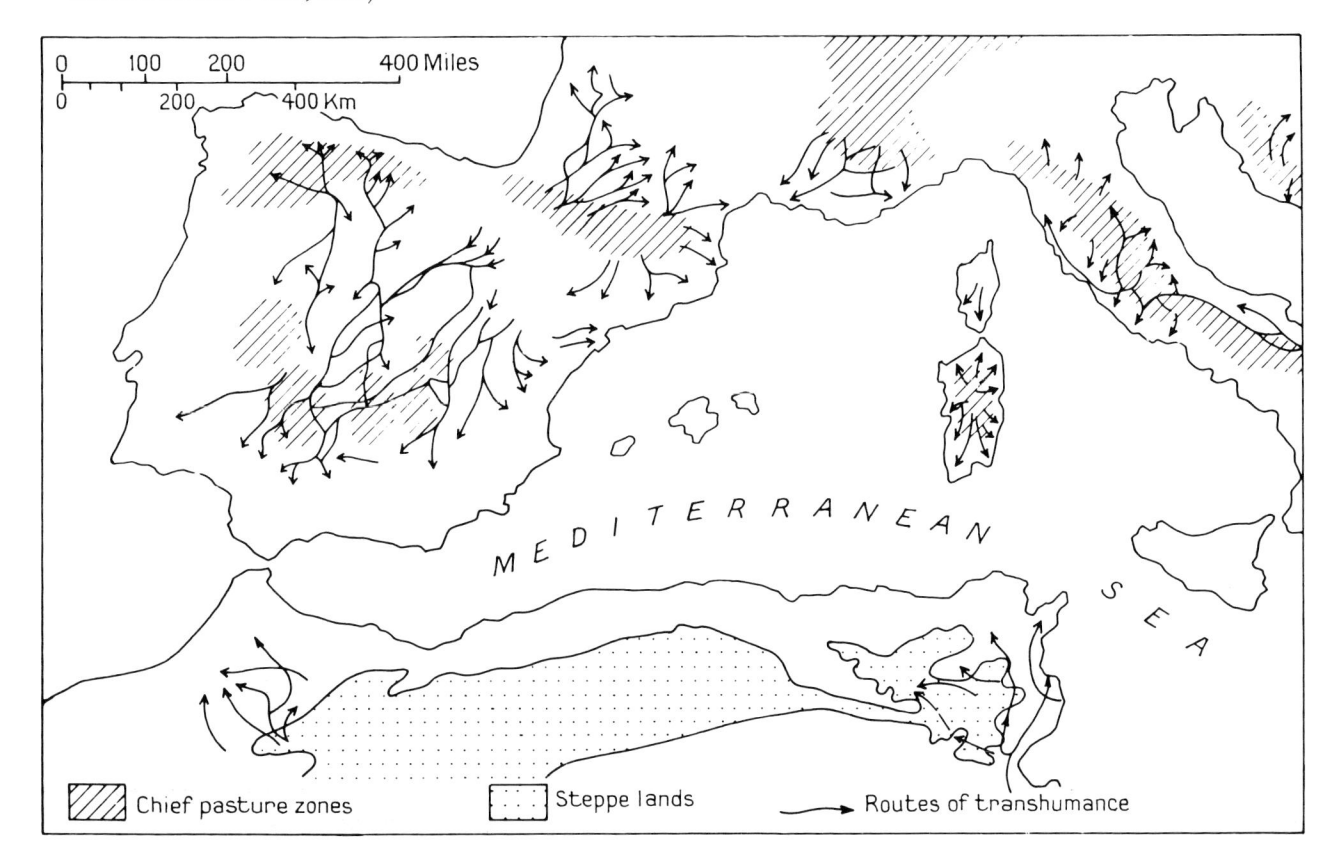

chestnut zone halfway between the mountain top grasslands and the coastal plains in Corsica, afford clearer examples. Another flourishing example concerns the stony expanses of the Crau in Provence which form traditional pasture for sheep from late November to mid-February. These flocks then spend five or six weeks on cultivated meadows before being moved by rail to the foothills whence in late spring they begin a normal transhumance up to the high alpine meadows. Today the irrigation of nearly 8000 hectares of the Crau has greatly increased its stock-carrying capacity. The pastoralists sell off their surplus lambs at sales at Arles in the spring whereas transhumant flocks and herds in central Europe are thinned out at Michaelmas before winter begins.

An ancient form of double transhumance of sheep survives in Bosnia and Herzegovina in the Dinaric Alps south of Sarajevo (fig. 29). The dry summers of the lower slopes near the Adriatic cause transhumance to the mountain pastures of Bosnia, especially near Mount Bjelasnica. In this area the Vlahs or Vlach people have existed for many centuries first as nomadic and later as transhumant pastoralists. Since the Middle Ages many of them have become assimilated with the surrounding Slavs and they have lost their language but not their traditional pastoral way of life. In 1961 only about 8000 Vlahs remained in Macedonia and 1000 in Serbia. Today, the transhumant pastoralists of Bosnia and Herzegovina consist mainly of Slavonic-speaking Muslims (converted during the Turkish occupation) and of lesser numbers of Serbs and Croats. They depend almost entirely on a small breed of sheep kept chiefly for milk. A ewe, with twice-daily milking, will in 4 months on the highland pastures, produce about 3 or 4 kilograms of cheese, together with the usual by-products of cream and a kind of butter. The main villages are at about 750 m altitude on the slopes facing toward the Adriatic between the neighbourhoods of Mostar (on the Neretva) and Dubrovnik. During the cool season (mid-December to early February) the shepherds take the flocks *down* to pastures at 600 m and below. These pastures are thin (mainly matgrass – *Nardus*) and cannot be used for hay. In February the flocks return to the pastures round the permanent villages where they stay till the end of March when the stock, shepherds and their families migrate upward in stages to pastures above the village, and by early June are grazing above the treeline, often 65 km away. Here the families dwell in hut-villages and may undertake a small amount of cultivation of potatoes and vegetables. Dogs are kept to ward off wolves and, owing to the scarcity of water in a karst terrain, much use is made of snow accumulated in natural hollows and sinkholes. By early October the stock and families are back on the pastures around the permanent village. It will be noticed that in this economy haymaking is not possible on the lowlands and in winter stock are not stall-fed, as in the Alps proper. A different condition prevails on the inland slopes of the Bjelasnica mountain group, facing the Sarajevo area, where a central European type of climate prevails. Here the valley pastures can be cut for hay in summer, and the livestock (mainly cattle) that do not move down to them in winter are kept in huts in the permanent villages. Moreover the journeys up to the Alpine pastures are quite short. The climatic statistics for Split, Bjelasnica and Sarajevo given in Chapter 21 illustrate clearly the differences in the summer rainfall and the reason why snow lingers long on the mountain summits.

It must be admitted that transhumance and the picturesque ceremonies and customs associated with it attract more attention than the numbers involved merit. Today a relatively small proportion of the livestock of southern Europe leave the arable stubble, cultivated fields and meadows for distant pastures. Yet it is important to notice that in European lands with a Mediterranean climate the role of livestock in the arable farming has not increased appreciably since medieval times whereas in central Europe it has been greatly strengthened by the addition of maize, sugar-beet, other roots, lucerne and other green forage crops. Animal manures and leguminous crops are the great natural fertilizers and it is the relative lack of these on Mediterranean farms that partly explains the mediocre yields of some of the staple cereal crops. The recent expansion in irrigation could lead to large increases in the area (hectarage) under maize, alfalfa (lucerne) and other leguminous fodders. But in many districts cash crops such as cotton and tobacco are preferred. Specialization in leguminous forage crops is already appearing in new huertas where the population density is not excessive, as in the Ebro basin and Pontine marshes. But in parts the farmers still prefer to produce lucerne for sale to cow keepers in and near large towns, for example Barcelona, where the beasts may be kept in stables. There is always the hope that irrigated areas would, in times of fodder shortage, act as suppliers and safety valves for large flocks and herds on the same methods used by the Mormons so successfully in semi-arid Utah. But the fact remains that the sheep and goat are ideal for the present Mediterranean land use and that enormous financial resources will be needed before the vast expanses of bushgrowth, stubble and fallow can be put to more profitable uses than pastoralism. In many ways too the olive is the Mediterranean cow.

It soon becomes obvious to a traveller that the typical Mediterranean arable farmer spends much of his time among fruit trees whereas the typical farmer in central and Alpine Europe spends much of his time with livestock. Polyculture in a Mediterranean climate is rarely intended for animals whereas in a central European climate it is normally intended for them. In Mediterranean summer-dry countries the shepherds and herders tend to be specialists whereas in lands with some summer rain the farmers can combine animals, crops and manuring much more satisfactorily. The amount of time devoted to livestock is, however, also related to the amount of enclosure.

In arable tracts where fields and strips of different crops are quite unenclosed – as is common – the movements of grazing animals are carefully controlled to keep to the stubble or fallow or non-crop patches. Thus on the Greek ploughlands the slow, orderly march of sheep and goats over the harvested cereal stubble usually ends in a night corral. Everywhere in unenclosed arable tracts of southern Europe it is common to tether the larger beasts when not in use. Almost throughout the Balkans, and particularly in rural lowland Yugoslavia, an essential daily chore in summer is the leading of a few animals for a leisurely stroll to graze the patches not under crops. The task falls particularly on the very young and very old. In the cool of a summer's evening pigs may be paraded slowly with complete orderliness by an aged woman or several massive oxen may be led by a tiny child. Probably nowhere else in western Europe are the links between livestock and people closer. In contrast, where fences become common, as in parts of Switzerland (and of Austria), the farmers no longer need to stand and watch their livestock grazing.

ARABLE AND TREE CROPS
Cereals

Discussions of cereal production in Europe are usually biased by the traditional view that the Mediterranean climate is perfect for cereals. This, however, is quite a misconception: the climate is ideal only for the ripening and harvesting of a few cereals. Normally it is not ideal for heavy growth and high yields for which the central and western European climates far excel it. Moreover it is quite unsuited to maize and to rice, the heaviest-yielding cereal. In Mediterranean lands the yields of cereals are liable to be very low in dry years and great variations in production occur. It must be admitted that throughout lowland southern Europe the precipitation is often inadequate for maximum crop growth and that the natural deficiency for crops is usually less in the Po basin and in the Danubian lands than in the Mediterranean peninsulas (see table, Chapter 5). What is more, throughout the Danubian floodplains and on broad-floored polja either river water or groundwater is readily available in summer and today there is a great deal of supplementary watering by sprinkler and by surface flow. The fact that wheat and barley ripen in May or early June in the Mediterranean lands, or a month or two earlier than farther north, is a doubtful blessing as they cannot be immediately followed by another crop unless irrigation is available.

For cereals generally, if France is excluded, the leading southern European producers are Italy and Yugoslavia with about 14 million tons each. As the table shows, they are followed in output by Spain with 9 million tons. The small production in Albania and Switzerland reflects their territorial smallness, their mountainous nature and a national predilection for pastoralism.

For wheat alone, if France is again excluded, the chief south European producers are Italy, Spain and Yugoslavia. The difference in natural productivity and general farming practices of the various countries may be judged roughly from the yields which average 1·1 tons per ha in Spain, the driest of the Mediterranean peninsulas, 1·9 tons in Greece, and 2·2 tons in Italy, the wettest of the Mediterranean peninsulas. Yields per hectare are still higher in Yugoslavia (2·5 tons) where on the Danubian plains late spring and early summer rains, rich loess and alluvial soils, and heavy manuring (partly from leguminous rotation crops) are excellent for cereals. The very low yields in

CEREAL AREA AND PRODUCTION, 1966 (in thousand hectares and thousand metric tons)

Country	All Cereals Area	All Cereals Prod	Wheat Area	Wheat Prod	Maize Area	Maize Prod	Barley Area	Barley Prod
Albania	336	332	130	115	160	165	9	9
Greece	1 641	3 200	1 018	1 959	142	320	321	639
Yugoslavia	5 234	13 938	1 833	4 603	2 500	7 980	394	714
Italy	5 991	14 379	4 274	9 400	988	3 510	179	253
Portugal	1 642	1,288	523	342	473	565	111	49
Spain	6 956	9 192	4 190	4 812	482	1 154	1 338	2 006
France (all)	8 896	26 731	3 992	11 297	964	4 336	2 642	7 421
Switzerland	176	579	111	358	4	19	32	107
Malta	4	4	2	2	–	–	2	2
European *Total*	72 190	173 065	27 942	62 600	11 066	33 132	14 746	39 662
Austria	901	2 649	314	897	55	275	230	706
Romania	6 821	13 901	3 304	5 065	3 288	8 022	246	483
Bulgaria	2 355	6 818	1 142	3 193	574	2 207	416	1 064

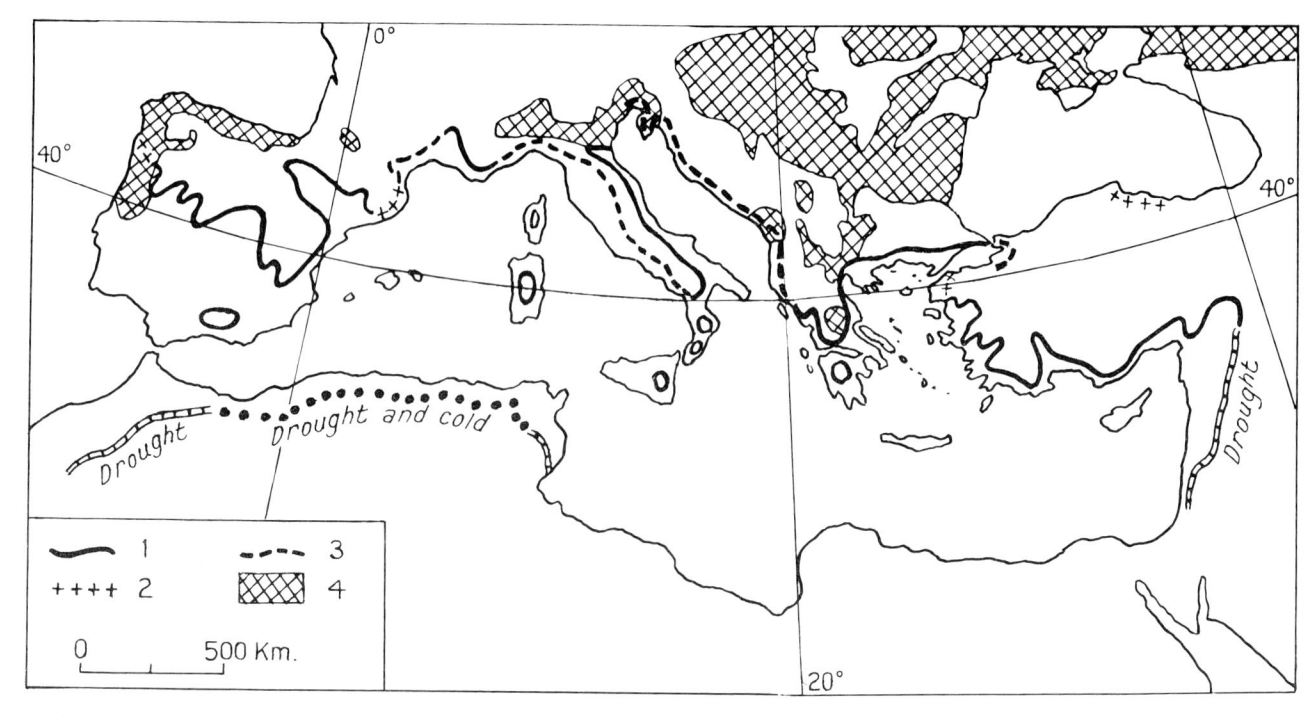

30 *Distribution of main areas of olive and maize cultivation in southern Europe*
  Here the successful cultivation of the olive for fruit is restricted mainly by
  1 *winter cold*
  2 *summer humidity*
  3 *combination of summer humidity and winter cold*
  4 *shows main maize areas: these coincide closely with areas with rain in the summer*
  (olive limits after P. Birot)

Albania and Portugal reflect inadequate manuring and, especially in Portugal, the practice of growing cereals beneath tree crops. The rolling oak-dotted wheatfields of southern Portugal contrast vividly with the open treeless expanses of the Danubian plains and of the Spanish Meseta.

For maize (corn) production the superior quality of the central European type of climate is always evident (fig. 30). By far the chief producer in southern Europe is Yugoslavia where maize is easily the chief cereal. The great Danubian cornbelt, which extends also for hundreds of miles up the wide flat valleys of the large tributaries such as the Sava and Drava, is reminiscent, except for human settlement and sunflowers, of the American cornbelt. Vast areas are planted almost entirely with corn, which in parts alternate with patches of lucerne (alfalfa) and sunflower. This Danubian pattern of cropping extends right into southern Austria but as a rule maize cultivation becomes unimportant in Switzerland where the summer heat and sunshine cannot always ensure maturity. The large production in France comes mainly from Aquitaine and that of Italy from the Po plain, each with appreciable summer rainfall. Similarly in Spain and Portugal the main producers are the peasant farmers on the minifundia of the humid northwest. In fact, maize was first introduced into Europe in Andalusia (in 1493) and by the end of the sixteenth century had dominated the agricultural economy of northern Portugal. The small extent of maize in arid Spain and in Greece is due to the certainty of summer drought. Even in the Balkans the crops, where possible, are given a little supplementary watering in dry years. Here in parts some cornfields near main roads are under-sown with melons which are harvested before the corn. Almost everywhere maize is predominantly a fodder crop, particularly for pigs. Its human use has decreased considerably, although it is still consumed in small quantities in numerous forms and the aroma of roasted corncobs enriches the sidewalks of some streets in Athens.

Of the less important cereals, barley is the chief. It is more resistant than wheat to drought and has been cultivated longer by man. It forms the second cereal in importance in Spain, Greece and Switzerland, where, and to a lesser extent elsewhere, it provides a fine animal fodder and also the basis of the brewing industry. Beer and light ales are drunk widely throughout southern Europe and brewing is common in the larger cities even in wine-producing districts. The industry tends, however, to be larger and more traditional outside the Mediterranean countries.

The oat and certain other lesser cereals provide fodder and locally may be used for human consumption. They are often grown on the bleaker cultivable areas and poorer soils. Thus rye comes in widely on schistose and other acidic soils on mountains at moderate altitudes, especially in northern Portugal, a country in which it is the third cereal for production. The oat, a fairly recent introduction, forms a winter crop in the Mediterranean and a spring-sown crop in central Europe. It becomes significant locally in distinctly inclement and rather sterile areas especially in Italy (359 000 ha and 477 000 tons in 1966), Spain (469 000 ha; 462 000 tons) Yugoslavia (320 000 ha; 386 000 tons), and to a lesser extent in Greece and Portugal.

Rice, which demands alluvial soils that can be kept flooded during early growth, is an obvious alien and was discussed in Chapter 5. Just over 90 per cent of the European production outside the U.S.S.R. comes from southern Europe and 40 per cent from Italy alone, mainly in the Po plains near Turin. Spain and Portugal follow in importance in non-Soviet Europe but today the brilliant green patches of paddy stand out also in the summer landscapes of the Camargue and of many estuarine and riverine tracts in the Mediterranean and on the middle and lower Danube where ample floodwater is available (see table in Chapter 5). Rice growing tends to be organized in relatively large units as it involves much expenditure of capital and labour. It is often government-sponsored and its spread is encouraged by high yields (at least double that of wheat), by national desires to cut down food imports and by the freedom today from malaria which in the past caused paddy cultivation to be prohibited or unpopular. The flat fields are usually devoid of habitations and most of the tasks, especially harvest, are done by a seasonal influx of workers.

Vegetables and industrial crops

Although the broad food-producing potential of a country can be judged from its cereals and livestock, which provide the basal foodstuffs, many other crops contribute to the national diet and national economy. Some of these are largely or partly for export and industrial uses.

Most culinary vegetables will not grow under natural conditions in Mediterranean climates in summer and in central European climates in winter. In the former, however, with irrigation, three harvests – spring, summer and fall – can be obtained of crops such as potatoes and tomatoes although the hot season crop is usually the most prolific. Throughout Mediterranean countries potatoes are mainly a cash crop consumed 'new'. Throughout Alpine and Danube lands they are mainly stored for winter use and large quantities are also sent to starch and alcohol factories. The exports from, for example, Malta and Crete, are new potatoes that arrive in Britain before the native crop is on the market.

Sugar-beet is not naturally suited to a Mediterranean climate although, with irrigation, its yields and sugar content are relatively high. In contrast the large production in the moister central European climate, particularly in northern Italy, has a lower sugar content but the luxuriant flesh and leaf provide also good cattle feed.

Tomatoes, fresh as salad, form a favourite food in most European countries, particularly the Mediterranean. Processing, including paste, takes the surplus and utilizes about 15 per cent of the total in Greece and probably more in Italy. The summer crop is usually irrigated. The German Federal Republic, France and the United Kingdom are large importers (total 564 000 tons from all sources in 1966) and Switzerland and Austria import small quantities. Sales to these from the Mediterranean are increasing, especially with improvements in motorways. Another noted Mediterranean vegetable is the onion, of which Spain (106 000 tons) is the third world exporter (after the U.A.R. and the Netherlands).

Among the non-cereal food crops of central Europe is the sunflower which beautifies large expanses in the Danube plains. The oil from its seeds is used domestically in the same way as olive oil is used farther south but the plant also provides excellent feed for poultry and other livestock.

Cotton and tobacco are outstanding among the purely industrial crops. Cotton is confined to the lowlands of the Mediterranean and lower Danube where summers are hot and irrigation water available. The chief producers are Spain (121 000 tons), particularly from the black earths of Andalusia, and Greece (88 000 tons), especially from the drained lake floor near Levadia.

Tobacco plays an appreciable role in the economy of several south European countries, and especially of the Balkan lands. It needs much labour as the leaves are picked successively only when they reach an adequate size. Large amounts of Turkish tobacco, mainly cigarettes and cigars, are produced in Greece, were there are about 120 000 ha under the crop, the surplus of which provides over one-third of the value of the national exports. The other countries, except Switzerland and Portugal, can supply all or most of their domestic needs.

Tree crops

Tree fruits or horticulture form a prime feature of the agriculture of Mediterranean countries and of the Danubian lands in Yugoslavia. Only a few fruits are restricted to the warmer south.

The crops that occur throughout the south European lowlands include the apple, pear, cherry, peach, apricot and plum. For apples, France with 3 million tons annually, leads the world and a considerable proportion comes from the south. Italy (2·16 million tons) is third in the world (after the U.S.) but is by far the world's chief exporter and sells abroad about one-fifth of its crop, mainly to western Europe. For cherries, Italy leads southern Europe and is second in the world to Federal Germany. For peaches (see table) it is supreme in Europe (60 per cent total) and exports one-fifth of its crop. France, on the other hand, consumes most of its own crop, which comes mainly from the Midi. During recent years the Greeks have been planting about 350 000 new peach trees annually and their present crop (105 000 tons) will probably be doubled by 1975. For apricots, Spain and Italy are outstanding. For plums, the Danube countries strongly dominate, and Romania and Yugoslavia easily lead the world. This characteristic, and the excellence of natural conditions for horticulture in the Danube plains, are seldom stressed enough. In Yugoslavia orchards cover 2 per cent of the state and are commonest in the interior hilly country. About half the fruit trees are plums (*sljiva*) which are largely consumed fresh or used in making plum brandy (*sljivovica*), the traditional national drink. In spring from a distance the

EUROPEAN PRODUCTION OF CERTAIN FRUITS, 1966, EXCLUDING U.S.S.R.

(thousand metric tons)

	Plums, fresh and dried	Peaches		Apricots	
			1968		1968
Albania	8	–	–	–	–
Austria	75	8	8	19	21
Belgium	5	3	5	–	–
Bulgaria	314	127	163	39	43
Czechoslovakia	144	6	6	22	18
Denmark	15	–	–	–	–
France	130	318	615	45	132
East Germany	144	4	6	–	–
West Germany*	581	27	57	2	7
Greece	24	105	168	27	36
Hungary	260	68	81	81	51
Italy	140	1423	1315	77	109
Luxembourg	4	–	–	–	–
Netherlands	9	2	1	–	–
Norway	14	–	–	–	–
Poland	179	–	–	–	–
Portugal	26	40	31	8	8
Romania	837	8	24	38	50
Spain	62	139	211	151	186
Sweden	18	–	–	–	–
Switzerland	43	–	–	6	10
U.K.	46	–	–	–	–
Yugoslavia	723	47	49	22	16
European total	3789	2325	2740	537	687
World total	4621	5022	5791	1036	1230

* includes West Berlin

white blossoms resemble a vast snow cover. In some ways in lands once under Turkish control the plum plays the role of the grape in Mediterranean countries. Muslims, forbidden by their religion to drink the juice of the grape, were not prohibited from drinking plum juice.

The *vine* is the chief tree crop common to southern European countries. It acquired an eternal kudos in classical times when the ideal of material bliss seemed to be wine within and olive without (as a cosmetic). It is hardy and will grow almost anywhere on the lowlands of western and central Europe, including England. Maps in geographical texts showing 'the northern limit of the vine' presumably mean the northern limit of commercial vineyards today. The juice of the grape becomes rapidly more acidic where summers are moist and cloudy and the roots are often waterlogged. Being a fast-growing climber the plant needs either severe cutting back or, if allowed to spread, heavy pruning and ample support. It thrives best on well-drained sunny habitats such as limestone scarps, loess, deep sands and flat alluvial floors with a low watertable.

Nearly two-thirds of the world's total area under vineyards is in Europe. For extent of vineyards, Spain, Italy and France lead the world, while for wine production Italy and France are supreme and Spain is a close third. The statistics however diminish the vine's areal significance as it is often grown mixed with other crops. Thus in Italy in 1966, vines growing alone covered 1 145 000 ha whereas mixed cultivation of vine and another crop or crops occupied a further 2 126 000 ha and produced one-quarter of the wine production. The next rank of European producers – five countries with over 200 000 ha under vines – deserve some mention. Portugal, if its national size is considered, will be seen to have remarkably extensive vineyards. It produced over 2 million tons of grapes in 1965, including port which comes from a specially defined tract on rather infertile terraced slopes overlooking the Douro near the Spanish frontier. However, here and elsewhere it is perhaps necessary to notice that the vine tends to attract more publicity than the humdrum cereal crops. The com-

NON-SOVIET EUROPE: VINES; AREA AND PRODUCTION OF GRAPES AND OF WINE, 1966

(thousand hectares and thousand metric tons)

		Production		
	Area	Grapes	Grapes for wine	Wine
Albania	13	43	7	4
Austria	45	186	186	131
Belgium	>1	12	–	–
Bulgaria	194	1 081	711	427
Czechoslovakia	25	59	54	31
France	1 400	9 624	9 294	6 225
West Germany	83	616	616	442
Greece	230	1 439	602	384
Hungary	245	569	503	337
Italy	1 644	10 239	9 271	6 514
Luxembourg	1	18	18	13
Malta	1	3	3	2
Netherlands	>1	5	–	–
Portugal	346	1 242*	1 191	893
Romania	326	954	880	620
Spain	1 720	5 037	4 761	3 240
Switzerland	12	110	108	77
Yugoslavia	259	1 230	976	569
Total	6 480	32 486	29 200	19 924
World total	10 110	50 956	38 823	27 418

* 1966 was a very poor year. 2 019 in 1965; 1 600 in 1968

bined vineyards of all Europe could almost be placed in the cereal fields of Spain alone.

The production of table grapes and of dried grapes is a secondary aspect of viticulture with considerable importance locally. Sales of table grapes have increased rapidly in the last decade, the prime markets being Germany and the U.S.S.R. For exports Italy has long taken first place (240 000 tons in 1966) but in spite of doubling its sales since the late fifties, it has been equalled or excelled recently by Bulgaria (260 000 tons in 1966) which sells mainly to COMECON states. Spain, France and Greece also export appreciable quantities of table grapes. For dried grapes, apart from lesser quantities of large Muscat seeded grapes from Málaga, Greece almost monopolizes the European output. It specializes in a small seedless raisin dried usually from the Black Corinth grape and known universally as the Zante or Greek currant. Its output of 1·8 million tons (21 per cent of world supply) is exceeded only by that of California and Turkey which produce mainly a medium-sized variety called sultanas.

Tree crops restricted in Europe to the Mediterranean climate

Tree-fruit crops that in Europe are monopolized by the Mediterranean lands include the fig which will stand fairly cold winters but has not yet achieved much popularity elsewhere. European countries produce 62 per cent of the world's supply and nearly all the remainder comes from Asiatic and African Mediterranean states. For production Portugal is easily first in the world, Italy second, Spain (after Turkey) fourth and Greece fifth. Portugal produces 365 000 tons of fresh figs annually or nearly one-quarter of the world output. Its main centre is Torres Novas just north of the Tagus. The Algarve is also famous for fig cakes and figs stuffed with almonds.

Citrus fruits are far more important (see table below). The fruits, and often the trees themselves, will not stand frosts much below about −3°C and the commercial European crop comes from Mediterranean areas normally with a relatively mild winter. The production of grapefruit is negligible, while that of lemons is dominated by southern Italy (76 per cent European total) followed by Greece and Spain. The orange and tangerine (mandarin) are more tolerant of frost and grow more widely except in the Midi and Albania. The great producers are Spain (55 per cent European total), Italy (32 per cent) and Greece (10 per cent). The bitter or Seville orange was introduced into the Near East by the Arabs before the ninth century AD and was taken by them to Spain. By the twelfth century it was common in the eastern Mediterranean and the crusaders brought it back to Italy and the Midi. It is used for marmalade, a common breakfast conserve in the British Isles (in Danubian countries replaced by plum preserve). The sweet orange and tangerine were introduced much later, the former probably in the fifteenth century by Genoese

CITRUS FRUITS AND FIGS, 1966

(Production in thousand tons)

	Oranges and tangerines	Grapefruit	Lemons, limes, etc	Figs Fresh	Dried
Albania	2			12	–
France	4			4	
Greece	480		162	110	24
Italy	1 370		659	241	29
Portugal	143	2	13	365	23
Spain	2 329	4	92	158	10
Yugoslavia	–	–	–	26	3
European total	4 328	6	926	918	89
Cyprus	62	36	17	1	
Israel	682	185	38	1	
Jordan	45		12	16	3
Lebanon	181		69	25	
Syria	5		3	54	7
Turkey	360	4	84	215	54
Algeria	321	5	15	46	4
Libya	13		1	5	
Morocco	596	16	9	59	20
Tunisia	85	1		14	24
U.A.R.	409		85	4	
World total	25 203	2 725	3 107	1 473	198

merchants trading with the Near East. The tree yields good crops for 50 to 100 years.

Citrus fruits are valued for their content of vitamin C, of which a person ideally needs about 70 mg daily. This is derived mainly from fresh fruit, vegetables and milk. In fact, black currants, peppers, mustard and cress, strawberries and raw broccoli, brussels sprouts and cauliflower all contain more vitamin C per gramme than do citrus fruits (0·5 mg/g). But many of the former are not always available or are too dear or are cooked before being eaten whereby their vitamin content is about halved and so falls below that of the orange and lemon, which form a hygienic, easily handled, relatively cheap fruit of considerable dietary value. The crop is harvested from mid-October to late May, and is sold mainly fresh and partly as juices. The fact that it ripens during the cooler months emphasizes more than does any other local crop, except perhaps the olive, the sunniness of the cool season. In the citrus-fruit trade man is exploiting commercially both the gloomy winters of northwest Europe and the sunny winters of the Mediterranean south. The two chief European producers tend to specialize in different crops as the following table of exports in 1966 shows.

80 Part I: General Patterns

	(*in thousand metric tons*)			
	Oranges, tangerines	Lemons, limes	Grapefruit	Total citrus
Italy	151	349	1	501
Spain	1267	35	3	1305
All Mediterranean*	2772	508	142	3422

* Europe, Asia and Africa.

Spain and Italy sell mainly to western Europe while Greece, where with the aid of heavy government subsidies citrus output has quadrupled since 1951, directs its exports mainly to east European markets.

The *olive* is the unique fruit of the Mediterranean climate and 98 per cent of the world production comes from Mediterranean countries. Europe alone produces about 75 per cent of the world output of olive oil, Spain (33 per cent world total), Italy (26 per cent) and Greece (14 per cent) being the chief producers. Portugal is usually fourth, or in some years fifth after Turkey. The native wild olive yields small quantities only of very acidic oil. A cultivated or improved variety was being grown in Crete by 3500 B C and was carried by the Phoenicians to north-west Africa and so to Spain. In Roman times it had an unrivalled reputation both as oil and pickled. The triad of

31 *Distribution of olive cultivation and olive oil production in Greece* (simplified from *Economic and Social Atlas of Greece*, ed. B. Kayser, K. Thompson *et al.* 1964)

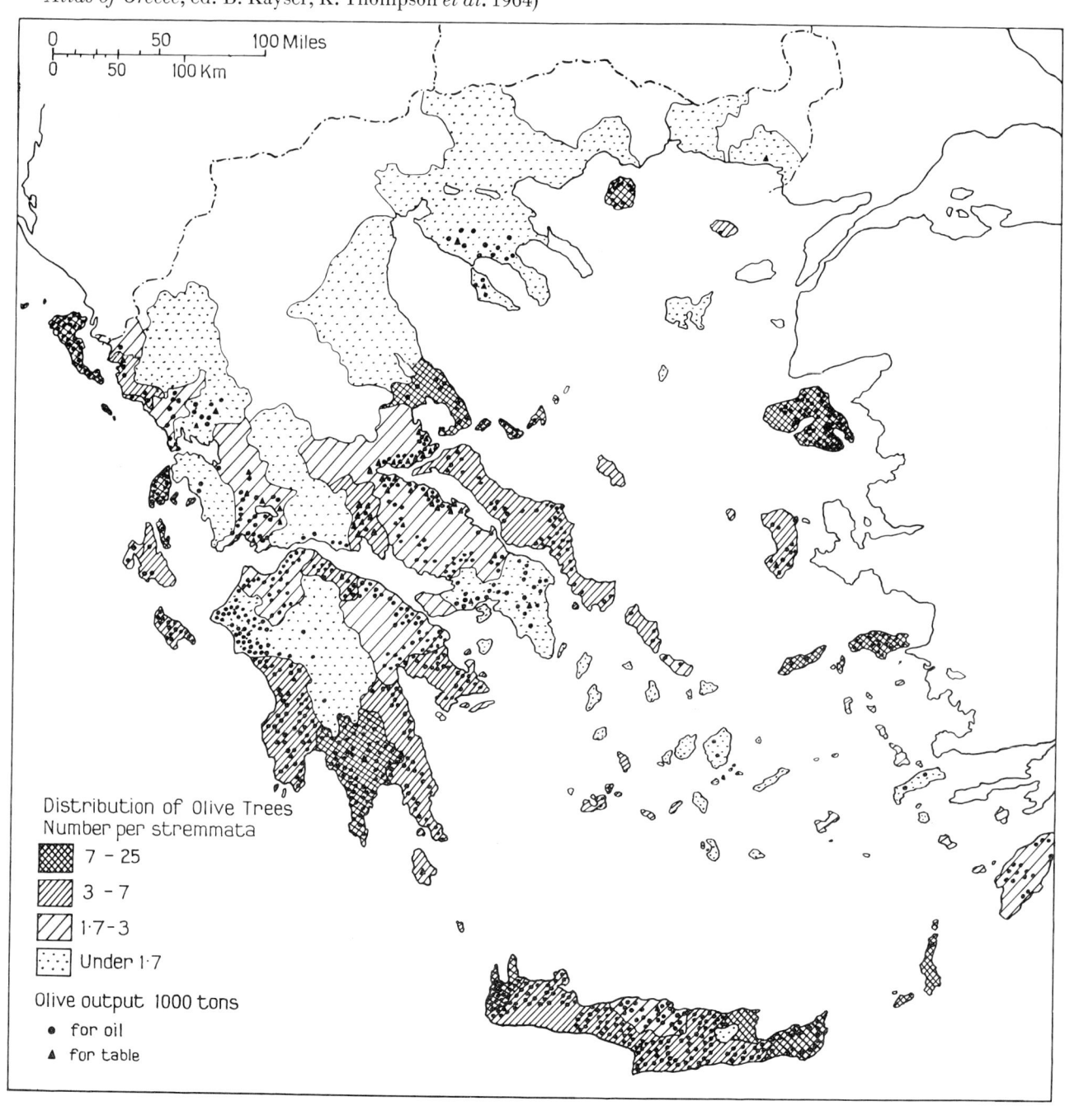

VIII *Olive groves at Cazorla, Jaen, southern Spain*

OLIVES, 1966

	Area (thousand ha)		Oil production (thousand tons)
	Total	For oil	
Albania	52	50	4
France	5	4	1
Greece	970	925	180
Italy	1802	1749	358[1]
Portugal	219[2]	218	38[2]
Spain	2224	2195	464
Yugoslavia	30	26	5
European total	5302	5166	1050
Cyprus	17	15	3
Israel	11	4	3
Jordan	33	30	5
Lebanon	29	25	6
Syria	116	90	24
Turkey	841	748	155
Algeria	140	75	15
Libya	100	25	18
Morocco	135	90	17
Tunisia	120	92	22
U.A.R.	12		
World total	7021	6435	1317

[1] Was 594 000 tons in 1967, and 425 000 tons in 1968
[2] Was 736 000 ha and 107 000 tons of oil in 1963 and 516 000 ha and 72 000 tons of oil in 1965; 53 000 tons of oil in 1968

Mediterranean cultivation became wheat or cereal on the richer lowland, the olive on the rocky screes and basal slopes and the vine on the lower slopes above the olive groves.

The olive is an evergreen tree that for good fruiting needs careful cutting and pruning to promote new shoots and to ensure that the foliage does not develop excessively. It will withstand appreciable frost (down to about $-10°$ C) but the problem is not so much growing the tree as obtaining a fruit rich in edible oil. Where summers are moist and cloudy the fruit either fails to form or becomes small and too acidic to be edible. The most suitable climate has abundant rain in late winter and early spring followed by hot dry sunny weather. The relatively sunny autumns and winters are also an advantage as the harvest is protracted and, for example, in southeastern Spain may hardly have begun in December. The tree will fill odd corners and line rocky paths as well as forming carefully spaced grey-coated regiments on gentle hillslopes and plains. Today in drier areas it often covers deep soils in alluviated basins which also support a ground crop, usually wheat.

The varieties of olive and methods of cultivation show slight differences. In Spain olives cover more than 2 million ha and abound especially in the semi-arid southeast where extensive new groves have recently been planted. In Portugal the olive does well everywhere except on the mountains and yields are exceptionally high on the fertile plains of Santarém. In Italy the olive is grown on about 7 per cent of the national territory and is predominantly a hillside crop intended for oil. Pure olive groves occupy about 900 000 ha and are characteristic, for example, of Puglia and Calabria. Olives in mixed cultivation occupy a further 1 350 000 ha and are seen to perfection in Sicily. In Greece oil production also predominates but the export of table olives is important (fig. 31). The large, dried, black fruit, renowned internationally as the 'Greek olive', is processed in oil or salted.

The yield of olive oil, as with that of most unirrigated Mediterranean crops, varies greatly from season to season. In Spain, for example, the output was 637 000 tons in 1963 and 314 000 tons in 1965. In some areas, especially in Portugal, excessive humidity in late spring may hinder flowering or the setting of the fruit. Sometimes the early spring rains may be inadequate, or exceptional rain in early summer may spoil the oil content. In Portugal recent yields have declined from 107 000 tons of oil in 1963 to an average of 61 000 tons from 1965 to 1968, while in Greece they have risen from 115 000 tons average in 1948–52 to 251 000 tons in 1963 and an annual average of 185 000 tons from 1965 to 1968. In the former the area under productive olive groves has declined drastically, while in Greece, as figure 31 indicates, there has been much new planting and the number of olive trees has increased from 60 million to 80 million since 1938. As the trees bear no fruit for 5 or 6 years and do not yield well until 10 years old, plantations take some time to come into full bearing. The demand for oil seems insatiable, and the olive remains the true sign of the Mediterranean climate and the unique feature of its agrarian economy. A single olive tree is preserved symbolically on the Acropolis.

9 Settlers and Settlement

The European continent has neither a political nor any apparent cultural unity. It is divided into many countries differentiated by language, culture, religion, race and history but these separate facets of nationality do not necessarily coincide for any one nation-state and their boundaries have fluctuated throughout history. To understand the European complexity of languages and cultures we need to penetrate through recorded history to archaeology and to the distribution of *homo sapiens* on the European stage during and after the closing phases of the last main glacial phase.

Race in Europe

Anthropologists classify race on the basis of such physical traits as the colour of the skin, the colour and texture of the hair, the colour and shape of the eyes, stature and skull measurements, although it is recognized that these are adaptive to environment and that some are recessive. Latterly blood groupings have also been studied. Their findings would appear to bear out the results of archaeological researches and that Europeans may broadly be classified under three racial groups:

1 the Nordic type of northern Europe with white skins, blue eyes, golden hair, large tall frames and long heads (dolichocephalic)

2 the Alpine type, with light brown skins, brown hair and brown eyes, tall frames and broad heads (brachiocephalic)

3 the Mediterranean type with dark skins and dark eyes, short slender frames and long heads

It must be stressed however that over the centuries there has been a long cross-breeding of these three types and pure breeds are seldom found. In fact the mingling of all these characteristics is the most marked feature of the European people.

The earliest inhabitants that concern us were the groups who moved slowly north in the wake of the retreating ice as the climate ameliorated. There appear to be survivals of these Palaeolithic men in parts of west–central France and in central Wales.

The Mediterranean race moved on to this backcloth about 4000 B C. Entering from the southeast they settled around the shores of the eastern Mediterranean and moved slowly into the western basin and thence finally filtered on to the Atlantic shores.

At about the same time, or rather earlier, as the Alpine and Baltic ice decayed, the 'Alpine' race advanced from the east across central Europe extending their settlements along the line of the Danube and other lowlands as far as the Pyrenees and Cantabrian mountains. The pattern of their distribution forms a broad wedge with its apex towards the west, and shows three sub-branches, the Dinaric–Alpine who diverged south into the northwest of the Balkan peninsula and Bulgaria; the south Slavs who are present today in Yugoslavia, and the Slavs who penetrated into Poland, Silesia and the Carpathians.

The Nordic people came from the Russian plains into Europe much later than the Mediterranean and Alpine races and settled around the shores of the Baltic Sea. Since then there have been many minor incursions of peoples into Europe mainly from the east. These include the Magyars who settled in Hungary, the Lapps who settled in northern Scandinavia and the Finns who migrated from the northern Russian forests and tundra. The Bulgars, who originated in the same general area as the Finns and Lapps, moved south into Bulgaria in the seventh century A D. Much later invasions were made by the Moors from North Africa who swept over Spain, much of southern France, Sicily, Malta and the south Aegean islands, and the Turks who overran the whole Balkan peninsula after the fall of Constantinople in 1453.

The picture that emerges is of successive waves of immigrant people from the east and southeast leaving vestiges of older inhabitants surviving in remote corners of the continent. Where uniformity of physical characteristics exists in Europe today it is the result of inbreeding due to an isolated situation. No homogeneous race exists in any European country and every nation includes a great variety of racial types. In Europe the word race has little real meaning.

Languages of Europe (fig. 32)

The distribution and structure of European languages have altered greatly within historic times but all the fundamental language groups are thought to have been already in evidence in the early Iron Age. It seems best to assume that they largely stemmed from a common tongue, the Indo–European, which was first established over Europe in the second millenium B C, the only time when Europe had a single language. It is thought to have spread westward from a source north of the Caucasus and diffused into Europe along the loess corridor. There are ancient survivals of languages unrelated to Indo–European but most have not been written down and since they are mutually unintelligible add no significant information to language studies.

At the present time there are more different languages in Europe than in any other continent except Asia. A suggested reason for this is that in times past communications were so difficult that differences in speech became sharply differentiated and finally resulted in the parting of tongues. It should be stressed that language means speech and not writing. Even today pronunciation and the meaning and use of words change from one generation to the next and within a comparatively short time in the past significant divergences in speech have developed.

During the Middle Ages and Renaissance, Latin was the language of diplomacy and the common language of Christendom but after the Reformation Latin fell into disfavour and subsequently language differences have helped nationalism and even a narrower regionalism to develop.

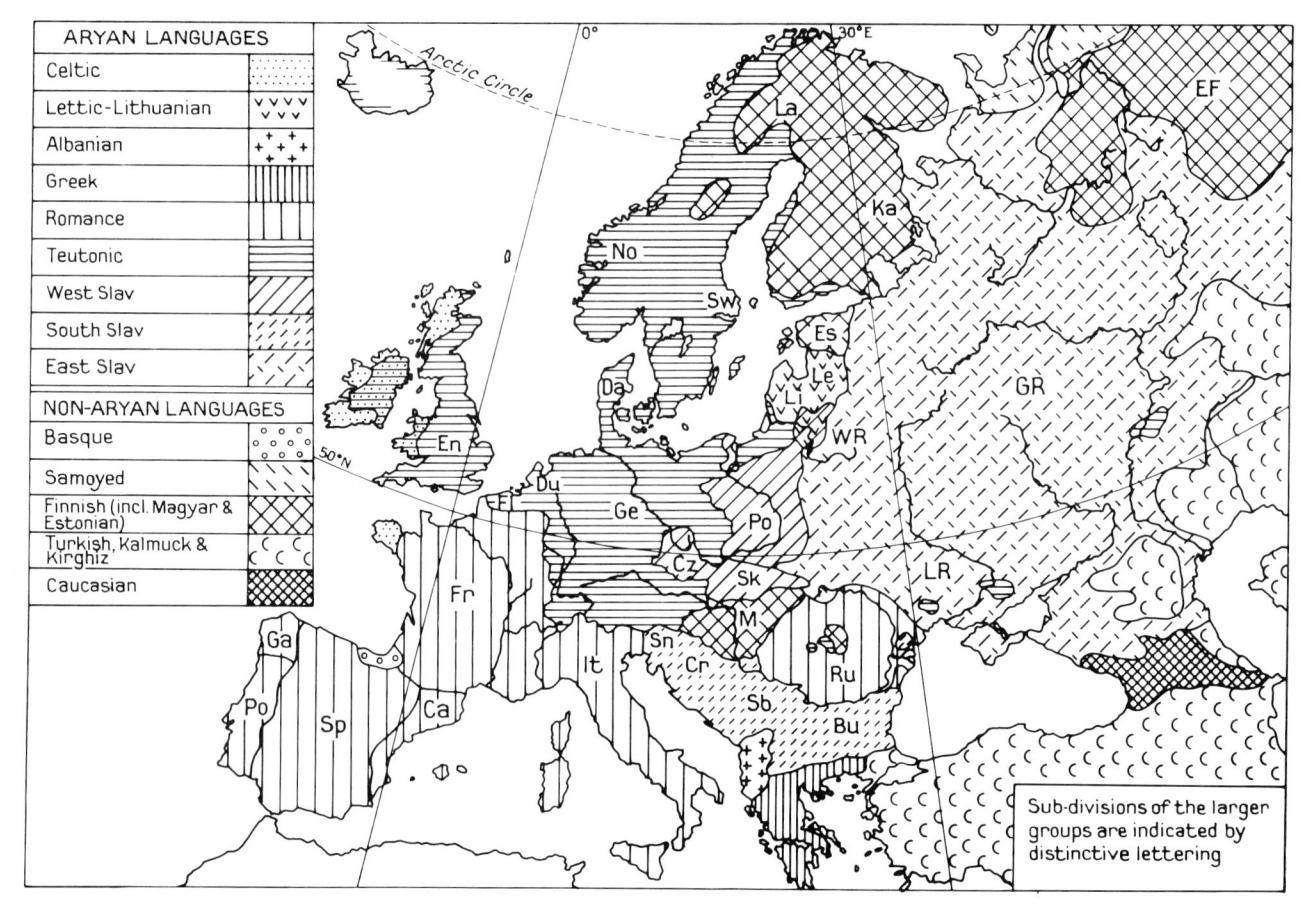

32 *Distribution of languages in Europe* (after J. F. Unstead)

There are four main families of speech present in Europe today:

1 Indo–European   3 Semitic
2 Ural–Altaic   4 Basque

The Indo–European has the greatest following and is split up into the most numerous languages while the last two have the least progeny. The Indo–European has eight different language groups descended from it: namely Celtic, Romance, Germanic, Baltic, Latvian and Lithuanian, Slavonic, Hellenic (Greek), and Thraco–Illyrian (Albanian). The following table shows the different spoken languages developed from the four main Indo–European groups.

Celtic	Romance or Latin	Germanic	Slavonic
Gaelic	Italian[1]	German, High	Russian, Great
Erse		German, Low	Russian, Little
Welsh	French	Dutch	Ukrainian[2]
Breton	Walloon	Flemish	Bulgarian
Cornish	Provençal	Friesian	Serbo–Croat
Manx	Spanish	English	Dalmatian
	Portuguese	Danish	Slovene
	Galician	Norwegian	Czech[3]
	Catalan	Swedish	Slovak
	Romanian	Icelandic	Polish
	Romansch	Faeroese	Serbian

[1] Plus Sardinian Dalmatian dialect.
[2] Including Ruthenian.
[3] Including Moravian.

The Ural–Altaic group covers the languages mainly of the Asiatic U.S.S.R. but there are outliers of these in the extreme north of Europe, in Hungary (Magyar) and in the Balkan Peninsula (Turkish). The Semitic language is represented in Europe by the speech of the Maltese; Basque has retained its identity in northern Spain, and in southwestern France in the Pyrenees (fig. 32).

While speech diverged from a common tongue by reason of isolation and the laws of speech itself, events like conquests and improved communications caused some languages to expand and others to decline. Where two languages came into contact usually the speech that prevailed was that of the people with the higher material civilization. This explains the spread in the western European world of Latin, from which French, Provençal, Italian, Catalan, Spanish and Portuguese have developed. There are also localities in Switzerland where Ladin and Romansch are spoken and islands of Latin speech in Romania. It is noticeable that the Romance languages, or those derived from Latin, do not extend to the farthest limits of the Roman empire. Subsequent invaders of the frontier zones contributed their language instead. Thus England lost Latin and took up Anglo-Saxon. Slavonic languages are now spread over the Balkan peninsula except in isolated pockets and islands, such as Vlach in northwest Greece and Wallachia (Romania), and Albanian, and Greek in the south.

In the Germanic languages, German, Dutch, Flemish and English form one group today, German becoming the main language of Germany, Austria and much of Switzerland, while English became later mixed with Norman–

French words. In the second Germanic group are the languages spoken by the Scandinavian countries, Norwegian, Swedish, Danish and Icelandic and these are mutually comprehensible.

The Slavonic languages are represented in south and central Europe by Czech and Slovak, which belong to the west Slavonic group and exist as pockets, surrounded by other languages; and by Bulgarian, Serbo–Croat and Slovene belonging to the south Slavonic language group. Macedonia is one of the frontier areas of language where Greek, Romance, Slavonic and Germanic tongues are all spoken. Albanian is the only surviving language of the Thraco–Illyrian group and was first written down in the late nineteenth century as also were Bulgarian and Serbo–Croat.

Today the linguistic map of Europe cannot be drawn with definite boundaries between language groups. The major tongues are parted by transitional areas where there are mixtures and every major language has alien minorities or 'islands' and 'outliers' of different languages spoken within their frontiers. In some cases language differences are divided by physical features such as mountain ranges, swamps, rivers, or even forests now cleared. Today physical features no longer form a barrier to communications and as these multiply and standards of living level out between nations so languages should decrease in number and international understanding increase, while it must be borne in mind that literature is not the prerogative of any one language. It is difficult to summarize so complex a distribution. But it seems obvious that, in matters of peoples and language, Europe must be considered a mere appendage of Asia. It has become a crazy jigsaw of 25 or 30 political states wherein racial traits, language and nationality nowhere coincide. The geographical personality of peninsular Europe embraces Mediterranean, Alpine and Northwestern segments of which only the two former are discussed in this book. The civilizations of these met and fused west of the Rhine along the line of the Rhône–Saône corridor. They developed along two great areas of traffic, the Mediterranean sea and the loessic Danube valley, separated by the great alpine chains with intervening links across them via the Morava–Vardar valleys, over the Postojna Gap and Pear Tree Pass, the Alpine passes, and along the Rhône corridor. The importance of these historic passageways can hardly be overestimated.

Peopling of southern Europe

As already noticed the number of nation-states and the number of different languages spoken make Europe the most politically fragmented of all the continents. The explanations for this are partly physical – a peninsula of smaller peninsulas, and so on – and partly historical. In the historical and social development, there are, according to C. T. Smith in his *Historical Geography of Western Europe before 1800*, five outstanding epochs during which a new impulse and direction was given to population and settlement. Briefly they are:

1. the long period of prehistory extending from the initial settlement by Palaeolithic man to the Iron Age
2. the age of Greek and Roman civilizations
3. the great age of migrations or Völkerwanderung
4. the expansion of Europe or the age of the mediaeval city states
5. the rise of nation-states

Unfortunately these historic periods do not have fixed boundaries in space or time so that, although they are dealt with separately in the short descriptions which follow, it must be recognized that there is considerable overlapping and interlocking.

Prehistoric settlement

If we neglect early man and Neanderthal man, and so delete a long span of prehistory, we can begin this survey 30 000 to 40 000 years ago when, during the final stages of the Würm glaciation, people capable of making elaborate tools appeared in Europe. These ancestors of Mediterranean man, as judged from the Grimaldi remains on the riviera, were not very tall, of slender build and with long narrow skulls. They were skilled craftsmen in flint, making sharp cutting edges, and combining flint with bone, wood, horn or antlers in their varied implements. Their armoury consisted of bows and arrows, spears and spear-throwers and barbed harpoons. At this time wooded steppe and tundra had established itself following the retreat of the ice in Scandinavia and on the Alpine chains; coniferous forest covered central Spain and northern Italy and the shores of the Mediterranean, while the warmer tips of the three southern peninsulas were under deciduous woodland (see Chapter 6).

In the sheltered valleys of the Alpine fold system from the eastern Alps west to the Pyrenees and Cantabrians most of the animal remains of this period are of reindeer and it is surmised therefore that, like the Lapps and Samoyeds today, these people were semi-nomadic followers of the reindeer moving to mountain pastures in summer, and retreating with the herds to lower ground in winter, when they settled in caves and practised, as in Lascaux, their ceremonials.

The forested margins of the Mediterranean were perforce settled by more isolated communities living in forest clearings and, their food and resources being more varied, their tools likewise are more sophisticated, smaller and more delicate. These people are thought to have originated in southwest Asia and to have migrated into Europe along the historic routeways of the Danube valley and the loess corridor from steppelands north of the Black Sea. In time they moved along the foot of the Alpine chains to the Cantabrian mountains and Atlantic shores.

With the final retreat and decay of the ice sheets the Stone Age hunters followed the tundra and the reindeer north, and the Mediterranean forest dwellers also migrated slowly north with the advancing forest. Arid conditions began to appear in north Africa and southwest Asia and provided, according to some archaeologists, the essential stimulus to the domestication or cultivation of the cereals that before grew wild and in greater profusion in all the Middle East and especially in the Jordan valley. The domestication of sheep, goats, and cattle also originated in the Middle East, probably through grazing and feeding on the fields of what must have been long stubble. Thus cereal cultivation and stock keeping were introduced into Europe by neolithic farmers from southwest Asia along with pottery, weaving and implements of husbandry. Early arrivals reached the Franchthi cave near Nafplion in the Gulf of Argolis in southern Greece by about 6000 BC.

Movement westward from the Middle East into the Mediterranean areas was easy since the soil and climatic conditions scarcely varied and so the slow tide of shifting agriculturalists was unhindered. Farther north, in and along the mountain chains, settlement sites needed to be selected more judiciously, and in any case were everywhere kept south of the great belt of boreal coniferous forests. The areas favoured for farming and settlement were gravel river-terraces, patches of loess and limestone areas not difficult to clear. In this way neolithic farmers settled the main and tributary valleys of the Danube. They practised a slash and burn technique that involved migrating after a few years and abandoning their fields to the regeneration of woodland.

The artefacts this civilization left behind show that these neolithic farmers stemmed in the main from two sources and followed two directions into Europe. First, from Syria and Anatolia they spread to the shores of the Mediterranean both in the eastern and western basins, being particularly well developed in Almería where they specialized in stock breeding and also grew cereals and pulses. From the Mediterranean coast of France and Spain they thrust north through the Carcassonne gap and the Rhône–Saône corridor. Mention must also be made of the Illyrians who were settled on the northeastern shores of the Adriatic at the beginning of the second millenium BC. bringing with them a Bronze Age culture that later developed into an Iron Age. Many traces of their piled *gomila* and fortified towns surrounded by concentric rings of dry walling stand on the crest of the mountains that run along this coast. A wealth of discoveries in materials, weapons, tools, coloured ceramics, ornaments, coins and massive sculptures indicate a highly developed culture with community life and a seafaring trade among the islands and straits of the Adriatic as the engraving of a boat on the side of a clay pot indicates.

The second great routeway used from at least 4000 BC onwards was the line of the Danube valley and the loess corridor leading west from the Black Sea to central France. The Rhône–Saône corridor and the middle Rhine valley were at the meeting-point of these two highways of migration. Similarly on the east, the sea route threading the islands and straits of the Dalmatian coast continued from the head of the Adriatic overland via the Postojna pass and linked the eastern Mediterranean basin to the lower Danube lands. Still farther east the Hellespont led to and from the historic hinterland of the Black Sea.

The rise of Greek civilization

Long before the neolithic farmers had started their attack on the northern forest the Minoan civilization had begun to flower in Crete and the higher civilisations of Sumer and Egypt advanced by this stepping stone in the Mediterranean area. Crete was well situated for trading links with the Nile delta and also with the valleys of the Tigris and Euphrates via the Fertile Crescent as well as with the islands of the Aegean and Adriatic Seas and these connexions were of great importance. But trade was also pushed by land via the Morava–Vardar valleys to Illyria and the Danube valley lands, as far afield as the rich metalliferous districts of Bohemia, Moravia and Transylvania. Gold, silver and tin were carried thence to the eastern Mediterranean, and beads, spices and other luxuries sent in return. North of the Mediterranean, trading was carried on throughout the area between the Po and the lower Rhine and along the whole Danube basin. Archaeological research supports these facts and they are confirmed by anthropological findings; in all these areas there has been in the past an incursion of tall, broad-headed people.

No doubt exists concerning the contacts made soon after 2000 BC by the Minoan civilization with the eastern and western Mediterranean basins and so by both sea and land the first hint of urban life was ushered into Europe. On the mainland it started at Mycenae in Greece about the year 1800 BC. This 'palace' civilization displayed a people using bronze and copper weapons, beautifully moulded gold and silver eating and drinking vessels and ornaments, ivory toilet articles and jewellery. They drove light two-wheeled chariots similar to those then used in the Near East.

Some centuries later, following the sack and burning of Knossos, the great city of Crete, the people of Mycenae took over its trading connexions and under the legendary King Theseus all Attica was temporarily united. *Poleis*, or city states, were established on the mainland and islands of the Aegean, Adriatic and Tyrrhenian seas as well as trading ports on shores and islands as far west as Sicily in the Mediterranean and even perhaps on the Atlantic and North Sea shores. Aristotle could rightly describe the Greeks of his day as political communities. Each *polis* that became a permanent centre of administration eventually grew into a town, generally walled.

From the geographical point of view the outstanding contribution of the Greeks to European civilization was the introduction of urban life and new crops to the whole Mediterranean world. The Greek city state was an independent or self-governing territory with clearly defined limits wherein settlement clustered around a fortified core, the 'polis', surrounded by the agricultural lands on which most of the people lived. By the year 600 BC there were two hundred such city states and Greek history is the record of these independent *poleis* scattered over the shores and islands of the Mediterranean, each with a small tract of dependent territory. Although Athens and Corinth, the two largest *poleis*, had populations of 300 000 and 90 000 respectively, the city states were generally minute by modern standards. Corinth controlled an area about as large as a minor English county.

Undoubtedly the urban aspect of civilization in ancient Greece has been overemphasized by historians. Thucydides tells us in 431 BC, at the time of the greatest expansion of the Athenian empire and greatest development of Athenian crafts, the majority of Athenians were living in the country.

The factors inherent in the varied landscapes of Greece and of the islands of the Aegean and Adriatic have been brought forward to account for the rise of the *polis* or city state. Certainly the islands had mountain slopes that yielded timber for construction, shipbuilding and fuel as well as pastures for flocks and herds, while their lowlands were suitable for cultivation. Some islands supported several city states; Crete had forty-three, and the island of Lesbos six. On the other hand in the Peloponnesus, Arcadia was a federation of several highland tracts, and the single city state of Sparta was very large, so that simple geographical explanations are insufficient to explain their growth.

In fact, Sparta was a *polis* in the widest sense (political community) from about 800 BC but although it gained control of all the southern Peloponnesus it remained a conglomeration of four villages without a surrounding wall for several centuries.

The suggestion that the warm, sunny Mediterranean climate favoured lengthy open air discussions is, assuming independent and progressive thought, probably nearer the truth. Certainly self-sufficiency of economy and society was the aim of them all. Plato estimated the ideal community at 5000, not counting slaves and women and children, and Aristotle was of the opinion that all citizens should be able to recognize each other. All business as well as justice was administered at the central 'polis' and the early Greek cities were therefore agglomerations of farm and artisan premises clustered round the 'polis', the whole being sited with more regard for productive farmland than for external trade and communications. The city states retained this character until overpopulation threatened their living standards and led between 750 and 550 BC to the dispersal of Greek colonies and the introduction of urban living into the western Mediterranean basin. Trading links between these cities were necessary from the beginning and led to the growth of merchant shipping.

It was fortuitous that of all the world's seas the Mediterranean has the best claim to be a nursery of seamen. Many of the city states had established themselves in the limited pockets of level land confined between the mountain foot and the sea. The mountain slopes afforded ship timbers at no great distance from the shore and their summits and outlines gave points and landmarks to steer by. The coastline is fretted with innumerable small indentations that provided sheltered havens and anchorages, and the headlands enclosing the bays gave safe sailing 'legs' and led from point to point around the whole sea. The islands provided similar landfalls and made summer daytime voyaging in the Mediterranean a relatively safe and easy matter. The day trips were assisted by the regular land and sea breezes but later when longer sea voyages were undertaken the prevailing northwest Etesian winds of summer gave early colonisers and navigators a wind to steer by and a steady course. As discussed in Chapter 3, the waters themselves are almost tideless and free from dangerous tidal currents and ocean drifts except in the narrows of the Strait of Messina and the Hellespont, which meant that small boats could be safely launched and beached at any time.

It was usual practice in these colonizing ventures to send out a balanced and viable community under a leader. The site would be surveyed and selected with regard to fertile soils, timber, water supply and defence. The objective was self-sufficiency, trading being of minor interest except at certain emporia. However, the pattern of trade, when it did develop later was typical of colonies everywhere: surplus agricultural produce was exchanged with the parent city, in return for manufactured goods in leather, metal, wool and pottery, though the colony was entirely independent from its inception. It is obvious, however, that the more daughter cities a mother city nourished the more its own industries flourished and the less dependent it became on home-produced food supplies. Inversely a daughter city gradually found it more profitable to specialize in certain commodities and its individual self-sufficiency was at an end. Instead of selfsufficient units a sophisticated and far-ranging trade system evolved with cities on the Black Sea sending hides, fish, furs, wax, honey and grain, and the western Mediterranean cities exporting bronze, copper, tin, silver and amber. Some of the sites of these early cities survive particularly in Magna Graecia which was colonized from Euboea and was the area of closest settlement, although the best-known colony is probably Massilia (Marseille). They were numerous and important along the eastern Adriatic where such cities as Issa and Pharos were established. On the island of Vis the Greeks from Lesbos founded in 392 BC the colony of Lissa but there were many others. Along this coast the first European vineyards were planted by the Greeks and the first olive groves. Abundant ruins and monuments bear witness to their long history on these eastern Adriatic shores.

Although largely confined to the Mediterranean seaboard the trading effects of the city states, especially in the great wave of colonization between 750 and 550 BC stretched from the Mediterranean to the shores of the North Sea, Black Sea and Baltic. During the later years of the movement the traders from the Greek colonial cities were coming into contact more and more with similar city states established mainly along the southern Mediterranean coast by the Phoenicians.

The Phoenicians

The Phoenicians are believed to have originated in the land of Canaan and to have migrated to the Levant coast where they established their trading cities. Their recorded history dates from 1600 B C when Egyptian power was extended over Syria. Writings on papyrus mention the Phoenician towns, and letters exist sent to the Pharaoh from the Egyptian governor of the Levant.

The Phoenicians had established themselves in island cities offshore where, secure from attack, they were ideally situated for maritime ventures. Their water supply came in some localities from submarine springs capped by leaden hemispheres and carried by leathern hoses into boats moored overhead. The Phoenician cities extended from Eleutherius in the north to Mount Carmel in the south and included besides Tyre and Sidon, the best known, also Beirut, Acre and at least ten other sites along the coast. From 1376 to 1366 B C the Egyptians along the Syrian and Lebanese coasts were harassed by raids from the Hittites and from desert nomads and were finally driven out. For a period between the twelfth and eighth century B C, before the westward drive of the Hittites destroyed their cities, the Phoenicians were a free people, and it was during these centuries that they established colonies, trading posts and merchant quarters throughout the 'known world'. Superb seamen, sailing in beautifully designed, well-equipped boats, they amassed a great knowledge of currents, winds and channels and gradually seized the carrying trade of the Mediterranean world.

They established colonies and trading posts first in the eastern Mediterranean, on Cyprus, some of the Aegean islands, and on the isthmus of Corinth but later extended their trade links into the western basin with colonies in Corsica, Sardinia, Sicily and Malta and along the coast of north Africa and southern Spain. Of them all, Carthage was the most important and remains the best known. Although the Phoenician cities never formed a federation, the colonies were never independent and remained closely tied to the parent city by trade links.

The Phoenician galleys picked up cargoes of timber and metals from Cyprus, embroidery and dyestuffs from Babylon and perfumes and spices from Arabia. Southern Spain through the city of Cádiz sent back rich metal ores and it was said that even the anchors of ships returning from Spain were made of silver. The Phoenician trade routes overland were equally significant, penetrating up the Nile valley as well as into the cities of the Tigris–Euphrates basin. In all the great cities of the ancient world they had their trading quarter. To them, a purely commercial people, we probably owe our alphabet.

The network of trade routes was designed for traffic in metals and so routes led from the lower Danube overland to the rich ore fields of Transylvania, and from the head of the Adriatic by the Postojna pass overland to lower Austria, Bohemia and Slovakia. Copper-working in the Salzburg area and metal-working in Bohemia–Slovakia dates from about 1500 B C.

Customs and skills also percolated by these trade routeways throughout the Danube valley and the inhabitants of this central European area (identified as *Celts*) began to advance in bronze technology and to produce tools, vessels and forgings of various kinds that enabled them to fix their agriculture. With permanent fields, mixed farming arose and with it permanent village sites, and an indigenous flowering of their civilization. Thus, while Greek and Phoenician cities were dominating the trade of the Mediterranean and establishing city states around its shores, the Celts were dominating the settlement and development of central Europe.

In Switzerland, and particularly along the Danube valley, the Celts built fortified townships girdled with high walls and towers that sheltered religious as well as mercantile practices. They minted coins and manufactured the tools of peace as well as of war. Beyond the town walls the farmlands extended in time to the edge of the coniferous forest. On the west the Celts expanded southward until they came into contact with the Phoenician cities in Catalonia; in the east they spread over the Hungarian puztas and penetrated to the Black Sea coast and south into the Peloponnesus where they made contact with the Greeks. They dominated the whole Alpine area and spilled through the Alpine passes into the Po basin and in 386 B C sacked the city of Rome. They had by this time made considerable inroads on the forests. They had their own notable towns and villages and were quarrying metal ores, but their major contributions to later civilizations were in the fields of language and race, as the Roman conquest, which wiped out so much of their culture, was unable to eliminate these.

Iron working, as distinct from copper and bronze, had meantime in Europe originated on the shores of the Black Sea and for a long time the smelting and forging processes were an industrial secret of the Hittites among whom it was used for making ploughshares and sickles. In the ninth and eighth centuries B C the Cimmerians and the Scyths, warlike and pastoral nomadic peoples who ranged Asia between the Black Sea and China, had begun to push westward and to filter by the age-old routes through and round the Carpathians into the middle Danube valley or along the loess corridor. They carried with them the knowledge and use of iron, which had reached central Europe by the seventh century B C. Thus it is generally assumed that while the 'Iron Age' was spread throughout the Mediterranean lands by the Greeks from about 1100 B C, the actual forging and smelting techniques developed later, mainly in the Alpine and Danube areas. This westward thrust of peoples also brought the Assyrians to the shores of the Levant, and they sacked and largely destroyed the Phoenician cities there. Phoenician power and influence were assumed for a time by Carthage in the western Mediterranean and lasted until the Punic wars with the Romans.

The Roman Empire

The Roman period that emerged out of the Iron Age and lapsed eventually into the Dark Ages, left on the landscape an enduring legacy of towns, town sites, villages, farms and field-patterns together with a road system and a wealth of extant place-names. For the only time in history, Mediterranean Europe and the lands forming the Rhine and Danube drainage basins were under a single control.

There seems little in the Roman environment that can explain satisfactorily the phenomenon of the Roman empire. The only certainty is that in the hills and plains of Latium an Iron Age pastoral society was in contact with Greek city states and Greek civilization on the south and with an advanced Etruscan culture to the north. The Etruscans had migrated from the Near East in the ninth century BC and had established city states in present-day Etruria. Their utensils and tools were made of bronze and iron but they also worked in gold, silver and ivory. Their metallurgical activities, based on local ores, were 'the most intense in all the central Mediterranean' (R. Bloch, p. 121). They never federated and their cities were conquered piecemeal by the Romans who nevertheless preserved much of their civilization. They were also advanced agriculturally having arable fields and many techniques of land drainage and water control. Their aqueducts and underground drainage channels (*cuniculi*) occur throughout northern Italy. Their influence extended from the plain of Lombardy to Magna Graecia and they had trade links by sea with other Mediterranean shores, especially in eastern Corsica. From about 616 to 510 BC the Tarquins of Tuscany or Etruria held sway over Rome itself and even when this domination ceased the Etruscans continued to control most of Campania for more than another century. The Romans did not finally subdue Etruria until about 280 BC.

The position of the Romans or Latium people between two advanced cultures was probably more important than the local situation of Rome but the site has proved enduring. Routeways naturally ran north–south each side of the Apennine spine and on the western route at the lowest crossing-place of the Tiber the city of Rome had its origins where an island in the middle of the river and high ground on both banks favoured the building of a bridge. The river was navigable and led inland to easy passes through the mountains into fertile mountain basins such as that of the Arno and the Liri. The estuary of the Tiber provided a safe harbour. At the bridgepoint were several steep-sided volcanic hills that formed useful defensive sites for the original cluster of Iron Age villages. By 650 BC Rome had grown into a small town and by 475 BC was no mean city with temples and public works within a walled perimeter ten kilometres long. It had become an important religious and trading centre and had already absorbed Magna Graecia on the south.

During these centuries the Etruscans farther north in Lombardy had finally succumbed to the Celtic invasions that penetrated Italy by the Alpine passes. Later the Celts moving through the Bologna gap crossed the Apennines and sacked Rome in 386 BC. The citizens rebuilt the walls and reorganized their army. Instead of a cavalry manned by patricians, they built up a well-equipped infantry drawn from the plebs and trained in hoplite warfare (heavy infantry in massed phalanx), an idea probably taken over from the Etruscans. Such an army actuated by technical intelligence as well as bravery was an innovation in the Mediterranean world, particularly one which concentrated on holding military strongpoints by a system of a strategically planned military roads along which supplies and reinforcements could move swiftly. Victories were consolidated by the planting of coloniae of veterans in newly acquired territories. By 257 BC the Italian peninsula was secure and Rome thus lay between the Greek maritime power in the eastern Mediterranean basin and the greater threat of the Carthaginian power in the western basin.

The Carthaginian influence extended through north Africa, into Spain and parts of Corsica, Sardinia and Sicily. Yet by 146 BC the Romans had destroyed Carthage and had gained naval supremacy in the western Mediterranean with a strong foothold on all its coastlands. Southern Spain was occupied and the coastal settlements linked by a military road that ran in an arc from Cartagena to Seville and Cádiz and had extensions into the mining areas of the Sierra Morena. The remainder of the Iberian peninsula was not brought under control until AD 19.

Southern France (Narbonensis) was formally annexed in 58 BC after an appeal by the Greeks of Massilia for protection against piracy. This annexation had the advantage of protecting both the Via Domitia (built 120–117 BC) into Spain and the route north into the Rhône–Saône corridor. The port of Massilia was bypassed by the Roman military highway into Spain and dwindled in importance. Its function was assumed by the port of Arles that grew up where the Via Domitia crossed the Rhône at the head of the delta from where it was linked to the sea by the Marius canal. Narbonne, on the military road into Spain, also became a busy port on this coast despite navigation difficulties through a lagoon.

Even during the Punic wars the conquest of Illyria had begun. The Alpine fold system from the Cantabrians to the Balkans appears at first sight to be a superb natural defensive barrier but from earliest times it had been penetrated by routeways like the Carcassonne gap, the Rhône–Saône corridor, the Alpine passes, the Morava–Vardar gap, the line of the Danube waterways and the entrance to the Danube plains at the head of the Adriatic. It was soon a matter of expediency to push the frontiers of the Roman empire back beyond the Alpine watersheds and foothills to

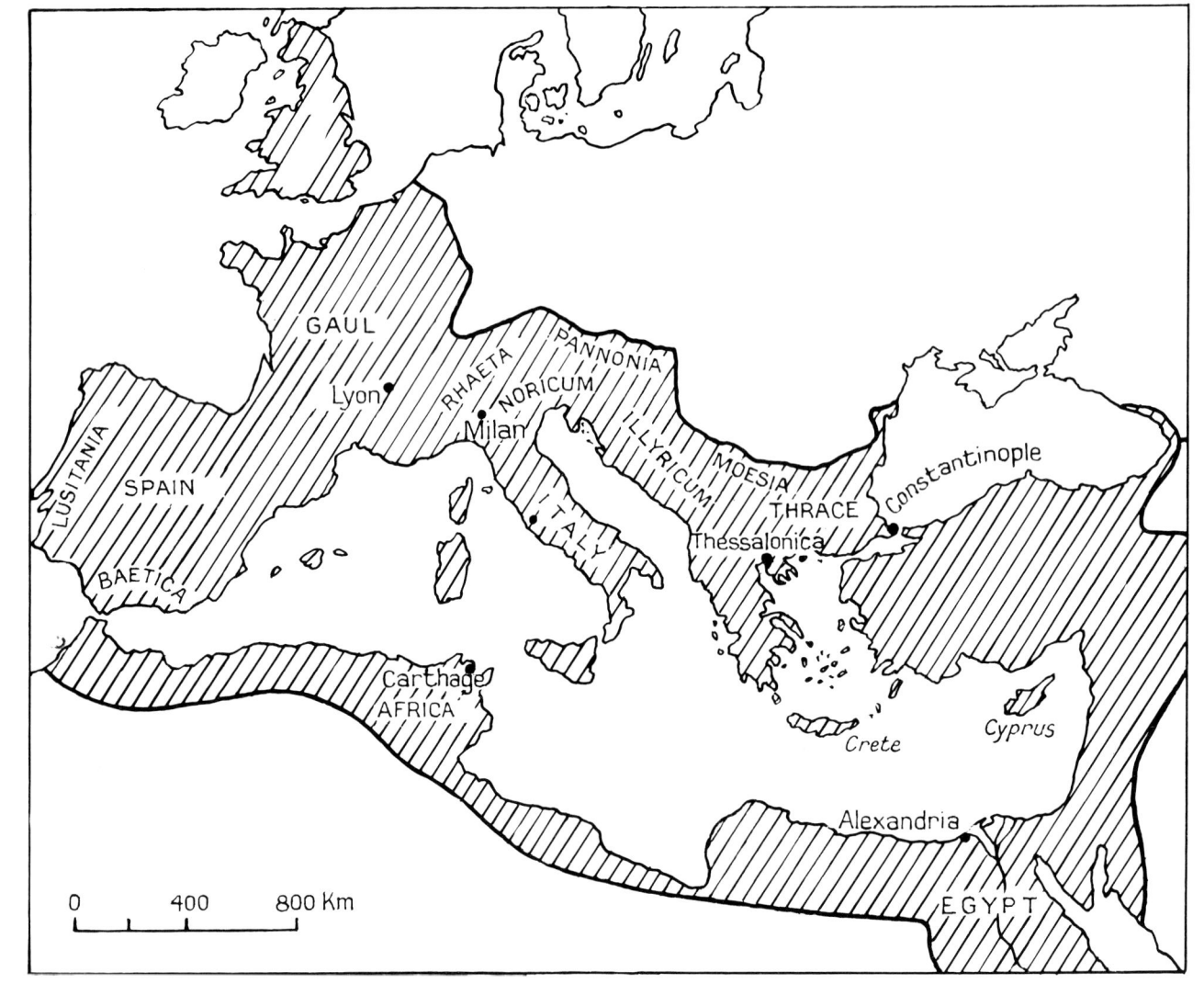

33 *The Roman empire in the fourth century* AD

the Rhine and the Danube (fig. 33). The Roman advance north of the river line is not the chief concern of this book but this initial frontier held until AD 406.

In the eastern Mediterranean basin, Greek power remained paramount until the second century BC, the city states acting as middlemen through which the trade in gold, silver and bronze goods, textiles, glass and ceramics flowed via the Black Sea and the Levant from Mesopotamia, India and the Far East. But in time the Romans acquired control of this traffic also and their empire extended over the Balkan peninsula, the Middle East, Lower Egypt and Libya. From the Middle East, knowledge and techniques of irrigation were introduced by the Romans into the Atlas lands and into coastal Spain.

Better farming methods and higher food production were spread throughout the Mediterranean world. The more significant agrarian improvements included more frequent ploughing of the soil, and the growing of fodder crops to ensure milk and meat supplies. It is now generally believed that, as a consequence, grain yields were as high in Roman times as at any time up to the eighteenth century. New varieties of crops and new breeds and types of stock were also introduced. Apricots, melons and lemons were introduced from the Far East and sesamum, clover, alfalfa, cherries, peaches, plums, figs, quinces, almonds, chestnuts, walnuts, radishes, flax and beetroots from the Middle East. Ducks, geese and some cattle were imported from China and the domestic cat from Egypt. The Greeks had already established viticulture around Massilia and it had spread throughout the Midi but during the Roman period vinegrowing spread north up the Rhône valley to the Côte d'Or and beyond. Andalusia first became of outstanding agricultural importance under the Romans and produced the finest quality olive oil, wine and grain as well as honey, wax, oakgalls, oysters, fish, wool and cloth.

The four main factors which led to this burgeoning of life and prosperity under the Romans were:
1 the existence of a large urban market centre of one million people in Rome
2 the military presence which protected and necessitated the smaller market centres throughout the empire
3 the superb civil administration that perfected the functioning of urban life and, with the military protection, made possible the Pax Romana
4 the great network of metalled roads, equivalent to the new railway network of the nineteenth century.

Colonies of Roman veterans and the spread of the villa population intensified agricultural development in the Mediterranean lands where, conditions being similar, farming practice was not pushing against new frontiers. In the Adriatic bays the remains of villas of Roman veterans are everywhere.

The Romans were ruthless and thorough in their search for metals but as their mining and smelting techniques were primitive they exploited only the most accessible and richest ores. They needed precious metals for their currency and to purchase goods from the East. Their uses of common metals were much the same as in Victorian Britain. Improvements in forged steel and the making of Toledo blades date from Roman times. Mining therefore grew into an important industry and in the second century BC there are records of Italian miners emigrating to the Sierra Morena and of gold rushes into Styria. Eventually, because transport was costly, metals began to be refined and manufactured near the mines and metallurgical developments, and settlements expanded in metalliferous areas.

Together with protection, the Roman army brought the direct benefit of a highway system planned to feed and serve important points and military bases. The road system linked the colonial towns together and fostered civilian trades and occupations.

Agriculture was stimulated by the demands made on the surrounding countryside for food and fodder for the military bases and these early acquired market functions. Of great importance also were the technical skills necessary for the building of the towns, aqueducts, viaducts, bridges and public buildings so that the Roman occupation must have stimulated growth in many ways.

The most important features that remained in civilian areas after the departure of the armies were undoubtedly towns, roads and urban life. By AD 15, there were 350 towns in Italy alone and a similar number in Spain. They were always sited with due regard to the Roman road network and were at important road and river junctions or crossings. Lyon and Châlons-sur-Saône were early Roman river ports. Many of them grew up on old Celtic town sites, as at Aix-en-Provence, and became Roman centres of administration and markets. Here and elsewhere the Romans seem often to have taken advantage of existing tribal centres. Other towns were new *coloniae*, planned in the formal Roman manner with two main highways crossing at right angles, and minor roads cutting the four quarters into blocks. Spaces at the major intersection were reserved for the public buildings of the forum and temples and so on. Most of the Roman coloniae were designed to house 6000 veterans, as at Narbonne, Arles and Orange but those established to relieve overcrowding and poverty were built to accommodate 60 000 persons as at Nîmes and Vienne.

Many old Celtic sites became 'expanded' towns by means of an extension of their urban function such as a

IX *Roman centuriation in the Po Valley near Cesina*

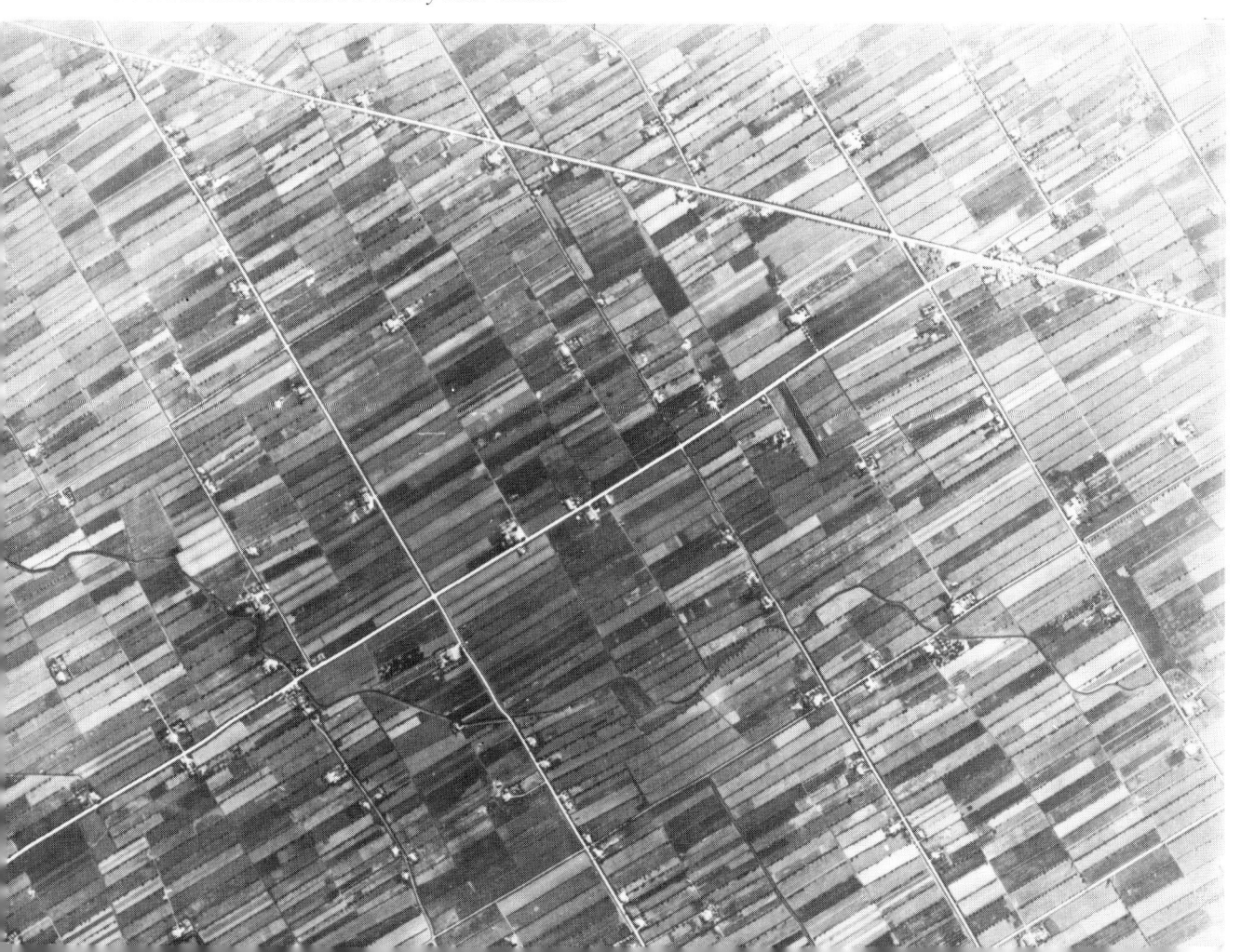

x *Aerial view of Roman arena and theatre at Arles*

bridge, a port or market and this extension would be carefully planned. Towns of this type are especially common in the frontier zone along the Rhine–Danube river line; such are Linz and Vienna, which lie outside our main field of inquiry.

So many of the Roman towns and ports have survived because they were sited to command strategic locations such as road and river crossings – bridges rather than fords – and were integral parts of the communication system. Even where the town has been destroyed, the site has survived for this reason, as for example at Dijon. Many Roman towns retained or increased their importance because, following the conversion of the Emperor Constantine to Christianity, they were made the seat of a bishopric and so retained their prestige through mediaeval times. The protection afforded by the presence of a bishop drew people to the shelter of the cathedral cities and in time the presence of saintly relics drew them also on pilgrimages. The great mediaeval fairs grew up, held at the festivals associated with the patron saints of the various shrines, some of which continue today.

Rural settlement under the Roman empire was of two kinds, the villa and the native village. The villa, or country estate of the Roman notables, like later country estates in England were selfsupporting communities with their own forges, carpenters' sheds, wheelwrights, weaving sheds, pigeon cotes, kilns and storage barns. They were generally built round a rectangular courtyard or two but varied much in size and wealth. They are abundant in the south of France and a cluster of them grew up about Lyon. At one time correlated with the limits of the vine and the olive, they came to be established like plantations in any productive corner of the empire, in which case the owner usually lived in the nearby town and drew the income from his lands which were run for him by a factor or estate agent. With the decline of empire, probably most of them eventually suffered this fate and became centres for the processing or semi-processing of surplus agricultural produce for sale in the towns.

A certain number of the villas can be traced along the frontier zones where veteran soldiers were settled, such as along the line of the Alps and the Danube. In these cases land grants were made by the state on a rentier system, the land being worked by native labour. These villas, after the retreat of the Roman armies, developed into the manorial system with a feudal economy and later became in some cases the nuclei of parishes. Around the coloniae and around some villas the agricultural land was parcelled out to veterans of the Roman army by the Roman surveyors into a regular chessboard pattern known as *centuriation*. Air photographs reveal traces of these centuriated fields in the Po basin, in southern France and in the Rhône–Saône corridor.

The second main type of rural settlement, the native village and huts is less impressive. In the countryside around the villas lived the native peasants poorly housed with few material possessions. Around Nîmes, Aix-en-Provence, Marseille and Narbonne and in some Alpine valleys nucleated villages of Roman age have been unearthed but normally the dwellings of the natives were scattered.

In considering the amount of land brought into continuous cultivation under the Romans it must be stressed that they were far more successful in the arid or semi-arid areas of the Mediterranean than elsewhere. In the deciduous forests the farms were established on gravel terraces along the river valleys, in limestone areas or on alluvium or loess patches. The dense oak forests of the Alps, Carpathians, in trans-Danubia or Transylvania were virtually still intact by the end of classical times. Little or no impact was made on the coniferous forest to the north. The shallow Roman ploughs were not efficient enough to have caused severe soil erosion except on steep slopes in the Mediterranean area. A study of place-names indicates a continuity of settlement from classical times to the present so that it is more rational to assume soil wastage and erosion from the later Middle Ages and the time of the Italian city state than from Roman farm practice.

A lasting legacy of the Romans to the visible landscape was the superb road system, much of which has been preserved. Even the minor roads may still be traced in secondary roads, and in field and parish boundaries and are retained also in some of the sheep and salt tracks.

In southern Europe the major Roman road system was retained and improved up to the present day. The Via Domitia still runs through the Midi linking Italy with Spain; the Pear Tree Pass is to have an international autostrada built over it; the Via Emilia still proclaims the first Roman frontier guarding the Bologna gap. In similar fashion, if we may digress a little, the Danube river line was paralleled by a frontier trunk road with feeder roads focusing upon strong points, a system echoed today in the 'Peace and Friendship Highway' that links Belgrade and the Morava–Vardar route with the Danube countries. Before the end of the Roman era all the more important Alpine passes had been brought into use.

The Völkerwanderung

By 114 BC the Goths had established themselves along the Rhine–Danube frontier, with trading links and settlement in depth, and from then on the Roman territory was continuously being eroded up to the fifth century AD. In AD 395, owing to the constant threat of invasion from the Germanic peoples along the northern frontier, the Roman empire for purposes of defence was divided along a line that followed the Dalmatian coast into a western and an eastern empire.

This division gives an opening date to the Völkerwanderung but the movement had begun long before. North of the Rhine–Danube frontier, the river Oder

34 *The Roman empire and the Barbarian kingdoms in* AD *528*

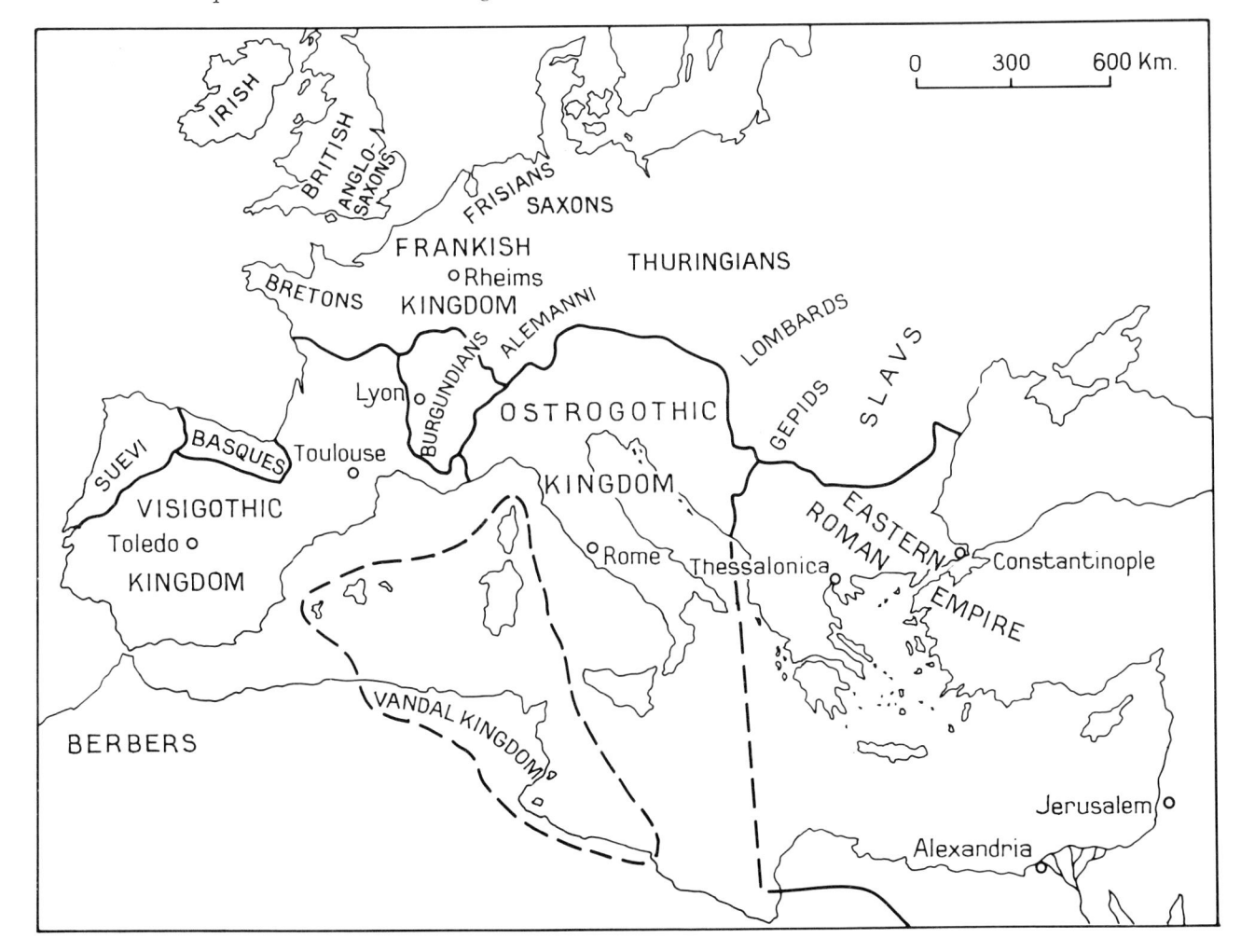

divided the Germanic peoples into those west of it who were settled on their *agri decumates* and those east of it who were pastoral nomads. East of the Vistula and ranging between the Black Sea region and China were warlike peoples known generally as Huns. The Huns were pastoral and nomadic and when mounted were highly mobile and aggressive. Repulsed on the east in the third and fourth centuries AD, they were thrusting with their herds west and dislodged the settled Germanic tribes west of the Oder. At this date the Bulgars settled in Bulgaria, arriving there from the north Russian plain, and the Magyars arrived in Hungary. In AD 410 the Goths sacked Rome itself and proceeded to establish themselves and their settlements widely throughout the empire. The Vandals reached Andalusia; the Franks settled in northern France, the Burgundians in the Saône valley, and the Lombards in northern Italy (fig. 34). But, on the other hand, Gothic settlement in southern France was sporadic and isolated and continuity with classical times was never broken. A rough boundary line may be drawn in France between the two areas of settlement, corresponding to that between the Romanesque architecture of the south and the Gothic of the north.

Along the southern shores of the Mediterranean a second migration was developing with the Arab advance. On the eastern frontiers the Roman empire was being harassed continually by a nomadic warrior tribe called the Saraceni, who eventually moved into the Mediterranean. Of all the peoples who migrated into this region the Arabs were the most politically mature, and their attainments in knowledge were far in advance of that acquired by the Slav invaders north of the Alps. The Arabs aimed to extend Islam and because they were familiar with the desert environment were able to spread their faith and customs throughout all the arid and semi-arid lands fringing the Mediterranean.

At first their power was entirely land-based and the cities of Damascus, Baghdad and Cairo stood well back from the sea. The trade and commerce of the eastern Mediterranean basin remained under the control of the Roman Byzantine empire. It held the island bases of the Aegean, controlled the Bosporus and the entrance to the Black Sea trade, patrolled off the Levant and Adriatic coast, and was thus able to command the supplies of timber and iron.

The Arabs therefore moved west along the coast of north Africa and launched an encircling movement around the western Mediterranean basin. One arm of their offensive was launched through the Iberian peninsula, the other through Sicily towards the Italian peninsula. They swept through Spain in AD 711–18 and established themselves in forward posts in southern France whence they could make forays up the Rhône valley or into Aquitaine (fig. 35). From this period date the charming fortified hilltop towns, like Eze, La Cadière and Le Castellet, that dot the coastlands of Provence and Liguria.

The Moors in Spain

The cultural impact of the Moors in Spain may be judged from the fact that, besides countless place-names in the peninsula that have been 'arabized', over 3000 Arabic names remain in use in Iberia. Such names include La Mancha (the tableland), Almadén (the mine), and the prefix Guadi in the names of rivers which stands for wadi (river). These names are most numerous in the south and east of Iberia where the conquest lasted for five centuries, and especially in Granada where the Moors remained for 800 years. Their influence on settlement sites and on agricultural techniques is probably more apparent in present-day landscape than their great contributions to western scientific knowledge and civilization.

That influence was greatest in the basin of the Guadalquivir and in the lowlands of Cartagena, Valencia, Almería and Málaga where they extended the knowledge and techniques of irrigation. The terminology of irrigation still in use here is Arabic in origin.

They introduced irrigation systems into the Guadalquivir valley and into Málaga and Almería where they founded the town in AD 825. The Turia canals, still in existence, of Alicante and Almería were built under the Moors in AD 911–76. Around Zaragoza in the Ebro valley, 10 000 hectares were irrigated under their direction, and in La Mancha and parts of Granada they developed the noria for use where water has to be continuously drawn by animal or wind power from wells. Elsewhere they developed and extended the existing irrigation systems introduced much earlier by the Romans and perfected a method of conserving water in the side canals.

It thus became possible to cultivate new food crops which they introduced. Rice was grown in Valencia; sugarcane on newly irrigated lands in Granada and Valencia, and bananas and cotton in the Guadalquivir valley around Seville. Lucerne was reintroduced indicating an understanding of basic soil maintenance. In the irrigated huertas on the estates of wealthy Arab landowners or merchants,

35 *The empire of the Caliphs about* AD 737

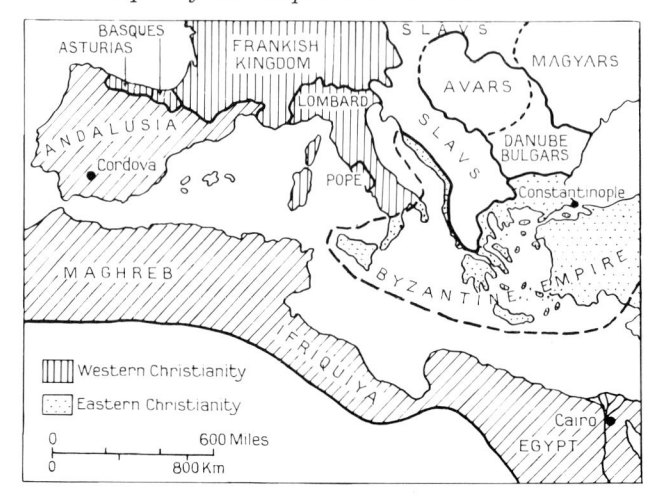

XI *Moorish architecture in Spain: a patio in the Alhambra, Granada*

intensive horticultural methods were employed in the cultivation of oranges, lemons, apricots, peaches and strawberries, but it is doubtful if there was much commerce in these luxury crops. Then as now swift transport and communications must have been difficult or expensive. On land which was beyond the reach of irrigation water, the olive was an important crop and its area of cultivation was extended. From China, by way of Syria, the mulberry tree and silkworm were introduced. Silk weaving was established in Cordoba, Almería and Valencia. The carob nut was introduced into southern Spain and Portugal, and date palms to the Elche region. On the Meseta, Arab influences transformed the pastoral economy, since they brought in new strains of horses and sheep and new methods of sheep handling. By these means they laid the foundation of the later Spanish wealth in wool and sheep.

The Arabs displayed an equal skill in metalwork and in exploiting with energy the mineral resources. The steel of Toledo, and the silver and copper of Cordoba were renowned; for the first time mercury was mined at Almadén, and tin in Galicia and the Algarve. Art and craft industries which survive to this day were established. Cordoba, the Moorish capital, made silk cloth, pottery, glass, leather goods, and gold and silver ware. At Játiva near Valencia the Moors first introduced to Europe the Chinese art of papermaking.

Their influence on settlement is still apparent in the hilltop and island sites capable of defence and in the haphazard arrangement of the dwellings within the built-up areas. Their intangible bequests in navigation, cartography, mathematics and astronomy are less easy to assess.

The conquest of Sicily took far longer than that of southern Spain but in 831 Palermo fell and in 902 the Saracens had mastery of the island. Sicily became a granary for them as it had been for the Romans and Greeks, and the new colonists developed pastoral farming in the interior so that it became famous for its mules and pack animals. The coastal plains came under irrigation and intensive cultivation, growing rice, cotton, oranges, sugar, dates and mulberries for sericulture. Palermo became an Arabic university city and one of the first European centres for silk weaving whence the craft spread to Italy and southern France. Eventually the Saracens extended their sway over Pantellaria, Malta, Sardinia and the Balearics in the western basin and over Cyprus, Rhodes and Crete in the eastern basin.

Under the Arabs trade moved freely since it was never held in dishonour as a profession. Their trading links were the lifelines by which the civilization of Byzantium and of the Orient continued to pass to the western Mediterranean and to Europe generally. Their caravans went overland to Samarkand and Tashkent and they had sea links with India and the Malay peninsula.

In 1065 Palermo and Syracuse were captured by Norman invaders and from then on Sicily became the main trading link between the Christian and Mohammedan worlds in the Mediterranean. It is interesting to note that the Normans proceeded to extend their domain over what had formerly been Magna Graecia, and Italy was partitioned from that date until modern times (fig. 36).

The age of medieval city states and of forest clearance
From the eleventh to the fourteenth centuries, at least

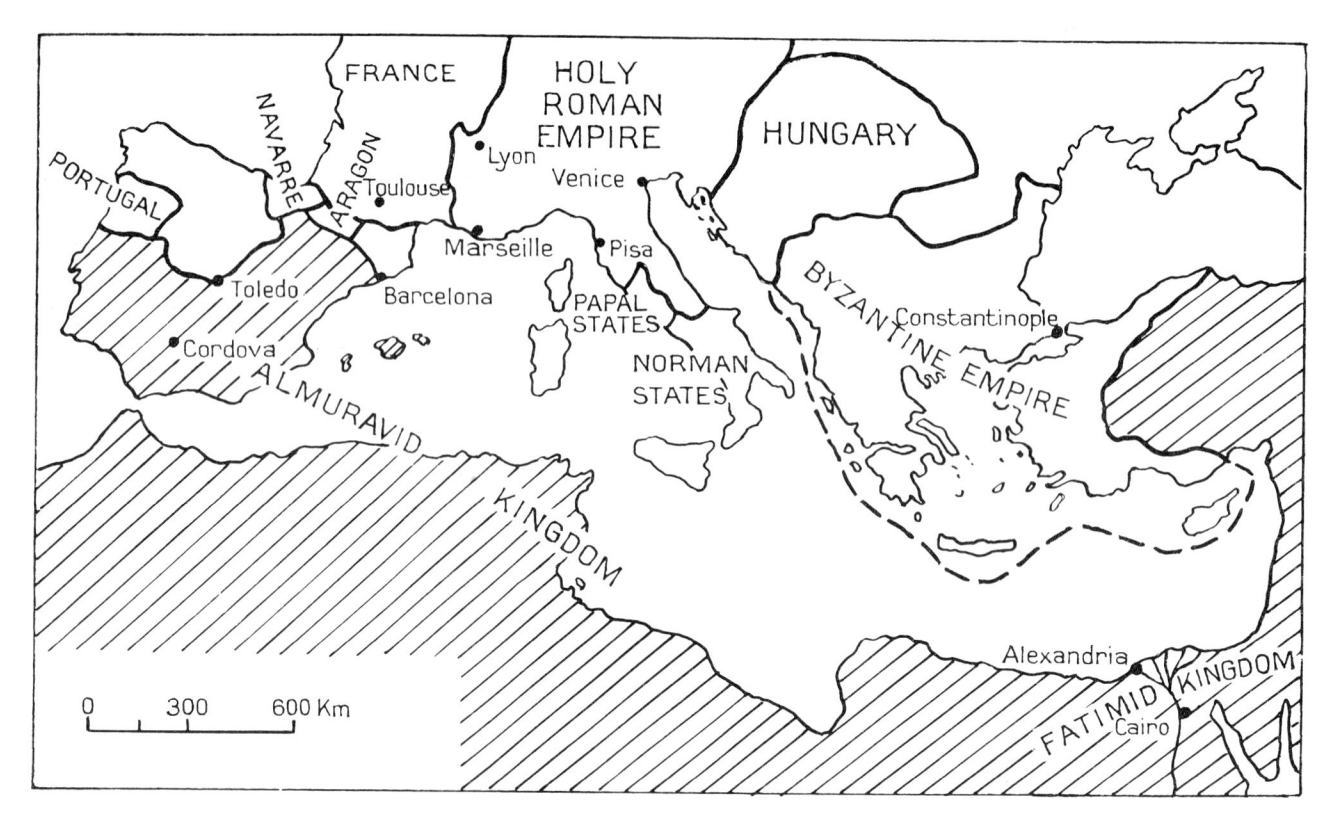

36 *Southern Europe and the Mediterranean lands about* AD *1150*

three great developments became apparent in peninsular Europe:

1 the *Reconquista* in Spain was a period of resettlement assisted by new religious orders, especially the Cistercians, Franciscans and Dominicans

2 in the mountain valleys and foothills north of the Alps there was extensive forest clearance for settlement and three of the Swiss cantons became a federation in 1291

3 a remarkable increase in productivity and trade encouraged the growth of great commercial cities such as Venice, Genoa, Pisa, Ravenna, Florence, Milan, Lucca, Siena and Dubrovnik.

During and after the decline of the Roman empire the general nature and direction of trade continued unchanged. Far Eastern produce still travelled to Rome, Marseille, Arles and Narbonne and return traffic in slaves, timber and cloth was carried by the old sea lanes to the Levant or to the Black Sea shores or by overland routes to Samarkand and Tashkent. But in the western Mediterranean, until the Arabs gained maritime supremacy, trade was harassed by pirates and traffic concentrated increasingly on the eastern basin. Active trade was controlled by the Byzantine empire through the ports of Amalfi, Bari, Naples, Gaeta, Venice and harbours on the Adriatic, but finally Venice took precedence of them all and after a struggle with Genoa took command of the sea trade of the eastern Mediterranean.

Venice, at the head of the Adriatic, found itself at this time between the eastern and western halves of the Roman empire and grew early as a distributing point between the two. The river Po was navigable up to Pavia whence Alpine passes were attainable, while a land route due north of Venice led to Vienna and via the Postojna pass to the lower Danube valley highway. The north Italian plain already formed a productive hinterland. The multi-island site selected for the city of Venice was safe from surprise attack by land but vulnerable to blockade of food supplies. The lagoons behind the sand bars (*lidi*) formed natural salt-evaporation pans and provided a useful item of trade (fig. 75). The Venetians ranged far afield for timber and soon became the leading shipbuilders and cargo carriers of the Byzantine empire. By the year AD 1000 they had their own quarter, with preferential tariffs, in Constantinople. In 1092 they were given freedom to trade in all Byzantine ports. Economic necessity led them to trade with the Saracens and they became a vital link between the Levant (the entrepôt of goods from China, Malay and India) and the western Mediterranean. They continued also the trade routes from Tashkent and Samarkand along the Danube valley and via the Postojna pass to the north Adriatic shores.

For the next two centuries the great crusades and the rise of the Papacy favoured the growth of cities on the pilgrim ways to the Holy Land and Rome. As a result of the first three crusades Venice and Genoa and Pisa acquired prefential trading rights and properties in many Levant cities. For the fourth crusade Venice furnished transport for 4500 horses, 9000 knights, 20 000 foot soldiers and provisions for one year. In return the Venetians received the Cyclades, Sporades, and the shores of the Black Sea, Sea of Marmara and of Thessaly. They purchased Crete and the islands and eastern shores of the Adriatic and so having control of sea routes to and from the Levant 'held the gorgeous East in fee'. Despite long and bitter struggles with Genoa and Pisa, Venice remained mistress of the eastern Mediterranean until the achievement of the Cape route to India by da Gama in 1498–9 broke her trading monopoly, and caused her citizens to hold a day of public mourning.

Other towns were also gaining in importance and prosperity, especially Genoa and Pisa, that acquired in addition to preferential rights in the conquered Levant towns, trading depots in Elba, Corsica and Sardinia. Many mainland cities also flourished, and Florence, Siena, Bologna and Milan rose to preeminence. Situated amid prosperous farmland, they early developed industries based on agricultural surpluses that attracted trade so that they too became entrepôts for the expensive eastern luxury trades. Banking and financial facilities automatically followed. Development of this kind was not unique to northern Italy. Several towns similarly situated north of the Alps enjoyed a similar flowering, but by the eleventh century Italian merchant bankers had houses everywhere in European towns particularly where fairs took place. The Italians had the financial expertise to gain control of the business organization and administration and were advancing credits, collecting market dues, and so on. Furthermore they employed their own carrier teams to haul the goods over the Alpine passes and their own boats and boatmen for water haulage. Gradually the wealthier towns accumulated highly skilled and specialized industries, such as the making of arms and armour, cloth dyeing and fulling, and leather craft.

The decline of these cities can be imputed to various causes such as overpopulation, the Black Death and the long costly struggle with the Ottoman empire in the Aegean, but surely the chief is that by 1498 the Portuguese had reached India by sea. In 1504 Venetian galleys in Beirut and Alexandria found no spices awaiting shipment and by 1510 the Portuguese in the Indian ocean had acquired a stranglehold on the spice trade. After 1571 the Venetians no longer traded with the Levant and the merchants there facing bankruptcy could not prevent the Ottoman empire spreading through Syria and Egypt.

According to tradition the Ottoman Turks were descended from nomadic tribes driven from their homes in central Asia by Mongol invaders. From the early ninth century there are records of Moslem fighters for the faith sworn to continuous battle with the infidel. These warlike tribes established themselves and their dynasties in the lands they had thus cleansed. By the thirteenth century they had arrived at the eastern frontiers of Byzantium and by 1340 the Byzantine possessions in Asia Minor were reduced to a few coastal towns near Constantinople (fig. 90). Five years later the first Ottoman army set foot on the Balkan peninsula and one hundred years later Constantinople fell to the Turks who then ruled the whole Balkan area. Their suzerainty stretched from the walls of Vienna in the northwest to Moldavia in the northeast and controlled the overland route and the Black Sea route to the caravan trade of central Asia (fig. 91). By sea the Ottoman Turks acquired footholds in the islands of the Aegean and, under Suleiman the Magnificent, their admirals harassed the Venetian navies and added the Barbary States and Malta to the Ottoman empire.

Administration under the Turks with its religious overtones was strictly controlled but peaceful. It was in the hands equally of the military and the religious, the latter propounding the law, the former enforcing it, while a civil service composed of Christian slaves administered it. Revenues were divided into seven parts, one of which was allotted to the building and upkeep of schools, hospitals and mosques. In 1571 the Turkish fleet was defeated at the second battle of Lepanto and the empire declined from that date, weakened by almost constant warfare. Although the Turkish suzerainty over the Balkans lasted a considerable time its lasting effects are not very evident. Certainly in the towns the mosques and the special workshop quarters are still to be seen as well as the houses built round an enclosed courtyard demonstrating the Moslem's preference for privacy in his domestic life. They introduced special types of metal work, such as silver and gold filigree and the manufacture of woollen and felt carpets and hangings. They also made a great difference to the clothing habits in various areas. The Albanians still prefer to wear the felt trousers, introduced by Turkish merchants, and many women went trousered and veiled until recent times. Numerous new plants were introduced and new products stimulated. Coffee arrived in Vienna under the Turks, along with sweetmeats of various kinds made from honey, mastic gum and various aromatic flavourings. In northern Bosnia where their settlement was densest they discouraged the keeping of pigs and the development of vineyards and instead fostered milk production and it is likely that the wide areas of plum orchards and the making of *slivovitz* is a direct result of Turkish rule.

The towns became nucleated defensive points but the native population was largely dispersed as, for example, are pockets of Serbian people in Croatia today. By and large the effects of the Turkish occupation on the Balkan peoples seem small except perhaps in their music, dancing, folklore and costume which are tinged with orientalism.

The making of the modern nation states

The making of the modern nation states in Europe dates back to the eighth century AD and can only be dealt with individually. Thus the history of the formation of each nation is given as an introduction to the geographical description of that state but we hope the reader will agree that modern Europe cannot be understood without recourse to its deeper historical roots described above.

10 International and Economic Patterns

The agricultural survey made in Chapters 7 and 8 showed wide differences in the agriculture and land use of the various states. A few were farmed intensively but many were farmed extensively, using a large labour force in not the most profitable way. An agrarian revolution which includes better crops, rational rotations and much mechanization has swept through southern Europe since 1945. At the present time the EEC (Common Market) is trying to coordinate and improve the agricultural production of its community.

National populations, densities, and development problems

Whether a country has an agricultural surplus or not depends largely on the number of its inhabitants, its cultivated area and the intensity of its land use. It may, however, specialize in crops for sale, such as tobacco, citrus fruit and cotton, as a profitable means of purchasing foodstuffs. The total population of a country is a guide at least to the local market it provides and its density is a clue to the utilization of its agricultural and industrial wealth. In areas where people live by supplying services or on capital from overseas investments, shipping, and sale of manufactured goods, the population is likely to be large. The table below relates to the larger countries of southern Europe and has an appendix of the smaller states of less than 500 sq km.

Apart from the small states, some of which are densely built over, the most densely populated countries in southern Europe are Italy, Switzerland and Portugal in that order. For comparison the density of population in Eire is 41, in Bulgaria 74, in Austria 87, Romania 81, in Denmark 111, the United Kingdom 224, Belgium 312 and the Netherlands 371 per sq km.

However, overpopulation and underdevelopment can only be judged on the standard of living, which is very difficult to assess in simple terms. In Greece, Spain and southern Italy the average personal incomes are only one-third those of the non-Mediterranean Common Market countries. Here the climate and historical inheritance of agricultural soils and resources are seldom favourable and agriculture does not produce funds sufficient to finance industries, indeed often the proceeds cannot pay for large-scale land-reclamation schemes. In southern Italy by 1950 the devastation of a fierce war heightened the already marked lack of adequate industrial development. Personal annual incomes averaged less than $4 a week, about one person in eight was unemployed and a large proportion of the workers were employed only part-time. Since 1950 the Cassa per il Mezzogiorno, an Italian government agency for the south, has encouraged agricultural developments and expansion of services to such an extent that the Mezzogiorno's share in the national agricultural production has increased from 30 to 40 per cent. To stimulate industrial

	Area (sq km)	Population (est.) 1966 (thousands)	Density (per sq km)	1968 (thousands)	Density (per sq km)
Albania	28 748	1 914	67	2 019	70
Southern France	136 620	10 300	76	10 600	78
Greece	130 918	8 614	65	8 803	67
Italy	301 224	51 962	172	52 750	175
Portugal*	88 420	9 335	102	8 907	101
Spain	504 879	31 871	63	32 411	64
Switzerland	41 288	5 999	147	6 147	149
Yugoslavia	255 804	19 735	77	20 186	79
Andorra	465	14	31	18	39
Gibraltar	6·5	25	4 167	25	4 167
Liechtenstein	157	20	127	21	134
Malta	316	317	1 005	319	1 009
Monaco	1·5	23	15 436	23	15 436
San Marino	61	18	291	18	291
Vatican City	0·4	1	2 500	1	2 500
Total	1 488 908·4	140 148	94	142 808	96
Total non-Soviet Europe	4 930 000	447 880	91	454 324	92
U.S.S.R.	22 389 470	233 105	10	237 808	11

* includes Atlantic Islands of Madeira and Azores

growth, state corporations were forced to locate a large proportion of their investment in the south and these firms were given loans, grants and fiscal reliefs. These methods and schemes were, it must be stressed, Italian but Italy as part of the EEC could draw heavily on the European Investment Bank which was specially formed to supply funds to backward districts of the Common Market. The EIB has spent two-thirds of its grants in the Mezzogiorno. The general schemes and results are discussed further in Chapters 17–19.

Spain, unlike Italy, is not a member of the Common Market and unlike southern Italy had considerable industrial development in the early twentieth century. Progress was checked and reversed by the disastrous civil war after which most countries ostracized Spain. Eventually in 1953 Spain made an agreement with the United States whereby the latter provided economic and financial aid in return for air and naval bases in Spain. In 1956 Spain was admitted to the United Nations and foreign trade and exchanges became easier. Since the 1960s, industry has been helped by a national policy of regional development whereby, in seven specified industrial areas, sites, easy loans and a heavy reduction of taxes are granted to new industries. A great deal of the capital comes from tourism (see p. 110) but some help, including technical aid, has come from international organizations such as OECD. Spain, it appears, would like to become an associate member of the European Common Market.

Greece has had problems very similar to those of Spain, such as lack of rainfall over large areas, mountanous terrain, the extensiveness of soil erosion, and a disastrous civil war (1944–9). During and after the civil war, massive United States aid largely offset the lack of surplus foreign exchange. But investment declined when the U.S. aid ceased in 1951. The main trouble was that the exports were so much lower in value than the imports (machinery, etc.) needed to develop the country. However, the growth of tourism and of ship owning, by 4 and 5 times respectively between 1953 and 1964, and large increases in emigrants' remittances home largely offset the trade deficit. In recent years, with the help of a state organization for industrial development, and with favourable terms for foreign investors and any other factory builders, many large new industrial concerns have been established. Since late 1962, when the country became an associate member of the Common Market, it has been able to draw on large sums from the European Investment Bank, for schemes such as the irrigation of the plain of Salonika. The problems that remain include overproduction of wheat, lack of adequate agricultural mechanization, lack of power, and the need to amalgamate or improve the numerous artisan concerns. One-third of the industrial production comes from 75 per cent of firms that employ less than 10 workers but form about half the total industrial labour force.

The development problems of Switzerland contrast sharply with those of the Mediterranean lands. It is already highly industrialized and as a longstanding neutral politically has avoided recently both foreign and civil wars. Its main problems are to obtain an adequate power supply and sufficient force of skilled workers, and to reorganize its agriculture and modernize its road system. Because of the attractiveness of its political neutrality and financial security it has had to curb the extent of foreign investments in its own industrial and business concerns.

The general condition of Yugoslavia is still more exceptional in southern Europe as it extends from the Mediterranean coast into the lower Danube plains. Today it is not a member of the COMECON countries (not discussed in this volume) but we must notice that all its Soviet-orientated neighbours have few or no problems of private ownership and having ruthlessly swept away, if they wished, traditional holdings and family concerns, can indulge freely in rational five-year plans. All, however, aim, as in the west, at reducing the vast numbers formerly engaged in agriculture by increasing mechanization and yields, and all aim to absorb the labour thus set free in greatly increased industrialization.

Industrial patterns and problems

The bases of modern industry are power and energy, raw materials and an industrially minded society. A characteristic of southern Europe is its poverty in solid fuels. Here only Spain, with about 12·5 million tons a year and, to a lesser degree Yugoslavia (1·3 million tons a year), have appreciable resources of bituminous coal. In Yugoslavia there are large resources of brown coal and lignite, which go mainly for thermal electricity generation and distillation, but generally speaking for any solid fuels southern Europe is poorly endowed compared with central and northwestern Europe. For crude petroleum and national gas the Alpine synclines in Italy and Albania yield small but useful quantities, and the outskirts of the Carpathian arc in Danubian countries have oil and gasfields of great use locally, but here by far the chief resources are in Romania, a COMECON country.

In southern Europe hydro-electricity has long been a chief source of power. This is the true home of hydro-electric development and of many inventions concerned with it. As the table (page 100) shows, water power is the prime supplier of electricity in Switzerland, Portugal, Spain and Albania. It provides more than half the electric power used in Yugoslavia and will soon provide three-quarters. It forms almost half the electricity production in Italy and France but if southern France alone were considered its proportion there would be over 90 per cent. So important is it and its projected schemes which are just beginning to gain impetus in the Balkans and on the lower Danube that it is described in detail below.

WATER POWER

The direct use of moving water in grist mills was common in Roman times and increased in popularity up to the early nineteenth century for driving almost any form of machinery. Later it was slowly replaced by steam power in areas with good coal supplies but its direct use still persists in a small way in many countries. On the Danube floating grist mills linked by a pontoon to the bank are still in operation.

During the second half of the nineteenth century the inhabitants of Alpine Europe played a notable part in the development of hydro-electricity. The first sizeable application of water power to generate electricity for industrial purposes was probably that of Aristide Bergès in the French Alps in 1869. He used a stream dropping 485 m into the Grésivaudan valley to produce electricity to drive machinery in his woodpulp and paper mill at Lancey, sixteen kilometres from Grenoble. In lower Austria a power plant with a capacity of 10 kW was set up in 1873. By 1882 there was a hydro-electric station (of 900 h.p.) on the Valserine in the French Alps and two years later Marcel Deprez began to transmit electrical current over appreciable distances from hydro-stations here.

The people of southern Europe were interested in hydraulic power partly because they were seriously short of coal for steam engines. They also happened to have numerous natural waterfalls and rapids that were easy to use, particularly where streams in hanging valleys plunged down the sheer walls of glaciated troughs. The increasing demand for electricity for domestic and industrial purposes was augmented by the electrification of railways. Today with the building of higher dams, hydro-electric power has become a major economic and social influence throughout the Alpine lands in Europe. However, its importance has also increased in Mediterranean lands generally, where large storage dams built mainly for irrigation water can be partly used for power generation. In the Pyrenees and Alps the storage reservoir is most needed against shortage of runoff in winter although an abnormally dry spell in later summer or early autumn may cause equal difficulties. In Mediterranean lands the season of shortage is nearly always summer.

For purposes of comparison the adjoining table gives for each country of southern Europe and of the Danube lands the actual production of electricity per country in million kWh and the percentage of the total national output that is supplied by hydraulic generators. In southern Europe the percentage ranges from 98 per cent in Switzerland and 95 per cent in Portugal to 49 per cent in Italy and 30 per cent in Greece. These large proportions contrast vividly with those of countries in the middle and lower Danube basin where, for example, Hungary and Romania derive less than 1 per cent of their electricity production from water power. The former has few falls and steep gradients while the latter is rich in alternative sources of power. It must be noticed however that international exchanges of electrical power by means of high-voltage grids is common today. In eastern Europe, the unified energy system or power grid of COMECON (the economic cooperation of the Soviet bloc in East-Europe) has a total installed capacity of over 35 million kW or 35 000 MW. But here Yugoslavia shares a large power scheme with Romania. Farther west, Switzerland and Austria have long exchanged power with western Germany and each other and in many other countries shared hydro-stations have been built near international boundaries, and electricity is bought and sold.

Already nuclear power has made a small but significant impact on the energy supplies of southern Europe. Both Spain and Switzerland have nuclear power stations. France in 1966 produced a total of 1395 million kWh from two nuclear stations, including that at Marcoule in the south; by 1968 this had increased to 3539 million kWh. In Italy, 8 nuclear stations, one of which has an annual output of 1000 million kWh, produced 3863 million kWh. Six of these stations are in peninsular Italy where hydro-electricity is less abundant. Here too the Italians have developed geothermal heat from *soffioni* in the volcanic areas. The chief stations are in southern Tuscany at Larderello near Pisa (in operation since 1894), Monterotondo and Sasso Pisano. Together the geothermal stations generate 2633 million kWh of electricity annually. As yet Italy has no rival for this kind of power in Europe and indeed its only notable rival is in the North Island of New Zealand (1268 million kWh). Thus the chief competitors of hydro-electric power

ELECTRICITY PRODUCTION IN SOUTHERN EUROPE, THE DANUBE LANDS AND CERTAIN OTHER COUNTRIES

(in million kWh)

Country	Hydro-electricity 1964	1965	1966	Total thermal and hydro	Per cent hydro (1966)
Albania	203			288	70
France	34 715	46 429	51 695	106 111	49
Greece	749	759	1 700	5 700	30
Italy	39 328	43 008	44 321	89 993	49
Portugal	4 220	3 983	5 306	5 591	95
Spain	20 646	19 687	27 176	37 466	73
Switzerland	22 663	24 015	27 444	27 962	98
Yugoslavia	7 575	8 985	9 880	17 174	58
Austria	13 179	16 083	17 331	23 817	73
Bulgaria	1 471	2 000	2 010	11 757	17
Czechoslovakia	2 727	4 456	4 256	36 528	12
Hungary	74	75	100	11 855	0.9
Romania	585	1 005	1 035	20 806	0.5
U.K.	1 760	4 625	4 461	190 000	2
U.S.A.	180 299	197 001	197 900	1 249 400	16
U.S.S.R.	76 600	81 431	90 900	509 800	18

Some common energy conversions:
 1 000 kWh = 0·4 ton coal
 1 000 thermal units of gas = 0·15 ton coal
 1 ton crude oil = 1·4 tons coal

today are coal, petroleum and natural gas. Most of the countries of southern Europe are, as we have seen, seriously short of coal and of petroleum; and few have been fortunate enough to find natural gas on a large scale. In any event, the growing national demands on energy have meant that the development of hydro-electricity has increased rapidly in the last twenty years. Some countries have already utilized most of their more economical hydro-sites.

In the Danube basin the chief hydro-electric possibilities are on the upper course, on the large tributaries from the Alps, and on the lower course at the Iron Gate. In the last-named locality, in Yugoslavia and Romania, the Sip–Gura–Vaii barrage across the Danube (opened in 1972) is the greatest single hydro-site in western Europe. A barrage 444 m long gives a fall of 26 m increasing to 34 m at low water. Twelve turbines with a total capacity of 2·1 million kW produce about 10 500 million kWh annually.

In the Balkan countries generally, hydro-electricity has a future if loans are available for construction. In Greece it forms only 17 per cent of the national output but in Yugoslavia it already provides 58 per cent of the electricity production or 21 per cent of the national energy consumption. The latter country has utilized only one-sixth of its hydraulic potential and the Iron Gate scheme alone will yield 7000 million kWh for Yugoslav use.

In the western countries of southern Europe the Alps and Pyrenees play a dominant or vital part in the power supplies of Switzerland, Italy, France and Spain. In Switzerland, hydro-electricity from Alpine streams supplies almost all the energy used for lighting, domestic purposes, railways and manufacturing industries, except those of a heavy type which depend mainly on power from imported coal. The chief railway lines, such as the Saint-Gotthard and Simplon, have their own hydraulic stations near their summit levels. Most of the Swiss hydro-sites are in the Alps but a secondary cluster lies on the Rhine above Basle where the potential sites include the falls of Schaffhausen. These lower stations, as at Eglisau, provide power mainly for electro–chemical and metallurgical factories. In 1958, the Swiss produced 16 700 million kWh from more than 6000 hydro-electric plants. This represented about 30 per cent of the total national consumption of energy. Since then the interchange of power with other countries, especially western Germany, has increased and conjoint schemes have been completed on, for example, the international boundary with Austria and Italy. Remarkable progress has been made in Switzerland with new installations and the enlargement of old. By 1966 the hydro-electricity output exceeded 27 000 million kWh and was over 98 per cent of the total electricity production. The raising of the height of dams, particularly by means of reinforced concrete, has opened up great new possibilities. Thus the recent constructions include a dam 255 m high across the Drance de Bagnes at Mauvoisin and another with an eventual height of 308 m at Grand Dixence, both on left-bank tributaries of the Upper Rhône in the Matterhorn area.

In Italy the value of Alpine streams in supplying power to a Mediterranean country very poor in coal is seen in a wider setting. The north predominates both in intensive agriculture and in manufacturing largely because of the Po drainage system. The Italians under the pressure of a dense population have expanded industries rapidly and have exploited any new forms of power. Yet in spite of recent developments of natural gas (equivalent to over 7 million tons of coal annually) they still import over 70 per cent of their energy requirements. Hydro-electric production has risen markedly but today its proportion of the national energy budget has declined, as the following table shows.

ITALY: PRIMARY SOURCES OF THE TOTAL NATIONAL ENERGY CONSUMPTION

	Percentage					
	1900	1930	1950	1961	1964	1968
Coal	97	69	40	18	11·3	9·3
Hydro	1	22	36	27	15·6	11·9
Petroleum	2	9	22	43·5	63·6	69·5
Natural gas	0	>1	2	11·5	9·5	9·3

In 1961, water power supplied about two-thirds of the electricity produced, whereas in 1966 when the hydro-electric output reached a record of nearly 45 000 million kWh it supplied about one-half of the total production. The continued increase in water-power installation says much for the skill of Italian engineers who have also built giant dams in many other parts of the world. In Italy probably 80 per cent of the potential water power has already been utilized. Two-thirds of the hydro-stations, producing over 77 per cent of the national hydro-electricity output, are in the north. The chief installations in the peninsula have installed capacities of about 200 000 kW each. The more important are San Giacomo and two others in the Abruzzi, Galleto near Terni on a tributary of the Tiber, and Mucone in Calabria. The reservoirs in the south have to be relatively capacious and are integrated in a grid with ample thermal (coal, nuclear or geothermal) supplies in the drier months. In July 1968 large offshore deposits of natural gas were struck on Italy's continental shelf in the Adriatic near Ravenna but multi-purpose reservoir building is likely to continue unabated in peninsular and insular Italy if only for flood control and irrigation.

In France, the southern half of the country produces almost all the hydro-electricity whereas the northern half has the main coalfields and most of the thermal stations. The south transmits power northward especially in summer and the north sends power southward at times of lower hydro output as sometimes happens, for example, in late autumn. France, like most European countries, is short of energy resources and imports a large part of her

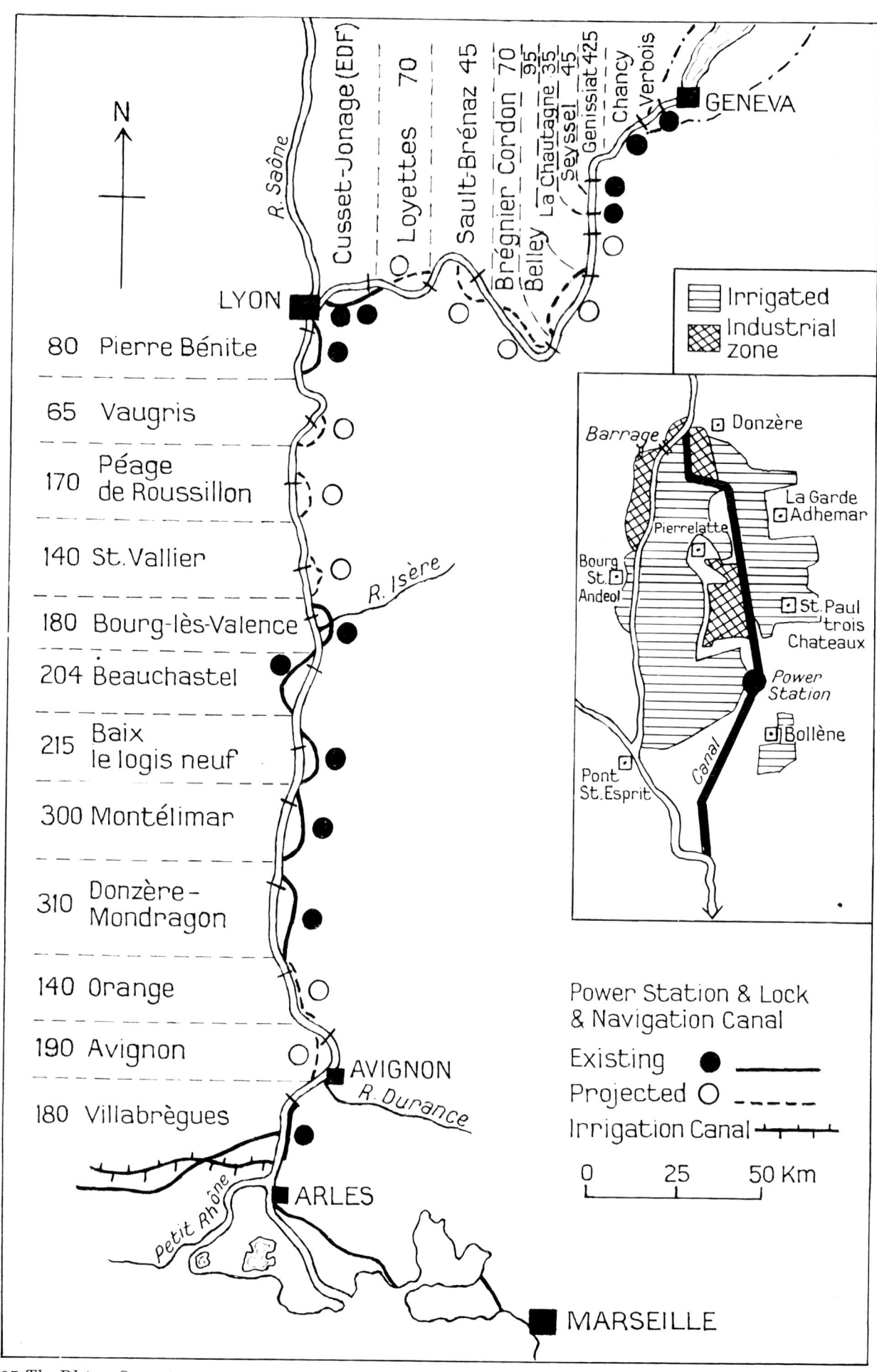

37 *The Rhône–Saône improvement scheme, with (inset) details of the Donzère–Mondragon project*
Statistics give installed capacity in thousand kW (after *La Documentation française*, 1968 et al.)

raw energy producers, especially petroleum. In addition, transmission lines link the country at 36 points with the west European grid and about 1000 million kWh of electricity are imported annually. The complementary geographical distribution of northern coalfields and southern waterfalls greatly encouraged the development of hydro-electricity. The balance between thermal and hydro power has varied considerably over the decades. Up to 1934, thermal electricity predominated but from 1935 to 1960, except in years of low rainfall, the hydro output was the greater. Since 1960, in spite of the opening of many large hydro plants, the thermal electricity production has exceeded that of water power. The primary sources of French power production have changed greatly since 1957 as the following table of percentage total energy output shows.

Power source	1957	1966	1968
Coal	65·8	35·4	30·7
Petroleum	25·3	49·4	55·0
Hydro	8·1	11·4	9·7
Natural gas	0·8	3·8	4·6

The result has been an increasing dependence on imported energy: in 1952 the French produced two-thirds of their power consumption while in 1968 the proportion had fallen to 37 per cent.

Yet the importance of water power remains great. In 1966, the 1650 hydro-electric plants in France produced 52 000 million kWh of electricity or the equivalent of 21 million tons of coal. The 60 thermal stations produced slightly more electricity. By 1986 it is esimated that the power installed at electricity plants (in megawatts or 1000 kw) will be, hydro (20 000 MW), nuclear (30 000 MW) and thermal (40 000 MW). The reasons for the relative decline of water power are that by then almost all the more desirable hydro-sites will already have been used; that the average cost of building a thermal station is less; and that thermal electricity production is cheapened by replacing coal with liquid fuels and gas. However, water power will retain its advantages of being non-pollutant, of regulating river flow without lessening it, and of occupying sites that usually are not otherwise highly productive.

The French hydro-electric output comes mainly from the Alps (about 65 per cent total, including Rhône and Rhine), the Massif Central (20 per cent total) and the Pyrenees (15 per cent total). In the French Pyrenees there are more than 100 stations but few produce over 250 million kWh annually. The supply provides local markets and allows the electrification of the main trans-Pyrenean railways. In the French Alps the development increased from ten hydro plants in 1896, to 57 by 1914 and to 118 (of over 1000 kW capacity) by 1946. At this stage the improvement of the Rhône began to materialize. The company formed in 1933 for this purpose (*Compagnie Nationale du Rhône*) was a form of public corporation, under the advice and financial backing of the state. Its main aims were hydro-electricity generation, irrigation, navigation and flood control. However, hydro-electric power has predominated and, as often happens in these multi-purpose schemes, has been the prime profit earner. The programme provided for the construction between Lake Geneva and the Mediterranean of 26 hydro stations that would use a total fall of 350 m with an annual potential of 16 000 million kWh. The middle Rhône below Lake Geneva was tackled first. At Génissiat about 50 kilometres downstream from Geneva where the Rhône breaches the southern edge of the highest ridges of the Jura, a dam, 80 m high, was built across the river. The power house (opened in 1946) has an installed capacity of 335 000 kW with an annual output of 1500 million kWh. By 1956, a smaller plant had been opened at Seyssel farther downstream and several other stations were planned for this stretch of the Rhône (fig. 37).

The 333-km stretch of the lower Rhône below its junction with the Saône offered great possibilities as the volume is much larger and the configuration of the valley is an alternation of gorges and small plains. Here twelve large installations are projected in all. The 150-km stretch downstream from Valence near the junction of the Isère to Beaucaire near the delta head, alone affords sites for eight large stations, with a combined potential of 10 000 million kWh. The first scheme completed here (in 1952) involved the damming of the Rhône near its exit from the Donzère gorge and the excavation of a derivation canal from above the dam to Mondragon. Eighteen kilometres along this canal a power station, using a fall of 26 m has an installed capacity of 300 000 kW and an annual output of 2000 million kWh. Typical of these large schemes was the building of a fine navigation lock and of several housing estates for workers at the power station and at neighbouring factories (fig. 37).

More recently several other large hydro-electric stations with lateral canals have been built on the lower Rhône, including one at Vallebregues near Beaucaire downstream of Donzère, and four in succession upstream of it near Montélimar, Baix, Beauchastel and Valence respectively. All these schemes function nearly 7000 hours a year and their steady power output has been an attraction for modern industrialists. For example, at L'Ardoise a new electro–metallurgical concern produces over half the French output of chromium plate and one-fifth of the national output of stainless steel. Some distance upstream near Lyon, the Pierre–Bénite scheme, with an annual output of 500 million kWh, was opened recently. It involved digging a long canal for navigation and drainage and the 15 million tons of earth moved was used for the creation of an industrial site of 500 ha and a railway marshalling yard at Sibelin and for part of the Lyon-Vienne motorway. By 1969, the CNR hydro-stations were producing 10 600 million kWh of electricity annually and within a

XII *Picos del Sil, Orense, Spain: hydro-electric dam*

few years the few remaining projects, including four large centrals on the lower course, will raise the total output to nearly 16 000 million kWh.

Since 1958, French engineers have also undertaken a large improvement scheme on the lower Durance, where five main hydro-stations and a barrage across the deep gorge of the Verdon at Castillon, were already producing 1000 million kWh annually (fig. 58). Here the output was reckoned to be one-tenth of the potential but the difficulties included a marked summer low water and a deep alluvial infill in the valleys. In 1960, a large earthen dam, 650 m wide at the base, was built across the Durance at Serre–Ponçon. It created a reservoir of 1200 million cub m capacity and the power house, of 300 000 kW, generates 700 million kWh annually. A lateral canal with a flow of 250 cub m/sec has been constructed alongside the lower Durance as far as Mallmort, where it turns south away from the Durance valley and enters the Étang de Berre near Saint-Chamas. Five hydro-stations have already been installed on this canal and the whole Durance scheme will eventually yield 6000 million kWh, in addition to irrigating extensive areas. The plans here include the Canal de Provence, which leads southward from a reservoir on the Verdon and will irrigate 60 000 ha as well as improving the water supply of Toulon and Marseille.

In Iberia the main demand for electricity has been in the major cities and in the industrial towns of Catalonia (fig. 38). Throughout the peninsula there is a close connexion between flood control and water storage and in all the drier parts the needs of irrigation are important or dominant. Most of the rivers in the south and east have a marked summer low water and need large reservoirs to ensure a modest flow all the year. The regime of Pyrenean streams is much more regular although those rising at moderate altitudes may dwindle after prolonged drought. Thus, during a prolonged period when rainfall was below normal from autumn 1948 to 1949 many of the industries dependent on hydro-electricity in the Barcelona area could only work one day a week. In arid Iberia, intense evaporation from reservoirs and the high sediment load of streams add to the difficulties of irregular seasonal regimes and large variations in annual flow but these problems have not deterred the Spaniards from building reservoirs. Since the late eighteenth century they have built storage lakes of a capacity quite exceptional for Europe at the time, as for example, at Valdeinferno and at Puentes (65 million cub m capacity). At Puentes a tremendous flood in April 1902 made a bridge-shaped breach in the base of the barrage which, however, survived as a whole. The reservoir at Tremp on the Noguera Pallaresa was at the time of its construction the fourth largest in the world.

Since 1950 the Spaniards have again invested heavily in dam-building and most of the schemes have a power house. The main reservoirs are in the basins of the Tagus, Duero, Ebro, Guadalquivir and Guadiana in that order of importance. They include many with over 1000 million cub m capacity as at Cijara (1760 million cub m) on the middle Guadiana, Buendía (1571) on the upper Tagus, Mequinenza (1530) on the lower Ebro, Ricobayo (1196) on the lower Esla and Iznájar (1110) on the Genil. The production of hydro-electricity has risen almost directly in proportion to the increase in water storage, from 7500 million kWh in 1950 to 18 000 million kWh in 1960 and about 30 000 million kWh in 1970. Nearly three-quarters of the Spanish production of electricity comes from water power.

The distribution of hydro-sites in Spain reflects the frequency of rapid streams and deep gorges as well as the abundance of runoff and consumer demand (fig. 38). Thus, the traditional industrial area of Catalonia has long made use of water power especially for driving textile machinery direct. Today these textiles continue to flourish and direct water drive has largely been replaced by electricity from local stations and from hydro-sites farther west and thermal plants near Badalona. Hydro-electricity generation in Spain is remarkably widespread. The chief area is on the lower flanks of the snowy central Pyrenees in Lerida and Huesca which supplies about 20 per cent of the national output. Throughout Cantabrian Spain small stations are common, the largest being at Grandas de Salime on the river Navia. In Galicia, the deeply incised valleys of the middle Minho provide sites for eight stations. In Old Castile on the lower Esla and the international stretch of the Douro a series of dams and storage reservoirs are equipped with six stations. The lower Esla on the level Meseta needs large storage lakes whereas the international Douro traverses a deep gorge where the storage is small but the steep drop (3·75 m per km) allows 3 stations to produce a total of 2750 million kWh from an installed capacity of 546 100 kW. In New Castile, the stimulus of the metropolitan market has encouraged the building of large reservoirs with power plants on the Tagus and of smaller stations on the steep headstreams of its tributaries in the Central Sierras north of Madrid.

Farther south the upper Guadalquivir has numerous gorges where ten dams have been built partly for power supply. Similarly sixteen small plants have been installed on the rapid streams on the flanks of the Betic Cordillera and especially near the snowy Sierra Nevada. In addition, each of the major east-coast streams now has a string of four or five dams with small power plants in the deep gorges of their mountain headstreams.

The Ebro is exceptional as its volume is exceptionally large (615 cub m/sec) and its chief gorge is at its seaward end. Here at Mequinenza, Seros and Flix are three moderate-sized hydro-stations, the first having a large reservoir. The Spaniards have probably already utilized about 50 per

38 *Distribution of electricity generation and transmission lines in Spain* (after *Atlas Nacional de España*)

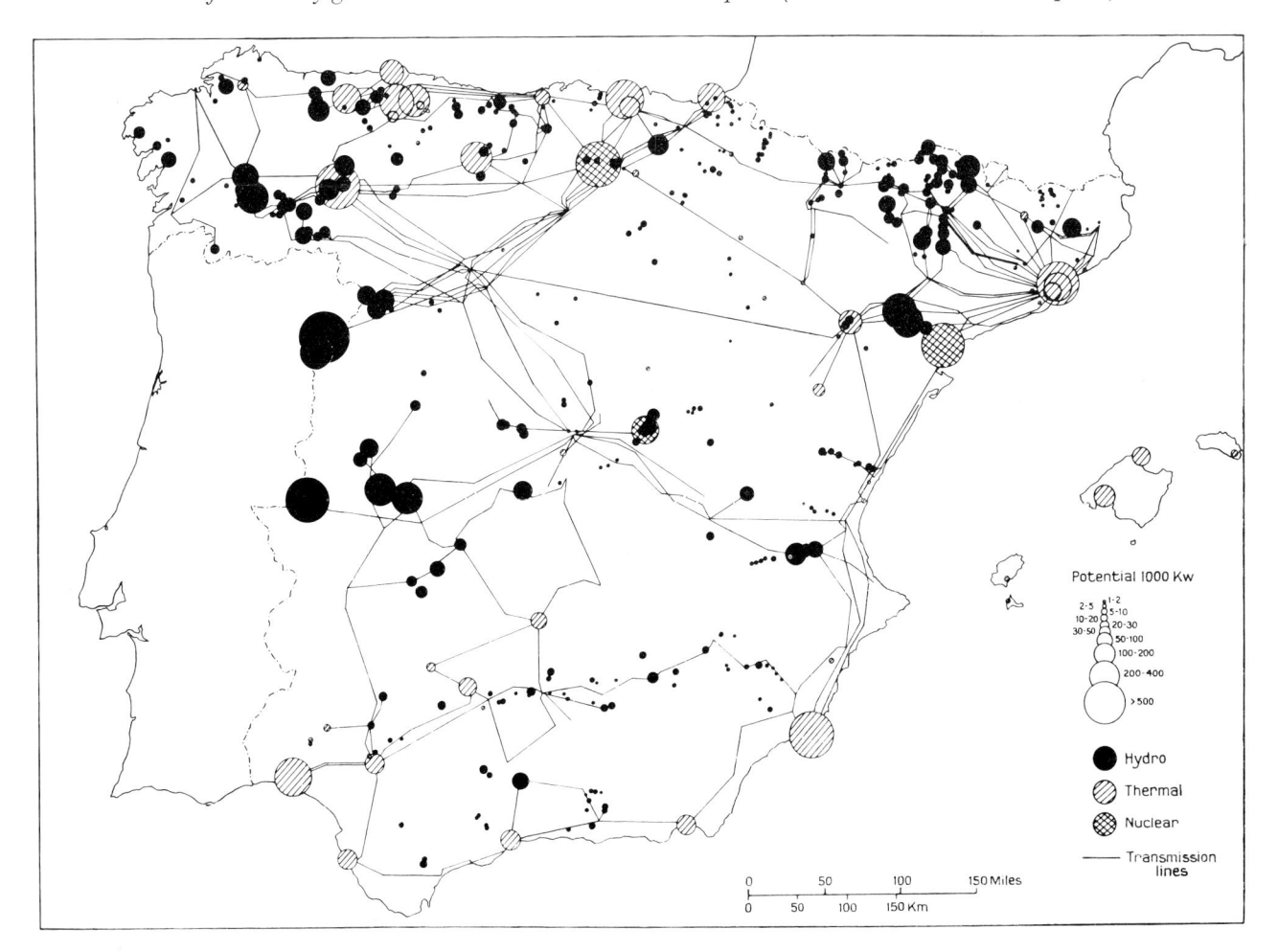

cent of their hydraulic potential and the utilization of the remainder will be still more difficult and expensive.

The Portuguese have also, on a smaller scale, rapidly expanded their hydro-electric production from 1000 million kWh in 1950, to 3200 million in 1960, and about 7000 million kWh in 1970. Here too, deep gorges and steep gradients abound, particularly on the north-bank tributaries of the Douro and Tagus. The present installations closely reflect the needs of the two major cities, Lisbon and Oporto. Near Lisbon a series of barrages and large reservoirs on the Zêzere river which rises in the rainy Serra da Estrêla, are equipped with seven hydro-plants that produce about half of the national output. Most of the remainder comes from four stations on the river Tâmega which enters the Douro not far upstream of Oporto. Nearly all Portugal's electricity is hydraulic and a considerable potential remains unexploited. The lower Douro and its steep northern tributaries alone could provide a further 5000 million kWh and a large scheme is already nearing completion here (see Chapter 13).

Industrial development

The industrial development of a state may be judged fairly accurately on its output of manufactured goods, on the number of persons employed in industries and the proportion of the gross national product formed by the industrial sector. As it would be impossible in a short space to discuss all manufactures, we have selected textile looms in cotton and wool, cement, crude steel and sulphuric acid as representative of modern industrial complexes.

Moreover, we have added these statistics for all adjoining and Danubian countries and for the United Kingdom for purposes of comparison and as some guide to probable interchanges of goods. In the statistics for wool, ordinary and automatic looms are not distinguished. All production statistics are given in thousand metric tons and relate mainly to the year 1966 or 1965; looms are given in thousands to the nearest thousand. If statistics had been given for a longer term they would show that textiles, especially woollen, have not greatly increased their production since 1948 while cotton textiles have also grown slowly except perhaps in Czechoslovakia and the Balkan countries. In the latter the raw cotton crop has expanded and so stimulated home manufacture. The great change, however, has been the partial replacement of ordinary looms by automatic machines, particularly in the countries with a shortage of skilled industrial workers.

Cement output is given as it gives some clue to the building programme including roads and other public works. The crude steel output can be used in engineering either in making machines and so on, or in construction work. However, in nearly all southern European countries it is too precious to be used lavishly on construction work except for railways where the scrap is easily recoverable. The output in crude steel has leapt upward since 1948 except in a few countries such as Greece and Albania. Thus the production in Czechoslovakia has increased nearly four times and that of Romania sevenfold in the last twenty years, partly because the COMECON states have laid great stress on the development of heavy industries. The table

SOME INDICES OF INDUSTRIAL DEVELOPMENT, 1966 (absolute statistics, in thousands)

	Looms Cotton Ord	Auto	Wool	Cement	Crude Steel Output	Apparent consumption	Sulphuric acid	Industrial Employment Number	Per cent national total	Per cent value G.N.D.
Albania				135				130	17	
France	21	59	13	23 438	19 585	17 143	3 072	7 652	39	41
Greece	4	3		3 587	210	839	138	698	19	27
Italy	9	71	22	22 374	13 639	14 175	3 369	7 996	40	44
Portugal	22	14	4	1 720	269	695	412	962	28	43
Spain	44	22	10	11 834	3 750	6 720	1 710	4 044	34	36
Switzerland	2	11	2	4 326	428	1 990	194	1 267	51	52
Yugoslavia	6	20		3 232	1 867	2 783	542	1 620	23	38
Austria	1	7	2	4 501	3 193	2 065	193	1 356	40	52
Bulgaria	6	6		2 851	699	1 691	353	1 097	25	46
Czechoslovakia	7	27		6 130	9 128	7 758	982	3 090	50	66
Hungary	11	4	2	2 601	2 649	2 377	418	1 885	37	64
Romania	11	5		5 886	3 670	4 084	619	2 281	27	48
U.K.	69	46	34	16 751	24 705	21 247	3 168	7 898	31	47

gives the apparent or probable consumption of steel for each country so that it can be seen at a glance which exports and which imports steel. It, and the table already given of the national population, also allow the amount of steel used per person to be evaluated; this ranges from 545 kilogrammes in Czechoslovakia to 97 kilogrammes in Greece and 75 in Portugal.

Mention must be made here of great modern advances in the techniques of making crude steel, which in conjunction with international planning by the ECSC or EEC (Common Market) has caused the steel industry to move to the coast in EEC countries. Crude-steel manufacturing is switching steadily to oxygen processes, which in 1969 accounted for 30 per cent of the Common Market steel-making capacity and by 1971 are expected to account for over 40 per cent. We mention this because the change in Italy especially concerns us. Here on the coastal areas about 45 per cent of the steel production is by oxygen processes such as L.D.* and Kaldo, about 45 per cent by open hearth, and 10 per cent by electric furnaces, whereas in inland Italy the steel comes from electric furnaces (70 per cent) and open hearth (30 per cent). The new plants in the south include tube concerns at Chieti, Brindisi and Palermo; ingot and steel works at Naples, Bari, Catania and Sant'Antioco (Sardinia), and a large integrated works at Taranto. Nothing could suit southern Europe better than being able to do without coke supplies in which it is so obviously lacking. However, vast quantities of water are still required and at Taranto a powerful desalinization plant has been installed.

The table opposite shows clearly enough the great national value of manufacturing, especially when compared with the relatively small labour force. This labour force, moreover, also includes construction and building, mining, electricity, gas and water concerns, and small domestic manufactures, that together in most countries occupy in themselves between 5 and 10 per cent of the national workers. The reader who consults the tables of employment in agriculture (Chapter 7) and in industry might well wonder what the rest of the civilian labour force does. Some, unfortunately, are unemployed and in for example southern Italy may comprise up to 5 per cent of the available workers but normally nearly all the remainder are engaged in *services* such as transport, communications, finance, insurance, administration, civil defence, social aids, education and so on. These functions increase as a national economy becomes more complex. Thus in Yugoslavia, services will occupy 23 per cent of the labour force in 1975 against 16 per cent in 1960 and in Spain 35 per cent against 27 per cent in the same period. Today services

* Linz–Donauwitz are the Austrian metallurgical centres where the oxygen process was invented. It does not need coke.

39 *Main railways and chief route and commercial centres of Alpine and Danubian Europe*

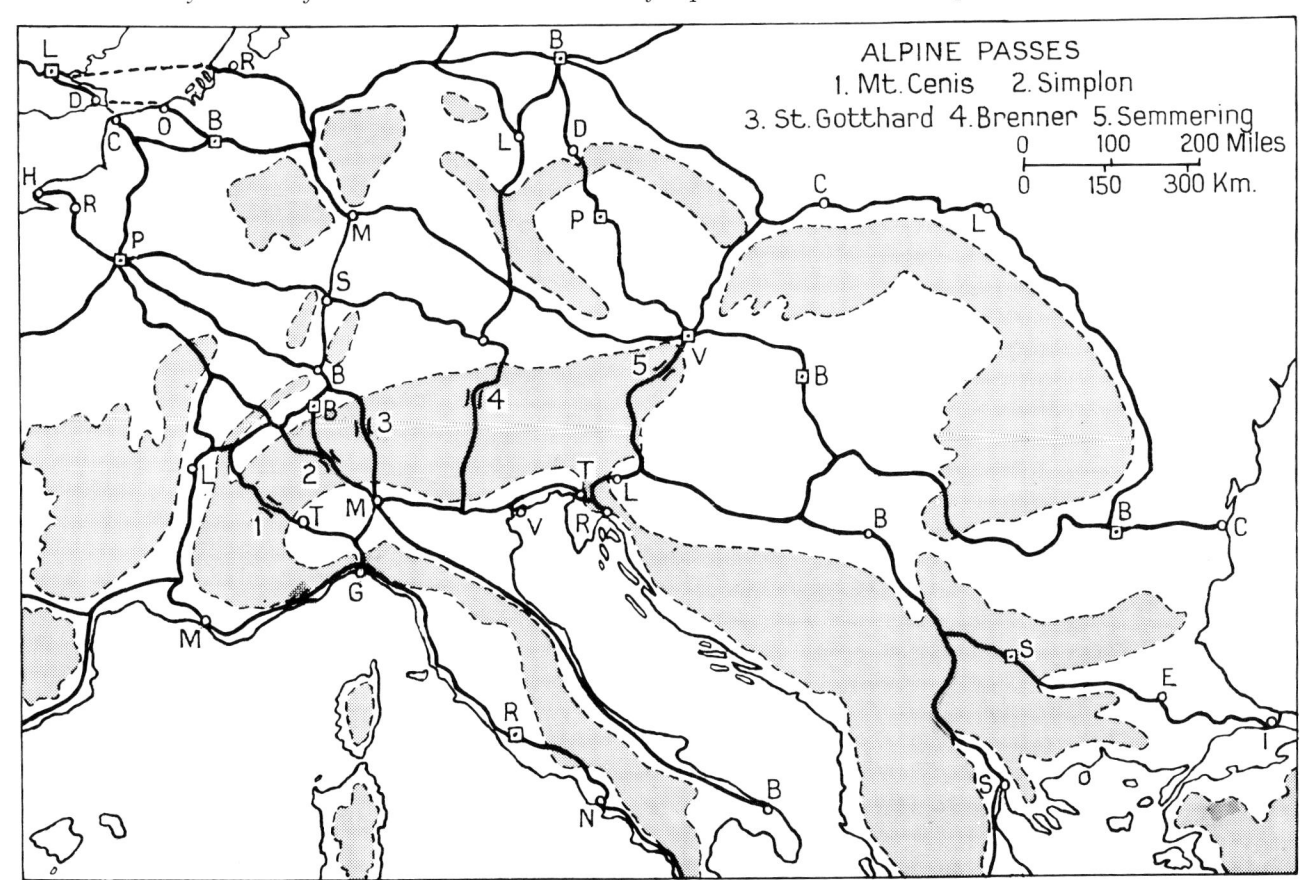

occupy about 30 to 40 per cent of the total working population in most countries of southern Europe, except in the Balkans where the proportion is appreciably lower.

International communications and transport

There is no truly international river highway in southern Europe comparable with the Danube and the Rhine in central Europe. The only significant riverway of the south, apart from the Swiss Rhine, is the Rhône, which will be of more international use when the Moselle–Saône canal is enlarged and modernized, a task which should be completed in the 1970s.

In regard to modern highways and railways the main international routes often follow much the same course and in the past the railway usually pierced the summits by tunnelling while the road went over the top. However, this is changing rapidly. On many main routes it is now possible to use car trains for the mountain crossing especially in winter. Also on some of the higher road passes the snowiest summits are now being avoided by the use of road tunnels as that beneath the Great St Bernard Pass.

The Pyrenees do not form a great international obstruction because they can be circumvented at each end (fig. 57). The Alps proper, however, are a very notable obstacle. In

40 *Crude petroleum pipelines in Western Europe*

Statistics show crude oil transported by pipeline in 1968 for lines then completed. The pipeline from Ragusa to Augusta in Sicily (400 000 tons) is omitted (after EEC, Energy Statistics Yearbook, 1969)

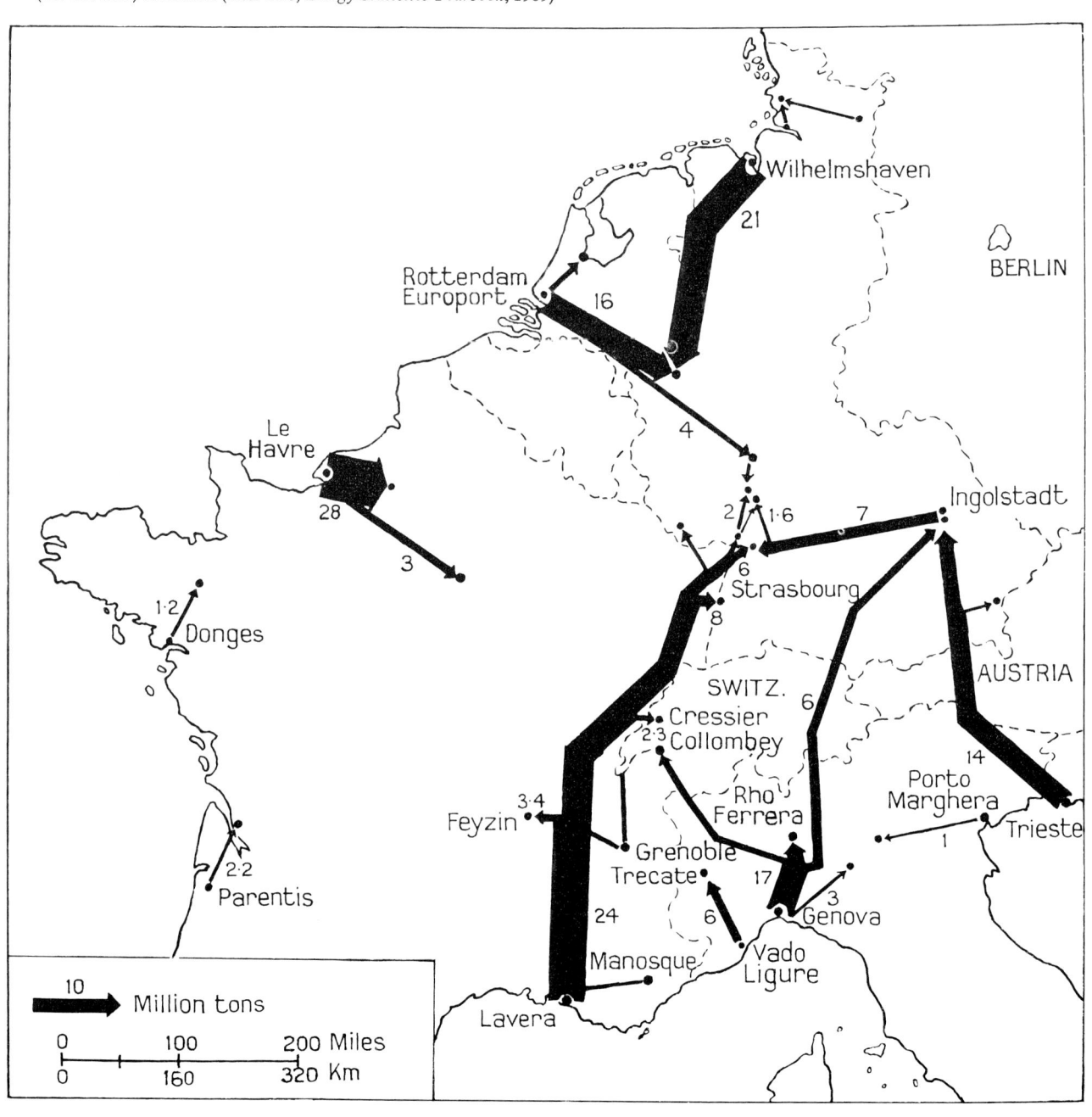

the south where they form more or less one massive range between France and Italy the prime route is along the riviera coast, where today Italy is building a fine but costly autostrada to relieve the coastal road. The main international road passes are the Little St Bernard (2188 m); Great St Bernard (2472 m); Chamonix–Courmayeur by a fine tunnel under Mont Blanc; Mont Cenis (2083 m); Mont Genèvre (1854 m) between Briançon and Cesana; and the Col de Larche between Barcelonnette and Cuneo. The chief railway uses the Fréjus or Mont Cenis tunnel. About 14 km long it has a summit height of 1294 m and was the first of the great trans-Alpine tunnels to be constructed (1857–70). In Switzerland the Alps bifurcate into two main ranges which have encouraged the rise of an independent state. Here the main railway routes are the Simplon, with two main tunnels, and the famous Saint-Gotthard tunnel. Figure 39 shows the general position of these and gives some details of the river valleys they utilize. Farther east in Austria the Brenner Pass (1371 m) forms the lowest road and rail open-air crossing of the Alps. It is the great routeway between central Europe and Italy. In Austria the Alps diverge into numerous ranges and the southern branch is prolonged southeastward into the Dinaric Alps. However, in the neighbourhood of Trieste and Ljubljana is a remarkable narrowing and lowering of this limestone range. Here the Postojna pass is of exceptional importance but has been bedevilled recently by political troubles. From Ljubljana routes go eastward to Belgrade and northward to Vienna and Germany, crossing the high northern Alpine range at the Semmering Pass (985 m).

The transport of liquids and gases by pipe spread rapidly in the late nineteenth century but has reached its greatest use in the last fifty years. The national pipeline grids for natural gas, crude petroleum and petroleum products, especially in Italy and southern France, are constantly being extended. However, the pipeline grid is today truly international (fig. 40). In 1967, a trans-Alpine pipeline for crude oil was opened, and a 540 km gas pipeline from the Dashava area in the U.S.S.R. over the Carpathian mountains to Shale near Bratislava in Czechoslovakia came into operation. The latter has an annual capacity of 2000 million cu m of natural gas and has recently been extended into Austria.

In western Europe there is now a crude oil pipeline grid joining Lavéra near Marseille to Lyon and to Strasbourg where it connects across the Rhine to Karlsruhe and so to a long central European grid. In addition, a crude petroleum pipeline, mentioned above, crosses the Alps from northeast Italy to Nuremberg, with a branch to the Vienna neighbourhood. The east European pipeline grid was based on the 'Friendship' crude-oil pipeline from the Urals oilfield to East Germany with branches to Hungary and Czechoslovakia.

The more recent natural-gas pipeline grids in several southern countries reflect the large new demand for power resources especially in Italy. The crude-oil movement from the Mediterranean coasts reflects their proximity, when the Suez canal is open, to the oilfields of the middle East. To this good fortune has recently been added a most favourable location between inland western Europe and the great oilfields of north Africa. Thus to the exchange of electrical current is today added the trans-frontier transport of petroleum products and natural gas. No doubt in the future many other chemical products will also be carried by this method in Europe.

TOURISM

One form of transport, aviation, depends largely on

INTERNATIONAL TOURIST TRAVEL, 1966   (in 10 000s: all numbers under 5000 are omitted)

	A	B	Cz	F	WG	Gr	H	It	P	Ro	Sp	Sw	UK	US	Ys	Total
Austria					423							19	47	44		696
Bulgaria					10										20	148
Czechoslovakia	28			22			87									352
France	14			175				133			79	72	170	104		1180
W. Germany	27			65				31				36	68	112		676
Greece				9	11			6					11	20	7	100
Hungary	12		61												12	160
Italy	262			469	551	22					20	367	184	125	122	2678
Portugal				18	8						84		25	23		193
Romania	1	2		1	9											29
Spain				752	126			24	107			24	138	43		1584
Switzerland	13			100	141			57					75	72		594
U.K.				42	29			10						75		318
Yugoslavia	48		25	30	66		11	42					25	11		344

passengers for its profits and causes little decrease of traffic on railways and certainly none on roads because the total number of tourists shows a steady increase. The private motor car and motor coach, aided by the autostrada, autobahn and Friendship highway, has given a new dimension to tourism. Today countries such as Spain, Greece, Italy and Switzerland rank earnings from visitors as a major source of income. The reasons for tourism are always obvious to geographers who by profession are interested in scenery and travel. The most popular attractions are sunshine, snow, mountains, warm seas for bathing and cultural achievements.

The chart or table (page 109) expresses the general movement in 1966 in units of ten thousand persons. It ignores internal tourism which is very considerable but does not add to nor subtract from the national wealth. It refers to the countries of southern Europe described in this volume together with the Danube countries and, for obvious reasons, the Federal Republic of Germany, the United Kingdom and the United States. Because we are concerned mainly with tourism in southern Europe the table has weaknesses. Thus it hides, for example, the fact that most visitors to Bulgaria are Turks (371 000), followed by Yugoslavs, East Germans (115 000) and Poles (102 000); and that in Czechoslovakia the greatest number of visitors came from Poland (907 000) and East Germany (130 000). But it was felt that to add these and various northwest European countries would make the general picture too complicated and might obscure the tremendous tourist industry of Italy and Spain or, to take a smaller territory, of Switzerland. The significance of this great source of revenue is discussed again in more detail under the various countries of southern Europe. The following table provides a fascinating summary of the economics of tourism in southern and western Europe for 1968. It shows clearly why most governments are strongly encouraging the building of new hotels and restaurants. The table records, first the total national receipts from foreign tourism and the percentage they form of each country's exports both of goods and of financial and other transactions. Secondly, it gives the total amount each nation spends on tourism abroad and the percentage that forms of the value of the total national imports of all kinds, visible and invisible. Thirdly, the numbers employed in catering for tourism, both foreign and domestic, are given. Here the statistics relate to hotels and restaurants except those for Italy, Portugal, Spain and Switzerland where employees in restaurants are not included. As cafés and bars abound also in these four countries, and, indeed, form an important and popular part of their social facilities, their numbers employed in tourism are seriously understated. The table also shows the number of foreign visitors recorded either in registered hotel accommodation or at the frontiers of each country. It has a notable omission that concerns both Italy and Yugoslavia where in 1968, in addition to the total shown, over 16 million foreign excursionists made a fleeting visit. The table should, for completeness, also have given the number of nights spent by tourists in each country but the statistics for these are unreliable except for fully registered hotels and lodgings. The statistics for Italy in 1968 were 11·2 million registered foreign guests staying a total of 61

SOME ASPECTS OF TOURISM, 1968
(Values in u.s. $ million; visitors in millions, recorded either at frontiers or at registered tourist accommodation)

Country	Total receipts	Per cent total exports	Total expenditure	Per cent total imports	Employees	Foreign visitors
France	954	6·5	1099	7	523	10·8
Greece	120	13	42	3	?	1.7
Italy	1475	10	363	3	309	11·2*
Portugal	219	23	67	5	23	1.1
Spain	1213	36	102	2·5	88	19·2
Switzerland	724	12	320	4	July 75 / Nov. 42	6·0
Yugoslavia	187	10	51	2	126	3·9*
Austria	687	23	256	8	87	7·3
Belgium and Luxembourg	274	3	398	5	est. 37	
W. Germany	906	3	1580	6	389	6·6
Netherlands	342	3	458	4	20	2·1
Sweden	109	2	313	5		
U.K.	677	3	650	3	543	4·8

* excludes 16·2 million foreign excursionists in Italy and 17·7 million in Yugoslavia

million nights. The figures for Spain (6·9 million registered visitors staying a total of 39 million nights) show how incomplete registration is, or how incompletely the registered accommodation represents the tourist influx compared for example with conditions in Switzerland, where 6 million visitors, staying a total of 19 million nights, were registered in hotels.

MAJOR POLITICAL PATTERNS AND GROUPINGS

The political affinities of the European Mediterranean and Alpine states differ appreciably. Italy is a full member and Greece an associate member of the European Economic Community (EEC) or the Common Market. This community since 1967 has shared a common European council, common court of justice and statistical and information services with two other communities, the European Coal and Steel Community (ECSC) and the European Atomic Energy Community (Euratom). The reasons for this become clear from their history. In 1951, following a suggestion by Robert Schumann that the French and West German coal and steel industries should be placed under one authority open to all European nations, a European Coal and Steel Community was formed and came into force in the next year. In 1955 at a conference at Messina, the idea of a European Economic Community and of a European Atomic Energy Community was favourably received and both treaties were signed at Rome in 1957 and came into force on 1 January 1958. In subsequent years internal tariffs between the signatory countries were seven times reduced by 10 per cent. In 1961 Greece was allowed to become an associate member and applications for membership of the EEC were received from the U.K., Denmark, Eire, Austria, Sweden and Switzerland. In 1962 Norway applied for membership. In that year a Common Market was started for grains and in following years a common policy came into operation for rice, dairy produce, beef, sugar, vegetable fats and oils, and fruit and vegetables. July 1968 was the time set for the final removal of internal customs duties. The United Kingdom, Denmark and Eire were formally admitted in January 1973, but as yet the application of other European countries for full membership has not been decided. So of the countries in southern Europe only Italy is a full member and Greece an associate member. The area and population of the Common Market countries in 1966 are listed below (in the adjoining column).

All the countries in this list are members of the North Atlantic Treaty Organization (NATO) a body set up with the United States and Canada for mutual defence in 1949. France, however, withdrew her forces in 1966.

Portugal, Austria and Switzerland are members of the European Free Trade Association (EFTA), with its headquarters at Geneva. EFTA originated in or just before 1958 when the creation of the Common Market revealed differences in the ideas of west European countries as to how a

	Area (1000 sq km)	Population (millions)
West Germany	248	59
Belgium	30	9
France	551	50
Italy	301	52
Luxembourg	3	0·4
Netherlands	33	13
European Community	1167	182

free trade community should be built up. It came into force in 1960 and subsequently led to many mutual tariff reductions and to elimination by 1967. Finland became an associate member. The area and population of all the members in 1966 were as follows:

	Area (1000 sq km)	Population (millions)
Austria	84	7
Denmark	43	5
Norway	324	4
Portugal	92	9
Sweden	450	8
Switzerland	41	6
U.K.	244	55
Finland	337	5
EFTA	1615	99

It will be noticed that all the full members have applied for membership of the Common Market with a view to forming one large combined free trade area; also noticeable is that of the above countries, Denmark, the U.K., Portugal and Norway belong to NATO, the other four countries being neutral. However, as noted, since January 1973 the United Kingdom and Denmark have belonged to the Common Market so the statistics of 1966 have been greatly altered and the population of EFTA reduced by nearly two-thirds.

A rather wider community for mutual economic aid and progress was formed in 1948 as the Organisation for European Economic Cooperation. In its early stages this body was concerned with the distribution of the American Marshall Aid and was later widened by the inclusion of Spain and West Germany and closer association with the United States and Canada. The growth of the EEC and EFTA did not entirely supersede the OEEC which in 1961 became the Organization for Economic Cooperation and Development. Today OECD includes all European countries, except those adhering to COMECON, as well as the United States, Japan and Turkey. The EEC, ECSC and Euratom all take part in the work of this broader organization, which publishes among other things many surveys of great value to geographers.

PERCENTAGE IN 1964 OF:

	Origin of imports				Destination of exports			
	EEC	EFTA	U.S.	Rest	EEC	EFTA	U.S.	Rest
France	37	12	11	40	39	17	5	39
W. Germany	35	18	14	33	37	27	7	29
Italy	33	14	14	40	38	18	9	35
Greece	42	20	11	27	38	13	15	35
Austria	59	14	5	22	48	18	4	30
Portugal	33	21	11	35	21	25	11	44
Switzerland	62	15	9	15	41	18	9	32
Spain	36	18	16	30	39	25	10	26
Czechoslovakia	8	5	1	86	8	4	1	87
Hungary	11	9	3	77	11	9	–	80
Romania	20	6	–	74	14	5	–	81
Bulgaria	1	–	–	99	1	–	–	99
Yugoslavia	25	8	13	54	26	9	6	59
Albania	4	–	–	96	6	–	–	94
U.S.S.R.	5	4	2	88	6	5	–	89
U.S.	15	11	–	74	17	10	–	73
U.K.	17	11	12	61	21	13	9	58

The middle and lower Danube countries have a mutual grouping or Council for Mutual Economic Aid (COMECON) which was founded in 1949 to develop resources and trade within themselves and with the U.S.S.R. The main plans have concerned coordinated trade and development in machinery, power resources and agriculture. They included in 1958 the decision to build an oil pipeline from the U.S.S.R. to Hungary, Czechoslovakia, Poland and East Germany. These coordinated plans for increasing regional specialization and generally integrating the economies have already been discussed for as far ahead as 1975. In 1962 Albania withdrew from it and in 1965, Yugoslavia was admitted as an observer member. The active full members, in addition to the U.S.S.R. (22·4 million sq km; 234 million people) and the Mongolian People's Republic (1·6 million sq km; 1 million people) are as follows:

	Cz	E. Germ	Pol	Hung	Rom	Bul
Area (1000 sq km)	128	108	313	93	238	111
Population (1966 in m.)	14	17	32	10	19	8

By far the major part of the trade of the COMECON countries is with each other, and particularly with the U.S.S.R. They also have a treaty for mutual defence which is governed by the Warsaw Pact.

The Danube countries are united in a Commission founded in 1948, with its headquarters at Budapest. The members are Austria, Bulgaria, Czechoslovakia, Hungary, Romania, U.S.S.R. and Yugoslavia. The Commission administers and advises on the flow of shipping on the river and improvements in the navigation.

The major political and economic groupings described above are not rigidly watertight. COMECON countries do some trade with the Common Market and EFTA as the table above shows, and there are many trading agreements between eastern and western European states. However, the mutual organization of industries and resources is necessarily restricted to within each group and it will probably be some time yet before even western Europe is firmly united in a single Common Market. The unity or federation of western or peninsular Europe has always been a problem. Often in the past there has been partial unification by conquest, as under the Romans and for or against Napoleon, but these communities have soon broken up once military force or threat was removed. The present international communities have arisen largely in time of peace and may well be more enduring than some of their predecessors. The differences between west and east Europe are largely ideological and reflect also the pull of Asia on the one hand and of North America on the other. The strange thing in Europe is that federations such as the Swiss set a model for the world while the lack of federations of the larger political units provide an example of just how the conomic life of a continent should *not* be managed.

In recent years Spain has shown a desire to become an associate member of the EEC. At the same time Albania retains a close alignment with communist China, and Yugoslavia maintains excellent relations with both western and Danubian Europe.

II
Western Peninsular and Alpine Lands

11 Spain: Basal Developments and Patterns

The growth of the nation-state

Ancient man moved into Cantabrian Spain from the Dordogne valley of France and formed there an area of settlement and development called the Biscayan of which the Basque people are thought to be a survival. At the same time Berbers from north Africa moved into the Almería district and spread themselves over the nearby islands of the western Mediterranean basin to form the early Catalan empire.

The grave structures of Iberia, the underground chambers of Malta, the nuraghi of Sardinia and the talayots of Majorca display certain resemblances and date from this period. The Iberians whom the Romans absorbed into their empire were descended from these Almerians. They were described as being short, dark, alert, and energetic with a pronounced lower lip and high cheekbones. They had already developed a considerable traffic in copper and silver within the ambit of the western Mediterranean basin.

Along the southern and eastern shores of Spain before the Roman conquest, Phoenicians, mainly Carthaginians, had established trading depots, and had thrust westward through the Pillars of Hercules and established the port of Cádiz near the mouth of the Guadalquivir. Few material traces of their sites remain.

The Greeks also established colonies along the Mediterranean coast of Spain, of which the finest remains, including art and sculpture, are at Denia (Hemeroskopean), near Málaga (Mainake) and Ampurias (Emporean). With the overthrow of Carthaginian power in the western Mediterranean during the Punic Wars, Rome found herself in 237 BC mistress of the southeast coastlands of Spain, the metalliferous area of the Sierra Morena and the Guadalquivir basin.

These were immediately interlinked by a military road system while the rest of the peninsula was more slowly brought under Roman control. It was AD 19 before Asturias and Galicia became part of the Roman empire. By then Baetica, the earliest possessions, had become a public province under a civil administration although the main part of the peninsula was still under military control with planned garrison and administrative centres such as Emerita Augusta (Mérida). Each new territorial expansion was accompanied by the planting of colonies of veterans with a watchdog function. All had the typical rectangular streetplan and many of them were imposing structures built in the imperial Roman style, linked by fine highways and strong bridges, where trade and technical development were fostered. Spain was a fine example of Roman colonial achievement and by the time of Vespasian the Roman franchise had been conferred on 350 towns in the peninsula. Rome in return had sole access to abundant mineral wealth in gold, silver, copper, lead and tin from the Sierra Morena.

With the decline of Roman imperial power, invasions from the north commenced once more, starting in AD 406 when the Vandals appeared south of the Pyrenees. Marauding in bands, they plundered far and wide without being sufficiently numerous to establish a rule of their own over a fixed territory. Simultaneously Galicia and Asturias, the least Romanized of the provinces, early established their independence of Rome and were soon followed by mountainous societies elsewhere. The Iberian peninsula now began to display those divisive elements which had been submerged by Roman law and order, although there still remained the unifying influence of Christianity which had spread through the empire after the conversion of Constantine. In AD 494 the Visigoths moved south into Spain from Aquitaine and, establishing their capital at Toledo, sequestered two-thirds of all the cultivated land to their own use. For a while the Hispano–Roman people and the Visigoths remained separate and distinct peoples. The Visigoths had an elective monarchy which had the effect of fomenting strife and intrigue between rival heirs but finally in 654 a legal code for the whole country was drawn up.

In 711 the Moors invading from north Africa began to undermine the authority of the Visigothic kingdom. They promised protection to the Jews, freedom to slaves, and preferment to discontented Visigoth nobles, and so by diplomacy and resounding victories the occupation of Spain was speedily accomplished. The *Reconquista*, or recapture from the Moors, was germinated by two factors, the legal code of 654 and the two independent mountain states of Asturias and Galicia. Its completion took seven centuries, despite the fact that the Moors were never united under one caliph as each successive Berber invasion professed allegiance to its own leader. Concerted action on the part of the separate northern Visigothic kingdoms was likewise difficult to achieve especially as over the centuries the Islamic and Christian societies of Spain had become inextricably mixed in an amicable coexistence (*convivencia*). It was not until religious fanaticism was aroused that Islamic rule was finally expelled from the peninsula.

The Reconquista

As already noticed resistance was never entirely eliminated in northern Spain. We may begin for convenience with Catalonia where during the reign of Charlemagne, the Franks expelled the Moors and occupied Barcelona (AD 801), where they set up an independent Visigothic kingdom with different laws and customs from the rest of Christian Spain. The resistance from northwest Spain was, however, earlier and more expansive (fig. 41). In the early eighth century the Visigoths of Asturias began to elect one of their own nobles to be king. Later Galicia joined Asturias and together they established their capital at Oviedo whence their armies pushed south to the plains. The discovery of the tomb of St James the Apostle at Campostella gave them a national symbol and by 910 the Asturian kingdom extended southward to Coimbra and eastward to Burgos, with a new capital at León. About this time the Basque kingdom of Navarre began to expand and

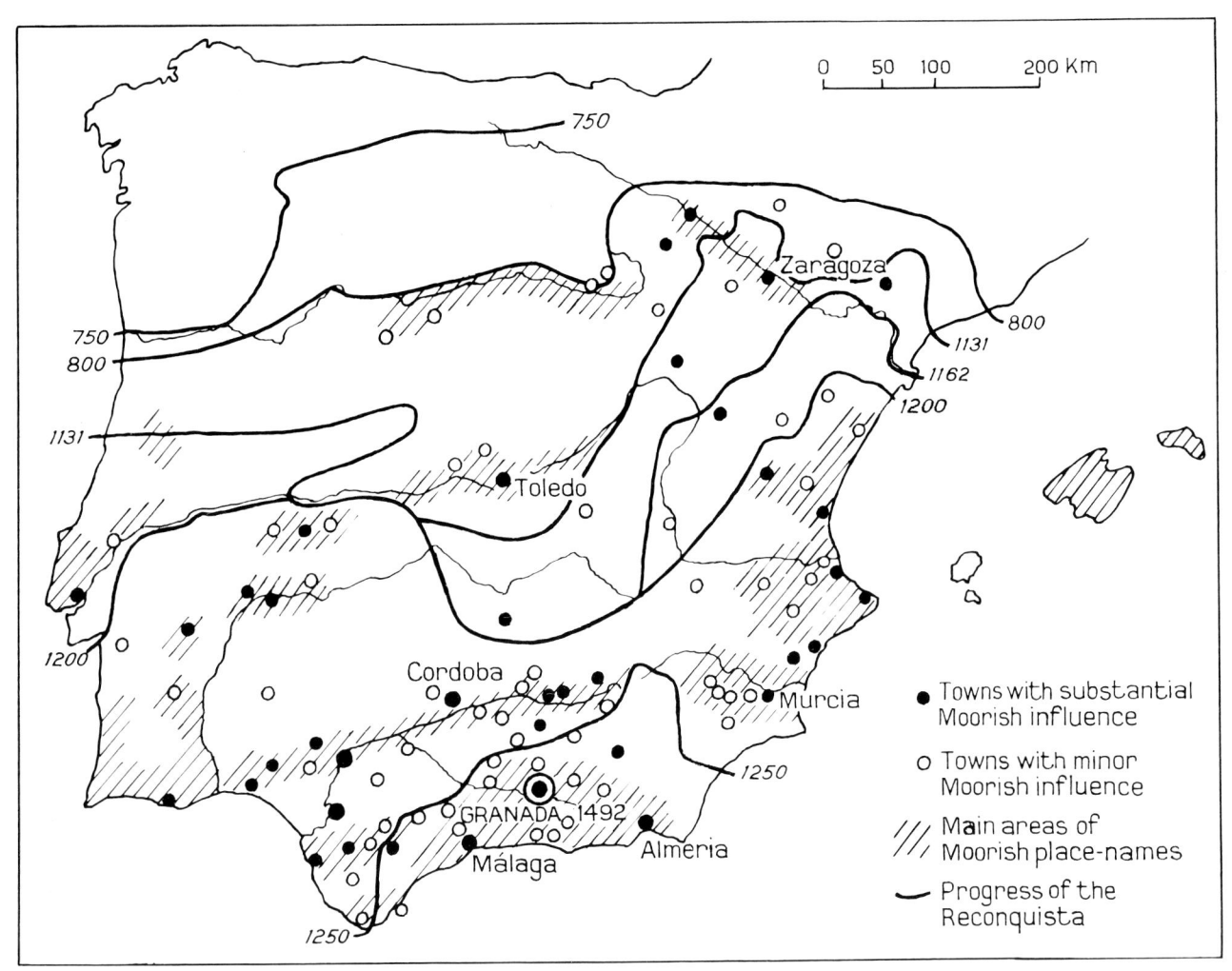

41 *Iberia: progress of Reconquista and Moorish impact on towns and place-names* (after Lautensach and C. T. Smith)

by AD 1000, Sancho III of Navarre had become master of Castile, Aragón, and of much of León. He declared himself 'Emperor of the Spains'. His successor pushed the reconquest south to the Tagus and Mondego. Under Alphonso VI (1065–1109) Toledo was permanently reoccupied and Valencia (1094) was won back by el Cid. At first Alphonso styled himself 'The Emperor of the Two Religions' and did not attempt to expel the Moors living under his jurisdiction. Indeed, since agriculture and manufacturing were dependent on them, it would have been folly to do so. For a time a Hispano–Moorish civilization under Christian authority flourished. Later, however, religious feelings became inflamed and the *convivencia* was doomed. The fanatical expulsion of the Moors and the civil strife among the kingdoms ruined the Spanish economy.

In 1139 Portugal declared its independence, and Aragón and Castile broke away from the suzerainty of Navarre. Aragón proceeded to gain ascendancy over Sicily and southern Italy and finally in 1443 its ruler became King of Naples with his capital in that city. When in 1469 Ferdinand of Aragón married Isabella of Castile they controlled a larger area of the peninsula than had been united for centuries because by then the kingdom of Aragón, in addition to southern Italy, included Valencia, Catalonia and the Balearics. With this marriage of two intelligent and able rulers, the unity of Spain was finally established. As territories were acquired from the Moors they planted new towns surrounded by their own allotted lands to attract settlers. Given special trading privileges, their citizens also had control over their own municipal affairs as well as representation in the national cortes.

The civil administration of Spain as a whole was put into the hands of trained administrators or ecclesiastics and gradually the power of the nobles was lessened. Later, however, the representation of the towns was reduced and the alcabala imposed (sales tax of one-tenth price), which gave the crown an income and almost absolute authority.

Modern Spain

After the conquest of Andalusia and Granada (1492), Ferdinand and Isabella proceeded to make Spain the most powerful country in Europe. The discovery and conquest of the major part of the New World was probably the most significant event in Spanish history, and is still reflected in its present-day geography. Isabella made the trade with the New World a monopoly of the port of Seville, organized and controlled by the Casa de Contratación which governed the licensing, loading and cargoes of all ships trading to the New World. The Casa also established a research centre for navigation studies that became one of the most important institutes in Europe for mathematical and scientific research. The Spanish empire is now but a fragment of its former extent but Spain still ranks as one of the universal powers by virtue of the wide spread of the

Spanish language effected at this time. Approximately 85 million people speak Spanish. As Portuguese is spoken by another 53 millions, and the two languages are mutually intelligible, it is estimated that about 150 million people speak one or other of the Hispanic languages making them second to English as a universal tongue. Relationships between Latin America and the Iberian peninsula remain close and friendly.

On the death of Ferdinand, the whole of Spain passed into the possession of Charles V who inherited besides Castile and the Aragón kingdom in Italy and Sicily, the Netherlands and the central European possessions of his grandfather Maximilian. Under his son Philip II (husband of Mary Tudor of England) the whole of the Iberian peninsula was united for a time under one control, and he held, besides the Spanish possessions in the Americas, a large part of central Europe. He was the most powerful monarch in the world and Spain entered into a golden age becoming the cultural and intellectual centre of the West with Madrid as its splendid capital.

During the centuries following Philip II's accession (1556), Spain has been involved in a series of ruinous wars. Periods of recovery and progress have all too often been offset by costly military campaigns and disastrous civil strife. At the close of the Thirty Years' War (1649), the nation was left with a corrupt administration, a bankrupt treasury and a heavily taxed people. In 1659, Roussillon finally went to France. Fresh efforts at economic rehabilitation were largely ruined by the War of the Spanish Succession at the end of which the Treaty of Utrecht (1713) awarded Gibraltar and Minorca to the British, Spain's Italian and Flemish possessions to Austria, and Sicily to Savoy. Then the nation, ruled by Philip of Anjou (grandson of Louis XIV), was forced back in Europe to within its Iberian frontiers. Nevertheless, Spain remained a great colonial power and made remarkable progress at home in the eighteenth century. Between 1700 and 1750 the population rose from $5\frac{1}{2}$ to $7\frac{1}{2}$ million and the Spanish court remained the magnet for scholars, artists and craftsmen. A decree of 1773 proclaimed the dignity of labour even for noblemen, and reforms in dress, education and law were put forward. Land was distributed among the peasants and colonization of derelict areas effected. In one instance, 3000 Bavarian miners were settled in the Sierra Morena. New industries under foreign instruction were initiated, new highways built and canals dug.

The Seven Years' War and the Napoleonic invasion (1808–14) brought this recovery to a halt. During the early nineteenth century the Spanish colonies in the New World broke away from the mother country with the exception of Cuba and Puerto Rico which, together with the Philippines and a few other Pacific islands were sold to the u.s. for $20 million in 1898. Spanish Sahara, and several small northwest African territories are all that remain of the former empire. The Canaries are an integral part of metropolitan Spain.

The nineteenth century in Spain was marred by dissension over the constitution and civil wars over the succession to sovereignty. Progress in industries and economic reforms was slow and the pace almost inevitably quickened in the twentieth century. The First World War saw the nationalization of the railways and the modernization of the Basque iron and steel industry.

Under the reforms of Primo Rivera (1923–30) many public works were undertaken, such as a roadbuilding programme, financial reform, railway electrification, fostering of industry, agrarian credit and especially the *Confederación del Ebro* the first of the great composite water schemes. Sadly, the whole reform movement was disrupted by the Civil War of 1936–9 which caused immense destruction and cost one million lives. It ended in the authoritarian regime of General Franco, who faced acute economic reconstruction problems.

A succession of droughts, poor harvests and the refusal of foreign credits caused shortages of food, water, gas, electricity, raw materials and industrial and agricultural machinery and made the nation's economic recovery painfully slow. During the Second World War, Spain remained neutral and at its close was, for various reasons, ostracized by the United Nations. She was thus deprived of Marshall Aid and of financial loans from the United States. However, after agreements made in late 1953 to allow the establishment of American naval bases at Cádiz and Cartagena, and air bases near Madrid, Zaragoza (Saragossa) and Seville as part of the Nato strategic plan, large economic aid was made available. This has led in recent years to great material progress.

THE PHYSICAL SETTING
Structure and relief

The Iberian peninsula is the largest in southern Europe and stands at one of the Old World's great crossways between the Atlantic and the Mediterranean, and between northwest Africa and southwest Europe. Of the three great southern peninsulas it is the least European in appearance. In essence, a fortress land, the greater part is occupied by the Meseta at 600 to 1000 m above sea level, an ancient earthblock into which granites have been extensively intruded. Structurally it is a detached fragment of the Hercynian mountains which unlike some other portions of the chain elsewhere remained largely unsubmerged. Recent marine deposits are therefore absent over large areas of the central plateau but instead have accumulated thickly around its margins. These during the alpine orogeny were subjected to intense pressure and folded against the plateau edge causing widespread disruption to the ancient block, manifested in huge fractures, upheavals, broken surfaces and the metamorphosis of the older strata into crystalline schists, gneiss and marble. In particular the whole plateau was tilted so that it is now higher on the east than on the west.

Huge upfaulted mountain blocks traverse the plateau

from west to east, the highest being the Central Sierras. These disjointed massifs, the Sierra de Gata (1867 m), de Gredos (2592 m) and de Guadarrama (2430 m), arranged en échelon, divide the Meseta into two main basins which themselves contain endless variations on the plateau theme from paramos to tabuleiro, mesa to butte, cuesta to badland, with everywhere piedmont gravel screes blurring the hard outlines. The northern basin in Old Castile centres mainly on an old lake floor drained by the Duero (Douro) and its tributaries. The southern depression in New Castile, consists of the drainage basins of the Tagus and of the Guadiana that are separated by the Montes de Toledo which merge eastward into the vast high plains of La Mancha (fig. 46).

The Meseta is in parts further hemmed in and cut off from the Atlantic to the north and the Mediterranean to the south by great mountain chains. On the south the Baetic or Andalusian Cordillera abuts ridge upon ridge on to the plateau and attains 3481 m in the Sierra Nevada. Down-faulted between this mountain barrier and the southern front of the Meseta is the large wedge-shaped trough of Andalusia. On the north of the Meseta the Cantabrian Mountains (1800 m average elevation) flank the Bay of Biscay and are continued eastward into the lofty Pyrenees (average elevation 2000 m) and possibly also in the Iberic mountains. Between the two last-named ranges is the downfaulted Ebro trough. The Meseta edge in the Iberic mountains gives a deceptive impression of easy accessibility. These mountains are not ridged like the Pyrenees but they are high and bleak, with summits at over 2000 m, and in fact form the main water-divide between Atlantic and Mediterranean rivers. Moreover in the east they come close to the sea and leave space only for a narrow discontinuous strip of crescentic coastal plains.

Consequently the main drainage of Iberia is westward to the Atlantic and is gathered up by the four great rivers Douro, Tajo (Tagus), Guadiana and Guadalquivir. Only the Ebro among large rivers drains east to the Mediterranean. At first sight these valleys would appear to lead into the heart of Spain and to provide useful routeways. But Portugal lies athwart the mouths and lower courses of three of the rivers and occupies their coastal lowlands. The fourth, the Guadalquivir, does not give easy access to the Meseta. On the Spanish tableland the rivers have cut deep gorges and fall from various plateau levels by rapids that, coupled with the very irregular regime, make east–west water travel impossible and cause impediments to north–south land communications. The characteristic Spanish Meseta river possesses a narrow, steep-sided valley and, except in the rainy northwest, little water in all the drier months.

In conclusion, although Spain has ocean on seven-eighths of its frontiers it has few easy points of accessibility to and from the inland plateau. It has practically no navigable rivers apart from the Guadalquivir up which ocean-going vessels can ascend for 80 kilometres. It is also over long stretches further isolated by a coastline that is rocky, steep and unindented, or backed inland by steep gradients. This makes the exceptions all the more important. Along the Bay of Biscay many small havens suitable for fishing and coastal craft feed the two ports of Santander and San Sebastián. In Galicia, Pontevedra, Vigo, Ferrol and Corunna have fine natural harbours on a ria coastline. In the southwest the Guadalquivir enters the sea through low-lying coastal plains, sea marshes, lagoons and sand dunes, but one fine natural harbour in Cádiz Bay provides shelter for shipping. The remarkable harbour of Algeciras Bay is shared by Gibraltar which has been a British possession since 1704. On the east coast, Almería, Valencia and Barcelona are not well endowed with natural sheltered anchorages but Cartagena has a fine natural harbour and has become Spain's chief naval port.

Climate

Since the Iberian Peninsula lies between 36°N and 44°N theoretically the northern area should be influenced mainly by westerly air streams throughout the year (western margin cool temperate) and its southern parts should experience moist westerlies during the winter and dry northerly airflow during the summer (western margin warm temperate). In practice the airmasses experienced in southern Spain may be strongly influenced by the close proximity of the Sahara. As described in Chapter 4, during the cool season and early summer the passage of cyclonic depressions westward into the Mediterranean either from northwest Africa or more rarely through the Strait of Gibraltar often brings a Saharan airmass to southeastern Spain where the scorching dusty wind is called the leveche.

Another important modification of the Spanish climate is due to the compact shape of the peninsula, reinforced by its high altitudes and coastal mountain barriers. This introduces a continental influence which confines the mild purely Mediterranean type of climate to southern coastal areas. The effect is strongest in winter when frontal weather associated with a moist westerly airflow is dominant. Then a local high pressure occasionally forms over the cold plateau and tends to fend off maritime influences which, in any case, are further restricted to the coastal areas by high relief barriers. In fact, the Iberian peninsula is almost a continent in miniature with a western margin cool temperate climate or Biscayan type in the north and northwest, a western margin warm temperate or Mediterranean type in the south and northeast, and a modified more continental type of these climates in the interior.

The type or types affected by continental modifications cover the larger part of Spain and the purer types are practically confined to the coastlands. Upon the interior plateau owing to height, distance from the sea and the warding-off of maritime influences the winters are rather cold in the north and cool in the south where the chilly weather decreases in length to a mere eight weeks in

Extremadura. Large tracts in the basin of Old Castile may be further chilled by the downhill gravitation of cold air (temperature inversion) from upland slopes during calm weather in winter but the climate is rarely cold on continental standards and days with snowfall are relatively infrequent. It is always wise to keep an eye on what the instruments rather than what the people say on climate. For each country we have given detailed lists of statistics which reveal the various climatic differences within the country itself and allow comparison between the various countries. Further information on many weather elements, for example sunshine, is given in Chapter 4.

These statistics give a fairly true idea of the climate an inhabitant experiences, except for wind strength and for heat on sunny days when, as temperatures are recorded in the shade, they understate the feeling of warmth experienced by bodies in the bright sunshine. In the cool season the bright sunny spells bring a human desire to avoid the cold shade while in the high-sun season the shadow is most comforting.

Throughout Spain summers are hot, except in the north and northwest. In lowland Andalusia, July and August are hotter than at the Equator. The warmer Mediterranean coastlands have very mild winters and hot summers that often on the flatter coastal areas are tempered in daytime by sea breezes.

The climatic statistics have been carefully selected to show the temperature and precipitation regimes of the various regions. The amount and efficiency of the rainfall is the climatic element of greatest importance over the greater part of Spain and the country can be further subdivided into a *pluviose* northwest which has a sufficiency of rainfall (over about 600 mm annually) and a much larger *arid* or *semi-arid* southeast with less than 600 mm. Normally the annual precipitation declines in amount from northwest to southeast. Pontevedra facing the Atlantic winds in Galicia receives 1500 mm a year, Almería diametrically opposite on the southeast coast a mere 250 mm. Naturally this progression is interrupted by relief elements, and rainshadow areas and enclosed basins are abnormally dry as Zaragoza (295 mm) will illustrate. Such regional variations abound. What cannot be obscured is that except on the higher mountains, the south and southeast are arid or semi-arid, the rainfall being both insufficient and inefficient as it falls spasmodically, irregularly and in heavy storms (see Chapter 4 and number of raindays). Here and in the Ebro trough and drier parts of Old Castile the high summer temperatures then create

SPAIN: CLIMATIC STATISTICS, MEAN MONTHLY TEMPERATURES (centigrade)

	Altitude if over 100 m	J	F	M	A	M	J	Jy	A	S	O	N	D	Absolute Max	Absolute Min	Annual days with frost*
Bilbao		8.7	10.4	11.1	13.3	15.8	18.8	20.8	21.3	19.6	15.7	11.8	8.9	43.9	−7.8	9
Pontevedra		8.3	9.4	11.1	12.8	15.6	18.3	20.0	20.0	18.9	14.4	11.1	9.4	40.0	−6.7	?2
San Fernando (Cádiz)		11.1	12.0	13.2	15.2	17.6	20.7	23.1	23.6	21.4	18.0	13.8	11.3	41.3	−1.9	0.5
Gibraltar		12.8	13.3	13.9	16.1	18.3	21.1	22.8	23.9	22.2	18.9	15.6	13.3	37.8	−1.1	0
Málaga		12.3	13.3	14.7	16.9	19.3	23.0	25.8	26.1	23.6	19.5	15.9	13.1	43.3	−1.1	0.2
Cartagena		10.5	11.4	13.0	15.2	17.6	21.3	23.9	24.4	22.2	18.8	14.8	11.4	40.0	−2.2	?1
Valencia		9.9	11.4	12.7	15.2	18.0	21.2	24.2	24.8	22.4	18.5	14.3	10.6	42.8	−7.8	1
Barcelona		8.7	10.3	11.4	14.0	17.2	20.8	24.0	24.1	21.6	17.1	12.9	9.3	37.2	−9.4	2
Palma		10.7	11.8	13.1	15.3	18.5	22.2	25.5	25.8	23.8	19.2	14.4	11.3	38.9	−4.0	1.6
Huesca	540	5.6	8.3	10.6	13.3	17.8	21.1	24.4	23.9	20.6	14.4	9.4	5.6	38.8	−13.2	34
Zaragoza	237	5.5	8.3	10.6	13.6	17.6	21.4	24.6	24.7	20.8	14.6	9.4	5.7	44.4	−15.0	11.5
Burgos	887	2.4	4.4	6.1	8.9	11.7	15.6	18.3	18.9	15.6	10.6	6.1	2.8	38.3	−20.0	79
Valladolid	715	3.3	6.1	7.4	10.3	14.0	17.7	21.4	21.3	17.5	12.0	7.1	3.8	40.0	−15.0	44
Madrid	667	4.5	6.4	8.1	11.9	15.6	20.6	24.7	24.6	19.3	13.1	8.5	5.0	44.4	−13.0	29
Ciudad Real	628	5.6	7.8	10.6	13.3	16.1	19.4	24.4	23.9	19.4	14.4	8.9	6.1	44.2	−9.4	45
Badajoz	186	8.9	10.0	13.3	16.1	18.9	23.3	26.1	24.4	22.8	18.3	13.3	9.4	47.0	−5.0	?1.9
Albacete	686	4.5	6.8	8.5	11.7	15.3	20.1	24.2	24.3	20.1	14.2	9.1	4.8	40.3	−22.5	70.5
Córdoba	120	8.9	10.6	12.8	17.2	20.0	25.6	28.3	28.3	24.4	18.3	14.4	11.1	44.6	−6.0	29.5
Sevilla		10.0	12.2	14.4	17.2	20.6	24.4	27.8	28.9	25.0	19.4	14.4	10.6	50.0	−5.6	1.5

* statistics relate only to 1967 and 1968

MEAN MONTHLY PRECIPITATION (in millimetres)

	J	F	M	A	M	J	Jy	A	S	O	N	D	Mean annual Total	Raindays	Days with snowfall
Bilbao	116	95	109	111	77	84	62	55	89	135	143	136	1230	164	4·6
Pontevedra	160	157	170	122	114	58	48	41	180	79	163	193	1501	149	0
San Fernando (Cádiz)	89	82	96	71	49	11	2	4	31	86	115	124	760	71	0
Gibraltar	125	113	118	68	43	12	1	3	35	81	162	138	909	84	0
Málaga	80	56	78	61	23	9	2	4	33	76	90	87	607	49	0
Cartagena	45	32	41	20	24	17	2	6	33	37	45	34	340	42	0
Valencia	32	40	42	30	36	20	13	9	80	67	56	43	472	56	0
Barcelona	35	38	44	46	33	33	27	38	84	72	44	28	526	72	3·1
Palma	43	39	39	36	31	20	9	17	44	77	50	55	465	73	0·9
Huesca	36	43	51	56	96	33	18	91	46	33	25	41	569	74	2·7
Zaragoza	17	19	23	29	38	29	15	15	24	33	30	20	295	66	2·1
Burgos	43	48	51	58	66	53	20	18	46	51	53	43	551	103	18·7
Valladolid	20	22	26	27	40	29	9	9	33	31	36	23	308	71	6·9
Madrid	36	36	43	49	42	37	11	13	42	43	51	38	444	94	3·8
Ciudad Real	36	36	46	61	43	18	3	5	28	43	36	38	391	98	3·5
Badajoz	48	46	69	43	38	13	3	5	23	43	58	56	444	65	0·4
Albacete	23	21	38	41	49	36	11	15	41	39	32	28	381	61	3
Córdoba	66	53	102	53	46	10	0	5	25	61	53	79	554		
Sevilla	59	53	65	44	41	14	1	2	18	53	74	69	500	67	0·2

desertic conditions, giving rise to the saying that 'Africa begins at the Pyrenees'.

Vegetation

The Iberian peninsula is richer in number and variety of plant species than any other region of equal size in Europe. This wealth of plant life is the result of many and varied local climates, different exposures, soils and altitudes within the country, and also to its position as a land bridge between west–central Europe and northwest Africa. Moreover as a consequence of its colonial expansion many new plants were introduced into Spain. The total recorded number of species is today 5660, of which ¼ are thought to be endemic, ⅖ of central European or Alpine origin, ⅕ from the Mediterranean area and $\frac{1}{12}$ from Africa.

They may be grouped into two main physiological types:
1 moisture-loving or mesophytic plants common to central and western Europe which are present in pluviose Spain and on higher moister mountains elsewhere in the peninsula
2 drought-loving or xerophytic plants, including the typical Mediterranean flora which is found over the remainder of Spain.

There are wide variations to be found in both these main types. The Central Sierras are covered with arid steppe species and species endemic to Spain while in the southeast the vegetation displays the semi-desert characteristics of African varieties. There are also broad transition zones between the two types. North of the Tagus, central European vegetation climbs up on to the Meseta from the west whereas south of the river the xerophytes of the plateau spread towards the lowlands of southern Portugal.

Within the broad distributional limits of the two main vegetation types local factors introduce boundaries for different species. The Serranía de Cuenca and the Montes de Toledo are the southeastern limit of the birch and the Sierra de Guadarrama of the beech. The evergreen oak (ilex) is not found at the higher levels nor in the rainy northwest and in these areas the cluster pine, olive and fig are also absent. Broadly speaking xerophytic plants fade out north of the Tagus valley and increase southeast of it making the northern plateau a broad transition zone between central European and Mediterranean associations. This changeover is noticeably most gradual in northern Portugal and in Catalonia north of Barcelona.

More details of the plant cover and botanical composition of the floristic areas, together with the Latin names of the main species are given in Chapter 6. When viewed in a broader geographical sense, the vegetation falls into three main kinds, forest, matorral and steppe.

Forest or woodland. Nearly one-quarter of Spain is classed officially as woodland but of this the major part (7 million ha) is used for grazing and only one-tenth of the country is true forest. Deciduous forests of oak, ash, birch,

SPAIN: FORESTRY DETAILS, 1967

By species		Area (thousands of hectares)
Treed Woodland	Coniferous	6 442
	Deciduous	6 914
	Mixed	905
Non-Treed Woodland		12 812
Total		27 073
By use and ownership		
Public	State	755
	Communes	6 018
Private		20 300
By height of growth		
Tall (monte alto)		8,220
Medium and short (monte medio y bajo)		3 368
Matorral, pasture, etc		15 485
Number of trees felled		24,146 000

Timber Production		(thousand cub m)
Coniferous		
chief species:	Pines	3 485
	Silver Fir	33
	Total	3 525
Deciduous		
chief species:	Oak	108
	Beech	206
	Chestnut	87
	Eucalypt	563
	Poplar (Chopos)	250
	Total	1 353
Grand total		4 878

Forestry Products	Amount	Value (million pesetas)
Leña (firewood)	16 256 000 estéreos	200
Resin	43 785 tons	97
Cork	52 334 tons	247
	(97 255 in 1960)	
Esparto	35 480 tons	20
Grazing or pasture*	20 459 000 ha	6 361
Timber	4 878 000 cub m	3 151
Others		2 291
Total		12 367

* including natural meadows and grassland

beech and sweet chestnut are confined to the north and northwest, the only parts of the peninsula where natural forest communities are extensive. Beech (nearly 600 000 ha) and deciduous oak (*Q. robur*; 350 000 ha) are the commonest trees, but Spanish chestnut, poplar and eucalyptus, the latter two recent introductions for wood-pulp, are also abundant. Navarre has some of the finest forests in Spain including extensive plantations of poplar, eucalyptus and cypress.

Elsewhere in the peninsula dense forests are mainly confined to the steeper valley sides and to the rainier slopes of the mountains. The Sierra de Cazorla, the source of the Guadalquivir, about 100 km east of Jaén, is clad in dense pinewoods, the home of deer, wild pig and Spanish mountain goat (moufflon). However, the typical trees of arid Spain, the Aleppo and maritime pines and the ilex and cork oaks, seldom form dense or particularly lofty growth. The evergreen holm oak or ilex covers all told over 3 million ha and is valued for its heavy crop of acorns. The cork oak (*Quercus suber*) covers over 500 000 ha in southern Spain but like the ilex its woods have the appearance of parkland with widely spaced trees. Pines occupy a total of over 4 million ha, by far the chief species being the Aleppo pine (*P. halepensis*) and the maritime pine (*P. pinaster*). The umbrella or stone pine (*P. pinea*) is fairly common especially near the coast and Spanish fir occurs often in the Baetic Cordillera. Various other conifers are also grown in recent plantations. These pines frequently grow isolated or in small clumps and occasionally in sizeable pinares or compact groupings of a single species.

The present tree growth is all that remains of widespread forests cleared mainly for fuel (domestic and industrial), timber, and for cultivation. Once the tall trees were felled, soil erosion, dry summers, burning and pastoral activities greatly hindered their revival. As discussed in Chapter 7 (Forestry), fuelwood is in great demand, and this can be supplied more quickly by bushes. Deforestation in Spain is as elsewhere followed by a secondary growth of shrubs and bushes, which if left alone, would slowly revert to tall tree growth, but over large areas constant cutting, burning, and above all its use for grazing, particularly by goats and cattle, maintain its degenerate condition. This is one reason why *matorral*, an expanse of bushes and shrubs, covers about two-fifths of Spain.

Matorral, parts of which may be called *monte bajo* (low woods) in distinction to *monte alto* (high forest), spreads over vast areas on, for example, the Sierra Morena and Montes de Toledo. Normally it is composed mainly of plants of the cistacae, labiatae, leguminosae and ericaceae families, many of which are oleaginous and aromatic evergreens with small leathery and spiny leaves. They flower in spring and autumn on the Meseta and are in bloom virtually throughout the year on the Mediterranean coast, so providing Spanish apiarists with a fine crop of honey and much beeswax (9000 tons of honey and 500 tons of wax in 1968). The superabundance of matorral reflects both the

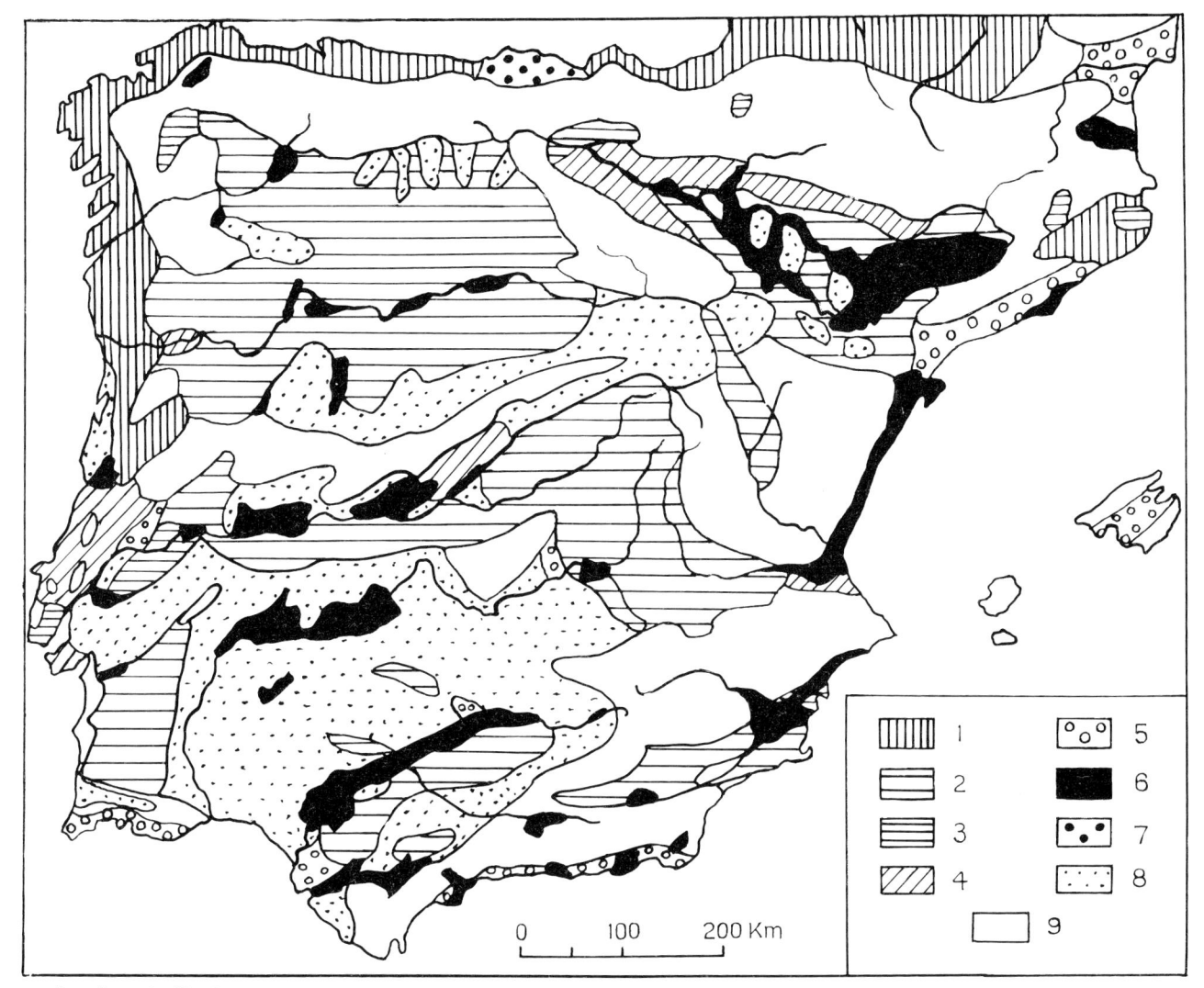

42 *Land use in Iberia* (modified from P. Birot)

1 *polyculture of non-Mediterranean type, based mainly on maize*
2 *wheat, barley and fallow*
3 *arable without fallow*
4 *dry Mediterranean polyculture*
5 *Mediterranean tree cultivation*
6 *huerta and irrigated cropland, including rice*
7 *specialized intensive stock raising*
8 *upland areas with a small amount of cropland*
9 *highlands with forest and scrub; cropland absent or restricted to isolated patches*

pastoral and agrarian economy as well as the low rainfall, bright, hot sunshine and high evaporation rates of the climate.

Over some areas discernible plant associations have developed in it but in others one species predominates and the type acquires a special name. Matorral in which cistus dominates is called jaral, a particularly abundant type on the southern Meseta. Other dominant species give rise to the terms *tomillares* (thyme), *espliegares* (lavender), *romillares* (rosemary) and *retamales* (broom). These and gorse are present in the matorral almost everywhere at least in small numbers. Other plants that occur frequently but rarely form single species scrub are myrtle, mastic, wormwood, sage, sweet marjoram, arbutus, tree heath, and stunted varieties of kermes and scarlet oak. These are described in more detail in Chapter 6.

The height and density of matorral varies greatly. In the driest habitats, bushes may be loosely spaced but in others they may intermingle to form impenetrable thickets. The shrubs may reach a height of 3 to 5 m but more commonly are half that height. During spring, tulips, lilies and asphodel brighten the ground flora and in autumn the meadow saffron. The bushgrowth is highly flammable and its main economic uses are for grazing, charcoal-making and firewood.

Steppe. Matorral grades through many forms of impoverished scrubland into steppe, an expanse of bare soil and rock with a sparse covering of low, isolated bushes and of tussock grasses. This thin plant cover develops on the driest areas and on porous, infertile or saline soils with excessive evaporation. Until recently it extended over about $3\frac{1}{2}$ million ha or nearly 7 per cent of Spain and where water is not available for irrigation it still occupies large areas in La Mancha, southeastern Spain and the Ebro trough. The sparse vegetation includes purslain, glasswort and various medicinal herbs while wormwood is common on the Ebro steppe and the dwarf palm in southeastern Spain. The chief bunch grasses are a coarse esparto (*Lygeum spartum*) and alfa, the latter, which covers about 500 000 ha, being the chief source in Spain of esparto. The commercial harvest of esparto has declined since 1959 from 97 000 tons to under 30 000 tons.

LAND USE AND ITS PROBLEMS

Outside the pluviose zone, aridity imposes tremendous burdens upon agricultural development in Spain. To the insufficiency of the rainfall is added the intense nature of its fall sometimes on warm and sun-baked ground which may be bare in fallow or partly bare under growing crops. The result often is that much moisture is evaporated, and more rushes off the surface, without penetrating it, dredging out the finer soil particles in its flow, and leaving behind impoverished soils. The rivers become suddenly engorged and temporarily rush headlong in their narrow defiles, spreading a waste of stones and pebbles on their banks wherever they overflow.

Being thin and poorly developed the soils lack cohesion and wind erosion also plays havoc with cultivated soil surfaces, and there are many landscapes reduced to rocky and stony expanses that have lost their soil covering in this way. Alkaline soils are common and there is a danger in more arid areas of soil being rendered useless by salination. Furthermore the fact that the rivers are deeply incised and short of water for most of the year means that irrigation projects need expensive dams and other engineering to provide water storage and to raise the water to the level of the fields. Except in the rainier north and northwest the eye seeks in vain over most of Spain for the rain-soaked, soil-mantled plains and valleys, and the gentle slopes of the more normal erosional processes. Areas suitable for cultivation and close settlement are isolated from each other by wide areas where only a meagre living can be wrung from the soil, or pastoral steppe prevails (fig. 42).

These natural difficulties are enhanced by the common methods of land ownership. In arid Spain much of the land is held in large estates: in Extremadura, Andalusia and La Mancha 2½ million ha are in the ownership of 7000 persons so that many peasants are detached from the soil and in rural areas like La Mancha can only expect seasonal employment for 150 days a year, or else, as sharecroppers (*braceros*) have only a tenuous grip on a livelihood. In fact about 30 000 farms occupy nearly half the agricultural land in Spain and the 5000 largest holdings cover 12 million hectares. These vast estates are as a rule understaffed and not infrequently include extensive tracts of grazing and bushgrowth. Hence their productive value is much less than their mere size indicates. The richer arable land and especially the highly productive irrigated areas have a more balanced spatial distribution among holders, as the table opposite shows.

In contrast to latifundia, many regions of Spain have minifundia or microfundia in which the farms are too small to provide an adequate task and income for their owners or tenants. Most of these tiny farms are inevitably overstaffed and much of their labour is part-time. Moreover, they are usually divided into numerous plots which hinder or prohibit the use of machinery. In Spain about 10 per cent of all agricultural land (which includes bushgrowth) is shared between 75 per cent of the holdings whereas

SPAIN: NUMBER OF FARMS AND AGRICULTURAL AREA, 1962

Size (hectares)	Farms Thousands	Per cent total	Agricultural area Thousand ha	Per cent total
No land	151	5·0	–	–
under 2	1 218	40·5	892	2·0
2–5	620	20·6	1 989	4·4
5–10	417	13·9	2 928	6·6
10–20	301	10·0	4 192	9·4
20–50	196	6·5	5 872	13·2
50–100	52	1·7	3 511	7·9
100–200	24	0·8	3 377	7·5
200–500	17	0·6	5 287	11·8
500–1 000	7	0·2	4 671	10·5
over 1 000	5	0·2	11 932	26·7
Total	3 008	100·0	44 650	100·0

SPAIN: PERCENTAGE DISTRIBUTION BY FARMS OF AGRICULTURAL AREA AND OF ARABLE AND IRRIGATED LAND, 1962

Farm size (hectares)	Total Agricultural area Farms	Area	Arable and orchards Farms	Area	Irrigated land Farms	Area
under 5	64·4	6·4	64·4	11·0	64·1	29·9
5–20	25·1	16·0	25·6	27·2	26·5	33·4
20–200	9·5	28·6	9·3	43·6	8·8	25·5
over 200	1·0	49·0	0·7	18·2	0·6	11·1
Total	100·0	100·0	100·0	100·0	100·0	100·0
Actual numbers (in thousands)	3 008	44 650	2 747	19 442	1 478	2 034

similar statistics for Portugal are 65 per cent of the farms and for Italy 60 per cent. Fragmentation, as noticed above, increases the economic problems in the minifundia areas. In Galicia and León fields average less than 0·5 ha and much time is lost in moving from one plot to another as well as much space lost in fences. In the better-watered lands of the northwest, holdings may be only 2 to 8 hectares but will be subdivided into more than a dozen plots. The problem of minifundia and fragmentation is less felt in the Basque lands and in Catalonia where the inheritance system (*mayorazgo*) gives the farm to the eldest son. Nor is it really serious in Andalusia where fragmentation of holdings is much less and the plots average nearly 4 hectares in the east and over 10 hectares in the west. But the fact remains that in Spain although about 72 per cent of the holdings are owner-occupied (against 12 per cent tenancy and 7 per cent share-cropping) minifundia is a greater problem than latifundia. The main progress since 1954, apart from irrigation land settlements, has been cooperative farming and concentration of holdings in the north with financial aid from the authorities. Each of these groups which in 1963 already numbered 2500, embracing 50 000 members and over half a million hectares, forms jointly a number of holdings but in practice most of the members prefer to contribute only their land and find work elsewhere. Since 1961 the requests for government aid towards concentration of holdings have increased rapidly and about half have been realized. Between 1961 and 1969 *concentración parcelaria* was applied to nearly 2½ million ha, which represents about one-quarter of the total area that needed it. However, no less than 84 per cent of this concentration of holdings lies in the provinces of Old Castile, León and in the northern strip of New Castile. Relatively little has yet been done in the wetter north and Galicia, very little in Aragón and Andalusia and nothing in Catalonia, the Levant and the Balearic Islands.

This grouping into contiguous holdings favours mechanization and ensures greater agricultural output per worker. The importance of maintaining productivity with a smaller labour force and improved methods is obvious. It sets free labour for factories and services and in recent years the number of agriculturalists has been decreasing and that of manufacturers increasing by about 50 000 annually. In 1965–6, on an average, 4 million people, or 33 per cent of a national working force of 12 million, were employed in agriculture, forestry and fishing; these sectors supplied 27 per cent of the total domestic income (Gross National Product) and formed nearly half of the total exports; and, although the Spanish people spent over one-third of their national income on food, most of their requirements were produced by their own farmers. However, since 1966 the rise of manufacturing has been exceptional. In 1968 out of a labour force of 12·5 million, agriculture employed 31 per cent and manufacturing 26 per cent. No doubt in the early 1970s the balance will for the first time in Spain's long history be tipped on the side of factories. In 1969 Spain's exports of manufactured goods rose by one-quarter, and agriculture's share of the GNP dropped to about 17 per cent while that of manufacturing and construction rose to 40 per cent. But if industrial growth means for agriculture more funds, better seed, more fertilizer, more machines and continued expansion of irrigation, farming will always play a much larger part in the life and economy of Spain than it does in those of the highly industrialized countries of western Europe.

Up to 1931 Spain was governed in the interests of the landowners which was not necessarily to the best agricultural advantage of the nation. Much of the better quality land was sacrificed to sheep farming and any legislation to alter this was always resisted. Despite a thousand years of irrigation practice and skill in dry farming introduced into Spain by the Moors, agricultural techniques generally are backward and crop yields are among the lowest in Europe. It is only since the Civil War that there have been heroic attempts to better agricultural conditions. In 1939 the establishment of the National Institute for Land Settlement, with powers to buy and divide large estates, to select certain areas for irrigation schemes, to establish cooperatives and provide loans and technical training, was a turning point in the economic and social development of Spain.

Animal husbandry

Largely because of its aridity and high relief, one-third of Spain is rough grazing and waste. However, the vast expanses of dry thin steppe and of matorral actually cause no less than 40 per cent of the state to be used mainly or solely for grazing. Pastoral farming therefore plays a very important role in the Spanish agricultural economy and Spain has more sheep (20 million in 1962; 17 million in 1969) than any other southern European country and is second to Greece for goats (2·5 million in 1968). Statistics for these and for crops will be found in Chapter 8. Sheep of the merino breed, native to Spain, are the most numerous of the farm animals and are kept mainly in the steppe lands of the Meseta, especially in Old and New Castile and in upper Extremadura. The wool is lustrous and of fine quality and 30 000 tons or 11 per cent of the European clip is produced here. Although the merino has been exported all over the world with great success, in its homeland the breed has deteriorated. The industry is in the hands of the Mesta, a federation of sheep owners who still control many of the ancient sheep roads, the *cañadas*, up into the summer pastures on the Central Sierras and Iberic mountains, although today the government offers cheap rates for moving sheep by rail (fig. 43).

Where the land is too rocky or steep for sheep and in the drier areas of matorral on the Sierra Morena and Baetic cordillera herds of goats thrive. They are a symbol of an impoverished countryside but provide welcome meat, mohair and milk (see Chapter 8). Cattle, apart from working oxen, are largely confined to areas where water and meadows are abundant. They are most numerous in the humid north where, with pigs, they form part of a system of mixed farming. Half of Spain's cattle (total 4 million in 1968) are here and most of the dairy cows. Their forage is supplemented by root crops and that of the swine by potatoes and forest pannage. Pork is the chief Spanish meat and pigs (6·7 million was the national total in 1968) are kept in large numbers also throughout Extremadura and the Central Sierras wherever acorns and chestnuts are plentiful. Fighting bulls are bred on the marshes near the mouth of the Guadalquivir and on similar ranchland near Seville and Salamanca.

Farming generally in Spain is mainly unmechanized and a great variety of working animals are kept. The country has more mules and asses (donkeys) than any other in Europe (see Chapter 8). These, together with horses and working oxen, impose their demands on the supply of grass, fodder and foliage. Overstocking is a definite farm-

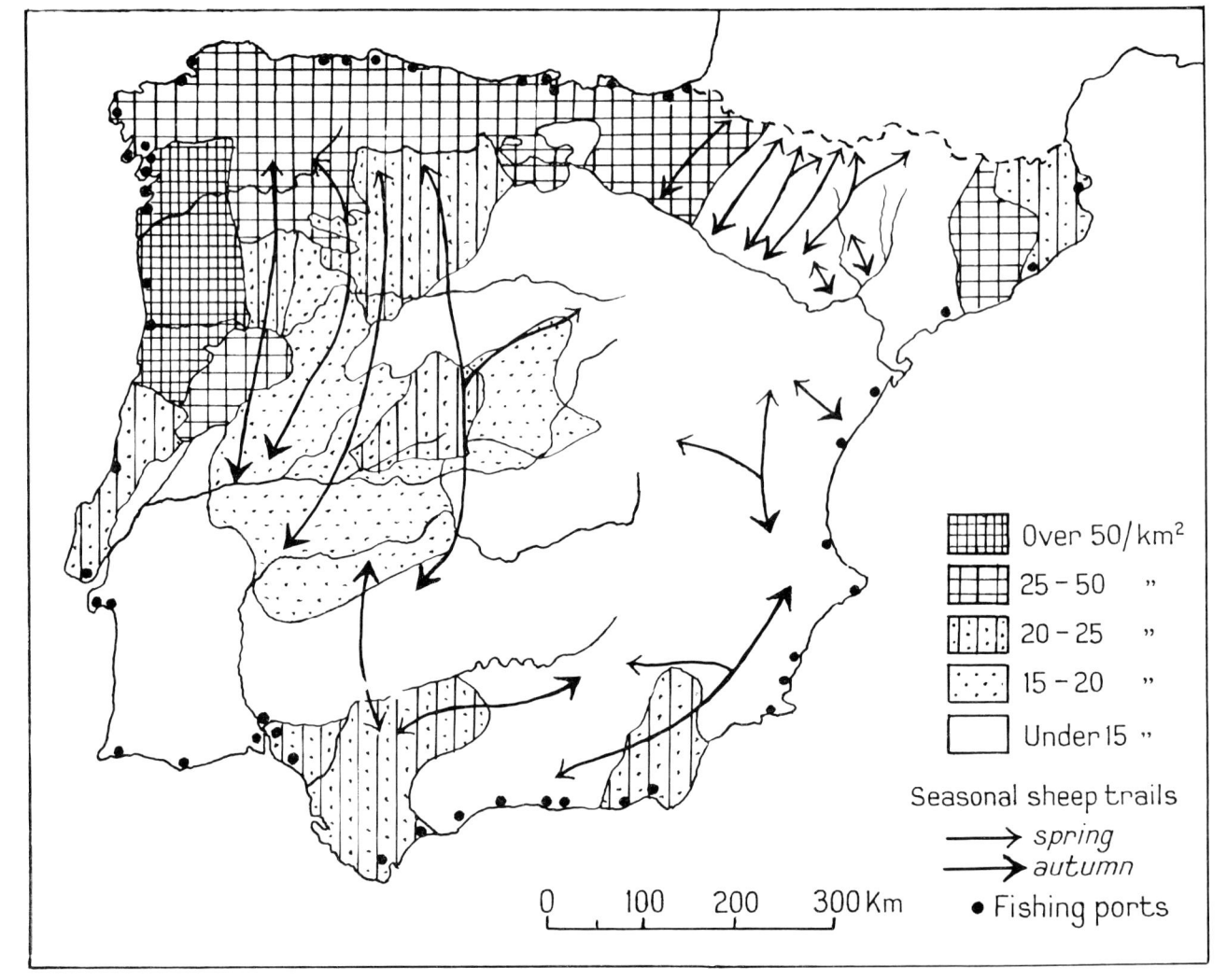

43 *Iberia: density of stock and direction of main stock routes.*
  Black dots show chief fishing ports   Blank areas have less than 15 farm animals per square kilometre

ing hazard and further impoverishes an overgrazed countryside by hastening the pattern of soil erosion begun, often centuries ago, by deforestation. The recent increase in mechanization is helping the transport and fodder problem but will not lessen the risks of soil erosion.

Arable farming

The dominant crops are the classical triad of wheat, olive and vine, each normally grown as a monoculture, with the addition of typical 'modern', Mediterranean crops of citrus and vegetables on irrigated areas. The area, production and relative importance of these and other crops are related to those of other southern European countries in Chapter 8.

Basal cereal foods occupy one-third of the 21 million ha. of cultivated land and are worth nearly 30 per cent of the total agricultural production. Wheat is by far the chief temperate cereal and is particularly extensive in Old Castile and Extremadura but the yields are low, averaging 11·6 kg/ha in 1966 and 13·4 kg/ha in 1968. Dry farming methods (*secano*) are essential, wheat being followed by a pulse or a fallow grazed by sheep. Probably one-quarter of the cultivated land lies fallow every year. The higher and drier districts produce barley, oats and rye. Barley (18 kg/ha) gives higher grain yields than wheat and is easily the second most important cereal in Spain, being useful for fodder and, when of good quality, for brewing. Maize, the third Spanish cereal for area and output, is the prime cereal of the pluviose north where it gives a yield double that of wheat in arid Spain. It is grown both for grain and fodder, often on tiny farms of under 1 ha, especially in Pontevedra. Rice is a minor but prolific crop on patches of the Ebro lowlands and coastal plains in Valencia and Murcia. From 62 000 ha about 375 000 tons are obtained making Spain the second European rice producer after Italy.

Orchard and garden crops cover between 5 and 6 million ha. Olives occupy 2·1 million ha of which 80 per cent are in Andalusia, usually on large estates under absentee landlords (fig. 44). In 1966, Spain, with 464 000 tons of oil, was the world's chief producer. Nearly all the olives are processed for oil.

Vines occupy about 1·7 million ha or 8 per cent of the cultivated area. The provinces of Valencia and Tarragona each grow one-quarter of the national vineyards. Spain, with 3 million tons average 1964–7, is the third wine-producer in Europe and is renowned among others for sherry (Xerez; Jerez). Much wine is sold to France for blending and, in addition about 6 per cent of the grape harvest is exported fresh or dried.

Citrus fruits, stone fruits and vegetables come next in importance. For oranges Spain, with about 2 million tons from 120,000 ha, leads European countries, and is third in the world for output (after the U.S. and Brazil) and second for exports. The bulk of the bitter or Seville orange crop, grown in Andalusia, is sold to Britain for marmalade. The exports of all oranges, sweet and bitter, and tangerines in 1965 (in 1000 metric tons) went to the following destinations:

West Germany	430	Switzerland	35
France	270	Norway	22
U.K.	103	East Germany	18
Netherlands	89	Denmark	17
Belgium	75	U.S.S.R.	13
Sweden	50	Others	31
		Total	1153

Of the drupes (stone fruits), apricots and peaches are the chief, indeed for the former Spain is first in Europe and second to the U.S. in the world for production (see table, Chapter 8). Vegetables, as with citrus and other more delicate fruits, are grown on *huertas* (irrigated market gardens) and *vegas* (irrigated orchards with a ground crop between the rows of fruit trees). Two or more crops can be harvested annually, among them being onions of which Spain, after the U.A.R. and the Netherlands, is the third world exporter (106 000 tons).

The many industrial crops include cotton, hemp, flax, sugar-beet and esparto. The last is harvested wild on the arid steppes, particularly in Almería, whereas cotton and sugar-beet are widely grown on huertas. The cultivation of irrigated cotton, particularly in the lower Guadalquivir valley, has made great progress recently and in 1966 covered 225 000 ha with an annual fibre yield of 90 000 tons or nearly half the total production in non-Soviet Europe. It benefits from a heavy protective import tax and by Government purchase of the crop.

Irrigation was introduced into Spain by the Moors along with crops such as the date palm, sugar-beet, orange, almond and cotton, but the great composite multi-purpose irrigation and land development schemes are a feature of

44 *Iberia: general limit of the safe cultivation of date palm, olive, and orange*

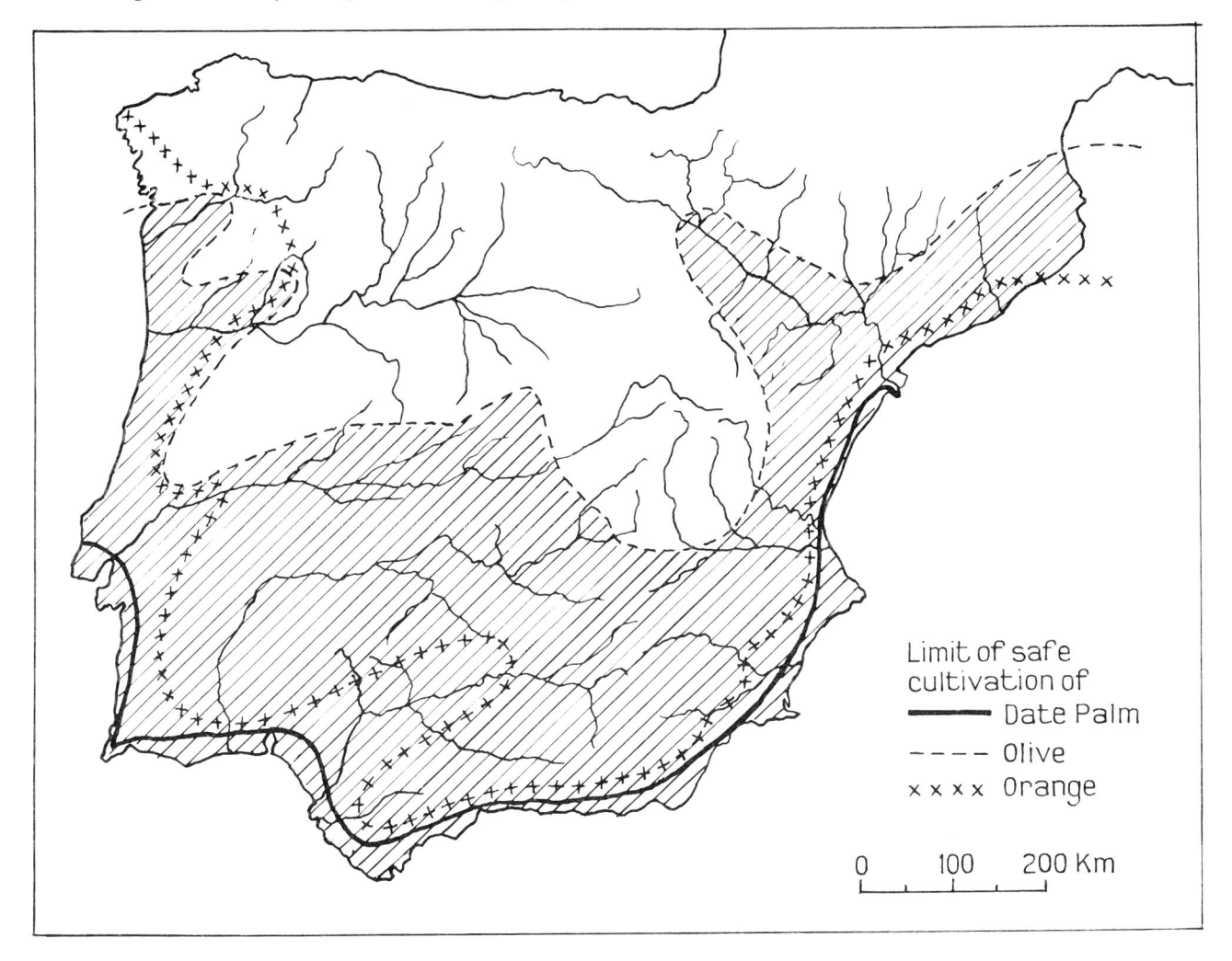

the country in this century when the *Institutó Nacional de Colonización* (INC) was established. Its achievements are among the most important European developments in this field, and have aimed rather optimistically at bringing a further 70 000 ha a year under irrigation. The INC, more recently in coordination with the Director General of Hydraulic Works, transforms *secano* into *regadío* (irrigated); drains salt marshes and swamps; undertakes reafforestation; plans village and farm communities; and organizes farm cooperatives. The landless peasants are settled on farms of their own from 5 to 10 ha in size, with all kinds of financial and technical assistance so that eventually they become peasant proprietors with rising living standards, unable to alienate any of their land but amortizing their loans within 40 to 50 years. It is a planning achievement unique in Europe. By 1962 no less than 347 437 ha had been irrigated, 225 agricultural communities established with schools, clinics, electricity and clean drinking water, roads, bodegas, oil-presses, etc., and 350 000 farmers settled in farms of their own out of an estimated 2 million redundant agricultural labourers. Between 1956 and 1968, the government-planned schemes brought nearly 450 000 ha under irrigation and in the last year, with 74 000 ha newly watered, actually achieved their very high annual target. The ultimate irrigable area is estimated at 4 357 929 ha, or 22 per cent of the cultivable land, or 9 per cent of the state. The greatest possibilities lie in the basins of the Ebro, Tagus, Guadiana and Guadalquivir but extension of irrigation is also possible in the Mediterranean coastlands of Valencia, Murcia and Almería and on various isolated areas of the Meseta. In some of the irrigated districts, increased water storage would permit a second or even a third crop to be grown where only one is now attempted, and in all districts *regadío* yields are at least double those of *secano*.

Among the largest continuous tracts irrigated to date, by old and by new projects, is that around Lérida (165 000 ha) in the valley of the Segre which has been served by the canal de Aragón y Cataluña since 1910. Nearby a further 38 000 ha are irrigated on the riverine lands of the Gállego near Zaragoza. These are connected by the Canal de Las Bárdenas to large irrigation schemes in the basins of the Arba and Aragón, both tributaries entering the north bank, the latter from the high Pyrenees. Most other of the north-bank tributaries of the Ebro are now also interconnected by canals and the recent constructions include new dams on the Gállego at La Sotonera and on the Cinca at El Mediano to feed the second largest irrigation canal in Europe. In addition several reservoirs have been constructed on the Aguas, Jalón and other rivers entering the Ebro from the Iberic mountains to the south (fig. 48). The schemes on the Ebro itself begin with a large dam at Reinosa. About 190 km farther downstream near Logroño the diversion of large quantities of water into lateral canals on the south or right bank commences. Here the Canal de Ladoso, 125 km long, was completed in 1915.

Near Tudela, the Tauste canal, begun in the thirteenth century and completed in 1775, waters the north side of the river valley and some way downstream the Canal Imperial de Aragón (begun in 1521 and finished in 1796) waters the southern side for over 100 km to Pina well below Zaragoza. On the lower course a large new dam built by the National Colonization Institute crosses the Ebro at Mequinenza. Thus the irrigated tracts on the main river now extend with few breaks from near Logroño to the gorge near its mouth, where the delta has been drained and the coastal plain as far as Cape de la Nao put under huerta or vega.

Farther south the huertas around the towns of Elche, noted for its date-palm groves, Murcia and Lorca are supplied with water from the Segura and Sangonera rivers, and even the upland basin of Granada has irrigation water brought to it from dams high up on the headstreams of the Genil river.

Andalusia is also the scene of extensive INC activities. One project is concerned with the reclamation of Las Marismas, the marshes at the mouth of the Guadalquivir river, but other sizeable schemes have developed the whole length of the river such as at Genil, Bembézar Rumblar and Guadales Bajo (fig. 49). Comprehensive plans also exist for the Guadiana, Tagus and Duero basins in Spain and many of the dams and reservoirs are already complete, for example the Montijo and Orellana on the Guadiana, Borbollon and Rosarito on the Tagus, and Villalazana on the Duero. For each area, the INC and the Director-General of Hydraulic Works, draw up a general regional plan which includes not only the hydraulic engineering (estimated to cost 25 per cent of the total) but also borings, soil analyses, settlements, communications and power, reafforestation, crop experiments, new industrial plant, credit facilities, and technical assistance for the farmers.

Away from these composite and settlement schemes, irrigation remains small-scale and primitive. Earth dams and wells (*norias*) are common. In the basins and alluvial depressions well and artesian water is generally and readily available but is not always potable. Electric and petrol pumps, animal and wind power are all in use to raise water for general purposes.

Power resources

Attempts to solve the shortage of water for irrigation have also partly remedied the second of Spain's basal deficiences, that of power. Spain produces only 13 million tons of coal a year, mostly in Asturias and Oviedo where the Nalon valley (Mieres, Langreo, Lena, Labiana) supplies two-thirds of the bituminous output and 2·8 million tons of anthracite. The bituminous coal is of poor calorific quality and mining is difficult as the intensely folded seams vary in thickness and often dip steeply and peter out at faults. Gijón is the chief distributing centre. Coal also occurs locally in Andalusia and León but is little worked. South of Barcelona about 2·7 million tons of lignite are produced each year.

To supplement the paucity and awkward national situation of the coal output, hydro-electricity has been widely developed, particularly in the pluviose north where rainfall is heavier and river flow more regular. This remarkable development is discussed in detail in Chapter 10 and illustrated in figure 38. The larger schemes are connected with dams for water shortage, irrigation and flood control but in the pluviose north power has sometimes been the prime objective, especially in the textile districts of Catalonia, or, to give further examples, on the Navia river to serve the iron and steel industries of Oviedo, and at the small hydro-electric schemes for the paper and textile industries of Bilbao, San Sebastián, Tolosa and Hernani. Since 1950 the Spanish hydro-electricity output has quadrupled from 7500 million to over 30 000 million kWh and water power now supplies 73 per cent of the national output of electricity.

In 1964 petroleum deposits were struck 64 km north of the city of Burgos but they have proved disappointingly small. Much larger deposits have recently been struck off the mouth of the Ebro. Elsewhere some use is being made of nuclear power (fig. 38).

Mineral ores and metallurgy

Deposits of iron ore are widespread in Spain, the total reserves being estimated at between 350 million and 600 million tons. The present annual output is nearly 6 million tons (2·6 million tons iron content) and comes mainly from Viscaya, Santander and Granada with lesser amounts from León, Almería, Teruel and Guadalajara. In Viscaya, although there are restrictions on the export of the best ores, the seams are showing signs of being exhausted, and Granada is Spain's major producing area at the present time. Teruel and Guadalajara ores are phosphoric and are largely exported via Sagunto. The chief ore-exporting port in Spain and also in southern Europe is Bilbao which, with Oviedo, Gijón, La Felguera, Mieres and Avilés, has integrated iron and steel plants, including a government-sponsored project of $1\frac{1}{2}$ million tons capacity a year or over half Spain's total production of steel. Situated on the Nervión estuary, Bilbao can import foreign scrap iron as well as Basque iron ore and coking coal cheaply by sea and can send by-products by coastal shipping to Gijón and Oviedo for use in chemical fertiliser and glass works. Elsewhere the iron and steel industry is on a small scale with an absence of combines and mergers. Typical of these smaller enterprises are the two blast furnaces at Sagunto and the iron-smelting concern at Zaragoza which uses high quality ore from Tierga. If these various plants were combined and modernized the present (1967) Spanish output of 2·7 million tons of pig iron and ferro-alloys and 4,330,000 tons of crude steel could be appreciably increased.

Dependent upon the foundries are several art and craft and small-scale metalwork industries such as the metal ware factories of Málaga and the small arms trade of Toledo. Fortunately Spain possesses important non-ferrous ores that widen its metallurgical products. Mining for copper in the Riotinto district of the Sierra Morena, for lead and silver at Linares and Peñarroya, and for mercury (40 per cent world supply) at Almadén have in some areas been carried on for centuries. The refined output of copper, lead and zinc each exceeded 50 000 tons in 1966. To these must be added small quantities of manganese, titanium, sulphur, etc. and, from quarries in Catalonia, no less than $2\frac{1}{2}$ million tons of potassium salts. The output of the last-named increased to $3\frac{1}{2}$ million tons in 1968 and became, after coal, mercury and iron ore, the fourth most valuable mining industry in Spain. In recent years mining has occupied about 140 000 persons and metallurgy nearly 200 000.

Light industries

Other Spanish manufactures have largely grown out of the agricultural background. Based first on direct water-power and later on hydro-electricity, 90 per cent of Spain's textile industries are in the northeast and are concerned with cotton, wool, silk, jute and linen. Barcelona is the industrial, financial and commercial centre with a wide ring of satellite manufacturing towns such as Mataró, Badalona, Manresa and Sabadell. The branches of textiles show less geographical separation than in Britain but the manufacturing processes are partly specialized. Manresa, Ribas and Baga primarily undertake spinning (of cotton) while Granollers, Igualada, Sabadell and Tarrasa primarily do weaving, the first two of cotton and the others of wool. Cotton manufacturing is easily dominant and with 3400 factories and over 200 000 employees forms the chief single manufacturing industry of Spain. Barcelona, in addition to being Spain's major seaport, has chemical, leather, paper and printing works and most of the engineering factories associated with the textile manufactures and electrical equipment, diesel engines, rolling stock and motor vehicles.

Industries concerned with food processing, such as sugar refining, flour milling, fish canning and oil pressing also tend to be concentrated in the northeast although market-orientated goods, such as electrical appliances, are also important around Madrid and the other larger cities. The processing and conserving of fish is naturally restricted to the fishing ports (fig. 43). The flourishing Spanish fishing industry which far excels that of any other south European country for catch and for processing is described in Chapter 3. In 1966 Spain produced 136 000 tons of frozen fish (mainly Atlantic cod), 175 000 tons of cured fish (mainly cod and haddock), 137 000 tons of canned fish (largely sardine, anchovy, tunny, bonito, etc.) and 129 000 tons of industrialized fish matter. The great good fortune in having so long a coastline on the Atlantic is evident. There are said to be 679 fish-canning factories in Spain.

The absence of large industrialized tracts and of extensive coalfields has, with more positive attractions, a helpful effect on tourism, one of Spain's greatest industries. In 1968, about 19 million foreign visitors came here in search mainly of sunshine, winter warmth, clean beaches, historic buildings, works of art, and picturesque countryside. The coasts of Málaga, Murcia, and the Costa Brava are especially famous and have recently experienced a great expansion of building, particularly of flats and apartments for letting and for foreign residents. Nearly half of the visitors come from France, over 2 million from Great Britain, and well over 1 million each come annually from West Germany and Portugal. Tourists find the sunshine and aridity of the southeast especially attractive.

Communications

Outstanding features of the Spanish communication system are the lack of continuous links along long stretches of the mountainous coasts and the marked focusing of radial routes on the centre at Madrid. The capital is linked by first-class roads to the coast at Corunna, Lisbon, Cádiz, Valencia, Barcelona and Irún and good roads link up the other major cities but away from these most local roads are not macadamized. As yet only 2 million motor vehicles are in use and no Spanish firm makes cars from its own components. The railways likewise converge on Madrid with minor regional networks around the chief inland towns and the ports. But one-quarter of the system is narrow gauge and three-quarters broad and neither fits the European standard width. Some of the railways need modernization but on the major routes services are good, punctual, relatively cheap and very popular. Airways are like other communications and the 40 civil airports communicate with each other via Madrid.

There is virtually no river traffic in Spain except on the Guadalquivir where ocean-going vessels can reach Seville and small craft Córdoba; and on the Ebro where sea-going ships can reach Tortosa. In contrast, coastal shipping is quite important (2000 vessels = 1½ million gross tons).

There are 230 ports in Spain and recent government

SPAIN: EXTERIOR AND COASTAL (CABOTAJE) SHIPPING MOVEMENTS, 1967

Port	Total entering (exterior and coastal) Number of ships	Passengers (thousands)	Cargo (thousand tons)	Total leaving (exterior and coastal) Passengers (thousands)	Cargo (thousand tons)	Total coastal traffic only Passengers (thousands)	Cargo (thousand tons)
Algeciras	2 499	294	3 491	311	2 881	598	1 464
Alicante	1 764	52	722	51	318	100	604
Almería	828	7	377	6	1 980	13	1 577
Avilés	3 078	–	3 513	–	2 080	–	3 524
Barcelona	7 338	237	5 725	250	1 518	417	4 496
Bilbao	7 454	39	6 774	36	2 058	–	5 667
Cádiz	1 796	20	786	20	350	39	711
Cartagena	2 623	–	7 526	–	4 935	–	4 891
Castellón	591	–	2 818	–	2 519	–	777
Corunna	1 415	6	3 035	2	2 048	1	2 141
Ferrol, El	544	–	230	–	202	–	202
Gijón-Musel	3 565	–	1 888	–	1 842	–	2 984
Huelva	1 335	–	2 502	–	4 931	–	1 529
Málaga	2 442	95	3 185	92	374	185	887
Palma	4 643	296	959	280	138	576	1,027
Pasajes	2 904	–	2 518	–	538	–	1 904
San Esteban	767	–	2	–	383	–	385
Santander	2 060	1	2 449	1	448	–	1 691
Sevilla	1 668	–	1 535	4	592	4	1 067
Tarragona	2 482	–	3 422	–	1 326	–	2 223
Valencia	4 319	72	2 670	92	1 103	163	1 808
Vigo	1 527	13	727	17	503	2	672
Others	14,017	178	2 682	186	6 511	362	7 342
Total	71 659	1 315	59 554	1 318	39 578	2 459	49 537

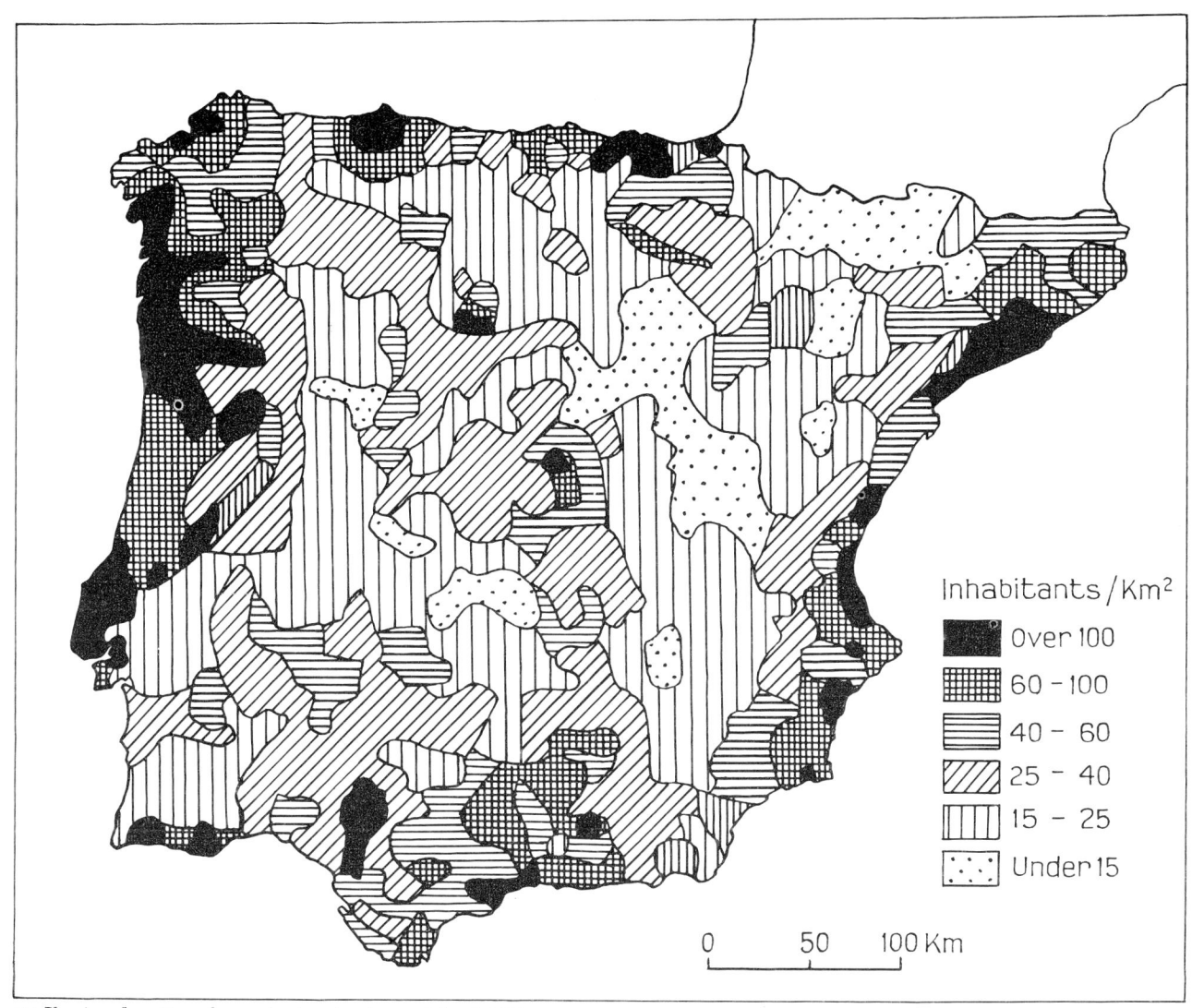

45 *Iberia: density of population* (estimated for 1969)

schemes include the modernization and deepening of several of the more important. In 1940 the ports handled 20 million tons of cargo; by 1963 the amount had risen to 70 million tons and by 1969 had increased still more rapidly to over 100 million tons. Container wharves are being built at Barcelona and near Cádiz and a giant-tanker terminal at Gijón–El Musel. The table gives shipping movement details for 1967; first, of all traffic external and coastwise, and second, in a separate column, for total coastal traffic only. It reveals that 50 per cent of all the national cargo trade is *cabotaje* or coastal and reminds us that the Spanish peninsula is a high plateau with a coastline of 2234 km on the Atlantic and 1670 km on the Mediterranean. The table shows clearly, e.g., the busy packet-port traffic with the Balearic islands and northwest Africa; the market pull of Málaga; the great ore exports from Huelva and some of the Cantabrian ports; and the continued all-round dominance of Bilbao and Barcelona. More surprisingly it shows that Cartagena was for cargo handled the first port of Spain – the effect of thermal power stations and of foreign bases (U.S.), especially on its imports.

Human and political patterns

The population of Roman Spain in the first century AD has been estimated at 6–7 million. Probably by the end of the empire it was 9 million. Wars and pestilence, economic decay, and emigration to the New World partly account for the slow growth of population to just over 10 million in 1787. Thereafter expansion quickened somewhat with a total of 15 million in 1860 and 18 million in 1897.

The population increases that accompanied the Industrial Revolution in other parts of western Europe did not occur in Spain but in the last century, 1860–1960, the population almost doubled to nearly 30 million. The estimated population for 1969 shows an increase of about 2 million to over 32 million indicating that the rate of growth for the next hundred years may be maintained.

The average density of the population is 64/sq km but this obscures a very unequal distribution and a rather high number of small regional capitals. The overall distribution shows a concentration of population mainly on the periphery of the peninsula along the Cantabrian and Galician coasts on the north and the crescentic coastal plains fronting the Mediterranean on the east and south. Here, as figure 45 shows, the population densities drop sharply on the north coastlands from high densities near the sea to low densities a short distance inland, and on the Mediterranean lowlands from the irrigated tracts outwards to the unirrigated areas. Other districts of dense population occur around Madrid, on the irrigated lands of the Ebro and Guadalquivir, and around the mining centres of the Sierra Morena and Sierra Nevada. The high population density on the Cantabrian coast can be attributed to the increased rainfall which gives greater agricultural possibilities, to the mineral wealth of Asturias and the Basque provinces, the rich fisheries, and the numerous harbours for small craft along the north coast and the magnificent ria anchorages in Galicia. The whole coast is easily accessible to western Europe and this has further encouraged port development, associated manufactures and urban development. Galicia has the highest rural population density in Spain with 130/sq km diminishing to 46/sq km inland. With similar population nuclei in the ports of Bilbao, Santander and San Sebastián, Asturias and the Basque Provinces also display the rapid decrease in population density inland with 180/sq km at the coast dropping to 33/sq km within a short distance of the sea. Coal mining in Asturias and iron ore and copper mining in the Basque Provinces also partially account for the higher densities. The ports that have risen in importance have access to easy passes across the Cantabrian mountains: Oviedo and Gijón have access to León by the pass of Pajares; and Santander leads to Reinosa and Valladolid by the valley of the Besaya. Bilbao and San Sebastián command the railway routes across the eastern Cantabrians to Burgos (also a noted coastal resort).

On the Mediterranean coastlands, although the two excellent harbours of Barcelona and Cartagena afford some explanation for the belt of high population density along this coast, the main reasons are the intensive agriculture in the huertas of Valencia, Murcia, Almería and Catalonia and the great expansion of modern tourism and the recent influx of international residents. On the irrigated huertas of Murcia densities of 200 to 400/sq km are attained while beyond the reach of irrigated water the land supports a mere 4 to 8 persons per square kilometre. In Valencia likewise densities of 580/sq km drop to under 20 away from the irrigated areas, and – today – away from the sea.

Availability of irrigation water also explains the increase in population densities of the Ebro basin and in the interior in parts of Andalusia and on deltaic fans at the foot of mountains. The Ebro valley displays densities of 50–80/sq km upon the irrigated mixed crop zone, above which the density drops to 20–30/sq km. Industrialization and marketing, and social and administrative functions largely explain the high concentration of population around Barcelona and Madrid. In Catalonia the population has increased sevenfold since the sixteenth century largely owing to the growth of manufacturing and irrigation. Barcelona province has 290/sq km with 1·7 million in the regional capital. Madrid, the national capital, has over 2·6 million inhabitants and favourable tax rebates for factories established in it favour its further industrial growth.

The agricultural areas of the interior in Old and New Castile and Extremadura have only 20 to 40 persons per square kilometre and the average density drops to under 8 persons on the arid, saline and eroded districts. These low densities of population on such vast expanses of rural interior Spain do not prevent the existence of sizeable towns. Following the Reconquest as each kingdom was regained the old Moorish trading centres were made regional centres with administrative functions. For example, Salamanca was put in control of 105 villages and small towns and 408 hamlets. The historical divisions of the Spanish kingdoms led to the growth of a large number of regional or provincial capitals. Today in Spain there are 48 provincial capitals, such as León (86,000 population), Salamanca (101 000) and Granada (162,000). Together they have 10·5 million inhabitants. The growth of other market centres, of ports and of manufacturing and mining centres all fostered urban growth but industrialization was weak compared with that in, e.g., Great Britain and Germany. In 1950 of the 225 towns with 10 000–100 000 inhabitants in Spain 180 were dominantly agriculatural, 19 were mainly mining and only 46 were largely manufacturing. Subsequently, these smaller towns have changed little, while the few larger cities, or the 20 with over 100 000 people, have widened their already diverse functions. However, the distribution of cities is more stable than their industrial structure and size.

The classification by number of inhabitants is given in the table, which reflects the continued importance of the provincial capitals. Today 34 cities in Spain have over 100 000 inhabitants and of these only 5 (Madrid, Valladolid, Badajoz, Salamanca, Burgos) are on the Meseta. No

SPAIN: CLASSIFICATION OF MUNICIPALITIES (*Municipios*) BY NUMBER OF INHABITANTS, 1960

Number of inhabitants	Municipios	Total population (in thousands)
Under 1 001	5 265	2 308
1 001–5 000	2 847	6 396
5 001–10 000	599	4 206
10 001–20 000	241	3 241
20 001–30 000	74	1 757
30 001–50 000	31	1 191
50 001–100 000	32	2 234
100 001–500 000	21	3 833
Over 500 000	3	4 323
Total	9 113	29 496

less than 24 are on or near the coast and the other 5 (Zaragoza, Córdoba, Granada, Pamplona, Vitoria) are peripheral to the Meseta block. Of the 13 cities in Spain with over 200 000 inhabitants only 2, Madrid and Valladolid, are on the vast interior plateau. Valladolid has only just reached this rank whereas Madrid has long been the great exception: a large metropolis, ideally situated to bind up the peripheral forces of the naturally more productive coastlands; a fascinating city with superb metropolitan qualities.

12 Spain: Major Geographical Regions

A popular pattern of the major regions of the Iberian peninsula is shown in figure 46. It consists of four major types, the peripheral mountains in the north and southeast; the interior plateau or Meseta; the lowland troughs of the Ebro and the Guadalquivir; and the Mediterranean coastlands. To them are added the Balearic Islands and Gibraltar while the semi-independent Andorra must be distinguished in the Pyrenees. It will be seen that these divisions are not based solely on structure, or else Galicia, with its pluviose, deeply dissected landscape, would be grouped with the Meseta rather than with the Cantabrians which it resembles so closely. They are distinguished by their general geographical characteristics except in the case of Gibraltar and Andorra where some political separation prevails. Their physical and human traits are mentioned frequently in Chapter 11 and have also been broadly discussed in the earlier systematic chapters.

The northern mountain regions

This area includes Galicia, the Cantabrian mountains, the Basque country and the Pyrenees and is essentially pluviose Spain. Its unity stems from the all-the-year rainfall which gives it a green landscape with abundant trees, deciduous and coniferous, rich meadows and plenty of surface water. Structurally it is less unified as the west consists of granite–schist plateaux and the east of relatively young fold mountains but the general topographic impression is mountainous, often deceptively so. Historically and socially it differs from interior and southern Spain in being fundamentally mainly Celtic or pre-Roman and distinctly non-Moorish. No area of Spain is freer of the controls of water supply unless the early settlement, as shown by innumerable Iron Age camps or *castros* on the hilltops in Galicia and Asturias, were partly intended to avoid water.

Galicia is a low plateau of granite and schist similar to

46 *Geographical regions of the Iberian peninsula* (after J. F. Unstead)

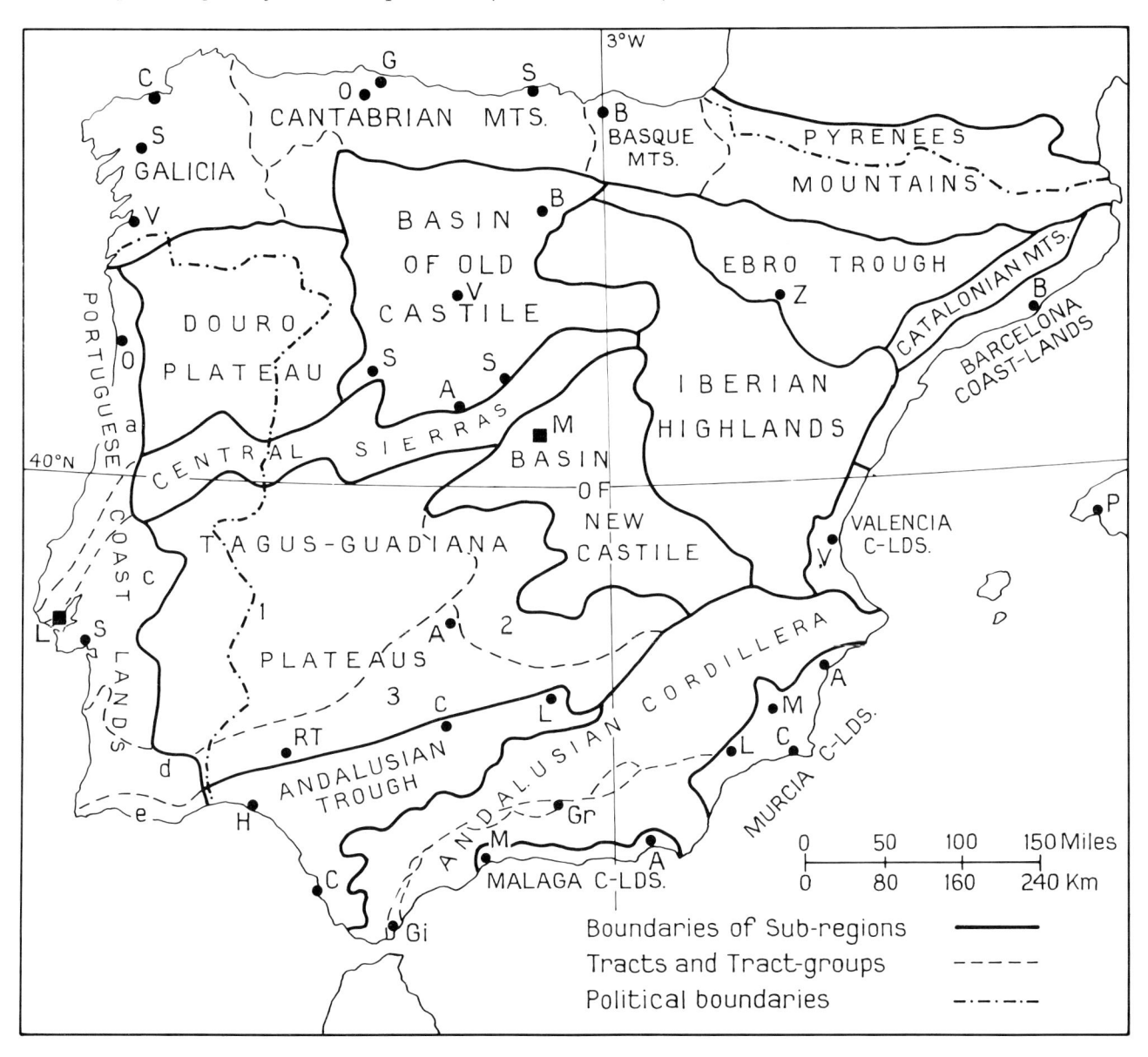

Brittany. This Hercynian block has been fractured and peneplaned and in a few interior basins has been floored by later deposits. The rivers escape from these basins and upper peneplain levels in deeply incised valleys, of which that of the upper Sil forms an indispensable routeway for road and railway to the Meseta. The plateau lies mainly below 500 m and most of the interior is flat or undulating. The Atlantic maritime influence is strong climatically and socially, as would be expected of a land with nearly 500 km of coastline and more than 80 ports and fishing villages. The southwestern coast has been shaped by river erosion, faulting and a rising sea level into a magnificent series of ria harbours. The drainage is mainly to the Minho, one of the most copious and regular of Spanish rivers and an important source of hydro-electricity (see Chapters 10 and 11).

On the plains and valley-floors below about 400 m the chief crops are maize and other cereals, potatoes, sugar-beet, beans and various leguminous fodders. These are cultivated intensively by hand usually on tiny farms (0·2–0·8 ha) that are subdivided into small fields enclosed by stone walls. In Galicia about three-quarters of the plots are under 1 ha. Apple trees abound and cider is a common drink. These, other temperate fruits and vines often line the enclosures and, with the more natural treegrowth on the hills, give a *bocage* aspect. Above 400 or 500 m where soils are thinner and rain is often excessive, stock rearing prevails and large expanses are clothed in heather (*Erica*) and especially in gorse, the latter being valued for fodder, bedding for stock, and charcoal. Permanent meadows are not common as the cultivated plots go right up to the gorselands and woods. Pigs are bred but cows are more important (one-third the national total) and some are used both for milking and farm work. The intensive maize–livestock economy is seen to perfection in Pontevedra province where the animal density exceeds 50 per square kilometre of agricultural land, by far the highest in Spain. The small coastal plains are mainly under market gardens for early crops of vegetables and potatoes. Fishing forms an important occupation, particularly longshore for sardines, anchovy and tunny, which are canned locally, and in distant waters for Atlantic cod, haddock and skate. The ports of Vigo and Corunna land and handle the bulk of the Galician landings which in 1968 amounted to nearly 450 000 tons or 43 per cent of the national total. Vigo is easily Spain's leading fishing port and with a fleet of 570 motor-driven vessels (averaging 26 tons) dealt with 16 per cent of all mainland landings. Corunna, with a much smaller fishing fleet (75 motor-driven vessels, averaging 35 tons), deals with about 10 per cent of national landings and was in 1968 exceeded in importance only by Pasajes and Vigo. Each of the smaller Galician ports, such as Marín, has a swarm of fishing boats and fishermen, often part-time. Octopus is one of the favourite items in the local diet.

The rural settlement is exceptionally dense in the west and quite dense in the drier Sil basin. Galicia (1900 sq km) is a non-industrialized land almost the size of Wales but its total population is slightly greater (2·7 million against 2·6 million). In rural districts the people live close together in separate houses and tiny hamlets mostly of less than twelve houses. The urban life centres on the ports and provincial capitals but includes Santiago, a great pilgrimage centre founded in the ninth century around the shrine of St James. The chief cities are Corunna (La Coruña: 218 000) and Vigo (186 000), the latter being one of the Spanish ports that grew rapidly with the overseas trade expansion in the sixteenth century. Both are important for fish-processing, ship-repairing, shipbuilding and as calling places for liners on the north Europe–South America routes.

El Ferrol (86 000) has a fine almost landlocked harbour and was created as a naval base in the eighteenth century. Today it is known as El Ferrol del Caudillo, because it was the birthplace of General Franco, the leader of modern Spain. The largest inland town is Orense (73 000), the market centre for the most extensive agricultural lowland on the Minho. The great development of hydro-electricity in this basin may help to stimulate new local industries and lessen the steady emigration abroad from parts of the overpopulated countryside.

The Cantabrian mountains. In the west this region closely resembles Galicia and consists largely of crystalline mountains but towards the east the rocks become mainly sedimentary although still largely 'Hercynian' in age. The change occurs mainly in Asturias east of Oviedo where the Cantabrian cordillera consists of three east–west structural belts: a lofty southern belt (the Cantabrian Mountains), a coastal range, and between them a longitudinal depression. The southern belt or Cantabrians proper is a high, powerful barrier developed on crystalline rocks near the Meseta and on carboniferous limestones at its seaward flanks. It exceeds 2000 m at several points near Galicia, and rises to 2417 m in the Peña Ubiña in the centre and culminates in the east in the Picos de Europa (2642 m), a heavily glaciated limestone mass, now a national park. The only easy pass across this range is at Pajares (1363 m) used by the highway and railway (now electrified) from Gijón to Madrid. When this road was built in the sixteenth century it was so costly that Charles V asked if it were paved with silver. The valley-slopes of some of the high mountains are finely wooded but large areas are flat-topped expanses of gorse, heather and poor pastures used in summer by transhumant flocks.

The northern structural belt is a coastal strip or range, the seaward foot of which is cut into a series of raised platforms (*rasas*), especially at 60 to 70 m. As far east as the busy fishing port of Avilés these coastlands are crystalline but eastward they consist of a range of younger sedimentary rocks on which, however, the inlets and most of the towns remain small. The dominant port is Gijón (149 000), the outlet for the nearby coalfields and a notable centre for metallurgical, glass and ceramic industries.

Between the coastal range and the high Cantabrians is an interior longitudinal depression that stretches from Oviedo to Langreo and Cangas. It is structural (fault) in origin and more recently has been floored by Tertiary deposits that have weathered into a hilly topography, well wooded and in parts richly cultivated with maize, beans, potatoes, vegetables and apples. The Carboniferous coal seams buried in the floor and sides of this trough form the richest coalfields in Spain. The chief mining centres are Langreo and Mieres, each with a dense suburban mining population and the latter also with interests in iron ore and chemicals. Oviedo (145 000) in the west of the trough is the ancient provincial capital and the nucleus of the modern industrialized area. In 1857 it had only 14 000 inhabitants and made little growth until the railway came thirty years later. It has now developed large industries concerned with coal mining, armaments and heavy chemicals. Agriculture has also been stimulated and the sales of farm products, especially milk, to the cities is well organized. Most of the farms are owner-occupied and average about 1 ha although many are less. This intensive farming coupled with the industrial development creates a very dense settlement, which exceeds 250 per square kilometre around Oviedo. The national government plans to make Gijón suitable for the anchorage of giant tankers.

East of the Picos de Europa in Santander province the Cantabrians become definitely 'Alpine' in age and structure. The rocks are Secondary (Jurassic) or Tertiary, the relief is lower, and except near Reinosa there are no workable coal measures. The landscape consists mainly of east–west folds broken by transverse fractures, and the coastal range and interior depression of Asturias are both absent. The coastal lands are noted for milk production. The chief city is Santander (140 000), a port with notable wool textile, metallurgical, chemical, engineering and paper-making industries. Within its ambit, Torrelavega manufactures rayon and chemicals and Reinosa high-grade steels. The lower Cantabrian relief makes road and railway communications with Old Castile less difficult than farther west.

The Basque mountain region links the Cantabrians with the Pyrenees and consists of folds of moderate height oriented almost west to east. It affords a relatively easy passageway from France to Old Castile, although quite mountainous locally. Its distinctiveness lies in its inhabitants, the Basque people, who number just over 1 million and speak *euskari*, a tongue once spoken widely throughout the western Pyrenees but now mainly confined in Spain to the provinces of Guipúzcoa, Viscaya or Biscay, Álava and northern Navarra. The language and people also extend into southwestern France. In Las Vascongadas or the Basque region *euskari* has lost much of its former dominance in the industrial cities but retains its traditional hold in rural districts. The political organization is also in a sense unique in Spain as the three provinces together cover only 7100 sq km, Viscaya alone being only 170 sq km. The Basques are distinguished by qualities of industry and energy as well as by their retention of an ancient language, old customs, and a sense of unity. For a mountainous country the area is densely populated and the two coastal provinces are among the most densely peopled in Spain. It will be seen, however, that they are favoured with a reasonable or ample rainfall and with rich metalliferous deposits.

Álava is drier than the coastal provinces and has rather colder winters. Its settlements too are more nucleated and the largest, Vitoria, does not exceed 120 000 inhabitants although it has almost doubled since 1950. On the coastlands, industry, mining, commerce, farming and tourism are important and of long standing. Metal work was stimulated early by local copper and iron ores and in Tudor England a bilbo (after Bilbao) was a sword of tempered steel, and perhaps also an iron shackle for slaves and prisoners. During the second half of the nineteenth century, the high-quality iron ores attracted a large metallurgical industry to the lower Nervión valley near Bilbao. This town was carefully planned as a port for Castile in the early fourteenth century but its great growth has been during the last hundred years. From 18 000 inhabitants in 1860 it grew to 83 000 in 1900, 230 000 in 1950 and 390 000 in 1968. Today Baracaldo, one of its suburbs, has over 100 000 people and probably another 100 000 live in the suburban fringe. The planned core of the city (Siete Calles or Seven Streets) is on the Nervión 13 kilometres from the sea and its quays can take ships of 4000 tons. Larger vessels are accommodated at the mouth of the estuary at Portugalete, which was the original port before the founding of Bilbao, with many special privileges, about 1310. The modern industrial complex specializes in steel products such as shipbuilding, agricultural machinery, diesel and railway engines, but the manufacturing complex is widened by other traditional manufactures including paper, wool textiles and heavy chemicals. The paper industry, as with textiles, spreads into nearby towns and depends partly on imports and largely on plantations of conifer and eucalypt in the Basque lands. Bilbao is in most years the chief Spanish port both for exports and imports (see table, p. 128), and has a growing car-ferry, passenger service with northwestern Europe.

A rapid growth of industry and population has occurred also in coastal towns farther east, as at Irún, Pasajes and especially at San Sebastián (160 000) on the transcontinental railway from Irún. Here the medieval fortress-town and fishing port had only 10 000 inhabitants when the railway came in 1864. Thirty years later when the Regent Maria Christina built a summer palace overlooking the fine bay, the town had trebled in population. It soon became a noted international summer seaside resort and acquired in its port area east (windward) of the town, cement, paper, food-processing and textile industries. The many smaller towns nearby have also benefited from tourism and new industries: Lasarte manufactures rubber goods, Villafranca has metallurgy, and farther inland Vitoria makes fertilizers and agricultural machinery. Recently Pasajes has dominated the Cantabrian fishing industry and has become the second most important Spanish port both for value and weight of fish landings (11 per cent national total in 1968).

The Pyrenean region. The Pyrenees stretch between France and Spain for 400 km and cover a total area of 55 000 sq km, of which two-thirds are on the southern slopes in Spain. The cordillera varies in width from 25 or 30 km in the west, to about 130 km in the centre and 10 km in the east, and has an average height of about 1200 m. It is a natural region but not a geographical unity because of the international boundary along its crest although transhumant shepherds fraternize on its alpine pastures. The French side is discussed in detail in Chapter 14. Structurally it is composed of three major longitudinal belts, a central axis of fractured crystalline rocks, flanked to north and to south by a broad belt of folded and faulted younger sedimentary strata, which forms the highest crests such as Monte Perdido or Perdu (3355 m) and Pico de Aneto (3404 m). The upper slopes have been severely glaciated and show magnificent cirques and hanging valleys as at Gavarnie where the waterfall down the cirque headwall drops 422 m. Today the permanent snowline is at about 2800 m but the snowcaps are small and, like the glacial lakes, insignificant compared with those of the Alps.

The structure and landscapes of the cordillera differ appreciably from north to south and west to east. The northern slopes in France (see Chapter 14) have few longitudinal depressions whereas the Spanish or southern slopes although also mainly of Cretaceous and Tertiary sedimentaries are eroded into irregular, tilted plateaux with sizeable longitudinal valleys overlooked by steep escarpments that have been notched by erosion into sierras. The Spanish side too is very much drier. Similarly, both the western and eastern parts consist largely of contorted and fractured sedimentary strata, often limestones, but the Atlantic west is lower and wetter, and carries much meadow and temperate forest. The eastern or Mediterranean part although high is narrow and carries a typically Mediterranean vegetation, poor in grasses. In each part, the chief road and railway skirt the coastline (fig. 57). The western section is crossed by several roads, including that through the Roncesvalles pass (1057 m) while the eastern has a high motor road from Perpignan to Figueras.

Between these two parts, is the central Pyrenees, a high, wide belt of formidable mountains with a broad crystalline core (mainly granites and schists) and lofty sedimentary flanks. It stretches for about 270 km from Canfranc to Puigcerda and has no passes under 1800 m. It is traversed by only three international roads – at Bonaigua, Portalet and the Viella road tunnel. But at each end of it, where the relief sinks to the sedimentary mountains, railway routes tunnel below the passes connecting Spanish and French valleys. The western line, opened in 1928, joins Zaragoza and Jaca to Pau via the Canfranc or Somport tunnel (marked S on map). The eastern line runs northward from Barcelona to Puigcerda and Pau by means of the Puymorens tunnel (also marked on map). At Puigcerda it is joined by a French road and railway, over the Col de Perche, from Perpignan. Throughout the length of the central Pyrenees the French slopes, open to moist Atlantic airmasses, remain green all the year while the Spanish slopes in the Aragón syncline soon become parched in summer.

The land use and vegetation of the Pyrenees varies appreciably with the climatic gradations noted above but their dominant trait is their altitudinal zoning (see Chapter 6). The agriculture is typical of pluviose Spain in the west with its maize, apples and meadows, but in the east takes on a Mediterranean tinge. It survives in patches up to about 1000 m above which forest gradually takes over. Above the tree line at 1700 or 1900 m, alpine pastures occur, especially in the west, and are grazed by transhumant flocks in summer. Today on many of the rivers a ladder of hydro-electric stations shows clearly enough the future fate of them all. Tourism is becoming important in winter as well as in summer and includes potholing in some of the deepest caves in Europe. Several local centres are important for gathering up these activities, among them Pamplona (125 000) in the western Pyrenean foothills.

The international boundary along the Pyrenees, demarcated in 1856 to 1868 is often hailed as an ideal mountain frontier. But, in fact, it posed many problems. The watersheds of its rivers and the crest or summit-line of the mountain range seldom coincide, and internationally it is preferable, if at all possible, for a nation to have control of the headstreams of rivers draining from mountains in their territory. In the eastern Pyrenees at Llivia on a headstream of the Segre is a small patch of Spain entirely surrounded by France. More unusual is the semi-independent community of Andorra, perched high up against the Pyrenean watershed.

Andorra. This *seigneurie* or semi-independent principality lies at the eastern end of the lofty central Pyrenees and consists of a cluster of six valleys (The Valls) that drain to the Valira a tributary of the Segre. The total area is 495 sq km and the total population about 15 000 of whom 5000 are Andorrans. The community is Catalan in origin but in 1278 its seignorial rights were shared between the Spanish Bishop of Urgel and the French Count of Foix. The latter's rights passed to the French crown and today the Andorrans pay a very small sum every two years to France and to the Spanish Bishop of Urgel. They are virtually selfgoverning. Physically the country consists of deep glaciated valleys ending in large steep-sided cirques, with a lowest point of about 1000 m and a highest of over 2900 m. Northward it rises to the watershed where the only motor road crosses into France by the Embalira pass at just over 2400 m; southward it drops towards the Segre valley which is easily reached by the same highway. In the cold season when the northward road is closed by snow (usually mid-November to early May), all road communications must go southward. The climate is severe even in the main and tributary valleys, being cool in summer and cold in winter. The whole area lies in a rain shadow, and above 1400 m suffers from summer drought.

The high meadows need additional water to get a hay crop and in the main Valira valley an overall irrigation scheme is being pushed forward. Availability of water, levelness, and maximum exposure to sunshine for early snow-melt decide the value of agricultural land, which is limited to stretches near the capital town, Andorra la Vella. Most of the cropland and many of the tiny meadows are supported by terracing, the best soils being given over to tobacco and the remainder to barley, rye, potatoes and vegetables. Wheat will not ripen at this altitude, and barley, replanted immediately after harvest, is quite often a failure. Arable is subordinate to the need for gathering the maximum amount of winter feed for the sheep and cattle. Every possible scrap of meadow grass is cut and carried, sometimes from seemingly impossible heights and slopes. The coarser summer pastures on the broad shoulder of the mountains are held in common and the sheep and cattle farming is transhumant, the families moving to temporary upland summer villages for 6 to 9 months annually. These temporary villages, like the permanent winter ones, are nucleated and so sited as to gain the maximum sunshine. At the end of summer, the sheep (estimated at over 25 000) are sent to winter pastures on the French and Spanish lowlands after large numbers have been sold off at autumn fairs. The cattle (about 3000) are wintered locally. Horses, mules and goats also find sustenance on the lower slopes and there is a mink farm. By far the chief cash crop is tobacco which is of good quality and processed in local factories. It finds a ready market in France and Spain and from tourists generally.

The four main sources of income are tourism and the sales of duty-free goods especially tobacco, spirits and perfumes, of water and hydro-electricity to Barcelona province, and of radio transmission rights. There are several small villages, some with growing ski-sports and camping sites, but the main cluster of dwellings stretches along the chief highway from Escaldes, a spa with warm sulphur springs and a hydro-electricity station, to Andorra la Vella. Here, where the main valleys meet, the elongated capital town has about 2500 permanent residents and presents an animated scene, especially in summer, with tourists busy acquiring duty-free goods. About 800 000 visitors come annually and the small principality seems to have a bright future in spite of the customs posts on its frontiers.

The Meseta

This Hercynian block occupies half of Spain and forms a plateau core over 600 m in average height. It lies within 'arid' Spain with annual rainfalls of less than 600 mm and includes districts with less than 300 mm. Its various major subdivisions are based mainly on the effects of differences in altitude and aridity (see fig. 46). The *Basin of Old Castile*, drained by the Douro, is mainly under extensive wheat farming making it the granary of Spain. Much of the grain is still grown under the *secano* system, a rudimentary dry-farming method with the minimum of machinery, although mechanization is increasing. After the cereal harvest, the land lies fallow or else is put under fodder for the work animals. This often involves the land being held in large estates divided among sharecroppers (*braceros*). Valladolid (203 000) acts as the flour-milling and agricultural market centre for the area. A beginning has been made in building a diversified economy by the construction of several barrages on the Douro and its tributaries for hydro-electricity for the smelting of aluminium, for smelting ferro-tungsten at Medina del Campo and for a car-assembly works at Valladolid. Salamanca (118 000), a picturesque university town, and Burgos (104 000) are noted tourist and provincial centres.

South, beyond the Central Sierras, the heat and drought increase and the arid plateaus and basins of the Tagus and Guadiana are given over largely to pastoralism. In Extremadura and elsewhere to the west where the pastures are better quality, about 1 million transhumant merino sheep are reared, and the wide stretches dotted with ilex provide acorns for over 1 million pigs as well as occasional ploughland for the cultivation of cereals, especially barley. Toward the east the climate is more arid and La Mancha or the Basin of New Castile has large areas of esparto steppe and poor matorral more suited to goats. Wherever water is available as in the valley bottoms of the Tagus and Guadiana there are emerald green patches of irrigated cultivation, under olives, vines, stone fruits, vegetables, sugar-beet and fodder, which supply food industries in towns like Toledo (42 000), at a bridgepoint over the Tagus, Badajoz (111 000) and Mérida. Increasingly these rivers are being controlled by projects under the INC and, for example, already three large multi-purpose dams are in operation on the upper Tagus and two more on its middle course.

The upturned edge of the plateau in the south is the Sierra Morena which contains much mineral wealth. Lead is mined at Linares, mercury at Almadén, copper at Tharsis and Riotinto and coal at Bélmez and Puertollano. This mineral wealth has been developed by foreign capital and is exported through the port of Huelva.

Details of the many ancient cities on the Meseta will be found in guidebooks and encyclopaedias but here we wish to make special mention of the former and present national capitals, Toledo and Madrid. Toledo stands on a steep-sided meander spur surrounded except on the north by the Tagus and became very early a Celtiberian tribal stronghold. The Romans captured it in 193 BC and later made it a colonia and a regional capital. It assumed, and has retained ever since, a dominant importance in Spanish church affairs from the first introduction of Christianity into the peninsula. It was a provincial capital under the Visigoths and under the Moors (712–1035) and then became head of an independent state. In 1087 Alphonso made it capital of the kingdom of León and Castile and it acquired remarkable wealth with the aid of a large Jewish colony who together with the Spanish citizens adopted much Moorish culture and were known as the Mozarabs. Toledo probably had a population of 60 000 in 1561 when Madrid replaced it as the national capital. However, its decline was slow and in 1575 the arrival of El Greco (*ob.* 1614) added to its artistic treasures. Today its narrow winding streets, lined with tall massive houses, are free of wheeled traffic and its fine Moorish gateways, city hall, alcázar, cathedral and numerous other churches and edifices form an enduring testimony to the grandeur of medieval Spain. It still makes swords as it did two thousand years ago but its truly significant industry is tourism. Toledo vies with Venice as the greatest of the cities that have survived almost intact from the golden age of southern Europe.

47 *Regional setting Madrid* (after P. P. Courtenay)
  Blank areas are mainly 400 to 600 m
  1 *Alcalá*
  2 *Aranjuez*
  3 *Ocaña*
  4 *Torrijos*
  5 *Escorial*
  6 *Segovia*
  7 *Ávila*

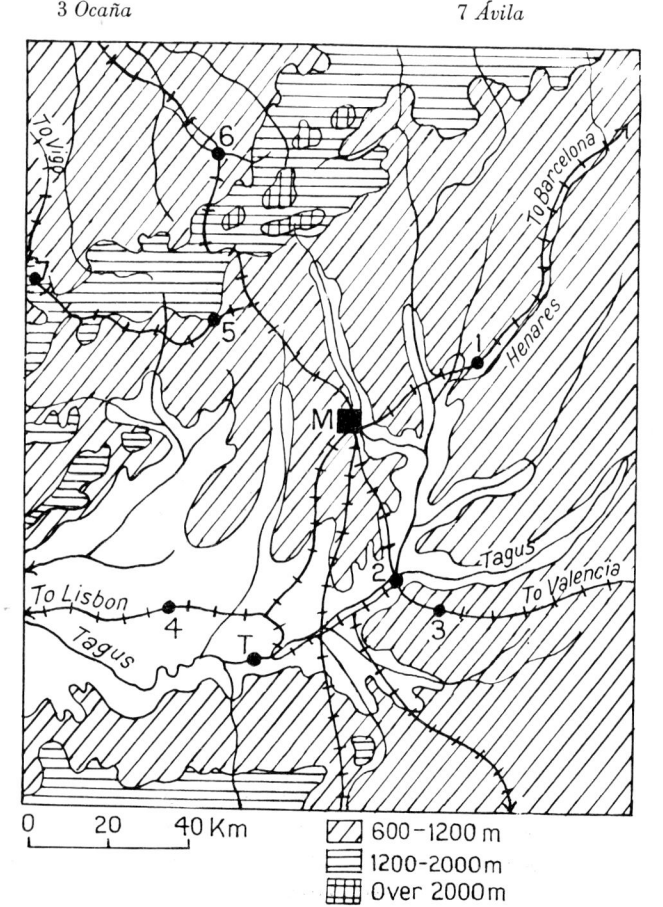

Madrid (3 million), the capital of a highly centralized government and the cultural centre of Spain, is at the core of the Meseta on the Manzanares, a headstream of the Jarama tributary of the Tagus, and in the midst of semi-arid, rather unproductive countryside (fig. 47). It was an arbitrarily but carefully selected national capital, and has been made the focus of road, rail and air networks. Passageways across the broken plateau and through the Central Sierras have always been of crucial importance. Three passes, commanded by the towns of Plasencia, Ávila and Somosierra, lead north across the Central Sierras into the northern basin, and other important routes run via the Sila valley to Corunna, the Jalón valley to the Ebro basin, and southward to Andalusia (Andalucía).

Despite the unpropitious circumstances of soil and climate in the vicinity of Madrid and the absence of raw materials except coal from Puertollano, the district has become the third most densely peopled area in Spain and the city itself houses about 9 per cent of the total Spanish population. It has therefore acquired the usual market-oriented metropolitan industries, such as printing, publishing, clothing, furniture, leather and food manufactures as well as light metal and electrical industries.

The city is either Roman or Visigothic in origin and was throughout its early history of far less importance than the easily fortified town of Toledo on the river Tagus 75 km to the south.

Under the Moorish occupation it had a castle or alcázar and good drinking water was conveyed to it by a system of underground galleries leading from the Sierra de Guadarrama. In 1085 the fortress was captured from the Moors. In 1369 the first cortes of Castile was held here and another stimulus to its growth came when Charles V visited it for health reasons in 1524 for which the supply of clean drinking water may have been responsible. Some of his immediate successors to the crown of Castile also stayed here often and in 1561 Philip II made it the permanent seat of the Spanish government, a decision in which the interests of the unity of the country were paramount. During the next forty years although emigration to the New World was denuding much of Spain of population, Madrid in the dusty centre of the country continued to grow and by 1600 had 40 000 inhabitants.

In the seventeenth century the architectural additions included a new city wall, the magnificent Plaza Mayor (1623; and still surviving) and the royal summer palace (since destroyed) and the lovely park and gardens of Buen Retiro. The university and churches were greatly enlarged and by 1700 the civic population exceeded 100 000. After that year with the coming to power of Philip V and the Bourbon dynasty much of the town was rebuilt in stone; scores of fine public buildings were erected; the Palacio Real replaced the moorish castle; public amenities were provided; agriculture was fostered in the neighbourhood and horticulture on the royal estates, and crafts in metal, weaving and pottery were established.

XIII *Aerial view of Madrid, showing district near the Retiro Park*

During the mid and late nineteenth century Madrid acquired new amenities such as gas and better water supplies, and railways gave it a greater centrality. By 1900 it had grown to a city of 540 000 people and the expansion continued unabated to 750 000 by 1920. The growth of the population and built-up area in the early nineteenth century was unparalleled in the city's history. Large suburbs developed to the north and south; the Manzanares was canalized and its banks reclaimed and built on; the metro or underground railway was inaugurated in 1928 and assisted the expansion of garden suburbs; many new parks and fine public buildings were erected including the new ministry quarters and the general post office on the Calle de Alcalá. Then disaster struck. From November 1936 to March 1939 Madrid withstood a murderous siege and was the first great city to be heavily bombed from aeroplanes.

The slow and costly reconstruction of devastated districts provided opportunities for new zonification and, for example, a fine university quarter has been constructed. At the same time public motor transport has encouraged the spread of new suburbs both industrial and residential. An aqueduct from the Sierra de Guadarrama gives a good water supply and hydro and thermal stations provide adequate electricity. Although many basement workshops remain in several areas, industry has concentrated in the southeastern quarter near the terminal of the Barcelona-Madrid railway. Owing to its centrality with regard to land and air communications Madrid has become the largest redistributing centre in Spain for all kinds of products and notably for the fish and farm produce of the north and northwest and for the manufactured goods of the Barcelona district. National administrative and economic control is tightly concentrated upon the capital, which has acquired economic power partly because of its political and financial supremacy.

Since 1953 Madrid and the Spanish economy generally has undergone a minor industrial revolution, partly with the aid of massive U.S. financial largesse. The city has grown to a population of 3 million. As an industrial centre it is second in Spain only to Barcelona. Lighter manufacturing industries, such as motor vehicles, electric equipment and household appliances, chemicals and food processing are the strongest sectors. Whereas once industrialists were encouraged to settle in the capital, today the Ministry of Industry tends to limit permissions and attempts, by incentives of cheap land, tax exemptions and so on, to persuade factories to move out to less congested zones as at Guadalajara and Alcázar de San Juan. Most factories in Madrid are less than twenty years old and the population has doubled in that time. It is the headquarters of numerous leading firms especially of petroleum companies, public utilities, chemicals, construction and finance.

Tourism is also a prime industry; the Prado is a mecca for art lovers, and within striking distance are Toledo, Aranjuez, Segovia and El Escorial, the eighth wonder of the world. Modern rebuilding has created fine libraries and museums; the city is rich in parks, handsome squares and boulevards and the complex near the main post office ranks, with the railway station at Milan, as one of the superb achievements of early twentieth-century city reconstruction. As in other capitals, the route to the airport has rapidly become a shop window for international firms. The poorer residential quarters are being rebuilt in a moderately high architecture to match the pleasant tall building of the past. The metro is the fastest and by far the cheapest – and least pretentious – in the world and its seats reserved for *caballeros mutilados* constantly remind one of the last civil war. Spain has made a remarkable recovery and nowhere more than at the metropolis. Madrid has long been a great capital; its main problem now will be not to allow spatial and population expansion to overwhelm its civic amenities. Unfortunately it has no successful example to follow in cities of comparable size anywhere in the world.

The *Central Sierras* are less an obstacle to routes north from Madrid than would be imagined from their great height (maximum 2592 m in the Sierra de Gredos) as, although they cross the Meseta from southwest to northeast for 400 km, they consist of a broken alignment of horsts. The direct route from Madrid to Burgos crosses them at 1454 m at the Somosierra pass. Their precipitation is fairly heavy and supplies ample water for numerous reservoirs on the headstreams of tributaries of the Douro (Duero) and Tagus (Tajo). Some of these reservoirs, as at Granadilla on the south-flowing Alagón, are very large and are fed partly by melt water from snow which on the higher summits often lingers until early summer and provides winter sports for the citizens of Madrid. The former extensive forests of deciduous forest and pine have been largely cleared but patches remain including the pinewoods around the Escorial, the magnificent summer palace of Philip II on the slopes of the Sierra de Guadarrama. This vast treasure house and pantheon is 43 km from Madrid and is a world-famous magnet for tourists.

Today many of the small towns on the Central Sierras have some influx of people in summer who wish to escape the heat. The chief natural resources are the farmlands for vine, olive, citrus and vegetables in the lower, warmest valleys and the pasturage for migrant flocks above the treeline. The population is very sparse but outstrips the economic and social facilities. The effect of the Central Sierras as a climatic and vegetation divide may be judged roughly from figures 42 and 44.

The Iberian highlands

This region, sometimes called the Iberic mountains, comprises a group of plateaus raised appreciably above the basins of Old and New Castile and descending by steep step-faults to the Ebro trough. The northwest part rises to 2262 m in the Sierra de la Demanda and 2313 m in the Sierra del Moncayo but the wider southern part is lower, drier and warmer. The north around Soria is renowned locally for its snow and cold. The rivers are deeply incised but whereas those flowing to the Ebro may, like the Jalón, provide routes and widen occasionally into productive basins, those in the south, like the Júcar, Guadalaviar and Turia, flowing to the Mediterranean, have gorges of little use even for roads and consequently Valencia and the neighbouring Mediterranean coastlands lack direct communication inland. The highest summits of the Iberian highlands are clothed with pinewoods and pastures but matorral prevails on the lower slopes and on limestones. Population densities are 5 to 15 per square kilometre except where irrigation occurs. Soria (22 000) is the smallest provincial capital of Spain but the *pantano* (reservoir) here and those on many of the Iberic mountain rivers show that water is a valuable resource in a society determined to use it.

The Ebro trough

This trough, 500 km long by 80 to 160 km wide, occupies a tectonic depression between the Meseta and the Pyrenees which was later invaded by the sea. The marine sediments then laid down were later, after emergence, overlain by alluvium and riverine deposits. Subsequently uplift caused the river to be incised and the landscape today is of low plateau interfluves separated by narrow valleys that sometimes open out into enclosed basins. The Ebro itself flows through an upper and a lower basin entirely surrounded by mountains through which it cuts its way to the sea. The trough, thus excluded from maritime influences, is blighted by aridity. Zaragoza in the centre of the drainage basin receives only 300 mm of rain a year. Being cut off from the sea, its winters are colder than on the Mediterranean lowlands while its summers tend to be hotter. The heat and aridity create large tracts of alkaline soils and brackish underground water. The natural vegetation is woodland in the upper valley and along the foothills where ground water is more abundant, dwindling to matorral and dry steppe towards the centre of the basin. Under natural conditions it would supply only pasture for sheep, but today the basin displays all the signs of an intensive commercial agriculture, modern industrialization and urbanization. This is entirely due to the harnessing of the rivers and the tremendous irrigation works. All the tributaries of the Ebro below Tudela have been controlled

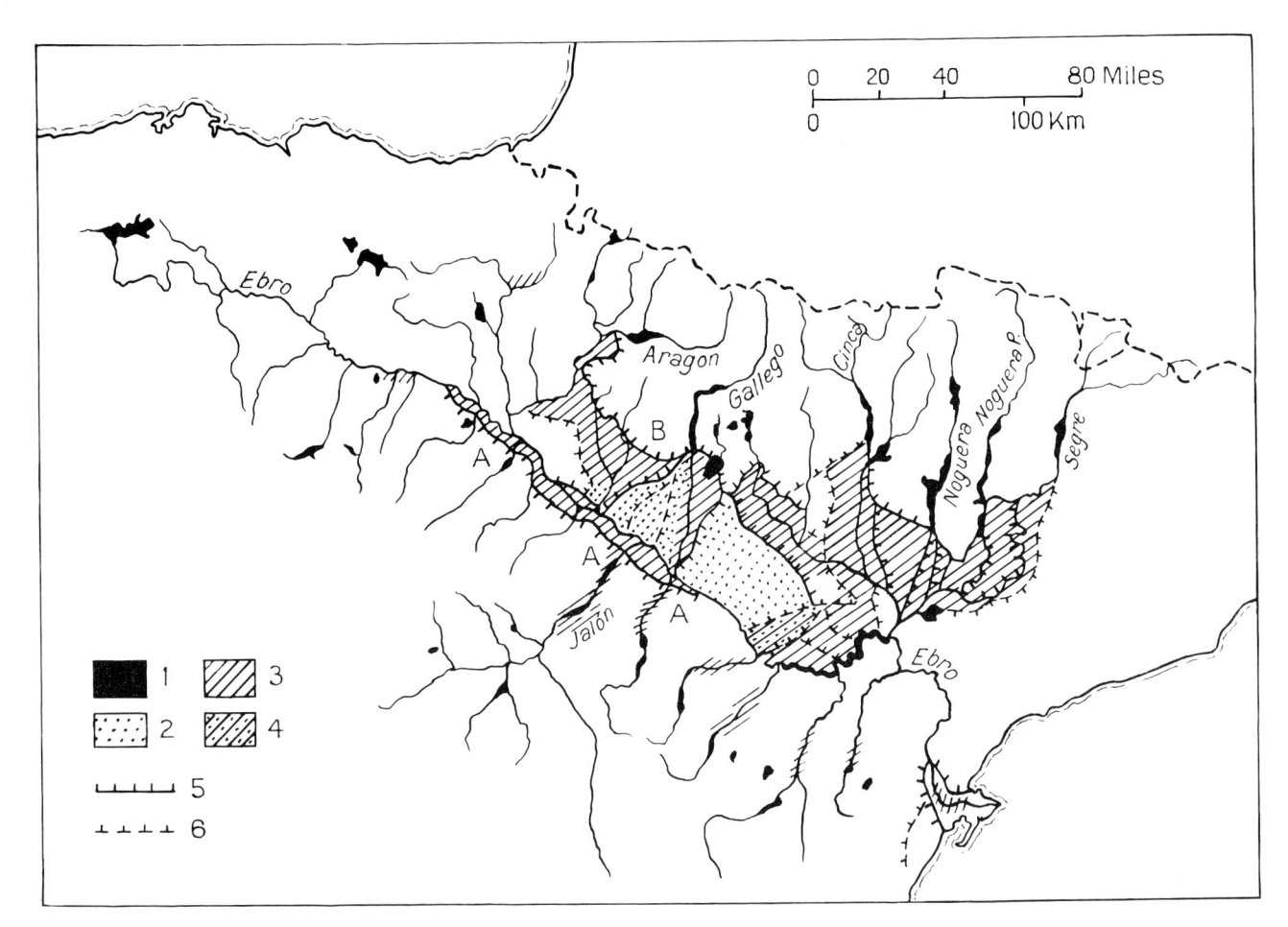

48 *Irrigated tracts in the Ebro trough*
 1 *main reservoirs and barrages, including that of Mequinenza on the Ebro*
 2 *areas most affected by gypsum*
 3 *irrigated*
 4 *irrigation being completed*
 5 *completed irrigation canals*
 6 *irrigation canal under construction*
 A *canals de Ladoso, Tauste, and Imperial de Aragôn*
 B *canal de Las Bárdenas*

by multi-purpose barrages and 22 per cent of Spain's hydro-electricity is developed here. The water is then fed to canals and so on to the huertas and vegas. The chief right-bank rivers are the Jalón, Huerva, Aguas, Martín and Guadalope; all have reservoirs and irrigation tracts and the first two also feed the Canal Imperial de Aragón. The Pyrenean tributaries irrigate the Llanos de Urgel, Los Monegros, La Noguera and Las Bárdenas; and irrigated lands extend almost the whole length of the main river from Logroño to the delta (fig. 48). The crops grown show that selfsufficiency is the first aim, with cereals, sugar-beet and alfalfa taking precedence over market-garden crops and fruit. The farming is mixed with animal husbandry, pigs and poultry on the lowlands and cattle and sheep in the tributary valleys. A few districts show a regional specialization, as in the Logroño neighbourhood which is largely under vines, but large estates under pastoralism are conspicuously absent, the typical farms being of 8–12 hectares, under protective landholding legislation. Zaragoza (430 000) developed as a route centre commanding the passageways along the Ebro valley, the Jalón valley on to the Meseta and the Gállego valley across the Pyrenees. Its industries are predictably largely food processing, handling sugar, olive oil, fruit, flour, wine, and other regional products. Recently the manufacture of vehicles, electric and railway equipment and machine tools has developed, using hydro-electric power generated in the basin. Logroño (69 000) and Lérida (79 000) also have food-processing industries, and are provincial capitals.

The Catalonian mountains

The *Cordillera Costero Catalana*, as the Spanish call it, is dealt with here as it lies across the seaward end of the Ebro trough (fig. 46). It is, however, more a natural than a geographical division as it is closely united economically with the coastlands. From a physical viewpoint Catalonia naturally falls into a coastal strip, an interior corridor (the Vallés) and an interior *Serralada*, the last being the Catalonian cordillera. These mountains rise to 1712 m in the Montseny block, 1236 m at Montserrat, and occasionally exceed 1000 m in the south near the Ebro. They are, however, in the north crossed by rivers such as the Ter and Llobregat from the Pyrenees which expand into wide green basins at Vich and Manresa respectively. South of the Montserrat block, the coastal streams rise on the cordillera itself and have a less regular regime. The Catalonian mountains generally are much better watered than the Ebro plains partly because they often receive moist northeast winds, a continuation of the mistral across the Gulf of Lions. The climate is also milder in winter and Montseny, for example, has extensive forests of mixed central European and Mediterranean species, with the ilex predominating. Farther south Mediterranean vegetation becomes more dominant. The agriculture and horticulture of the middle and lower slopes are concerned with typical Mediterranean products. The existence of some coal and salt and the abundance of water power (today largely hydroelectric) from rivers like the Llobregat with a fairly regular regime have been fully exploited by the enterprising and active Catalans. As described in Chapter 11, Manresa, Igualada and several other towns are notable textile centres and this industrial growth together with good farming has created areas of relatively dense settlement between the main upland blocks. The important potash deposits at Suria near a tributary of the Llobregat feed chemical factories near Barcelona.

The Andalusian trough

This tectonic trough, drained by the Guadalquivir, is floored with marine and river clays, interrupted in its interior parts by isolated limestone outcrops and covered on much of its seaward end with an expanse of marsh, lake and water channels, known as Las Marismas. The drainage pattern is markedly asymmetrical. From Andújar downstream nearly to Seville the Guadalquivir has been pushed northward close to the foot of the Sierra Morena by the action of copious rivers, such as the Genil, that derive their water from the heavy rains and snows of the Baetic Cordillera. This has resulted in a plain 40 to 70 km wide between the left bank of the river and the Baetic foothills.

This wedge-shaped Andalusian lowland faces the Atlantic between Huelva and Cádiz and in the cooler seasons is wide open to Atlantic airflow. It enjoys winters that are markedly warmer and wetter than those of New Castile but the higher rainfall tends to be offset by the more southerly latitude. Its summers are roasting, dry and dusty and consequently the vegetation resembles that of the Ebro trough, being xerophilous in nature and steppe-like in appearance. The winter warmth, as the climatic statistics show, is favourable for tourism and the spring weather is ideal.

Under the Roman and Moorish occupations Andalusia was highly productive and quite populous. Nearly all of its present towns and cities are Roman in origin. Today, with notable exceptions, it is largely a region of latifundia under absentee landlords. The crops are tended by a poor peasantry using backward methods of tillage, little machinery, and showing little knowledge of intensive cultivation. Between Seville and Córdoba is one of the exceptions and here the more fertile loams and black earths (tirs) yield rich harvests of wheat, citrus and vegetables with vines and olives on the lower hillslopes. Another fertile area lies south of Seville where large areas of Las Marismas have been reclaimed and planted with rice, sugar-beet, chick-peas and other cash crops (fig. 49). But up to 1950 the Plain largely remained under improved pastures and crude arable. Vast areas of olive and wheat occur on the flat southern interfluves away from the rivers. The population is concentrated mainly in villages and small towns but isolated farmsteads (*cortijos*) are relatively common on the drier upper terraces. The farming generally is extensive, unmechanised and poor in yields. The peasantry till the soil for very low wages and do not have permanent employment. The *tercio* system is used, wheat being followed by fodder and, in the third year, by fallow. The fact that about half the total land lies fallow each year is some indication of the low soil fertility. In some districts, as around Jaén and Seville, the vine is grown extensively as a monoculture and bulls are bred for the corrida.

The potentiality for irrigation is tremendous and not surprisingly the lower Guadalquivir valley was one of the larger areas designated for development under the plans of the INC. Here less than 15 000 ha were irrigated in 1918 but from 1950 onwards the cash cropping of cotton rapidly expanded. Today over 200 000 ha are under middle-quality American varieties often grown as a monoculture for year after year on the same irrigated land. Cotton picking requires much hand labour and has done much to ease the agricultural unemployment. The irrigation areas of the Plain are shown in figure 49, and discussed also in Chapter 5.

The chief Andalusian city is Sevilla or Seville (610 000), which can, with the aid of a short stretch of canal, be reached by ocean-going vessels up the Guadalquivir. It exports mainly agricultural produce and mineral ores. Its industries are concerned chiefly with food processing (flour, wine and olive oil), textiles, ceramics and chemicals, but probably the chief is tourism. It was successively the Roman capital (Hispalis) of southern Spain, the capital of a Moorish kingdom and the commercial centre of trade with Hispanic America. It has too been remarkably fortunate in escaping the rebuilding characteristic of city cores in the twentieth century. Córdoba (226 000), head of ancient navigation on the Guadalquivir, is also an important provincial centre with food-processing industries but here again tourism may well be the first money-earner. Cádiz (732 000) is the chief coastal settlement and occupies the tip of a peninsula jutting north on the west side of a fine bay. It is a Spanish naval base with arsenals and shipyards nearby at La Carraca, supplied by heavy engineering industries at San Fernando and Matagorda. A large container port is being planned for this neighbourhood which has long been famous for the export of sherry, a wine named from the nearby town of Jerez de la Frontera (148 000). At the opposite or western end of the coast of the Andalusian Plain the river Odiel forms a small estuary on which is situated Huelva (93 000), the outlet for the Riotinto copper mines and other mineral ores of the Sierra Morena. It is also a notable fishing port, being fourth in importance in Spain and handling 5 per cent of the total national catch.

The Andalusian or Baetic cordillera

These southern fold ranges consist of two main mountain chains divided, except in the northeast, by an interior longitudinal fault zone. The northern range is of moderate

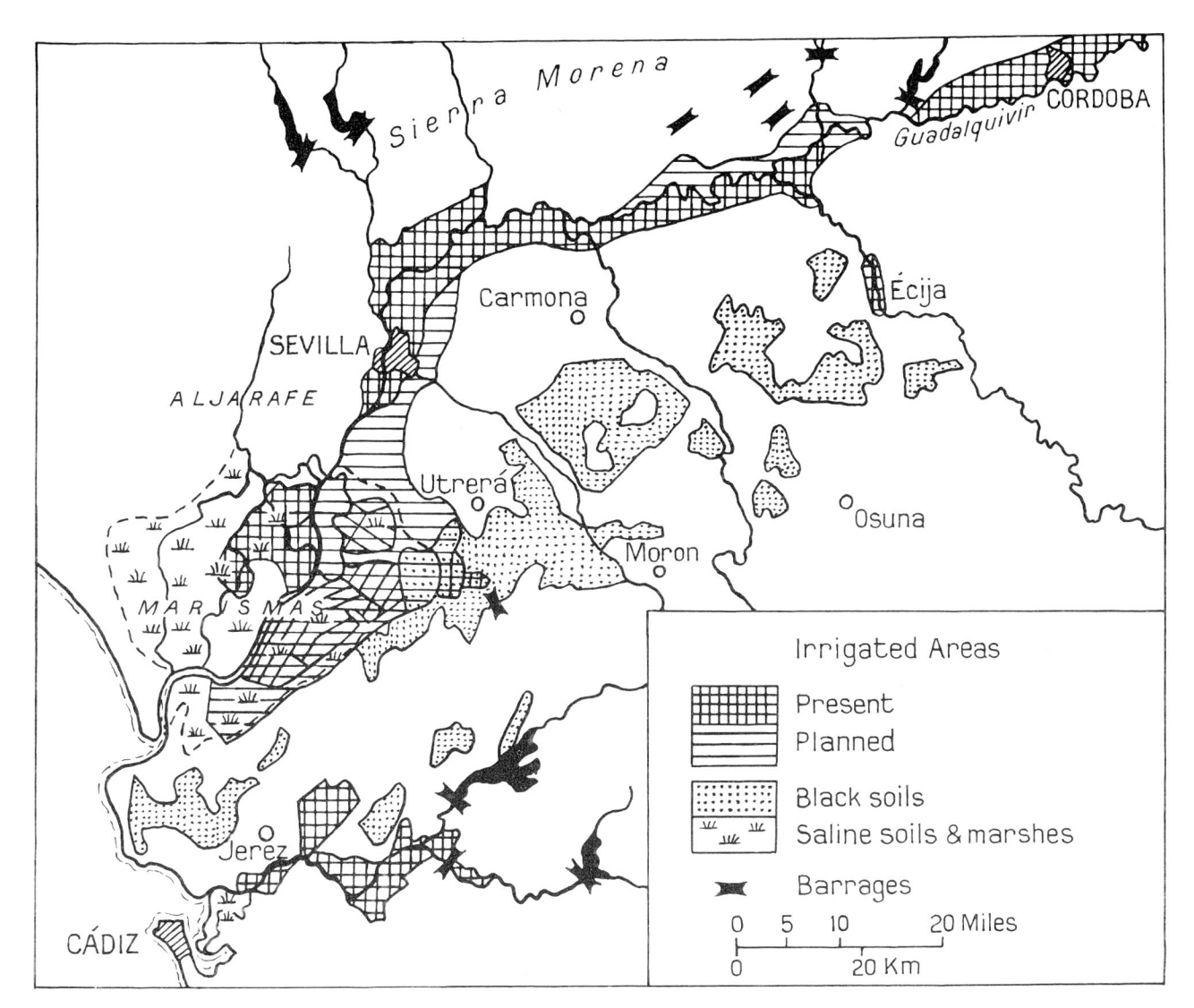

49 Irrigated areas in the middle and lower Guadalquivir valley

height and is formed mainly of limestones and flysch. In parts of the west its rainfall is sufficient for wheat, vine and olive but eastward the slopes become more barren and matorral is common. Some of the limestone tracts show typical karstic features.

The southern range has undergone considerable faulting and some vulcanism. It consists largely of schists and other crystalline rocks and is much higher than the northern range. It culminates in the Sierra Nevada where the Cerro Mulhacén (3481 m) and Pico de la Veleta (3470 m) are the highest points of Spain. The upper slopes here are glaciated and still keep snow patches until late summer. The abundant runoff of the range as a whole feeds many head streams that eventually supply water to vegas in the mountains, as at Guadix and Baza, or in the interior fault-zone, as at Granada, or on the coastal lowlands, as at Málaga. The lower valleys on the southern side of the cordillera are productive where floodplains exist, and olives persist to over 1000 m, chestnut, mulberry and walnut to 1600 m and rye and potatoes to about 2500 m. Much of the former forest has gone but fir and pine are common on some sierras.

The narrow interior fault-zone is important as it gives opportunity for communication and its basins are usually densely peopled vegas. It is drained by the upper courses of several rivers, notably the Guadiaro, Guadalhorce and Genil, and where these valleys widen irrigation gives fine crops of cereals, fruits and fodder. Such vegas exist in the Guadalhorce valley near Antequera and on the Genil near Loja and Granada. Granada (168 000) overlooks a wide trough, at 800–1100 m above sea level, where the Genil valley is lined with shallow river terraces. The chief crops of its vega are cereals, sugar-beet (refined locally) and fodder. It is a great tourist centre as would be expected of a city that was the last Moorish capital and bulwark in Spain and has the Alhambra, a hill-top miracle of Moorish architecture and gardens.

Where the longitudinal fault-belt is broken at the Strait of Gibraltar it descends into the large bay of Algeciras, named after the town (75 000) on its western side, a busy fishing port with frequent ferry connexions with north Africa at Ceuta and Tangiers. The eastern side of the bay is formed by the steep-sided limestone ridge of the Rock of Gibraltar.

Gibraltar

The history of this Rock reflects that of numerous peninsulas and offshore islands round the shores of the Old

XIV *Gibraltar from the south*

50 *Gibraltar*

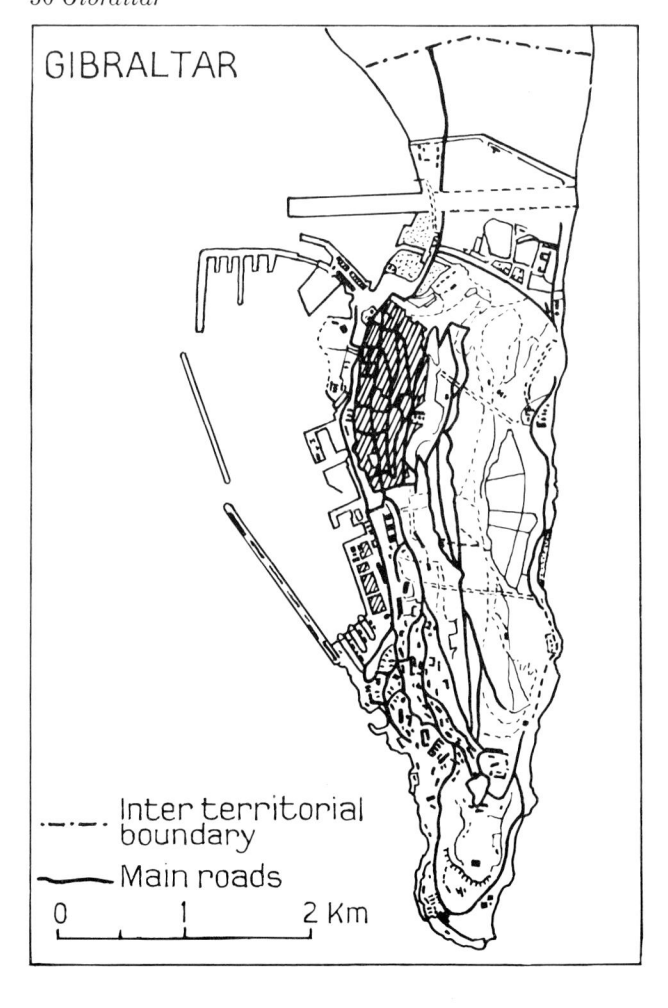

World in that it has been captured and held in turn by different maritime powers throughout historical times. Cave dwellings with rock carvings reveal the presence of Palæolithic man and it is known that Greeks and Romans also lived here. In AD 711 it was occupied by Tariq ibn Ziyad, a Berber leader, who went on to defeat the Goths at the present site of Jerez. The name Gibraltar is a corruption of the arabic Jebel Tariq, meaning Tariq's mountain. Apart from a short Spanish occupation in the early fourteenth century, it remained Moorish until 1462 when Spaniards captured it. In 1704 the British navy took it and it has remained under British rule since. Twice during war in the twentieth century it has provided them with a naval base for the protection of shipping in the Strait, and for ship repairing and refuelling. Between 1966 and 1969 the Spanish government prohibited the entry of a large number of Spanish workers that commuted to it daily and eventually completely closed the frontier to all land traffic.

This was a serious economic blow to the colony as the Rock has no natural resources and no agricultural land. It consists mainly of a narrow, low, sandy isthmus and a steep-sided rock, in all about 7 km long and with a maximum height of 426 m and a total area of 55 ha. An airport, extended into the bay by debris got from tunnelling in the Rock, occupies much of the sandy isthmus (fig. 50). The main buildings of the town occupy the northwest side of the Rock where a fine port, also partly constructed on material got by tunnelling, protrudes into Algeciras bay. Buildings are also scattered on the south side of the Rock where two wide natural (wave-cut) platforms occur, but the east side is precipitous. The upper slopes of the ridge are under water-catchment schemes and matorral, mainly of wild olive where the denizens include Barbary apes, the only wild monkeys in Europe. Today the colony's water supply is enlarged by several diesel-driven seawater distillation plants.

The people (25 000 in 1966) are almost entirely dependent on the ship-repair depot, the fine port, the military garrison and tourism. A ferry service is run to Tangier and there are other regular connexions with Morocco. The peninsula enjoys a long dry summer with an average of 10 hours of sunshine daily and mild winters with 4 to 6 hours of sunshine daily. It is a favourite port of call on Mediterranean cruises and before the closing of the Spanish frontier had a large car-ferry service with Morocco. The Gibraltarians are bilingual in Spanish and English and have undertaken a considerable building programme to encourage tourism. They would, of course, benefit from the reopening to traffic of the Suez Canal.

The Mediterranean coastlands

These are the coastal strips bordering the Mediterranean that are sufficiently low-lying to have Mediterranean products and a character of their own. Most are alluvial plains built up from sediments deposited by rivers and longshore currents in the almost tideless sea. They are absent where the lofty Andalusian fold mountains pass out to sea as a bold rocky coast. But it is usual to include with them the hilly coastal strip of Catalonia, which is often cliffed in the Costa brava.

The *Barcelona or Catalonian coastlands* include an interior corridor and a purely coastal strip. The former is called the Vallés in the north and the Penadés south of the Llobregat. It is under typical Mediterranean polyculture interspersed with rich market gardens and near Barcelona has industrial towns such as Sabadell (146 000), Tarrasa (128 000) and Granollers. In the far south the Penadés merges into the plain of Tarragona.

The coastal strip is hilly and cliffed in the north (Costa brava) but south of Blanes has a narrow coastal plain with market gardens and many industrial towns, notably Badalona (139 000), a rapidly growing suburb of Barcelona. From the mouth of the Besós river to the Llobregat delta the houses of Barcelona and its continuation in Hospitalet (207 000) line the coast. The Llobregat delta is rich huerta and to the south, beyond a short stretch of limestone garrigue (Costa de Garraf), the plain of Tarragona forms an extensive lowland mainly under olives, vines, hazel nuts, almonds and carobs. These Barcelona coastlands end in the Ebro delta where the older upper huertas have scattered cottages and the extensive newly reclaimed seaward flats are ricefields devoid of habitations.

Barcelona (1 790 000) is the most important manufacturing city of Spain and industrialization is the prime characteristic of all its vicinity. Its port or harbour is entirely artificial and lies almost halfway between the mouths of the Besós and Llobregat rivers, each of which has excessive deltaic deposition. Fuel supplies and raw materials are easily imported to supplement the Catalan development of hydro-electricity (see Chapter 10) and the local mineral resources. The city is fortunate in having a considerable population of skilled artisans and a large professional middle class. It has had two great periods of growth. During the first, from about 1300 to 1500, it was a powerful commercial emporium, and capital of a Catalan kingdom. However, it was debarred from 1492 to 1778 from free trade with the Americas (a monopoly of Seville) and did not begin to expand rapidly again until late in the Industrial Revolution. Towards the close of the eighteenth century textiles based on water power were introduced and with other industries gradually caused a modern expansion. Since 1850 its population of 175 000 has increased tenfold. Today it has all the functions of a great modern city – cultural, financial, commercial, industrial and administrative. Its main industries are textiles, engineering, foodstuffs and, above all, engineering including railway equipment, electrical equipment, diesel engines, automobiles (from purchased components) and ships. It is the second or third port of Spain and monopolizes Catalonia's trade, and has a busy packet-ferry traffic with the larger Balearic islands.

The Valencia coastlands. This region is an almost unbroken lowland floored by relatively recent river alluvium and experiencing a marked summer drought. It is virtually one great huerta that has been under cultivation for two thousand years. It is crossed by the rivers Mijares, Turia, Júcar, Palancia and Serpis, the first three draining limestones and therefore yielding their flow rather more uniformly. They are further regulated by dams where they debouch on to the plain. Near Valencia the waters of the Júcar are diverted into the Acequia Real which runs from north to south the whole length of the Valencian huerta, but their use, owing to the low rainfall, has to be rigidly controlled and the Tribunal de los Aguas meets weekly before the doors of Valencia cathedral.

In this stretch the coast is fringed with dunes which have cut off a lagoon, the Albufera de Valencia, from the sea. These are mostly clothed in pines but the marshy areas behind them have been turned into ricefields using the lagoon for drainage. Inland from the ricefields are the huerta holdings growing crops of cereals, fodder and tobacco for home consumption, and oranges and other fruits, tomatoes, artichokes, onions and cauliflowers as cash crops. Where the plain meets the mountains are discontinuous terraced slopes under vines, olives, almonds and citrus. Oranges form the chief cash crop and are harvested from December to April.

The whole area is closely peopled, supporting densities of 200 to over 400 per square kilometre even in rural districts where houses are dotted thickly among the irrigated plots. The activities of this populace focus mainly on Valencia (614 000), situated on the Guadalaviar river five kilometres

xv *Barcelona*

from the sea and forced to use El Grao as its outport. It is by far the chief town and industrial centre of the huerta and has considerable textile manufactures, especially silk, and a variety of smaller industries such as shipbuilding, pottery, glass, fruit packing and chemicals. It imports iron ore from Teruel for its engineering industry.

However, the chief metallurgical concern on this stretch of coast is 30 km north at Sagunto (30 000) where a modern port and blast furnaces were opened in 1917. The ore comes via the Aragón railway from the Sierra Menera near Teruel. Sagunto was an ancient Greek or Greek–Iberian town founded on a hill near the mouth of the Palancia river by colonists from Zacynthus. Later it joined the Romans and was besieged and taken by the Carthaginians under Hannibal in 219 BC so starting the second Punic War. The Romans rebuilt the town and it has extensive Roman remains overlooking the present settlement on the plain. But Sagunto never recovered its early importance because in 138 BC Valencia was founded by Junius Brutus and populated with captured Lusitanian soldiers who were given Latin rights or franchise. Valencia was on the Turia, a much more copious river than the Palancia, and in this part of Spain potential must be judged on the amount of water and plain available for irrigation. The alluvial plains having been built up by the rivers bear a close relation to their volume.

In the north the large orange crop from the plain near Castellón (84 000), capital of that province, is exported from its own outport.

The Murcian coastlands. From cape de la Nao south to cape de Gata the coastal relief changes. Whereas in the provinces of Valencia and Castellón the coastal plain is a flat strip at the faulted edge of an upland block, south of cape de la Nao the folds of the Andalusian cordillera run out obliquely to the sea and as a result the coastal plain is frequently interrupted and, where present, stretches irregular distances inland up lowlands between mountain folds rather than along the coast. The rivers first encounter lowland some distance inland and most of the irrigated tracts are a considerable distance from the sea. The coastlands are often semi-desert plains and hills almost bare except for agave, opuntia and esparto. On large areas the only crops that survive the drought are the olive, vine and carob but wherever water is available rich market gardens and cash crops, including banana, date and sugar-cane flourish. The rainfall is less than in Valencia and its occurrence very variable, floods being as much a menace as drought. Safe irrigation demands elaborate engineering works and is limited mainly to the Segura basin. The traditional organization of the water supplies also differs from that in Valencia. In Murcia, where greater barrages and more costly canals are usually necessary, the owners or tenants of the land purchase their water needs from those who financed the constructions.

It is necessary to describe the inland huertas first. These mainly align the Segura, the headstreams of which rise far inland on the western ranges of the Andalusian cordillera and so provide much water. More than a dozen dams store sufficient water to irrigate the huertas downstream. There are upland irrigated tracts as around Yecla but the main coastal huerta begins on the lower course near Cieza and widens into a plain 25 km long by 6 km wide near Murcia city. Here two canals, under water laws dating back to the Moorish occupation, distribute water to several hundred small settlements. The farmers grow a wide variety of fruits and vegetables, particularly tomatoes, peppers and melons. The dense scattering of detached dwellings surrounds a built-up core (Murcia city) in which live about one-quarter of the municipal population of 275 000 persons. From the Murcia huerta, irrigated lands extend up the tributary valley of the Sangonera or Guadalentin to Lorca where 11 000 ha are irrigated mainly for cereals, and downstream along the main valley to Orihuela. To the northeast, outside the Segura basin, Elche (101 000) uses the waters of the Vinalapó river for its date palms.

The actual coastlands in Alicante and Murcia, as already noticed, are often hilly and mainly arid. In the north the steep coastal slopes are terraced for olive, vine, almond and carob. A coastal plain appears at Alicante (136 000), the provincial capital which also acts as the port for agricultural produce and as a metallurgical centre for the silver, lead and iron ore mined in the inland sierras. South of this vega, aridity prevails with much dry, dusty steppe that supports mainly goats and esparto grass except where water is raised from solitary wells. Along some of the coastal margin there are lagoons and marshes, including the large Mar Menor, that could be reclaimed for agriculture. At present several lagoons form evaporating pans for salt for the local chemical industry while the sands are in places used for glass manufacture. The chief town in the south is Cartagena (144 000), the Spanish naval base for the Mediterranean. It has a sheltered harbour that appealed to the Romans (*Carthago Nova* of the third century) and as a naval base was largely rebuilt by Philip II in the sixteenth century and enlarged two centuries later. Today it is also a metallurgical centre and exports mineral ores as well as agricultural produce. With the stimulus of foreign (U.S.) bases in its hinterland it became Spain's first port for cargo in 1967, being ahead of Bilbao for weight of imports and second to it for exports.

These coastlands will be transformed agriculturally in the near future. The schemes under way include a long canal to irrigate 50 000 ha of the coastal plain near Cartagena and another to run from Orihuela on the Segura to Alicante. The latter, the *Saladares de Alicante* scheme, has been nearly completed by the I N C.

The Málaga coastlands or Mediterranean Andalusia. These 300 km of coastland consist partly of relatively small lowlands and partly of the terraced south-facing slopes of the Andalusian cordillera (fig. 46). Here the annual rainfall is much greater than on the eastern coast, and the lofty highlands, especially the snowy Sierra Nevada, provide more numerous streams. The summers, however, are among the driest and longest in Spain and the Costa del Sol is justly famed for its sunshine. The coastal plains are restricted in size and disconnected physically although linked by a coastal road. All are where streams from the high sierra have built up deltaic flats. In classical times many of the settlements on these rivers were ports but today they are surrounded by cultivated alluvium, much of it deposited since the inland mountain slopes were exposed to erosion by deforestation. Many of the vegas are very densely peopled. Almería (103 000) in the north was a great Moorish port in the eleventh century and today exports mineral ores, table grapes and other agricultural produce. The Motril vega goes in more for sugar-cane while that at Vélez Málaga on the Vélez delta is equally subtropical and palm-clad. The largest vega on this coast is at Málaga at the mouth of the Guadalhorce valley which provides a railway routeway to Córdoba. Here the terraced vineyards on the hillsides overlook a typical huerta, abounding in orange groves. The city, which dates back to Phoenician times, and is sited some way north of the Guadalhorce mouth, has today more than 350 000 inhabitants. It expanded rapidly in the nineteenth century as a port for African trade and as a winter resort of the grandest kind. Its port facilities were greatly improved and industries such as cotton textiles, sugar-refining, smelting and engineering grew up. The expansion has been phenomenal during the last four decades, when its population has doubled. The conjunction of lofty mountains, warm sea, abundant sunshine, almost rainless summers, balmy winters, and of tropical and desert vegetation make tourism and residential functions outstandingly important. Its population is becoming steadily international.

The Balearic Islands

At cape or Cabó de la Nao the Andalusian cordillera runs out to the sea and continues as a submarine sill for 80 kilometres until part emerges as Ibiza the westernmost of the Balearic Islands. This archipelago consists of 5 larger islands and 11 rocky islets which are spread out over a length of 330 km and together occupy a land area of 5014 sq km. Majorca (Mallorca) and to a lesser extent Ibiza are the only islands that repeat the southwest–northeast fold pattern of the Andalusian mountains. Majorca has three structural strips, a northwestern which is a complex belt partly of Jurassic limestone that rises to 1445 m but has occasional small basins floored with terra rossa; a southeastern ridge also largely of limestone; and between the two a lowland corridor of horizontal recent marine sediments of great fertility. Minorca (Menorca) is completely low-lying although it has Hercynian crystalline outcrops as well as limestone plains.

The climate is typically Mediterranean as the statistics for Palma show. The higher parts are mainly garrigue or matorral which affords grazing for sheep and goats. The better-watered hillsides are terraced and planted with carob, vines and olive on their upper slopes and market-gardening crops on their lower levels. At the foot of slopes where the soil is thicker and ground water more abundant, huertas abound while the rest of the plains are under wheat, barley, beans and cash crops. Generally the citrus and vine are less important than almonds, which are exported in large quantities, and apricots and carobs.

Settlement on the islands began early and the ancient remains include remarkable structures of the later Bronze Age from about 1000 B C to Roman times. These are mainly temples; *naus* or *navetas*, a type of boat-shaped or keel-shaped burial structures; and numerous *talayots* or conical towers, built of stone blocks, that resenble the *nuraghi* of Sardinia and, no doubt, like them were fortified dwellings. The later history recalls many occupations, especially Spanish. Minorca belonged to the British from 1708 to 1756, to the French for the next seven years and alternately to the British and Spanish until 1802 when it went finally to Spain. These changes are reflected in the fine architecture of Port Mahón.

Today rural settlement is usually in large agricultural villages situated some way inland and in small ports on coastal inlets. Non-agricultural industries are few owing to the lack of raw materials and power. A little lignite, cement from local limestone and small metal workings are present but food processing and a variety of textiles at Palma, tourism, and a wide range of handicrafts are dominant. The capital of Minorca is Port Mahón (20 000) which has a commodious harbour. It and Ibiza (13 000) the chief town of the island of that name, were occupied by the Carthaginians. On Majorca, Manacor is an agricultural market centre with a tourist industry concerned partly with fine limestone caves and underground lakes. By far the chief city is Palma (204 000) one of the fastest growing towns of Spain. It has a fine harbour with regular sailings to Alicante, Valencia and Barcelona, and exports almonds, grain, other agricultural products and textiles. It is a regular port of call for Mediterranean cruises with benefit to its great variety of local handicrafts. Regular air services are maintained with Barcelona and Alicante and, with numerous charter flights, largely account for the recent growth of tourism and the spate of new building. The total resident population of the islands was 366 000 in 1936, 420 000 in 1950 and over 500 000 in 1968. Already the density of 100 per square kilometre exceeds that of most provinces on the mainland but it gives little idea of the populousness of these sunny islands. Palma de Mallorca is by far the busiest passenger airport in Spain, and the number of visitors arriving at it has risen from 241 000 in 1961 to 1 228 000 in 1968. With recent developments of 'package tours' its popularity will increase and it has a bright future if its amenities are carefully maintained.

13 Portugal

Historical development of the state

Portugal first tasted independence when it severed its connexions with León in the twelfth century. Before that date its history was part of Spain's although ethnic, geographical and cultural differences were already apparent.

The Lusitani tribe of northwestern Iberia fiercely resisted the Roman invading armies but finally they were defeated. Their name Lusitania, however, survived in this westernmost province of the Roman empire which was wider than present-day Portugal and had its administrative centre in Mérida farther east. Lusitania was independent for a time during the Visigothic invasions but was later absorbed into a larger Visigothic territory. It was again submerged during the initial Moorish conquest of Iberia but since there was little or no Arab settlement in the cool, damp northwest corner of the peninsula, after fifty years the land lying between the Douro and Minho rivers became part of the Christian kingdom of León and Asturias. Known as Portucallis it was a hereditary county ruled by a governor owing fealty to León. In 1064 the *Reconquista* pushed its southern frontier in Portugal down to the Mondego River and a capital city was established at Coimbra (fig. 41). The whole territory was later made the dowry of Teresa, daughter of Alphonso IV of León, on her marriage to Odo, Duke of Burgundy, but it was still held as vassalage of León. Their son Afonso Henriques known as Count of Portugal in 1114, ruled over Galicia. By 1139 he had separated Portugal from Galicia and had pushed his southern frontier to the Tagus with the aid of Flemish, French, German and English crusaders who were on their way to the Holy Land. The first bishop appointed to the restored see was an Englishman, Gilbert of Hastings. Afonso Henriques was assisted by the Knights Templar in establishing order and peace along his new frontiers while the Cistercians from their new monastery in Alcobaça acted as architects and developed agriculture in central Portugal. The independence of Portugal was recognized by the King of León in 1143 and four years later Lisbon was captured from the Moors.

Afonso's son Sancho I continued the development of Portugal by enfranchising many communities and embodying their trading rights in charters and markets thereby attracting settlement. His successor called together the first cortes of nobles and prelates at Coimbra in 1191 to establish the nature and rights of landholding thereby curbing the temporal power of the church. The next king, Sancho II, extended Portugal's frontier south once more to include the Alentejo and the Algarve. His heir, Afonso III, moved the capital south from Coimbra to Lisbon and, in 1254, summoned the commoners to be represented for the first time in the cortes.

One of the greatest of the succeeding Portuguese Kings was Denis, the farmer king (1275–1325) who established a university at Coimbra, encouraged a national literature, started a mint, and began the planting of pine trees on migrating dunes near Leiria to supply timber for his ships and to protect the nearby agricultural land. He engaged a Genoese admiral to build his navy and had begun to lay the basis of a foreign trade. After Denis, the Portuguese rulers for the next half century were embroiled in quarrels with Castile but reasserted their independence at the Battle of Aljubarrota (1386) which they won decisively with the aid of English bowmen. The Anglo-Portuguese Alliance dates from this battle and the treaty was cemented by the marriage of the Portuguese King John (João I) with Philippa, daughter of John of Gaunt. There followed a substantial increase in trade and with the capture of Ceuta in northwest Africa (1415) the great age of Portuguese discovery and exploration began. Their aims included the finding of the lost Christian kingdom of Prester John and of a new way to the spice trade of the Far East.

Under the guidance of Prince Henry the Navigator, national enterprises were set in motion. The country was in a superb situation to undertake the circumnavigation of Africa and quite early the Azores and Madeira were colonized and sugar exported from them to Europe. By the time of Prince Henry's death (1460) the newly designed Portuguese caravels had reached Sierra Leone. In 1482 Diego Cam explored the mouth of the Congo and in 1488 Bartholomew Diaz rounded the Cape. Ten years later Vasco da Gama reached India and arrived back in 1499 with a rich cargo of oriental merchandise. The dominance of the Mediterranean ports and of the Near East trade routes over the spice trade was now broken for ever.

Meantime under Spanish patronage Columbus had reached the New World in 1492, and the Tordesillas line was drawn dividing it between Spain (to the west) and Portugal (to the east). Cabral landed on the coast of Brazil in 1500 and from then until 1506 Portugal was establishing her empire over Madagascar, Ascension, St Helena, Tristan da Cunha, and the coast of India and of east Africa. Fifty years later they were in possession of Ormuz in the Persian Gulf, Goa, Malacca, the Moluccas and Macao and had reached Canton.

For the next hundred years by this system of strong land points and trading posts based on sea power, the spice trade of Europe with the Indian ocean and south China seas was secured to Portugal until the Dutch broke their monopoly in the seventeenth century.

Portugal's golden age began with the discoveries of metals and diamonds in the Minas Geraes in Brazil. The crown claimed a 'royal fifth' of all incomes derived from Brazil and this paid for the building of the many academies, libraries and palaces erected at home. During the same period peace was made with Madrid and Portuguese independence permanently assured. There followed the treaty of Methuen (1703) with the English which laid the foundations of the port wine trade whereby port wine was exchanged for English woollens.

From 1750 Portugal was served by a great administrator, Pombal, who reformed the sugar, spice and diamond trades, and set up national industries in silk, cutlery and hats, which survive to this day. He formed companies for the development of the wine industry, the tunny and sardine fisheries of the Algarve and for the Brazil trade, and initiated a system of universal education. In 1910 Portugal became a republic following the establishment of universal suffrage, and a year later compulsory education was enforced. From 1951 to 1968 the country was governed by Dr Salazar who devised a democratic constitution in which the elected members of the cortes are representatives of all the trades and professions in Portugal. The country still retains much of its vast colonial empire. Ninety per cent today consists of Portuguese West Africa, Angola, Guinea and Mozambique, and the remainder, of the Cape Verdes, São Thomé, Príncipe, Macao, and part of Timor. Portuguese India (Goa, Damão, Diu) was seized and forcibly incorporated into the republic of India in December 1961.

General patterns

The Iberian peninsula which is so enigmatic a part of Europe plays a far more remarkable role in the southern hemisphere. The Spanish language has become the mother tongue of Latin America while the trading relationships of of Portugal with its overseas empire are world-wide extending from its former colony of Brazil, with which it still has close social ties, to Africa (Angola and Mozambique) Timor and Macao.

The Portuguese homeland, 89 000 square kilometres in area, is a wedge-shaped country on the southwest of the Iberian peninsula, 580 km long and 225 km at its greatest width in the north and half that width in the south. The name itself is probably derived from the Latin *Terra portucallis*, the Roman settlement on the north bank of the Douro, a term which came to be applied later to the whole area between the Minho and the Douro.

Structure and relief

The Romans, in describing the country as a portcullis were apt in their recognition of the parallel ranges of mountains that traverse the country from east to west and of the great barrier wall of the plateau edge that crosses these from north to south. Physically Portugal begins on the western edge of the Meseta where the ancient Hercynian block starts to crumble away and to fragment towards the Atlantic. The residual ranges of the Spanish plateau continue west into Portugal and jut out as rocky headlands into the Atlantic. The Spanish Sierra de Gata merges in this way into the Serra da Estrêla, the highest of the Portuguese ranges at 1991 m, the Sierra de Guadalupe becomes the Serra d'Ossa (900 m) and the Sierra Morena ends seaward as the Serra do Malhão and the Serra de Monchique (902 m). Between these ranges are basins of four large rivers, the Minho, Douro, Tagus and Guadiana which have their middle and upper courses in Spain, and four small entirely Portuguese rivers, the Zézere and Sorraia tributaries of the Tagus and the Mondego and Sado which together form a belt of coastal lowlands lying at the Atlantic edge of the Meseta. Neither of the two larger rivers, the Douro and Tagus, is navigable until after their descent from the plateau in gorges and rapids, and the navigable reaches of all the rivers are in Portugal. The geographical entity of Portugal is thus a coastal area traversed by navigable waterways and linked by an ocean frontage and highway.

Additional Spanish landforms repeated in Portugal are the downwarpings of the plateau rocks which have later become filled with Tertiary sediments to form fertile and productive lowland areas in both countries. This feature which dominates Old Castile continues into Portugal as the lower Mondego valley. The river Tagus occupies a similar trough (Estremadura in Portugal and Extremadura in Spain), the Sado basin can be traced in New Castile while the southern Algarve is but a continuation of Andalusia. Nevertheless the frontier between the two countries is natural and decisive and has been described as a desert of rock boulders at the edge of the massive plateau.

Climate

A second key to the geographic identity of Portugal lies in its unique maritime variant of the Mediterranean climate. Average temperatures are fairly typical on the coastal strip, being between 10°C for January and 20°C for July in the north and increasing to 11°C for January and 24°C for July in the south, but a short distance inland a more interior character prevails. In the north in January inland stations are at least 3° or 4° colder than the coast and will be below freezing at times while in summer temperatures will be 5° higher on average, reaching at times maxima of 38°C and rarely of 40°C. Conditions along the coast are always pleasant due to the proximity of the sea and the fanning effect of the sea breezes, and the shelter from the hot, dusty winds of the Meseta afforded by the inland mountain ranges. The table of climatic statistics shows, apart from the Serra da Estrêla, the general features of the climate of the more densely populated areas.

The westerly airflow with its various fronts during the cool season, brings to Portugal a heavier rainfall than is usual due to the varied and high relief although the rainfall is concentrated to a certain extent in the valley funnels opening to meet the prevailing rainbearing winds. The effects of the higher relief on climatic phenomena are readily apparent, for example in the more extreme character of the climate of the higher slopes and summits which are bleak and exposed; in the mountain and valley winds such as the *nortada* of the Lisbon summer; the rainshadow areas and the drier interior basins; the belts of maximum precipitation; the importance of a southern aspect; and especially in the increase of precipitation with altitude on slopes facing moisture-laden winds. Parts of the Serra da Estrêla in central Portugal receive over 2750 mm of rainfall annually and even the southern slopes draining to the Tagus measure a useful 750 mm. Precipi-

PORTUGAL: CLIMATIC STATISTICS, MEAN MONTHLY TEMPERATURES (centigrade)

	Altitude (if over 100 m)	J	F	M	A	M	J	Jy	A	S	O	N	D	Absolute Max	Absolute Min	Annual days with frost
Oporto		9·3	10·2	11·4	14·1	15·9	19·3	20·2	20·7	19·2	16·0	12·1	9·2	40·1	−4·4	2·8
Coimbra	140	9·2	10·7	11·9	13·9	16·5	19·6	21·3	21·6	19·9	15·8	12·3	9·5	38·2	−0·6	0
Lisbon		10·3	11·3	12·6	14·2	16·6	19·3	21·4	22·1	20·3	16·9	13·7	10·5	40·3	−1·7	0
Évora	321	8·6	9·9	11·4	13·8	16·4	20·0	23·0	23·2	20·8	16·5	12·4	9·6			1·4
Serra da Estrêla	1441	1·0	1·9	2·6	4·1	7·8	12·4	15·6	16·0	12·6	7·7	4·6	2·3			

MEAN MONTHLY PRECIPITATION (millimetres)

	J	F	M	A	M	J	Jy	A	S	O	N	D	Mean Annual Total	Raindays	Days with snowfall
Oporto	146	126	126	126	88	41	24	21	70	127	166	128	1210	113	2·6
Coimbra	91	83	97	106	82	45	19	15	57	88	111	89	894	139	0
Lisbon	89	88	87	75	50	22	5	5	37	75	116	98	755	116	0
Évora	63	83	74	51	47	31	4	5	29	75	99	84	650	92	0·1
Serra da Estrêla	323	277	357	303	227	115	57	33	167	360	411	321	2951		

tation in the Minho to the north averages 1000 mm, most of which falls normally between October and May. Inland this increases with altitude to 1300 mm apart from in the occasional enclosed and drier mountain basin.

Although the incidence and seasonal distribution of the rainfall remains much the same, the amount dwindles considerably towards the south and the summer drought intensifies and lengthens, as the climatic statistics for Évora demonstrate. Eventually in the Algarve it averages 500 mm to 625 mm and semi-arid conditions begin to appear. Along with the decrease in amount is a decrease in the efficiency of the fall, and evaporation and runoff become excessive. Rivers dwindle to very low water in the height of summer and flood disastrously after the onset of the winter rains. Flood control and irrigation are both necessary and government schemes are on foot and in operation for the regulation of all Portuguese rivers and in particular the Tagus, Mondego and Sorraia.

Vegetation and land use
Forestry

The vegetation associated with this peculiar Portuguese variant of a Mediterranean climate also helps to distinguish Portugal from Spain. Where ridges of the Meseta extend into Portugal the vegetation is coarse grass and steppe but, where these ridges approach the sea and maritime airflow, tree growth flourishes and, unlike most Mediterranean countries, Portugal still possesses a generous forest cover; over one-fifth (28 per cent) of the country is wooded and if orchards and olive groves, almond and carob cover are counted, then over one-third of the country is under trees.

The forest cover is remarkable in that it contains both tropical and temperate species and their distributions change over quite short distances. The Serra de Sintra (northwest of Lisbon) supports some of the most luxuriant vegetation in the world where a Mediterranean climate acts in conjunction with a rich volcanic soil. Portugal is at the crossroads of the world of trees; here African palms and jacaranda meet European chestnut, ilex, lime, elm and poplars, and the araucaria and maple of the New World mingle with the magnolia of the Old. The abundance of succulent plants, such as cacti, and of tree ferns also reflects the maritime exploits of Portugal. The famous national forest of Bussaco is just such a medley and includes in addition cedars of Lebanon and Australian blue gums.

The nature of the vegetational distribution contrasts with that of arid Spain as also does the careful management of the forests. The increased rainfall in Portugal fosters trees and the relatively mild winters permit their growth to an altitude of 1300 m but farmers in addition find in their woods an alternative source of income. Even the deciduous trees are trimmed for firewood and the lower branches are grazed by stock so that often oak woods resemble orchards. Nearly half the trees are pines, in particular the cluster pine, one-quarter are cork oaks, and most of the remainder ilex, deciduous oak and sweet chestnut. The pines chiefly occupy sandy areas along the coast and mountain slopes up to 1000 m on the serras. The oaks are mainly in Ribatejo and Alentejo, about half being *Quercus suber*, the cork-producing variety.

Most of the forests are privately owned but the state actively pursues an afforestation policy particularly over areas of dune, serra and waste. By 1938, over 12 000 ha of

XVI *The port wine country near Lamêgo in Alto Douro, Portugal*

dunes and a slightly greater area of mountain had been planted with pine. Since then about 10 000 ha have been planted or replanted annually and in normal years planting exceeds cutting. Forestry is taught in all the agricultural colleges and trees are provided free for public planting.

By far the most valuable forest product is cork of which Portugal produces half the world's supply (fig. 19). The trees are pruned to 13–15 m high and limited to three or four main limbs and each limb to two or three small branches so that long, wide lengths of cork can be cut. The cork oak flourishes especially south of the Serra da Estrêla where the summer drought is longer and more severe. The main cork-producing areas are the Tagus valley above Santarém, the Sorraia basin, the flat country between Setúbal and Évora, and the Serra de Grândola east of Sines. Usually the cork-oak stands are *montados* or cultivated lands but in the Serra de Grândola some of the woods are thick and rise above an undergrowth of evergreen shrubs. There have been great developments in western Alentejo in recent years owing to the investments by large companies in cork-oak estates. Nearly 600 000 ha are now under cork-oak woods in Portugal which is easily the world's chief producing area, averaging about 200 000 tons of cork annually in recent years. The bark is first cut when the tree is fifteen years old and then every subsequent ten years. Usually one hectare will support 175 trees and yields when mature about $2\frac{1}{2}$ tons of bark per cutting. The trees will go on producing for 150 years and the older woods, which regenerate naturally, are thinned out for charcoal. The cutting is done in the summer months and the cork sold in sheets, or as shavings and granules. The shavings are widely used for fruit packing and as insulating material, the granules for the making of linoleum and floor coverings but the moulded uses are many and ingenious. Over 300 factories are engaged in processing cork in Portalegre, Évora and Beja and the Portuguese output of cork products has in recent years averaged 350 000 tons. They provide about 18 per cent of Portugal's exports. The tree also provides acorns and the first stripping yields tannin for local pigskin industries.

The pine forests of northern Portugal and of the sand-dune littorals of Alentejo follow the cork oak in importance as a forest resource. The plantations cover 1·3 million ha and are one-third larger than those of France. The thinnings and prunings are used for fuel and considerable areas also provide grazing. The chief economic uses, however, are timber for pitprops, building, telegraph poles and railway sleepers. The exports are mainly of mine timber and amount to 180 000 tons annually. These pinewoods are also an important source of resin and turpentine especially on the sandy littoral west of Leiria where the trees are tapped regularly as on a rubber plantation. More than 100 factories in coastal Portugal deal with the distillation of the resin and between them produce in a good year over 90 000 tons of pitch and resin and 17 000 tons of turpentine, the bulk of which is exported (fig. 20).

The sweet chestnut flourishes mainly on the interior serras north of the Tagus, where it supplies staves for wine barrels, vine props, and fruit for human consumption. It and the trees mainly for paper-pulp are mentioned again in Chapters 7 and 8 where the great value of pannage for pigs is also stressed. The ilex covers vast areas in eastern Alentejo and the acorns from it and the cork oak (which necessarily yields a much smaller crop) feed about half Portugal's pigs.

Cultivated tree crops and vines

Cultivated tree crops and vines play an important part in the Portuguese agrarian economy and a still larger part in its landscape. The world and European importance of the country for these crops is discussed in Chaper 8, and here they will be described mainly from a national viewpoint. As would be expected the vine and olive are easily the chief although Portugal leads the world only for figs.

Vines. Portugal is one of Europe's great vineyards and the area under vines (about 4 per cent total land area or 356 000 ha) is only exceeded in France, Italy and Spain and its production of wine makes it fifth in the world. Port wine is one of the most valuable exports of the country and the Douro valley has been famous for port since 1654 when it became an 'English vineyard'. Vines grow almost anywhere in Portugal but especially on warm, well-drained soils on south-facing slopes (fig. 51). In normal years ninety per cent of the crop grows north of the Tagus and the remainder is mainly to be found around Setúbal, with smaller quantites around Évora and between Lagos and Tavira in the Algarve. These southern areas produce liqueurs and muscatels.

The Lisbon peninsula south of Santarém accounts for nearly half the wine output and most of the sunnier slopes near Lisbon, Santarém, Bombarral, Torres Vedras, Mafra and Sintra (Cintra) are under extensive vineyards. The majority produce white wine, some of which is distilled into brandy for addition to port. Some localities here and farther north have acquired reputations for high class wine and are legally recognized viticultural regions.

The vineyards of the Mondego valley producing the celebrated Dão wines have rigorously controlled limits from Viseu in the north to Gouveia and Oliveira do Hospital in the south. These fine table wines have to mature and although grown on small individual holdings are marketed by cooperatives.

The famous vineyards of the Douro valley which produce port wine commence 90 km upstream from Oporto and extend for 100 km on both sides of the river from Barqueiros to the Spanish frontier (fig. 51). Its legal limits are set by Vila Real, Alfandega da Fé, Barca d'Alva, Castelo Rodrigo, Meda Pasquiera and Lamego. Wine production within this area is controlled by a large-scale organization and is made into port at Vila Nova near Oporto. Only persons registered with the Association of Exporters of Douro Wines may export the finished product which must carry a certificate of origin. The whole of this middle section of the Douro and its tributaries is under vines which cover 26 250 ha or nearly 8 per cent of the total Portuguese vineyards. The yield is heaviest around Régoa where two-thirds of the crop is harvested.

The Douro valley is floored with schistose soils which are harder and more compact than elsewhere in Portugal but apparently being rich in clay and lime are fertile, friable, warm up quickly and are thought to give a special quality to the grape. The climatic conditions are also particularly favourable with a hot dry ripening period from June to October. The vineyards are held in walled estates (*quintas*) which include the holding of a proprietor and those dependent on him, such as bailiff, steward, accountant, household manager as well as the labourers. Their maintenance and management calls for specialized and incessant work. All the slopes are terraced and held by retaining walls. Each terrace is usually made wide enough for about nine rows of vines, and each stock is planted in a hole about one metre deep. The vines grow up the slopes and so receive the maximum sunshine. In the third year they are arched over and tied to a supporting stake or stakes made usually of a granite splinter but on many quintas wires are strung between the granite uprights. The plants are sprayed for rust and mildew and in September and October the grapes are harvested by large numbers of pickers, many of whom are migrants. The pressing is still done by treading by the men who work all night in shifts, the women picking all day, and the men carrying the laden baskets to the vats. The grape juice is then, after treatment with brandy, sent down to Vila Nova in flat-bottomed boats (*rabelos*) on the Douro, or by rail. At Vila Nova more brandy is added and the wines carefully blended and stored for 15 to 20 years. Many of the large 'caves' belong to English wine merchants and the whole industry has been developed under English management and capital since 1703. About half a million hectolitres of wine are produced here each year. The exports in 1966 were 314 000 hectolitres.

The Lower Douro and the Minho valley produce the well-known *vinhos verdes* which are green in colour and sparkling and piquant, but poor travellers. The wine is the result of a terrain deficient in lime and a climate which is rather more humid than other Portuguese vinegrowing areas. The vines are grown on a lofty trellis work and up every tree, and drainage ditches are dug to mitigate the effects of too much ground water. They are produced on tiny family farms usually in association with minute plots of cereals. The grapes are harvested in mid-September before being properly ripe and trodden for juice in large stone troughs. The best quality vinho verde comes from the Minho valley, and from the vineyards adjoining the port wine region. The output is four times as large as that of port.

The total Portuguese wine harvest varies appreciably with the weather and has in recent years ranged from nearly 15 million to 9 million hectolitres. These annual variations are still more characteristic of the olive, the chief rival of the vine.

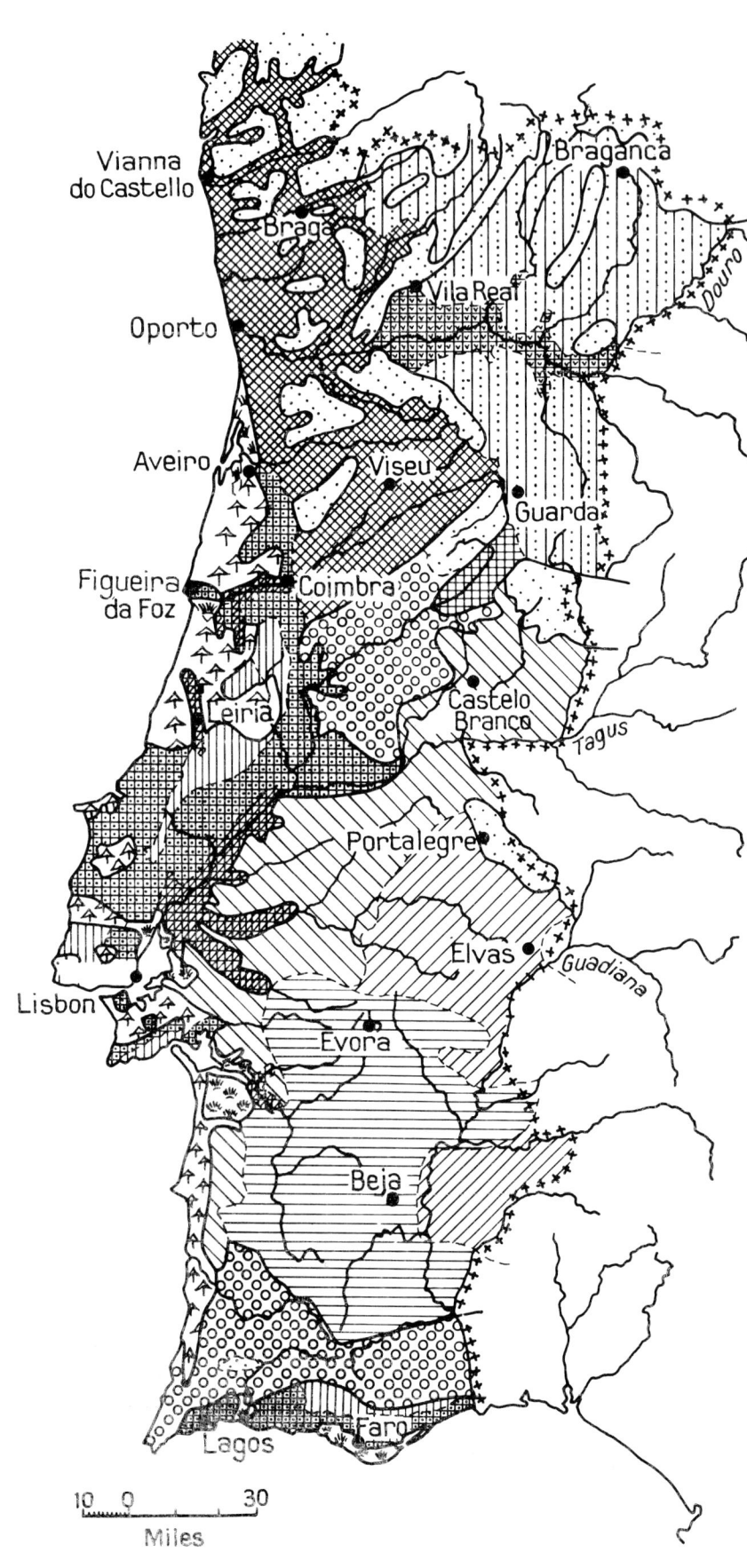

152   Part II: Western Peninsular and Alpine Lands

1. *Intensive cultivation of maize, beans, rye, and vine. Forests of pine and deciduous oak.*
2. *Olive and vine in lowlands; potatoes and cereals (mainly rye) on uplands. Woods of sweet chestnut and evergreen oak.*
3. *Predominantly vine-growing, with fruits and rye in west.*
4. *Intensively cultivated alluvial soils. Much rice, maize, beans, and oats.*
5. *Intensive cultivation of cereals, maize, and fruit. Large vineyards and olive groves. Pines near coast and cork oak inland.*
6. *Chalk hill-lands; stunted oak-growth with small cultivation of cereals and some fruit. Fig and carob in Algarve foothills.*
7. *Highland of granite and gneiss; oak and stunted shrub-growth in west, sweet chestnut and low growth of lavender, sage, and steppe in east.*
8. *Schistose mountains with large expanses of heath and* cistus *matorral*
9. *Sandy areas; largely under pine and heath with sand dunes near coast.*
10. *Salt marsh and unutilized coastal areas.*
11. *Extensive cultivation of wheat, beans, oats, rye, and maize, much olive, vine important locally. Large areas under cork oak and matorral*
12. *Plateau and uplands of north Alentejo. Cork oak and evergreen oak are abundant. Extensive cultivation of cereals and beans, olives frequent.*
13. *The lowlands of south Alentejo. Much wheat, olive, and beans. Cork oak and evergreen oak abound; heath and rough pasture common on foothills to west and south.*

51 *Vegetation and land utilization of Portugal* (after Lautensach)

Olive. Olive groves occupy 9 per cent of the total arable land or 3·7 per cent of the total land area of Portugal. The trees yield on average about 700 000 hectolitres of oil annually but the yield has varied recently from 1 million to 414 000 hectolitres. Olive oil presses only exist on the larger estates. The tree will grow in dry areas and on extremely poor soils but the best crops come from the fertile plains around Santarém, where it grows more freely and taller than in any other Mediterranean area (see Chapter 8).

The chief producing areas are in central and eastern Alentejo, northern Santarém and northern Estremadura, the slopes of Beira Alta and the upper Douro valley while extensive olive groves surround most of the towns in the valley of the Tagus. The oil is of a distinctive flavour, but the quality varies, and so, for the sardine-packing industry a few thousand tons of Spanish olive oil has to be imported annually. The home crop is almost all consumed in Portugal except for about 3000 tons exported to Brazil and to the Portuguese colonies.

Other tree crops. Fruits other than grape and olive are grown on 320 000 ha throughout Portugal either planted in orchards intercropped with vegetables or with the trees outlining the fields. They include stone varieties, such as peaches, apricots, plums and cherries including morellos, and pipped varieties such as apples, quinces, medlars, and the finest pears in Europe. Oranges which are also grown tend to be more numerous on the west, while dates and bananas (for the home market) are limited to the Algarve. The production of figs is quite remarkable (see Chapter 8). They grow in small clumps in the Douro valley and cover large areas farther south near Torres Novas and in the Algarve. The coarser varieties are fed to pigs but many are used for distilling alcohol and the higher class fruits for processing, drying and preserves, as for example at Faro and Lagos.

Of the nut tree crops the almond is common from Coimbra southwards but is most numerous in the Algarve. About 2000 tons are exported annually. The carob (20 000 ha) also is most abundant in the Algarve and is used partly for export and partly for human and domestic consumption at home.

The dried fruit and nut industry is strictly controlled by government regulations and for export the products have to reach a high standard of excellence. Food industries based on fruits are widespread throughout Portugal. Lisbon, Alcobaça and Caldas da Rainha have quince marmalade factories; Elvas is famous for dried and conserved plums (and apricots); raisins are produced at Alpiarca in Santarém province; citrons and mandarin oranges are packed at Setúbal; pears are handled at Alcobaça, Aljubarrota and Viseu, and peaches at Sedouim. Most of these soft-fruit products go to the Portuguese colonies and Brazil, except the figs, which like the nuts, find a ready market in the Christmas-keeping countries of western Europe.

Arable crops

Land devoted to purely arable crops covers about 30 per cent of Portugal and is given over mainly to cereals and vegetables. The chief cereals in order of output are wheat, maize, rye, oats, rice and barley, but wheat and maize have by far the largest acreage and yield (see Chapter 8). These cereals are grown throughout Portugal but the climatic differences between the maritime coast and the drier interior and between the cool north and warmer south favour certain distributions. For example, maize predominates in the Minho and Douro area; Beira Alta and Beira Baixa are better suited to rye, and the Alentejo to wheat and oats.

Cereal yields are very low compared with those of northern Europe and fluctuate so much that the country's chief need is to increase the cereal yield per hectare and to stabilize crop production. Portugal in normal years produces enough wheat to be selfsupporting but in bad years enough only for about half the domestic needs.

About three-quarters of the wheat is grown south of the Tagus in the 'bread basket' of Portugal, a belt of rolling country from Portalegre to Beja (fig. 51). The bulk of the remainder comes from the northern bank of the Tagus in Estremadura between Santarém and Lisbon. Wheat is a common crop on the lower slopes of the interior serras of Tras-os-Montes, Beira Alta and Beira Baixa around the towns of Bragança, Guarda and Castelo Branco but is practically absent along the coast. To ensure against a fluctuating national output, grain elevators have been built in various parts to store surpluses of good years to offset the lean years.

Maize is the predominant cereal on the coastal plain north of the Mondego valley where the climate is more humid and soils are deeper and more fertile. Here three-quarters of the national crop is produced. Maize is still important in the Tagus basin around Santarém but south of the Tagus it plays little part in the agricultural economy. In dry years, when the cereal harvest may be seriously short of moisture, the maize crop is irrigated wherever possible and consequently its annual yields are relatively steady. The following statistics in 1000 hectares and 1000 tons demonstrate a year (1966) that climatically was very unfavourable to wheat, rye, oats and barley.

	1965		1966		1967	
	Area	Yield	Area	Yield	Area	Yield
Wheat	628	612	523	312	586	637
Maize	484	459	473	565	436	577

The drier parts of north and northeastern Portugal produce three-quarters of the rye crop. It is also important in the neighbourhood of Portalegre south of the Tagus but elsewhere is restricted mainly to the higher, drier and more exposed serras. About 240 000 hectares in all are under rye, producing on an average 180 000 tons a year and supplying 10 per cent of the population with their main bread grain. Barley and oat crops tend to coincide with wheatlands and therefore occur mainly in the south where Alentejo produces 90 per cent of the oat harvest, and most of the barley.

Rice has been grown in Portugal for 700 years and is a favourite dish of the people but up till 1940 only half the domestic rice needs were met from homegrown supplies. Since that date under rice-growing cooperatives a surplus of rice has been produced annually.

The rice fields are restricted to floodable alluvial soil in, for example, the basins of the Vouga and Mondego rivers around Aveiro and Figueira da Foz. The Vouga district accounts for 4000 tons and the Mondego district for 10 000 tons. Rice growing has also been extended farther to the south around Leiria, but the chief rice-growing areas today are in the valley of the Sorraia and along the Tagus and into the Setúbal peninsula where nearly three-quarters of the Portuguese rice is produced. The yields are high (179 000 tons from 38 000 hectares in 1964 and 1969) and moreover the crop covers land that would otherwise be almost useless, and provides a great deal of seasonal employment. The bulk of the rice is decorticated in several large factories in and near Lisbon, and the remainder in small factories, windmills and watermills near the paddy.

The agriculture of Portugal is biased toward wheat and vine rather than on salad and vegetable crops but the government has made great efforts to stimulate horticulture and market gardening. The chief vegetable crops are grown in polycultures with orchards and tree crops or with vines. They are temperate vegetables, such as peas, beans, potatoes and carrots but in addition melons, pumpkins, tomatoes, peppers, and cucumbers are grown and a proliferation of pulse crops, such as haricot and butter beans, broad beans and chick peas. Potatoes, which occupy 100 000 hectares and yield 1 million tons annually, are especially important in the wetter north, but the most remarkable vegetable crop is the bean family and especially French and haricot beans north of the Tagus and broad beans south of it. No country in Europe has such a large area (over 400 000 hectares) of haricots although the Romanians and Yugoslavs grow them widely mixed with other crops. Other species of cultivated beans and pulse are also popular in Portugal. These crops are leguminous and nitrogen-fixing and their products are rich in protein. Almost every meal in Portugal is in more senses than one a bean feast.

Pastures and animal husbandry

Natural pastures cover only about 6 per cent of Portugal and play a relatively small role in its agricultural economy even for a Mediterranean country. In the interior serras of northern Portugal and in the drier areas of the Tagus, matorral, heath, steppe and semi-arid pastures provide transhumant grazing for horses, mules, donkeys, goats, sheep and cattle. Yet livestock play a considerable part in Portuguese agriculture because the farming is not mechanized, and maize, oats and acorns are used for animal fodder, and stubble for grazing. There are all told about 1 million cattle, 5 million sheep, 600 000 goats, $1\frac{1}{2}$ million pigs and nearly $\frac{1}{2}$ million donkeys, mules and horses. The sheep and goats provide the dairy produce, the ox is the draught and plough animal, the donkey the beast of burden, and the pig the source of meat.

Cattle are most numerous in the north, sheep and goats in the south, while pigs are ubiquitous. The bulk of the livestock breeders are peasant proprietors who cannot afford to improve their stock. For this reason the government has set up animal research stations with the object of improving the strains.

For animal husbandry or livestock farming five areas are outstanding.

1 The coastal lowlands and valleys north of the Tagus where half the country's cattle and one-third of the pigs are bred. The cattle consist mainly of working oxen but dairy cattle are found in the river valleys and in the vicinity of the towns, such as Oporto and Braga. The Minho valley shelters the native Barrosa breed and the Vouga valley the Aroucesa. During the summer they are pastured in the Mondego valley. Butter is made in Oporto and Aveiro but cheese is usually made of goats' milk.

2 The uplands of Tras-os-Montes and Beira support one-third of the sheep and one-quarter of the goats of Portugal. The sheep are grazed on stubble and on fallow lands to manure them for the following year, so in general are kept on the higher ground in summer and brought down to the cultivated valley fields in the autumn. Their chief product is milk for cheese which is usually produced cooperatively.

3 Central Portugal, particularly the Tagus plain. For livestock farming this area is transitional between the pluviose north and the drier south and has a wide variety of farm animals. It is the chief breeding region in Portugal for donkeys and horses. Frisian cattle are kept in the Lisbon neighbourhood being fed on peas, fig leaves and ensilage, often in stalls. About one million crossbreed sheep feed on the autumn stubble and the Tagus pastures, and Castelo Branco province is a leading producer of merino type wool. Farther east, and on the Serra da Estrêla in summer, sheep and goats are pastured for milk and cheese production, the chief markets for which are at Lisbon and Tomar.

4 The Alentejo. The dry summers of this large region limit livestock keeping largely to sheep who graze the stubble, fallow and mountain matorral, and pigs that find abundant pannage from the oak trees. One-third of Portugal's sheep are bred here and Alentejo leads for cheese production and for wool (40 per cent national total). Unfortunately many of the fleeces are grey or black. About one-quarter of the nation's pigs are kept here, every large estate having a sizeable herd which after harvest, or all the year on fallow, roams beneath the fruit trees and oaks. In addition Alentejo is the chief Portuguese breeding-ground for the mule, a useful, sterile beast which stands up well to the hot climate.

5 The Algarve. The climate of this semi-arid region restricts the common livestock mainly to sheep and goats, both of which find rough grazing on the mountains, and to pigs which can forage beneath fruit trees and oaks or anywhere else. A few mules and donkeys serve as working beasts but horses and cattle are notably absent.

Intensification of agricultural production

The economy of Portugal is dominantly agro–forestal. About one-quarter of the nation's income (gross domestic product) comes from agriculture and over 40 per cent of the working population are engaged in it. Yet farm facilities are in the main primitive, yields are low and efficient marketing methods and adequate transport, fertilizers, machinery and food-processing plants are all too often lacking. The north shows abundant signs of rural overpopulation with the resultant splintering of the farmland into tiny, uneconomic units, while in the south the land holdings are largely in latifundia cultivated under sharecropping systems and with seasonal unemployment.

In the lower parts of the north the average farm is under 2 ha and is usually managed by the owner with the unpaid help of his family. In the south farms average over 20 ha in size and are 36 ha in Alentejo. Here a relatively few big landowners employ a relatively large paid labour force and in Beja, for example, hired workers, fulltime and casual, comprise 86 per cent of the agricultural employees. This farm-size distribution is obviously connected with natural conditions, especially water supply, climate and topography but it was strengthened by historical events in which, as no doubt also in Spain, the Moorish occupation drove the pre-existing population northward to the moister and more mountainous tracts. Later to encourage reconquest the Christian kings granted to the nobles large estates to take and to hold much as happened in the marchlands between England and Wales.

The systems and size of land holdings are of great significance in modern Portugal. Of 800 000 farms nearly two-thirds are owner-occupied and most of the remainder have some form of sharecropping tenancy. About half of the total farms have less than 1 ha of arable land and cover only 4·3 per cent of the arable area (fig. 52). About 88 per cent of all the farms are under 5 ha and occupy together only 23 per cent of the total arable land. But many of these small properties are further subdivided and nearly 40 per cent of all farms in the humid north have more than 6 land parcels. Here minifundia are supreme.

Yet latifundia form a more serious problem, in fact more serious even than in Spain. Less than 1 per cent of all farms occupy half of the arable land and 0·2 per cent occupy 39 per cent of the agricultural area. As said, large estates predominate in the south except in coastal Algarve. So also do large fields, and on the Alentejo uplands about 90 per cent of all farms have less than 6 land parcels.

The table of agricultural statistics, although relating to some years back, is still reliable and presents a statistical picture of the northern minifundia and southern latifundia. The districts of Lisbon, Santarém and Castelo Branco include parts of the Central Serra. The first belongs to the pluviose north; the second, Santarém, has exceptionally wide floodplains; Castelo Branco is clearly transitional and although north of the Tagus is mainly south of the Central Serra and has on its lowland a cereal–cork oak economy reminiscent of that of Alentejo. This table can be applied to all the regional accounts of Portugal and confirms statistically the visitor's impression of leafy miniscule plots in the north redolent of earthy smells, and of sweeping sun-drenched parklands in the south resonant with the sounds of animals seeking forage and shade.

The pressure of population on the land is grievous in the north while in the south the land system oppresses the people. The great hope is for better yields, for more irrigation in the drier and less fertile tracts, and for extension of arable land. Portuguese authorities claim that although 15 per cent of the country consists of heath, matorral, swamp, marsh or dune, much of this is capable of reclamation and only 5 per cent of the country is classed as worthless. At the present time, however, one-fifth of the country makes little contribution to the agricultural production.

To within the last few decades, irrigation has been more related to the ease of obtaining water supplies than to the need to offset drought. In 1963 about 620 000 ha had some form of irrigation, and of this 516 000 ha were for arable and horticultural crops; 65 000 for other orchards and vineyards; 17 000 for permanent grass; and 22 000 for olives. The table shows the percentage of the farmland irrigated in each district in the early 1950s. Clearly the northern farmers made much use of supplementary watering whereas in the south, if the Tagus floodplains and coastal Algarve are omitted, irrigation was relatively unimportant. Since then several irrigation schemes have been undertaken. Some are truly multi-purpose. In the north they always involve flood control and valley-floor drainage because of the heavy winter rainfall; in the south irrigation and flood control become the prime objectives. Schemes covering 100 000 ha have already been approved by the government in the valleys of the Sorraia, Mondego, lower Tagus and Idanha. In the first three areas, power

52 *Farm sizes in Portugal*
  Total number of farms was 801 200, and total arable area 4 115 000 hectares

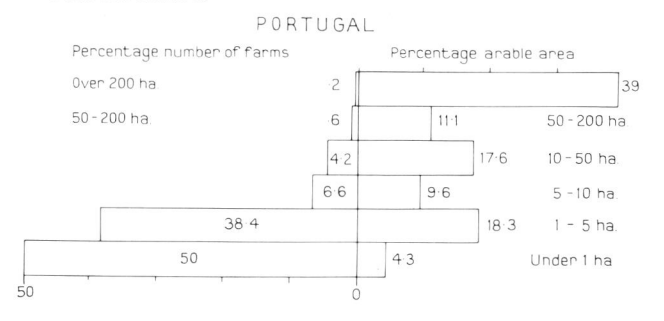

Percentage number of farms	Farm size	Percentage arable area
2	Over 200 ha	39
6	50 - 200 ha	11·1
4·2	10 - 50 ha	17·6
6·6	5 - 10 ha	9·6
38·4	1 - 5 ha	18·3
50	Under 1 ha	4·3

PORTUGAL: AGRICULTURAL WORKERS AND FARM HOLDINGS (by districts)

District	Agricultural workers[1]	Per cent total working population	Number of farms[2]	Farmland per farm (hectares)	Cattle	Sheep	Goats	Pigs	Per cent farmland irrigated average 1951–6
Viana	60	60	46	1·4	2·4	4·0	5·3	1·4	68
Braga	86	41	56	1·8	3·5	4·0	4·0	1·7	79
Porto	73	16	59	1·6	3·3	3·0	2·4	1·8	75
Vila Real	87	75	47	3·4	3·0	10·8	6·9	2·0	20
Bragança	62	75	37	12·6	3·5	79·7	11·8	2·2	2
Aveiro	65	35	62	1·5	2·4	3·1	3·1	2·1	57
Coimbra	77	50	76	1·9	1·8	4·5	2·5	1·7	37
Leiria	80	55	66	2·5	2·0	3·7	2·4	2·2	17
Viseu	117	70	89	2·2	2·2	6·1	2·4	1·8	40
Guarda	65	68	54	6·4	2·1	34·0	3·7	1·6	11
Santarém	95	55	67	5·5	4·0	8·8	2·7	2·1	11
Lisboa	77	13	49	4·0	3·4	15·9	3·6	1·8	4
Setúbal	46	30	14	19·4	5·4	57·1	7·1	7·8	6
Castelo Branco	65	57	47	7·7	2·4	32·4	6·0	1·9	10
Portalegre	49	65	16	21·6	9·1	121·4	11·5	10·7	4
Évora	53	61	11	36·6	13·3	131·1	9·5	30·5	2
Beja	77	71	20	36·2	6·9	77·6	10·4	18·1	1
Faro	63	50	38	9·3	3·0	9·2	4·5	2·2	5
Portugal	1297	42	854	5·7	2·9	13·2	3·8	2·6	

[1] Active agricultural population in 1960, in thousands
[2] Number of farms in 1952–4, in thousands

stations, drainage, irrigation and flood control were involved while the last was purely an irrigation project.

These schemes in fact were not designed to reclaim new land so much as to intensify production in land already under cultivation. They were intended to increase the area under maize, rice and fruits and to improve the grazing. Two are already completed: that designed to control flooding and to irrigate the lower Sado valley in western Alentejo by means of a dam across the Rio Xarrama in Vale de Gaio, and the Idanha scheme which dams the Rio Ponsul at Cabeco Monteiro. The large schemes on the Sorraia, Mondego and Tagus, although implemented, are not yet complete but will affect 55 000 ha as well as increasing power supplies. These land improvement and reclamation schemes when in full operation will provide employment for an extra 100 000 workers. Two small additional government schemes are the drainage and irrigation of 450 ha on the south bank of the Tagus east of Abrantes, and the flood control of the river Alcoa. The privately owned Companhia das Lezirias has undertaken the drainage and irrigation of 520 hectares of low-lying floodable land near Salvaterra in Santarém.

Fishing

An alternative source of food supplies and employment lies in the fishing so opportunely presented along 850 km of Atlantic coastline. About 36 000 persons are engaged full-time at fishing for a livelihood, exclusive of those who fish as a part-time occupation, or who earn a living in the sardine canneries. In 1969 the full-time fishermen operated the industry in 3000 motor boats and 13 000 sailing and rowing vessels representing a total of 178 000 tons.

In a normal year half the value of the catch is provided by sardines followed by hake (10 per cent), cod (9 per cent), mackerel, bream and tunny (3 per cent). The sardines also form the greatest weight of fish caught.

The Portuguese fisheries are classed under inland, shore, coastal and deep-sea. The inland fisheries are confined to the rivers of the north and centre. The famous trout rivers are the Cávado, Mondego, and Zézere, but trout are also caught in the tributaries of the Minho as well as in the mountain streams of Tras-os-Montes and Beira. The Minho is a salmon river (300–400 is the catch in a normal run). Shad and lampreys are taken by seine nets in the Tagus and the lower courses of the northern rivers.

The shore fisheries are conducted in the coastal lagoons which in many cases are freshwater and yield the same types of fish as the Portuguese rivers. Cuttle fish and octopus are caught in considerable quantities and sardines and small fry are netted at various times, the seines being dragged by men and oxen. This is especially important at Palheiros de Mira where sardine fishing goes on from May to December and then the men take temporary agricultural work in the Alentejo. The shore fisheries also are responsible for the large harvest of shellfish which forms 2·5 per cent of the total catch. They include mussels, cockles, crawfish, crabs, lobsters, shrimps and oysters which are packed for export, at, for example, Viana, Lisbon, Peniche and Sines. Portugal also has some of the world's finest oyster beds of which the largest, in the estuary of the Tagus, covers 22 000 hectares.

The longshore fisheries are usually divided into coastal or inshore (within territorial waters, or 20 km off the coast) and deep-sea, in the more distant waters. The continental shelf off Portugal is very narrow and natural harbours are few but this is compensated for by the great variety of fish caught. The most important are the sardine and the tunny. Sardines have been landed in Portugal for 800 years but their importance dates from 1880 when the French founded a sardine cannery at Setúbal. Today sardines are landed at 28 ports and fishing villages in Portugal and 14 of these depend almost entirely upon this fish. The main ports are Setúbal, Matosinhos, Olhão, Portimão, Peniche, Lisbon, Lagos, Aveiro and Vila Real de Santo António. All kinds of craft and all kinds of nets are in use in the sardine fishing which is done mainly at night. The annual catch averages 130 000 tons and is sold at the ports to local canneries, where the fish are beheaded, gutted, graded, soaked in brine, cooked, dipped in boiling oil and finally canned and sterilized. Portugal is the world's leading exporter of tinned sardines selling 60 000 tons a year (cf. Spain, 16 000 tons). It is, after cork and wine, the most valuable single export and, like wine, the grading and quality are carefully controlled; and the fisheries conserved during certain seasons of the year. Other species of fish caught are mainly consumed in Portugal but the tunny and tunas are landed in sufficient numbers to leave a surplus for export. The tunny migrate along the coast of the Algarve from May till August every year, and are mainly caught in nets that stretch sometimes for more than three kilometres from the shore. The fishing and conserving of tunny are largely confined to the Algarve, the chief ports being Vila Real de Santo António, Tavira, Cezimbra, Setúbal, Faro and Lagos. About 10 000 tons of canned tunny are exported every year but much of this consists of species of tuna from farther out in the Atlantic (see Chapter 3).

The deep-sea fisheries have long been important to Portugal. Portuguese fishermen have been sailing to the Grand Banks, Newfoundland, since the early sixteenth century and in 1550 about 150 boats used to set out every year from Aveiro. This port and Figueira da Foz are still the most important ports engaging in cod fishing, although Lisbon, Oporto and Viana do Castelo send out fleets today. In 1968 about 10 000 fishermen were employed in the Portuguese cod industry and the catch was 80 000 tons. About 50 000 tons of cod, hake and haddock are dried or cured each year and dried cod (*bacalhau*) is almost a staple Portuguese diet. Other details of the Portuguese fish catch, which exceeded 500 000 tons in 1966, are given in Chapter 3. The associated industry of boatbuilding affords additional employment at each port, as the fishing fleet needs replacements of about 800 small boats annually.

Power and mineral resources

The economy of Portugal could be transformed if adequate power resources were available. Coal is present in a few places but the deposits are small and variable in quality. The two chief fields are near Vila da Ingreja and Pejão in the Douro valley about 30 km from Oporto. These produce all told 420 000 tons annually or 95 per cent of the Portuguese output. A little lignite is obtained in the Leiria district and small quantities of bituminous coal in the Alcáçovas valley in Setúbal province. However, in a normal year, the bulk of Portugal's coal supplies are derived from imports (1 million tons), mainly from Britain, West Germany and the United States.

The generation of hydro-electricity is hindered by the irregular regimes of the rivers, and south of the Tagus also by the very low water in summer. Nevertheless the government has encouraged water-power generation wherever possible and has constructed numerous storage dams and supplemented the hydro-stations by thermal plants and by a nuclear generator at Sacavem north of Lisbon. The largest thermal stations are on the banks of the lower Tagus near Lisbon and near Setúbal and Oporto, the chief power-consuming areas.

In 1966, of the installed capacity of 2·3 million kW about 1·8 million kW were water-powered (see Chapter 10). In 1968 of the total national electricity output of 6215 million kWh, 84 per cent was generated by water power. There are numerous small hydro-electric plants but the chief stations are on the Zézere, a right-bank tributary of the Tagus near Lisbon, and on the Tâmega, a right-bank tributary of the Douro near Oporto. These two rivers are stepped with barrages and the Zézere has a notable series of large reservoirs. Other less elaborate schemes exist, for example, on the Lima in north Portugal and on the Rio Ponsul in Castelo Branco province. The large barrages and lakes in Alentejo on the Sado and Sorraia rivers have an appreciable potential but are intended primarily for irrigation and flood control. The latest project is on the Douro just above Oporto where the river has cut a deep trench through crystalline rocks. Here in 1971, a dam 55 m high was completed at Carrapatello and when full will allow navigation upstream almost to the Spanish frontier, and will, it is anticipated, provide cheap transport for iron ore from Moncorvo and coal from Pejão. The installed turbine capacity of 200 000 kW will be a welcome addition to the power supplies of the Oporto region.

As yet the only finds of oil or natural gas in Portugal have been small quantities near Coimbra so the domestic supplies are imported. The metallic mineral resources of the country do little financially to offset the general lack of power supplies. However, the geological surveys are by no means completed and the poor communications in remote areas militate against the commercial exploitation of most of the known deposits. Small quantities of tin ores and of high-grade haematite are mined in Bragança district near Tôrre de Moncorvo in the Douro valley. Tin ores are widespread in the Central Serra and mined in small quantities around Guarda, Castelo Branco and Viseu. Wolfram, for which Portugal was for long Europe's chief producer, also comes, now in very small quantities, mainly from the northern parts of the Serra da Estrêla around Panasquera near Guarda and from the plateau north of the Douro around Borralha near Vila Real. The chief mineral wealth of southern Portugal is copper ore, an extension of the richly mineralized Sierra Morena in Spain. Mining is most active around Aljustrel and Beja, and the production, with the aid of foreign capital, has risen to about 600 000 tons of ore. The bulk is exported through Setúbal but small amounts are used at home in the manufacture of heavy acids.

Manufacturing industries

In 1960, manufacturing provided about 38 per cent of the national income (gross domestic product) of Portugal and employed 21 per cent of the working population. The metallurgical industries based on the mineral and power resources described above are small and of national importance only. Yet a great effort has been made to supply the bulk of the country's iron and steel requirements. In 1961 an integrated iron and steel industry, *Siderurgia Nacional*, was established at Seixal opposite Lisbon on the banks of the Tagus where the tributary Coina enters and provides extra water frontage. The plan was to build 4 blast furnaces and produce 1 million tons of pig iron a year by 1972. At present in production there is 1 blast furnace (annual capacity of 229,000 tons of pig iron), 1 Linz–Donauwitz converter (annual capacity of 235 000 tons of steel) and 1 electric furnace capable of producing 40 000 tons of finished steel products annually.

In 1961 the pig iron produced amounted to 86 000 tons, the steel to 92 000 tons. In 1965 the second converter was built and the first plate mill assembled and between 50 per cent and 80 per cent of all Portuguese imports of iron and steel were replaced by home products. In 1966 the home production of steel ingots was nearly 260 000 tons.

The reserves of iron ore, although of 42 per cent iron content, are only sufficient to last a few more years and, moreover, the coke has to be imported from the U.S. Power for the furnaces is mainly hydro but the uneven regimes of the rivers has necessitated the building of a thermal station on the Seixal site. The water consumption of the project is also very high, and Seixal, which uses more water than the city of Lisbon, takes water from the Tagus and from deep wells. Labour is cheap and plentiful and the technical training and assistance were provided by the foreign firms that built the converters. The industry is well sited near the main market, Lisbon, and is admirably placed with regard to cheap water transport to other coastal situations but both raw materials and finished goods have to be discharged from freighter to barge and vice versa to enter or leave the Coina river where the works have their wharves.

Despite the great cost to the country of establishing and fostering the industry the government are convinced that it is worthwhile since besides saving foreign exchange it is valuable as a growth industry for attracting auxiliary manufactures, providing technical instruction and raising the standard of living.

Older iron and steel industries exist in Lisbon and Oporto for the making of armaments, agricultural equipment, tools, hardware and so on. A recent development at Lisbon is the manufacture of textile machinery. The Seixal plant also supplies steel to two small shipbuilding yards.

The Portuguese chemical industry is based partly on salt produced in coastal evaporating pans. There is a modern well-equipped factory on the south bank of the Tagus at Barreiro, which produces sulphuric acid, superphosphates (fertilizer) and copper sulphate for vine spraying.

The pottery industry also depends partly on home-produced raw materials as kaolin is quarried (42 000 tons in 1968) in several places around the granite masses of the Douro plateau. Glazed ceramic tiles (*azulejos*), first introduced into Portugal by the Moors, are still made on the outskirts of Lisbon. Most of the larger towns produce artistic pottery and there are more important manufactures of industrial and electrical porcelain. A notable glassworks, based on local supplies of silica sands, has been established at Leiria for the making of bottles and flasks.

The remainder of Portuguese manufacturing and processing industry largely concerns primary products of farming, forestry and fishing. The factories have a strong family bias and are concerned primarily with consumer goods such as textiles, leather, food and paper. Many have a local arts and craft flavour and have not wrenched the employees out of the agricultural society. Most belong to the twentieth century and have not greatly stimulated rapid urbanization. By far the chief are textiles taken as a whole and including garment manufacture. There are over 700 small factories divided among the chief towns and these are supplemented by domestic work. For example, the villagers of Arraiolos near Évora make carpets in their homes.

Cotton manufacture leads the textile branches and 200 small spinning and weaving mills, centred mainly on

Oporto, provide, with the aid of imported colonial cotton, surplus yarn and fabrics to send back to the colonies. Guimarâes is noted for thread. Woollen textiles are widespread and employ 8000 operatives, mainly at Oporto, Coimbra, Lisbon, Évora and Portalegre. The long coarse wool (*churra*) of the north goes chiefly for rugs, carpets and mattresses while the finer merino type of the south goes largely for woollen cloths. Linen is manufactured mainly at Guimarâes, jute at Lisbon and garments are made up chiefly at Lisbon and Oporto. The exports of textile yarn and fabrics together far exceed those of any other single product and go largely to the colonies.

The chief food exports are fish and fish preparations which consist chiefly of sardines (nearly 2 million tins annually) processed at Setúbal, Matosinhos, Olhão, Portimão, Vila Real de Santo António and Lagos. Tunny preserving is almost confined to Lagos, Portimão and Albufeira in the Algarve.

The forestry products have already been discussed in Chapter 7. The output of paper pulp (217 000 tons in 1966) has doubled in the last five years, and is now a leading export. Cork and cork products rival wine as the chief

53 *Portugal: provinces and provincial capitals*

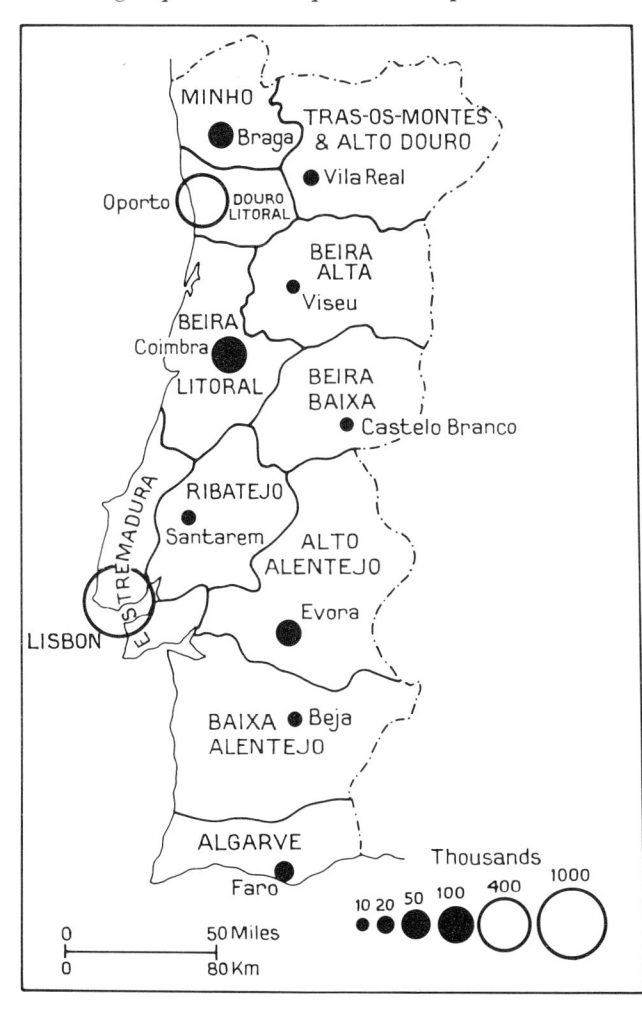

PORTUGAL: AREA AND POPULATION IN DISTRICTS

District	Area (sq km)	Population (in thousands) 1960	1968	Density (per sq km) 1968
Viana do Castelo	2 108	278	286	136
Braga	2 730	597	659	241
Porto	2 282	1 193	1 353	593
Vila Real	4 239	325	341	81
Bragança	6 545	233	245	38
Aveiro	2 708	525	578	214
Coimbra	3 956	434	444	112
Leiria	3 516	405	426	121
Viseu	5 019	482	490	98
Guarda	5 496	283	272	50
Santarém	6 689	462	479	72
Lisboa	2 762	1 383	1 565	567
Setúbal	5 152	377	434	84
Castelo Branco	6 704	317	321	48
Portalegre	5 889	188	185	31
Évora	7 393	220	226	31
Beja	10 240	277	275	27
Faro	5 072	315	328	65
Total	88 500	8 293	8 907	101

export to foreign countries and very seldom fail to exceed it.

Main towns and population pattern

Urbanization has been slow in Portugal where in spite of a national population of about 9 million, only 2 towns exceed 100 000 and only another 4 are over 40 000. In 1833, the country was divided administratively into 18 provinces and a century later, with the aid of a geographical adviser, these were readjusted into the following 11 provinces, the capitals of which are given in brackets and shown in figure 53.

Minho (Braga)	Beira Baixa (Castelo Branco)
Douro Litoral (Oporto)	Estremadura (Lisbon)
Tras-os-Montes and	Ribatejo (Santarém)
Alto Douro (Vila Real)	Alto Alentejo (Évora)
Beira Litoral (Coïmbra)	Baixo Alentejo (Beja)
Beira Alta (Viseu)	Algarve (Faro)

The employment afforded by the provincial functions and marketing may be judged from the population, in 1966, of Évora (26 000) and Faro (20 000). The chief towns are near the western seaboard where manufactures and commerce have also developed.

Lisbon (pop. 1 564 000 in the administrative district) is the capital city and financial centre. An ancient foundation,

XVII *Central Lisbon from the air showing late eighteenth-century planning*

it grew on a site of seven hills, 15 km up the Tagus on the north bank where a deep channel leads into a vast tidal lake which is in process of being silted up by the Tagus and the Sorraia. The estuary is one of the finest natural harbours in the world, capable of accommodating the largest vessels and is well-equipped with shipbuilding and repair yards and docks for over 9 km of its length. The river, 2 000 m wide, is bridged here by the longest suspension bridge in Europe and the fourth longest in the world.

The city was shattered by the earthquake of 1775 and was subsequently beautifully planned and rebuilt by Pombal. It stands in a commanding situation in the midst of a fertile agricultural hinterland and has developed in the factory suburbs of Belém, Barreiro and Alferrarede a wide variety of industries. Today small family businesses are merging into large combines as the industries become more sophisticated in lay-out and output. The chief exports are cork, wine and petroleum products, the last-named based on large imports for local refineries.

Lisbon is also a port of call for boats and passengers on the Atlantic run and its excellent sea connections, as the capital of a large overseas empire, have made it the most important collecting and distributing centre in Portugal, as well as giving it a large entrepôt trade in oil seeds, tomatoes, and in hides and skins. It was an obvious choice for a NATO port. It is in addition one of the home ports for the cod-fishing fleet to the Grand Banks and is served by the airport of Portela.

Oporto, the second city of Portugal, stands on the north bank of the Douro river about 5½ km from the sea. It has a population of 340 000 with a total of 800 000 persons in its vicinity and 1 353 000 in its administrative district. It is Roman in origin, having grown up on the site of the old *Porto Cale*. Owing to the dangerous sandbar at the mouth

PORTUGAL: CHIEF PORTS, COMMERCIAL SHIPPING MOVEMENTS, 1968

Port	Entering Ships (number)	G.T. (thousands)	Passengers (number)	Cargo (thousands of tons)	Leaving Passengers (number) In transit	Disembarked	Cargo (thousands of tons)
Lisbon	5 777	28 432	84 403	5 995	262 854	75 950	2 016
Douro (Oporto)	654	373	4	306	10	8	118
Leixões	3 075	5 898	869	1 629	4 373	638	1 039
Setúbal	1 226	1 330	16	325	284	2	801
Portugal	11 310	52 868	85 294	8 255	267 528	76 598	3 975

The above table excludes tourist vessels which at Lisbon totalled 115 ships of an aggregate 2 280 000 gross tons; and at Leixões 114 vessels

SOME MAIN ITEMS OF TRADE, 1968 (by weight, thousand tons)

	In				Out		
Item	Lisbon	Leixões	National Total	Item	Lisbon	Leixões	National Total
Sugar	89	91	181	Sand	22	–	22
Cotton fibre	–	86	86	Cotton yarn and fabrics	–	41	43
Ground nuts	113	–	113	Granite	–	100	101
Machinery	20	25	47	Machinery	4	–	4
Coal	412	94	651	Paper pulp	21	70	197
Cement	113	7	302	Cement	86	34	235
Iron bars or plate	57	113	209	Iron bars and rods, etc.	104	42	147
Phosphates	185	–	289	Superphosphates	54	61	115
Fuel oil	268	249	518	Fuel oil	149	–	150
Diesel oil	290	184	479	Diesel oil	179	–	179
Gasoline	176	115	291	Gasoline	129	–	129
Timber	58	120	179	Timber and wood products	51	188	241
Minerals	17	2	19	Minerals*	9	55	68
Maize	354	28	382	Maize	9	–	9
Iron ore	127	–	127	Iron ore	3	37	41
Crude mineral oil	1770	–	1770	Slate	–	17	17
Petrol	167	30	198	Petrol	109	–	109
Chemicals	58	35	99	Chemicals	10	1	11
Wheat	63	30	93	Tomatoes	119	–	119
Wine (ordinary)	6	114	120	Wine (ordinary)	122	114	237
Cork	5	–	5	Cork	105	33	138
				Port wine	–	38	42
				Fish conserves	–	34	37
	5995	1629	8255		2016	1039	3975

* minerals, excluding marble and alabaster (88 000 tons in 1968) nearly all from Lisbon

of the Douro, an artificial harbour has been established beyond the river entrance at Leixões to handle larger vessels and to act as the anchorage for a fishing fleet. Oporto, including Vila Nova de Gaia (50 000), its suburb across the Douro, is the headquarters of the wine trade, and port wine is the chief export, but the city is also the largest industrial centre in the north, and the second largest in Portugal. Its industries include iron foundring, textile machinery, electrical equipment, tyres, fish canning and processing (especially at Matosinhos), textiles, pottery and tobacco. The accompanying table reveals a host of details as well as many fascinating differences between the trade of Oporto and of Lisbon.

Setúbal (50 000) has grown up on flat alluvial land at the mouth of the Sado river and sheltered by the cliffs of the Serro da Arrábida on the north. The very fine harbour which is 8 km long is set on a lagoon formed by a wide sandspit that shelters the port. It is the third most populous town of Portugal and the most important fishing centre, but suffers from proximity to Lisbon. Its chief industries are sardine canning, and the manufacture of cement and artificial fertilizers. It is the chief port for the products of Alentejo, including pyrites from Aljustrel and Louzal.

Coimbra (about 50 000) is sited on a low hill above the Mondego and has spread from the north bank to the south bank of the river. An ancient city, it contains Portugal's oldest university (founded in 1290) and is the intellectual centre of the country. Its various industries include ceramics and wool textiles, the latter scattered in scores of small factories. Coimbra has its own airport and is on the main railway from Lisbon to Oporto and the junction of a branch line from the port of Figueira da Foz at the mouth of the Mondego. Braga (about 44 000) is the provincial capital of the densely peopled Minho for which it acts as a market centre and as a small industrial nucleus with manufactures of silk, hats and metal goods.

These and other main towns are linked by broad-gauge

railways, the chief of which joins Lisbon and Oporto. The main roads are excellent but other road networks are poor or locally almost absent. In 1968 only 623 000 motor vehicles were registered in Portugal, including tractors.

The density of population increases generally from south to north and from the sea coast towards the interior. About 80 per cent of the population live north of the Serra da Estrêla or in pluviose Portugal. This is despite the fact that northern Portugal has half its area above 400 m while southern Portugal has little land above that height. In the more humid north all the coastlands support well over 100 persons per square kilometre, by far the largest extent of such a density in Iberia. The dense, scattered rural population resembles that of coastal Galicia. On the bleaker plateau the people tend to cluster more into nucleated villages. Southern Portugal is part of 'arid Iberia' and is thinly peopled. Commonly the houses cluster together in tightly packed towns and villages but special mention must be made of coastal Algarve which, like the Sintra peninsula near Lisbon, has experienced a remarkable spread of bungalows and villas in the last few years. This sunny Mediterranean strip with its warm winters and exotic vegetation is also rapidly becoming internationalized.

Geographical regions

The geographical or physico–geographical regions of Portugal are shown in figure 46. In the following description, however, the coastlands are not dealt with together as a separate unit as they are economically bound up with the interior and those in the north bear little resemblance to to those in the south. The real geographical division of Portugal is between the pluviose north and semi-arid south; this change occurs south of the central mountain ranges.

Northern Portugal

This region stretches from the frontier with Galicia southward to Cape Carvoeiro and the Serra da Estrêla. Its northernmost part, the Minho and Douro, is the historical core of Portugal from which population and political power have expanded south. Here along a fault-lined coast there is only a very narrow coastal plain and the whole area belongs structurally to the Douro plateau. However, climatically the coastlands (Minho and Douro) differ markedly from the interior uplands (Tras-on-Montes and Alto Douro).

Minho and coastal Douro

This dissected plateau between the Minho and lower Douro consists of ancient rocks that have been repeatedly peneplaned and the scenery displays tabular residual mountains divided by steep-sided valleys which, like the Lima, Cávado and Tâmega, thread northeast–southwest fracture lines. The valleys are sometimes terraced naturally and artificially while the foot of slopes everywhere is blurred by granitic debris that yields seepage springs. These valleys are open to westerly airflow with its benign Atlantic influence and are mild and humid with 1000 mm of rain annually. Their mouths open out on a narrow coastal plain which, however, is insignificant compared with that south of the Douro. It is a land of peasant proprietors with tiny farms of 1½ to 2 ha, practically self-supporting and even producing small surpluses of butter. Owing to inheritance, splintering of the land holdings is commonplace. It is quite usual in the division of agricultural property to divide up single trees, members of the family being allotted the fruit of certain branches. Pressure on the land is very great and, in fact, has had the effect of intensifying the farm management. Crop rotation consists of maize (often irrigated), followed by beans and melons, that are succeeded by irrigated sown grass, so ensuring several crops of hay from a single sowing as fodder for the dairy kine and draught oxen. Cabbage and potatoes follow the hay crop in the rotation. Fallow is unknown and soil fertility is maintained by green and farmyard manuring and the addition of the silt-laden irrigation water. Fruit trees are grown around the fields and land too steep for the plough is usually under olives and vines.

Tras-os-Montes and Alto Douro

Inland, from the densely populated area described above, Tras-os-Montes and Alto Douro form a mountainous country across which two roads lead into Spain. Patches of oak and chestnut woodland remain but most of the plateau has long been cleared and replaced naturally by heath that provides summer pasture for sheep. The deforestation also led to heavy soil erosion and to combat this there has been much new planting of pines. The annual rainfall averages only 500 mm and, as the winters are bleak and summers hot and dusty, the landscape resembles that of the adjacent Spanish Meseta. The soils likewise are thin and poor being developed on impermeable schists that on the steep slopes drain rapidly, dry out quickly and soon develop gullies. The commonest farming practice is to alternate rye with fallow and to leave the higher rough pastures for sheep and goats. The farms average about 3·4 ha in the district of Vila Real and 13 ha in Bragança.

South of the river Douro, a relief section from west to east across northern Portugal would show the existence of a notable *Coastal Plain*, which in parts widens to 50 km (marked (a) on map). Here the lower floodplains of several purely Portuguese rivers conjoin and with the aid of longshore currents and onshore prevailing winds have created a fascinating combination of alluvial riverflats, coastal lagoons, longshore bars and spits, and large expanses of sand-dunes. South of Espinho the coastlands are ill-drained and waterlogged, fringed seaward by lagoons and sandbars. Most of the lagoons have been converted into salt-evaporation pans, fish farms for crab, lobster, prawn and even for algae for fertilizer. Inland the flat polder-like areas, where drained, grow rice; on the seaward edge pines have been planted to stabilize the dunes and to provide resin and pulp. South of Aveiro the sands become very extensive and the timber, turpentine and resin industry more widespread, together with glassmaking. Away from the sea the river alluvium produces maize, fruit and truck crops except on the ill-drained valley floors that are grazed by cattle. Most of the settlements are fishing villages set along the coast in each bay and harbour. The larger such as Aveiro, with paper-pulp industries, and Figueira da Foz are important local ports. Generally the rural settlement is dispersed and in the wetter areas aligns the roadways in true ribbon fashion.

Inland the plateau south of the river Douro (part of the Douro Plateau in figure 46) in Beira Alta is cooler and slightly more humid than the coastlands. Rye and oats for fodder are widely grown and the land use deteriorates eastward into rough pastures, heath and woods, with cultivation confined to the warmer valleys and irrigable patches. A few such favoured valleys may be under maize followed in the rotation by grass. The richer district around Coimbra supports grain and fruit, with vines on the steeper slopes.

The most distinctive part of the Douro plateau lies to the northeast on the Douro gorge near the frontier with Spain. Here in the port wine district the mica–schist soils are thought to give a special quality to the grapes. The steep sides of the main valley and those of the right bank or south-facing tributaries (in Alto Douro) have been terraced and closely planted with vines. Further details of this legally defined wine-producing area are given in the earlier part of this chapter.

The central serra

At the southern end of the Douro plateau the land between the Mondego and Zézere rivers rises to the Serra da Estrêla, the highest mountains in Portugal. The height (maximum 1991 m) is sufficient to cause heavy rain and snow, and the ample runoff feeds large reservoirs with hydro-electric stations at Castelo Bode, Bouçã and Cabrill on the Zézere. The mountain summits, once partly forested, are now mantled with heath grazed in summer by sheep. The lower, more sheltered slopes have chestnut and pinewoods while the warmest south-facing valleys produce good wine.

At its southwestern or seaward end the Serra da Estrêla grades into a lower hill range of younger rocks (marked by pecked lines on the map). The metaphoric and granitic outcrops are here replaced mainly by Secondary limestones and sandstone. The limestones in parts give rise to rather barren karst scenery but the sandstones are more productive and carry woods of chestnut, pine and oak as well as citrus groves. The Sintra (Cintra) Hills overlooking Lisbon have in the twentieth century, when Portugal remained neutral in the Second World War, become aglow with the delightful villas and bungalows of an international populace. Here the only extensive area of volcanic rocks in Portugal adds to the scenic variety which is further enriched with castles and the old royal palace at Cintra to create one of the most picturesque tracts in Europe.

Southern Portugal

South of the Serra da Estrêla and its seaward continuation described above lie the lowlands of Estremadura and Ribatejo which are the floodplains (*lezirias*) of the Tagus and Sorraia (marked (c) on map). These link up southward with the plains of the Sado and together form the true coastlands. The greater part is floored with river alluvium and displays flat, fertile fields, many of which are flooded annually in spring, thus renewing their fertility. Intensive mixed farming is common, with heavy crops of wheat, maize, rice, and fodder for dairy cattle and pigs. The stretch between Santarém and Vila Franca specializes in the breeding of fighting bulls. Owing to the rainshadow effect of the Sintra peninsula the rainfall is often inadequate for crops and much supplementary irrigation water is taken from the Tagus and Sorraia. Recent government schemes have created vast reservoirs on the Sorraia and Zézere and plans are afoot for the main river near the frontier. In Ribatejo the Tagus has some natural terraces and much land is devoted to arboriculture and vineyards. The whole society centres on Lisbon which provides the best single market in Portugal.

Farther south on the coastlands the rainfall decreases rapidly and five months suffer drought. Soils become thinner and excessive evaporation develops a tendency toward salinity. The country in its natural state is mainly rough pasture and habitations are few. This monotonous pattern is broken near the Sado which waters paddy and has a large lagoon with extensive salt pans. Setúbal controls the local fishing and marketing and industrial activities.

Alentejo uplands

South and southeast of the Tagus–Sado lowlands stretches a vast rolling plateau with a semi-arid climate. Cork-oak woods are common in the west and widely spaced ilex in the east but the general impression is of heath and garrigue or *campos* grazed down by sheep, donkeys and goats. Cereals are planted at intervals to alternate with the grazing, the crops being taken by the sharecropper while the pastures belong to the estate owner. The *montado* system works alongside the *campos* method, and results in a cultivated woodland mainly of cork oak and ilex to produce bark, timber and pannage. Beneath and between the trees cereals are grown as a winter crop and are alternated with long periods of fallow. Each system is a variant of dry farming so necessary here in view of the long drought and difficulties of irrigation in a land where most streams are deeply incised and are dry in summer. Exceptions, of course, occur as for example the Évora district and the Guadiana valley which produce wheat, grapes, olives and figs. Elvas is famous for plums; Beja, in the centre of extensive *alcornales* (oakwoods), handles cork, and pig and sheep products; Évora (16 000) is a provincial market centre with food-processing and woollen textile industries. The population tends to cluster in compact groups and some of the towns are walled.

The South Portugal uplands

The South Portugal Uplands (marked (d) on map), an extension of the Spanish Sierra Morena, are variously called the Serra de Monchique, da Mesquita and da Malhão. They form the chief metal-mining region of Portugal but otherwise are poor, rugged hilly country with much matorral or garrigue.

Their poverty and thin scatter of population contrasts vividly with the populousness of the *Algarve Lowland* (marked (e) on map), a narrow coastal strip of alluvium resting on a Tertiary sedimentary base and backed inland by limestone scarps. Along its eastern half from Faro to Vila Real it is fringed by a line of sandy islands but on the west rises to coastal platforms and cliffs at Sagres and Cape St Vincent. The long, hot, dry, sunny summers and mild sunny winters, the flat-roofed Moorish architecture, the palms and cacti, the steep-sided wadis and even the natives recall north Africa. The land, given water, yields abundantly of typical Mediterranean and tropical fruits. The Portuguese aspect enters especially in the tunny fisheries and the many small fishing villages dotted along the coast from Tavira to Sagres and Cape St Vincent, once the headquarters of Prince Henry the Navigator and today a symbol of the greatest surviving colonial power. In recent decades thousands of new residences have been built and this narrow, dry coastal strip, for widely different reasons, will soon rival the Minho in density of population. Here, indeed, is Africa in Europe. The modern settlement is in parts on well-organized estates and in others spreads as a proliferation of detached dwellings seeking sea views and access to sandy beaches. When the old Vila Real, which had been slowly washed away by the sea in medieval times, was replaced by a new town by Pombal in 1774 he gave it a plan and architecture like that of the *Cidade Baixa* in Lisbon. Today on this sunny coast the visual effect of both planned and unplanned is charming.

14 Southern France: the Mountains

In western Europe the Central Massif, Vosges and Black Forest Hercynian blocks acted as rigid buffers against which younger sediments were compressed and rucked up during the long Alpine orogeny. In France the folds of the Alps proper are represented by what appears at first sight to be a structurally continuous loop of mountains swinging through the Jura, the Savoie and Dauphiné Alps and the Maritime Alps of Provence, thereby linking the Swiss Alps with the Italian Apennines. However, the chains which run from east to west in coastal Provence are part of an earlier Alpine folding that extended, as the trend lines show, into the Pyrenees (fig. 4). Where these Maritime Alps come into contact with the later north–south folds of the French Alps farther north there is great structural complexity.

The alpine fold areas of France are usually subdivided geographically into: the Jura; the French Alps of Savoie and Dauphiné; the Maritime, Provençal, or Mediterranean Alps; and the Pyrenees. These divisions are outlined roughly in figures 46 and 62.

THE JURA

These mountains, which are between 1000 m and 1500 m in average height, overlook the wide floodplain of the river Saône and swing in a great arc for 250 km between the Rhône and the Rhine. Festoon-like they are widest in the centre where they attain a width of 25 km and contract at either end to 10 km. Built of enormous thicknesses of limestone to which they have given their name, they were ridged up into fold mountains during the Alpine orogeny by pressures mainly from the southeast exerted against the rigid Hercynian blocks of the Central Massif, Vosges and Black Forest. The mountain-building forces were neither so strong nor so prolonged as in the Alps to the south, and the folding of the Jura is far less complex. The first folds that rise above the floor of the Swiss Mittelland are high and dip steeply but thereafter the folds sink in height gradually and decline in amplitude one after the other in parallel lines until the folding peters out altogether and the horizontal strata form a level plateau. A distinction is therefore possible between the folded Jura and the plateau Jura (*Jura plissé* and *Jura tabulaire*).

The Folded Jura is a classic example of anticlinal ridges (*monts*) and synclinal valleys (fig. 54). Some 160 anticlines have been counted in the central Jura alone. The synclinal, longitudinal valleys (called *vaux*; singular *val*) are followed by the rivers which escape their confines by gorges or gaps called *cluses* across the monts. A fine example of such is at Pontarlier where the river Doubs makes a right-angled bend to breach the anticline or *mont*. A typical espalier drainage pattern results with its inevitable river captures. The river Doubs again displays evidence of these, and has truncated one of the headstreams of the Ain.

Already erosion is beginning to carve out a relief pattern that is a reversal of the structure. The summits of many anticlines have been breached and excavated by mountain torrents into hollows (*combes*) which have steep, inward-facing, bare limestone scarps, that have been incised by the streams on the downhill side into deep gorges through which the torrents rush to join the synclinal master rivers. One of the steep scarps, the Crêt de la Neige, is in fact the highest point on the Jura (1723 m) and typically rises close to the Col de la Faucille (1323 m), the easiest pass across the mountains. The presence of surface drainage on the limestones can be attributed largely to patches of boulder clays left behind by the Quaternary glaciation. These deposits are found in patches on the ridges and plateau surfaces but cover the floors of the synclinal valleys and depressions rather more thickly. Lakes, such as the Lacs du Val and de l'Abbaye, are held up by these clayey deposits, and may be strung along a river's course or sited in the cluses. Elsewhere water escapes by subterranean flow to emerge from the mouths of caves as swollen springs (*grosses sources*) usually at the mountain foot.

The Plateau Jura suffered sufficient pressure to fracture it into blocks without over-much tilting the strata. Today tablelands at 1000 m, 600 m and 500 m approximately drop towards the Saône valley ending in a steep wall which declines in height southwards from 900 m to about 250 m. In some places the mountain scarp is almost unbreached, in others it is fretted into embayments and spurs. Over large areas of the plateau, karstic phenomena and topography prevail with sinks, dolines, uvulas, solution hollows and barren slopes and rock pavements but, whether surface or subterranean, the drainage goes ultimately to the Rhône or the Saône via in some cases the Ain and the Doubs. The stepped, lower western fringe of the plateau, dropping to the Saône trough is sometimes called the Tabular Margin (fig. 62).

54 *Characteristic structure and landforms of the French folded Jura* (after D. V. Ager and B. D. Evamy)
a *box-fold*
b *molasse-floored syncline*
c *cluse*
d *box-fold with faulted flank*
Locally minor anticlines develop on the flanks of major box-folds.

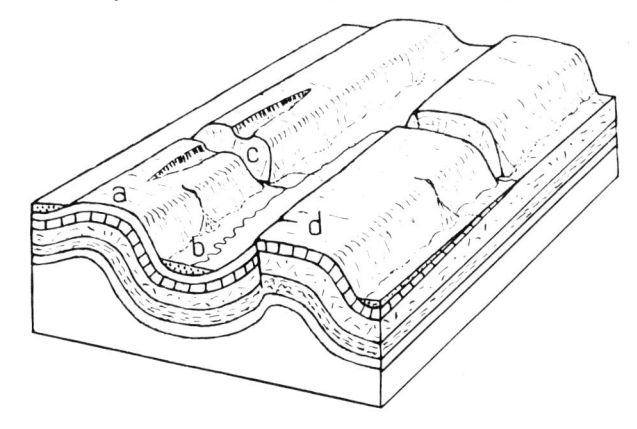

Climate

The Jura mountains lie in that part of France which has continental tendencies in its climate, with rather cold winters and warm summers. Spring and autumn are of short duration and the plunge from winter into summer and back again is fairly permanent. In summer much of the rainfall is in the form of thunderstorms. However, this climatic pattern is modified by altitude and its concomitants, aspect and the direction of the ranges. Temperatures decrease with altitude more rapidly on north-facing than on south-facing slopes; and more rapidly in the northern Jura than in the southern. Likewise temperature inversions, especially in winter are more frequent in the northern Jura. Temperatures are on average below freezing for 10 weeks on the plateau Jura in winter and for $3\frac{1}{2}$ months on the higher folded Jura but aspect is of great importance in determining the length of winter and the number of hours in the day when the thermometer rises above freezing point. Winters in the Jura are in fact more severe than in parts of the Swiss Alps as evidenced by the upward limit of tree growth which is often 600 m lower in the Jura. Summer temperatures hover around 19°C but the air being clear and dry, insolation is intense and there are long hours of sunshine. In favoured localities vines will ripen and maize reach maturity. Precipitation is heavy, being about 1000 mm on the plateau and increasing to 1750 mm on the folded Jura. During the cold season much of it falls as snow which covers the floors of the higher valleys to a depth of 3 m for about 6 weeks and remains frozen there for as long as 4 months at 1000 m. On occasion the mountain roads and passes are blocked but nor for long.

Vegetation and land use

The original cover of the Jura region is likely to have been forest and today one-third of the region is wooded. The Jura foot bears deciduous growth of plane, maple, beech, oak, and mountain ash and towards the south groves of sweet chestnut. The plateau has been reafforested in parts with conifers, while beyond, the mountain slopes are muffled in trees which rise in dark masses above steep white limestone cliffs. The pastoral floors of the valleys are sufficiently sylvan to merit the description of wooded meadowland. Above 1300 m and the limit of tree growth are the mountain pastures or *alpages*.

The agricultural response to the mountain environment shows an emphasis on dairy pursuits. The plateau is largely under rough grazing and the alpages are often owned in common by the village communes. Dairying has developed on cooperative methods which both streamline production and distribution and engage in technical research into all aspects of the industry. Out of about 500 000 cattle of the Tarentaise breed, three-quarters are milch kine. The table of statistics (right) underestimates this as it relates only to

SOUTHERN FRANCE: MOUNTAINS

Département	Area (sq km)	Population 1968 (thousands)	Density per sq km	Capital	Population 1968 (thousands)	Arable	Permanent Pasture	Fruit and vine	Agricultural, not cultivated	Non-agricultural	Woods and forests
Jura											
Jura	5008	234	47	Lons	22	98	166	4	18	33	186
Doubs	5228	426	82	Besançon	120	135	174	1	1	20	195
Ain	5756	339	59	Bourg	40	166	173	5	48	41	135
Alps											
Haute Savoie	4391	379	48	Chambéry	54	62	153	1	15	88	142
Savoie	6036	289	86	Annecy	57	22	175	4	65	197	162
Isère	7474	769	103	Grenoble	166	191	182	17	128	80	212
Hautes Alpes	5520	92	15	Digne	16	41	208	5	26	142	150
Basses Alpes	6944	105	17	Gap	25	80	213	8	143	32	223
Alpes Maritimes	4294	722	168	Nice	325	18	126	6	70	56	152
Var	5999	556	93	Draguignan	20	29	54	73	109	39	296
Pyrénées											
Basses Pyrénées	7629	509	67	Pau	76	170	156	11	201	76	149
Hautes Pyrénées	4507	226	50	Tarbes	59	72	137	4	75	69	96
Ariège	4890	139	28	Foix	10	71	169	3	80	41	130
Pyrénées–Orientales	4086	282	69	Perpignan	104	22	68	74	70	73	102

N.B. Haute Garonne is omitted as only a minor part of it is in the Pyrenees. Vaucluse, part of which is in the French Alps, has been included in the separate table of statistics in Chapter 15 on the lowlands of southern France (p. 179). Pyrénées–Orientales includes lowland Roussillon.

adult cows that were actually yielding milk. The valley floors are given over to sown grasses, hay and fodder crops, the cattle being moved off them in the spring to the alpages for the summer, whence they return in autumn to more sheltered conditions. Liquid milk is despatched fresh to Montbéliard, Besançon, and as far as Lyon and Paris. Some even goes to Geneva for processing in chocolate manufacture. The bulk of the liquid milk is however made into dairy products, butter, cream and all kinds of cheese but especially gruyère, either in local dairies, or in cheese factories in towns like Pontarlier, Dôle and Lons.

Certain localities display a specialized agriculture. South-facing slopes are often planted in orchards, apple, plum, cherry, peach and apricot for sale fresh or for making into liqueur. Around Arbois is the largest concentration of vineyards in the Jura today. Vines grew formerly much more extensively but were reduced by phylloxera and have never recovered the position they once held here. Arbois was the home of Louis Pasteur and it is rather fitting that the Jura countryside should have posed him with the problems to which he owes much of his fame, the fermentation of grapes and anthrax in sheep. Arbois is one of the smaller classified areas for table wines and Château Chalon, much prized by the French, is made in the district. Truck farming makes its appearance in specialized localities but owing to distances from large population centres tends to depend on one crop, for example peas near Frasne, cauliflowers around Hautepierre, herbs near Pontarlier, and flowers between Culoz and Seyssel. Elsewhere and in general the farming is subsistent in character and alternates with forestry and chalet industries between summer and winter to supply a livelihood.

Water power and industrial growth

Before 1946, when hydro-electricity was nationalized in France, a few small power stations were in operation in the valleys of the Ain, Doubs and upper Rhône. Since then, France's largest hydro-electric station has been built on the Rhône at Génissiat by the Compagnie Nationale du Rhône. The barrage at Génissiat was completed in 1948 and today impounds a stretch of water that reaches almost to Bellegarde. The scheme is described further in Chapter 10. The station is harnessed to the French grid, and with the completion of a secondary hydro-site at Seyssel, a chain of 6 power stations now extends the length of the French Alpine section of the Rhône as far as Cusset and 5 more are planned for this same section.

As the table of statistics (left) shows, the indigenous industries of the Jura have been rejuvenated and developed by this assured electricity supply. The dairy industry for example has been greatly extended to include factory-made cheeses, butter, yoghurt and casein. Milk surpluses have led to bakelite industries and some centres now have manufactures in consumer objects such as combs, mirrors, fountain-pen cases and electric fittings.

Simple market towns have become small industrial centres with alternative employment, as has happened for example at Moirans, Arinthod and Longchaumois. Other food-processing industries include liqueur making, brewing and the manufacture of chocolate.

An old-established local iron industry has burgeoned into steel manufacture with small foundries at Pontarlier and Bellegarde, rolling mills at Syam, wire-drawing at Lods and Vuillafans, and nail and bolt manufacture at Ornans, Clerval, Jougne and L'Ile-sur-Doubs. Larger scale engineering industries have grown up at the edge of the Jura, notably at Montbéliard which manufactures Peugeot cars as well as aircraft engines, bicycles, agricultural machinery and machine tools.

The watch and clock making industry from across the Swiss frontier has become established at Besançon (113 000) which is the chief French centre for the industry, and in a ring of smaller towns in the Bienne valley such as Morez, Morbier and Perrigny. Altogether 14 000 workers are engaged in this trade which has strong business links with Geneva. Subsidiary industries like jewellery and glassmaking have also been established.

A small but flourishing chemical industry based on salt deposits mined on the northern flanks of the Jura has led to brine pumping at Poligny and salt refining at Dôle. For centuries the forests have provided winter employment and wood and timber industries are still active particularly in the valley of the Bienne. The textile industry, once based

Livestock 1966 (thousands)				High Tension Electricity Consumption (in million kWh)				
Bovine		Sheep	Goats	Railways	Mines	Electro-chemical and metallurgical	Other manu-factures*	Total
In milk	Total							
92	187	41	2	19	4	–	961	984
107	238	15	1	9	2	–	762	773
131	265	59	21	90	2	662	337	1091
83	138	11	6	31	3	355	515	904
57	100	21	11	67	7	2967	1366	4407
137	261	52	53	69	35	920	2568	3591
17	34	224	6	1	2	372	96	417
5	10	256	13	1	1	464	666	1132
4	7	89	4	5	12	–	261	278
1	1	117	2	23	16	–	170	209
135	271	355	2	51	6	1660	829	2545
70	110	128	1	28	8	1084	668	1788
42	89	140	3	10	13	822	215	1060
7	11	69	3	9	13	–	49	70

* includes paper, textiles, some metallurgical manufactures etc

on local wool and hair supplies, has extended into specialized branches for the large factories of Lyon and Mulhouse. Silk and silk thread is spun at Argis, Saint-Rambert, Ambérieu and Pont d'Ain, silk material is woven at Nantua, Bellegarde and Maillat. Linen is made at Belley, and lace and net at Ambérieu and Nantua. Similarly rayon, hosiery and small-wares are manufactured at Besançon which was the first French town to establish a factory for synthetic fibres. Altogether approximately 160 000 workers are engaged in industry either whole or part-time in the Jura.

Settlement and communications

Population is fairly evenly dispersed in the valleys of the folded Jura but is sparse over large areas of the plateau and here tends to be concentrated in the towns. The three largest urban centres are at Montbéliard (24 000), Besançon (113 000) and Lons-le-Saunier (15 000). The first-named has 190 000 people in its conurbation and the second 150 000.

Important trans-continental railways cross the Jura into Switzerland en route for Italy, or Austria and the Balkans. They were difficult and expensive to build and involved cuttings, tunnels and viaducts but they carry a heavy tourist traffic, while three mainroads carry the motor traffic by three mountain passes southwards.

THE FRENCH ALPS

The French Alps are treated here in one section but in such a way that the differences between the Savoie–Dauphiné Alps and the Provençal Alps can be clearly seen. Each illustrates the same geographical facets of a lofty mountain environment, although the former bears strong resemblances to Alpine Switzerland and the latter to the Mediterranean world. The statistical table (pp. 166–7) shows clearly many of the vital differences in the geography of these two broad divisions and of the various départements that comprise them.

Structure and relief

The Savoie and Dauphiné Alps swing as a crescent 330 km long from Lake Geneva to the Durance valley. This mountain arc is highest in its medial strip and declines in altitude on its flanks which are penetrated by valleys and embayments that lead to passes across the watershed which practically coincides with the frontier between France, Switzerland and Italy. The whole crescent is from 150 to 170 km wide and consists mainly of four north–south relief-structural zones.

On the east are the High Alps, the highest range of the system with peaks exceeding 3000 m and rarely 4000 m. The main massifs, such as Mont Blanc (4807 m), Pelvoux and Mercantour lie along the frontier and reveal the crystalline core which consists of Hercynian fragments overwhelmed and incorporated in the newer sedimentary folds. Other types of crystalline rocks present are batholiths formed at the time of the upheavals and pressure-induced metamorphic rocks (*schistes lustrés*), each later revealed by denudation of overlying *nappes*, as near the Great St Bernard and in the Graian Alps.

The High Alps are bounded on their northern and western rim by a longitudinal trench or sub-High Alpine depression which can be traced from Chamonix to Grenoble and includes the Grésivaudan in the Isère valley. This long trench is followed by the lower Drac and the Vale of Chamonix and continues into Switzerland and Austria where it is drained by the upper Rhône, Upper Rhine, Inn and Salzach. In France for 180 km it separates the High Alps from the Pre-Alps with a trough locally several kilometres wide and everywhere about a thousand metres below the mountain tops on either side. It seems due mainly to the erosion by rivers of a relatively soft cover of clays and shales. It carries routes into the heart of the mountains via the Maurienne and Tarentaise to the Mont Cenis and Great St Bernard passes (see figure 39) and has turned the Alps as a barrier into what Napoleon called 'a splendid traitor'.

The Pre-Alps extend from Lake Geneva to the river Durance and are built of limestone folded into parallel ranges that reach an altitude about half that of the High Alps. These mountains repeat the same wide-spaced, espalier drainage-pattern of the folded Jura, in which rivers follow the longitudinal valleys and escape from them by a sharp right-angled bend in a transverse valley (*cluse*). In the Pre-Alps there are four important *cluses* leading from the longitudinal furrow into the Rhône valley. The most northerly is used by the river Arve, the most southerly by the Isère where it leaves the Grésivaudan, and the two central ones are occupied by glacial lakes Annecy and Le Bourget respectively. These *cluses* divide the Pre-Alps into separate mountain masses of Vercors, Grande Chartreuse, Bauges and Genevois. As in the Jura, river capture is common. The Isère, to name an obvious example, is a master river that has reduced the streams flowing into lakes Annecy and Le Bourget into truncated or beheaded misfits; and turned the passes into *les cluses mortes*. Here, as in the Jura also, much of the landscape is developed on permeable limestone ranges that front the prevailing westerly rainbearing winds. Precipitation is abundant and the subterranean drainage that results is both magnificently complex and labyrinthine. In the Gouffre de Berger in the Vercors massif the deepest known continuous cavern system in the world at 1130 m has been excavated by the river Germe which issues in a *grosse source* at the foot of limestone cliffs eight kilometres northeast of Grenoble. Similarly the Grande Chartreuse is honeycombed by an incredible subterranean system with spectacular waterfalls.

At the foot of the Pre-Alps is the region known as the Alpine Foreland that extends from about the Isère to the Rhône and beyond (marked Rhône–Rhine plateau in figure 62). It is covered with thick fluvio–glacial deposits: moraines and outwash material sorted and re-deposited by glacial meltwater and later terraced by rivers with glacial regimes. The hidden basement consists of little disturbed sedimentary rocks.

Throughout the Savoie and Dauphiné Alps erosion has been, and still is, severe. Glaciation, past and present, has set its stamp on the landscape. In the High Alps snowfields still cover 520 square kilometres of the Mont Blanc massif and give rise to glaciers like the Mer de Glace and the Chamonix but much of the glaciation is the result of a former more extensive icesheet. Most of the valleys have been excavated by tongues of ice into U-shaped, overdeepened troughs that outline the separate mountain massifs. The long profiles of these valleys are characterized by treads ending in rock sills marked by rapids or waterfalls and known as *verrous* which form useful dam sites today. At Les Bossons the Chamonix glacier overhangs such a tread. The valley floors are wide and flat with high precipitous sidewalls, that are often overhung with tributaries that pitch their water headlong into the main valleys. Often the upper valley slopes end in broad shoulders that are well seen again in the Val de Chamonix. These upper valleyside flattenings are called *replats* by geomorphologists and, *alpages* by the herders who use them for grazing. Beyond the limit downstream reached by the valley glaciers, deposition is widespread and the morainic material has been partly dispersed by meltwater and fluvial action into spreads of gravel, banks of sand and pebbles, deltas and fans of debris. Among the crests of the ridges and at the summits of the mountains, cirques and arêtes, cols and horns are everywhere apparent, attacked today by frost action so that needle-like pinnacles, bare rock slopes and screes abound. Because of heavy precipitation normal surface erosion also proceeds most rapidly below the permanent snowline and avalanches and rock slides of late winter are followed by torrential floods and landslips of spring and early summer.

The Maritime Alps

Folded at the same time as the Pyrenees, the Alps of Provence have the same east to west axis and are probably a detached continuation of them. Isolated fragments do exist near the Rhône in the neighbourhood of Arles, including the Chaîne des Alpilles, but the main chains stretch from east of the Rhône–Durance floodplain to the Italian frontier and beyond. These constitute a confused mass of broken ranges thrust up like blocks with steep scarp edges and gently sloping lofty plateau-summits (*plans*) that isolate numerous lowland basins from each other. Made of limestones, these fragmented mountains are barren and rocky with a thin, poor soil cover and widely spaced surface streams, that normally flow in steep, wild gorges like that at Verdon but there are occasionally more open valleys floored with alluvium, for example those of the Argens and Gapeau. Over large areas the drainage is underground as in the plateau of Vaucluse but the surface often lacks typical karstic features. The disorderly array of massifs persists to the coast which is strewn with rocky offshore islands and presents steep cliffs usually free of marine aggradation since the prevailing longshore currents sweep the sediments westward to the coasts of Languedoc on the other side of the Gulf of Lions. The cliffs form corniches and the coastal inlets calanques. Where rivers enter these inlets as in the case of the Argens, Gapeau, Siagne, Gisele and Arc, small alluvial plains have accumulated. These tiny plains continue wide-spaced and infrequent along the coast to Menton and behind them the cliffs rise to 300 m and higher forming the sheltered Côte d'Azur. Between Hyères and Cannes however two highland massifs of red porphyry, Maures and Esterel, interpose their rugged relief into the landscape and jutting out to sea separate the Côte d'Azur from Bas Provence. These with Corsica and Sardinia are remnants of a Hercynian block that foundered and forms today the floor of the western Mediterranean basin (see Chapter 2).

Climate

Strong differences between the Maritime Alps and the Savoie–Dauphiné Alps are also evident in climate and vegetation. The Maritime Alps lie within the latitudes of the Mediterranean summer drought while the French Alps to the north enter into the sphere of interior continental influences and all-the-year rainfall. Basically the French Alps have a climate transitional in character. A continental influence of cold winters and warm summers, marked by thundershower rainfall and of a rapid changeover of the seasons, is profoundly modified by altitude. In the first place, the mountains running from north to south present high barriers to the westerly rainbearing airflow and receive ample precipitation throughout the year. Secondly, this rainfall is redistributed so that western slopes and ranges receive the brunt of the fall, and the lee slopes and longitudinal valleys have much less. The yearly falls of 1500 mm on the Pre-Alps dwindle to half that amount on the High Alps except on exposed peaks like Mont Blanc (1100 mm). Average falls in the Isère valley amount to 900 mm but where several valleys meet and the landscape is more exposed as at Grenoble the rainfall increases to 950 mm. Thirdly, since here in the French Alps the normally cool winters are increased in severity by altitude much of the cold-season precipitation falls as snow. The whole mountain area lies under snow for two months in the west and south and six months in the east. The high mountain passes are blocked by snow from October to May.

Altitude also has an important effect on the temperatures. The overall decrease is particularly noticeable in winter, and is greater in the north and east and on shaded sides than in the south and west and on sunny slopes. To the general temperature decrease with altitude must be added the prevalence of cold temperature inversions at night particularly in enclosed valleys and basins during the winter and spring. Aspect assumes a great significance in mountainous regions since the south-facing slopes

SOUTHERN FRANCE: CLIMATIC STATISTICS, MEAN MONTHLY TEMPERATURES (centigrade)

	Altitude (if over 100 m)	J	F	M	A	M	J	Jy	A	S	O	N	D	Absolute Max	Absolute Min	Annual days with frost
Pic du Midi (Pyrenees)	2859	−7·9	−7·6	−7·1	−5·2	−1·7	2·8	6·5	6·5	3·2	−0·8	−4·6	−6·9	17·6	−25·4	268·5
Perpignan		6·7	8·0	10·0	13·1	16·2	19·9	22·7	22·2	19·2	14·7	10·3	7·4	34·8	−4·6	16·9
Montpellier		5·0	6·7	8·9	12·8	16·1	20·0	22·8	22·2	18·3	13·9	8·9	5·6	37·1	−8·8	43·9
Marseille		6·6	7·4	9·2	12·6	16·2	19·8	22·4	21·7	18·9	14·6	9·9	6·8	38·3	−11·7	26·2
Nice	340	6·5	7·0	8·7	11·8	15·1	19·0	21·9	21·9	19·0	14·6	9·8	7·1	35·0	−5·0	3·0
Avignon		3·9	6·1	9·4	13·3	17·2	21·7	23·9	22·8	18·9	13·9	8·3	4·4			
Lyon	175	1·7	3·9	6·7	11·1	14·4	17·8	20·0	19·4	16·1	11·1	5·6	1·7	34·6	−12·4	64·5
Chamonix	1040	−5·8	−2·3	2·2	7·0	11·0	15·2	16·8	16·1	12·3	6·5	0·8	−3·9			
Briançon	1298	−2·2	−1·1	1·1	5·6	10·0	13·9	16·7	16·1	13·1	7·2	1·7	−1·1			

MEAN MONTHLY PRECIPITATION (millimetres)

	J	F	M	A	M	J	Jy	A	S	O	N	D	Mean Annual Total	Raindays	Days with snowfall
Pic du Midi	182	164	160	171	142	127	71	73	109	115	137	162	1631	181	117·9
Perpignan	58	47	49	47	57	40	20	27	44	59	57	41	554	86	4·0
Montpellier	79	69	61	58	71	46	23	51	76	104	86	61	785	90	
Marseille	47	38	40	43	46	26	14	23	60	99	79	50	574	81	2·5
Nice	60	63	66	60	68	49	16	26	64	147	125	73	828	81	4·2
Avignon	38	38	41	51	61	51	28	51	74	89	66	46	645		
Lyon	36	38	48	61	81	84	74	79	76	97	66	43	795	148	22·6
Chamonix	58	69	63	66	86	117	117	130	112	112	86	86	1120		
Briançon	33	36	43	51	58	53	41	43	53	76	58	46	592		

(*l'endroit*) have warmer temperatures and more sunshine than slopes facing north (*l'envers*). Equally the sunny slopes also have shorter spells of frost and longer periods when the ground is free from snow cover. Typical average winter temperatures are −5°C at Chamonix in the High Alps, −0·5° at Annecy in the Grésivaudan and 4° in the Durance valley. These low figures however give no hint of the clear, dry sparkling air of high mountains during anticyclonic weather in winter and of the long hours of sunshine despite cold air temperatures. Similarly, summer temperatures of over 16°C mask high sunshine amounts and absence of humidity, as well as intense insolation in sunny areas. Mountain and valley winds are of local importance and form on the whole an unpleasant feature of the climate. Normally in calm clear weather warm air from the valleys blows up slopes during daylight, and cold air moves downhill at night.

The Maritime Alps enjoy the mild winters and hot summers of the Mediterranean and the small plains at the mountain foot sheltered by the barren limestone ridges behind them have the clemency of winter and the heat and sunshine of summer intensified. Nice for example, on the warm Mediterranean, has mean January temperatures of 8°C and average July temperatures of 23° which again gives only a hint of the scorching noonday heat and of the brilliance of the sunshine. Summer everywhere is a time of drought, dust and glare; mid-September to mid-May is the season of rain with just over 800 mm at the coast diminishing to 250 mm within a short distance inland. Generally at these temperatures, the rainfall is insufficient, especially as it occurs in sudden heavy showers, liable to cause soil-wash and floods. The Durance in flood has laid waste much agricultural effort in this manner.

A local wind of much importance is the cold mistral, 'the masterful', which blows when a cold airmass over northern Europe moves down the Rhône corridor, usually in the rear of a depression that is travelling eastwards in the Mediterranean. Crops exposed to the mistral are often withered by cold and windbreaks are planted to protect them and dwellings and open stretches of road. Further details of its climatic nature are given in Chapter 4.

Vegetation and land use
Because of the heavy precipitation on the western parts

of the French Alps approximately half the surface is under forest. Deciduous trees climb to about 1000 m and above that height conifers extend to the treeline. Farther east deciduous trees become increasingly rare and the conifer forests consist more and more of single species and eventually mainly of spruce. There has been much deforestation in the past and the present tree growth is largely new plantations for timber for industrial purposes or as avalanche breaks or as protection against soil erosion on steep slopes.

The valley floors everywhere have been cleared to provide meadows and arable land, and at the upper limits of tree growth to increase the area of mountain grazing or alpage. In the Genevois massif today only the steepest slopes have been left under forest and the same is probably true for the more accessible mountains elsewhere but over the whole of the French Alps of Dauphiné and Savoie nearly one-quarter of the land is forested and, just over one-quarter under permanent grass.

The aspect of the Maritime Alps is very different. Large areas lie under Mediterranean maquis or garrigue of laurel, myrtle, box and kermes oak, dwindling inland with diminishing rainfall to a shrub growth of rosemary, lavender, thyme and broom. In neither is there much herbaceous groundcover and each is aromatic and oleaginous and a desperate fire hazard during the baking summer. Where this has been cleared by cutting or overgrazing or by fire, soil erosion is inevitable and there are many such bare and rocky hillsides that once had both soil and vegetation cover. Where rainfall is more abundant or where soil has accumulated to some depth patches of tree growth of typical Mediterranean species occur albeit sporadically (see Chapter 6). In Provence the Esterel and Maures massifs are covered in this way. In the Maritime Alps generally about 40 per cent is classed as woodland but much of this is scattered brush and shrub and does not compare with woodland classed as such in the French Alps. A further one-third is designated as pasture but again this is poor quality grazing not to be equated to true meadow or grassland. The département of Var is truly Mediterranean (see table, p. 166); permanent pastures are few and fruits extensive; over half the land is classed as forest, a proportion exceeded in France only in Les Landes.

Pastoral farming

The primary agricultural response is pastoral although everywhere throughout both mountain regions this is declining in importance relative to the modern industrial growth. In the French Alps most of the half million cattle are milch kine of well-known French breeds. The Swiss or Jura pattern of transhumance is repeated here with the cattle moving off the valley floors where they have spent the winter, stall-fed in byres, up the mountains to the alpages for the first flush of spring grass. The valley floors are then put under hay and forage crops. Vines appear on the sunny valley sides above the danger of cold-air drainage and severe frost. Some favoured localities have fruit orchards as well which has led to the establishment of apiaries.

As the season advances and the snows retreat, alpine pastures are exposed and the cattle are driven higher still. In autumn they are back on the alpages and finally reach the valley floors in time to graze the last growth of grass in the hayfields. The winter feed, hay and roots, has all been gathered in and the cattle are stall-fed during the winter months. Many variations on this theme are practised. In some localities, families move with their herds to summer quarters up the mountains; elsewhere the cattle are grazed for the valley farmer by other herders while he takes to the plough for the summer and retrieves his animals in the autumn. Normally two crops of hay can, with the aid of irrigation, be got off the valley fields. Irrigation is usually a cooperative venture and is well exemplified in the Grésivaudan. Similarly the marketing of the milk is increasingly under cooperatives. Liquid milk is sold directly off the farms to cheese-making (gruyère) establishments and processing factories or else for consumption to the growing urban populations especially around Grenoble and Lyon.

In the Maritime Alps the economy is different although the change to it is gradual. The numbers of dairy cattle decrease to one-thirtieth of those in the Savoie–Dauphiné Alps, their place in the agrarian economy being taken by about half a million merino sheep. These transhumant flocks pasture in winter on the Crau, a vast expanse of Quaternary gravel brought down by a proto-Durance, and in the lower Durance valley. After lambing in spring the sheep move up to the pastures on the limestone plateau (see Chapter 15). Many of the lambs are slaughtered then and the meat sold in nearby towns and the skins sent to Grenoble for gloving. The surplus ewe milk is made into Roquefort cheese.

Wherever possible every parcel of land, especially on the coast, is cultivated for cereals, vegetables and tree fruits. Digne, for example, is a noted centre for plums and prunes, and Sisteron for almonds. The Grasse district was devoted to perfume-yielding flowers but this trade is now dominated by synthetic products. Many of the interior basins are entirely given over to viticulture with olives on the drier slopes. Both harvests, of olives and grapes for *vin ordinaire*, are run by cooperatives, notably at Brignoles and Aix, which maintain standards and by efficient marketing assure the growers an income. The nut harvest, especially chestnut, almond and walnut is also increasingly handled by cooperatives. The market-garden products near coastal resorts include asparagus and, near Nice, cut flowers that are flown north all the year.

As elsewhere, the pattern of life in the French Alps is changing and today only one-third of the labour force is engaged in agriculture. The rest are mainly in factories made possible by the development of hydro-electric power.

Water power and industrial development

The French Alps with their snowfields and glaciers for assured runoff, their abundance of waterfalls, deep valleys and dam sites for reservoirs, are ideal for hydro-electric power generation. In fact, a French engineer Aristide Bergés was the first man to develop electricity from water power for his wood-pulp and paper factory near Grenoble in the Grésivaudan (see Chapter 5). Before 1939, much water-power had been developed in this region, but it was small in scale, limited to single valleys and consumed locally in the electro–chemical and electro–metallurgical industries in and around Grenoble. Since the development of long-distance transmission of electricity much larger schemes have been undertaken, designed to fit in with the national grid.

Among these schemes, of which there are 10 with more than 150 stations, are the Sept Laux; the Bissorte barrage, which was constructed across the lip of a hanging valley with a drop of 1144 m; the Chambon scheme which uses the hanging valley of the Romanche for a storage reservoir; and the Barrage de Sautet in the upper Drac valley. The most important schemes, apart from the Rhône, are developed in the Isère valley where a barrage at Tignes supplies power stations at Les Brévières and Malgovert. Also on the Isère four underground power stations have been constructed at Randens and another at Pisançon. Schemes are also projected and several completed in the valleys of the Arc, Arve and Romanche. By 1956, 60 per cent of the hydro-electricity and 34 per cent of France's total power was being produced in the French Alps; these figures include the power produced by the Génissiat dam on the middle Rhône (see Chapters 10 and 15).

A beginning has also been made on the more difficult task of providing adequate hydro-electric power in the Maritime Alps where the river regimes are less regular, irrigation more essential and porous limestones more common. The Durance, which rises near the Franco–Italian frontier, flows over impermeable rocks and therefore has provided more suitable dam sites. Its regime before regulation was highly variable with low water in summer and winter and high water in spring and autumn. Disastrous droughts were followed by equally disastrous floods. A complete hydrological survey of the Durance drainage basin was carried out in 1955, and plans prepared to prevent flooding, to provide irrigation and drinking water, and electric power (fig. 58). A large earth barrage has been erected at Serre Ponçon below the confluence of the Ubaye with the Durance and the reservoir behind it forms a lake 18 kilometres long. A power station erected at the dam provides 700 million kWh a year and 10 smaller stations line the river between the dam and the Verdon confluence, where the Castillon barrage controls the flow of water in the huge Verdon gorge. Below that confluence the Durance water is deflected into a derivation canal and is carried via the old Durance channel over the Col de Lamamon into the Étang de Berre and so to the open sea by means of the Rove canal tunnel. Five large power stations utilize the water of this derivation canal and 20 others are planned for the upper and middle course of the river. When complete the Durance system will add 24 per cent to the total French hydro-electric power output, 30 per cent more land will be brought under irrigation in the departments of Vaucluse and Bouches du Rhône, and Marseille will have an assured water supply.

Already high-voltage transmission lines link these hydro-sites with the main grid to Paris, Marseille and Lyon. Eventually they will be grouped under the Génissiat and Donzère–Mondragon power schemes.

Industrial development

Industrial development in the French Alps shows a long history of domestic and small factory industries based on local deposits of iron or precious metals, for example silver at Argentière, or upon forest products which supported cabinet making and wood carving, or upon agricultural surpluses of which the glovemaking of Grenoble is the best example. Those small local industries survived until the railways introduced cheap, mass-produced goods but subsequently there has been a great resurgence of industry in the area due to abundant, cheap hydro-electric power and of improved communications. The food, wood-pulp, paper, leather and glovemaking industries have increased

55 *The tourist industry of the northern French Alps* (after P. and G. Veyret, *Les Alpes*)

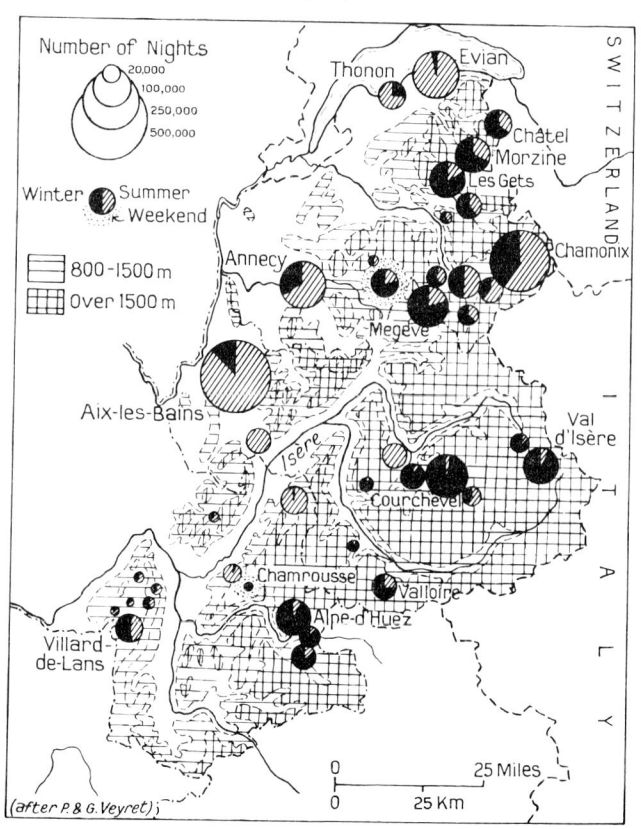

in size and scale. The significant changes, however, are the establishment of electro–metallurgical and electro–chemical industries and the concentration of the bulk of all industry in and around Grenoble (233 243) in the Grésivaudan. The table of the consumption of high-voltage electricity (p. 167) provides details and interesting sidelights on this distribution.

The electro–metallurgical industries are mainly concerned with the production of aluminium and the making of steel alloys. Ninety per cent of French bauxite, or 2 million tons a year, is mined in the Argens valley around Brignoles and Le Luc. This is first processed at La Barasse, Gardanne and Saint-Louis des Aygalades and then refined into aluminium in many old metallurgical centres including Saint-Jean de Maurienne, and Chedde. The total output amounts to approximately 150 000 tons a year. Special steel alloys, also using electric furnaces, and involving chrome, copper, cobalt and magnesium are smelted at Chambéry and Grenoble. Engineering based on these metallurgical products centres on Grenoble and is specifically designed to manufacture turbines and power-house equipment. Other heavy industries that rely on the abundance of cheap hydro-electric power are cement-making at Grenoble and chemicals in the Grésivaudan.

Tourism

Another outcome of improved communications is the remarkable growth of the tourist industry which is the mainstay of Provence. At Nice the Promenade des Anglais was built in 1824 but the great expansion here and elsewhere on this coast came with the opening of the main railway from Paris in 1865. Another rapid growth was associated with the motor car and regular air services. Air flights to Nice, the Blue Train from Paris, the Route Napoléon through Grenoble, Sisteron and Digne, and a new motorway along the coastlands funnel 1 million visitors annually to the French riviera resorts. Inland provincial mediaeval towns like Gap, Briançon and Digne and the charming old fortified hilltop towns such as La Cadière and Le Castellet share in this prosperity and popularity. Fourteen car–sleeper trains run weekly between Paris and Avignon which is a convenient starting point for the Côte d'Azur, the Alps and Roussillon. Another two car–sleeper trains run weekly from Paris to Saint-Raphael and another to Nice, which is now the fifth largest city in France.

As the summer season along the coast merges into its cooler season, a winter season opens also in the High Alps with the coming of snow for winter sport. Grenoble then becomes the 'gateway to the Alps' and towns such as Aix les Bains (one of the oldest spas in Europe) and Chamonix become skiing metropoli (see fig. 55). Winter sports are very popular in France and over 200 resorts participate in the sport, 40 being fully equipped. The new resorts include Bourg–Saint-Maurice and scores of others in the French Alps. Grenoble was the centre for the 1967 Olympic winter sports. Cable cars and télérifics which in the warm season carry milk from the alpine pastures now carry tourists and skiers in winter. The summer too brings vacation visitors to the mountains where national parks are now being created.

Urban growth

The most important town in the whole alpine region is Grenoble which grew up in the wide, fertile valley of the Grésivaudan on the left bank of the Isère river (fig. 56). The Grésivaudan itself suffers no marked effect of shadow and is followed for its entire length by a string of manufacturing towns, creating a conurbation of 405 000 people which focuses on Grenoble. The city also controls valley routes leading into the High Alps and is on the important railway line to the Mont Cenis pass and Turin. With a population of about 162 000 it has administrative, financial and university functions, and is besides a large tourist centre with 100 000 visitors a year and important manufactures as described above. The other notable city, besides Nice (323 000) and other famous resorts, is Toulon which owes its importance largely to its naval base. Its industries are concerned with shipbuilding and maintenance, including merchant vessels and fishing boats, and it has a population of 176 000 (350 000 in neighbourhood) which makes it the twelfth largest city in France. On the eastern end of this rocky coastline, upon the populous Côte d'Azur is the tiny principality of Monaco.

56 *The site, growth and industrial layout of Grenoble (after P. & G. Veyret et al.)*

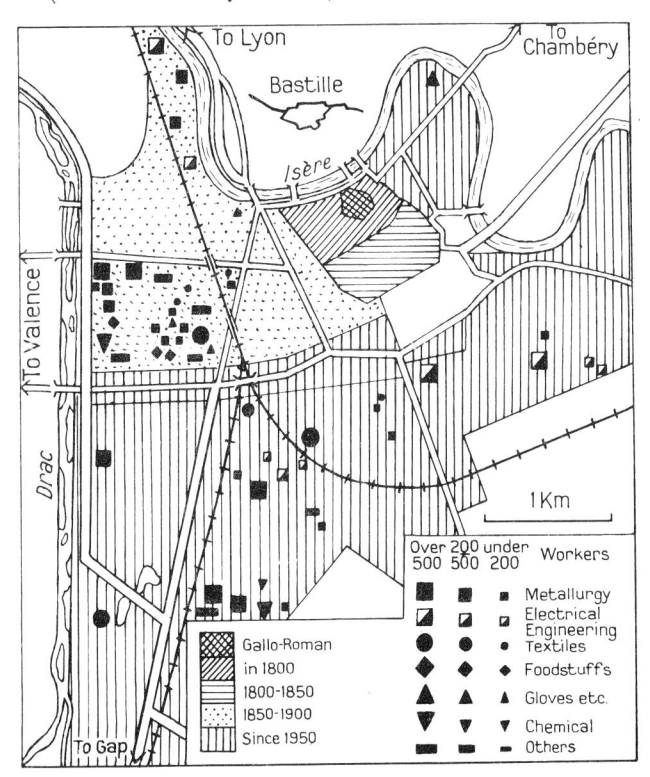

XVIII *General view of the principality of Monaco showing harbour and Monte Carlo*

MONACO

This sovereign independent state under French protection is situated on the coast of the French riviera 15 km east of Nice and not far from the Franco–Italian frontier. A micro-state of 150 hectares and 23 000 people, it was established on a terrace overlooking the Mediterranean and backed landward by high mountain walls. Offshore it has an anchorage with a sizeable harbour and fairly deep water (8 m) along its quays. Landward access to it is by winding mountain road.

The settlement dates back to Phoenician times but its modern independence stems from the Treaty of Paris in 1861. The principality consists of three communes: Monaco-ville (1860 population), which is the ancient settlement and retains its mediaeval ramparts around the later royal palace and cathedral; La Condamine (9858), a winter resort west of the bay of Monaco, with manufactures of perfume and liqueurs; and Monte Carlo (8838), a playground of Europe with a casino and luxury hotels. Taxes on gaming pay for all the improvements and services in the principality which also has an income from the sales of postage stamps. The native Monegasque population of 2700 pay no taxes and are not allowed to gamble. Tourists number about 650 000 annually and provide the chief means of employment. The attractions include a motor-car rally and a world-famous oceanographical museum and botanical gardens.

THE PYRENEES
Structure and relief

Extending for 400 km across the neck of land between the Bay of Biscay and the Gulf of Lions and with a greatest width of 130 km, the Pyrenees form one of France's 'natural' frontiers, although two-thirds of the Pyrenean area are in Spain.

The Pyrenees with an average height of 1200 m, are lower than the Alps. They are highest and widest (80 km) in the centre where they culminate in the Pic d'Aneto at 3412 m. To west and east they decline in height and width, and in the east attenuate to a mere 8 km and a single range, the Monts d'Albères, that ends at the coast at Cape Cerbère. In long sections of the central Pyrenees the crestline does not coincide with the main watershed and the international boundary often follows neither. As a result the head valley of the Upper Garonne, the Val d'Aran, lies in Spain although accessible by road from the south only since 1925; and the upper Segre valley, with the small Spanish enclave of Llivia, is in France being reached by a neutral roadway 5 km long. Some of the summit passes on routes across the mountains are either in France or in Spain and not at the international boundary. The chief international and Spanish passes are shown opposite; the other main summit cols entirely in France on routes leading direct or indirect across the central Pyrenees are the Col de Perche (1577 m) on the Perpignan–Bourg Madame–Puigcerda

Southern France: the Mountains 175

route; and the Col de Puymorens (1931 m) on the route from Pamiers to Puigcerda and also giving access to Andorra via the Puerto d'Envalira. Within the ranges on the steep French flanks the longitudinal valleys are weakly developed and communications between the various valley-incisions are difficult and expensive. The climb to the high crest of the main massifs also poses steep gradients and serious engineering problems, and the main movements of goods and people have invariably flowed round either end of the range. In 1968 of the 11·5 million tourists entering Spain from the French side of the frontier, 4·8 million crossed or circumvented the western end of the mountains and 4·2 million the eastern end. Had Napoleon seen figure 57 probably he would not have said 'Il n'y a plus de Pyrénées'! To relieve the long queues at the customs posts at the coastal ends of the Pyrenees a new international road tunnel has been opened between the Neste d'Aure valley at Aragnouet in France and Bielsa on a heastream of the Cinca in Huesca province. With a length of 3 km and a summit height of 1827 m it provides a more direct route from Toulouse to Zaragoza and Madrid. As usual it will also open up possibilities of winter sports and of summer camping and climbing.

The ranges themselves belong to an earlier Alpine mountain-building movement, and are of great complexity. A Hercynian crystalline core was involved in the folding, suffering much contortion and fracturing in the process. It and the longitudinal structural zones of the Pyrenees and the various geographical sections into which the whole range is usually divided are described in Chapter 12.

Near the eastern end the Canigou (2785 m) is interesting as it is a high, detached Hercynian mass that faces the

57 *Main routes and passenger traffic into Spain across the Pyrenees in 1968*
  1 *Port Bou railway (878 000 persons)*
  2 *Col de Perthus (3 336 000 persons)*
  3 *Envalira pass in Andorra*
  4 *Col de Puymorens in France*
  5 *Puigcerda (697 000), road and railway reached from Perpignan by the Col de la Perche*
  6 *Puerto de la Bonaigua and upper Garonne, with road tunnel in Spain*
  7 *new road tunnel from Aragnouet (Y) to Bielsa*
  8 *Col du Pourtalet (80 000)*
  9 *Somport at Canfranc, road and rail (178 000)*
  10 *Roncesvalles or Puerto de Ibañeta*
  11 *small arrow, Dancharinea (150 000)*
  12 *Irun, where widest arrow (2 421 000 persons) is main highway, next widest is Behobia route (1 848 000) up Bidasoa valley and over the Puerto de Velate (868 m) to Pamplona, and narrow arrow to west is main railway (577 000 persons)*

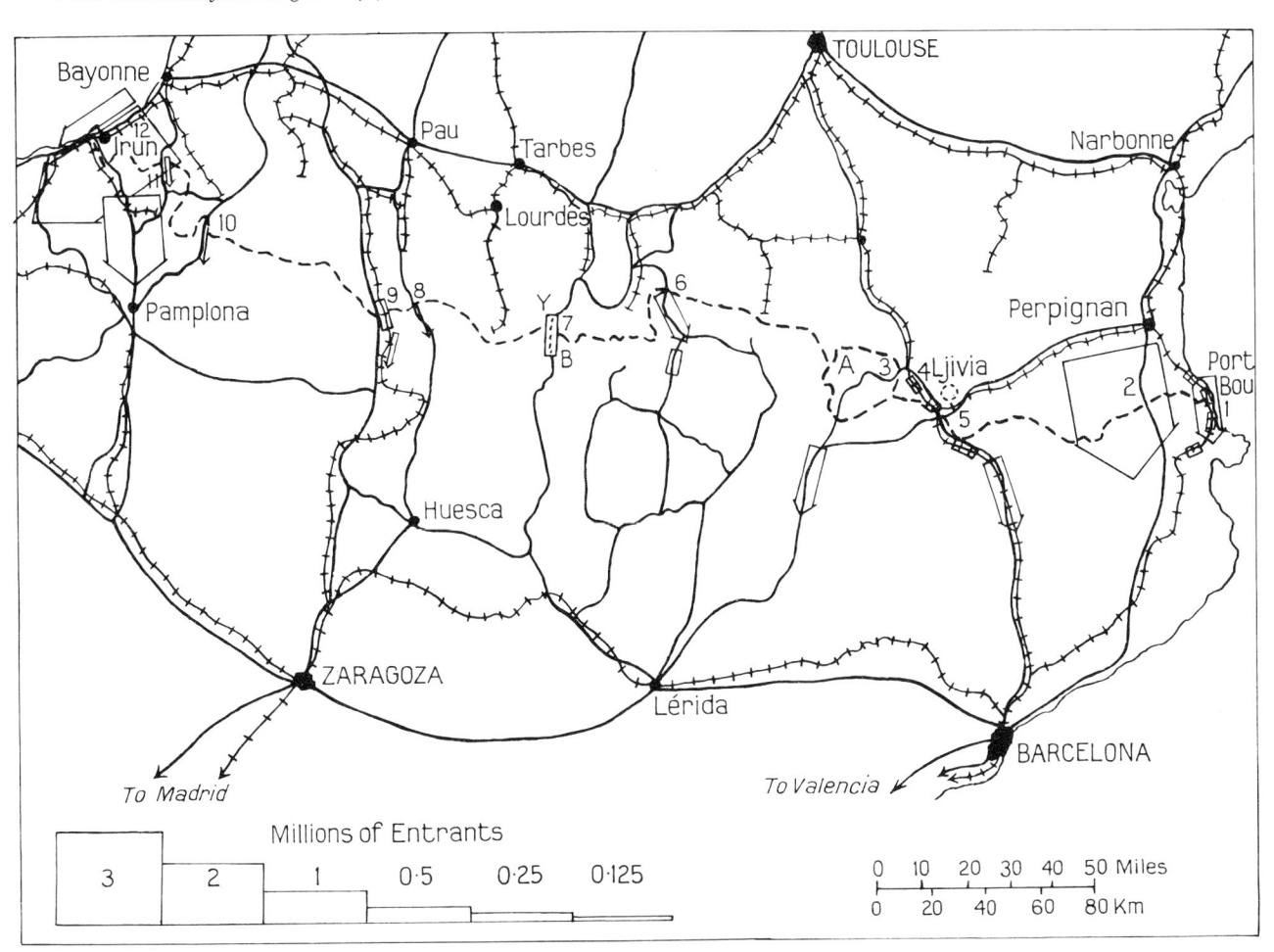

similar detached fragments of Esterel and Maures across the Gulf of Lions.

The glaciation and its scenic effects, particularly the fine cirques, have also been described elsewhere. About 34 square kilometres of ice and snowfield exist today on the northern slopes of the central Pyrenees, the largest patches being on the Maladetta and the largest single glacier on the shaded slope of the Pic de Vignemale. Fine crescentic terminal moraines survive in some of the glaciated valleys.

Drainage and fluvial deposition

The runoff of the Pyrenees goes mainly into France in spite of the smaller Pyrenean area on the French side of the watershed. The chief rivers are the Garonne and its tributary the Ariège, the Adour and its tributary the Gave, and the Aude. This asymmetry of runoff is highly fortunate for France and depends basically on the greater wetness of the northern slopes. Other characteristics of the drainage are the marked infrequency of longitudinal valleys and the fine subterranean systems. The only notable longitudinal river courses are on the Ariège in France, on the Aran in Spain and on the Têt, Agly and Tech in the east. The subterranean drainage on the limestone mountains includes several remarkable series of caves, waterfalls and lakes. The Garonne has its source underground on the south side of the watershed and the *Union Pyrénéenne Electrique* has tapped this underground water near the Val d'Aran to obtain a fall of nearly 1000 m. The many karstic potholes include the Gouffre de Pierre Saint-Martin, 346 m deep and leading by a string of seven enormous caverns to a depth of 728 m.

At least two other aspects of Pyrenean drainage on the Frency side deserve mention. In the central Pyrenees a number of springs are sulphurous or chalybeate and have encouraged the creation of numerous spas or health resorts. Also on the northern foot of the mountain mass there are remarkable deltaic fans of gravel and sand deposited in Pliocene times by the ancestors of present Pyrenean rivers. These include the plateau of Lannemezan which has a typical simple consequent drainage over the gently sloping surface of the gravels. These old deposition fans are, however, mainly outside the Pyrenees and belong to Aquitaine.

Climate and vegetation

As these too have already been described (Chapters 6 & 12) little will be added here more than a few details relevant to the French side of the range. This is the rainier flank especially when Atlantic depressions move inland in autumn eastward across the plain of Aquitaine. The Pyrenees tend to deflect these 'lows' through the gap of Carcassonne into Languedoc and in so doing may receive heavy rainfall on their northern slopes.

The severity of climate and abundant precipitation, often as snow, on the higher summits are shown in the statistics for the Pic du Midi (p. 170). If these are compared with the statistics for Chamonix and Briançon in the French Alps it will be seen, even if allowance is made for altitude differences, how fortunate the French are in the runoff of their Pyrenean streams.

The vegetation, as already noticed, changes from central European plant associations in the wetter west to Mediterranean species in the east. In the west the Irati forest still has about 2000 hectares of tall treegrowth. In the east, kermes and cork oak gradually become dominant on the lower slopes and grade upwards into sweet chestnut, rhododendron, pine, birch and juniper. Maquis and garrigue surround these eastern woodlands while in the west bracken and fern dispute the woodland clearings with meadow grasses. Much natural regeneration of forest is prevented by the grazing of sheep and goats, especially in the east, where at the end of dry summers goats often have to survive on tree foliage and vine prunings. Great improvements could be made from a farming point of view if extensive liming were financially possible, as lime would discourage the large areas of dwarf rhododendron in the east and of bracken and fern in the west.

Land use and farming

The emphasis everywhere in the French Pyrenees is on pastoralism, but statistics are not readily available for the relative extent of the Pyrenean pastures and crops as some of the départements, especially Haute Garonne, are partly or largely in the lowlands. In the départements of Hautes Pyrénées and Ariège, which consist largely of the Pyrenean range and foothills, about one-sixth of the land is arable, one-quarter to one-third permanent pasture and one-quarter wooded. In the Pyrenees on the French side there are about 520 000 cattle and nearly 750 000 sheep and both are most numerous in the west and central Pyrenees. The sheep are of various breeds kept for wool, milk (for Roquefort cheese), meat and hides. The cattle are mainly dairy breeds for supplying fresh milk to towns and for cheesemaking. In autumn the surplus stock is sold for veal or as store cattle to farmers in Aquitaine. Many of the bullocks are still reared as draught animals. The goats, about 60 000 in all, are kept chiefly on the garrigue mountain slopes in the east. In parts, there is also the breeding of horses for working animals and of mules for export to Spain.

The arable farming in valleys and wherever soils and temperature allow, is geared to stock-feeding except on warm sunny slopes, which are usually under temperate fruits, nuts and vines, and near towns, where market gardening pays best. In all higher districts the pastoral activities are typically transhumant. With the recent increase of tourism, cultivation and dairying have intensified and cooperatives increasingly handle the supply and marketing problems. The eastern Pyrenees show the effect of heightened summer drought by larger areas of vine and the appearance on the lower slopes of Canigou of the olive and citrus. Here, too, however, the cooperative system leads occasionally to specialization and the upper Tech valley, for example, has extensive cherry orchards.

Power supplies and industrial growth

The French Pyrenees have a large hydro-electric potential. As early as 1901 high-voltage transmission lines carried power from the Aude valley to Carcassonne and Narbonne but, as in the French Alps, most of the development here up to 1939 was on a small scale. The need to electrify the French railways in the absence of coal supplies was a great stimulus to hydro-electrical development in the Pyrenees and by 1939 the two railways across (or round) the mountains and many of their connexions were electrified. After 1945, French power resources were nationalized and the whole power network revitalized. By 1955 no less than 96 hydro-sites were in operation although none equalled in output the greater schemes of the French Alps. Today the Pyrenees produce one-eighth of French hydro-electricity and supply railways, local industries and the national grid.

The various local industries have expanded under the stimulus of an assured power supply. The mineral resources include talc as at Trimouns and Céret in the east (one-tenth of the world output); and bauxite at Péreille and Roquefixade that is made into aluminium at Beyrède in the Aure (Neste) valley and elsewhere. More important is the rich haematite ore (about 250 000 tons annually with 55 per cent iron content) mined in the Ariège valley at Vicdessos, Rabat, and in the Canigou. Today in the French Pyrenees alongside the small indigenous craft (iron-work, etc.) industries and traditional textile factories there have arisen elaborate modern electro–chemical and electro–metallurgical concerns in an integrated iron and steel and engineering industry stimulated, as in the French Alps, by hydro-electric power. The blast furnaces at Tarascon are still coke-fired but Pamiers has electric furnaces and the two towns together produce 20 000 tons of high-grade steel a year. Electrically operated ovens produce special steels at Rebouc, Saint-Antoine and Mercus. Electro–chemical and electro–metallurgical industries have developed in numerous other small towns dispersed through the larger Pyrenean valleys, such as chemicals at Boussens, explosives at Marignac in the Garonne valley and carborundum abrasives at Sarrancolin in the Aure valley. However significant concentrations of metallurgical and engineering plants occur around Tarbes (insulators, military equipment, etc.), Lourdes and Bagnères de Bigorre. Aircraft are assembled at Ossun, reflecting a similar concentration around Grenoble in the French Alps.

Population

Some of these manufacturing towns could more properly be included with Aquitaine but they owe their prosperity to the Pyrenees, for it is a characteristic of the French side of the range that the main clusters of people have been swept with the drainage down to the peripheral slopes, and in places right on to the plain. The regional centres such as Pau, Pamiers and Lourdes, overlook the plain of Aquitaine. They are at the junction of routes and at the entrance to a valley, and a few are on routes across the Pyrenees. The trans-Pyrenean railways and roads are described in Chapter 4 (under snowfall) and in Chapter 12 (under Spain). These routes mainly affect towns on the Atlantic and the Mediterranean coastal end of the mountain barrier, especially as there is a change of railway gauge at the boundary. Many French towns probably benefit from the fact that the central Pyrenean roads are closed by snow in winter when snow sports, notably at Cauterets, are of considerable importance. The creation of national parks and of nature reserves on the mountains has stimulated summer tourism and the spas have long been popular. Over fifty spas have thermal and medicinal springs, among the best known being those at Ax-les-Thermes, Bagnères de Bigorre, Bagnères de Luchon, Amélie-les-Bains and Le Boulou. Multitudes of pilgrims visit Lourdes in the Gave de Pau and an airport has been built there. In the meanwhile there has been a general drift of population in the Pyrenees away from isolated valleys and difficult terrain towards the new employment in the expanding towns at the mountain edge. Some more mountainous tracts that could be sparsely inhabited are now quite empty of permanent residents and possible problems of rural overpopulation and lack of social services have been solved in the process. Much of the mountain isolation must be attributed to the way in which so many headstreams and local tracks end in steep-walled cirques, and to the rarity of linking longitudinal valleys. Significantly the main highway from Bayonne to Perpignan near either end of the Pyrenees lies 40 km north of the frontier and watershed.

It would be quite misleading to close any account of the highlands of southern France on a note of emigration and depopulation. Tremendous efforts have been made to attract immigrants, especially from abroad. Of the départements concerned in our survey only Ariège decreased in population between 1946 and 1968 (146 000 to 138 000). In every other département, even in Hautes Alpes, the development of hydro-electricity, of manufacturing and of tourism has caused a steady population growth. During that period all told over 1 million new residents were added, the absolute increase being greatest in Alpes Maritimes (269 000), Isère (226 000) and Var (185 000). When viewed in the light of dwindling mountain economies in so many parts of the world, it is little short of a miracle.

15 Southern France: the Lowlands

The lowlands of southern France described in this volume consist of the Rhône–Saône valley terminating in the delta of the Camargue and the coastal plains of Languedoc and Roussillon. The latter will be described first, as geographically they are a littoral appendage of the great Saône–Rhône depression.

LANGUEDOC AND ROUSSILLON

These lowlands in recent geological times were overwhelmed by a marine transgression to a height of about 52 m above present sea level. The water penetrated far inland up the estuaries of the Aude and Hérault and laid down the sediments now forming the Garrigue and the low plateaux of Bas Languedoc. It lapped against the foot of the hills today forming the gap of Carcassonne (191 m) and of the Corbières ranges in the south. Out of this shallow sea, hills such as the headlands of Cap Leucate and d'Agde jutted as islands. Into it rivers emptied their sediments which spread out in huge gravelly deltaic fans that today stand at 30 to 55 m above sea level. Notable examples occur in the lowland east of Les Aspres in Roussillon, the Grès district of Montpellier and the low plateau, 32 km wide, forming the Costière de Saint-Gilles northwest of Arles.

During the Quaternary period world sea level oscillated periodically by 100–200 metres and has almost certainly risen by about 70 metres in post-Quaternary times but it seems to have been fairly stable since the beginning of the Roman era. In Languedoc, river and coastal deposition was then already gaining on the sea. Salt-water retreat was slow and water was trapped on the land surface as shallow lakes, for example, the Roman *lacus Narbonenses*, which had on their seaward edges salt marshes and lagoons or *étangs* of which 3200 sq km still exist.

The soft limestone areas like the Garrigue were easily eroded by rivers and have stupendous gorges. The sediments accumulated at their mouths as low fans, now the low plateaux that can be traced from the Rhône to the Étang de Thau and round the south side of the Corbières massif. Between them and the margin of the sea the emerging coastal plains were converted by rivers into level alluvial floodplains. On the coast, deltaic river deposition, longshore drift and lagoon-filling extended the land margin and the sea which once carried crusaders from Aigues-Mortes is now six kilometres from that medieval port. The longshore drift is westward and sediments brought down by the Var and Rhône are drifted along the shores of Languedoc and Roussillon smoothing out the coastline and building up bay bars and beaches. The long loops of sand dune are today usually fixed by plantings of marram grass and their natural movement inland is checked by a causeway which carries the coastal railway. The river mouths are kept clear by dyking, which also assists the drainage and reclamation of the salt marshes behind the dunes.

Beyond this coastline of étangs, dunes and salt marsh are the alluvial lowlands of Roussillon and Languedoc. Here four distinct types of landform can be seen: the reclaimed salt marshes or *souillères*; the river floodplains of the Aude, Agde, Herault, etc.; the deltaic gravel fans and low piedmont plateaux; and the Garrigue or Cévennes foothills covered by aromatic shrub that has given its name to similar vegetation elsewhere. Roussillon is a separate, semicircular plain lying in the extreme south between the high country of Corbières and Albères and backed by the Pyrenees, but it displays on a smaller scale the same landform elements met with in Bas-Languedoc.

Climate

Over the whole lowland the typical Mediterranean summer-dry climate and the mistral exerts a strong unifying influence. The climatic statistics for Perpignan and Montpellier, more or less speak for themselves, and the mistral and various weather types are discussed in Chapter 4. Cold airflow from the north, or mistral, brings frequent slight frosts during the mild winters but for delicate crops they can usually be warded off by matting covers or cloches or Dutch frames. At Perpignan frosts are much less frequent than near the Rhône corridor and crops rarely need protection at night. Autumn is the rainiest season and the annual amounts decrease rapidly southward from 785 mm near Montpellier, to 550 mm near Perpignan and 445 mm at Salses, the driest locality in France.

As in all areas of Mediterranean climate, local winds are very important. They are usually associated with the passage eastwards of depressions in the Mediterranean area, and are therefore more frequent in autumn, winter and early spring. The front of the depression brings tropical airflow from the south, which having passed over the Gulf of Lions is rainbearing, and is called expressively the *marin* or *l'autan*. In the rear of the depressions, the Polar or Arctic airflow from a northerly quarter is cold and dry and is called locally the *cers* and is in fact normally a mistral. Land and sea breezes too have descriptive names. The fresh wind off the sea in spring is called the *grec*; when it blows inland charged with moisture during the summer bringing oppressive weather its name changes to *labech*. Cold airs which drain down off the Pyrenees when the lowlands heat up are called *tramontane roussillonnaise*.

Two important limiting climatic–pedological factors must be stressed in considering the land use of these lowlands. First, aridity increases southwards; and secondly, large areas are floored with thirsty sands and gravels which do not hold moisture, and with alluvium, which dries out rapidly and cracks in summer.

Agriculture

Under natural conditions the Mediterranean triad of wheat, olive and vine is almost imposed upon the area. It was one of the granaries of Imperial Rome and today about one-sixth of the ploughland is under wheat. This crop is not usually successful where the rainfall is less than 600 mm and so it is absent from Roussillon and the

SOUTHERN FRANCE: THE LOWLANDS

Département	Area (sq km)	Population 1968 Number (thousands)	Density per sq km	Capital	Number (thousands)	Arable	Permanent grazing	Fruit and vine	Agricultural not cultivated	Non-agricultural	Woods and forest	Bovine *In milk	Bovine Total	Sheep	Goats
Haute Saône	5 343	214	40	Vesoul	18	109	158	1	39	26	202	109	193	41	1
Côte d'Or	8 765	421	48	Dijon	151	266	166	10	76	62	292	96	227	88	3
Saône et Loire	8 565	550	64	Mâcon	35	205	358	13	16	67	197	236	518	119	37
Rhône	3 215	1 326	412	Lyon	535	55	97	22	24	22	180	58	94	28	17
Drôme	6 525	343	53	Valence	64	171	76	40	25	129	213	32	57	177	59
Vaucluse	3 566	354	99	Avignon	89	67	10	65	80	25	106	2	3	97	7
Bouches du Rhône	5 112	1 470	288	Marseille	894	92	66	53	64	146	104	5	12	280	1
Gard	5 848	479	82	Nîmes	130	68	132	120	71	53	163	3	6	113	27
Hérault	6 113	591	97	Montpellier	167	56	105	181	74	55	142	2	3	88	5
Aude	6 232	278	45	Carcassonne	46	144	132	120	101	51	86	14	25	72	2
All France	533 998	49 779	92			18 331	13 632	1 749	3 693	4 881	12 485	10 694	21 184	9 186	1 017

* 2 years or more old in milk

southern half of Languedoc; in the north it is grown sporadically on local patches of clay or wherever a deep tilth can be provided by terracing the lower hillsides. It is in any case grown in rotation either with fallow in the drier areas or where irrigation water is available with lucerne or some other green fodder crop. Because of the marginal rainfall conditions and the bright sunshine it is quick-ripening, hard wheat with a high protein content.

The greater part of the remaining agricultural land, or nearly two-thirds of the total cropland, is under the vine and this is an outstanding area of monoculture (see table at top of the page). Introduced first by the Phoenicians in 600 B C, the vine has been grown here since classical times, and has almost ousted every other crop. The vines occupy the barren slopes of the Garrigue, the coastal lowlands, valley sides and recently reclaimed land so that obviously soil and moisture conditions are not crucial. Nevertheless the best dessert grapes, and wine grapes do tend to favour distinct areas. The quality of both table grapes and the flavour of the wine largely depend on the amount of warmth and sunshine between the setting of the fruit and its final ripening. In this the nature of the soil and the slope have a significant part to play. The stocks are planted about 1½ metres apart and in the Midi the plants are kept low with a straight stem and two side-branches springing from the top. Every year pruning removes about ninety per cent of the new growth, the flowers are thinned out, and so are the berries, particularly of the dessert variety. In spring the ground is carefully cleared of weeds to give every available scrap of moisture to the vines and to allow the soil to reflect its warmth to the fruit.

Single-crop economies (monoculture), though efficient in the purely mechanical methods of cultivation, in marketing and in the processing of the produce, are particularly vulnerable to attack by disease, both of the soil and the crop unless strict preventative methods are adopted. Increased humidity predisposes towards mildews and fungus growth; arid conditions towards blight and insect infestation. Vine stocks imported from America in 1863 brought with them phylloxera and mildew. This caused such devastation in the vineyards of France that 1 million hectares of vines were ruined and the vineyards were reduced in a few years to one-fifth of their former extent.

Immune American vine stocks were imported and the old French and other European stocks gradually replaced by them. The replanting and reorganization of the industry were achieved through the cooperative system. The first of these was started in the Biterrois district in 1901 but today there are over 500 cooperatives concerned with viticulture with 190,000 grower–members, representing 90 per cent of all vinegrowers. Only the viticulturalists that produce fine table wines have little need of a cooperative. For the small growers producing 'vin ordinaire', cooperatives handle, and own, all the marketing and processing in *caves co-opératives*, and have extended their control to bottles, corks, labels, barrels and the necessary insecticides and sprays. These last are essential since no means of combating the pests and mildews have been discovered except spraying with copper sulphate solution every few weeks. Probably owing to the increased security following the establishment of the cooperatives there are still here about 200 000 farmers and 1 million persons dependent mainly on the vine. Languedoc and Roussillon produce 40 per cent of the total French wine output or 25 million hectolitres or 9 per cent world total; the bulk of it is good but

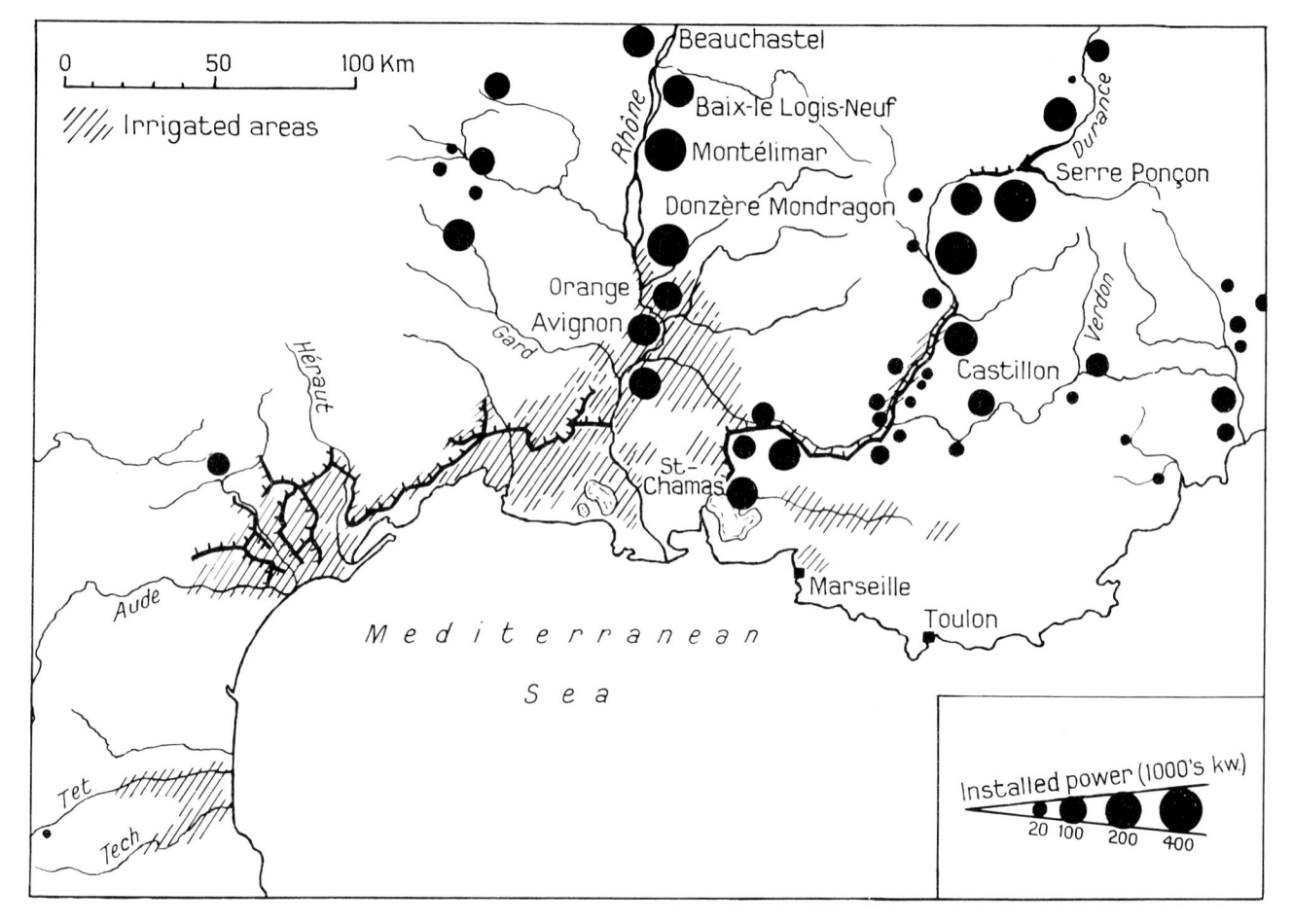

58 *Hydro-electric and main irrigation developments in Mediterranean France* (after P. Carrère and R. Dugrand)

undistinguished red wine. The region faces competition from the import of large quantites of cheap wines from north Africa, and also discouragement from the government who see in this monoculture not only a health hazard but also an extravagant use of a valuable agricultural resource.

Near the towns market gardening makes an appearance and it is hoped to increase the scope and scale of this by a unified system of irrigation for the whole of Languedoc and Roussillon. This scheme includes a master channel, the canal du Languedoc which runs from the Rhône north of Arles to the river Aude, and a series of feeder canals leading off from it (see figure 58 and Chapter 5). The valley of the Aude has already a prosperous intensive cultivation of temperate cereals in rotation with fodder crops and roots, alongside orchards intercropped with vegetables and stocked with bees. This arable farming further supports and is supported by a well-balanced animal husbandry associated with dairy cattle, pigs and poultry.

Market gardening also flourishes in Roussillon where an efficient irrigation system uses river water. As the market gardening has prospered due to increased demands from the growing urban and industrial populations of the north, so has the demand for increased water supplies. Wells have been sunk to tap the underground water supplies in the gravel pediments and diesel and wind pumps are a frequent sight. These irrigated lands, 18 000 ha all told, form strips alongside the rivers Agly, Têt and Tech. Here the mild winters and very early springs encourage the production of early vegetables and salad crops that are grown under orchards of peach, apricot and plum before the flush of leaves shadows the ground. Figs are grown against limestone slopes with a southerly aspect and vines climb the lower valley sides. Asparagus and artichokes flourish on the reclaimed coastal salt marshes and on the irrigated flats table grapes yield good profits. The crops are marketed through cooperative schemes which also handle packaging and at Perpignan (102 000), the most important distributing point, there are cold-storage depots alongside the railway yards. This town acts as the administrative, financial and market centre for Roussillon.

Olives make their appearance usually where the site is too arid for other crops, and occasionally on steep slopes where they help to prevent soil erosion, and alongside the roads and field boundaries. They are especially numerous in the valleys of the Garrigue which drain to the Hérault. The total French output of olives is 13 000 tons p.a. making France the eighth olive-oil producing country in the world (see Chapter 8). Other typical Mediterranean tree crops are the walnut and mulberry, the latter being sufficient to support a small silk industry at Nîmes and Perpignan.

The more barren areas of the country such as the salt marshes and the Garrigue are only suitable for the pasturing of sheep and goats. These number approximately 300 000 and 40 000, respectively. During the parching summer months they have to be moved to cooler pastures on the Massif Central. The meat finds a ready sale in north African countries and the salt marshes are supposed to lend a specially sweet flavour to the mutton. The raw materials find their way to the towns, the leather and wool

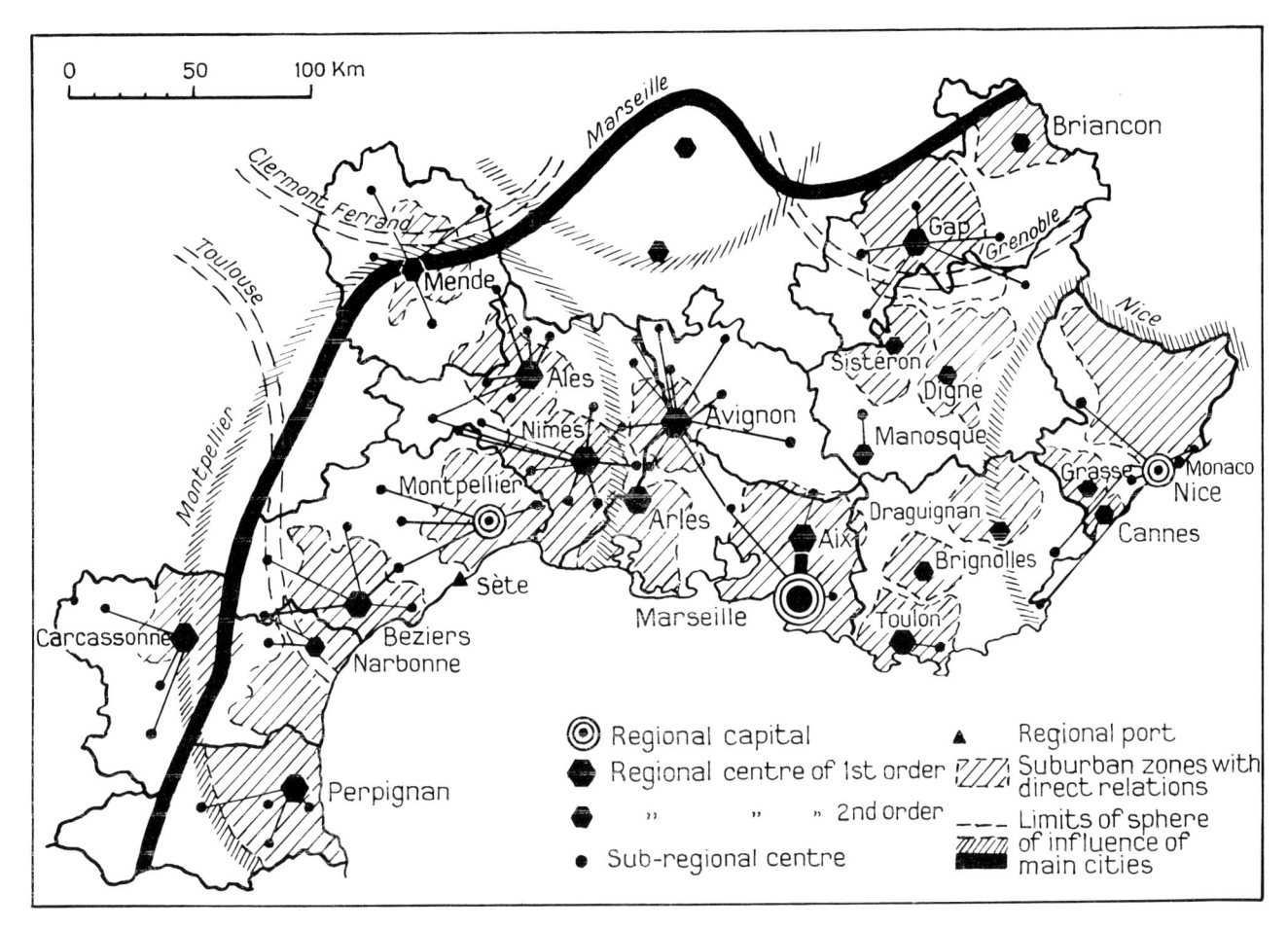

59 *The urban network of Mediterranean France* (after R. Dugrand)

going to Nîmes for shoes and cloth and the milk products to cooperative creameries for making into Roquefort cheese.

A small fishing industry provides an alternative occupation for 1900 fishermen during the summer before the vine harvest begins. A number of small fishing villages along the coast land about 11 000 tons of fish, mainly tunny, sardines and anchovies. A beginning has been made in fish-farming for eels, oysters, lobsters and crayfish. Further details will be found in Chapter 3

Urban and industrial growth

Evaporation pans along the coast provide salt for chemical industries as well as for food-processing concerns such as pickling olives at Montpellier. A petroleum refinery at Frontignan handles 1 million tons of crude oil annually and has brought modern chemical industries to the nearby port of Sète, which was founded in 1666. This, the largest coastal town (41 000), also has cement works, and the manufacture of agricultural equipment chiefly for viticulture. The port handles about $3\frac{1}{2}$ million tons of goods a year and is for cargo the seventh port of France. It trades mostly with Algeria, importing crude mineral oil, phosphates, sulphur, wire, copper sulphate and wool in exchange for refined petroleum products, other chemicals, fertilizers, meat and wine. Its small fish-preserving industry depends on the local sardine catch. Sète is also an inland port being the terminal of both the canal du Midi and the Canal du Rhône–Sète. This latter canal uses the seaward side of a series of étangs in order to get a sufficient depth of water and is mainly used for the transport of refined oil to the Rhône valley. The Canal du Midi is only navigable by shallow barges which carry mainly wine, articles for the wine trade and vineyards, petroleum and cement.

Montpellier (168 000), the largest town in Languedoc, trades mainly in wine but food industries, pharmaceutical manufactures, including soap and perfume, are gaining in importance. It is the administrative and commercial centre of Languedoc and has a flourishing university.

Nîmes (130 000) stands on a low spur of the Garrigue overlooking the lower Rhône plain. It dates from Roman times when it was an important spa and still possesses beautiful Roman remains, a fine amphitheatre, temple, theatre, and so on. Today an important nodal centre for road and rail it has developed textile industries, the manufacture of leather goods and agricultural machinery and has railway repair shops. Other towns of local importance in Languedoc are Béziers and Narbonne both of which are concerned with all aspects of the wine industry as well as being agricultural market centres (fig. 59).

Mention must be made of the French government's enlightened scheme for the tourist development of the Languedoc–Roussillon coastline. Nearly 30 000 ha of coastland from Nîmes to Perpignan are protected from uncontrolled speculation and uninhibited development. Planned communications, full service facilities, including yachting basins, and building sites are made available for villas, camping sites and shopping arcades which also receive government financial assistance with the construction, provided they conform to the regional plan. This warm, sunny, lagoon-lined coastline may well prove a glorious exception to the haphazard development of so many of Europe's fine beaches.

By 1971 the value of the government planning efforts for Languedoc were evident. State assistance for industrial development has led to the establishment of major plants at, for example, Frontignan for petroleum refining and for fertilizers; at Montpellier for the manufacturing of computers; at Narbonne for sulphur refining; at Béziers for metal containers, and at Nîmes and Vauvert for food processing and preserving. The growth of industrial production in Languedoc actually exceeded the national rate. Among the expanding sectors was the mining of bauxite at Villeyrac, Bédarieux and Cazouls in Hérault. This is partly reduced at Salindres but for final smelting goes to Alpine hydro-sites. In 1967 of the French bauxite production of 2 813 000 tons, about 700 000 tons came from Languedoc and the remainder from Var.

The undertakings of the Compagnie Nationale d'Aménagement du Bas Rhône–Languedoc have helped significantly to convert the monoculture of vines into a more diversified irrigated polyculture and to add new agricultural tracts. The value of apples, apricots and table grapes has doubled since 1954 and the proportion of the value of the region's total agricultural produce supplied by the vine has dropped from 60 to 50 per cent. About 50 000 ha were irrigated by 1968 and considerable progress made in *remembrement* and in the construction of new farm buildings, cooperative factories and collecting centres. The efforts by government agencies include the introduction of many thousands of new workers, especially from Algeria and Spain, and this too has proved most successful. French Algerians have started new vineyards.

The progress made on the 150 km of sandy beach has also been remarkable. New coastal resorts have been built at Saint-Cyprien near Perpignan; at La Grande Motte on the Etang l'Or near Montpellier; and at Leucate–Barcarès on the Etang de Salses. Others also being developed are at Cap d'Agde; the mouth of the Aude; and at Gruissan on the Etang de Sigean. All told there will be new accommodation for 450 000 tourists and Languedoc will eventually become the California of France. It would indeed be an unimaginative society who could not capitalize intelligently on the sunshine and sands of the Gulf of Lions.

THE RHÔNE–SAÔNE VALLEY
Structure and relief

One of Europe's most important passageways follows the deep corridor that lies between the Central Massif and the mountains of the Jura, Savoie and Dauphiné and is drained southwards by the Saône and Rhône. From the Mediterranean this corridor leads either northwest over the low plateau of Langres to the North Sea or northeast by the Belfort Gap or Burgundian Gate to the Rhine. Today the roads and railways that follow this route are among the most heavily used in France.

The long narrow lowland is the result of subsidence along a well-marked fracture line, on the east of the Massif Central, of a portion of the Hercynian range. On the east the demarcation at the edge of the Alpine folds is not clearly defined but the rivers Isère and Drôme for part of their lower courses are thought to follow fracture lines. The surface of the sunken Hercynian mass lies at such widely differing depths that it appears to have been thoroughly shattered and fragmented. Some of the fragments even remain upstanding as horsts such as the plateau of Crémieu east of Lyon. As in the much larger sunken block of the Hungarian Plain and middle Danube, the subsequent history of the lowland is of deposition, lacustrine, marine, glacial and fluviatile, the last of which is still proceeding.

When the Pyrenean folds were continuous into Provence, it is believed, judging by the nature of sediments preserved on the floor of the corridor, that a large body of water was thus confined in the corridor and extended so far north as to occupy most of the valley of the Saône. When the central stretch of these Pyrenean folds collapsed and formed the Gulf of Lions the sea entered a long narrow inlet which reached as far as the 'sill of Vienne'. This is where the plateau of Crémieu approaches closest to the Central Massif and the present river is flowing in a narrow defile on a band of Hercynian rock linking the two.

The sea invaded the Swiss Mittelland, and limestone deposits of various kinds can be traced the length of the Rhône valley and on the plateau. Changes of sea level gradually drained this water off to the south to its present confines, and terraces cut in the valley sides attest to the continuously dropping base level.

North of the Vienne sill a freshwater lake covered the whole plain of Bresse which drained south by a notch in the sill into the marine gulf. A forerunner of the present river Doubs and other rivers emptied their load of pebbles, gravels and sediment into Lake Bresse and today the plain of Bresse displays all the characteristics of an old lake floor.

Rivers flowing down from the Alpine fold mountains east of the marine gulf spread delta fans of gravel and pebbles where they entered the sea. Such is the low gravel plateau of Bas Dauphiné and the stony area of the Crau. Gradually uplift south of Lyon caused the retreat of the sea and the formation of the Rhône valley.

During glacial times glaciers moved downhill from the Alps and formed a piedmont sheet of ice at the foot of the mountain which extended across the valley floor as far as the town of Bourg. As a result boulder clay is found widespread the length of the corridor, together with much ground moraine. Lines of terminal moraines can be traced alongside outwash areas of gravel and sand. In places too are patches of loess, blown and carried by the wind from exposed floodplains and outwash aprons of Pleistocene age.

Ever since, the streams issuing from the alpine glaciers

have added their burden of pebbles and stones to the valley floor where they have accumulated in wide spreads. In this way the upper Rhône has built up the Pays de Dombes where it joins the Saône. This process can best be discerned during the late summer months of low water when rivers like the Durance and the Drôme enter the Rhône between wide banks and shoals of pebbles while at the delta the finer mud and silt accumulate at contact with salt water.

Major geographical Regions

The Rhône–Saône valley can be divided by physical and human factors into four major regions:
1 the Saône basin
2 the Rhône valley from Lake Geneva to Lyon
3 the Rhône corridor
4 the Camargue or Rhône delta

The Saône basin

Les Monts Faucilles where the Saône headstreams rise west of the Belfort Gap are as far from the Mediterranean as the Tyne is from the Solent in England. This length and the southward direction make it one of the most important rivers in France. The highlands rise to 1000 m in places but their average elevation is about 500 m and when the Saône enters the Plain of Bresse it has fallen to half that height. In its subsequent long journey of 330 km over an old lake floor it is flowing at or very near its base level and exhibits all the traits of a slow meandering river. Powerful left-bank tributaries like the Ognon, Doubs, Loue, and Seille have thrust the river to the west side of the valley and it also receives from the Jura much subterranean drainage issuing as springs at the foot of the mountains. The plateau de Langres and the Côte d'Or also contribute such rivers as the Tille and the Grosne. The long gentle gradient of the lower Saône and the large amount of karstic subterranean water contributed to its volume gives the river a far more regular regime than that of the upper Rhône. The gravels brought down by the Rhône that form much of the Pays de Dombes mark the confluence of the two rivers.

The Rhône valley from Lake Geneva to Lyon

The Rhône enters France 20 km below Lake Geneva (Lac Léman) and is already a considerable river, having traversed central Switzerland for 170 km. The French Rhône has to negotiate the Jura plissé and the Jura tabulaire. The first obstacle it penetrates in a zig-zag fashion with long sluggish sections in the vaux where the channel is braided and meandering, and with swift gorge sections in the cluses, as at the defile de l'Écluse ten kilometres above Bellegarde and the defile de Saint-Pierre. The val section, below the cluse of Bellegarde is the site of the Génissiat dam and reservoir. From this point downstream the river is navigable for flat-bottomed boats. The Rhône crosses the Jura plateau in a northwest direction and, after skirting the north side of the Hercynian plateau of Crémieu, enters the Plain of Bresse where it is joined by the Ain. The floodplain of the Rhône in this section is 3 to 5 km wide and consists largely of reedbeds, marshes and gravel banks liable to inundation. To limit the danger of flooding a lateral canal has been built each side of the floodplain to take surplus water. After a journey in France of 200 km and a drop of 210 m the Rhône joins the Saône but the emerald waters of the latter can be distinguished from the milky water of the Rhône for some distance downstream of their confluence.

The Rhône corridor

This corridor section is the real funnel of ideas and commerce which has made the Rhône such an important river. It flows due south for 210 km in a valley filled with alluvium but at first its course still traverses the old lake floor of the Plain of Bresse until the river has crossed the 'sill of Vienne'. Thereafter the river flows through a series of seven plains or basins linked by gorge sections.

The plain of Vienne is succeeded by the plain of Saint-Rambert d'Albon and that by the plain of Valence which is considerably more ample than the first two since here the Drôme and the Isère enter the Rhône on the left bank. Leaving the plain of Valence the river enters the Cruas gorge which is so constricted that the river has been strongly dyked to make room for roads and railways to share the valley floor. The Montélimar basin follows and ends in the Donzère gorge. South of this is the Pierrelatte basin, the setting for the Donzère–Mondragon power scheme. South again the Mondragon gorge leads to the plain of Orange, and so to the gorge of Roquemaure where the Rhône enters the wide plain of Avignon. A few kilometres below Avignon the last tributary, the Durance, comes in from the Alps. Before reaching Arles the river bifurcates and its delta begins.

The Rhône delta

The Rhône in late summer is a diminished stream and increasingly threads its way between wide shoals of pebbles. In the cool season, below the junction of the Durance south of Avignon the river is over 150 m wide and has to be confined between artificial embankments.

Five kilometres north of Arles the river bifurcates at the apex of the delta and formerly bifurcated several more times before discharging by many mouths into the Gulf of Lions. Today the water has been confined to two channels, the *Grand Rhône* on the east which carries about four-fifths of the flow and empties into the Golfe du Fos, and the meandering *Petit Rhône* which enters the sea 40 km to the west.

Since the greater discharge is funnelled down the Grand Rhône, the bulk of the river-borne debris is also carried seawards by this channel. The longshore drift in the Gulf of Lions takes this material westward and deposits it as sand spits and bay bars, often redeposited by the wind as sand dunes, between the two mouths of the river.

Formerly the whole of the delta area, or Camargue,

between the two arms of the river and the sea, was a flat, monotonous expanse of reedbeds, salt-marshes, fens and shallow meres. Old river channels followed by lines of willows or poplars interrupted the general level of the surface which was also threaded with a maze of runnels that drained southwards into a series of étangs or meres. The largest of these is that of Vaccarès which with a depth of water no more than 1 m covers 150 sq km. These étangs increase in number and complexity towards the line of coastal dunes and drain through them by a series of narrow channels. Reclamation has altered the aspect of the northern part of the Camargue and will gradually change the greater part of it. The Ile du Plan du Bourg in the southeast has been set aside, however, as a nature reserve and is noted for its pink flamingoes, wild white horses and bulls. Recently its area has been extended and it has become a regional nature park.

The land lying east of the Grand Rhône consists of a 'dry' delta, called the Crau, made up of two vast gravel and pebble fans built formerly by the Rhône and the Durance. It slopes from about 70 m on the northeast down to the line of the Rhône. The Durance, which has abandoned its former course and now flows to the north of the Chaîne des Alpilles to join the Rhône near Avignon, no longer builds up the delta of the Crau. Today this is a vast sheet of waterworn pebbles of various sizes, much of it uncompacted but occasionally cemented into a hard dry surface. Everywhere water percolates rapidly downward leaving a thin and dusty soil except where saline marshes are held up on the surface by local impermeable hardpans.

Climate and land use

The climatic statistics for Lyon, Avignon and Montpellier illustrate the transition in climate that occurs in the 5° of latitude traversed by the Saône–Rhône valley. The Saône lowlands have a typical central European climate with least precipitation in winter and most in the warmer seasons. At Lyon there are on an average about 150 raindays a year, the lowest monthly occurrence being 10 in August. Snow falls on about 22 days in a normal year and the winter half-year is cool to cold with over 60 days with frost. Then when Polar or Arctic airmasses move down the valley, icy fogs may occur and ponds may freeze and ice form on the rivers. Spring frosts are often a menace to *primeurs*. Summers are decidedly warm but the heat is less prolonged than in the Midi. From at least Mâcon on the Saône south past Lyon to Valence in the Rhône corridor, this climate is essentially a valley climate as the general direction of surface winds is controlled by the topography. Lyon experiences virtually only a *vent du Nord* and a *vent du Midi*. The former is polar or arctic air that brings low temperatures and clear skies, and lower down the Rhône becomes the mistral; the *vent du Midi* is a maritime tropical or Mediterranean airmass that occasionally brings clear skies but normally gives wet, blustery weather.

Where this Lyonnais climate actually changes to a Mediterranean climate is uncertain, as the transition migrates with the seasons up and down the Rhône corridor. It is popularly put either at Valence – *à Valence le Midi commence* – or at the exit from the Donzère gorge. The great differences lie in the intense summer drought of the south (July with 5 days or less of small local showers); the decrease in raindays to well under 100; and the all-the-year increase in sunniness and temperature. The mild winters of the Midi make the cold northerly airflow, the mistral, all the more noticeable and the flat plains allow it to whistle along with unabated persistence. Consequently, as discussed in Chapter 4, the Midi has more days with frost than other Mediterranean coastlands. Its heaviest rains come in autumn, often about mid-September, with monsoon-like intensity.

Land use

These changes from central European to Mediterranean climate are naturally reflected in the land use which also reflects centuries of agricultural experience, regional and personal bias, and of varying response to local and national markets. Details are given in the table of statistics at the end of the chapter.

In the north where the Saône basin rises to the elevated Plateau de Langres and along the headstreams leading up to the Belfort Gap much of the land is still forested, as in the forests of Longchamps, Cîteaux and Chaux. Normally it is deciduous woodland with oak, ash and elm associations, but recent forestry operations have introduced formal coniferous plantations, especially where the natural pastures are not particularly lush. Coppice growth of alder and willow is found along all water courses or in low-lying marshy ground of which quite a surprising amount remains. Much of this upland is under permanent pasture for stock raising and beef production. The terraces alongside the rivers of limon-covered clay or gravel are cropped on the best mixed farming methods. Cash crops of tobacco, hops and mustard are followed by fodder crops of roots or green silage and eventually by wheat or maize. These terraces lie alongside water meadows that under irrigation in summer produce abundant hay crops, and support a fine dairy industry (see table, with cows in milk). Liquid milk is sold to the larger towns like Dijon, Besançon and Lyon as well as dairy produce and veal. Even the smaller market towns, like Auxonne and Chalon, have developed food-processing industries such as flour milling and sugar refining. Around the towns well-managed market gardens produce early potatoes, cauliflower, beans, carrots, onions and asparagus.

The Plain of Bresse to the south displays an intensification of this type of farming with an emphasis on cows, pigs and hens under a cooperative system which controls the sales of liquid milk and the milk-processing factories. Cheese and milk products are produced in sufficient quantity for export (*Bresse bleu*) and the skimmed milk and farm surpluses support a considerable pig population. Similarly the hens are reared under modern intensive battery methods. As in the northern part of the plain the river terraces are under constant rotation of root and green-fodder crops, wheat and maize and cash crops, but towards the west, facing the Jura foothills, the area under cereals tends to increase. Bourg is the chief agricultural market for the plain of Bresse, and its Bresse capons are known throughout France.

Pays de Dombes

In the midst of this richly productive region, the Pays de Dombes, between the confluence of the Rhône and Saône forms a poor landscape. It consists of moraines, drumlins, eskers, and gravel ridges interspersed with lakes and marshes. Despite drainage and reclamation that has been proceeding since the eighteenth century, water still covers 8000 ha. The centre is still ill-drained but drainage and reclamation have transformed a belt round the edge into fertile market gardens, dairy farms and piggeries. An increased planting of trees has resulted in the effective drainage of some waterlogged areas. For a time fish farming was a profitable venture but this gradually declined in favour of orchard and truck crops which provided a better return on investment. Duck farms on the damper areas, and sheep on the gravel for meat also reflect the proximity of the industrial populations of Lyon. The Dombes is now a protected area and a regional nature reserve.

The Côtes

The western side of the Saône valley displays another type of specialized land use developed over the centuries. These are the famous Côtes Bourguignonnes consisting of the Côte d'Or, the Chalonnais, the Mâconnais and the Beaujolais which stretch under vineyards the length of the Saône valley and for some distance along the Rhône valley as far as Trévoux. The slopes are steep and quite lofty, and are developed mainly on Jurassic limestones. The climate, soil, aspect and rainfall conditions are such that if the vines are to produce a suitable harvest they need to be tended with care. The result is that in this region some of the finest vineyards in France produce burgundies with household names, such as Pommard, Nuits Saint-Georges, Chambertin, Meursault, Montrachet and Mâcon. Each vineyard claims to have a unique soil, aspect and method which produces a unique wine but because of the borderline character of the area the quality and quantity at each harvest varies appreciably from year to year. In any case only a small amount of quality wines is produced annually and not every year is a vintage year. Grown as a monoculture the industry depends, as in the Midi, on the *caves co-opératives* to market their produce and to make the necessary bulk purchases of equipment. Occasionally the lower slopes of the Côtes display a different green in orchards of soft fruits and delicate stone fruits (apricots, peaches, cherries, etc.), that are used in the making of liqueurs. Fields of wheat, maize, sugar-beet and even hops make an occasional interruption where the valley floor widens. The table of land use (p. 179) shows clearly that orchards and vineyards occupy only a very small proportion of the farmland.

Beaune, Mâcon and Villefranche are the chief centres of the wine industry but Dijon is its largest administrative centre, for which function its fine nodal situation makes it admirably suitable. Since every main road and railway to Paris from the southeast crosses the Côte d'Or at Dijon it has also attracted to itself food industries based on the rich farming of the Saône valley and including confectionery, biscuits, liqueurs, mustard and tobacco.

The Couloir

From Lyon to Arles the Rhône threads its way through a series of seven gorges that unite the basins of Vienne, Saint-Rambert d'Albon, Valence, Montélimar, Pierrelatte, Orange, and Avignon. These towns are in each case the chief agricultural market centre and deal with food processing. In each of the basins, on the eastern side of the valley, is a pediment of gravel and pebbles brought down in flood by the Alpine tributaries of the Rhône and which forms an area of only moderate fertility at the foot of the mountains. Here are grown the main food crops of grain and fodder while the remainder of the basin floor is given over to the specialized cultivation of fruit and vines. For example the Vienne and Saint-Rambert d'Albon basins are renowned for fine peaches. Apricots, plum and other stone fruit orchards stretch as far south as Montélimar which is especially noted for pears. Before the first flush of leaves causes too much shade, market-garden crops are grown on the orchard floors. Early potatoes, strawberries and spring flowers are despatched to northern markets and, after the fruit harvest, melons are planted out to bring an early autumn harvest. Wherever shade trees are required and along the roadsides, mulberries are planted for the silkworm industry of the Lyon district. The hill slopes are vineclad and have more certainty of a fruitful harvest lying so much farther south, but few areas aspire to the production of fine table wines.

Mixed farming dominates the agrarian economy in the Bas Dauphiné between the Rhône and the Isère valleys. This is an area of piedmont debris brought down mainly by ice action but also by flood waters. The country is fairly high (about 600 m) and trenched by rivers that flow west to the Rhône. Being covered with clays that hold up lakes and marshes, it has experienced constant improvement. The heights are being reafforested and the lower parts drained for improved pastures and arable crops. The tract nearest the Rhône is gradually being covered with cereals, fodder drops, tobacco, flax and hemp and in specially favoured areas with market gardens, orchards and nut groves.

South of Montélimar in the basins of Pierrelatte, Orange and Avignon market gardening dominates. Cauliflowers

XIX *Aerial view of Lyon showing the Rhône*

are grown during the winter and, as the fields are cut, they are quickly planted with lettuce between the rows. They are intercropped with potatoes to be harvested in early March followed by artichokes, early peas, spinach, onions, carrots and finally tomatoes and peppers. The fields are continuously being tilled and planted but the parcels of land are scattered and large-scale improvements difficult. Added to the hazard of summer drought is the incidence of the mistral and every parcel of land has its windbreak. The slopes are under flower cultivation (carnations and lavender), orchards, nut groves and vineyards. *Vin ordinaire* is produced everywhere while the district near Avignon produces a quality wine, Châteauneuf du Pape, and the best French rosé wines. Considerable amounts of table grapes are grown and marketed locally. Avignon acts as a general market for agricultural produce and has food-processing industries.

The delta – the Camargue

The delta of the Rhône has long been under reclamation and much progress has been made in the last fifty years. On lands least liable to flooding, drought-resistant hard wheats are grown in rotation with fodder, sown grasses and market-garden crops but the land is cursed by aridity in summer and blasted by the cold mistral in winter and therefore yields are meagre. The dwellings lack doors and windows on the northern side and are protected by windbreaks. On the improved pastures Andalusian and Camarguais cattle are bred. The black fighting bulls of the Camargue are famous as also are their guardians who ride the white horses. On the salt pastures, merino sheep are bred for their wool and the flesh of mutton sheep grazed on salt meadows is highly esteemed.

Irrigated ricefields now cover 40 000 ha and a further 7300 ha are under the vine but during periods of heat and drought, salt from the subsoil is brought to the surface and salinization can occur unless great care is taken with the drainage. Plans for increasing the reclaimed areas depend on draining the large Étang de Vaccarès. South of this lake is an interconnected series of shallower meres which could also be reclaimed. At present they are only used for salt-evaporation pans or as fish farms and in part as a nature reserve for pink flamingoes and wild white horses.

The Crau

On this veritable desert of stones even the garrigue shrivels under the summer heat. Rain in autumn and winter nourishes a thin growth of grass which provides winter grazing for sheep that spend their summer on the alpine meadows as the old sheep roads testify. Following the increased wool prices since the Second World War great improvements have been made in the productivity of the sheep farming. The native Crau breeds were crossed with merino to improve the wool clip and better feed for the animals provided by irrigation. Canals now outline the area. A navigation canal runs from the Port de Bouc alongside the Rhône to Arles and this is fed by irrigation canals that run to north and east of the Crau. The Canal de Craponne follows the ancient course of the Durance south of the Chaîne des Alpilles and several branches lead off from it to the navigation canal while two other canals run along the eastern margin.

Cultivation has thus been made possible on 11 000 ha along the margins of the Crau and crops of hay, alfalfa and sainfoin provide sustenance for the lambing ewes in spring after the natural pastures have been exhausted. Rail transport has replaced the herding along the old stock routes and the district supports over $\frac{1}{4}$ million sheep every year. An annual sheep fair at Arles disposes of the surplus lambs and the older sheep culled from the flocks.

After the sheep have left for the mountains the cultivated areas are put under drought-resistant varieties of wheat, sown grasses and salad crops. Olives, almonds and vines have been established in a few localities and lavender for the perfume industry but this latter faces severe competition from the new synthetic pharmaceutical industries based on petroleum. The interior wastes of the Crau provide suitable ground for a car-racing circuit, a sewage disposal plant for Marseille and sites for military installations.

Power and industry in the Rhône–Saône valley

Power resources and modern industry in the Rhône–Saône valley have been up to this century mainly peripheral to the corridor. The Saône basin has always drawn on the coal supplies of the fields which lie along the eastern edge of the Central Massif, namely those of Burgundy and of Saint-Etienne, both of which have been essential to the industrial growth of Lyon and Dijon.

The coal derived from these fields is good quality but the mining and access are difficult. The Burgundian field produces about $2\frac{1}{4}$ million tons a year from at least four separate mining centres. The canal du Centre which links the upper Saône to the Upper Loire traverses the mining region, and a main railway line runs from it to Lyon. A thermal power station has been built at Epinac and from this town south along the canal and railway stretch miscellaneous metallurgical and specialized engineering industries. The Saint-Etienne field farther to the south and southwest of Lyon produces just over 2 million tons of high-quality coking and gas coals every year. Of this output one-quarter is consumed in industry in the city of Lyon and the bulk of the remainder is used in the making of coke and gas and in the development of thermal electricity. The largest nuclear power station in France is being built at Bugey 37 km upstream from Lyon. Steel making is concentrated in the valley towns leading to Lyon such as Givors and Rive de Gier.

Lyon, the administrative centre of this industrial complex, with 530 000 citizens and over 1 million in its conurbation, is the third city of France. It commands an important position at the meeting-place of river and land routes in a productive agricultural and industrial hinterland. Here, at the confluence of the Rhône and the Saône, is the focus for routes leading up the Rhône to Geneva, via the Belfort Gap to Basle, or the Dijon gap to Paris. The town also functions as a river port, the eighth in France, with wharves, sidings and port facilities along both rivers that handle every year about $1\frac{3}{4}$ million tons of freight.

The coal from Saint-Etienne and the abundant hydro-electricity power supplies of the corridor have established the city as an important industrial centre. The first group of industries is concerned with electrical and mechanical engineering and numerous branches of metallurgy, of which tinplate is especially important in that the food processing and canning industries rely on it. The second group is concerned with chemicals, including dyes for the silk industry. In addition, leather and leather products, vegetable-oil refining, foodstuffs, glassware and paper all reflect in some way the abundance and variety of the agricultural surpluses of the region.

About half the industrial workers of Lyon are engaged in the manufacture of textiles, chiefly silk, real or artificial. Sericulture has long been a subsidiary occupation of the farming communities in the district for which the mulberry trees provide food, but the bulk of the silk filament is now imported from Japan. The industry is organized in about 5000 small enterprises in and around the city which acts as its administrative and merchanting centre. Embroidered silks come mainly from the Jura towns, and cloth of gold, silver tissue and brocades from the towns of the Bas-Dauphiné. Eighty per cent of French silk is produced in the Lyon district which has long had almost a monopoly in embroidery silks and ribbons. The city has also attracted to itself the greater share of the French synthetic silk industry.

Dijon (151 000 and 200 000 in its conurbation) is the largest urban centre in the Saône valley. Its importance derives from its command of the route from the upper Saône and the Belfort Gap leading to Paris. The main railway lines and roads from south and southeast France converge upon it. Dijon also functions as a river port on the Burgundy canal that links the Saône navigation with the Seine system. Its industries are largely concerned with food, wines, liqueurs, confectionery, flour milling, biscuits and canned fruit and vegetables, as well as the famous Dijon mustard. A small metallurgical and engineering industry has been established here recently and motor bicycles, pedal bicycles, machine tools and aluminium articles are manufactured.

Chalon at the junction of the Canal du Centre with the Saône is a river port and the outlet for the Burgundian coalfield. It has developed food-processing industries, barge-building and engineering trades. Mâcon is a similar commercial centre with food-processing and light engineering industries.

The Rhône corridor south of Lyon and the Marseille district lack local coal supplies except for the small coal-

xx *General aerial view of Marseille showing old harbour in distance and in foreground new development*

field of Alès in the west and some lignite beds at Gardanne on the east. The lack of power has been rectified since 1933 by the harnessing of the master Rhône and many of its tributaries under the Compagnie Nationale du Rhône. This magnificent scheme, which is almost completed, is described in Chapter 10. It involves twenty major hydro-sites and nine navigation canals and has turned the area into an important exporter of power. New industrial areas and workers' towns have been created, with marshalling yards and motor highways, to take advantage of the abundance of power and in some instances to improve and make permanent the large villages built to house workers engaged on the construction of dams and other installations. Several old towns, such as Avignon (86 000, with 220 000 in conurbation), now have food-processing and artificial textile industries. All have been stimulated by rising standards of living and easier communications, and the great barrages have become tourist attractions in themselves.

The port complex of Marseille (Marseilles)

The deltaic mouths of the Rhône discouraged the development of a large port but just to the east of the Grand Rhône Marseille has grown to be the first port and second largest city (890 000, with over 1 million in conurbation) in France. Its importance stems from three urban functions:

1 as an outlet for the prosperous Rhône–Saône valley
2 as a Mediterranean port for France
3 as an industrial conurbation

The town is an ancient foundation dating from the time of the Greek city states. It grew up round the tiny harbour of Lacydon which is today the Vieux Port and the fishing port. The settlement known as Massilia was sited too far from the entrance to the Rhône corridor to rival nearer anchorages such as the port of Aigues Mortes. The city grew in importance with the rise of the French north African empire and the building of railway links to Lyon, Paris and Bordeaux. The great stimulus came with the cutting of the Suez Canal which converted the Mediterranean sea into a highway of world trade and lastly the growth in importance to world industry and transport of the Middle East oilfields underlined the necessity to France of having an ocean oil terminal on the Mediterranean.

The port site is sheltered on the north by the Chaîne de l'Estaque and the anchorage has deep water close inshore. Because it lies east of the delta and is not estuarine, it is free from silting and with the absence of tidal range ships can enter and leave harbour at any time. Its port facilities are continually being improved and extended. An artificial breakwater increases the sheltered anchorage along the coast and the large étangs west of Marseille have provided space and opportunity for much artificial harbour creation. The Étang de Berre and the Golfe de Fos have been incorporated into the port complex by the canal de Bouc which tunnels through the Chaîne de l'Estaque. When complete the industrial and port area of Marseille, Lavéra and Fos will be one of the largest of its kind in Europe. It will also be one of the world's largest ocean oil terminals, discharging petroleum in the Golfe de Fos which has been made capable of handling the largest tankers. Three oil refineries are already in operation at Lavéra on the Golfe de Fos and the storage capacity is to be increased to 7 million cubic metres or half that of Rotterdam. At present one-third of French petroleum requirements are off-loaded at Lavéra–Fos, amounting to 50 million tons a year, and 5 million tons of refined products are exported. The oil pipeline from Lavéra to Basle has recently been doubled in capacity. By 1975 the maritime traffic of the Marseille complex is estimated to reach 6 million tons of imports and $3\frac{1}{2}$ million tons of exports, irrespective of 70 million tons of oil. The new harbour of Fos, opened in December 1968, handles all oil shipments, specialized bulk cargoes and container traffic as it has a basin 700 m wide and 20 m deep and extensive areas of flat land for development. A large industrial complex will be constructed alongside with a wide variety of metallurgical, chemical, timber and food-processing industries. The town of Fos-sur-Mer is expected to grow from its present population of 3500 to 100 000 inhabitants.

Today the port complex of Marseille handles 40 per cent of French imports by weight (see table, p. 190). Some of these move on to the Rhône navigation system (see Chapter 5). The Marseille district is especially important for chemicals, light and heavy, based on oil imports and on local salt. North of the city are cement, brick and tile works using for fuel lignite from Gardanne. Some of the great deposits of bauxite in Var are reduced here to alumina. Other industries include vegetable-oil refining, sugar refining, margarine making, soap manufacture, flour milling, paper-making, textiles and a wide variety of other market-oriented manufactures. The making of locomotives and rolling stock, ship building and repairing, and marine and electrical engineering are also important. An airport and aircraft industry have been established at Marignane and these and other undertakings have involved the employment of large numbers of immigrant Italian workers for whom the municipal authorities have built large blocks of housing (*cités-ouvrières*) around the older part of the city

60 *Number of workers and daily commuters in the Marseille district* (after R. Dugrand)

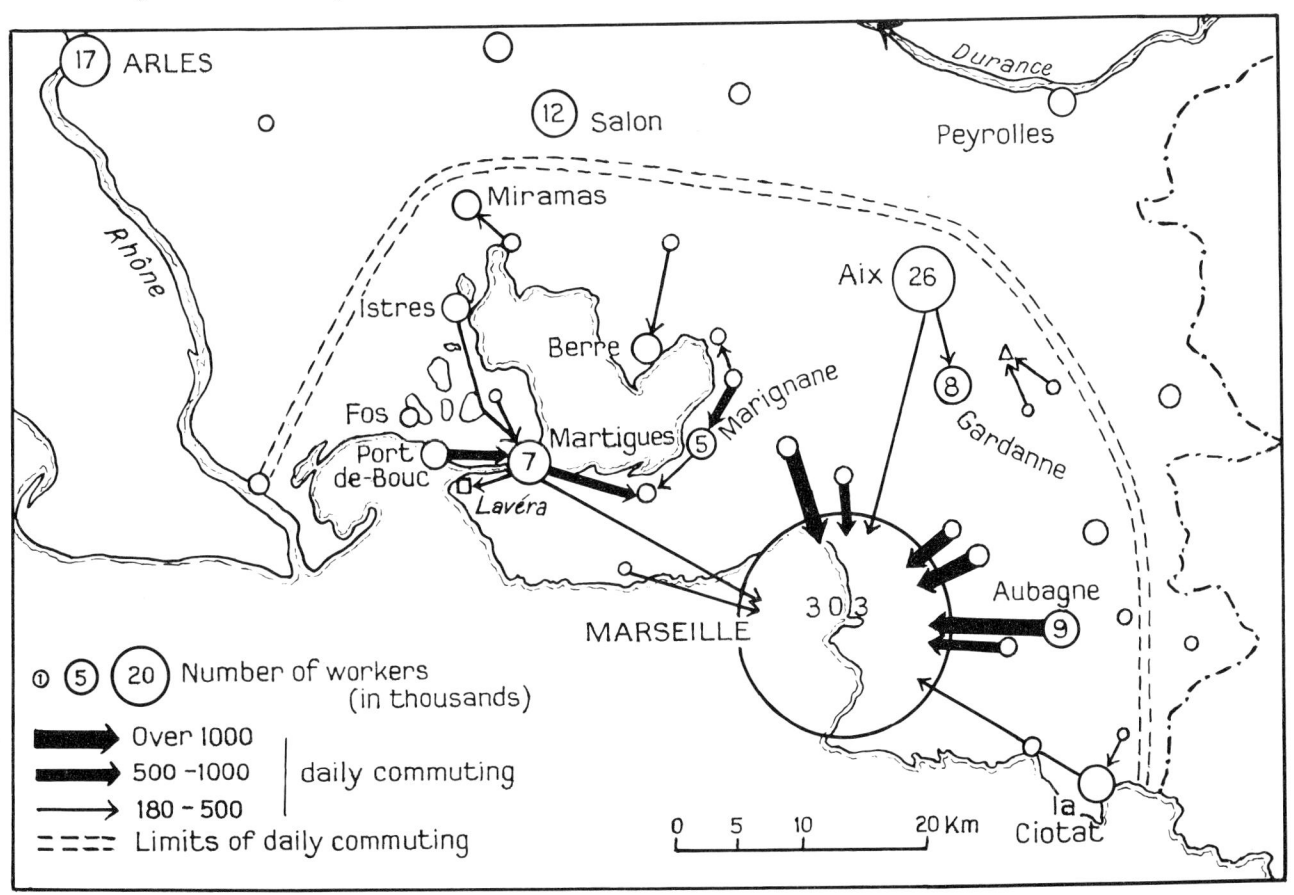

LEADING FRENCH PORTS, 1967, SHIPS ENTERED AND GOODS UNLOADED (thousand tons)

Port	Ships (number)	Net registered tonnage (thousands)	Agricultural products	Foodstuffs; fodder	Solid fuels	Crude petroleum	Refined petroleum	Minerals for metallurgy	Metallurgical products	Machinery and metal articles	Fertilizers	Other: chemicals, manufactures, etc	Total	Passengers
Marseille area	10 470	38 576	1 497	1 039	83	47,647	1 448	554	180	16		390	53 607	362 555
Le Havre	6 624	30 334	801	535	1 237	27 631	1 257	156	185	68	8		32 392	197 133
Dunkerque	5 471	12 480	477	410	108	5 163	1 527	4 550	36	51	139		12 933	103 403
Nantes area	3 571	5 693	190	397	272	6 047	1 292	1	132	1	265		8 810	137
Rouen area	5 034	7 042	544	433	1 511	731	548	96	77	18	915		5 894	1 967
Bordeaux area	2 838	5 519	269	546	86	2 622	716	139	39	3	269		4 929	2 059
Sète	1 899	2 966	140	299	–	1 761	559	29	2	2	383		3 308	345
National Total			5 454	4 329	4 822	91 634	11 736	6 329	808	1 888	3 409		134 809	2 851 820

GOODS LOADED (thousand tons)

Marseille area			180	497	87	–	4 781	88	174	246	31		7 633	367 725
Le Havre			241	289	317	12	2 206	1	74	126	17		3 802	200 827
Dunkerque			499	226	7	–	805	1	1 081	106	364		3 586	104 837
Nantes area			11	18	1	–	2 042	25	88	12	23		2 367	106
Rouen area			958	363	43	–	3 107	6	103	76	76		5 335	2 350
Bordeaux area			349	217	4	24	1 311	13	12	23	223		2 401	1 393
Sète*			59	118	12	–	339	14	2	14	303		940	694
National Total			3 457	1 934	523	67	14 630	469	1 974	622	1 359		31 018	2 927 432

* In 1967, Bayonne exported 1 193 000 tons of raw materials for the chemical industry and was France's seventh exporting port, Sète being eighth
The main items included in 'Others' are chemicals, paper-pulp, glass, various other manufactures, and raw materials and manufactures for the building industry

(fig. 60). Yet far from being congested the Marseille–Lavéra–Fos complex has at its disposal an abundance of poor quality land in the Camargue and the Crau. With an assured freshwater supply, ample power resources from the Rhône and Durance, a main motorway from Lille via Paris, and an enlarged Rhine–Rhône water link due for completion by 1976, the future of Marseille seems bright.

The table of statistics for the seven leading ports of France by weight of cargo handled gives the more significant items of trade only but it demonstrates vividly the important role that Marseille plays in French commerce. Even in passenger traffic it is exceeded only by Calais (1 722 000 passengers) and Boulogne (903 000) as it has a busy packet service with Corsica and numerous world-wide sailings. But Marseille – and to a lesser extent Sète – reflects above all the great change in modern commerce caused by vast international movements of crude petroleum and tremendous national developments of petroleum refining and petro–chemicals. The port was ideally situated to deal with the former French Mediterranean colonies and today not surprisingly it is the prime importer of north African crude oil

16 Switzerland and Liechtenstein

SWITZERLAND
Growth of the modern state

On 1 August 1291 the mountain cantons of Schwyz, Uri and Unterwalden formed a league for common defence against any attacker, and especially the Habsburg emperors. This nucleus, the core of modern Switzerland, was overwhelmingly German. Gradually the defensive league grew as other districts and towns federated with it although they did not always join with each other (fig. 61). After a great deal of fighting in which the league, or parts of it, helped nearby towns, it was joined by the towns of Lucerne (1332), Zurich (1351), Glarus, Zug (1352) and Berne (1353). Hence in the mid-fourteenth century the defensive confederation consisted of 8 cantons, and its associated territory had spread west to the edge of French-speaking Savoy. However, Glarus and Zug were not completely independent of the Austrian Habsburgs until 1394.

During the fifteenth century the league gained control by conquest of most of Aargau and in 1410 Uri took command of the whole of the Saint-Gotthard pass whereas previously the league had sat astride the northern side only of this important routeway. Other federated cantons also expanded and Appenzell, Sankt Gallen and upper Valais had close associations with the league but were not admitted to full membership. However, in 1481 the cantons of Fribourg and Solothurn were allowed to join. This was the beginning of the great age of Swiss mercenaries who fought with distinction throughout Europe.

In the early sixteenth century Basle, Schaffhausen (1501) and Appenzell (1513) were admitted to the league which comprised these 13 independent cantons until the French Revolution. The Reformation, which secured Calvinism, and the counter-Reformation, led to bitter civil wars but the Swiss mercenaries were engaged mainly abroad and in the Thirty Years' War (1618–48) fought on both sides and supplied both with arms. At the close of that war the cantons again agreed to defend their frontiers by a confederate force against any aggressor. This same period saw the rapid growth of domestic industries, especially textiles, carved wooden objects, musical boxes, and clocks and watches, partly for sale abroad.

Almost throughout the eighteenth century the defensive confederation remained a collection of separate cantons with widely different methods of government and there was virtually no Swiss nationality as such. However in 1761 the foundation of a Swiss society (*Helvetische Gessellshaft*) showed signs of coherence. In 1798, aided by the lack of local unity, Napoleon took over the league, enlarged the number of cantons from 13 to 23 and imposed on them a new centralized, democratic constitution. This

61 *Territorial growth of the Swiss confederation*

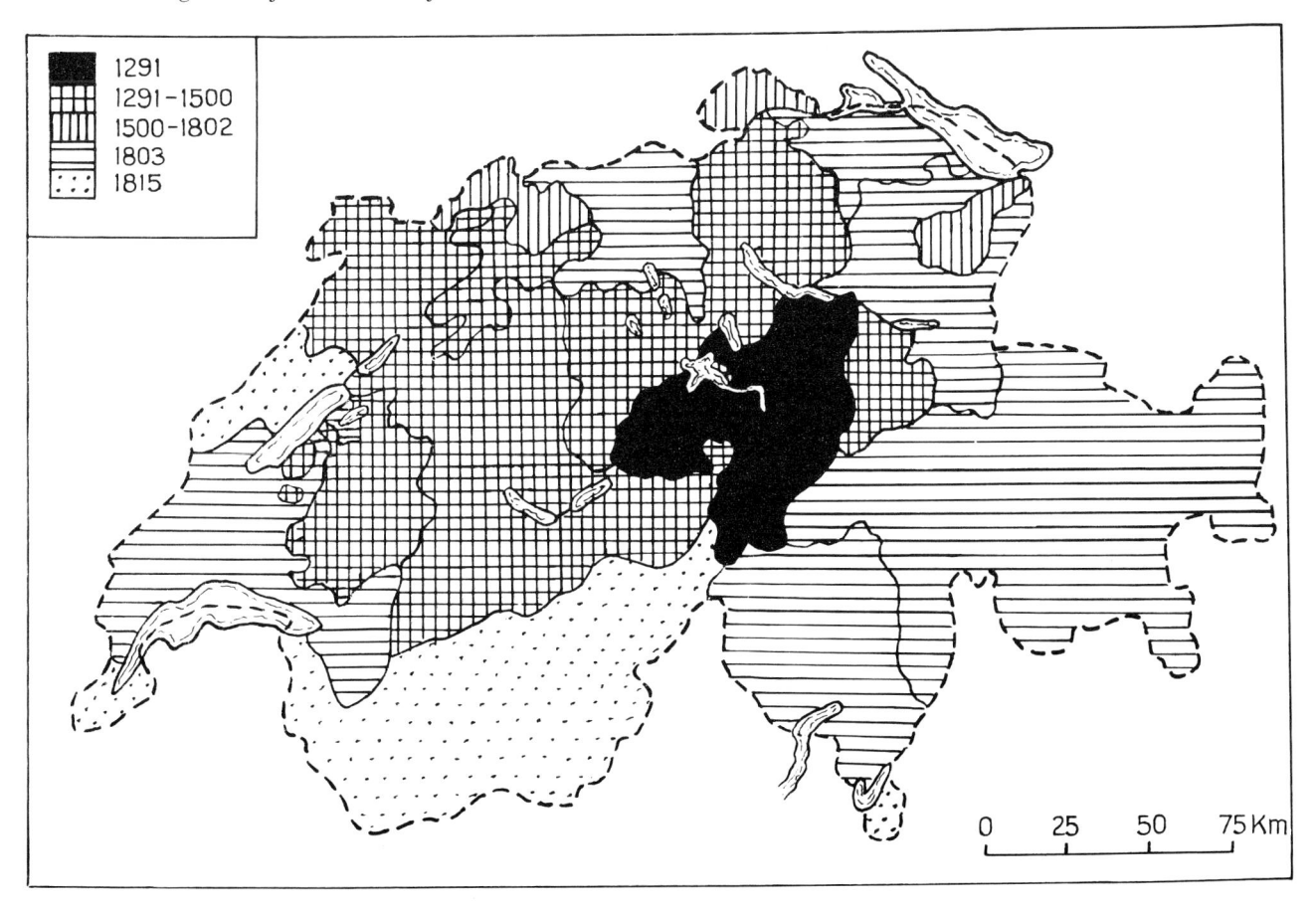

constitution and cantonal structure of the Helvetic Republic was reorganized in the next few years and the cantons were given much the same boundaries as today. Aargau and Vaud were separated out of Berne canton and Grisons, Sankt Gallen, Ticino and Thurgau became part of the confederation which now contained 19 cantons.

In the European peace conferences of 1814 and 1815, the cantons of Geneva, Valais and Neuchâtel (a possession of the King of Prussia until 1857) were united with the confederation and Swiss neutrality was guaranteed for ever by the leading European powers (fig. 61). A new constitution restored sovereignty to the cantons – now 22 in number – the conjoint army being virtually the only Federal concern. For the federal meetings, Lucerne, Zurich and Berne acted in turn as headquarters for two years. In 1848, after religious troubles, a new constitution, which greatly strengthened the central government, was introduced and, with modifications in 1874, still survives. At about the latter date the French franc and metric system of weights and measures were adopted.

Railways were not welcomed at first but by 1859 a dense network had been built. By 1882 the Saint-Gotthard railway had been completed and within a few decades the railways were nationalized. In the war of 1914–18 Switzerland remained neutral and, after it, the League of Nations had its headquarters at Geneva. During the 1939–45 war the federal republic again remained neutral and supplied war material to both sides. It was, however, a neutral who would have attacked bitterly any aggressor. As such it has not become a member of the United Nations nor of the European Common Market, although it is seeking a special arrangement with the latter.

In recent years it has been noted for its strong financial condition and its financial attractiveness for foreign investors. The territorial boundaries are interesting. Several parts of Switzerland lie north of the Rhine and the town of Constance is cut across by the Swiss–German boundary; in the south some small bits of Italy are entirely surrounded by Swiss territory. The peaceful conditions obtaining show that boundaries are less important than the attitudes of the nations on either side of them.

General Aspects

Switzerland is one of the smallest states of Europe with an area of 41 288 square kilometres and a maximum width of 330 km from east to west and 220 km from north to south. Oval in shape it sits astride two major chains of Alpine mountains and commands the nexus of routes leading to the Saint-Gotthard pass. In Switzerland the great land routes between the Baltic and the Mediterranean cross those from southwestern Europe to the Black Sea. The country occupies the main watershed of western Europe where the Rhône begins its journey to the Mediterranean, the Rhine to the north Sea, the Inn to the Danube and Black Sea, and the Ticino to the Po and Adriatic. It is therefore not surprising that Switzerland has for centuries commanded some of the most important routeways in Europe.

Switzerland and the Swiss are also unique in having elements that in other states are cited as divisive but here are given as evidence of social and political maturity. Three facets are of outstanding significance.

First, the difficult mountain terrain divides the valleys from each other yet the country is a federation of 22 cantons, three of which (Basle, Unterwald, Appenzell) are subdivided politically into two parts. Each canton plans its own development and manages its own legislature as well as participating in federal decisions.

Second, there is no unity of language. German is the mother tongue of 72 per cent of the population, French of just over 20 per cent, Italian of 6 per cent and Romansch 1 per cent. These languages have equal status and there are no linguistic divisions. The majority of the people normally speak German except in the five southwestern cantons, which are predominantly French-speaking and the Ticino where Italian is most used.

Third, there is complete religious toleration despite the fact that this is the country of the Calvin reformation. Protestants form 53 per cent of the population and Catholics 46 per cent.

Alongside the respect afforded to the identity of the individual and his canton, it is equally recognized that the Federation can handle best such general matters as national defence, communications and power supply. Because of Switzerland's neutrality and political maturity many world organizations have their headquarters at Geneva as well as numerous international business concerns.

Structure, relief and landforms

The general nature of the Alpine orogeny, so well studied in Switzerland, has been outlined in Chapter 1. Here little more will be attempted than to summarize the broad patterns of structure and geomorphology.

Four main lines of structure and folding can be distinguished in the Alps proper, each running almost west to east:

1. the Pre-Alps or Outer Ranges, with summits at about 2000 m, that stretch from Lake Geneva to Lake Thun
2. the High Calcareous Alps, with summits at about 3200 m, extending from the Dents du Midi to the Bernese Oberland and Glarner Alps
3. the Crystalline Core of Hercynian massifs, which includes the Mont Blanc (4807 m, just in France) and the Saint-Gotthard masses, Monte Rosa (4634 m), the Matterhorn (4481 m) and the Pennine Alps
4. the Root Belt or southern limestone zone, which is almost absent in Switzerland

These main highland ranges are divided by deep structural depressions which provide relatively easy passageways through the cordillera. The most easily recognized of these depressions is that followed by the upper Rhône, upper Reuss, and upper Rhine. North of the trough is the

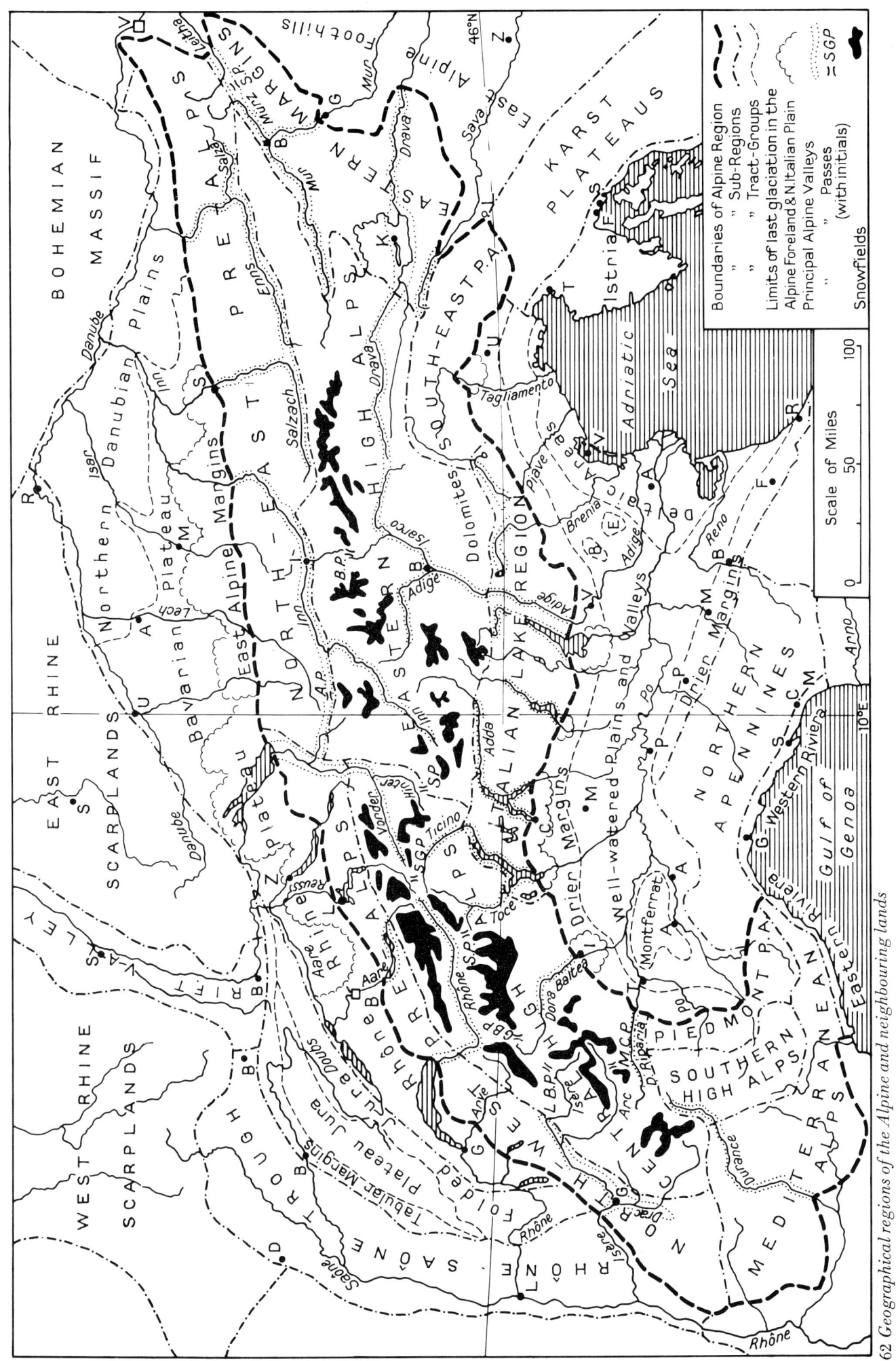

62 *Geographical regions of the Alpine and neighbouring lands*

line of the Calcareous Alps and to the south are the granite, gneiss and schist of the crystalline core.

As with fold mountains elsewhere the ranges display an arcuate pattern. The more pronounced Alpine mountain-building movements were spasmodic and were accompanied as well as followed by prolonged denudation that was intensified by isostatic readjustment or uplift, thus exposing the crystalline core of the mountains. Beyond a line joining Lake Zurich to Lake Como the eastern Alps being lower were far less denuded. The present height of the Alps is due to the excess of orogenic uplift and isostatic readjustment over denudation. But the sharp relief features, the knife-edged ridges, the pyramidal peaks, and the deeply cut valleys are due mainly to erosion although some steep scarps are the faces of faults.

Glaciation in the Quaternary was extensive and severe. The Alps were almost buried beneath ice sheets which covered all but the highest peaks, which were shattered by frost action at their summits to form the typical 'horns'. The legacy of the ice within the Alps proper lies in the over-deepened and 'stepped' glaciated valleys with hanging tributary valleys and truncated spurs; in the upper mountain sides scalloped with cirques; and on the periphery, where the ice tongues gouged out the flatter piedmont, in the large moraine-dammed lakes. The present glaciers and ice fields are small but by far the chief in western Europe.

A huge downfold in front of the Alps to the north can still be traced from Lake Geneva to the Vienna basin and today forms the Mittelland (fig. 62). This syncline was occupied by a branch of the former Mediterranean sea that dwindled away in shallow sheets of water and inland lakes leaving a cover of marine sediments behind. Meanwhile rubble, stones and gravel were crammed into this sub-Alpine trough from the slopes of the gradually rising and steepening Alps, forming molasse which today covers most of its surface.

These newer deposits have since been modelled by subaerial erosion, the deposits being sorted by river action and affected by later earth movements and more reeently deranged by ice. Glaciation appears to have here been of two kinds. In the earlier ice advances a piedmont ice-sheet was formed by the merging of glaciers spilling out from the foot of the Alps. This covered the surface of the Mittelland plateau and left behind immense and well-preserved morainic ridges, an abundance of ground moraine and drumlin fields, and the terminal lakes Geneva, Lucerne and Constance. Some of the smaller ice advances and, of course, the ice retreat of any glacial phase, took the form of glaciers, the chief effects of which were the deepening and widening of the valley floors, the deposition of the great lateral moraines that in some cases impound lakes, and the derangement of the drainage. Today the plateau, known variously as the Mittelland or the Rhône–Rhine plateau or the Swiss Foreland, slopes from an average height of 1400 m at the foot of the Alps to 400 m at the Jura foot. It presents an undulating aspect of flat-topped, steep-sided hills, divided by flat-floored, glaciated valleys with terraced sides. The drainage goes mainly to the Aare and so to the Rhine (fig. 62).

North of the Mittelland, the Jura extend between the Rhône and the Rhine as a simple fold system which represents the dying away of the effects of the Alpine orogeny. Rising abruptly from the Mittelland the first range is formed by the northern boundary wall of the tectonic trough underlying the Mittelland plateau. Other ranges rise parallel to it divided by longitudinal valleys with the height of land decreasing and the slopes becoming more gentle until the folding dies away altogether beyond the Franco–Swiss border and the plateau of the Jura is attained. This simple fold system called the *Jura plissé* is easily distinguished from the *Jura tabulaire* which lies in France. In the Swiss Jura anticlines normally form mountain ranges and synclines coincide with depressions or *vaux* (*val*, singular).

At their northwestern and northeastern extremities the Jura are drained direct by the Rhône and Aare but in the middle section of the mountains the rivers show three unusual characteristics. First, in their vaux courses they are sluggish as in reaches of the Birse, but where they pierce the ridges in gorge-like transverse gaps (*cluses*) they become rapids; secondly, this trellis river pattern lends itself to river capture, as is well exemplified by the Doubs; and, finally, since the mountains are largely built of the limestones to which they have given their name, many of the *vaux* have no apparent water outlet and drain by subterranean flow.

Climate

Switzerland extends through only 2° of latitude but because of its mountainous character has wide variations of climate. Its weather and climate are essentially central European with strong mountain or Alpine characteristics. Nowhere in Europe is the effect of altitude on climate seen to better advantage. The general decrease in temperature amounts to 0·75°C on north-facing slopes and 0·5°C on south-facing slopes for every 100 m of ascent. Above 1500 m, although shade temperatures are low, the marked decrease in absolute humidity and density of the air often makes the air feel dry, and in daylight when cloud cover is thin or absent the insolation at ground level is far more intense and richer in ultra-violet rays than at sea level. For this reason Switzerland's high-level sanatoria attract sufferers from lung diseases and equally the bright sunshine of calm clear days in the cold season is attractive for winter sports.

The high altitudes, and frequency of snowcover which favours above it local anticyclonic conditions and cold air, introduce another aspect of climate that profoundly affects settlement and cultivation. This is the tendency of mountain valleys and basins to suffer from temperature inversions on calm days. Then at night, cold air resting on the

CLIMATIC STATISTICS, MEAN MONTHLY TEMPERATURES (centigrade)

	Altitude (metres)	J	F	M	A	M	J	Jy	A	S	O	N	D	Absolute Max	Absolute Min	Annual days with frost
Neuchâtel	487	−1·0	1·1	4·1	9·0	12·9	16·6	18·8	17·8	14·7	8·7	4·1	0·0	33·9	−16·9	76
Basle	273	−0·1	1·9	4·8	9·6	13·3	17·1	19·0	18·3	14·9	9·6	4·5	0·6	34·7	−24·2	67
Berne	572	−1·9	0·3	3·5	8·3	12·1	15·8	17·9	17·0	13·8	8·4	3·1	−1·2	33·0	−19·8	86
Lausanne	553	−0·5	1·6	4·2	8·7	12·6	16·2	18·4	17·5	14·7	9·1	4·5	0·6	31·8	−15·9	70
Geneva	405	0·0	2·0	4·9	9·4	13·3	17·1	19·5	18·3	15·1	9·5	4·9	0·9	36·4	−16·7	69
Lugano	275	1·3	3·5	6·9	11·4	15·1	19·1	21·5	20·5	17·2	11·5	6·2	2·3	35·2	−11·0	53
Davos	1561	−7·4	−5·1	−2·6	2·3	6·7	10·2	12·1	11·2	8·4	3·3	−1·3	−6·1	28·9	−29·8	177
Pilatus	2068	−6·2	−5·7	−5·5	−2·0	1·7	5·3	8·1	7·9	6·3	1·3	−2·3	−5·4	22·6	−25·2	
Säntis	2500	−8·8	−8·7	−8·4	−4·7	−0·8	2·5	5·0	4·7	2·9	−1·7	−5·1	−8·1	18·0	−26·3	

MEAN MONTHLY PRECIPITATION (millimetres)

	J	F	M	A	M	J	Jy	A	S	O	N	D	Mean annual Total	Raindays	Days with snowfall
Neuchâtel	42	50	52	62	71	99	92	91	93	104	68	69	906	144	23
Basle	36	41	50	66	79	105	85	83	78	79	62	50	829	139	19
Berne	43	52	58	70	81	104	99	102	83	92	67	57	922	145	29
Lausanne	48	55	63	67	82	97	93	102	112	123	81	65	1004	141	23
Geneva	41	51	50	64	75	75	80	83	91	121	84	57	887	117	17
Lugano	63	72	118	159	184	185	168	195	183	214	141	78	1790	121	11
Davos	45	58	53	55	56	102	120	122	95	66	56	61	903	145	73
Pilatus	65	89	114	157	148	182	176	156	139	75	40	33	1395	167	103
Säntis	139	182	165	206	193	284	303	266	210	177	122	150	2432	188	149

upper mountain slopes gravitates down on to the valley floors which consequently are shunned by settlement and are usually kept for growing summer crops of hay. The houses cluster some way up the valley side so avoiding the coldest chill of winter nights and often also getting a few hours of extra sunshine by being above the ground fog then formed in the valley-bottoms. The deep recent incision of the rivers and the presence of high-level terraces and alluvial fans help to provide suitable sites for settlement.

In the mountains, aspect or orientation becomes of great importance. Settlement and cultivation cling to the warmer south or sun facing slopes and valleys, where the sun shines longer and more often. Slopes in the shadow are either under grass, forest or bare rock according to general altitude, and on the higher areas their temperatures remain below freezing-level for long periods. The gradient of the slope is also important in mountains since the angle of the sun's rays affects the intensity of insolation. Slopes almost at right angles to the noonday sun will on clear days in summer be extraordinarily warm. This explains the preference of peoples and crops in certain valleys for particular heights and shoulders. So effective is a south aspect that rye can be grown at 1800 m at Cresta above Saint-Moritz and on other favoured sites. The sunny and the shaded slopes have distinctive names: *sonnenseite* and *schattenseite* in German and *l'adret* and *l'ubac* in French.

The local variations due to altitude and topography cannot be allowed to obscure the general coolness and wetness of this central European climate. The long list of climatic statistics appended speak for themselves. Winters are cool to cold everywhere and even the Mittelland has frost on about 70 days annually. The summers are warm and begin to be hot in the south. On the high mountains the temperatures given were recorded in the shade and represent well the conditions of shadowed areas and at night. They give no idea of the burning heat and glare of the noonday sun.

Precipitation decreases from west to east and more abruptly in rainshadows everywhere. Mountain massifs exposed to westerly and northerly air flow are notably wet. Winter and spring, when much of the area is snowclad, are the drier seasons and summer the wettest season except in the far south where October is normally the rainiest month of the year. The cold season precipitation usually comes as snow which on the high Alps reaches 7 m or more in depth and lies for as much as six months, closing some railways and mountain passes from November to May. The higher parts have snowfalls throughout the year and those above about 2800 m in the west and 3200 m in the east rise above the permanent snowline. There are approximately 1200 separate névé fields and glaciers in the western Alps. The largest, the Aletsch, is 26 km long with a catchment area of 130 sq km; the longest is the Mer de Glâce (31 km) and the lowest descends to 1000 m near Grindelwald. The three main dispersal centres are the Mont Blanc massif, the Pennine Alps and the Bernese Oberland, which feeds the three largest Swiss glaciers.

A further climatic consideration in a mountain region is the influence both of the local topography and the general direction of the main ranges. The local topography leads to the formation of mountain and valley winds, of which the latter has already been discussed as a cold downward creep at night leading possibly to inversion of temperature. The former, or mountain wind, often causes convection and heavy rainfall on mountain tops in summer. The general direction of the main ranges greatly affects the possibility and frequency of transverse airflow. In the Swiss and Austrian Alps the ranges run roughly east–west while the chief paths of cyclonic depressions both over central Europe and the Mediterranean tend to be in an easterly direction but separated by the mountain cordillera. This broad situation, coupled with thermal differences between central Europe and the Mediterranean Sea, is likely to lead to airmass invasions from north to south and south to north across low parts of the Alps. The transmontane airflow is funnelled into north–south valleys and leads especially to the warm föhn winds (see Chapter 4). Depressions moving north of the Alpine system tend to induce air from south of the Alps to cross the cordillera. This airflow may be channelled into river valleys such as of the Aare and Reuss and with the typical mechanism of rapid heating during descent with an unsaturated, adiabatic lapse-rate, gives warm, dry weather. Similarly, in winter when a cold airmass resting on the north European plain expands southward into the rear of a 'low' moving eastward over the Mediterranean, a föhn-like wind may occur in valleys, such as those of the Ticino and Toce, on the south side of the Alps. The föhn, however, is felt most on the north flank as on the Mediterranean side it tends to be colder and, indeed, here is often known as the *bise*. At Martigny the föhn blows on average on 39 days annually and at Altdorf for 48. In the upper Rhine, Reuss and Aare valleys it also blows on an average for 48 days a year, and can raise the mean temperature of January to zero and of July to 21 °C. Even in the southern Alps in the Ticino it may bring in autumn tinder-dry weather that rapidly ripens the grapes. As explained in Chapter 4, the dryness and heat of the föhn seems often only explicable by assuming that some of the airflow has descended in lee-wave eddies from far above the mountain crests.

Vegetation and land use

Switzerland's vegetation is essentially central European with invasions of Mediterranean flora in southern valleys and on south-facing slopes. In such locations on the Mittelland in general and Rhône valley in particular, as well as in the Ticino, camelias, magnolias and even palms flourish. Up to about 550 m on northern slopes and 700 m on southern slopes all cultivable patches are under farm crops, including the vine. Above that height deciduous forest dominates with a preponderance of sweet chestnut on the *sonnenseite* and beech on the *schattenseite*. At 1750 m conifers begin to dominate, the spruce yielding to larch and pine. At their upper limits the pines decrease in size and are interspersed with dwarf species of willow, juniper, rhododendron, azalea and bilberry. The stunted tree growth eventually gives way to alpine meadows most of which lie above 2000 m and have a growing season of six to thirteen weeks. They are brightened with crocuses, anemones and narcissi in May; lilies, orchids, campanulas, larkspur, daisies, gentians and alpine roses in June; and finally with the meadow saffron or autumn crocus. At the upper limits of the pastures the cushion-type flowering plants give way to mosses, lichens and finally to snow and bare rock.

As has been demonstrated, Switzerland consists of three parallel belts of country running from northeast to southwest, namely the Alps which occupy three-fifths of the national territory, the Jura which takes up another one-tenth and the Mittelland between, which represents one-third of the state. This means that seven-tenths of the country is mountainous and since Switzerland is one of the smallest countries in Europe the land available for arable farming is very small. It has been estimated that of the entire state, although only 24 per cent is unproductive, less than 10 per cent is suitable for rotational arable crops and artificial meadows. In 1968 about 169 000 ha were under cereals of which 99 000 ha were for wheat and 30 000 ha for barley. Potatoes occupied 35 000 ha and the vine nearly 12 000 ha. In 1965 there were 162 414 farming units, excluding a few thousand horticultural smallholdings, and these owned just over 1 million ha excluding alpages which are, of course, difficult to disentangle from forest statistics. As the table shows about one-fifth of the farm holdings were less than 1 ha and seven-tenths were 10 ha or less. Even on the largest farms the cultivated area seldom exceeds 30 ha and in all Switzerland only 700 farms have more than 50 ha of ploughland. Yet the smallness of the average farm and the mountainous terrain does not preclude the use of farm machinery and over 70 000 four-wheel tractors and 100 000 two-wheel cultivators are in use. However, much of this machinery is used for hay cutting and carting. The long snow cover and low average temperatures ensure that the fodder crops dominate the farming system. Of the whole country, 43 per cent is under permanent meadow and pasture and a further 3 per cent under artificial leys. In addition much of the ploughland is devoted to animal forage. Hay is the chief crop everywhere and the amount of bread grain produced supplies only one-third of the home requirements.

Despite the small size of the farms, the emphasis on dairy farming for milk and cheese production and the relatively common use of farm machinery of all kinds ensure that less than 10 per cent of the total national labour force of nearly 3 million is engaged in agriculture. The table (p. 198) shows, however, that many Swiss people give some help to the farmers at harvest and seed time. The average milk production per cow is high, being 3570 kg/year, against 2080 in Italy, 1620 in Spain, 1200 in Greece, 3910 in U.K. and 4250 in the Netherlands. Everywhere the farmer's chief concern is to garner enough fodder and hay to keep up milk yields and to feed stock. About two-thirds of the cattle are in the Mittelland where they are stall-fed for most of the year and only graze the fields in autumn after the sixth or final mowing of the valley grasses. Most of the cattle are *tâchetée rouge* or *brune* and number in all 1 796 000. *Tâchetée noire* is common only in Fribourg canton and the Hérens type only in Valais. The chief fodder crops consumed are sown grasses, lucerne, colza, clover, maize and sugar-beet, besides which the cattle are also fed on every available mountain pasture. The remarkable number of pigs is naturally related to the milk production and to fodder crops.

In the Alps a transhumant system of dairy farming is practised (fig. 29). The cattle are stall-fed during the winter in the farm dwellings near the floor of the valley on hay laboriously cut, dried and carried during the summer months from minute fields on the valley sides. This hay is stored in huge lofts over the cattle stalls. As soon as the snow leaves the lower pastures the cows are led on to them while the fields are ploughed, manured and sown with feed crops. As the snow retreats up the mountains the cattle are herded after it, the lower slopes being termed the mayen and the highest pastures above the treeline the alps. The milk is used for the manufacture of Emmentaler, Gruyère and processed cheeses; all kinds of cream; evaporated, dried and condensed milk preparations; and in other food processes, especially chocolate preparations. Formerly in the remote valleys the cheeses were made on the alpine pastures and carried down to market at regular intervals. Latterly plastic pipelines convey the liquid milk to cooperative creameries for processing. With the first snows of autumn the herds retreat to lower and lower pastures down the mountainside until at the onset of winter they are back in their stalls again. In the Alps only about 2 per cent of the land is suitable for crops and the greater part of this is under fodder. More than one-quarter of the surface is barren and the remainder is forest and alpine pasture. Here livelihoods depend on the number of cattle that can be brought through the winter so that every wisp of hay is gathered and every possible patch cultivated. The Jura shows a similar pattern of beasts fed in stalls during winter and on mountain pastures in summer. Over 30 000 tons of cheese are produced in the Jura each year, manufactured at Winterthur, Emmental, Fribourg and Appenzell. Orbe and Stalden make condensed and powdered milk, and chocolate is manufactured at Vevey, Geneva, Neuchâtel and Kilchberg.

Arable farming occurs mainly on the Mittelland. Wheat is the chief cereal and accounts for half the grain grown. Crops are confined in any case to sunny slopes but in favourable areas climb with a succession of oats, potatoes and rye up to 2000 m. On the best soils, such as alluvium, glacial loams and loess in the valley floors, maize makes an appearance, being used principally for silage. A certain amount of specialization can be detected in other cultivated crops: for example, tobacco is grown in the southern cantons of Fribourg, Ticino and Vaud; vegetables in Vaud canton and around Zurich and Berne; peaches in the Ticino valley and apricots in Valais. Many of these products are processed in various ways, for example in the famous dried soups of Knorr and Maggi at Kemptthal and Thayngen.

In a small land-locked country like Switzerland maximum agricultural returns are essential and one of the main reforms necessary to achieve large-scale agricultural improvement is in *remaniement parcellaire*, or consolidation of the land holdings. Half of all Swiss farms are in a fragmented condition and morcellation is particularly acute in the Jura and in the southern alpine valleys. Plans to initiate this agrarian reform were set forth in 1918 but by 1942 only 50 000 ha had been regrouped. The desire for reform has to be expressed by the commune and regional plans have to be drawn up by the canton for the building of roads, industrial sites, tourist attractions, etc., before funds for the reorganization are made available by the federal government. About 30 000 ha a year are regrouped and to date 374 000 ha have been brought into land-reform schemes. Reorganization is completed in Schwyz and Geneva cantons but 435 000 ha of holdings are not yet consolidated. The largest areas remaining to be regrouped are in the cantons of Zurich, Vaud, Aargau, Berne and Graubünden.

Many of the difficulties are human. Attachment to particular parcels of land is real, particularly where the piece has special qualities. Differences in price obtained between building and pastoral land encourage some farmers not to give up their land in the hope of a better price. The mayen pastures held in common by the commune are generally made part of the tourist attractions and remain unenclosed, and in many cases there is a great difficulty in providing each farmer with a unit of from 10 to 15 ha. The success of regrouping and enlarging agricultural holdings, with consequent saving of labour, may be judged from the table of the cultivated extent of landholdings arranged in order of holding size for the years 1955 and 1965.

The accompanying tables of statistics also show the main elements of the agricultural production, which forms about

SWITZERLAND: LAND HOLDINGS AND AGRICULTURAL WORKERS

Size (hectares)	Number of holdings (thousands) 1955	1965	Per cent change 1955–65	Workers (thousands) 1955 Permanent	Occasional	1965 Permanent	Occasional
up to 1	42	30	−28	17	68	5	47
1–5	67	44	−34	101	81	30	73
5–10	53	40	−25	133	51	68	47
10–20	35	37	+6				
20–50	8	10	+24	170	50	127	58
over 50	0·5	0·7	+36				
Total	206	162		421	250	229	226

MAIN AGRICULTURAL PRODUCTS

Products or Crops (thousand tons)

	1965	1968
Wheat	371	409
Barley	105	112
Rye	66	64
Oats	30	31
Maize	18	24
Sugar-beet	362	453
Potatoes	1206	1098
Grapes	126	136
Milk	3117	3321
Butter	33	38
Cheese	77	85

Livestock (thousands)

	1965	1968
Cattle	1773	1849
Pigs	1672	1849
Sheep	249	298
Horses	73	59
Goats	89	est. 72
Poultry	5800	6200

10 per cent of the gross domestic product (GDP) at factor cost.

Power resources and industrial development

Half the total working population is engaged in industry (51 per cent, against just under 10 per cent in agriculture, and nearly 40 per cent in services) despite the barren mountain environment, land-locked situation and absence of industrial raw materials. The Swiss have always displayed genius in the utmost use of natural and human resources, and by doing so have built up one of the highest national living and educational standards and a worldwide trade. Lacking coal and iron-ore deposits, the economy concentrated on goods of high value and skilled workmanship. 'Condemned to superiority', the Swiss specialize in quality goods of high weight/value ratio because they cannot undertake bulk and mass production.

The country has, in fact, a few tiny basins of lignite and coal in Valais and at Sankt Gallen and in parts of the Mittelland which are mined and used in thermal power stations. But, although 2½ million tons of coal are imported each year from the Ruhr, by far the greater part (98 per cent) of the domestic electricity production is hydroelectric and the country has a close transmission grid (see Chapter 10 and figure 63). Power stations are designed to serve specific undertakings, for example that at Eglisau on the Rhine serves the aluminium works at Neuhausen, that at Visp serves the Simplon railway and that at Amsteg on the Reuss the Saint-Gotthard. In the Jura many small hydro-plants have been erected; in the Mittelland the major stations are on the Rhône at Chancy–Pougny and on the Rhine between Schaffhausen and Basle, while a huge underground station has been built in the Aare valley at Innertkirchen. The glaciated landscape of the Alpine region with its steep gradients, hanging valleys, U-shaped gorges, numerous lakes and heavy precipitation is almost ideal for hydro-electric development but the low March flow and high July runoff together with the variable daily regimes of Alpine tributaries, make storage reservoirs desirable. At the present time one-third of the power is developed in the Valais but new projects are planned for the upper waters of the Rhône, Reuss and Aare. The total potential output of 35 000 million kWh will nearly be reached by 1975. Already considerable power is exchanged with Germany but in the near future Switzerland will have to look increasingly beyond its own frontiers for power or to new power resources to maintain its policy of dispersal of industry and population. At present 40 per cent of the power is used in

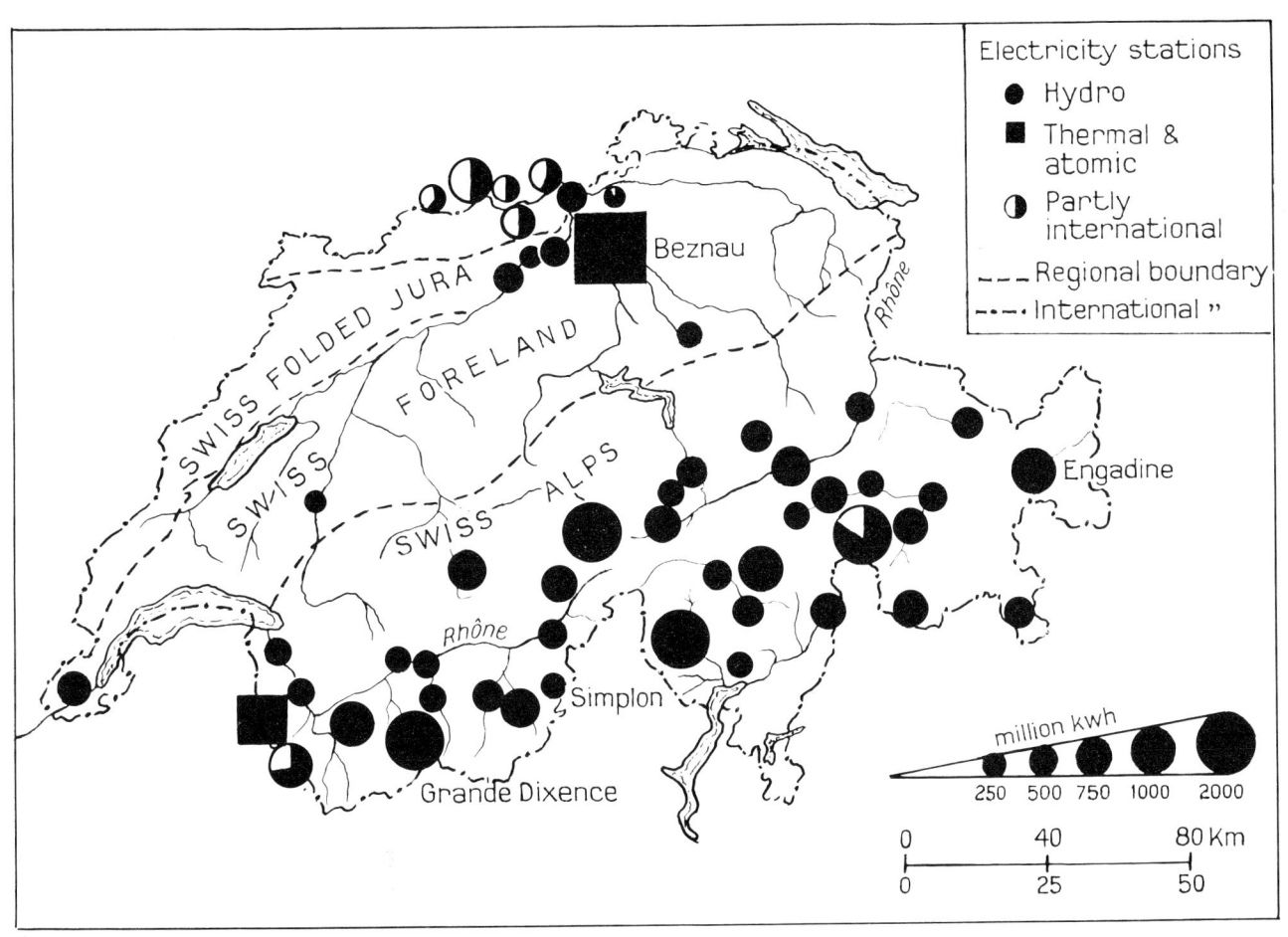

63 *Distribution of chief electricity generating stations in Switzerland* (after A. F. Mutton *et al.*)

industry, 40 per cent for domestic consumption, 8 per cent for transport, the railways being electrified, and 12 per cent is lost in transmission.

An oil pipeline links Basle and Aigle to Lavéra and Savona on the Mediterranean and so serves the Rhine navigation and road and air communications, and also forms, with locally mined salt, the basis for a chemical and pharmaceutical industry at Basle. Perfumes are made at Geneva as well as cosmetics, plastics, soap and insecticides. Electro–chemicals are centred in the Valais and electro–metallurgical industries below Schaffhausen. One-sixth of the country's exports are dyes and drugs. Up to 1967 decreasing amounts of iron ore and manganese ore were mined in Sankt Gallen canton, in the Jura at Delsberg–Choindez and at Frick near Olten but this production, amounting to 100 000 tons a year, was exported to the Ruhr for refining and 75 000 tons of refined metal were imported via the Rhine navigation. Domestic scrap is treated at Solothurn, Choindez and Lucerne all of which is converted into highly processed steel, together with $1\frac{1}{4}$ million tons of semi-finished steel from the Ruhr. The major steel centres are Biel, Winterthur and Schaffhausen, and there are minor concerns at Baden, Brugg and Oerlikon.

Switzerland also manufactures about 1 per cent of the world's aluminium, amounting to 77 000 tons a year and largely exported in the form of aluminium foil wrappings for cheese and food products. Neuhausen, Rheinfelden and Orsières are the refining towns while the rolling mills are situated at Kreuzlingen. Altogether about 120 000 people were engaged in the metallurgical industries in 1968.

The engineering industry which has developed from the metallurgical trades is significantly specialized and two-thirds of the products are exported. They consist of a wide variety of electric engines, motors, transformers, generators, turbines, cables and safety devices for electric installations as at Zurich, Baden, Basle, Geneva and Berne. A second group makes diesel locomotives, marine diesel engines and jet engines, as at Winterthur and Zurich. A third group is concerned with textile machinery and sewing machines, as at Oerlikon, Schaffhausen and Neuchâtel and a fourth with machine tools, precision instruments, calculating machines, papermaking and food-processing machinery. Many are internationally famous. This engineering industry (263 000 workers in 1968) centres on Zurich and stretches in a belt across the Mittelland from Sankt Gallen in the east to Basle in the west.

Metallurgy, engineering and machine making together employ about 40 per cent of the working population. As already noticed the products are characterized by their light weight compared with their value, their intricate or specialized nature and the large amount of labour and skill involved in their manufacture. This is well illustrated by the making of watches and clocks which engages 75 000 people in 1300 factories of which only about 60 make complete units, the rest supplying components. Geneva is the main centre, followed by Schaffhausen and the Jura towns such as Biel, Neuchâtel, Le Locle, La Chaux de Fónds and Saint-Imier. Nearly all the 70 million watches

XXI *Zürich, the old core as seen looking northwestward from the cathedral tower, showing the river Limmat soon after its issue from Lake Zürich. In left foreground is the Rathaus bridge just beyond which on the highest point of the left bank is the Lindenhof, the site of the Celtic and Roman settlements.*

and clocks made in Switzerland in 1970 were exported, and represented one-seventh of the national income. Jewellery and silver goods are also made in these towns and others, often as a subsidiary of the watch trade. Musical instruments are made at Le Locle, Berne, Basle and Zurich, and record players and cinematic material at Sainte-Croix, stimulated by the annual film festival at Montreux.

The watch and clock trade, of which Switzerland has long supplied 70 per cent of the world exports and 50 per cent of the world production, illustrates the country's ability to face growing competition. Since 1955 the Japanese have made great inroads in the world market for watches but Switzerland has almost doubled its output and still remains the world's chief producer and seller with 45 per cent of the world output.

Textiles form another important group of industries, which, with clothing and footwear, employs about 135 000 workers. The cotton spinning centres on the cantons of Aargau, and Thurgau, and its weaving, largely on automatic looms, on Basle. Woollen textiles and knitwear are made at Schaffhausen, Solothurn and Interlaken, and silk is spun and artificial fibres manufactured mainly at Zurich, Lucerne and Berne. Glarus in the Linth valley specializes in the dyeing and printing of cottons. Sankt Gallen has long been famous for embroidery and lace, now factory products employing only a few thousand workers. Typically it also makes synthetic fibres and linen goods, and has a technical school for fabric design. The prime urban market of Zurich dominates the clothing industry.

Some idea of the importance, relative and absolute, of the industrial concerns may be gleaned from the table of employment provided by the various economic groups. The decline in number of employees in some sections, such as textiles, has been remarkably small compared with that in most other countries. But it has been accompanied with a marked increase in output achieved largely by increased automation, a process at which the Swiss have few equals. The watch industry and the textile concerns are, for example, very highly automated. Yet there has been in recent decades a serious shortage of workers, and a good deal of labour, often skilled, has come in from adjoining countries. So great has been the desire to live in this highly educated, socially attractive state that restrictions have had to be put on immigration. Today about 18 per cent of the national labour force is foreign and in August 1968, a busy month, about 648 000 foreign workers had permits for temporary residence. Of these 409 000 were Italians, 81 000 Spaniards and 60 000 Germans. There is virtually no unemployment and in both agriculture and manufacture productivity per worker per hour rises steadily.

Tourism

This is a very important source of income. The magnificent scenery, the snow in winter and spring, mountaineering all the year and the high standards of living of a bi- or tri-lingual society who specialize in tourism, attracted

SWITZERLAND: EMPLOYMENT BY ECONOMIC GROUPS

		(thousands) 1965	1968
Agriculture and Forestry		252[1]	
Mines and Quarries		9	
Manufactures:	Food processing	128	
	Textiles	74	64
	Textile goods and footwear	93	68
	Wood and cork goods	73	
	Paper industry	22	
	Graphic arts	61	
	Leather	8	
	Rubber and plastics	14	
	Chemical industry	58	60
	Crude petrol refining	1	
	Clay and stone products	35	29
	Metal working	192	120
	Engineering	301	263
	Watchmaking	76	78
	Jewellery making	5	
	Musical instruments	1	
	Others		
	Total	1143	
Building and construction		321	
Electricity, gas and water		19	
Services:	Wholesale trade	108	
	Retail trade	211	
	Banking and insurance	63	
	Transport and postal	149	
	Hotels and restaurants	152	
	Medical	100	
	Others	92	
	Total	875	
Grand total[2]		2612	

[1] Excluding 268 000 part-time workers but including 593 full-time fishermen.
[2] The total working population is variously estimated at between 2 775 000 and 2 900 000 and the number of full-time persons employed in agriculture and forestry at 252 000 to 275 000.

6 million foreign visitors in 1968. These, irrespective of residents at health clinics and numerous political refugees, provide one-eighth of the Swiss revenue and, with railway freight and passenger charges, banking, insurance and other concealed imports, balance the national budget. The tourism, as with so many other aspects of Swiss industry, is superbly organized and employs more than 150 000 people. Switzerland has become the playground of western Europe. Not surprisingly Zurich is by far the chief centre for number of visitors and Basle and Berne are also important. In the mountains and on the mountain fringe the main tourist resorts are on lakes, the chief being Lucerne, Geneva, Lugano, Lausanne, Montreux, Interlaken and Locarno. Among the most popular ski and mountaineering resorts are Saint-Moritz, Davos and Zermatt.

Distribution of population

Three-quarters of the Swiss population live on the Mittelland and half in cities and towns of 10 000 or more inhabitants (fig. 64). The total population is 6·1 million which represents a density of about 150 per square kilometre. Of the ten towns with over 50 000 people, by far the chief is Zurich, with 673 000 persons in its conurbation. This city lies where the route from Basle to the Saint-Gotthard pass crosses the east–west routeway along the Swiss Mittelland. It grew up at the bridgepoint on the Limmat where that river leaves the northern end of Lake Zurich. The modern city has sent two long tentacles of settlement southward along both shores of the lake. It is the chief financial, industrial and marketing centre of Switzerland as well as an important educational and route centre. Its manufactures are especially concerned with textiles and engineering, including turbines, armaments, aero engines, electrical and railway equipment, and machine tools.

Basle (370 000 in conurbation) is the chief commercial centre of Switzerland and a leading industrial city (fig. 65). It stands at the southern extremity of the Rhine rift valley and at the head of large boat navigation on the river. In recent years the imports have reached 8 million tons annually and the exports about ½ million tons. The city has 5 kilometres of quays and large storage spaces particularly for the main imports of coal, coke and iron ore. Nearby are important salt deposits which have led to the establishment of large chemical industries, and dyestuffs are among the chief exports. The modern development in pharmaceutical and other chemicals has been stimulated by the petroleum pipeline from Lavéra. A further stimulus may come from a proposed pipeline for natural gas from the Netherlands. Basle is also a regional centre for engineering and textiles in the Swiss Rhineland.

Berne (258 000 in conurbation), the national capital, grew up within an incised meander of the river Aare. Its industries are concerned largely with clocks, watches and knitwear, and, as would be expected of a capital city, with printing and publishing.

Geneva (310 000 in conurbation) also has an attractive site, in this case where the Aare joins the Rhône just below the Rhône's exit from Lake Geneva. The town early became the chief centre of the Swiss watch, clock and jewellery trades. Banking and insurance are also important and the activities of the headquarters of numerous inter-

XXII *Aerial view of Basle (compare with figure 65)*

64 *Switzerland: cantons, cantonal capitals and other chief towns*
   The cantonal capitals are named except Winterthur

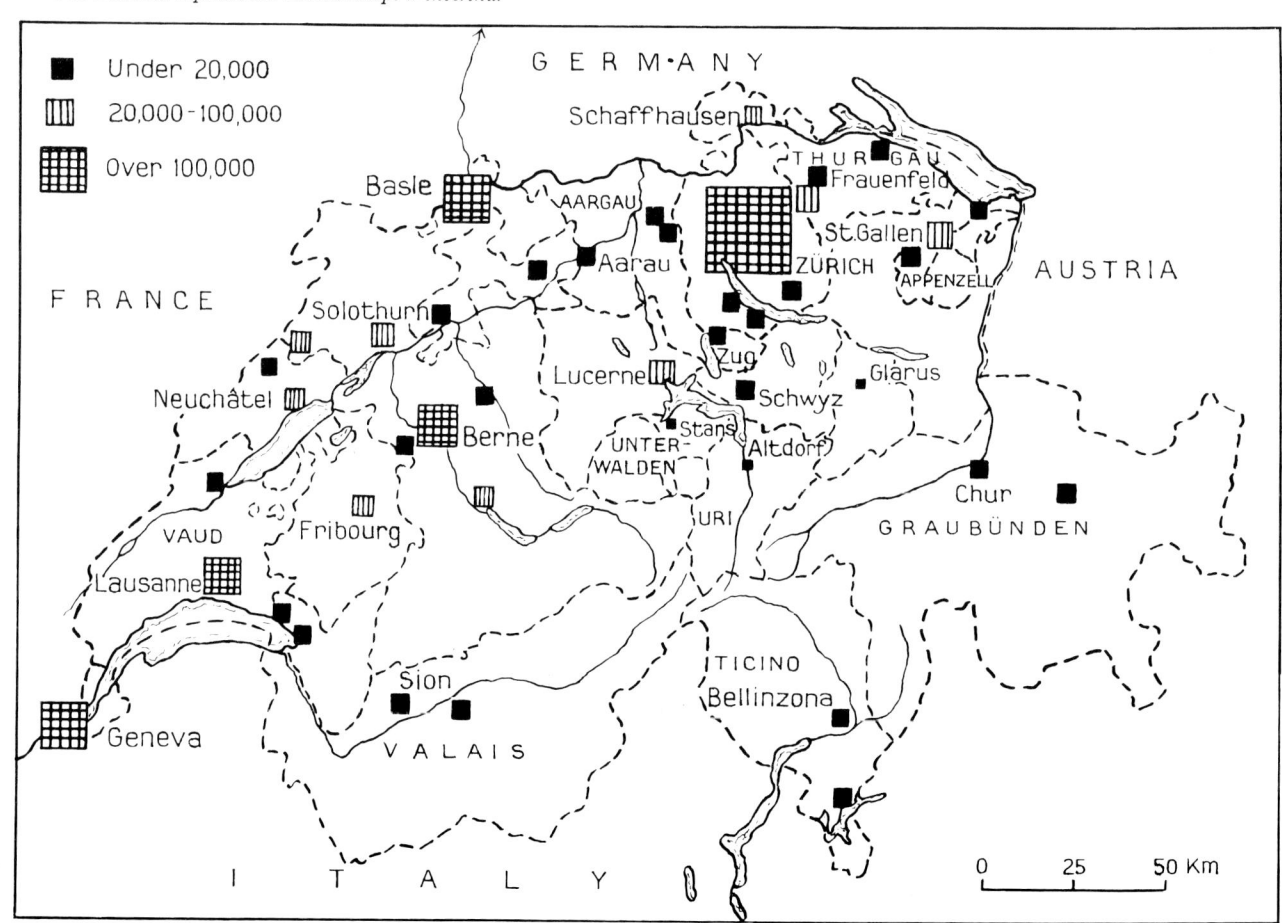

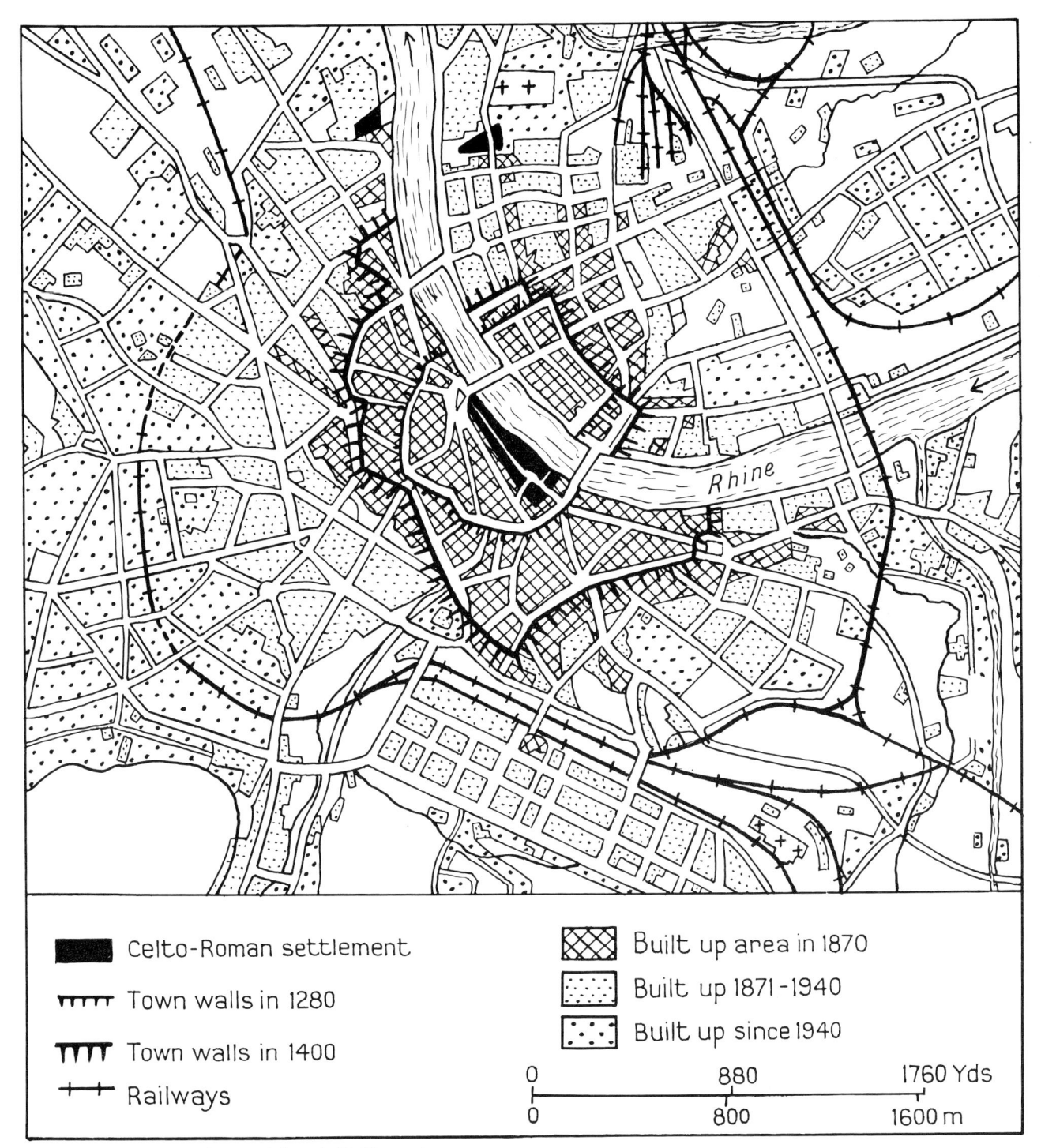

65 *Site and growth of Basle* (after A. F. Mutton)

national organizations such as the Red Cross, the International Labour Organization, various branches of Unesco, the World Health Organisation, the European Postal Union, EFTA and so on.

Lausanne (216 000 in conurbation) overlooks Lake Geneva on which it has an outport at Ouchy. It rose to importance partly because of its position on the Simplon–Lötschberg railway which stimulated tourism, food-processing industries and more recently electro–chemical and electro–magnetic and cutlery trades based mainly on the production of aluminium from French bauxite.

Lucerne, with 150 000 people in its vicinity, is a famous tourist resort on the lake of that name and at a bridgepoint over the Reuss. It has the good fortune to be on the Basle–Saint-Gotthard route and has become a centre for the making of jewellery.

The largest of the other towns are Winterthur (105 000 in conurbation), Biel (92 000) and Sankt Gallen (80 000), all notable manufacturing centres with large world-famous firms.

Communications

The Swiss Alps sit astride the direct routes from north-west Europe and the Baltic to northern Italy and the central Mediterranean (fig. 39). Of the many Alpine passes in Switzerland, the difficult Great St Bernard (2469 m) was

XXIII *Berne, capital of the Swiss Federation, founded as a fortress in 1191 on a spur in a loop of the river Aare. All the bridges shown, except the Mydegg bridge at the bottom right, are modern*

used in pre-Roman times although it had to wait for a carriage road until 1905 and for a road tunnel through the summit until 1964. The Saint-Gotthard (2108 m) has been in regular use in summer at least since a mule tract was built through it in the late thirteenth century. The Simplon pass (2005 m) became important after Napoleon had a roadway constructed through it in 1801. Today almost every possible Alpine pass carries a road and the lower passes are traversed by railways usually with the aid of summit tunnels. The transport system of Switzerland is one of the engineering marvels of the world and the Alps instead of being a serious barrier have become an international passageway. None the less all the higher road passes are closed by snow throughout the colder months and tunnelling still proceeds in order to give relatively low summit levels, with routes free longer of snow.

The route network as a whole was greatly aided by the longitudinal depression drained by the upper Rhône and Rhine, and by the remarkable divergence of four large rivers (Rhône, Rhine, Reuss and Ticino) from the Saint-Gotthard massif. The deep incision of glaciated valleys leading up to the watersheds encouraged roads and later railways to penetrate all the main Swiss valleys. But to connect these valleys by direct routes needed expensive engineering and largely for this reason the railways were nationalized (1898–1905) and federal funds have since developed a very efficient network.

The three chief railways leading from Switzerland into Italy are from west to east:

1 The Simplon, opened in 1906 with a 20 km summit tunnel (summit height 700 m), leading from Brig in the upper Rhône valley to the Toce valley and towns on Lake Maggiore. This route, however, involved another long summit tunnel, the Lötschberg, through the Bernese Oberland Alps north of the Rhône trench.

2 The oldest trans-Alpine railway, the Saint-Gotthard, has a summit tunnel or tunnels about 15 km long and a summit height of 1154 m. It was opened in 1882 and joins the Reuss valley to the Ticino valley and so to Bellinzona and Milan where it meets the Simplon line.

3 The Albula pass line (summit tunnel, 1820 m; length 6 km) from near Chur on the upper Rhine south to Tirano and the Po plain is of lesser importance except for famous winter sports resorts such as Saint-Moritz.

The Swiss railway network also includes a railway from Martigny up the Rhône to its source in the glacier near the Furka pass (above the Saint-Gotthard tunnel) and so eastward to the Rhine valley and Chur. This high-level line, however, unlike the main trans-Alpine lines, operates only in the summer.

There are nearly a score of frequented road passes over the two main ranges of the Swiss Alps, and in a few road and railway work together. Thus on the Saint-Gotthard route trains carry large numbers of tourists and their cars through the mountains all the year. Chapter 10 contains an outline of the relative position and importance of the Alpine passes generally. The busy transverse routes to

Italy and the central Mediterranean are outstandingly important but some notice should also be given to longitudinal routes leading into France on the west and into Austria on the east where, for example, the river Inn, which rises near Saint-Moritz, is followed by a road to Innsbruck. Today some of the internal road system is rather narrow and difficult on international standards but the costs of improvement in a mountainous terrain are enormous and probably the Swiss transport network will always remain famous for its railways rather than for its autobahn. Because some cantons have made more progress than others with their road-improvement schemes the stretches of autobahn are intermittent and it will be a long time before a national autobahn system is completed. Among the great new schemes are road tunnels under some of the Alpine passes. That under the Great St Bernard pass is 6 km long and was opened in 1964. In the following year the French opened a road tunnel (11 km long) under Mont Blanc. In 1967 the Swiss opened the San Bernardino tunnel (7 km long) and then commenced work on a tunnel, 16 km long, under the Saint-Gotthard, which, when completed in 1977, will be the longest road tunnel in the world.

Geographers will quite rightly continue to divide Switzerland into 'geographical regions', such as the Alps, Mittelland and Jura, but the real interest of the country lies in its self-imposed federation and coherence, in its national solidarity and international neutrality, and in the high quality of its social and economic systems.

LIECHTENSTEIN

The principality of Liechtenstein with an area of 158 square kilometres, and maximum lengths of 24 km from north to south and 9 km from east to west, is the smallest independent state in Europe. German-speaking it exists as a buffer state between Switzerland and Austria but more particularly because it commands the Gap of Ragaz that leads the valleys and passes of the Grisons to the Rhine crossing at Vaduz. The country has the Rhine for its western frontier and the crestline of the Vorarlberg as its boundary with Austria. Formerly the valley of the Rhine was marshy, ill-drained and liable to flood but this section has been carefully regulated and, with the aid of an irrigation canal, the land is now productive with cereals, fruit orchards and vineyards.

Most of the remainder of the country is mountainous rising in height to the south and culminating in the two peaks of Falknis (2564 m) and Naafkopf (2574 m). It is bisected into two highland blocks by the north-flowing Samina river.

Except for two small villages in the north, settlement still clings to the foothill country parallel to the Rhine and consists of five small towns including Vaduz the capital, with 3800 inhabitants. The total population is 21 000. All the settlements are linked by a road which runs from Ragaz in Switzerland to Feldkirch in Austria, and some by a railway, the Arlberg express route from Paris to Vienna, via Buchs in Switzerland. These communications have enabled a variety of small manufactures to be established such as precision instruments, ceramics, cotton textiles, false teeth, leather and embroidery. Thus since 1930 the economy has changed from predominantly agricultural to largely industrial and the manufacturing concerns employ 4500 workers including some immigrants from nearby countries. Tourism is increasingly important and the country also derives much revenue from the sale of postage stamps. The state, which first became independent in 1719, entered into a customs as well as a currency union with Switzerland in 1924. It is a constitutional monarchy and has no army. The absence of most taxes and the minimal nature of others have made Liechtenstein a headquarters for many companies and the state derives one-fifth of its income from this source.

17 Italy: Basal Patterns

Political development of the state

From AD 476 when the Goth Odovucar deposed the last Roman emperor in Rome, the Italians have lacked political unity until modern times. With the fall of Rome, Italian history becomes that of separate city states and petty kingdoms and emphasizes the physical differences in a country that never lent itself to easy unification. The Po basin is partly cut off from the western coastal plains by the great curve of the Apennines; in the peninsula the Apennine backbone fostered separate communities in mountain-girt basins and tended to divorce the Tyrrhenian from the Adriatic shores; the length of the peninsula and its island extension encouraged the development in the south of the Kingdom of the Two Sicilies. Geography segments the country into north, centre and south and these distinctions are repeatedly reflected in its history.

The Roman conquest (348–173 BC) was the first of the three major impulses which led to the modern unification of Italy, the other two being the Italian League of the City States, and the nineteenth-century *Risorgimento*. Under Roman rule the country was a confederacy under Roman protection and overall authority. The nature of the confederacy is significant as tribes and towns were not equally privileged and there are signs here of the seeds of regional and urban jealousies that delayed modern Italian unity for centuries. The remarkable urban growth and road network and their influence today have been touched on in Chapter 9. Here we shall only attempt to enlarge some aspects of particular importance to Italy.

The administration and defence of Italy was made possible by the Roman colonial policy. After 89 BC all inhabitants of Italy south of the Po were *de iure* Roman citizens, citizenship being granted in desperation to bring to an end incessant revolts. This discontent was finally appeased during the principate of the first Roman Emperor Augustus (ruled 27 BC to AD 14). However, the truly 'Roman' territory extended from coast to coast south of the Tuscan Apennines as far as the southern boundary of Campania and occupied about one-third of Italy. Here, Roman towns, *coloniae*, built on the usual gridiron pattern were set to guard the coasts of Latium and Campagna and the opposite Adriatic shore. Outside this core another line of Roman towns, including Bologna, Parma and Piacenza lay along the main Emilian way on the Po plain just north of the Apennines. Next in importance were the Latin towns established at strategic points along the military highways, such as Turin (Augusta Taurinorum), Cremona, Genoa and Brindisi. Another class of urban centre was the 'allied' town which was brought into the administrative net by means of establishing in it a Roman military garrison; Tarentum (Taranto), Capua and Rhegium (Reggio di Calabria) were of this order. Settlement of Roman veterans was also made on the 'public lands' acquired by conquest, and in villages in the hills and countryside newly brought under Roman protection. Around these new towns and in the new territories 'centuriation' or parcelling out of the agricultural land took place. These ploughlands were generally laid out geometrically and the rectangular field patterns may still be discerned especially in the southern part of the Po basin along the line of the Emilian way.

As time went on, it became apparent that there were flaws in the Roman colonial system. Many soldiers were not necessarily good farmers and preferred quick and easy booty from military expeditions to the hard toil of cultivating newly acquired land. Much land was vacated. Those who were good farmers found it difficult to make a livelihood in face, in the Roman area, of cheaply imported wheat from other parts of the empire, and elsewhere in competition with rich landowners working their large estates with slave labour. Many colonies in Italy dwindled away, and much public land donated to soldiers in this way was absorbed into the latifundia and tended to become cattle-grazing lands and subject to soil erosion.

By the time of Julius Caesar every town had an element of paupers in its population while large areas of the countryside were desolate and untilled. Under Caesar's agrarian reforms which treated the effects without removing the causes of distress, the latifundia were left intact though their owners were compelled to employ a certain number of free men on their estates and at the same time import duties afforded their crops some protection.

The key to the efficacy of the Roman administration was of course the system of military roads which were planned largely from the point of view of military strategy and therefore to conform to the needs of an efficient pattern of swift movement, and direct links between military strongpoints. From Rome highways radiated in most directions. Besides the roads running west on either side of the Tiber to Ostia the port at the mouth, the via Appia and the via Latina led south to the heel and toe of Italy to the towns of Brundisium (Brindisi), Tarentum (Taranto) and Rhegium en route for Sicily. Eastwards the viae Salaria, Caecilia and Flaminia linked Rome to the ports on the Adriatic, the last being extended to the Emilian way at Piacenza (Placentia) and the crossing of the river Po. The via Aurelia linked Rome with Pisa to the north, whence the via Augusta led to Genoa and southern France and the via Postumia to Cremona, Verona and Aquileia. The last-named, a port on the coastal flats of the north Adriatic, was destroyed by the Huns and is one of the few towns planned by the Romans in Italy that have been lost. Here only an eleventh-century basilica, built on Roman foundations, and extensive excavated ruins mark the site of a great port that flourished long before Venice was founded. Nearly all the other Roman towns in Italy have survived as living communities and retain much of their original streetplan as well as an architectural legacy.

In Cisalpine Gaul (the Po basin) the road pattern was conditioned by the existence of river crossings and alpine passes but was directed north to Rhaetica, east to Pannonia and southeast to Illyricum. The passes used by the Romans through the Alps were the Brenner, Septimer, Splugen, Mont Genèvre and Great and Little St Bernard.

When the western Roman emperor was deposed by Odovucar the Goth in AD 476, Italy suffered the same history as the other provinces in that empire. The Roman landowners had to yield one-third of their cultivated land to the invaders and while the civil administration remained Roman, the military authority was Goth. Three years later the Lombards invaded north Italy and a permanent settlement of Germanic people in the peninsula followed, spreading gradually over central and southern Italy. The territory was divided into dukedoms and Roman civilization was only spared by the fortunate circumstance that the Lombards became Christian and therefore open to the spread of Roman ideas.

The rise of the city states

The old Roman cities became capital cities of the new dukedoms, with the dukes having their official residences in them. Some cities were administered in the name of the king, some were the seats of the newly founded bishoprics but all served as administrative and market centres for the surrounding districts. Roman administration in the control of the eastern Roman emperor survived only in the Italian periphery in Istria, the Veneto, Ravenna (established as the new capital), the Pentapolis (Rimini to Ancona), Perugia, Calabria, Apulia, Naples and the Campania, Rome and the Campagna, Sicily, Sardinia and Corsica. Naples gradually assumed a semi-independent authority and in Rome the secular power of the papacy was established by the acquisition of large estates in central Italy and Sicily. The various bishops, as their lands and territories increased, asserted an ever-growing authority.

Further Lombard invasions from the north excited interference from the Franks also and in AD 800 Charlemagne conquered the Lombard Kingdom in northern Italy and made it part of the Holy Roman empire that had its capital north of the Alps, while extending its influence south over the papal states in central Italy. The partition of the Carolingian empire on the death of Charlemagne brought Italian territory into the long narrow Kingdom of Lothair between the East and West Franks that stretched from the North Sea down the Rhine to the Apennines.

Meantime in the south the Moors had begun their conquest of Sicily and southern Italy. Palermo was taken in 831, Messina in 843 and thereafter the war was carried to the Italian mainland, and towns along the Tyrrhenian coast as far north as Rome were threatened. From this date on approximately began the rise of the Italian city states. Even the most successful never succeeded in uniting Italy, nor in controlling the whole country nor in defending it against foreign aggression. The Hungarians overran the northeast plains of Italy and the Saracens were able to settle in the south, but where certain dukedoms were able to defend their territories they gained in power and authority while the Byzantine control remained dominant in other areas.

In 960, Otto the German King reestablished German control over Italy which lasted for two centuries but it was marked by frequent rebellions against the central authority, and against the overlords by the commonalty. In the Byzantine territories several cities had become prosperous through trade with the eastern empire. The greatest were Venice and Amalfi but others were Bari, Otranto and Salerno. Gradually other ports and route-centres like Pisa, Genoa, Pavia, Milan and Lucca became prosperous trading marts. Other small towns that were gradually growing in importance were the fortified *castelli* to which the people crowded for protection during Hungarian, Saracen or other raids. The very necessity for defensive measures laid the foundations for active participation in government to which in any case many were not strangers. Venice was electing its doges as early as the eighth century and Amalfi its magistrates in the ninth. Elsewhere the municipalities had dukes or bishops at their head with trade and manufactures and defence in the hands of the citizenry.

This communal movement was strongest in the central areas and the north in what had been the earliest Roman conquests. Gradually there developed city states enjoying complete selfgovernment, with municipal affairs in the hands of the merchants and landowners although conditions and practices varied between the various towns. The earliest of the city states were Lucca, Genoa and Pisa.

In Italy there never had grown up a marked division between townsman and countryman. Most of the townspeople had farms or villas in the countryside and the countryman interests in the town, so that gradually a city gained control over its surrounding area and began to absorb also the nearby smaller *castelli*. Unfortunately jealousy between the city states prevented federation. They would only combine for the destruction of another. Trade had grown enormously through the launching of the crusades and towns such as Genoa, Pisa and Venice on routes to the Holy Land had flourishing manufactures in textiles, leather and metals established to equip and supply crusading armies and their transports. Merchant and craft gilds grew up to organize the industries and the trade and the rivalry between the towns intensified. Pisa and Genoa were rivals in the trade to Sardinia; Genoa and Venice for the Levant trade. Inland, Milan was a commercial rival of Cremona and Pavia, while Florence competed with Siena and grew sufficiently wealthy to plan a port at Leghorn. The men who could wage the trade war most effectively gained leading positions and were appointed for life. So grew the fame and importance of the Visconti in Milan, the Este in Ferrara, the Gonzaga in Mantua and the Medici in Florence. The territories and trade of the stronger city states gained enormously at the expense of the smaller ones and the rivalries intensified into wars. The unification of Italy was also constantly being prevented by wars between the imperial authority and the nobility. At the same time the growth of manufacturing and trade led to the aristocratic assumption of power being challenged by the merchant and banker class. The Visconti lands extended from Genoa to Bologna and, including Siena, Pisa and Perugia, challenged the trading interests of Florence. Venice acquired a considerable empire widely scattered in the Adriatic, Aegean and Black Sea areas.

As a result of a common language and the realization of the threat to trade and living standards posed by constant wars, in 1455 the Italian League was founded and, freed from war, the arts and crafts of peace flowered in the Italian Renaissance. The picture painted by Botticelli in 1485 called today 'Pallas and the Centaur' really celebrates the peace achieved by Medici diplomacy showing wisdom (Minerva or Pallas, with the Medici crest on her gown) triumphing over brutish war policies.

This Italian League was, however, doomed to failure as Portuguese voyages round the Cape ruined the Italian trade with the Levant and, in any event, southern Italy had followed a history quite different from that of the north. Owing to Saracen invasions and settlement, it became the target of Norman crusaders and in 1054 the Norman settlement of southern Italy began. Eventually the Normans welded together Italian, Sicilian, Greek and Arabic elements into one territory known as the Kingdom of the Two Sicilies under one authoritarian and central power, which permitted no municipal share in administration. Even when Aragón was invited to take control, the government was still centralized and repressive with all control emanating from Naples. In central Italy, the Papal States lived under an equally centralized and authoritarian government.

The discovery of alternative trade routes to the spice trade in 1498 robbed the northern city states of their prosperity and eastern traffic. Later Antwerp and not Florence, became the central European banking city. With declining fortunes, and the Spanish and Austrian wars of succession, Italy was sacrificed to the maintenance of the balance of power in Europe. Only the tiny republic of San Marino retains today the form and constitution of a medieval city state.

The Risorgimento and final Unification of Italy

The age of enlightenment and reform slowly brought about an awakening of national consciousness in Italy which was quickened by the French occupation of the whole country under Napoleon Buonaparte. The country was mercilessly exploited in French interests but the unified government and the application of the Code Napoleon prepared the way for the Risorgimento under the inspiration and leadership of Cavour in the north and of Garibaldi in the south. In 1861 the parliament of a monarchical united Italy met at Turin. Unfortunately unification was followed by a general outbreak of brigandage in southern Italy which went on for decades. In 1870 Rome, a walled city under the Pope, was captured by government troops and united with Italy, the pope being allowed to retain ownership of the Vatican and Lateran palaces. In the following year the government, now at Florence, moved finally to Rome and began to modernize the city.

During the twentieth century Italy took part in the European scramble for colonies in Africa and in 1912 conquered Libya. In the 1914–18 war Italy joined the Allies and after the war obtained a frontier up to the Brenner pass, with many German-speaking peoples in the area that became the provinces of Trentino and Alto Adige. It also gained Trieste and Venezia Giulia and the Istrian Peninsula which had among its inhabitants 25 000 Slavs (fig. 67). These boundaries were to cause constant trouble. In 1920, after some fighting, Italy and Yugoslavia made a new agreement whereby Italy waived its claims to Dalmatia, except the town of Zara, and retained the islands of Lagosta and Pelagosa, while Fiume was to be an independent city.

Mussolini and a fascist regime prevailed in Italy from 1925 to his execution in 1945. Mussolini made determined but fruitless efforts to italianize the 200 000 German-speaking people in Alto Adige and the half-million Slavs living in Italy since the 1920 treaty. From 1935 to 1937 Italian troops ruthlessly conquered Abyssinia. During the Second World War Italy fought against Great Britain and France and in 1943 the Allies (in north Africa) invaded Sicily. In 1944, Rome was taken and Italy then became a co-belligerent of the Allies and in the following year north Italy was captured with the invaluable aid of Italian partisans. In the peace treaty of 1947, Yugoslavia received Fiume, Zara, Pola and Istria, thus the boundary here went almost back to the pre-1918 boundary as shown in figure 66. The fate of Trieste was not decided. Small parts of western Alpine Italy went to France, and the Dodecanese islands in the Aegean to Greece (see Chapter 23). These islands had been taken by Italy from the Turks in 1912. In 1954 Italy, now a republic, agreed with Yugoslavia to divide the territory of Trieste. The Italian share, which covered 234 sq km and largely comprised the city itself, had a population of about 310 000 including 63 000 Slovenes whose full rights were guaranteed.

In the economico–political sphere, Italy has looked northward. In 1952 it joined the European Coal and Steel Community and in 1958 became one of the founder members of the European Economic Community (the Common Market). Among the projects of the latter has been considerable financial aid for industrial and general economic development in southern Italy.

Today, for local government purposes the republic is divided into 19 regions (*regione*) of which 5 – Sicily, Sardinia, Val D'Aosta (which is partly French-speaking), Trentino–Alto Adige, Friuli–Venezia Giulia – have a considerable degree of regional autonomy, each with a regional council and regional executive junta. The other 14 regions have administrative functions only. Italy is also

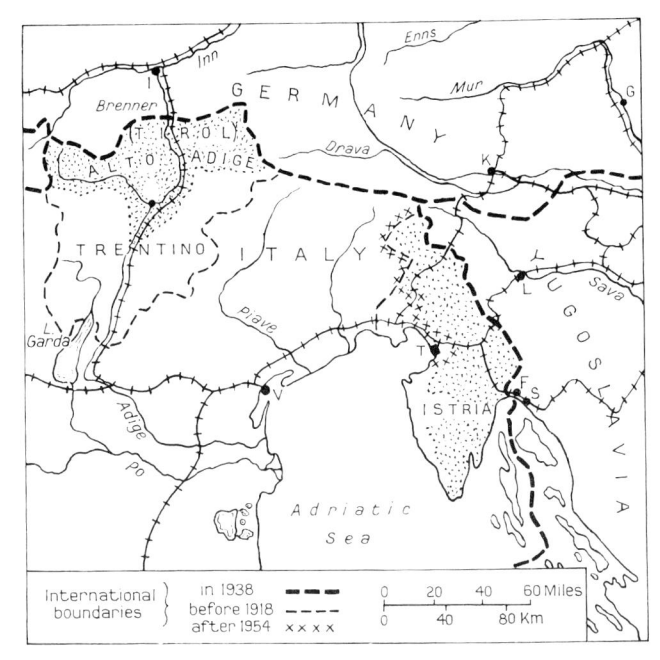

66 The northeast borderlands of Italy
F *is Fiume or Rijeka. Dotted area shows tracts of the Alto Adige where German is commonly spoken and of the northeast Adriatic formerly in Italy where Slovenes and Croats predominate.*

divided into 92 provinces, each with its council and executive junta. For example, Sicily has 9 provinces, Sardinia 3 and each great city is a province. 3 provinces, Milan, Rome and Naples, have over 2 million inhabitants.

Structure and relief

Italy which divides the Inland Sea into two distinct basins is the most mediterranean in character of the three southern peninsulas of Europe. The western Mediterranean lies in a circle of fold mountains while the eastern Mediterranean is a jigsaw of interlocking seas and generally low shorelines. Italy has often been compared to a pierhead thrusting for 1200 km in a southeast direction between these two basins.

The forerunner of the Mediterranean was a sheet of water called Tethys that lay between the ancient shield of Africa and the Hercynian mountain chain in Europe. It occupied a geosyncline in which enormous thicknesses of sediment accumulated. These and no doubt subcrustal processes caused the floor to sag or ruck into folds and the deeper sediments to be metamorphosed by pressure and intrusions of igneous material. It is generally assumed that in early Tertiary times the African block moved towards the Hercynian range exerting great pressure upon it. Some of its area was dislocated and fractured into blocks, some foundered, and the floor of Tethys was later uplifted as great recumbent folds that, although by no means continuous today, form the Alpine chains. The portion of the Tethys sea in which they originated can be identified by the nature of their sediments. The Italian Lakes area, for example, from Lake Maggiore to the Venetian Alps, is recognized as deriving from the Tethys seafloor nearest Africa.

The Apennines emerged later and although they seem to be a simple mountain range have a long and complicated history, with two great periods of foldings thrusting towards the northeast, and at least one epeirogenetic or general uplift. The first foldings occurred in early Tertiary times and were at once subjected to a period of intense denudation so that by mid-Tertiary times the land was much reduced in height and eventually sank to be covered by the sea and laden with silts, clays, muds, sands and gravels of marine origin. Uplift as folds, foundering and renewed sedimentation were repeated in mid-Tertiary times transforming the earlier sediments into flysch but also emphasizing the folds that can be traced through Abruzzi, Tuscany and Liguria and the longitudinal valleys and basins of the western Apennines. The sea penetrated these lowlands to the foot of the Alps and, submerging all but the highest summits, turned southern Italy into an archipelago.

At the end of the Tertiary period an epeirogenic or general movement uplifted the land so that the recent marine deposits are found today at great heights, for example, at 1200 m in Calabria. This latest movement was accompanied by faulting especially along the west coast where the sea floor plunges to 2750 m and by its usual concomitant vulcanism for which the evidence remains today in the lava fields and crater lakes of the Roman countryside, in Campania with its cones, and the Campi Flegrei, the Lipari Islands and elsewhere, areas that became classic fields of study and gave the vocabulary of vulcanism to geological literature.

Italy consists therefore of three main structural elements: the Alpine folded zone, the later Apennine folds, and the north Italian geosyncline between them.

Alpine zone

On its landward side Italy has its frontier mainly near or on the lofty watershed of the Alps at 4000 m in places. Covered by a continuous icesheet during Quaternary times, the Alps fed ice tongues that moved south on to the north Italian plain down the valleys such as the Adige, Ticino and Dora Baltea, leaving a legacy of overdeepened and moraine-dammed lakes and expanses of outwash plain and debris.

North Italian plain

This occupies a geosyncline of which by far the larger portion drains today to the Po, which rises on Monte Viso at 1830 m above sea level and flows east for 670 km (500 km in direct line) to its delta in the Adriatic. Formerly the plain was a shallow extension of that sea which has been converted into dry land by the tremendous amount of fluviatile waste brought down by rivers from the Alps and Apennines. At the mountain foot where the rivers enter the plains, large alluvial cones and fans of debris have

XXIV *Po Delta, aerial view*

spread out to form a continuous piedmont zone, in which the material over the centuries has been graded from coarse pebbles and gravel round the perimeter to fine silt along the central axis of the basin. On the east is the Venetian plain, a broad alluvial flat adjoining the lagoon coast at the head of the Adriatic, that drains independently of the Po into the sea.

The whole north Italian plain is a fascinating study in deposition but a distinction has to be made between the origin of the deposits and the nature of its deposition in various parts of the plain. At the foot of the Alps between the Dora Riparia and the Adige huge moraines form amphitheatres to enclose the Italian lakes, such as Como, Garda and Maggiore, and to the south of these are typical outwash plains that have had their surface debris re-sorted by spring floods or past and present-day rivers. Gravel grades into silt and both have been formed into terraces at various heights along valleys of the alpine tributaries of the Po. The presence of impermeable clays coming to the surface in these outwash areas accounts for the zone of seepages and springs, the fontanili, which divides the coarser fluvio–glacial gravels from the finer and more fertile silts.

The course of the Po itself is influenced by the load-carrying capacity of its various feeders. In the west of the basin the swifter Alpine streams push the river to the south, but farther east the Apennine rivers cutting through softer sediments bring down by far the heavier load and deflect the Po to the north.

Alluvium has been deposited by all the tributaries in broad stretches downstream from Turin and the main stream is so laden that it tends by the deposition of its burden to deflect eastwards the lower courses of its tributaries, especially the Alpine. The Adige, in fact parallels the course of the Po for 64 km before entering the Adriatic by an independent mouth.

Below Piacenza despite much dyking and poldering, large areas of fen exist owing to the extreme difficulty of effecting drainage. In other areas, notably Emilia, the land is raised above flood level by *colmate naturali*. This method consists of enclosing a low-lying area and diverting the river in flood onto it, which then being heavily charged with silt gradually builds the land up above normal flood level. In all its lower course for at least 420 km, the Po is constrained by embankments, and finally enters the Adriatic by a delta today about 33 km in width from north to south. The whole coastline is in fact deltaic and it is difficult to sort out the courses of the various river mouths of the last two thousand years. The coastline is being steadily aggraded and is fringed by lagoons and sandspits that quickly become inactive. The Po della Maestra advances about 80 m a year and the Po di Goro by about 70 m a year and, although today the lagoon on which Venice is situated is 65 km long, it was formerly much more extensive providing a system of interior navigation that extended from Ravenna to Altrino. Ravenna once a Roman port is now 11 km inland. Many of the former water expanses have been converted into dry land and form part of the alluvial plain of Veneto.

The set of the Adriatic currents is anticlockwise and the material brought down by the rivers is very quickly redistributed laterally, so that rivers tend to enter the sea by their southern distributaries. Venice keeps open its three *portos* or channels into the Adriatic fairly easily because the waters and silt of the Brenta, Sila, Adige and Po are all diverted to the south of its lagoon.

The Apennines

The Apennines are considered to begin at the Cadibona pass in Liguria and at first run mainly west–east and tend to cut off the plain of Lombardy from peninsular Italy by a continuous sandstone barrier which though lower than the Alps is of considerable width. Averaging 700 to 1000 m in height they are higher in the east than in the west culminating in Monte Corno in the Gran Sasso at 2914 m but they are traversed by a number of passes below 600 m. The rivers flowing south in Liguria are torrents entering the sea by huge clefts. From Rimini to Pescara the Apennines run southeastwards, parallel to the coasts, but the watershed lies closest to the Adriatic and the highest peaks are also here.

It continues to be a mountain range of considerable width made up of minor ranges and longitudinal valleys, such as the Salto, Liri and Aterno, running parallel in a northwest–southeast direction. On the west the rivers break out of the mountains by transverse passes through the ranges which assist the penetration of the summit areas. The centre is cold mountain country, sometimes karstic, reaching in general to 1500 m.

The complicated structure is hard to categorize. West of a line from Rimini to Rome, sandstones, clays and conglomerates predominate, especially in the Arno basin (fig. 77). Here the relief is deeply dissected with rounded interfluves. East and south of that line the main backbone of the so-called central Apennines consists mainly of limestones, often arched and everywhere fractured. The mountain blocks and ranges and their peripheral fault scarps have a trend parallel to the peninsula. Some of the fault scarps are truly impressive particularly that which coincides with the eastern wall of the Monti Sibillini and stretches 200 km southward to the Gran Sasso. The limestones give a massive, tabular, rugged relief such as is developed in the Gran Sasso where hard Eocene dolomites cap softer Liassic and Cretaceous beds. Here, however, the upper 300 m of Monte Corno (2914 m) have been glaciated. Equally typical is the extensive limestone plateau near Aquila where karstic landforms abound and fracturing has created a longitudinal rift valley now floored with new sediments. In these central Apennines Tertiary sandstones and clays may locally be preserved at high altitudes (Monte Gorzano, 2455 m) but they are restricted to the east and become truly important only in the Molise which is deeply dissected by headstreams of rivers draining to the Adriatic and has an undulating skyline with summits at 800 to 1000 m.

At least three other aspects of the geomorphology of the peninsular stretch of the Apennines described above deserve emphasis. First, there are eighteen quite large semienclosed Pliocene lake basins, many of which are associated with fractures and a few with limestone solution. The chief lie in the drainage of the Arno (Val di Chiana, etc.) and Tiber and form flat, terrace-lined basins within a mountainous landscape. Second, the western side of the peninsula was invaded by the Pliocene sea and has a 50 km wide coastal lowland in Tuscany, Latium and Campania. Here, however, volcanic activity has greatly altered the relief and south of the Ombrone river the plain is repeatedly interrupted by undulating lava flows, small volcanic massifs and low volcanic cones, often with crater lakes. On this coast too, especially south of Rome, limestones frequently form cliffed peninsulas as at Sorrento and the isle of Capri. Third, the eastern Apennine flank or Adriatic coastland is very different, although also invaded by the Pliocene sea. Here vulcanism is almost absent, the mountains lie closer to the sea and consist largely of conglomerates, sandstones and clays. They are fringed on the coast from Rimini to Termoli by Pliocene and early Quaternary platforms, in all over 300 km long by 35 km wide, down which hundreds of small consequent streams drain directly to the Adriatic. These platforms commonly lie at 90 to 130 m above sea level. From the plain of Foggia, the Pliocene platforms and basins continue inland southeast to the Gulf of Taranto and so are here separated from the Adriatic by the flat-bedded limestone (Cretaceous–Pliocene) tablelands of the Gargano peninsula and southern Apulia, the heel of Italy.

The main structural lines of the Apennines run northwest to southeast as far as Potenza where, in a tortuous mass of high country and divagating rivers, they turn south to link up with the crystalline massifs of Calabria (fig. 78).

Besides the landforms associated with processes such

CLIMATIC STATISTICS, MEAN MONTHLY TEMPERATURES (centigrade)

	Altitude if over 100 metres	J	F	M	A	M	J	Jy	A	S	O	N	D	Absolute Max	Absolute Min	Annual days with frost
Cortina d'Ampezzo	1275	−2·2	−1·1	1·7	5·0	8·9	13·3	16·1	15·6	12·8	7·8	2·8	−1·7			
Turin	276	0·3	3·0	7·5	12·0	16·1	20·4	23·1	22·0	18·3	12·2	6·0	1·7	32·4	−8·4	73
Milan	147	1·0	4·0	8·2	13·0	17·2	21·7	24·4	23·2	19·4	13·1	6·7	2·6	37·8	−13·9	45
Mantua		1·2	4·1	8·3	13·1	17·5	22·0	24·7	23·8	19·9	13·8	7·4	2·7			58
Venice		2·6	4·6	7·9	12·7	17·2	21·2	24·1	23·2	19·5	14·1	8·0	4·0	36·1	−10	29
Trieste		4·1	5·2	8·3	12·4	16·7	20·7	23·4	22·7	19·1	14·4	9·3	6·0	34·4	−5·1	
Genoa		7·5	8·7	11·0	14·0	17·3	21·1	24·1	24·1	21·3	16·7	11·8	8·6	35·0	−8·3	3
Leghorn		7·1	8·4	10·6	13·9	17·5	21·6	24·4	24·1	21·2	16·4	11·6	8·2	36·7	−6·7	7
Florence		5·6	6·1	10·0	13·3	17·2	22·2	25·0	24·4	21·1	16·1	11·1	6·1	36·3	−5·4	24
Rome		6·7	8·1	10·4	13·8	17·7	21·8	24·8	24·3	21·2	16·3	11·3	7·9	42·2	−8·3	10
Naples	149	8·9	8·9	11·7	13·9	17·8	22·2	25·0	25·0	22·2	17·8	14·4	10·6	37·2	−4·4	2
Palermo		10·3	11·0	12·6	14·9	18·1	21·7	24·6	24·9	22·9	19·4	15·1	11·9	49·0	−1·7	2
Syracuse		10·6	11·1	12·8	14·4	17·8	22·2	25·0	25·6	23·3	20·0	15·0	12·2	40·6	0·0	0
Ancona		5·5	7·2	9·8	13·8	18·1	22·6	25·6	24·9	21·6	16·5	11·2	7·2	38·9	−6·1	20
Foggia		6·3	7·6	10·0	13·7	18·0	22·6	26·1	25·7	23·0	17·0	11·3	7·7			16
Bari		8·0	8·8	10·7	13·6	17·4	21·2	24·1	24·0	21·4	17·5	12·7	9·2			2
Lecce		8·9	9·5	11·4	14·3	18·1	22·4	25·2	25·1	22·3	18·5	13·6	10·4			3
Aquila	735	2·2	2·8	6·7	11·1	15·0	19·4	21·7	22·2	18·3	13·3	8·3	3·9			
Cagliari		9·4	10·2	12·0	14·2	17·6	21·6	24·8	24·6	22·3	18·3	14·1	10·6	37·2	−2·2	1
Sassari	224	8·5	9·4	10·9	13·5	16·9	21·1	24·2	24·2	21·5	17·1	13·1	9·9	37·0	−0·2	2
Ajaccio		9·4	10·0	10·5	13·8	16·6	20·5	23·3	23·8	21·1	17·2	13·3	10·5	37·2	−3·8	
Malta		12·8	12·8	13·9	15·8	18·8	22·7	25·7	26·2	24·5	21·6	17·8	14·4	40·6	1·1	0

as vulcanism, fracturing and Pliocene planation, peninsular Italy is noted for its recent marine terraces, its landslips and gully erosion. The recent marine terraces are seen to perfection around the coast near Brindisi and Taranto, and inland around Foggia. The landslips (*frane*) and gully erosion (*calanche*) are associated in peninsular Italy and in Sicily with thick Jurassic–Tertiary beds of gravels, sands and clay. Where subsurface drainage undermines a more impervious bed overlying it, collapse or landslipping is common. Sometimes as in the *scagliose* clays of the northern Apennines, subsurface flow down the bedding planes may, where the dip is downhill, lubricate the rocks so that they slip frequently. Calanche or deeply dissected badlands are due both to undermining by seepage and to severe gully erosion during torrential rains. Such badlands are common locally on the eastern flanks of the central Apennines in the Abruzzi–Molise as well as on the edges of the easily eroded Pliocene platform in, for example, Basilicata. The frane leave bare scars on hillslopes and debris fans at their foot. They are most severe on the edges of the Pliocene deposits between Foggia and the coast of the gulf of Taranto, and are also common on the edges of the older Apennine Tertiary clays and sandstones. Frane and calanche are, of course, products of the same process and may often be seen in combination as in the valley sides of the Adriatic rivers Sangro and Biferno in the Abruzzi–Molise.

The common occurrence of earthquakes in some localities is connected with slips and fractures that originate much deeper in the crust. These subterranean movements cause tremors at the surface most of which are merely disconcerting but the more violent of which do great destruction locally. In February 1971 the centre of the Etruscan city of Tuscania, founded about 500 BC, was devastated by earthquakes. The distribution of earth tremors is closely connected with the volcanic tracts, the old fractured blocks of the south, and the high backbone of the Apennines where apparently crustal movements are still active.

Climate

Italy, like Iberia, sits astride the latitudes where the Mediterranean summer drought climate merges into the central European climate with rainfall all the year. Full bases for simple climatic study are given in the detailed list of climatic statistics; and the general meteorological

MEAN MONTHLY PRECIPITATION (millimetres)

	J	F	M	A	M	J	Jy	A	S	O	N	D	Mean annual Total	Raindays	Days with snowfall
Cortina d'Ampezzo	51	46	81	140	132	130	152	117	117	119	117	61	1273		
Turin	56	40	58	112	121	107	59	67	72	91	65	40	902	115	9·8
Milan	60	66	78	97	106	82	70	69	91	125	103	83	1045	121	9·1
Mantua	36	39	44	59	75	65	47	44	59	81	63	45	667	106	
Venice	36	44	53	57	72	74	47	50	65	68	55	40	667	98	2·0
Trieste	57	59	72	73	94	104	93	93	101	120	91	87	1062	125	6·4
Genoa	105	111	114	111	89	90	50	48	125	173	189	118	1342	119	1·7
Leghorn	68	62	77	61	60	50	26	36	84	140	119	100	883	111	2·9
Florence	49	54	69	74	77	68	37	48	83	101	99	70	839	117	2·3
Rome	83	65	73	65	55	39	19	26	64	127	114	98	828	102	1·6
Naples	94	74	72	65	50	35	16	27	73	116	114	112	848	112	1·1
Palermo	86	70	69	49	29	16	6	11	46	81	85	95	643	111	0·8
Syracuse	93	60	40	36	19	5	6	8	51	83	127	98	626	76	0·1
Ancona	66	41	45	57	48	51	27	31	59	98	62	49	644	99	4·1
Foggia	49	32	34	42	44	31	15	30	35	54	57	42	473	79	2·8
Bari	65	55	40	48	47	26	17	31	52	67	68	68	590	104	
Lecce	64	54	52	55	41	24	12	15	54	85	80	82	618	99	
Aquila	55	50	55	66	61	54	31	37	57	84	79	55	694	116	17·0
Cagliari	47	35	50	42	29	19	2	4	31	60	75	55	449	89	
Sassari	60	49	52	57	44	27	6	14	38	86	100	72	612	107	2·2
Ajaccio	75	52	61	52	41	28	4	21	50	91	118	72	665	80	
Malta	82	56	38	22	10	2	1	3	32	73	91	94	504	77	

and climatic setting is given in Chapter 4. However, certain generalizations are also helpful.

First, Italy because of its southerly latitude is in summer one of the hottest countries in Europe (see mean summer and absolute maximum temperatures). The summer daytime heat is often tempered near the sea by sea breezes, near large lakes by lake breezes and, always, on mountains by low temperatures in the shade. For the same reason and because of the warmth of the Mediterranean sea, the coastal lowlands of the riviera and peninsular Italy are among the mildest in Europe in winter.

Second, the plain of the Po has a unique position, sheltered by high mountains on all sides except the east. It is truly central European except towards its southeastern borders.

Third, the Ligurian Sea, as explained in Chapter 4, is a notable source of cyclogenesis and the 'lows' or depressions formed and rejuvenated here move eastward so that its coastlands and those of the Tyrrhenian sea receive ample rainfall most of the year. However, the Adriatic shores of the Italian peninsula are in the rainshadow of this eastward-moving airflow and, although open to southerly air masses and to winds from the Balkans, are appreciably drier than the western coastlands. For example, the statistics show clear differences between the precipitation of Leghorn and Ancona, and Naples and Foggia.

Fourth, only the extreme south has a marked summer drought and a distinct winter maximum of rainfall. Elsewhere late autumn is the rainiest period except in the Po basin, especially around Turin where spring normally brings most rainfall.

Fifth, the airmasses associated with cyclonic depressions are typical of the north Mediterranean. The hot dry southerly airflow (cT), on the advancing side of a depression is called the sirocco but after a long sea traverse may reach Naples, for example, moist and muggy and is known as the sciroccosa. The rear of the depressions often brings cold dry northerly airflow (the bora) to the Adriatic but on the west shores of Italy these winds are called the Tramontana and have lost much of their keenness.

Sixth, the Apennines, although lower than the Alps, carry a mountain influence far southward. In fact they modify the kind of climate found on the adjacent coastlands and add in winter a barrier of cold between the two coasts. The summits of the Abruzzi have some of the coldest winters in Italy and carry snow from November to

May and in severe seasons its wolves may be driven by hunger to wander on the lower slopes.

It is possible to recognize in Italy four main climatic provinces.

1 *Alpine Italy*. Here to a marked degree the mountains make their own climate, with a bewildering complexity of sunny (*adretto*) and shaded (*opaco*) slopes, temperature inversions, föhn effects and so on. The maximum rainfall is definitely in summer and precipitation is heavy and on the higher areas falls as snow at least throughout the cooler months.

2 *The north Italian plain*. Here the winters are cool to cold and summers hot. As in the Alps the central European type of climate is associated with a minimum of precipitation in winter but the maximum is usually in spring and autumn with summer far wetter than winter. The regional variations include a distinctly warmer winter near the shores of the larger piedmont lakes, where in sunny areas and narrow strips exposed to lake warmth, citrus trees will grow; and the coastlands near Venice where, in spite of the bora, winters are warmer.

3 *Peninsular Italy* experiences mainly a summer-dry climate with a rainfall maximum in autumn. This Mediterranean type prevails on the Ligurian riviera and most of the Apennines, together with their adjacent coastal plains. The chief regional variations are concerned with altitude and with position on west or east coasts.

4 *Southern Italy and Sicily* have a true Mediterranean climate with a severe, prolonged summer drought and maximum rainfall in winter. The transition to this is vague and occurs somewhere between latitudes 41° and 40°S. In parts of the transition zone October may be the rainiest month but everywhere winter is the rainiest season.

Land use and agriculture

Italy has a total area of 301 230 square kilometres or slightly less than the British Isles. Of this area about 42 per cent is classified as arable, 17 per cent pasture, 9 per cent tree crops, 20 per cent forest, 3·4 per cent waste and 8·6 per cent built on and water. The forestry aspects have been dealt with in Chapter 7. Here attention will be given to arable farming and animal husbandry.

Although in recent years agricultural output has been expanding steadily, its expansion has not kept pace with that of industry, yet it remains of great importance. Only recently have the numbers employed in it dropped below 5 million and in 1964 out of a total labour force of nearly 20 million, 25 per cent were employed in agriculture against 42 per cent in industry and 33 per cent in services. Agriculture then supplied 14 per cent of the gross domestic product at factor cost (industry 44 per cent; services 42 per cent). In 1967, just over 4·6 million persons were engaged in agriculture.

A shortage of productive land is easily Italy's most intractable problem. Today about one-tenth of the country, including built-up areas, cannot be put to crop uses of any kind, but at least one-sixth of the present cultivated area consists of difficult land that has been reclaimed over the centuries. Many steep slopes have been terraced and hand cultivation is everywhere apparent rendering large areas of hill country productive. Under the *Bonificale integrale*, a government-sponsored land-reclamation plan, money has been supplied by the national ministries toward land improvements and the radical transformation of farms. In 1950 a new series of important land-reform laws were passed, and under them over 800 000 ha have been acquired for fundamental reorganization. By 1962, about 634 000 ha had been allocated to 114 000 peasant families, and nearly 100 new villages and scores of new farmhouses had been built. The methods differed slightly in different regions. In La Sila where the plans had already been prepared for some time, only estates of over 300 ha were affected. Here 12 new villages and many new farmhouses have been erected on 75 000 ha of expropriated estates. Elsewhere, under the *stralcio* law, in a number of regions in various parts of Italy, the large estates generally were broken up unless they were already exceptionally well farmed. The owners received fair compensation, a fact which induced many of them to part voluntarily with some or all of their estates. The main areas affected by the *stralcio* law were the Po delta (45 000 ha expropriated), the Tuscan maremma (178 000 ha), Apulia–Lucania (190 000 ha), Sicily (over 80 000 ha) and Sardinia (about 50 000 ha expropriated).

The method of settlement can be exemplified by the work of the state-controlled *Ente Maremma*, the authority that dealt with the extensive coastlands of Tuscany. The land was drained and prepared for tilling before being divided into holdings of 5 to 15 ha (average 8 ha) each with a pleasant house, farm buildings and a well. The occupants were picked by lot from a list of applicants and can, after a two-year probation, buy the freehold over a period of thirty years. Tools, seeds and other requisites are available from the Ente or from cooperatives.

Everywhere the result of such efforts is either to reclaim fresh land or to replace monoculture, especially of low-yielding cereals in dry districts, by polyculture. The average size of the holdings ranges from 4 ha in much of the south and Sicily to 6 ha in the Po delta and 13 ha in Sardinia. It is enough to give a reasonable chance of a marketable surplus which ideally can be disposed of to cooperatives.

However it must be stressed that these great efforts since 1950 are the continuation of policies that have been active since early times. For example, the period between 1920 and 1939 was especially important and its notable achievements included the final reclamation of about 80 000 ha of the Pontine Marshes south of Rome. Here several thousand peasants were settled on 10 to 12 ha farms in carefully planned villages and farmhouses with excellent road and other services. By 1950 land reclamation in some form had been applied to about 7 million ha in Italy, leaving, it was estimated, only 3 million ha still seriously in need of attention. But the problems of latifundia which called for *trasformazione fondiaria* were not really tackled until after 1950. Since then various government ministries have seen to land improvements throughout Italy while the *Cassa per il Mezzogiorno* has been mainly responsible for reclamation and reorganization in all the south, including Campania, Abruzzi, Molise and for some government assistance, in parts of Latium, Tuscany and Marche. The activities of the Cassa involve 41 per cent of Italian territory, with 20 million people (36 per cent national total) and producing 23 per cent of the gross national product. When it is realized that the great changes in farm reorganization since 1950 have affected less than 1 million ha or 5 per cent of the national agricultural land, the immense problem remaining becomes obvious.

In 1961, after a decade of active estate expropriation, there were in Italy 4·3 million farm holdings, of which one-third were under 1 hectare and nearly two-thirds covered less than 3 hectares each. On the other hand, as the table shows, 49 000 owners (1·1 per cent total) owned nearly 10 million ha or 36 per cent of the national territory occupied by farmers. This disparity of minifundia and latifundia is aggravated by the fragmentation of holdings. Although much reorganization and unification had already been done, in 1961 no less than 28 per cent of all farms, covering over half the total territory involved, were splintered into four or more separate parcels.

The farming problems are not helped by the practice of sharecropping which has decreased greatly in recent decades but still involves 337 000 farmers (8 per cent total) and over 3 million ha or 12 per cent of the total agricultural land. Under this system the sharecroppers or *mezzadria* provide only their labour and receive half the profits. In contrast, 8 out of every 10 Italian farmers work their own farms without hired labour of any kind.

The natural difficulties involved do not always decrease with time although modern techniques may lessen them. Soil erosion, an ancient problem, is only capable of solution with large-scale planning and heavy capital expenditure. Watersheds must, wherever vulnerable, be kept wooded. The soft sediments laid down so thickly in the Apennine folds, following deforestation or overgrazing during historical times, are easily destructible and slopes quickly lose their soil, silting up the rivers so that they flood the lowland areas and increase the coastal and former malarial marshes. Many soils in the arid south are thin and immature; lacking humus they need costly fertilizers or manures to be rendered productive. The only alternative is frequent fallow which is virtually a variant of shifting agriculture. On the other hand the alluvial plains of the Po basin and the volcanic areas are wonderfully fertile and on these, intensive mechanized farming is the rule.

Another difficulty is the drought that afflicts all Mediterranean peninsulas during summer and encourages monocultures and latifundia. This is particularly severe in southern Italy in Calabria, Sicily and Basilicata, where large land holdings are under vines, olives or wheat. Here in the past the land was worked by day labourers for absentee landlords, the workers often having to journey several kilometres to their work, and for their own sustenance to toil at tiny plots at exorbitant rents. This system still survives in parts where reorganization has not yet been effected and water supplies are not readily available. Elsewhere irrigation has played a large part in offsetting the effects of drought in plains and lowlands. Loans and technical education also lead to increased yields.

There remains, however, a tremendous pressure of population on the soil, especially in the south, which has been relieved to a certain extent up to the present by emigration (337 000 in 1960). The birthrate is notably higher in the poorest overpopulated districts that are most exclusively dependent upon agriculture. More efficient mechanized agriculture is increasing, particularly in the north but it needs fewer labourers and already has caused the unemployment of 2 million farm workers. The primitive sickles and hoes and the use of farm animals are all evidence of an inefficient use of farm labour and of the extreme poverty of the south, while the malarial conditions

ITALY: HOLDINGS, NUMBER AND AREA, 1961

Size (hectares)	Number (thousands)	Total area (thousand ha)	Percentage Number	Percentage Area
Under 1	1 401	710	32·8	2·7
1–3	1 296	2 460	30·3	9·2
3–5	567	2 241	13·2	8·4
5–10	561	3 976	13·1	15·0
10–20	288	4 008	6·7	15·1
20–50	117	3 493	2·8	13·2
Over 50	49	9 684	1·1	36·4
Total	4 279	26 572	100	100

ITALY: FRAGMENTATION OF HOLDINGS, 1961

No. of parcels	No. of holdings (thousands)	Area (thousand ha)	No. of parcels (thousands)
1*	1 754	8 635	1 754
2–3	1 318	6 677	3 140
4–5	524	3 449	2 304
6–10	441	3 665	3 296
11–15	123	1 343	1 578
16–20	52	727	934
Over 20	69	2 075	2 648
National total	4 279	26 572	15 654

* not fragmented

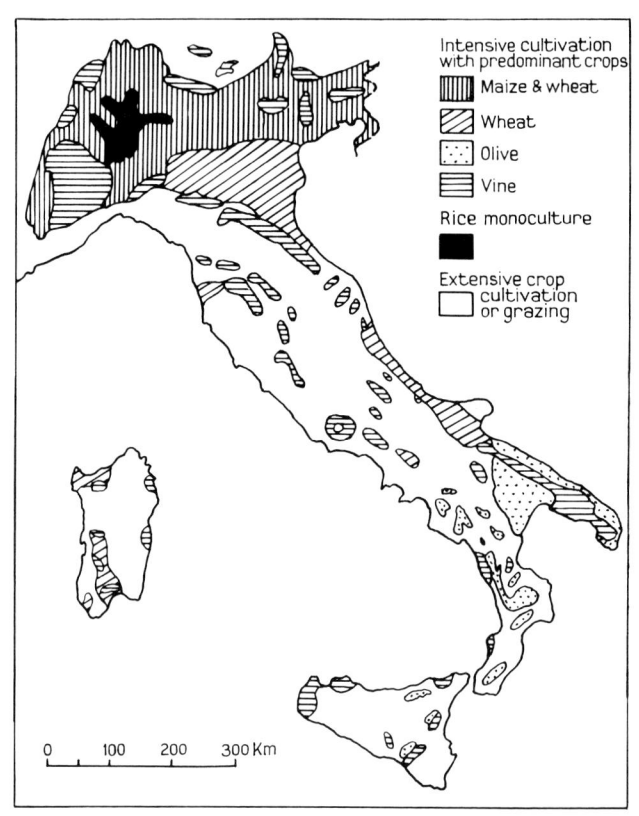

216　Part II: Western Peninsular and Alpine Lands

67 Some aspects of land use in Italy

used to mean that in addition to an expectation of employment for only 100 to 150 days a year, many were incapable through ill-health of doing a full day's work. So to the problems of unemployment and underemployment were added that of unemployability. Malaria has long been eliminated but its malignant effects have only just ceased to be significant.

A study of the crops grown indicates how the soil and climatic conditions are reflected in the harvests and in the variety of products.

Arable farming (fig. 67)

As the table of cereal area and production in Chapter 8 shows, cereals occupy nearly 6 million ha in Italy or 48 per cent of the total arable area irrespective of tree crops. The national cereal harvest of over 14 million tons exceeds that of any other country wholly in southern Europe. Wheat alone occupies over 70 per cent of the Italian cereal area and yields 9 to 10 million tons annually. Yet Italy is not selfsupporting in bread grains. Owing to the summer drought, the commonest wheat grown is the durum variety and is used in pasta, a staple diet of the Italians. The chief producing areas are the plains of Lombardy and of Emilia where owing to deep ploughing, seed selection and fertilizer the yield is double that of the second producing area Apulia and Sicily which, despite the low yield per hectare, produces one-seventh of the total crop. The table of details on agricultural production reveals a marked increase in the yield per hectare.

Maize is grown on one-sixth of the cereal area or on 1 million ha producing 3·9 million tons of corn which is used partly for making polenta, another peasant staple dish. It is grown in Campania, Veneto and Lombardy and yields are high but the quality poor. All other bread grains, oats, barley, and rye are confined to the hilly areas except rice which occupies 160 000 hectares in the Ticino and Po valleys. Introduced by the statesman Cavour and grown under irrigation it has one of the highest yields per hectare in the world but, like maize, the quality needs improvement. Of the 640 000 tons produced, about one-third is exported. As the table in Chapter 5 shows, Italy is by far the chief rice producer in Europe.

Vegetables

These form an important crop throughout Italy, the chief being the potato (350 000 ha, 3·8 million tons) which is widely grown in the cooler north in the Po valley, as well, for example, as on the Abruzzi (for a main crop) and in Campania for either an early or a main crop. On the warmer Mediterranean coastlands vegetables ripen all the year, especially tomatoes (133 000 ha, 3½ million tons) which are an important item of export. Peas and beans, melons, onions, garlic, cauliflowers and cabbage are also important locally.

The chief industrial crops are sugar-beet (300 000 ha) which flourishes especially in the irrigated districts of Veneto and Emilia, and fulfils the domestic requirement. Hemp, flax and tobacco are also grown and in the north, the mulberry for feeding silkworms.

Tree crops

As discussed in Chapter 8 where tables of areas and production are given, it is difficult in Mediterranean countries to separate arable and tree crops. In 1966 in Italy tree crops occupied 2·8 million ha or nearly one-tenth of the whole country. Italy is quite outstanding for horticulture, the chief tree-fruit crops by area being the vine, olive, almond, citrus and temperate fruits, especially peaches, apples, pears and cherries. The great increase recently in the output of these is shown in the accompanying table.

The vine covers about 1 145 000 ha as a monoculture and shares another 2 126 000 ha in mixed cultivation. Italy is the first world producer of grapes (over 10 million tons) and regularly since 1965 has just exceeded France for wine (6½ million tons). The output includes a few fine table wines, notably chianti, asti, orvieto, marsala and vermouth. The methods of cultivation are varied and ingenious and the crop is widespread both in the north and in the south. All sorts of factors favour viticulture but probably the dry, reflective limestone soils, the long hot sunny summers, local sulphur supplies for purification and the traditional flair for vine-tending are the chief.

ITALY: AGRICULTURAL DETAILS, AREA PLANTED AND YIELD OF IMPORTANT CROPS

	Average 1952–6			1968		
Crop	Area (thousand hectares)	Prod. (thousand tons)	Yield 100 kg/ha	Area (thousand hectares)	Prod. (thousand tons)	Yield 100 kg/ha
Wheat	4 790	8 481	17·7	4 275	9 590	22·4
Barley	246	285	11·6	175	258	14·7
Oats	446	537	12·0	323	390	12·1
Rice	167	855	51·3	156	639	41·1
Maize	1 262	3 037	24·1	967	3 988	41·3
All cereals	7 007	13 333		5 950	14 970	
Dry broad beans	553	433	7·8	375	325	8·7
Dry haricot beans	437	148	3·4	230	166	7·2
Potatoes	392	3 170	81·0	319	3 960	124·0
Sugar-beet	228	6 992	307·0	306	11 457	375·0
Tomatoes	93	1,477	158·0	129	3 258	253·0
Vines (grapes)	1,719*	8 587*		1 613*	10 298*	
Olives		1 399 (olive oil 273)			1 933 (olive oil 425)	
Oranges, etc		676			1 683	
Lemons, etc		343			826	
Apples		1 058			1 923	
Pears		408			1 369	
Peaches		419			1 315	
Figs, fresh		332			220	

* Amounts for vine cultivated alone reckoned at 1023:4405 for 1952–6 and 1159:7897 for 1968, which was a relatively poor year for yields, as also with rice. Grapes for wine reckoned at 7977 in 1952–6 and 9348 in 1968, and wine production at 5385 and 6524 respectively (all in thousands)

FARM MACHINERY (in thousands)

	1952–6	1968
Tractors	124	543
Garden tractors	4	130
Combined harvester–threshers	1	17

The vine produces more than 10 per cent of the national agricultural revenue and is the first love of most Italian farmers. It gives work directly and indirectly to nearly 4 million people and is grown to a certain extent on 2·5 million farms (against 1·5 million farms in France). In 1968–9, about 1·3 million Italian farmers specialized in the vine and another 1 million grew it in mixed cultivation (*cultura promiscua*). It is absent or insignificant only on the higher mountains, on heavy clayey soils, as in the hills of southern Tuscany, and where the water-table remains high as in the Tyrhennian maremma and in the northern plain north of the Po between Vercelli and Mantua. In the north it dominates on all hills and in the south its domination creeps down also to the coastal plains with free subsoil drainage and prolonged summer drought. The great majority of viticulturalists own less than 10 ha and produce, or at least declare that they have produced, less than 10 hectolitres a year. The most prolific specialized vineyards are in Puglia, Sicily, Piemonte and Veneto while the chief producers from *cultura promiscua* are in Emilia, Veneto and the northern half of the peninsula. Since 1929, the specialized vineyards have expanded from 935 000 ha to 1·2 million ha, while the area of farms on which vines are grown amid other crops has dimimished from 2·9 million to 1·9 million. At the same time Italy has become the world's leading producer of wine, and since 1967 an official system of labelling for certain brands has been started.

Olives occupy 930 000 ha as a monoculture and share about another 1·4 million ha with other crops, especially wheat. Italy alone produces one-quarter of the world's supply of olives and olive oil The crop is restricted by winter frost and warm-season rains to peninsular Italy where it is most common in the drier south and especially in Apulia with its expanses of dry limestone and protracted summer drought. Usually it is grown on large estates since

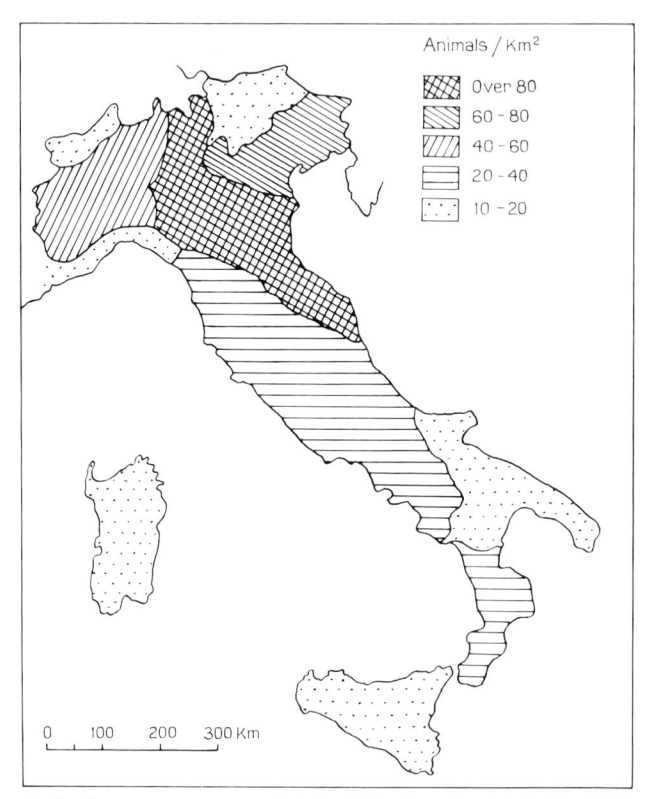

68 *Italy: some aspects of animal husbandry*

There seems no doubt that except in the plain of Lombardy and Emilia and a few other localities, animal husbandry is a rather neglected branch of the agrarian economy, although the output of milk, cheese and meat has increased markedly in the last decade (see table below).

The mountainous districts everywhere support sheep and goats especially on the macchia of Calabria, Apulia and Basilicata. They are used for both meat and milk and for making pecorino cheese. Altogether 8 million sheep and 1¼ million goats were kept in 1966. Although the very finest wool comes from Apulia the wool sheep are usually found in the north and sheep for hides in the south. The transhumant flocks of the central Apennines are becoming increasingly restricted in their movements and can probably only increase in numbers at the expense of goats as water supply and pastures are improved in the impoverished south. It is doubtful if peninsular Italy could support a high-class mutton breed under present conditions.

Italy's cattle (nearly 10 million in 1966) are concentrated mainly in the better-watered parts of the Po valley, the Veneto, and of the Alpine valleys such as Alto Adige, where excellent grass is available. A livestock industry comprising over 4 million pigs and 80 million poultry is

it takes about ten years for a tree to come into full bearing and only yields well every other year. The Naples district is famous for olive oil but the finest quality comes from Lucca. The yield varies considerably; the production of olives was 1 801 000 tons (357 000 tons of oil) in 1966, and 2 712 000 tons (594 000 tons of oil) in 1967.

Citrus fruits in Italy are grown mainly south of Naples. Sicily is the chief producer yielding two-thirds of the national harvest of lemons and half of the oranges. The total area under citrus is about 116 000 ha and the yield about 1¾ million tons, of which two-thirds consists of oranges. The rapid increase in production since the 1950s is shown in the table of agricultural details.

The remarkable extent of orchards for almonds, peaches, apples, pears, cherries and so on is discussed with tables in Chapter 8. Some idea of the importance of these may be gained by the fact that Italy produces 60 per cent of the European crop of peaches and is the world's chief exporter of apples and second exporter of cherries.

Animal husbandry (fig. 68)

Although pastures cover only 17 per cent of Italy the animal husbandry is less restricted because in peninsular Italy it makes extensive use of coppice, bushgrowth, macchia, and garrigue. Unfortunately much of the natural pasture is of poor quality, particularly in the south. Of the area of so-called pastures nearly two-thirds are in the south whereas nearly two-thirds of the cattle live in the north, an indication of the general poverty of southern pastures.

ITALY: LIVESTOCK NUMBERS AND PRODUCTION

		Number (in thousands)	
	Average:	1952–6	1968
Equines:	horses	660	320
	mules	388	223
	asses	696	325
Bovines:	total	8 721	9 539
	dairy cattle	2 732	3 430
	buffaloes	13	37
Ovines:	sheep	9 391	8 285
Caprines:	goats	2 462	1 124
Porcines:	pigs	3 990	6 185
Poultry		75 500	113 000

	Production (thousand tons)	
Average:	1952–6	1968
Prime meat*	691	1 245
Poultry, horse, offal, etc	218	614
Milk: total	8 622	10 767
of which cow	7 854	10 077
buffalo	11	28
sheep	484	515
goat	273	147
Butter	62	62
Cheese	306	430
Wool, greasy (clean)	14 (7)	12 (6)

* Prime meat is beef, veal, mutton, lamb and pork direct from carcasses

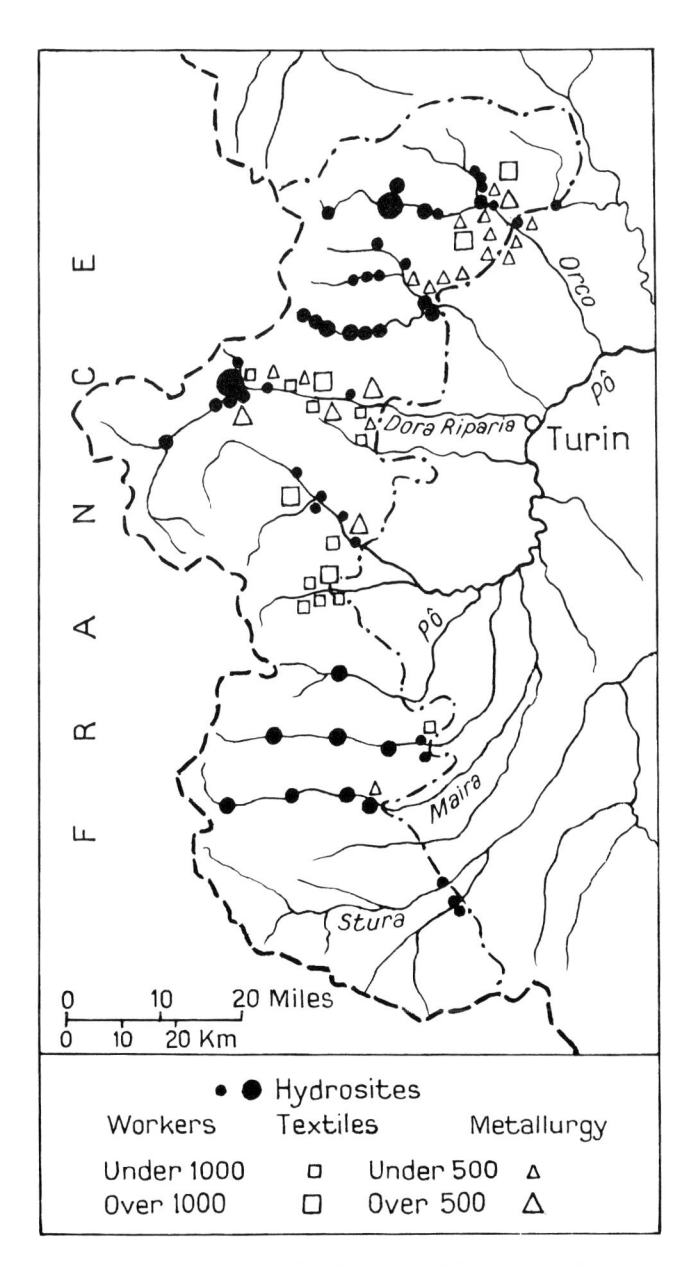

69 *Hydro-sites and chief industries of the west Italian Alps* (after R. Blanchard and P. and G. Veyret)

based on these northern herds and 250 000 tons of excellent cheese such as Parmesan, Bel Paese and Gorgonzola are produced annually. Fodder is always desperately short and there is little stockraising for beef, except in Emilia. An interesting development north of Naples involves the keeping of herds of buffalo which can endure the summer heat. From their rich milk a special soft cheese, mozzarella, is made locally.

Nearly 1 million working horses, mules and donkeys provide some of the labour force on Italian farms although in the south cattle are also used as work animals. Nevertheless the proportion of larger farm animals to the total population is only 1 to every 2 persons and, since the wild birds and smaller denizens of the countryside are always at risk for the peasant's pot, Italian farms do not usually present the same varied spectacle of life as those of northwest Europe.

Power resources and industrial development

It is recognized that, with 30 million people, peninsular and insular Italy are seriously overpopulated but the 26 million that live in the industrial north are almost as grave a problem. Industry affords a livelihood for 8 million persons or 40 per cent of the national labour force and this has only been made possible by developing hydro-electric power. Of Italy's annual power output in 1966 (90 000 million kWh from 27 million kW installed) no less than 49 per cent came from water, and the remainder mainly from coal, with smaller supplies from nuclear energy (642 000 kW installed) and earth heat (342 000 kW installed). The chief hydro-electric installations are in the Alps (fig. 69) and Alpine foothills where Trentino and Alto Adige produce half the national water power output. The second main concentration is on the central Apennines. Minor developments have been undertaken on the Apennines between Genoa and Pisa where the many plants are small, and in Calabria where four medium-sized and several small stations are in operation. The thermal stations show much the same distribution with outliers at Naples, Bari and in Sicily. Most of the nuclear power stations are between Rome and Naples. The chief geothermal stations are at Larderello south of Leghorn. The disparity in the distribution of power reflects the drought of southern Italy and the difficulty of using rivers with an extremely irregular regime. It also reflects an industrial tradition in the north and contact for centuries with industrially minded societies and markets.

As 75 per cent of Italy's water power potential is already harnessed, it was doubly fortunate that natural gas (methane) was discovered in 1946 below Caviaga thirty kilometres southeast of Milan. Subsequently many productive gas fields have also been found as, for example, at Cortemaggiore, Ripalta, Cornigliano, Bordolano and Ravenna (fig. 70). By 1957 a pipeline had been established on the north Italian plain carrying 13 million cubic metres of gas daily. Of this 71 per cent was used industrially; $11\frac{1}{2}$ per cent in power stations; 6 per cent in chemical plants; 3 per cent in road transport and $6\frac{1}{2}$ per cent domestically, including calor gas in containers for cooking and heating in 3 million homes. Since then the finds and the pipeline network have been widely extended. In 1959 Abruzzi Molise and Basilicata began production and in the next few years were joined successfully by submarine fields off Ravenna and large fields in Sicily and Puglia. Within the last decade several new gas fields have been struck in the Adriatic. The national output has risen from 6500 million cub m in 1960 to 8800 million cub m in 1966. The situation in 1968 was approximately as follows, the individual output of important fields being given in million cub m in italic type:

Area	Production million cub m	Chief fields
Piemonte	4	—
Lombardy	572	Bordolano *174*
		Sergnano *151*
Emilia Romagna	5402	Ravenna *1930*
		Minerbio *1402*
		Spilamberto *623*
		Corregio *348*
Marche	18	—
Tuscany	1	—
Abruzzi Molise	1031	S. Salvo Cupello *834*
Puglia	822	Roseto Montestillo *317*
Basilicata	569	Ferrandina–Grottole *508*
Sicily	1114	Gagliano *1002*
Submarine	880	Portocorsini Mare Est *607*
		Ravenna Mare *182*
National Total	10 413	

In this year the Netherlands produced 14 000 million cub m natural gas, the Federal Republic of Germany 6500 million and France 5700 million cub m. The remarkable Italian output is distributed mainly by a network of 3500 km of pipes, ranging from 305 mm to 762 mm in diameter. This is shown diagrammatically in figure 70. Natural gas now represents 15 per cent of the total national consumption of power and saves the import of at least

70 *Italy: natural gasfields and pipelines*

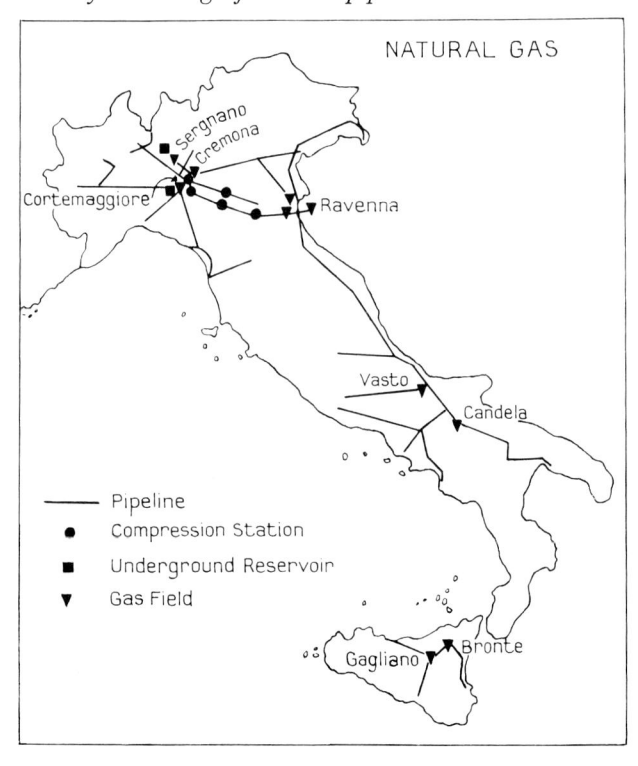

13 million tons of coal annually. In spite of greatly increased power needs, today only 12 million tons of coal need to be imported each year.

The country's fuel supplies have been further boosted recently by the development of a petroleum field at Ragusa, Gela and Fantanarossa in Sicily. The crude-oil output in the period 1964–6 averaged 2·25 million tons, and just exceeded 1·5 million tons in 1968, nearly all coming from Sicily.

The true value of the water power, natural gas and petroleum is seen when the national production of solid fuels is studied. The total coal output in 1968 was under ¼ million tons, partly from the Val d'Aosta, and the total lignite output was nearly 3 million tons, mainly from the Tiber valley near Terni in Umbria. However, from an Italian viewpoint, the whole energy problem needs to be seen against the following balance sheet, which includes the import of 92 million tons of crude petroleum in 1968. Two years later the imports of crude petroleum for refining in Italy reached 115 million tons. The increase is amazing; in 1950 the Italian petro–chemical concerns consumed about 5 million tons of crude oil.

ITALY ENERGY BALANCE 1968
(in thousand tons coal equivalent)

Domestic Production of primary sources:

Coal	209
Lignite	643
Natural gas	12 743
Petroleum	2 154
Primary electricity	15 616
National total	31 097

Imports

From EEC	4 117 (of which 3 382 coal)
From other countries	146 537 (of which 8 342 coal and 133 000 petroleum)

Italian mineral resources as a whole are very small, and the total production of iron ore has now dwindled to less than 800 000 tons, mainly from Elba whence it is shipped to Piombino on the mainland, and in small quantities from Cogne in the Val d'Aosta. Yet the industrial needs for steel are such that with the aid of large imports the output of pig iron and ferro-alloys reached 8 042 000 tons and the crude steel output, with the aid of scrap, 16 950 000 tons in 1968. Since 1947 the small iron and steel producing units in Turin, Genoa, Milan, Terni and Piombino have been concentrated, with government assistance, into larger, more economic iron and steel plants at Genoa, Piombino, Bagnoli, Aosta and Taranto (see pp. 239–40). The last-named will eventually be expanded from 4·5 million tons capacity in 1970 to 10·3 million tons.

The most important mineral exploited in Italy is bauxite from Monte Gargano in Apulia, and from sites in the Abruzzi, Campania and Latium, which yield all told ¼ million tons. In addition, nearly half the world's mercury is mined at Monte Amiata, south of Siena, and large quantities of crude sulphur (541 000 tons in 1968) are obtained from the volcanic areas of Sicily and near Naples. Small quantities of lead, zinc and manganese are also mined.

In spite of this unpromising raw material and power resource background Italy has accomplished an economic miracle and made itself the most highly industrialized country in the Mediterranean world. An important and growing chemical industry has been established for the production of artificial fibres, fertilizers, synthetic rubber and heavy acids at, for example, Milan, Venice, Ferrara, Naples and Terni. Ravenna has the largest synthetic rubber industry in Europe. Oil refineries operate at Genoa, Venice, Naples and Bari and upon these bases consumer industries concerned with textiles, metal goods, electrical and general engineering, and vehicles have been more securely established.

Probably over the whole country food processing occupies the greatest number of factories and every town has one or more *pasta* factories. But the engineering of high-quality metal goods, and the manufacture of textiles and of clothing and footwear each employs many more workers than food processing. Italy manufactures a wide variety of textiles and, despite home-produced wool, silk, flax and rayon, cotton is easily the chief with 960 factories mostly situated in and near Milan. In 1968, Italy had 81 000 cotton looms and of these over 70 000 were automatic, the greatest number in any state of non-Soviet Europe. Wool is the oldest textile industry and with 22 000 looms Italy equals the U.S.S.R. and is outnumbered only by the United Kingdom. Turin is the chief woollen centre. For silk (20 000 looms), Como, Bergamo and Brescia are famous but the manufacture of the natural fibre is delining in the face of Japanese competition and the increasing popularity of synthetic fibres. In fact Italy comes third after Germany and Japan in the manufacture of artificial fibres and its 30 factories produce 6 per cent of the world output. For these artificial yarns the chief manufacturing towns are also in the north Italian plain at and around, for example, Turin, Milan, Padua and Cremona. Jute imported from east Pakistan is manufactured at Turin, Naples, Lucca and Terni and hemp at Bologna, Ferrara and Modena. The whole of this textile industry, apart from jute, is still dominated by the northern towns as it was in the late eighteenth and early nineteenth century when a minor industrial revolution occurred with the increasing use of direct water power. The Italian textile factory concerns, as with those in Catalonia in Spain, were based on water wheels which were not replaced by hydro-electricity turbines until the twentieth century. Today the Italian consortiums are famous builders of dams and makers of hydro-plant machinery.

The engineering and metal industries of Italy are, for the most part, large combines producing high-quality goods. These branches of industry are today the largest employers of labour. They are concerned mainly either with locomotion, such as diesel locomotives, ships, cars and motor cycles (Fiat, Ferrari, Lambretta, etc.), or with smaller machinery also needing precision engineering, but requiring relatively small amounts of raw materials, such as calculating machines, Necchi sewing machines and Olivetti typewriters. Italian aluminium household goods also find world markets (Espresso coffee machines, etc.), for these bauxite is refined at Porto Marghera, Bolzana and Moro. The production of motor cars in 1968 was 1·6 million of which just over one-quarter were exported.

Another class of consumer goods developed out of local raw materials and native craftsmanship is concerned with ceramics, shoes, paper, printing and bookbinding, silverwork, jewellery and the making of muscial instruments. These various manufacturing concerns and the great new

ITALY: INDUSTRIAL AND COMMERCIAL CENSUSES
Numbers of workers (in thousands) of chief branches and sections engaged in industry and commerce

Branch		1951	1961
Connected with:	Agriculture	–	94
	Forestry	–	9
	Fishing	–	43
Extractive		119	104
Manufacturing:	*Total*	3498	4496
	Food and tobacco	413	424
	Textiles	651	599
	Clothing and shoes	412	513
	Skins and leather	39	50
	Wood products	294	381
	Metallurgy	145	192
	Chemical	149	220
	Rubber and paper	104	137
	Engineering	709	1143
	Non-metalliferous minerals	207	319
Building and construction		532	894
Electricity, gas and water works		93	116
Commerce:	*Total*	1803	2419
	Wholesale	244	377
	Retail	958	1333
	Hotels, etc	325	441
Transport and communications		579	743
Banking, insurance, etc		162	219
Services, Social, etc		208	324
National Total		6995	9463

NB There were 1 673 000 firms recorded in the 1951 census and 1 939 000 in 1961

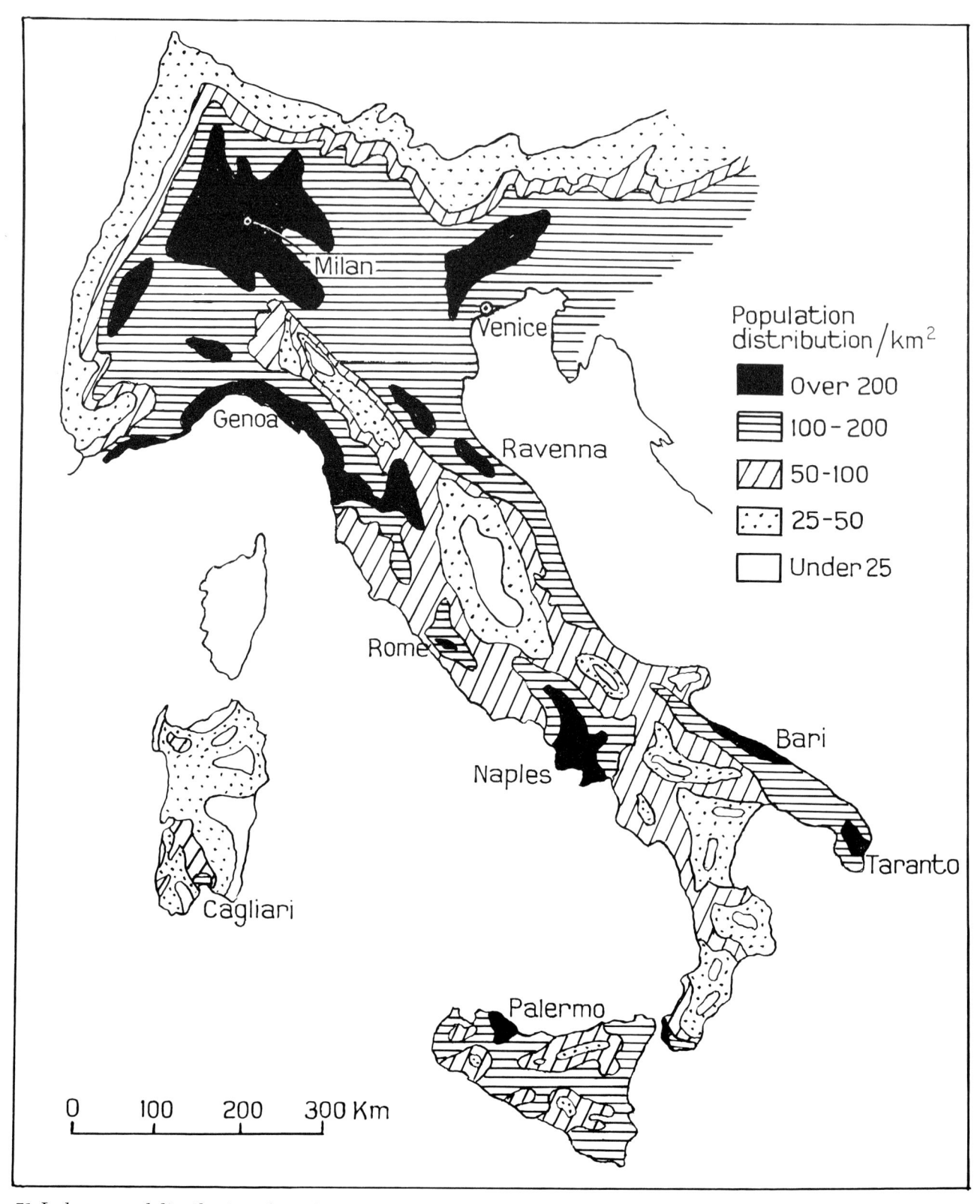

71 *Italy: general distribution of population* (estimated for 1969)

establishments for petrochemicals have expanded steadily in the last few years and manufacturing employment has risen by 2·5 per cent annually and the gross national product at 6 per cent as compared with 5 per cent in France and 3 per cent in the United Kingdom.

Tourism

The above-mentioned craft industries are bolstered by the important tourist traffic. Nearly 27 million people visited Italy in 1966, including 5·5 million from Federal Germany, 4·7 million from France, 3·7 million from Switzerland, 2·6 million from Austria, 1·8 million from the United Kingdom, 1·4 million from the Netherlands and 1·2 million from the United States (see tables pp. 109–10). These come for the artistic and historical treasures of Italy, including the Vatican city, and for the fine beaches, sunshine and scenery. The income from foreign visitors (922 000 million lire in 1968) more than balances the deficit on the external trade budget. The characteristic scene of non-industrialized urban Italy includes groups of tourists entering or leaving museums, churches and palaces. In summer, beaches such as those near Rimini are hidden under an endless strip of multicoloured sunshades; in winter the sunny coastal south comes into its own.

Distribution of population, towns and communications

The tourist traffic and internal economy and transport as a whole are greatly aided by an efficient communications system, particularly of roads. As already noticed the Italians make 1·6 million cars a year, a considerable proportion of which are used at home. About 8 million motor cars and 4 million motor cycles and light vans are licensed annually and these, and visitors' cars, are catered for by a road network that is remarkably efficient considering the mountainous terrain of so much of the country. The Italians maintain the Roman tradition of being fine highway builders. Their first autostrada was built in 1925 linking Como with Milan and by 1939 there were many stretches of major motorway such as those between Turin and Venice, Florence and Pisa, Naples and Pompeii and Rome to the nearby coast. Since 1945 plans have been put into operation to reconstruct this system, to widen its carriageways and extend it to cover the whole country. The progress has been remarkable and still continues. The chief autostrade at present in service (in addition to those given above) run from Milan to Rimini; Milan to Genoa; and from Bologna to Florence, Rome, Naples, Salerno and beyond. The last, the *autostrada del Sole*, has already reached beyond Salerno and is being extended to Reggio in Calabria and to Sicily by a bridge over the Strait of Messina. Other autostrade under construction include that from Genoa west to Savona and so along the riviera coast to relieve congestion on the winding coastal road; an extension of the autostrada del Sole northward from Bologna to Padua; and an extension of the autostrada north from Turin up the Dora Baltea valley as far eventually as Aosta where it will take traffic coming from the newly constructed summit tunnel under the Great St Bernard Pass and from the recently opened French tunnel beneath Mont Blanc on the road from Chamonix to Courmayeur.

This fine road network with its carefully planned circular bypasses round the great cities reflects the distribution of population and a desire to link the poorer south with the rest of the country. The distribution of population itself is clearly related to industrial development *and* the intensity of agriculture (fig. 71). Total population in 1966 was estimated at 53 million which gives an average density of nearly 175 persons per square kilometre. The largest areas of greatest density, or over 200 per sq km, are as follows: in the northern plain, Lombardy around Milan, Piemonte around Turin, Veneto around Padua, Emilia around Bologna and in a belt stretching thence southeast to Faenza. Here in the Po basin the chief industries and some of the best agricultural land in Italy, support over 22 million people.

On the northern coastal strip and the lower Arno basin a belt of dense population stretches from Florence and Leghorn (Livorno) along the *riviera di levante* to Genoa and thence, with small breaks, along the *riviera di ponente* to the French frontier

ITALY: COMMUNES WITH OVER 100 000 RESIDENTS

December 1968 (in thousands)	
Roma	2682
Milano	1691
Napoli	1271
Torino (Turin)	1153
Genova	843
Palermo	654
Bologna	489
Firenze (Florence)	458
Catania	409
Venezia (Venice)	368
Bari	349
Trieste	279
Messina	270
Verona	255
Padova (Padua)	224
Cagliari	220
Taranto	217
Brescia	204
Livorno (Leghorn)	173
Parma	170
Reggio di Calabria	165
Modena	165
Ferrara	157
Salerno	148
Foggia	138
Prato	137
Ravenna	130
La Spezia	129
Reggio nell'Emilia	127
Perugia	125
Bergamo	125
Rimini	114
Pescara	114
Vicenza	111
Ancona	108
Monza	105
Sassari	105
Terni	105
Bolzano	103
Forli	103
Piacenza	103
Pisa	103
Siracusa	102

ITALY: LEADING PORTS BY WEIGHT OF CARGO, 1969

	International						Cabotage					
	Ships		Cargo (thousand tons)		Passengers		Ships		Cargo (thousand tons)		Passengers	
Port	In	G.R.T. (thousand tons)	In	Out	In	Out	In	G.R.T. (thousand tons)	In	Out	In	Out
Genoa	6 548	29 384	39 399	2 808	143 620	176 099	3 837	8 094	6 503	1 077	311 148	263 251
Augusta	1 292	10 026	17 237	6 686	–	–	2 200	2 721	615	6 271	–	–
Trieste	2 737	10 771	20 020	929	30 895	35 546	3 341	1 883	1 928	952	84 166	80 809
Venice	3 699	9 853	11 071	1 487	121 915	112 046	3 292	4 463	7 703	561	20 746	14 833
Taranto	634	4 246	9 433	708	–	–	1 097	1 967	915	4 187	–	–
Naples	3 000	12 715	7 647	1 828	83 123	96 441	14 472	7 838	1 170	799	1 274,978	1 252 590
Porto Foxi*	597	5 694	8 462	5 795	–	–	–	–	–	–	–	–
Italy	39 402	137,460	182 634	32 341	901 594	796 864	230 517	88 391	39 615	39 508	9 036 203	8 945 472

* Near Sarroch southwest of Cagliari in Sardinia. A new petro–chemical refinery and port

In the south, the largest areas of dense settlement are around Naples, around Bari in Puglia and around Brindisi in the heel of Italy where intensive agriculture in warm, fertile lowlands is associated with industrial and commercial activities. The municipalities of Rome and Palermo are densely peopled but occupy relatively small areas.

The importance of urbanization in Italy may be judged from the fact that just over 25 per cent of the people live in cities with more than 100 000 citizens and another 35 per cent live in towns with between 10 000 and 100 000 inhabitants. Probably three-quarters of the total population could be classed as urban. Of the main conurbations, Rome (2 750 000), Milan (1 691 000), Naples (1 271 000), Turin (1 160 000) and Genoa (850 000) are outstanding but as the table shows there are nearly another 40 towns with over 100 000 people.

Rome acts as the administrative and cultural centre, Milan and Turin as the leading industrial and route centres, and Genoa as the great seaport. Like the landscape and the historic monuments, the ports are crucial to the country's economy. Since peninsular Italy forms a gigantic pier in the central Mediterranean it is ideally suited for the import of raw materials both for its own needs and for transit to the countries of central Europe. Foreign goods offloaded in Italian ports amounted to 85 per cent by weight of the country's total international cargo trade and its merchant fleet is now eighth in the world. The chief port, Genoa, handles 17 per cent of the national and 20 per cent of the international cargo trade. Here harbour developments to keep pace with the requirements of modern oil tankers and container shipping are being completed. Italy has a long coastline backed everywhere by high mountains and all its 144 ports are cramped for space. Plans have been realized for Genoa, Trieste and Augusta (Sicily) to have deep-water harbours for giant oil tankers. Panigallia near La Spezia has been developed to receive natural gas shipments from Libya, and Porto Marghera near Venice and Augusta for petro–chemical industries. The north Ligurian ports are being drawn into a huge commercial complex that includes Genoa, Voltri, Savona and Vado. Leghorn and Naples are also increasing their container traffic facilities. Taranto is developing fast with modern harbour installations and a large industrial complex financed partly by the Cassa per il Mezzogiorno and partly by special development funds of the European Economic Community.

Today the chief Italian imports by value are raw materials, particularly wool, wood and metals, for industry; foodstuffs, especially maize and meat; manufactured goods; and mineral fuels. The chief exports are machinery, fabrics, footwear and clothing, transport equipment, petroleum byproducts, and fruit and vegetables. In 1968 the imports came mainly from West Germany (19 per cent), the United States (12 per cent) and France (11 per cent), while the exports went largely to West Germany (19 per cent), France (13 per cent), the United States and the Benelux countries. The close economic ties with the Common Market fit in well with Italy's geographical position while the large trade with the United States partly reflects the needs of the millions of Italians who have emigrated there.

18 Mainland Italy: Regional Patterns

Mainland Italy has two fundamental climatic divisions, a northern where most rainfall falls in the warmer half of the year, and a southern where summers are distinctly dry. Relief, however, confuses rather than strengthens this division and, in fact, the southeast of the Po basin near Bologna is decidedly transitional. However, the dividing line between north and south is usually taken along the northern edge of the Apennines. Northern Italy falls geographically into an Alpine region and the north Italian plain, each of which is shown in figure 72.

Alpine region

The Alps sweep in a huge arc around the north of Italy. In the west the slope is steep up to the watershed and the mountains rise like a wall from the Po plain. Here the rivers cut deeply into the highland and plunge directly to the Po lowland with scarcely any development of longitudinal valleys. As many of the structural and erosional lines run east–west, aspect or orientation is of great importance. The north facing slopes are under forests of deciduous oak on the lower levels followed upward by sweet chestnut, beech and hornbeam and finally by conifers and on high mountains by alpine pastures (see Chaper 6). The south-facing slopes are under corn and vines on the valley floors and hardy cereals and hay at higher levels. The dwellers on the upper slopes raise cattle and produce mainly cheese. They live in typical Alpine wooden chalets, roofed with large stone slates and store the hay in wooden barns. These valleys also provide a refuge from the sweltering heat of the plains in summer and ski resorts in winter. The Mont Cenis and Great and Little St Bernard passes and tunnel, together with the French tunnel under Mont Blanc (opened in 1965), admit tourists from other countries. The Val d'Aosta and Piedmont account for nearly one-fifth of Italy's hydro-electric power and the great dams form an added tourist attraction. Aosta has chemical and steel plants and Perosa manufactures ball bearings. The former was founded as Augusta Praetoria Salassorum by Augustus about 24 B C and has remarkable Roman remains. Today it is only 33 km from the Mont Blanc tunnel and is a popular tourist resort.

The central stretch of the Italian Alps is characterized by north–south valleys that end at the piedmont in fine lakes, such as Maggiore, Lugano, Como and Garda. These inundated glacial troughs form a world of their own climatically, economically and socially. They are deep and modify the heat of summer and cold of winter on their shores. The citrus will grow here and other plants noticeably absent on the north Italian plain. The lakes regularize the downstream regime of their rivers and also prove useful for fishing and sailing, and have boat services. They are overlooked by a ribbon of villa settlement and where flat land occurs by large lakeside towns. The morainic arcs and outwash plains that stretch outside the southern end of the lakes are terraced for vines and orchards. The higher parts and the mountainous projections between the main lakes are either under forest or poor scrub. As the valleys and lakes face south they are especially favoured by sunshine yet they are open to southerly airflow and rains, and almost immune to cold-air drainage. Olives and orchards of apricot, peach and almond abound with maize, vines and fodder crops. The local hydro-electric power and excellent communications with the Po plain have fostered electro–chemical, light engineering and woollen industries that form useful complements to farming and tourism.

East of the Adige the Italian Alps broaden to 170 km and include the Dolomites, an assembly of steeply folded limestone ranges with crests fretted by erosion into turrets, bastions and pinnacles of various hues rising up from a base of barren scree and overlooking fertile basins and green wooded plateaux. Glacial lakes add their beauty to a unique landscape portrayed in many an Italian old master painting. Three-quarters of the Dolomites are forest, waste or summer pasture but the south-facing valleys support agriculture, including potatoes and hay on the higher slopes and maize, tobacco, vine, fruit and winter fodder for dairy cattle on the flatter valley floors. This region is the threshold of the Brenner pass and has in the Alto Adige a large German-speaking population (fig. 66). Each main valley is now a string of large hydro-sites, many being among the biggest in Italy. Bolzano and Trento, at the exit of valleys and on the main route to the Brenner Pass, have industries based on electro–chemicals, metallurgy, paper, cement and engineering as well as tourism.

North Italian plain

This lowland between the Alps and the Apennines occupies 16 per cent of Italy. It extends lengthwise for 400 km and has a width that varies from 80 km in the west to about 180 km near the Adriatic. Everywhere below 100 m except in the volcanic hills of Euganea and the limestone country of Monferrato, it forms the chief agricultural and industrial region of Italy. The plain is naturally well endowed with areas of rich soils and a long hot growing season with an appreciable rainfall that can be supplemented by extra water supplies to increase crop yields. Here are produced one-third of the wine, half of the wheat, three-quarters of the maize and sugar-beet, and nine-tenths of the rice and milk of Italy. The chief agrarian economy is intensive mixed farming with an emphasis on cereals (wheat, maize, rice and rye), dairy produce, vines, and tree and industrial crops, particularly sugar-beet, flax and hemp. Dairying is associated with rich fodder crops such as irrigated hay, alfalfa and clover and the common livestock include pigs and poultry. Certain areas lean more heavily on dairying and others on rice monoculture (fig. 67) or on tree crops and the plain can be divided into 'agricultural regions' but the real contrasts are between alluvial floodplain, the lower plain or river terraces (*bassa pianura*), and the upper plain of coarser fluvio–glacial deposits (*alta pianura*). The transition between plain and Alps is often

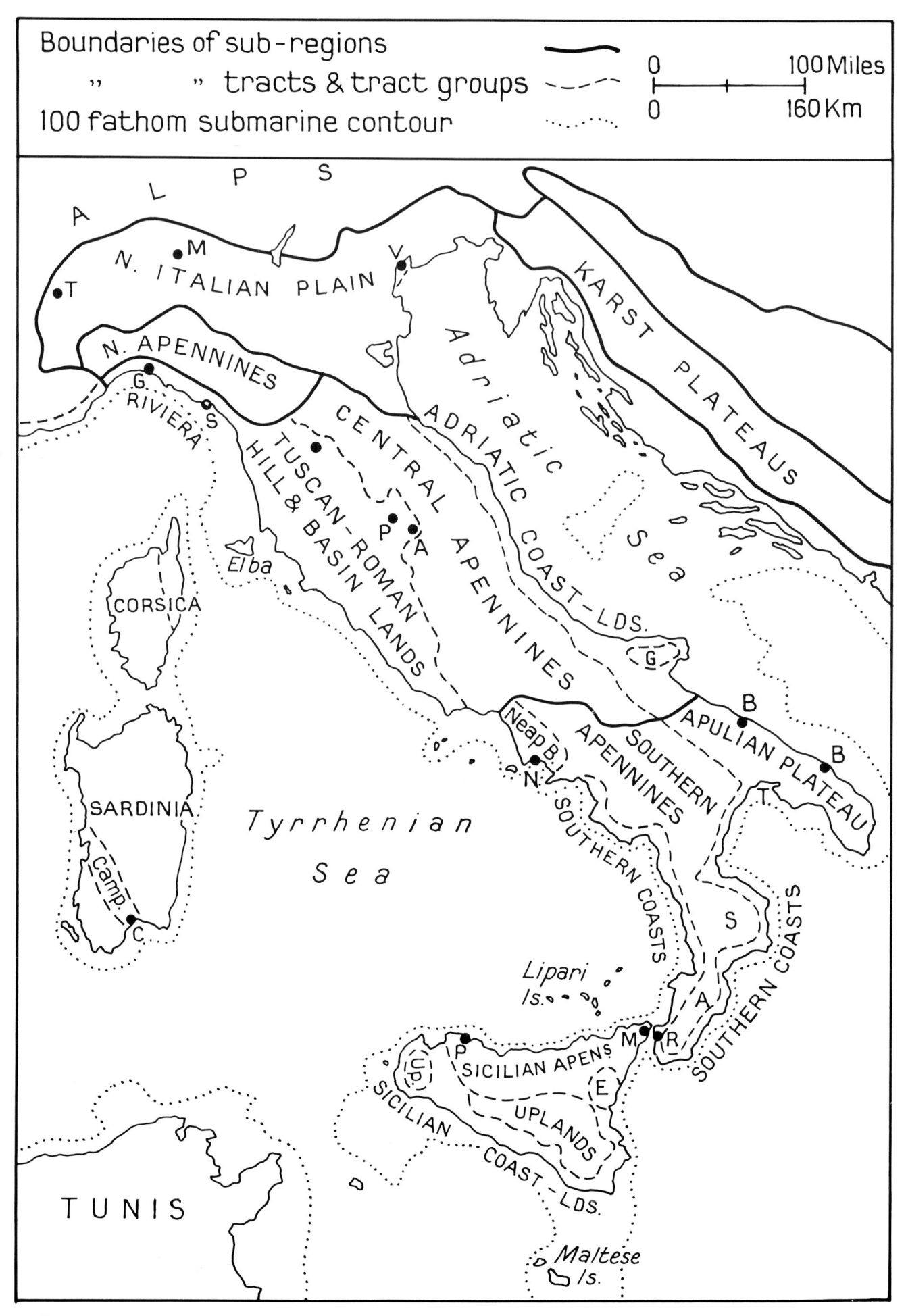

72 *Geographical regions of Italy* (after J. F. Unstead)

xxv *Rice seedlings being transplanted in the Vercelli district near Turin*

marked by glacial moraines at the valley mouths and solifluxion scree at the foot of mountain spurs. This sequence is strongest on the Alpine side of the Po lowland where between the coarse gravels of the upper plain and the finer alluvial deposits of the lower plain a wide belt of seepage, or *fontanili*, comes to the surface. The water problems clearly define these zones. The porous outer zone is relatively dry; the lower plain between the fontanili and the floodplain is irrigated; while the Po floodplain is dyked and drained against a superfluity of moisture. Wherever large capital expenditure has been undertaken, whether in drainage or in irrigation, the land is worked more economically in large estates and there is an emphasis on quick cash returns from dairy products, sugar-beet, flax and hemp or on large-scale cereal cultivation, wheat, maize or rice, with as much mechanical aid as possible supplemented by permanent and seasonal employment.

The floodplain

The valleys of the Po and of its chief tributaries, in their lower courses, are floored with recent alluvium which increasingly impedes the drainage towards the east and results in large tracts of bog or fen. Here saturation of surface soils is the main difficulty and as pumping machinery becomes more efficient reclamation is constantly being undertaken. Downstream from the Adige much land still remains to be reclaimed. Everywhere drainage is inevitably followed by the problem of the land shrinking and losing height and also of the loss by erosion of the extremely fine surface silts. *Colmate naturali* is an attempt to offset this. These polder lands produce on large estates sugar-beet, hemp or rice alternating with alfalfa and clovers in careful rotation. The green-fodder crops are given to stall-fed cattle while poultry are reared on maize meal and broken grain. The farming tends to become more mixed upstream and intensive dairying concentrates on the Po delta towards Venice.

Bassa pianura

Away from the river-bottom lands are the fine fluvio-glacial silts demarcated from the coarser detritus by the broad line of fontanili. This lower plain has been transformed by irrigation using both river and spring water. Two thousand of these fontanili exist in Lombardy alone and provide water for one-third of Italy's irrigated area. Here again large capital investments have retained the land in large commercial agricultural holdings with both permanent and seasonal employment. The land lying close to the river floodplain is under rice, maize, wheat, clover and lucerne. Away from the floodplain, the meadows irrigated by fontanili water, which remains at 10°C throughout the year, will yield normally 6 to 7 hay crops in a season. An intensive dairying industry has resulted, producing milk for the manufacture of butter and cheese under cooperative enterprises centred at Lodi, Cremona, Pavia, Piacenza and Parma. The typical cow–sow–hen routine is followed, with skimmed milk fed to pigs and maize meal to battery poultry. Also a considerable quantity of veal is released onto the meat market.

This is the largest continuous area of irrigated land in

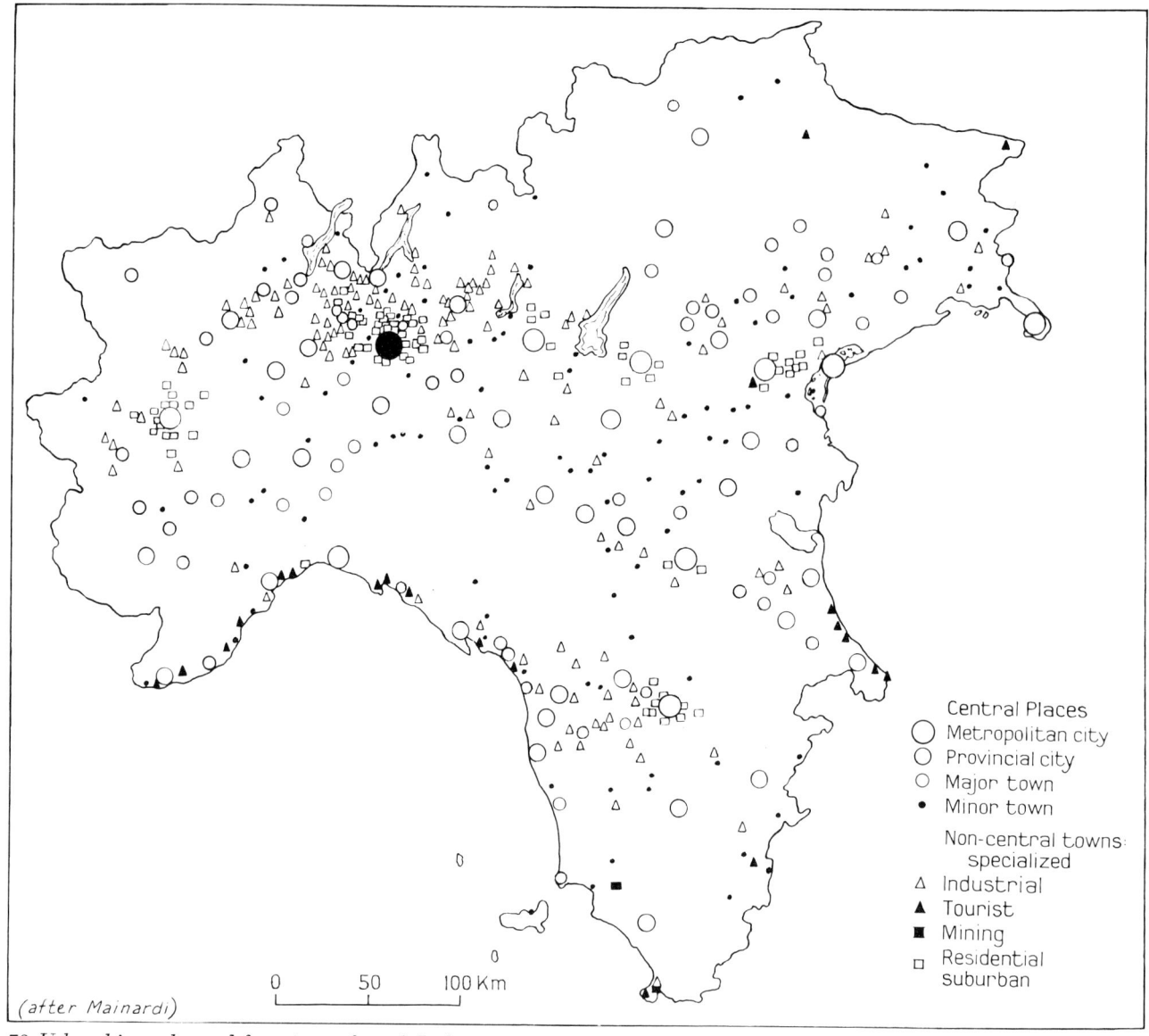

73 *Urban hierarchy and functions of north Italian towns*
Milan, the main metropolitan city, is shown as a black circle (after Roberto Mainardi and the Centro di Documentazione, Milan)

Europe and by far the chief in Italy, as has been discussed in Chapter 5. The fields are small, ditch-lined and outlined by pollarded mulberry trees linked by vines. The crops are cultivated in strips both to protect the fine soil from surface erosion and to have the land always in full production.

Variations from this general pattern, of course, exist. Mantua, for example, is the centre of a vine monoculture and wine industry. Hemp is almost a monoculture around Ferrara where the soils are lime-free and the water is soft. Again, 70 per cent of the area between the Po and Ticino around Vercelli is under rice. Using water from the Cavour canal, 500 000 hectares of land are carefully terraced and sloped to assist drainage. This canal, 85 km long, linking the upper Po to the Dora Baltea and Ticino, was built in the nineteenth century, and waters a district that has developed high-class dairying and produces the best Parmesan and Gorgonzola in Italy. The yields of rice are high but the quality is not outstanding. The cultivation of the crop is entirely mechanized except for planting, which is still done by girl seasonal workers.

Alta pianura
Beyond the reach of the fontanili, on the coarser gravelly deposits the rivers on the upper rim of plain of the basin are deeply incised and irrigation is almost impossible. Most of this 'dry' zone experiences over 1100 mm of rainfall a year but the very porous soils and intense evaporation during the hot summers so much offsets the rainfall that only the moister areas are cultivated and are under the usual cereals, with orchard crops and vines on the slopes and permanent pastures on the gravelly interfluves. These last can sometimes be indifferent *brughiere* (heathlike scrub). The Monza district under mulberries is one of very few tracts to strike an individual note.

The three land-use zones described above produce a large and varied agricultural harvest that provides the raw material for much Italian industry. Food processing is important in all the towns of the northern plain, for example Bologna is famous for its sausages, sauces and prepared meats; Milan manufactures pasta and confectionery; Vercelli handles rice and Asti wine. Lombardy produces the well-known Gorgonzola cheese. In addition the abundance of cheap hydro-electric power, the natural gas supplies, the skilled labour market, and the stimulus of the Common Market have all contributed to make the northern plain one of Europe's most important industrial

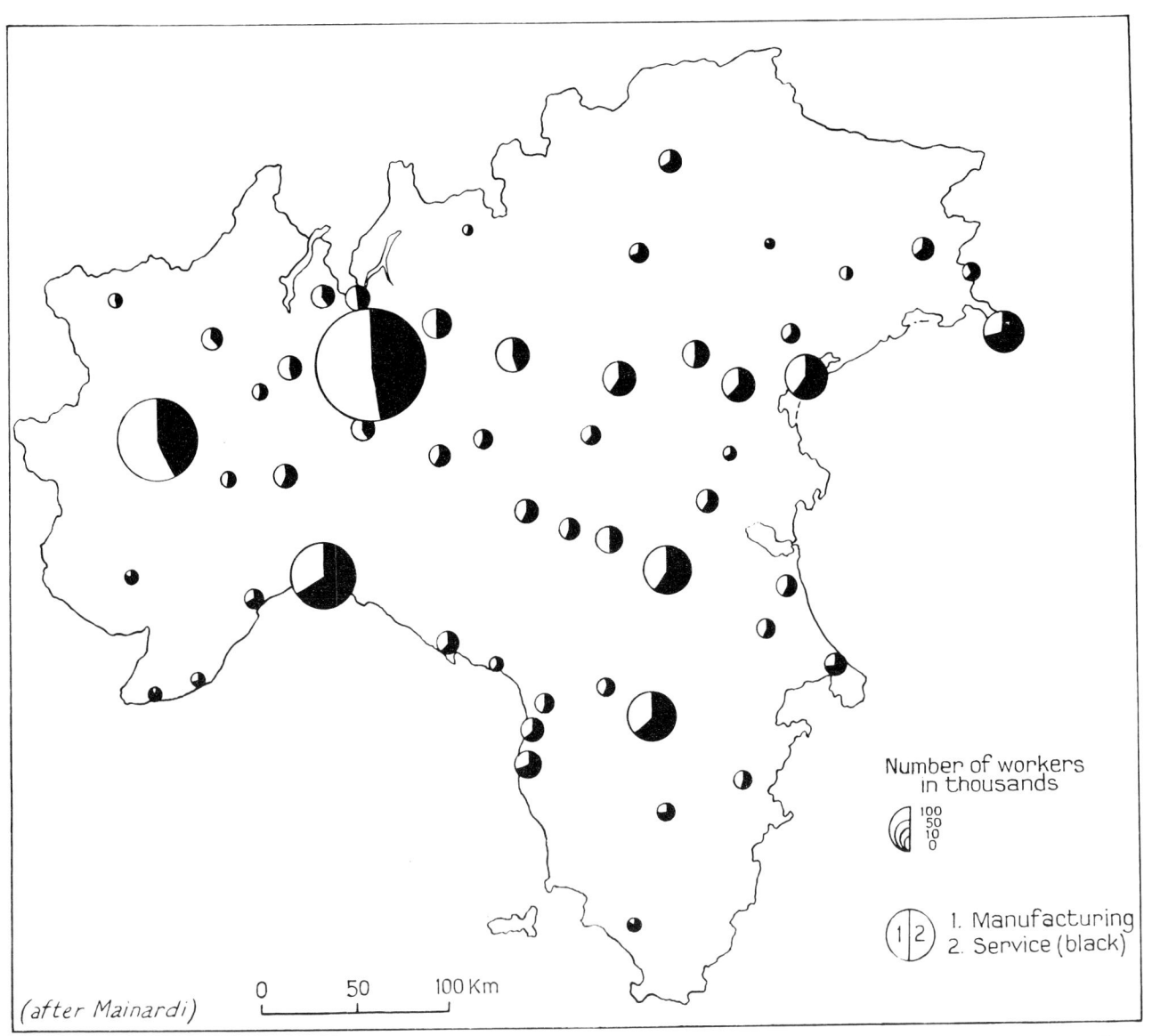

74 *Number of workers in manufacturing and in service industries in north Italian towns* (after Roberto Mainardi and the Centro di Documentazione, Milan)

areas. Owing to the nature of the power supplies, industry is dispersed and lacks a good deal of the pollution associated with coal-based manufactures, and the ancient urban pattern has not been dislocated. Heavy industry is established at Turin, Monza, Bergamo and Brescia with rolling mills at Cornigliano. Aluminium is smelted at Porto Marghera, Bolzano and Moro. The heavy chemical industry has also been established here using natural gas supplies at Bovisa near Milan. Ravenna is the largest centre for synthetic rubber in Europe.

Today the key industry in the north Italian plain, as in many highly industrialised countries, is car manufacturing since it provides a basic training in many skills and a natural progression from simple to complex engineering. At Turin, Fiat produces 85 per cent of Italy's cars, and the factories make fast touring cars as well as motor coaches, lorries, railcars, boats and aeroplanes. About 1 million units are made each year making it one of the most important car-manufacturing centres in Europe. The Mirafiori works of Fabbrica Italiana Automobili Torino alone employs 52 000 people or nearly one third of Turin's labour force and another one-third is dependent on components and accessories for it.

Other important firms are Lancia and Alfa-Romeo of Milan who specialize in hand-finished and sports cars. Coachwork and design are a special feature of Italian cars. At Modena, the firms of Maserati and Ferrari build racing cars; at Brescia the specialities are commercial vehicles and Lambretta motor cycles.

Electrical engineering has developed out of this background and usually produces high-quality goods with famous names, for example, Olivetti office equipment, Necchi sewing machines, as well as fine household equipment, such as espresso coffee-machines and kitchen ware. In the same way the chemical industry has led to the usual subsidiaries, for example, artificial fibres in Milan and its ring of satellite towns Monza, Saronno, Gallerate and Busto Arsizio, along with the necessary dyes and finishes, plastics, detergents and soap.

The older textiles tend to remain where they were first established, near the rivers for water power. The wool towns stand at the exit of rivers from the mountains close to sources of supply and abundant water, for example Biella and Bergamo. Como is still the centre of the silk industry near the mulberry country of Monza. The cotton towns are the newer industrial centres that have access to machinery and other fibres and good communications, as at Varese, Legnano, Busto and Gallarate. There is a constant home demand for cotton clothing in view of the hot climate, and also for sale to tourists. Hemp and flax industries have migrated, like cotton, to those towns where local chemical industries have assisted in their manufacturing processes.

Urbanization is not a modern development in Italy and the towns of the north Italian plain already established have absorbed the new industrial growth so that other factors have to be assumed for their position. One may discern a line of towns at the entrance to the valleys leading into the Alps and Apennines, for example, Turin, Como, Bergamo and Verona on the north, and Rimini, Bologna and Modena on the south and also on the old Via Emilia (figs. 73 and 74).

The middle Po encourages urban growth where it is joined by tributaries but below Piacenza flooding prevents this growth. Certain towns are ports on the Adriatic. Such were the now silted up sites of Adria and Ravenna but Venice remains a commercial artery. Finally, along the broad line of the fontanili, are important cities like Milan and Padua (Padova). Of them all, Milan, Turin, Bologna and Venice are oustanding. Milan (1 691 000) is the second largest Italian city and one of the most important industrial centres in Europe. It has attained this preeminence because it commands the Saint-Gotthard and Simplon routes across the Alps and has become the chief manufacturing and financial centre of Italy. With a ring of satellite towns it has a wide range of metallurgical, engineering, chemical and textile industries, managed mostly as small and medium-sized businesses. Locomotive engines of all kinds, cars (Alfa-Romeo), trains, aero and marine engines, domestic appliances, and agricultural machinery are its chief manufactures. An example of one of Milan's satellite towns is Lecco, an Alpine foothill town beside Lake Como, one hour away by rail. With a population of 45 000 it has 93 factories, 54 of which are concerned with metal. The factories tend to be small family businesses which have developed as auxiliaries of each other. There are 2 blast furnaces, 5 foundries and 10 factories producing metal; the remainder make iron bars, wire flex, wire netting, nails, screws, locks, chains, springs, etc. These in their turn have stimulated the building and construction industries and others concerned with telephones and electric parts.

The cluster of manufacturing and residential satellite towns around Milan is shown in figure 73. The city is interesting geographically for its circular plan. Beginning as a Celtic and then a Roman *castrum* of about 7 ha beneath the site of the present Piazza del Duomo, during the centuries it expanded outwards in concentric walled rings. The Romans called it Mediolanum or the central point and it remains the focus of routes across the Alps and so to Genoa, and along the Alpine foothills from Turin to Venice. It grew especially with the coming of railways and in the 1930s a redevelopment scheme included a remarkably fine railway station complex. Today Milan's pleasant high-rise suburbs are associated with a series of typically circular autostrade bypasses.

Turin (1 160 000) is likewise a route focus and lying at the junction of the Po with the Dora Riparia commands routes to the Mont Cenis and the Mont Genèvre passes. Its main industries are concerned with Fiat and Lancia cars and railway rolling-stock. Nearly 90 per cent of Italy's production of cars is made here, 130 000 workers being employed in this industry in and near the city. It is also the headquarters of many other famous firms including Olivetti (office machines) and the chief world producing centre for vermouth.

Bologna (490 000) commands the main route across the Apennines to Florence and also lies on the Via Emilia between Piacenza and Rimini, both ancient highways now paralleled by the Autostrada del Sole. It is therefore a busy rail and road junction and has attracted to itself miscellaneous industries like textiles, leather, machinery, glass and food processing, particularly salami. It is a famous university city and the centre of a rich agricultural district.

Venice (370 000) has grown up on numerous islands which are interconnected by more than four hundred bridges (figs 75 and 76). However, its most important link today is the 5 km causeway to Mestre where its numerous manufacturing concerns are located. These include petrochemicals, shipbuilding and the making of machinery, electrolytic aluminium, soap, textiles, leather and glass. Although the port, Porto Marghera, is busy (nearly 24 million tons cargo in 1970) and has regular sailings to Greece and the Levant, the main outlet for the rich north Italian plain today is ironically over the Bocchetta pass in its old rival Genoa.

75 *The site and setting of Venice*

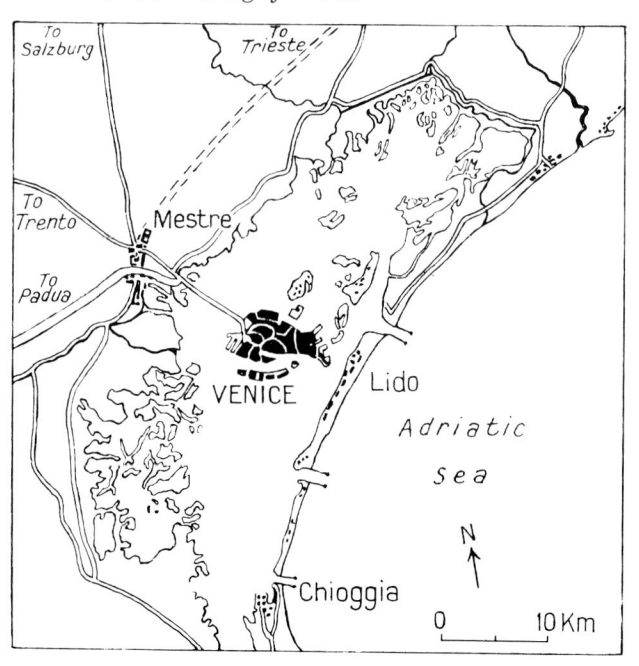

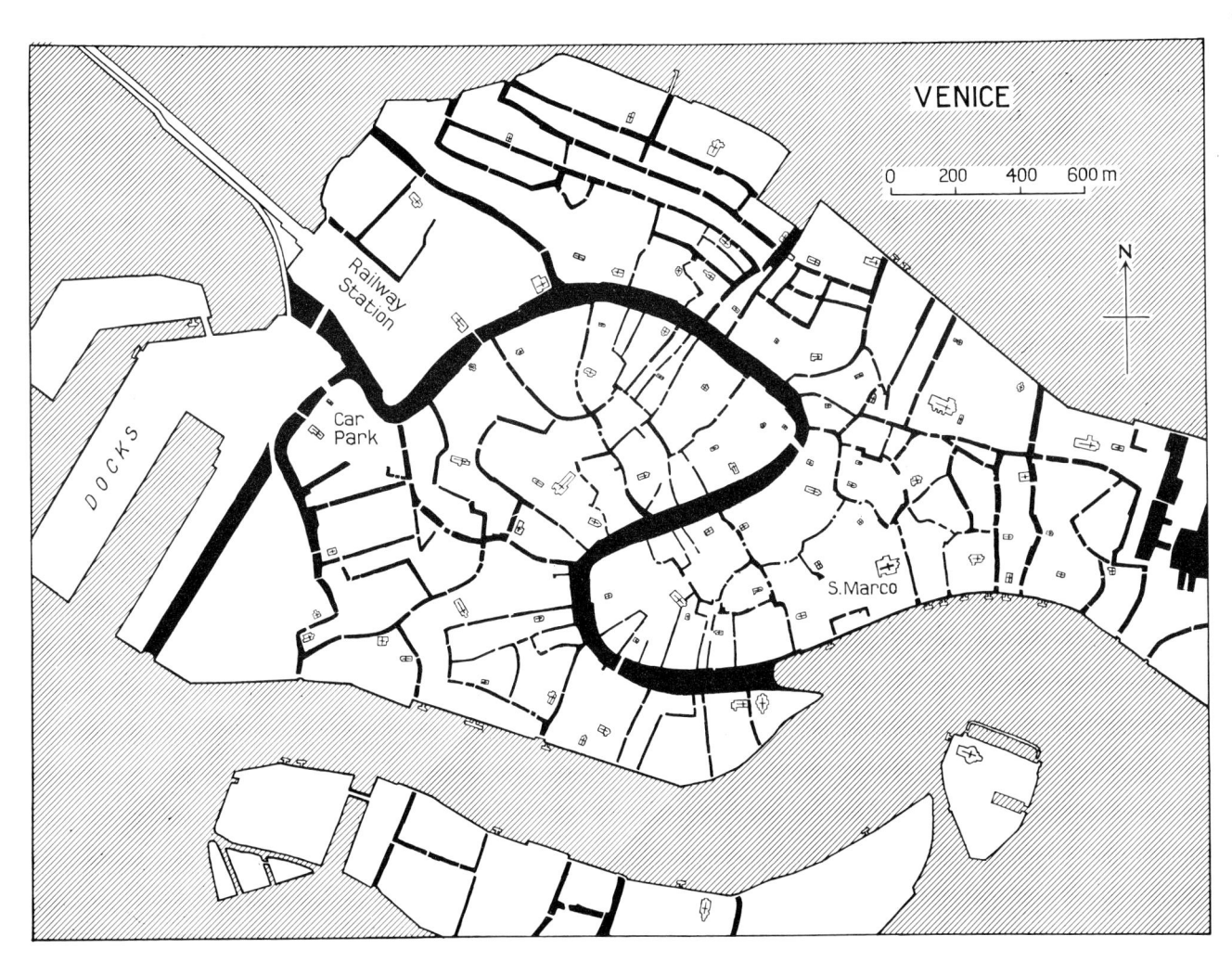

76 *The canals, bridges and churches of old Venice*

The state and fate of Venice are more than of regional importance. The town was founded beside the Rivo Alto (Rialto) in AD 811 by people from a village near the present Lido who were fleeing from the Franks. Twelve years later the bones of St Mark the Evangelist were brought here from Alexandria. When the town grew in size and wealth the new stone buildings were built on millions of wooden piles cut from the forests especially of the eastern Adriatic coastlands. The piles of oak, pine or larch about 3 m long were sunk vertically into the alluvium and upon them were placed, horizontally, thick baulks of timber so as to form a platform at a depth of 3 or 4 m. On this deep wooden platform the foundations of the buildings were constructed. The city reached its zenith of glory between about 1410 and 1571 but was weakened by Turkish advances in the Mediterranean (Constantinople fell in 1453), by the Portuguese sea trade with India after 1498 and by the discovery and colonial exploitation of Latin America after 1492. The 'most serene republic' of Venice ended in 1797, killed by the jealous machinations of Napoleon. Then it passed to either Austria or France until 1866 when it became part of a united Italy. The new port of Marghera with its adjacent industrial district and a garden city alongside the Mestre–Padua highway were developed from 1917 onwards, and now has 250 factories of various sizes.

Today the old city is an international magnet for tourists, and the Canal Grande is reckoned the finest street in the world. It is a marriage of man and the sea, fulfilled with the riches of commerce of seven or eight centuries. This beautiful urban museum is unique for its dependence on waterways and its freedom from motor-car traffic. Although having few trees it is built on a dense forest of timber and today its preservation poses a great problem. Slow subsidence of the built-up tracts combined no doubt with a slight rise in sea level threatens to engulf the city. The only permanent remedy seems to be to surround the lagoon on its seaward edge with dykes through which drainage to and from the Adriatic could be regulated with the aid of locks and pumping. At present three main channels lead into the sea (fig 75) and one can take tankers of 80 000 tons.

In fact, Venice is not the largest Italian seaport on the Adriatic, that status being held by Trieste (281 000) which handles a considerable traffic for Yugoslavia and Austria. It has civil and naval shipyards specializing in large vessels, and its main industries include metallurgy, diesel engine and turbine making, and petroleum refining. Trieste, however, can hardly be considered a port for the north Italian plain. In 1969, with 24 million tons, it was the third port for cargo in Italy but 70 per cent of its trade was crude oil mainly for the transalpine pipeline.

PENINSULAR ITALY

Two major structural elements make up peninsular Italy, the Apennine mountain chain and the lowland country on either flank. The 'wind-grieved' Apennines

xxvi *General aerial view of Genoa*

extend in a sinous curve for 1125 km from Ventimiglia to the toe of Italy and interpose their barrier between the Tyrrhenian and Adriatic seas. They also separate the two most productive and populous parts of the country, the western coastal plains of Tuscany, Latium and Campania from the Valle Padana of the north Italian plain.

Fortunately they are neither so high nor so rugged as to be impassable nor are they particularly wide. As figure 72 shows, they can be divided for convenience into three sections, north, central and south.

The northern Appenines
The northern Apennines extend from the Franco–Italian boundary to about the Reno valley which forms a routeway between Bologna and Pistoia in the Arno basin.

In Liguria the mountains run mainly from west to east and, swinging to the southeast, lie close to the sea as far south as Lerici, leaving a coastal plain so narrow that terracing is not only necessary for crops but also for dwellings, and settlements cling to every available piece of level land. Within an amphitheatre of mountains the streets of Genoa are linked by stairways and are so constructed as to be able to take cars and a cable railway. 'Genoa the Superb' has a population of 850 000 and is the fifth city in Italy and its most important port, handling trade for Austria, Switzerland and Germany as well as the plain of Lombardy. The port installations and harbour are constantly being modernized and plans are afoot for a new port at Voltri. The harbour within the shelter of a breakwater 5 km long, has five major basins and 30 km of quays. On an average the port is visited by 9000 ships annually and in 1970 handled 54 million tons of freight, mostly imported raw materials. The main iron and steel and engineering industries are concentrated in Savona west of the port and the naval dockyard and shipbuilding yards are at La Spezia on the east. The city is linked with Nice and Marseilles by a coastal railway that pierces the mountain ribs by seventy tunnels. A fine autostrada leads over the Apennines through the Bocchetta pass to Milan.

The coastlands here average 1250 mm of rain a year and, since northerly winds are fended off by the mountains, olives, vines and citrus fruits can be grown. Inland, the mountain slopes and hollows such as Vallombrosa that front the westerly winds are richly wooded. Chestnut and beech clothe the lower slopes and conifers the higher parts. In the deep narrow valleys at right-angles to the coast the villages cling to any patch of level land and the terraced fields are under vines, wheat, maize, sugar-beet, peaches (Oneglia) and vegetables. Latterly roses and carnations have been grown commercially and the Riviera di Ponente and di Levante are fretted with small bays and harbours that are becoming increasingly attractive to tourists and otherwise have small fishing interests.

Inland east of Genoa the Apennines increase in height and begin to trend from northwest to southeast. They are mantled in soft sandstone, clays and marls which weather deeply and make modern communication systems difficult to maintain. Landslides are frequent and lead to badland topography.

The central Apennines
Here the mountain range is higher than in the north and has a different drainage pattern. The short eastern flanks, south of the latitudes of Ravenna and Rimini, drain by numerous consequent rivers direct to the Adriatic while the long western slopes have developed a normal type of espalier pattern of fold-mountain drainage in which the rivers in the middle courses tend to follow longitudinal valleys between the folds and break across the uplands in narrow transverse valleys. This drainage partly reflects the different physical history of the Tyrrhenian and Adriatic Sea (see Chapter 2), the former being a great

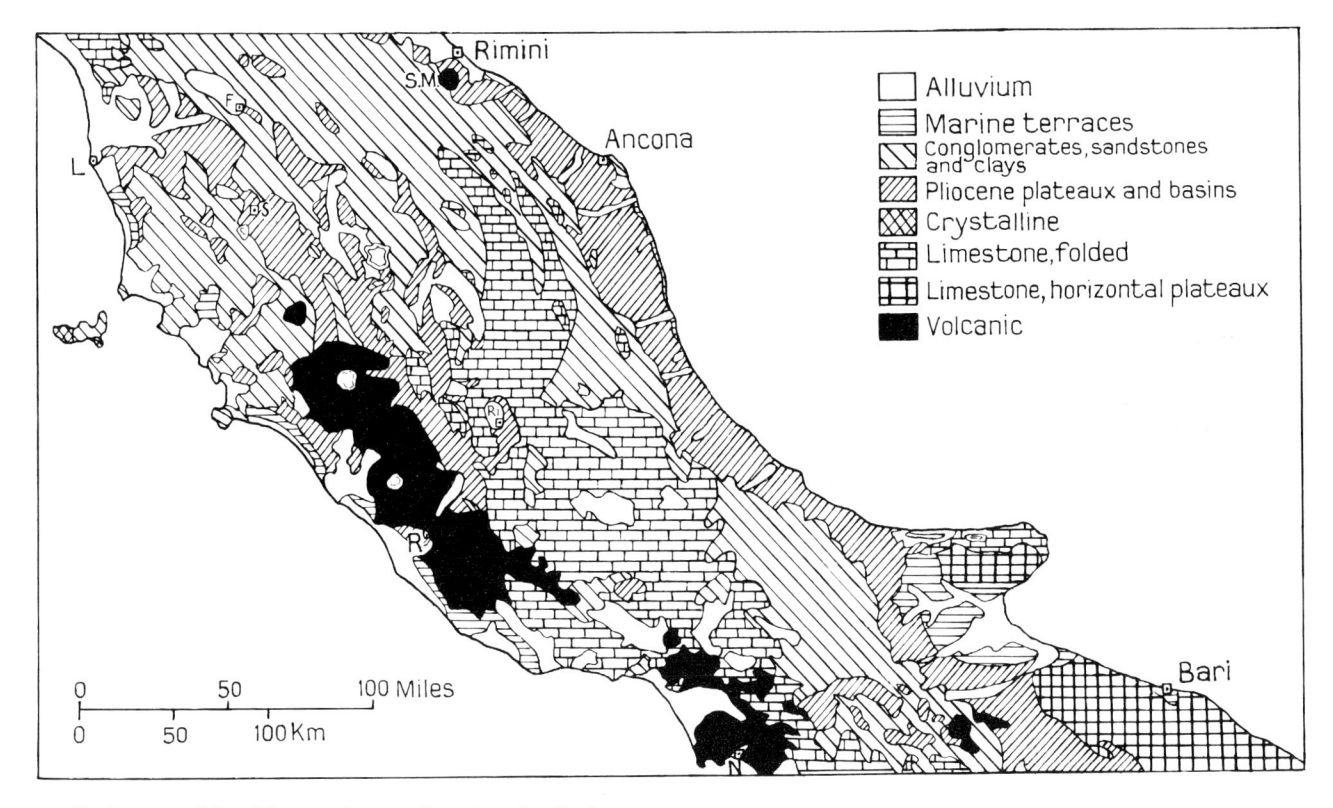

77 *Geology and landforms of central peninsular Italy*
   S.M. *marks San Marino*

78 *Geology and landforms of southern peninsular Italy and northeast Sicily*
   P *marks Potenza*

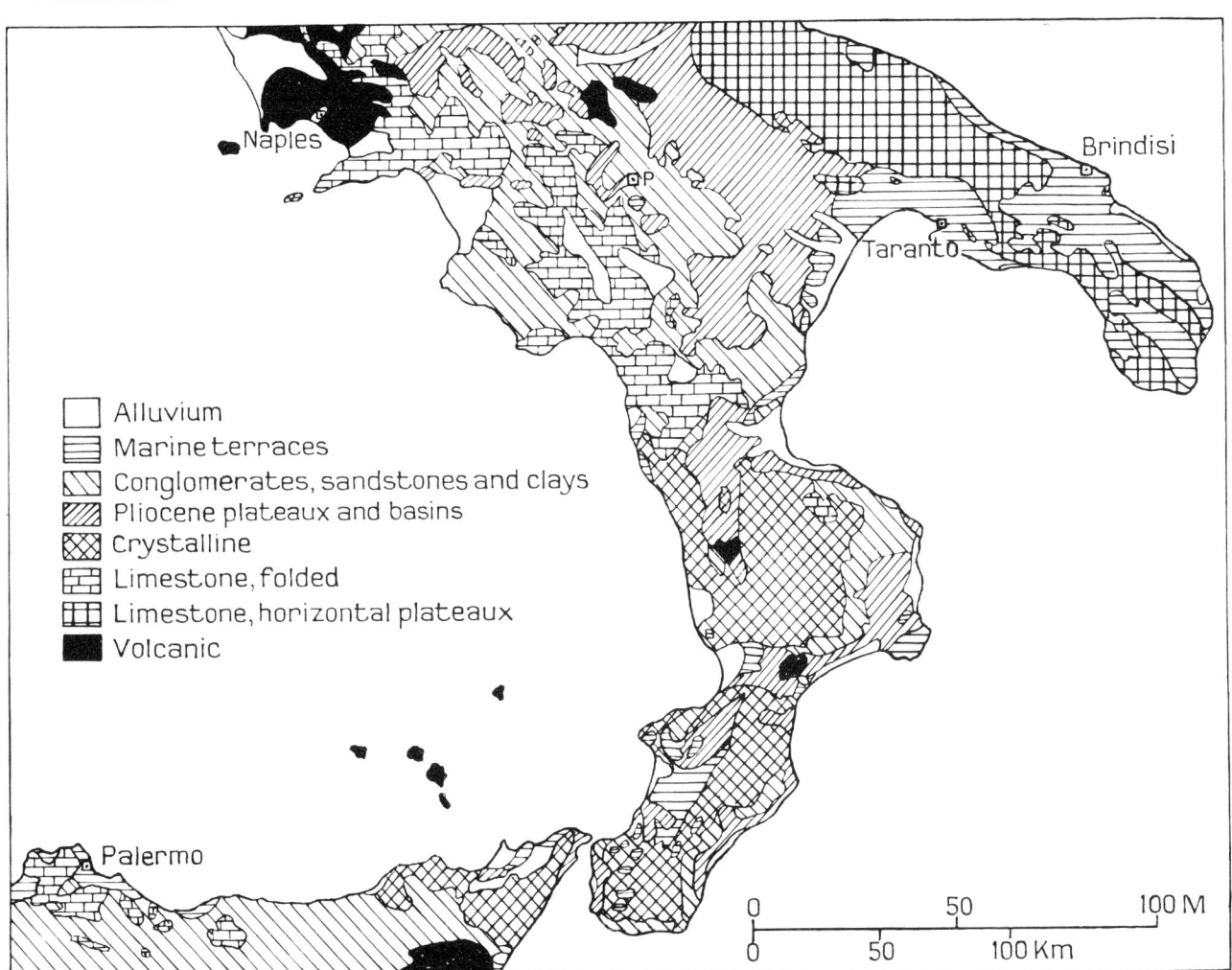

XXVII *The republic of San Marino, showing the town with its three citadels perched on Monte Titano (738 m). The abrupt scarp is a small outcrop of resistant Miocene limestone dipping steeply from left to right; the lowland consists of soft Pliocene clays*

subsident basin and the latter a shallow downfold. On the east much of the consequent drainage is over Pliocene platforms while on the west the lowlands are largely erosional and broken by fractures and vulcanism. The increased height of the central Apennines does not add much to their mountainous aspect since, except where glaciated on the higher summits, they are for the most part flat, rounded plateaux but the steep sides of these peneplains and the depth of the valleys make transverse communications difficult. As already described in Chapter 17, the dominant rock structure changes considerably from north to south. North of about Perugia the mountains consist largely of conglomerates, clays and sandstones liable to landslips whereas farther south these softer beds form only the upper eastern flanks and the main Apennine mass consists of limestones. These culminate in the Abruzzi in the Gran Sasso (Monte Corno, 2914 m) not far from Rome and 2000 m above Aquila, for which the climatic statistics are given in the table. This is an area of summer transhumance, of *frane* and of earthquake shocks. The Pliocene marine plateaux and Pliocene basins, much silted up in Quaternary times, are of great human significance (fig. 77).

The Southern Apennines

Here the summer drought is longer and more severe, the relief lower and more fragmented. Limestone plateaux and karstic landforms appear on the east coast in the Gargano peninsula and Apulia and are separated from the Apennines proper by a wide depression floored with Pliocene deposits liable to develop *frane* and *calanche*. The Apennines change direction near Potenza where a fractured, tortuous series of plateaux are formed against the crystalline masses of Calabria (fig. 78). The uplands apart from a few fertile valleys are less productive than in central Italy. The limestones are often under poor macchia or garrigue and the younger flysch deposits yield thin soils that are easily gullied. In Calabria the Sila mass (S on map) is crystalline, mainly granite, and has rounded summits rising to 1929 m while its southwestward continuation Aspromonte (A on map) is mainly of gneiss and has sharper watersheds rising steeply to 1956 m near Reggio. Here the highest parts are snow-covered in winter and under macchia but the middle and lower slopes are wooded, especially with sweet chestnut. The increased runoff from these impermeable surfaces has encouraged the building of dams for irrigation and for hydro-electric power, particularly in Calabria where five large and several small plants have been built.

Before leaving the Apennines, some mention should be made of the *Mezzogiorno* as used in modern development plans. Needless to say, it is not definable by strict geographical boundaries but includes the administrative region of Abruzzi–Molise and lies south of an approximate line from Rome to Ascoli and so embraces parts of the central Apennines as well as the southern Apennines shown in figure 72. Nor should a very small political subdivision of the Apennines be overlooked. In the northeast near Rimini and linked to it by an excellent highway, lies **San Marino**, the oldest independent republic in the world, dating back traditionally to the fourth century. It is a small (62 sq km) mountain fortress centred upon the volcanic eminence of Monte Titano (743 m) that drops almost sheer on its eastern face (fig. 77). The people, *Sanmarinesi*, live mainly by agriculture based on the cultivation of wheat, vine and olive and the rearing of sheep, goats and cattle. Small food processing industries have developed as well as the quarrying of road metal and marble, pottery making from local clays, and silk weaving. However, the main income of the republic is from tourism, including the sale of postcards and postage stamps, and from rents paid for offices of merchant-shipping firms who wish to sail under a flag of convenience. About 14 000 emigrés also invest their savings in the republic. The capital, San Marino, is a citadel on the mountain top and contains 4000 of the republic's 18 000 inhabitants. The country has been

79 *Distribution of types of rural settlement in Italy*

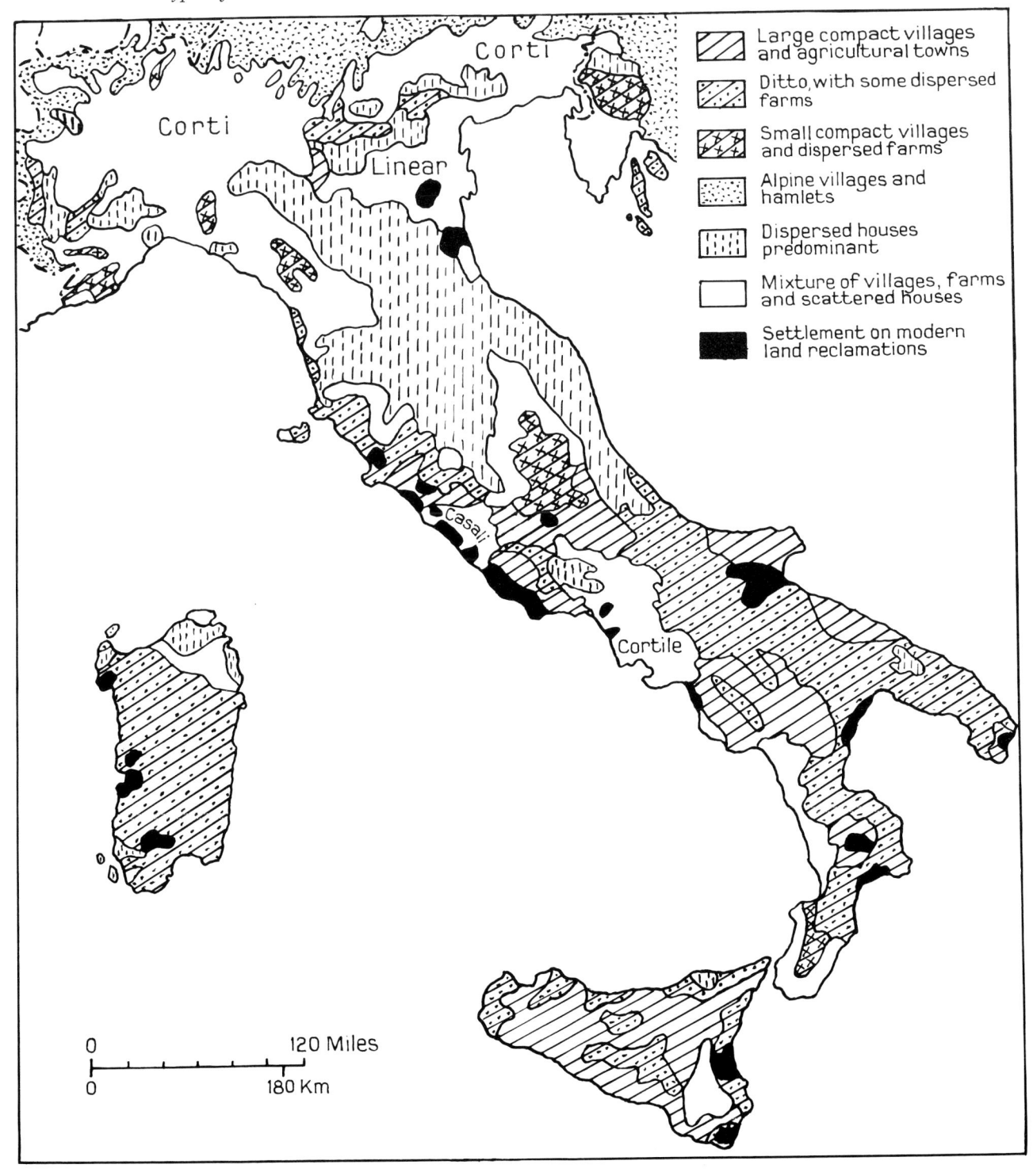

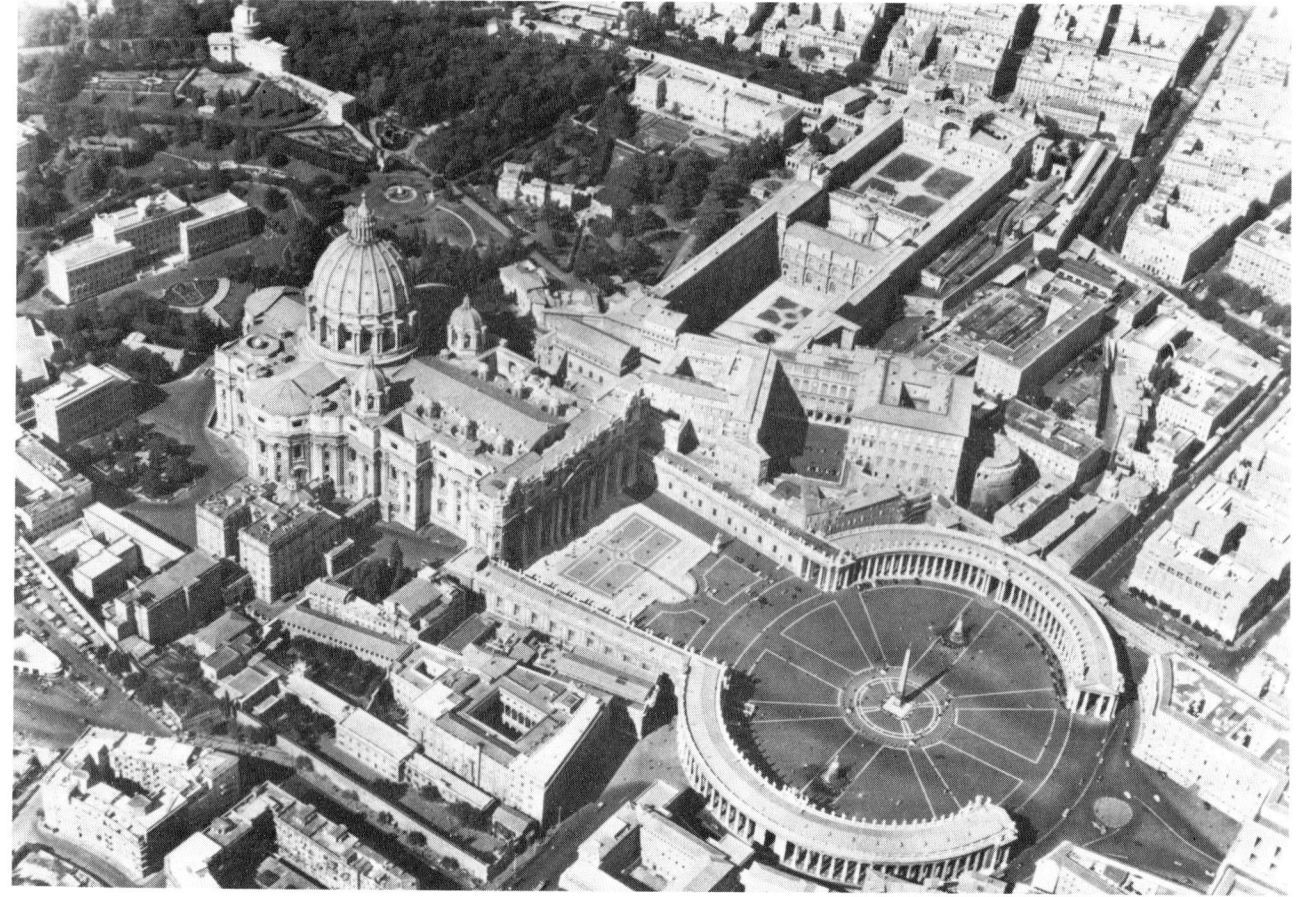

XXVIII *Rome: the Vatican City, the finest example of seventeenth-century town planning, undertaken under the direction of G. L. Bernini*

united with Italy in a customs union since 1862 but remained neutral in the Second World War.

The peninsular coastal lowlands

The lowlands and coastal hills flanking the central Apennines are, as already noted, wide on the west and narrow on the Adriatic, whereas in the southern Apennines the coastal plains are wider on the Adriatic side (figs. 72 and 77).

The *western lowlands* begin fragmentally in the riviera north of La Spezia but they widen out in the Arno basin where the Apennines swing away from the western Italian coast and enclose a lowland and foothill country of great complexity. These foothills consist of a series of basins best displayed between Pistoia and Cassino that were once occupied by arms of the Pliocene sea which retreated gradually leaving water trapped on their floors. Lake Trasimene is a shrunken remnant of such a water body. These inland seas degenerated into marshy flats since drained and linked by rivers such as the Arno, Chiani, Tiber and Salto, so forming important routeways through the mountains, notably the La Futa pass between Florence and Bologna, and the Val di Chiana route to Rome. These alluviated tracts are also very fertile and bear rich crops of cereals, vegetables, olives and vines. The ancient cities of Florence, Perugia, Siena and Assisi all grew up as the capitals of such basins. The 'burnt siena' of the artist is a reminder of the terra rossa of the Siena basin. The rich families like the Medici and Pazzi had farms in the country as well as business premises in the towns, which developed as agricultural markets, with domestic industries especially in leather and wool. Florentine cloth brought much wealth to its citizens while its richer inhabitants became famous as bankers. Florence (460 000) grew up where the old Roman Via Flaminia crossed the Arno at the site of the Ponte Vecchio. Like its nearby rivals it retains its wealth of craft industries in silver, jewellery, leather and embroidery and, as a gesture to the increasing influx of visitors, has added *haute couture*.

Between the basin and foothill country of the Apennines and the sea is a discontinuous coastal plain that was until recently malarial marshland partly brought about by the deforestation of the uplands and subsequent soil erosion and river silting. The Arno mouth entered the sea by the Maremma. Pisa the port 11 km from the mouth of the Arno has long since silted up and become a tourist centre. Most of its trade has been taken by Leghorn (Livorno; 174 000) which has also thereby acquired the twentieth-century industries of chemicals, soap and glass. Farther south the Tiber mouth was befouled by the marsh of the Roman Campagna while to the south lay the Pontine marshes and the plain of Pesto. These have all been reclaimed, drained and rendered productive. The land is now intensively farmed and is under market gardening and dairying with areas devoted to sheep grazing and to wheat. The reclaimed land has been apportioned among peasant proprietors, and agricultural loans and technical assistance made available to them. Model villages and townships

XXIX *Rome: Ancient Rome in the foreground, Medieval Rome to the left and Renaissance Rome to the right*

complete with social services, hospitals, schools and public buildings, are included in the reclamation as well, in parts, as many detached dwellings (fig. 79).

In contrast to these reclaimed lowlands are two areas of rather higher ground that lie along this coast. The first, the hill country of Tuscany, intervenes between the Maremma and the volcanic plateaux on either side of the Tiber. Tuscany shows great geological complexity which yields generally a poor farming endowment but which nevertheless is cultivated with cereals, vines and olives wherever possible, and these activities support small hill-top towns like Siena and Volterra. More significant, however, is the local mineral wealth. Manganese and lignite are mined near Siena, and mercury at Monte Amiata. There are widely dispersed pockets also of tin, copper, silver and zinc ores which have given rise to a metallurgical industry at Piombino on the coast where iron ore can be imported from Elba.

The second area of higher land is the low plateau of volcanic tuff and ash breached by the lower Tiber. Volcanic peaks rise from its surface, whose craters are now filled with water and form Lakes Albano, Bracciano, Bolsena and Vico. The district around Rome like Tuscany is unrewarding agriculturally and is scantily peopled but farther south the Campania around Naples, also volcanic in origin, is wonderfully rich and fertile. The mild winters ensure that cropping continues throughout the year, so to the winter cereals are added vines, olives, citrus and a profusion of other fruit as well as market-garden crops under irrigation during the summer.

Two important towns have grown upon these coastlands, Rome and Naples. Rome a city of 2¾ million inhabitants is the capital and larger city. Founded in 735 B C, it grew up on seven volcanic hills at a bridge point 24 km from the mouth of the Tiber. The city occupies a central position with regard to all Italy but has only grown to its present size since the unification of the country in the nineteenth century. Beyond servicing industries and the varied functions of a capital city it has no commercial importance. As the pilgrimage centre of the Roman Catholic church and the core of the ancient Roman empire and of modern Italy it is one of the most popular tourist attractions in the world. Rome airport is easily the busiest in the Mediterranean. Within Rome, on the right bank of the Tiber, is the **Vatican City State**, a tiny tract of 65 ha created in 1870 out of the last vestige of the large territorial possessions formerly held by the papacy. It has a resident population of about 900, two-thirds of whom are Vatican nationals. It is, however, by day packed tight with visitors. Internationally it forms a neutral and inviolable territory.

Naples (1 271 000) has a magnificent natural harbour that helps to make it Italy's fifth commercial (17·3 million tons cargo in 1969) and first passenger port. The city has a rich hinterland of productive soils supplying food and raw materials. Hydro-electric power has assisted in the establishment of food processing, cotton and jute textiles, metallurgical, engineering, shipbuilding and chemical

xxx *Naples, general view from the air*

manufactures but the city is overpopulated and presents a picture of 'mules with bells waiting at refreshment kiosks; strings of lemons hanging in the windows; hurdy-gurdies, shoeblacks dragging gilded stools, guttersnipes clinging to the footboards of trams; vendors of tortoises, glow-worms, catsmeat and peanuts'. In addition tourists make an important contribution to the city's economy, visiting Sorrento and Amalfi along the coast, the isle of Capri with its famous marine grottoes, and the ancient cities of Pompeii and Herculaneum which were buried by a disastrous eruption of Vesuvius in AD 79. Until that date the volcano had seemed extinct; its slopes were clothed in vines to their summits and forests grew over the crater where the while boar was hunted. Thereafter except for seven eruptions up to 1131, the volcano remained quiescent until 1631 and again cultivation had climbed to its summit. Eruptions have occurred periodically ever since, the last of a violent nature taking place in 1944. Further activity is expected at any time.

The Naples area has suffered from a depressed economy since about 1955 and to prevent further economic decline the authorities have set up here several large new concerns, including the headquarters of Aeritalia, the national aircraft industry, and a southern branch of Alfa-Romeo.

The *eastern lowlands*. The coastal plain of peninsular Italy bordering the Adriatic is very narrow as far south as Ancona, the only important harbour on this stretch of coast and a local centre for food processing, silk weaving, boatbuilding, fishing and the manufacture of accordions, paper and sailcloth. South of Ancona a coastal strip about 32 km wide stretches for nearly 250 km as a low platform dissected transversely by numerous parallel river trenches most of which are waterless during the height of summer. The low rainfall and mediocre soils support poor farming only, the typical Mediterranean triad of wheat, olive and vine being strongly dominant.

Between Monte Gargano and the Apennines is the plain of Tavoliere, 96 km long and 50 km wide, made up of clays overlain by sands and gravels. Waterlogged in winter it is arid in summer and until recently could only support sheep. The Apulian aqueduct, which reverses the waters of the river Sele and conducts them through the mountains to near Foggia, has transformed the area into a landscape of wheat fields and olive groves.

Olives also grow well in Apulia and, with oranges and lemons, along the Gulf of Taranto. Where terra rossa soils occur on the limestone plateau of Apulia, cultivation of a wide variety of crops is possible, for example wheat, olives, vines, mulberries, figs, almonds, flax and hemp; but over large areas natural conditions support only macchia which forms pasture for sheep and goats. Owing to the latifundia system of landholding there is little dispersed rural settlement and the population tends to concentrate in the towns (fig. 80).

Il Mezzogiorno

Italian economic development plans confirm the existence of a drier south, which includes nearly all the peninsula and the islands. But it was particularly south of a line from Gaeta to Ascoli and in Sicily and Sardinia that

XXXI *Trulli landscape in valle d'Idria near Martina Franca, Taranto*

the south was impoverished. The *Cassa per il Mezzogiorno* (The Southern Fund) was founded in 1950 to effect improvement here in social welfare by improving the agricultural and industrial economies. Since 1951 about £2500 million have been spent on these schemes.

Agricultural development was designed to proceed along three lines: land reform; irrigation; and the establishment of cooperative farming.

80 *Nucleated settlements in the Murge*

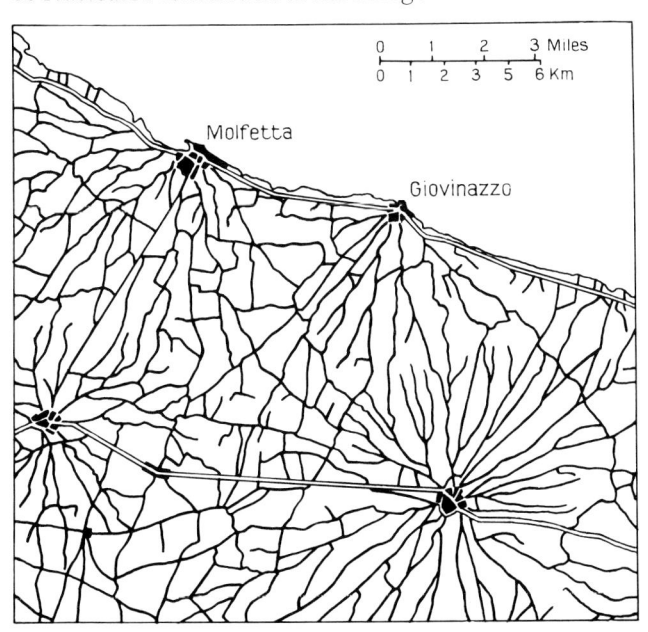

The agrarian reform includes the sharing out of uncultivated land on large estates, the increasing of the productivity of the cropland by irrigation schemes, and the introduction of new crops. These lines of approach are often mutually exclusive since most irrigation and drainage schemes are best carried out in large units, and it is only after large capital investment schemes and planning operations that the farm parcels can be allotted. Even then they are only economic in cooperative ventures. Land reclamation also includes the improvement of poor terrains and marshland around the Gulf of Taranto near the Sila massif and in Calabria.

The *Cassa* also envisages the setting up of industrial development zones around Bari, Brindisi and Taranto. Ever since 1957, not less than 60 per cent of investment to instal new industrial plant by government-controlled undertakings have by law to be established in the Mezzogiorno. This has resulted in basic industries in gas, oil, iron and steel, and mechanical engineering being founded in the south. The progress was assisted by two important developments, the discovery of natural gas in the Bassento valley 64 km west of Taranto at Ferrandina, and of considerable natural gas and oil deposits near Ragusa in Sicily (fig. 70).

The port of Taranto, a naval base and fishing port, in the instep of Italy was selected to be the site of the largest integrated iron and steel complex in Italy, and is today one of the largest in Europe, with new jetties, wharves and port equipment. Being on oil and natural gas grids, Taranto has now begun to generate its own growth and has assumed

XXXII *Vines in* cultura promiscua *near Perugia, Italy*

commercial importance. Its population of 219,000 is engaged mainly in iron and steel industries, cement manufacture, brick and tile making and brewing. A 'new residential town' is being planned, as well as an oil refinery.

A survey of the industrial potential of the tip of Italy was made by a study group of the EEC, the result of which was to attempt to set up groups of combined and integrated industries that establish their own momentum. In the present efficient state of the world's transport systems it is far more advantageous for industries to be sited near auxiliary industries than to be near raw materials. The 'growth' industries are those of agricultural and pumping machinery, excavators, cranes, machine-tools, elevators, domestic appliances and structural steel works. These are being established at Bari and Brindisi. Bari (349 000) formerly only refined and exported olive oil. Brindisi (90 000) is the terminus of the transcontinental rail and road systems.

Trade fairs are to be organised at Naples, Bari, Messina, Palermo and Cagliari and tourism further encouraged by increasing the availability of electricity, water and hotel accommodation. The whole scheme reflects the needs of an area where the average personal income was one-third that of the north. But it also reflects the oppression of a severe summer drought and of an unfortunate social history, and of the growing pull of coastal locations for heavy industries. As recently as 1962 in her *The Awakening of Southern Italy* (London; O.U.P. p. 64), Margaret Carlyle could write of Bivonghi, a characteristic hill village reached by a steep winding road constructed since 1954 from the coast near Stilo in Calabria.

> Bivonghi consists of a mass of houses piled one on top of the other, plastered against the rocks, with streets so narrow that they are only passable by donkeys or foot-passengers. . . . The majority of the (4000) inhabitants are agricultural day-labourers, who set off, men and women, early in the morning to walk seven or eight miles down the valley of desolation and over intervening hills to poor lands where they find work for 100 or 120 days in the year. . . . With conditions so poor in so many similar villages tremendous effort and large expenditure will be necessary for a long time to come if the people are to achieve a civilized way of life.

Some idea of the desperate need to find new occupations was provided during recent attempts to create a Calabrian regional province. The national government chose Catanzaro as the administrative capital, Cosenza as the university centre, and Reggio as the future site for a proposed steelworks. Reggio, by far the oldest and largest city (165 000) could not afford to be generous and its citizens rioted intermittently for seven months until they eventually acquired some of the regional administrative offices. Despite a large population the only sizeable manufacturing concern at Reggio was a small factory making railway coaches.

Since 1950 about 5 million people have left the south for other parts of Italy and 4 million have gone abroad. In spite of many financial incentives to investors, over 70 per cent of Italian industry and over 90 per cent of all firms employing more than 500 workers are in the north. Between 1951 and 1969 the number of new jobs created in the south numbered 1 million but that did not offset the 1 713 000 persons made redundant by agricultural improvements and reorganization. Today the south is advancing but it is not advancing at a greater rate than the north, where annual production growth has averaged 6 or 7 per cent for many years.

19 Islands of the Western Mediterranean

SICILY
Historical development

A day's journey by boat from the north African coast and separated from Italy by the Strait of Messina, Sicily stands guard between the western and eastern Mediterranean basins. It lies at the crossroads of Mediterranean trade routes and at the centre of the shifting currents of its various civilizations. The original neolithic inhabitants of Sicily are believed to have arrived from Italy in three groups, the Elymians, the Sicani and the Sicels, the last of which gave their name to the island. They are all thought to have spoken some from of Italian dialect. When the Phoenician traders first arrived in Sicily these natives were living for reasons of health and safety in hilltop towns surrounded by clearings of tilled fields and pastures in the midst of forest. Some of these sites have survived as towns to the present time, as for example, Centuripa and Enna. The Phoenicians, using anchorages and promontories, established trading bases around the coasts but when the Greek colonists arrived withdrew to the western shores where they founded the towns of Panormus (Palermo), Motya and Solocis.

Traders from Mycenae had been calling at Sicily since 1500 BC but the main wave of Greek colonists started in the eighth century BC, coming from Euboea to found Naxos and from Corinth to found Syracuse. They also occupied Corfu to protect the sea link between Syracuse and the mother city. The east coast was then rapidly settled and Catania, Megara and Taormina founded. The west coast was never colonized by the Greeks and the interior remained inhabited by the Sicels and Siceni who, by their trading contacts with the Greek colonies, became increasingly hellenized in their ideas, language and material culture.

The Greek colonies were successful from the beginning and were exceptionally prosperous, the buildings surviving from this period being among the finest monuments of the classical world. Vines and olives were introduced and the rearing of asses, mules, goats and horses. The economy was run on typically colonial lines, corn, hides, wool, metal and fish being exchanged for tools and implements, pottery, leather goods and textiles from Greece. By the fifth century the prosperity of the cities had led to the rise of petty dictatorships especially where the threat of Carthaginian aggression provided the need for a strong centralized control. The most powerful of these rulers was Gelon of Syracuse who dominated all the Greek cities in Sicily and extended his control even to those on the Italian mainland. Nevetheless he attracted to his court poets, philosophers, artists and scholars and was responsible for much fine city planning and building. This stimulated the Sicels to do likewise and they built the city of Palici on the plain of Catania and Cala Acta on the north coast.

In 409 BC, the Carthaginians under Hannibal began a series of inconclusive campaigns against the Greeks in Sicily and finally a boundary along the line of the Halycus river was accepted by the two peoples. In the peace which followed both sides were busy planting their territories with new towns and transplanting populations to them. The Greeks on the eastern half of the island began to penetrate up each side of the mainland of Italy, reaching as far north as Etruria on the Tyrrhenian shore and fanning out over the islands of the Adriatic, and in this way Syracuse became the greatest and most populous city in the Hellenic world, and the centre of Greek thought and learning.

Because of its power and influence it was soon involved in the first Punic war and Sicily became a battlefield between Carthaginian and Roman armies. The war ended by Carthage ceding in 241 BC all its Sicilian possessions to Rome, so that the island became one of the earliest Roman provinces. In the second Punic War the Greeks fought alongside the Carthaginians against Rome. Syracuse fell and was sacked and the whole of Sicily now came under Roman control. The Romans chose Syracuse as their capital and enfranchised a number of cities but the remainder paid tithes to Rome and the island became a granary of the empire. The interior was divided into latifundia worked by gangs of slaves, the earliest examples of plantation or collective farming practice in Europe. Roman colonies were planted at six cities, Panormus, Syracuse, Tauromenium, Thermae, Tyndaris and Catania but the major part of the island remained largely Greek. The introduction of Christianity following the conversion of Constantine led to the building of some of the finest early Christian churches and catacombs in the empire. From earliest times Sicily was a favourite place of retirement for wealthy Romans and, having become largely latinized, at the division of the empire it was made part of the empire of the west.

Intermittently invaded by Vandals during the decline of the western empire, it fell under Byzantine authority when the eastern emperor tried to reassert imperial control. Along with southern Italy it became a 'theme' of Byzantium, the papal estates in the island passing to the patriarch of Constantinople and the church there adopting the Greek rite. Sicily also bore the brunt of invasions toward Italy from Africa. In AD 740 the Saracen invasions began from north Africa, and Palermo fell in 831, Syracuse in 878, and Taormina in AD 902. The Arab conquest gave them control of Sicily, isolating it from the mainland and from Byzantium. In the following years large numbers of Arabs migrated to Sicily from north Africa and the island became part of Islam with scattered Christian communities.

The Arabs had great skills in agriculture, horticulture and in irrigation methods. Wheat growing still formed the mainstay of the agriculture but rice, cotton, dates, oranges and sugar-cane were probably introduced by the Arabs. Mulberry trees introduced by the Greeks were extended and silk manufacture established. Under the Arabs the island had 18 cities and 320 fortified towns. Palermo became famous for the manufacture of brocades, silk cloth,

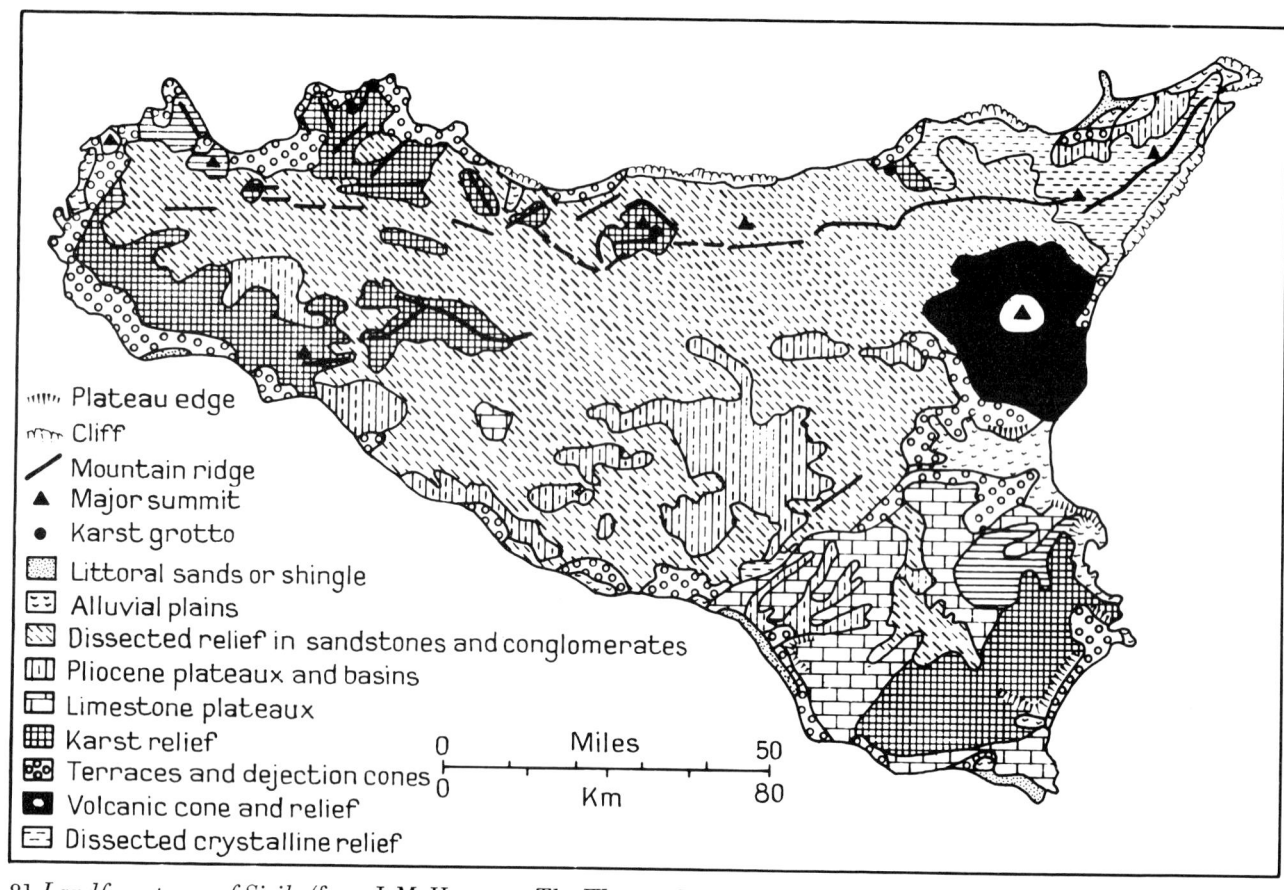

81 Landform types of Sicily (from J. M. Houston, *The Western Mediterranean World*, 1967)

leather, wood, gold and silver ware and in 970 had 200 mosques. Three new districts were built outside the old city walls, papyrus was grown for paper and rope making, and iron ore was mined outside the town at Balhara. Reliable water supplies for gardens and mills were obtained from the Oreto river. It was a period of great material prosperity yet the Norman crusade against Islam in Sicily in 1060 was welcomed as a liberating movement by the underprivileged Christian communities. The Norman leader Roger Guiscard was enfeoffed by the Pope with Sicily, Calabria and Apulia. Messina was captured by the Angevins in 1061 and Palermo eleven years later.

The Norman rule was tolerant in that Arab, Roman and Norman lived in peace together but it was feudal in administration and overcentralized and in 1282 the Norman sovereignty of the Kingdoms of Sicily was rejected in favour of the Aragonese. This proved just as oppressive but the island became increasingly latinized by settlement from the mainland, and the Sicilian parliament grew slowly in power and function until it gained control of finance.

From 1707 when the Spanish government ended, until 1861, when Italian independence was declared, Sicilian history shared in that of the mainland where quarrels of succession alternated with wars of aggression.

Invaded by the Allies during the Second World War, it is now a region of special development under the Cassa per il Mezzogiorno and may, once the autostrada del Sole is completed, be on the threshold of another golden age.

Structure and relief

Sicily covers 25 710 sq km and is separated from the mainland by the Strait of Messina, which is only 2 to 5 km wide, and from the African coast in Tunisia by the shallow Sicilian Strait which narrows to 145 km. Thus structurally the island is a broken link between peninsular Italy and northeastern Tunisia. The Apennine chain continues into Sicily and extends along the northern shore of the island as far as the valley of the river Torto (fig. 19). It can be traced for 160 km in the crystalline belt of the Peloritani mountains in the east and in the sandstones and limestones beyond of the Nebrodi mountains (1847 m) and Le Madonie ranges (1977 m).

The whole forms a plateau of between 900 and 1800 m with a steep but broken northern edge furrowed by gorges and flanked by aprons of debris, and a gentler south and southwest slope that sinks to an average height of 300 m. The character of the plateau changes just east of the Torto river where limestones and sandstones from an area of broken and tangled high relief, displayed either in smooth tabular masses or barren, much-dissected karst, that peters out near the sea to allow the formation of crescentic coastal lowlands such as the Conca d'Oro and the plains of Trapani and Alcamo (fig. 81). The mountain chain can be traced as far as the Egadi archipelago and as far south as Sciacca. The southwest of the island consists of low plateaux of chalk interbedded with clays, marls, and sands displaying slight uplift and shallow folds and again intricately dissected by south-flowing streams. The interior is a wild and barren country, associated with salt, gypsum and sulphur deposits but the coast is scalloped with small fertile plains 8 km in width or more that are spaced at intervals from Marsala all the way to Syracuse.

The Iblei mountains in the southeast likewise consist of sedimentary rocks, here interrupted by volcanic eruptions that have mantled the surface with lava flows, ash deposits and cones. The crustal instability can be traced from Monte Laurro through Etna to the volcanic islands of Lipari in the Tyrrhenian Sea with their seven volcanoes that include Vulcano and Stromboli. The volcanic cone of Monte Laurro (968 m) is quite dwarfed by that of Mount Etna (3269 m) which covers 1600 sq km and has a girth of 200 km. The mountain and surrounding area have been built up as the result of submarine eruptions in a gulf of the sea now occupied by the plain of Catania. There are records in ancient times of 135 eruptions, and of at least 3 disastrous eruptions in the Middle Ages, the worst being in 1669 when lava streams reached the sea devastating the city of Catania. In modern times the outstanding eruptions took place in 1910 when 23 new crafters appeared; in 1917 when a jet of lava squirted to a height of over 800 m; and in 1923 when the lava ejected was still hot 18 months later. Etna has erupted frequently since then and still smokes today from a huge, black, distorted cone. It was discharging violently in the spring of 1971. The lava fields at the summit contrast with the lush vegetation at the foot of the mountain, which is circumscribed by a curious furrow that receives the radial drainage of the cone and is followed for parts of their courses by the rivers Simeto and Alcantara.

The plain of Catania, which extends between Monte Laurro and Etna, is formed of the lower drainage basin of the Simeto and its tributary the Dittaino, and is the largest lowland in Sicily. In fact only 15 per cent of the island lies below 100 m and nearly two-thirds of it is hill country and low plateaux over 300 m. Because of this and the crystalline impermeable structure of the northern ranges, much of the stream runoff is rapid, and the limestone uplands covering nearly one-third of the island with their underground water supplies are of great importance. The steep slopes are also the reason for the large areas (40 per cent of the surface) which are liable to soil-slip. Another natural hazard is the instability of the volcanic districts.

Climate

Sicily, as the statistics for Palermo and Syracuse show, has a typical Mediterranean climate with a marked winter maximum and a prolonged summer minimum of rainfall. In fact, this means that the island suffers from drought for much of the year. The annual rainfall amounts to about 640 mm on northern coasts and 400 mm on southern but the intervening higher ground may receive as much as 800 mm and rainshadow areas very much less. The plains of the Conca d'Oro receive 640 mm; the plains of Catania on the other hand have 730 mm and Taormina 750 mm. The scanty warm-season rainfall is erratic, unreliable in amount and incidence, and torrential in fall, rushing quickly off the hot dry surface which dries out rapidly. Some of the rainfall is held temporarily as groundwater in the debris fans at the foot of slopes, and is tapped by wells but the greater part flows uselessly to the sea.

The drought is worsened by the frequent occurrence of the sirocco, the hot dry dusty wind which originates in north Africa at the front of a depression. Arriving along the south coast it is at times oppressive and humid but in its passage over the interior mountains it loses most of the moisture it acquired in its journey over the Sicilian Strait and descends the lee or northern slopes as an increasingly desiccating wind. Along the north coast the size and quality of the harvests are diminished by it.

The drought may also partly be attributed to the high temperatures and long periods of sunshine experienced throughout the year which greatly increase evaporation. A marked feature of the lowland climate is not simply the complete absence of winter frost but of winter cold which means that vegetation grows all the year. Neither Palermo nor Catania has any month with an average below 10°C and even Caltanissetta in the interior has an average January temperature of 7° and only 4 months below 10°C. Summer temperatures, despite the insularity, are high everywhere in Sicily. The hottest month on the coast averages 25°C and six or seven months average above 15°. The highest known temperatures include 49°C at Palermo, obviously during a strong descending sirocco (cT) airflow. Evaporation rates are high and the average annual potential evapotranspiration exceeds 1200 mm, which is usually about double the yearly rainfall.

Land use and agriculture

Woodland covers less than 4 per cent of the island, being restricted to the northern slopes of the Nebrodi and Madonie mountains. Macchia clothes the lower uncultivated slopes up to 500 m and includes dwarf palms as in southern Sardinia. Chestnuts begin to appear at about 500 m, sessile oaks at 610 m and beech at 1250 m. Papyrus, introduced by the Arabs, grows in the valley of the Cyane river, its only locale in Europe, and reeds tough enough to be used as vine supports line most river banks.

Despite the rugged terrain and drought, three-quarters of the island is cultivated and half the population is engaged in some form of agriculture. There are three main forms of land use:

1 The *latifundia* of the south and centre (fig. 82). Here mainly on hard, dry intractable soils liable to drought and seared at irregular intervals by the sirocco, are extensive wheatfields. The variety is drought-resistant and short-stemmed and gives yields only just over half the average for Italy but the grain is hard with a high protein content. Sicily produces one-tenth of the Italian wheat harvest and has no less than 1½ million ha, or over 60 per cent of its total cultivated land under a wheat rotation. Yet in any one year only about 650 000 ha are actually planted with wheat, a large proportion of the rest being under fallow and the remainder poor fodder. This is the region where much land reform is necessary. Water supplies need to come under

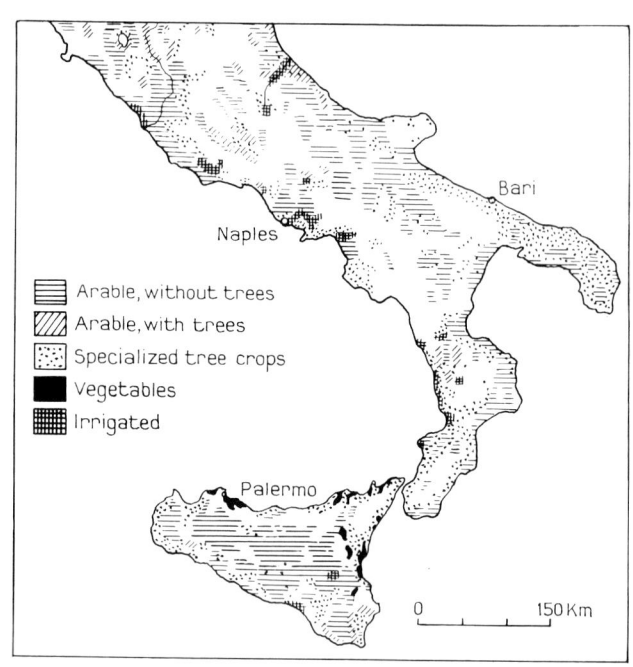

244 Part II: Western Peninsular and Alpine Lands

82 Land use in southern peninsular Italy and Sicily
(after P. Birot)

public ownership and communications in the rural districts need improving. Of the 550 000 agricultural holdings nearly half are 1 hectare or less but, on the other hand, 16 000 holdings are over 20 ha and cover all told over 1 million ha or nearly half the farmland. The major part of the latter consists of large underdeveloped estates. Already 61 000 ha have been expropriated, any intensively cultivated land being exempt, but much capital expenditure is necessary before the maximum land use is attained. Sheep rearing provides an alternative occupation on the higher and drier slopes and a certain amount of olive cultivation goes on. The donkey is the usual beast of burden and there are 142 000 of these animals, some of which are bred for sale elsewhere. Poor quality cattle (330 000) are kept on the fodder crops and nearly ¾ million sheep and goats on the natural and rough pastures.

2 *Huertas*. These are mainly in the coastal plains where water is more plentiful, either through a more abundant rainfall, from springs at the foot of the limestone, from wells in alluvial and debris fans, or by means of irrigation from the rivers. Where sufficient water is available, crops can be grown throughout the year. On the terraced slopes surrounding the plains, vines, olives and almond are planted. The level areas are under market-garden crops of green and salad vegetables and flowers, especially on the orchard floors, followed later in the year by cotton and forage. Two-thirds of Italy's lemons, or 400 000 tons, are grown in Sicily and half of the orange and tangerine crop. The citrus groves tend to encircle the towns and to be under irrigation. The plain of Catania with an area of 430 sq km has, by means of dams on the Salso and Simeto rivers, 16 000 ha of irrigated citrus orchard mainly around the towns of Lentini and Catania. The Conca d'Oro has its water supply safeguarded to a certain extent by the aqueduct leading to Palermo from the Oreto river but all the other coastal plains use simpler irrigation methods such as the water wheel turned by a mule. Thirty-eight per cent of Italian soft fruit is produced on Sicilian huertas mainly from the fertile coastal plains extending from Milazzo to Termini near Palermo. Properties in these areas tend to be small (3 ha) and the population density is accordingly higher than average.

3 *Tree crop areas*. The foothills surrounding the coastal plains up to 500 m, where soils are sufficiently deep and a certain amount of ground water is available, are given over to olive groves, and orchards of almond, cherry, peach, plum, nectarine and pistachio and, in the drier areas, carobs. The slopes of Etna and the hills surrounding the plain of Catania are especially fine examples of this type of land use. The southern slopes of Etna progress upward from citrus through vines, cherries, and apples at 610 m, above which pistachio orchards give way gradually to forest at about 1300 m, consisting of chestnut, oak, beech and birch. Above 2000 m the slopes are barren except for a vetch (*Astralagus Aetnensis*) which grows among the clinkers and on the slopes of secondary craters. On the newer lava slopes lower down the mountain, arboreal broom is today being planted because its exceptionally strong roots break up the lava. It is followed by figs, almonds, groundnuts and olives and, it is hoped, finally, by vines, but the process takes a century or so.

The hills overlooking the northern coastal plains are mainly under vines and olives as far west as Palermo. Sicily follows Sardinia as the second olive oil producing region in Italy, with 12 per cent of the national output. The Trapani district beyond Palermo specializes in vines for making marsala. In these areas too the population density is oppressive being in parts well over 300 per square kilometre.

Industrial development

Sicily is seriously overpopulated and the annual emigration is inadequate to relieve the pressure on the island's economy. Apart from intensifying agricultural production, the Cassa per il Mezzogiorno has made a start in establishing manufacturing industries. The only local mineral assets are widespread deposits of salt, gypsum, potash and sulphur for which until recently power for processing was absent. In 1953 the Ragusa oilfield started production and today has an annual output of 2 250 000 tons from the three drilling centres of Ragusa, Gela and Fortanarossa. A refinery has been established at Augusta, the island's largest port, which processes 10 million tons of crude petroleum annually, the bulk being imported. Petro–chemical concerns have also been built at Gela and Milazzo. In 1961 large supplies of natural gas were struck at Gagliano and are now distributed by a pipeline grid.

Caltanissetta, Agrigento and Catania produce 4 per cent,

or 150 000 tons, of the world's output of sulphur but the refining process is primitive, and the island's former world monopoly in sulphur has now passed to Louisiana. Along the coast of the plain of Trapani and in the Gulf of Catania some of Italy's largest salt evaporating pans operate and, on the bases of salt, sulphur, and mineral oil byproducts, a chemical fertilizer plant has been established at Porto Empedocle, and plastics, pharmaceuticals, dyestuffs and heavy acid industries at Palermo and Messina. Financial incentives have also been largely responsible, for example, for the building of a cement works at Augusta, a knitwear factory at Licata and a Fiat car assembly plant at Termini near Palermo.

Otherwise the main industrial development has been in food processing, such as olive oil production, wine making (marsala) and fruit and vegetable packing in most of the larger towns and especially at Messina and Catania. Leather industries have concentrated on Palermo and Catania.

The fishing industry, which engages some 10 000 boats, or one-quarter of the Italian fishing fleet, lands about one-fifth of the national catch, including half the landings of tunny. This has fostered a small shipbuilding and repair industry at Palermo and engineering and foundry plants at Catania and Messina.

Population and cities

The total population of Sicily is about 5 million and the average density nearly 200 per sq km which is slightly above the national average. This density and the marked urban growth is all the more remarkable when seen against the relative lack of industrial growth. Three of Italy's largest cities are in Sicily. The chief is Palermo (654 000), the regional capital and main port. Of Phoenician origin and with Saracenic and Moorish remains, it grew up on a tongue of land between two inlets, now built over. A new deep harbour, with extensive quays and dry docks for shipbuilding and repairs, has been excavated near Monte Pellegrino. Palermo is also an important tourist centre.

Catania (409 000) exemplifies the difficulties experienced by ancient foundations and historic settlements throughout western Europe. The increasing size and scale of modern integrated industries and of the communications which feed them cause these cities to burst out of their medieval confines and to destoy much of historical value in the process. Catania has important Greek and Roman remains among which is the third largest colosseum known. Strabo described it as the most important city on the island. Parts of it are at intervals destroyed by volcanic activity from Mount Etna and are then rebuilt. In 123 BC the town was so devastated that the inhabitants were relieved of all taxes for ten years; in AD 251 streams of lava from Mount Etna entered the colosseum; in 1170 the town was partly destroyed by earthquake and again in 1693.

Messina (270 000), the third Sicilian city by number of inhabitants, has a fine site around a semicircular harbour protected landward by mountains and seaward by a curving headland. This anchorage on the Strait of Messina has made it a trading port since earliest times and probably was also the reason for much of its disastrous history. In 1743, plague killed 40 000 people here; in 1783, earthquakes flattened much of the town; in 1854, nearly 15 000 citizens died of cholera; and in 1908, Messina had one of the most disastrous earthquakes in history when 84 000 lives were lost. Today it exports citrus and other fruits and food products and manufactures silk and rubber goods.

Of the other Sicilian cities most are of considerable tourist interest and several are noted winter resorts both for national and foreign visitors. In the west, Trapani (80 000), mentioned in the *Aeneid*, is a port and market centre for a fertile coastal plain. In the east, Siracusa or Syracuse (102 000) is a Greek foundation that grew up on an offshore island now connected by a mole to the mainland. Just north is Augusta, Sicily's main port and the second port for cargo in Italy. It handles nearly 11 per cent of the national foreign trade, mainly crude petroleum, with smaller quantities of wine, olive oil, cereals and citrus fruit. Having been turned into an ocean terminal for giant tankers, plans are afoot to give it a similar function for container traffic. On the more arid south coast, Agrigento, together with Porto Empedocle (50 000), has been concerned with the export of agricultural products, silks and sulphur since classical times.

The urban centres undoubtedly have an inflated population in regard to their functions. In rural districts most Sicilians live in large nucleated villages, or agro–townships, of about 5000 to 10 000 inhabitants at about 300 m altitude. Infant mortality, as throughout the Mezzogiorno, is high and reflects considerable poverty and overpopulation in spite of a large emigration which includes about 50 000 workers annually.

Communications are excellent in the coastlands but there is a dearth of good secondary and feeder networks on the plateaux. Inland rail and road transport focuses on Caltanissetta. Palermo and Messina have regular boat services to Naples, Genoa, Tunis and Tangiers, and Palermo and Catania have airports. From the western end of Sicily at Trapani a regular ferry service goes to the Egadi islands, which from May to June have an active tunny fishery, for local canneries.

THE LIPARI ISLANDS

Off the northwest coast of Sicily and close to the Italian mainland, is a group of seven volcanic islands and several small islets. Known today as the Lipari group, their main interest lies in their volcanic character, as they cover only 177 sq km and support a population of only 12 000 inhabitants.

Lipari, the largest island, is made of volcanic rock that rises vertically from the sea and in which bands of red obsidian can be discerned. Obsidian from Lipari has been found at Knossos but the chief quarrying activity at the present time is for pumice. Sulphur springs and vapour baths abound which with the abundance of marine life for underwater fishing and the warm sunshine would form a basis for a profitable tourist industry. Unfortunately fresh water is scarce and the land largely infertile. Every cultivable patch is under vines. The chief town is Lipari (4500), the capital for the whole group, which has ferry and hydrofoil services with Sicily and the Italian mainland.

Salina, 5 km northwest of Lipari, is made up of two extinct volcanic cones of which Monte Salvatore (962 m) is the highest point of the Lipari group. Stromboli, 'the lighthouse of the Mediterranean', lies 35 km northeast of Lipari. Here the volcanic cone erupts every two hours emptying lava and incandescent stones into the sea from a crevasse called the Sciarra del Fuoco, 'the pit of fire'. As elsewhere wherever the land is cultivable vines are grown and produce a dark sweet wine like madeira. Just south of Lipari a basalt needle, 75 m high, juts out of the sea and immediately south of it lies Vulcano, a wild, barren island with a rugged coastline fringed on the northwest with basalt reefs. The cliffs eject hot mud flows and springs and from the sides of the volcano sulphurous vapours escape from crevices and the crater still smokes. A tourist village has been built on Vulcano, which, as with the islands generally, where safe from eruptions, is wonderful for sailing and underwater fishing.

SARDINIA
Development of the cultural landscape

If we neglect the earliest settlements, the present landscape in Sardinia dates back to before the Bronze Age, when people using copper lived in hutted villages and buried their dead in tombs arranged in cemeteries today called 'witches' houses' that can be seen, for example, near Alghero. From this period also date the menhirs and dolmens and the unique stone platform of Monte d'Accoddi.

Cretan and Phoenician traders were attracted early to the island presumably for metals and the Bronze Age was a time of great prosperity judging from the number of buildings erected. The most striking are the *nuraghi* which abound throughout the island (fig. 83). These round structures consisted inside of a complex of bastions, turrets and cisterns connected by an interior staircase and corridors, as is well seen in the massive fortress at Nuroxi. Some nuraghi form part of defensive systems, while others in fortified villages resemble store rooms. Rectangular temples, ornamental wells and fountains and giant tombs also survive from this era.

In 500 BC the Carthaginians invaded Sardinia and left behind their various structures while the earlier inhabitants were restricted to the mountains. In 238 BC the Romans gained control and the island was made with Corsica a province of the Roman empire. For a long time it was treated as conquered territory and forced to send corn

83 *A Sardinian nuraghe*

and money taxes to Rome. Eventually in the time of Pliny, Cagliari was made a free city and *coloniae* were established at Porto Torres, Usellis and Cornus but the island was always considered a granary of the empire and a place of exile.

After AD 456, the occupation by the Vandals was a time of great cultural revival. Then 120 bishops expelled from north Africa settled here and contributed to the landscape the basilica at Cagliari and the monastery near by. The successive invasions by Goths, Lombards and Saracens are of more historical interest than geographical consequence. After the fall of the Byzantine Roman empire the Sardinians set up their own government and divided the island into four regions, with their administrative centres at Cagliari, Torres, Arborea and Gallura respectively, and the effect of these divisions is visible in the distribution of population today. The Arabs were only expelled from the island with the aid of troops from Pisa and Genoa and for a long period these city states and Marseille vied for local supremacy until Pisa gained the ascendancy. Under Pisan influence many churches were built and the fortifications of Cagliari constructed.

Between 1326 and 1478 the Aragonese conquered Sardinia and in so doing ruined the economy. In the early eighteenth century the island was annexed by Austria and later handed over to Savoy. Under a beneficent Savoyard administration, schools were built, the Italian language taught and agriculture and manufactures encouraged. Settlement by Piedmontese, Ligurians and Corsicans followed but the French invasions of 1793 halted progress. In 1861, Sardinia became part of the kingdom of Italy.

Occupation again occurred in the Second World War when the island was garrisoned by the Germans and air bases built. Since 1952 Sardinia has been an area of special

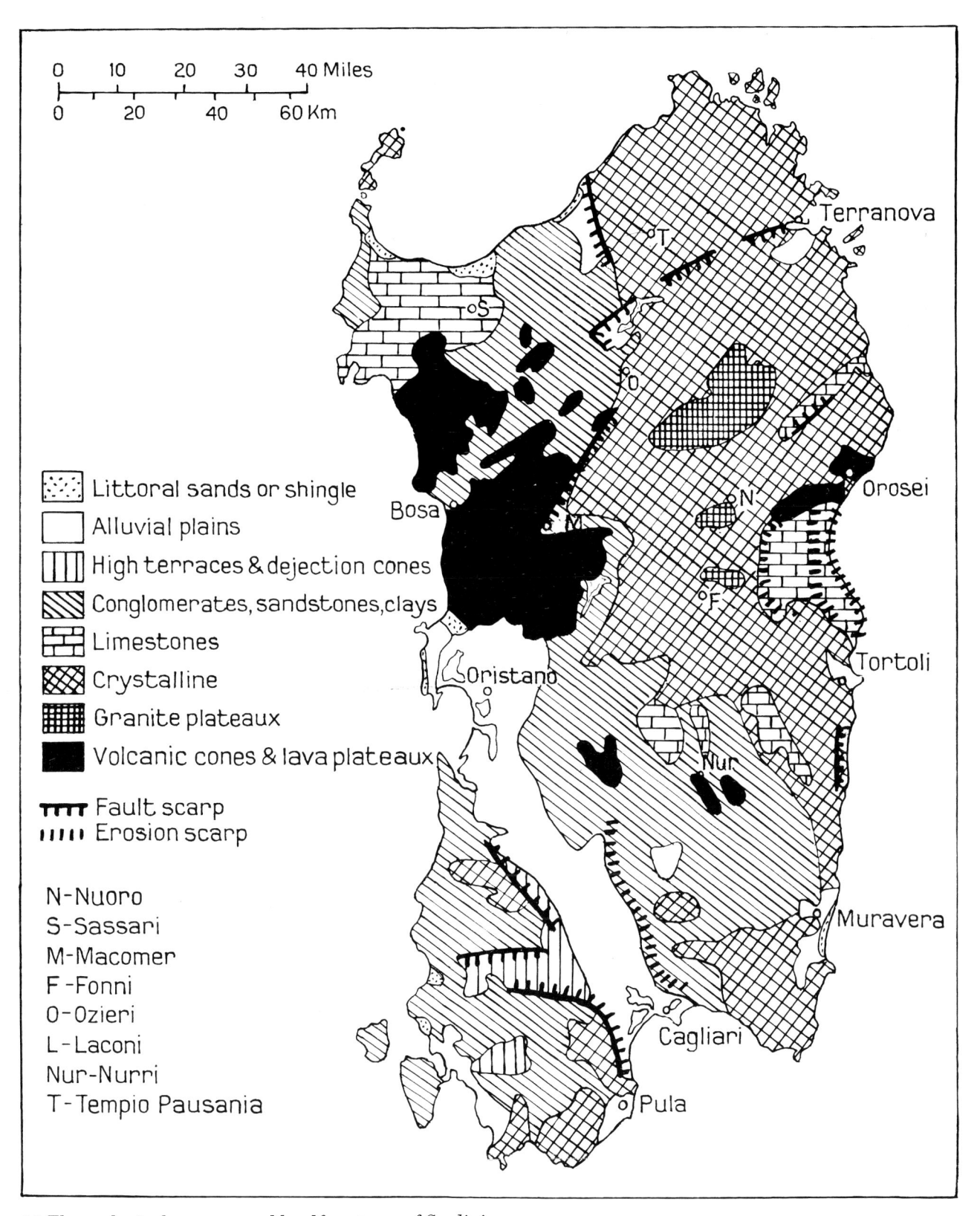

84 *The geological structure and landform types of Sardinia*

development within the aegis of the Cassa per il Mezzogiorno.

Structure and relief (fig. 84)

The island Hercynian horst of Sardinia lies about 260 km west of Naples and covers 24 000 sq km or three times the size of Corsica. Its maximum dimensions are 264 km from north to south by 96 km from east to west. Like Corsica, it is a remnant of a much larger crystalline block that has foundered to form the adjacent basins of the western Mediterranean.

This Sardinian fragment after prolonged denudation has been upthrust, fractured, tilted and much disrupted. These crustal movements were accompanied by vulcanism and earthquake activity so that lava today covers a considerable area in the west. Parts of the block suffered marine transgressions and show planed-off surfaces with a legacy of old erosion levels and sedimentary deposits that are mainly preserved in valleys and lowland pockets. Nine-tenths of the island is mountainous, including three main granitic massifs: the Eastern Highlands, the Nurra and the Iglesiente.

The *Eastern Highlands* extend from the Strait of Bonifacio in the north for the whole length of the island to Cape Carbonara in the south. They form a mosaic of continuous plateaux that lie at different heights but maintain an average elevation of 1000 m. In the northeast they are tilted below sea level to form a ria coastline and offshore islands that shelter the naval base of La Maddalena but south of Olbia the mountains rise precipitously at a fault shoreline that presents an almost unbroken wall of cliffs to the sea.

The Gallura mountain-block in the north (1362 m) is separated from the central massifs by a fault trench extending from the gulf of Terranova to the upper valley of the Coghinas river. It displays typical granite scenery of rounded massive mountains with bare summits. These domes exfoliate under weathering and the slopes are strewn with debris and the upland saddles with coarse debris. South of the Coghinas fault line the rolling granite plateaux culminate in Punta la Marmora (1834 m) in the Gennargentu massif. Some of the summits here have cappings of limestone which weather into mesa-like landforms especially west of the Gulf of Orosei. South of the Gennargentu block the granite is closely dissected by the Flumendosa river system and in the far south the barren granite dome of Sarrabus has some of the wildest scenery on the island. The summits here were glaciated during the Quaternary ice age.

On the west coast crystalline rocks are exposed in the extreme north in La Nurra and the extreme south in Iglesiente, each separated from the main eastern massifs by deep troughs. La Nurra is low (summit height 464 m) but the much larger Iglesiente often exceeds 1000 m.

Between these two horsts volcanic outbursts have overwhelmed the surface and large areas of sedimentary rocks blur the crystalline basement, forming low plateaux as in the Sassari district or mesa-like landforms in the volcanic flows behind Bosa and Alghero (fig. 84). Here an amazing cave, the Grotto Nettuno, in the cliffs of Cape Caccia has an inner lake and chambers rich in stalactites. The drainage of this central area is westward, notably by the Tirso, the longest Sardinian river, into the gulf of Oristano.

The lower Tirso opens out into lowlands which continue southward into the long fault trough of the Campidano and together they form the Arborea, the only large lowland in Sardinia. The Campidano rift valley, nearly 100 km long by 15 km wide, ends seaward in lagoons on the gulf of Cagliari. Once occupied by an arm of the sea it nowhere exceeds 100 m in altitude.

Climate

As the climatic statistics for Cagliari and Sassari show, Sardinia has a Mediterranean type of climate with a marked summer drought and most rainfall from October to December. The warm surrounding seas ensure mild winters on the coasts but on the lofty interior snow often falls in winter.

The main disadvantages of the climate are the low average rainfalls and the violence of the winds. The chief rain-yielding airmasses approach from the west and the west coast receives about 500 to 750 mm of precipitation annually. The normal annual rainfall increases to 1000 mm on the interior mountains and dwindles to about 500 mm on the east. In marked rainshadows the amount may decrease to about 400 mm. Consequently severe drought lasts for 4 to 5 months each year on the lowlands and the aridity may be unbroken for 3 months even on the mountains at 1000 m. Violent winds increase the evaporation rate and add their hazards to an effective adjustment to the climate. The squally *maestrale* from the northwest and the hot humid siroccos from the south are so frequent in early autumn and spring that reafforestation has to commence with seedling growth because transplanted saplings cannot withstand the blast.

Vegetation and land use

In Sardinia the altitudinal zoning of vegetation is less marked than in Corsica to the north. There is also a slight latitudinal change as compared with Sicily to the south, and subtropical drought-resistant plants like the dwarf palm here give way to evergreen (ilex) and cork oaks more typical of the north. Probably the drought accounts for the absence of beech and of highland species of pine. However, it is at once apparent that the natural vegetation of Sardinia has suffered severe destruction. Excessive deforestation in the past permitted soil erosion on such a scale that the bare rock has been exposed over large areas, while natural regeneration has been prevented by aridity, over-grazing, high winds and fire. Today forests have only survived in the Gallura and Gennargentu and even here much of the tree cover is cork oak that does not grow in close stands. Tempi Pausania is the centre of a cork industry. The destruction of the original forest cover has resulted in a woody growth over the uncultivated lowlands and the hillslopes below 900 m of evergreen shrubs like box and laurel. Above these are wide expanses of macchia of myrtle, heath, wild olive, juniper and tree paeony that degenerate with soil and rainfall conditions into garrigue of cistus, thyme, rosemary, sage and, in the south, euphorbia. Mat vegetation of alpine associations takes over at higher altitudes. Two-thirds of the island is estimated to be clothed in these kinds of natural growth, the proportions being as follows:

    130 000 ha of forest, of which two-thirds is cork oak
    180 000 ha of coppice
  1 135 000 ha of macchia and garrigue
    410 000 ha of unenclosed grazing

Agriculture

Despite the difficult soil and climatic conditions, half the working population of Sardinia are engaged in agricultural pursuits but most of these are pastoral as nearly half of the whole island is classed as rough pasture. Most of the arable land is only cultivated at irregular intervals and then generally to improve the grazing rather than to grow cereals. Pastoral occupations are pursued almost to the total exclusion of other agricultural pursuits and the herds include ¼ million cattle, 2¼ million sheep and ⅓ million goats, as well as 100 000 pigs. In fact, one-third of Italy's sheep and goats are to be found in Sardinia; the cattle, however, are poor dairy animals and are bred mainly for draught. The sheep are the milk yielders and large quantities of cheese are exported to the mainland along with mutton and hides. The wool is extremely coarse but the blankets, woven in Cagliari, are almost impervious to damp. The pastoral industry is carried on under feudal conditions or at best is transhumant tribalism. Ancient drove roads, called *drailles*, lead up to the plateau summits and the open pastures. Most of the sheep farmers own flocks of 50 to 300 head and some flocks are over 500 strong. The extent of communal grazing is also shown indirectly in the relatively few sheep farmers who keep less than 10 sheep. Whereas in the central Apennines 4 or 5 out of every 6 farmers with sheep keep less than 5 animals, in Sardinia only 1 in every 6 has less than 10 animals. The result is a large amount of surplus milk for cheesemaking.

The nucleated villages, lacking piped water and electric light, are situated at some distance inland away from the once malarial coasts. Around them are the open fields cultivated in strips, with vines on the hillslopes. About half the farm holdings are less than 3 ha including one-quarter that are less than 1 ha. A great many are remote from the habitations and are split into tiny fragments, often 3 to 10 in number, considerable distances apart. These patches are enclosed by stone walls (*tancas*) or by hedges of prickly pear. The few large estates are usually entirely under forest and macchia. The cultivated farmland of cereal, forage and tree crops occupies 32 per cent of the island but the small size and splintered character of the holdings and the poverty of the land indicate a primitive economy. Ploughing is usually effected by a wooden plough drawn by a bullock or an iron plough pulled by a horse.

Cereals occupy about 200 000 ha and sufficient wheat is grown for domestic needs although the yields are only half the Italian average. Better results could be achieved by the use of better seed, farm machinery and improved flour milling processes. Barley grown merely for fodder sometimes replaces wheat on the ploughland. Vines are grown around every village on slopes up to 1000 m but the output of wine is small. Olives grow wild and in the Sassari district are cultivated up to a height of 700 m but they occupy a very small proportion of the cultivated land, although the oil is of high quality and good flavour.

Improvements in land tenure and cultivation were instigated by the Fascists and continued under the Cassa per il Mezzogiorno. An important reclamation scheme has been developed in the Arborea–Campidano area which has the only mixed and intensive cultivation in the island. Up to 1948 when this scheme first came into operation 190 000 people in Sardinia were seriously affected by malaria and it was estimated that 2 million working days were lost annually because of it. Between 1940 and 1952 the main rivers of the island were hydraulically controlled so that the lower courses did not flood and the marshlands gradually diminished. Groves of eucalyptus trees were planted on these swamps to assist in drainage and to provide windbreaks. In this way artificial lakes have been formed on the river Coghinas, on the Flumendosa and its tributaries and on the Tirso where the artificial Lake Omodeo is one of the largest reservoirs in western Europe. The dams also provide electric power for the Cagliari district and irrigation water in the Campidano.

By 1952 malaria had been almost completely eliminated in Sardinia. Today reclamation schemes exist for numerous small plains and river basins involving all told 1 million ha of lowland and ½ million ha of mountains. But as yet only the 82 000 ha of the Campidano and lower Tirso have been brought under full cultivation. New settlements, Mussolinia and Terralba, have been established and 3000 peasant farmers from low-lying areas in the Veneto settled on farms varying in size up to 70 ha according to the quality of the landholding. The new settlers in all these schemes are obliged to form themselves into cooperatives for the first twenty years so that better seed and equipment will start a rising spiral of farm and living standards. The wheat yield in the reclaimed areas has already doubled, cattle breeds improved and vine stocks are producing a higher quality wine. The farming in the Fertilia area has been mechanized. Yet the fact remains that half the farmland of Campidano is still under pasture and on the other half cereals rotate with fodder crops. The normal arable crops are wheat, maize, rice, sugar-beet, beans and tobacco which shows a welcome beginning towards mixed farming. Oristano acts as the market town for the area. The sand dunes at the western end of Campidano have been stabilized while the hotter southern end is given over to salt evaporation pans.

Mining and manufacturing

Since the Bronze Age mining has provided an alternative employment to agriculture in Sardinia. The Iglesiente massif yields 80 per cent of Italian zinc ore (200 000 tons) nearly all the lead and small quantities of manganese, copper, nickel, tin and iron. Iron ore is also mined in the northwest in La Nurra. In addition two small Sardinian coalfields produced the bulk of Italy's output of coal. The Carbonia field in Iglesiente produces about ½ million tons annually and a small field near Tortoli about 1000 tons a week. Granite is quarried widely for road metal that is used throughout Italy.

Until recently the industrial growth has been insignificant, and mainly concerned with agricultural surpluses such as wool and olive oil. The non-ferrous metals are concentrated locally. Unlike Sicily, Sardinia is not self-governing in respect of education, public order, agriculture, commerce and industry. But recent progress has been considerable. The manufacture of artificial fertilizers has been established at Oschiri while at Cagliari, chemicals based on local salt, cement making, petroleum refining and miscellaneous engineering mark the beginning of a more balanced employment structure. At Porto Foxi a petro-chemical works has been opened which will eventually give work to over 10 000 people; the shortage of water has been overcome by a desalinization plant.

Population and communications

The total population of Sardinia is nearly $1\frac{1}{2}$ million and averages just over 60 persons per sq km, or one-third that of Sicily and lower than that of any other major Italian province. Town life is but weakly developed and the low standard of living is reflected in the high incidence of tuberculosis and infant mortality.

Communications in the island are improving. A standard-gauge railway links Cagliari to Porto Tórres, the north coast port for Sassari and for La Nurra iron ore. This small port also has tunny fishing and coral collecting and runs regular boat services to Genoa. Cagliari (211 000), the capital, is also linked by rail to Olbia, the packet port for sailing lines to Civitavecchia, and to Iglesias and the mining districts. Several main roads cross the island. Cagliari is remarkably large considering the total population of Sardinia. Its trade and new industries have been facilitated by the construction of an artificial harbour and regular sailings to Naples and Palermo. An air service to the mainland is also assisting the tourist industry which is increasing in spring and summer especially near the small picturesque ports of Alghero and Arbatax.

CORSICA
Historical development

The early history of Corsica shows that it shares many of its aspects with other islands of the western Mediterranean. The early settlers probably arrived from the Ligurian coastlands and, as in Sardinia, their remains consist of rock dwellings, cist graves and standing stones. A camp with massive stone ramparts may be of the same date as the Balearic talayots. Evidence of the Bronze Age dates from 700 to 600 B C.

In 535 B C the island was controlled by the Etruscans since it commanded their sea approaches and after 506 B C for a short period by the Greeks who founded a colony at Atalia. Carthage also tried to claim it but in the third century B C ceded their claims to Rome. In the Christian era, Corsica became a senatorial province and a place of exile for political offenders. Seneca spent eight years of exile here. Following the decline of the Roman empire, from A D 469 to 534 the Vandals dominated the island which provided them with timber for their ships. It then became part of the territory over which Byzantium tried to reestablish imperial authority but, although levying taxes on the Corsicans, the eastern empire was too distant and too weak to protect them, and they were plundered in turn by Goth and Lombard. In 713, the Saracens began to encroach on its territory and remained in partial possession at least until 930. Bonifacio was built as a fortress against them during this struggle. The tenth century was one of feudal anarchy and to limit this warmongering a national diet was assembled which succeeded in forming over the north of the island a republic of twelve autonomous parishes. However, it never succeeded in extending its government over the whole island nor in maintaining itself in face of constant external aggression.

In 1077 Corsicans were made subject to the Holy See under the trusteeship of the Bishop of Pisa and for a while Corsican life flourished. Unfortunately Genoa and Pisa were becoming intense trade rivals, and soon Corsica was divided between the bishops of Pisa and Genoa. In 1195 the Genoese by capturing a nest of pirates at Bonifacio gained their first foothold on the island, and for the next two centuries Genoa and Pisa each sought to control Corsica. To try to set more peaceful conditions, the pope gave Aragón sovereignty over the island and in 1325 the Aragonese so crushed the Pisan fleets that Pisa was no longer a power in Corsica.

The anarchy that followed persuaded the *Terra di Commune*, or government of northern Corsica, to offer the sovereignty of the island to Genoa, under certain conditions. Corsica would pay tribute to Genoa and be represented in its parliament; the commune of the Twelve would administer the north of the island and a newly founded commune of the Six the south. Sad events blighted the beginning of what the Corsicans thought would be a new era. The Black Death killed off two-thirds of the population and Aragón reasserted its claim to the government of the island. The Genoese founded the fortified towns of Bastia, Ajaccio, and Calvi and strengthened Bonifacio to resist the Spanish and took control of the Council of the Twelve. The Corsicans in an attempt to regain dominance of their domestic affairs then offered the government of the island to the Bank of San Giorgio, a commercial company in Genoa. The bank accepted and having driven the Spanish from the island, established a system of cruel and ruthless exploitations. No native institutions were allowed to operate, so justice lapsed and the vendetta took its place. In fact the bank fomented feuds to maintain itself in power and profit from the sale of arms. Defence was neglected and pirates rampaged unhindered round the coasts. Pestilence and floods, due to neglect, impoverished and barbarized the people. In 1559, after a short occupation by the Angevins, Corsica was taken over by the Genoese republic.

National liberation movements erupted finally but for

nearly three hundred years the Corsicans were excluded from office in their own island government. Defenceless and harassed by piracy, they withdrew from the coastal towns and villages and from the more fertile coastal lowlands. Decimated by pestilence and brutalized by the vendetta, they were mercilessly exploited by the Genoese. The population migrated in considerable numbers while the Genoese monopolized trade and stifled progress. In 1729 the Corsicans revolted and with the aid occasionally of French, British and Sardinian troops, kept up a spasmodic and relatively unsuccessful struggle. At the Treaty of Aix-la-Chapelle in 1748 Corsica was once more assigned to Genoa. Seven years later the Corsican patriot Paoli took up the cause of liberation. He united the Corsicans and, although not strong enough to eject the Genoese entrenched in the coastal towns, he established a wise and democratic government which was later adopted for the whole island. In 1768 the Genoese sold the sovereignty of Corsica to France. The French takeover was at once resented and resisted by Corsican patriots among whom was Carlo, the father of Napoleon Buonoparte, but a year later the French were masters of the island and the Corsicans in 1770 swore allegiance to Louis XV. For two years during the French Revolution the island was governed by the British under Sir George Elliot until in 1796 Napoleon took the island after token resistance by the British.

As a French department the Corsicans enjoyed a mediocre prosperity in the nineteenth century but largely because so many left the island for France and its colonies. Gradually rural depopulation became serious especially after 1860 when cheap wheat imports from the New World flooded European markets. Corsica did not keep pace with agrarian and industrial improvements elsewhere and suffered the inevitable rural decline. It is estimated that since about 1890 the land under cereals has decreased from 140 000 ha to about 7000 ha and the total cultivated from 200 000 ha to 22 000 ha.

During the 1930s the Italian fascists tried unsuccessfully to resuscitate rebellion aimed at making Corsica part of Italy. The island was occupied by Italian and German forces during the Second World War. In a decisive battle after the Italian armistice, French and Italians together drove out the Germans, who retreated to the Italian mainland. Subsequently the Italians retired to Sardinia leaving the Corsicans to enter once more into the possession of their poor, war-scarred island.

Structure and relief

Corsica lies about 170 km from Nice and 90 km from Leghorn while the shallow Strait of Bonifacio 11 km wide separates it from Sardinia to the south. It covers 8721 sq km and is at its greatest extent 180 km from north to south, and 87 km from east to west. In addition it lays claim to 43 archipelagos of smaller islets.

Structurally the island consists of three main elements, two mountain systems separated by a depression extending southeastward from the Ostriconi to the Solenzara rivers.

1 The high relief of the northeast consists of fold mountains of the alpine system, that display the same nappe formations and the same anticlinal ridges as at the Cape Corse peninsula, and synclinal valleys as at Saint Florent. The height of the mountains, and amplitude of the folds, die away northwards from 1500 to 1000 m. Low passes are frequent throughout this alpine area which is crossed transversely by the rivers Golo and Tavignano.

Along the east coast for 65 km rivers such as the Golo in the north and Tavignano in the south have built up the lowlying alluvial plain of Aleria to a width of 13 km. The seaward edge is marshy and fringed by lagoons and bay bars.

2 West and southwest of the depression are mountains of Hercynian origin which occupy two-thirds of the island. This is an upstanding remnant of a granitic block that foundered beneath the adjacent waters of the Mediterranean. Similar but smaller fragments remain in southern France in, for example, the Esterel and Maures massifs and the Canigou.

The main trend lines of the Corsican Hercynian massif run northeast–southwest and cause thirty or forty lofty ridges to run out to sea as steep headlands with deep intervening valleys. The ridges rise to a lofty crest line fronting the depression and culminate in high mountain peaks such as Monte Cinto (2710 m) and the glaciated Monte Rotondo (2625 m). These also mark the line of the island's main watershed.

Seawards the valleys and ridges plunge into a ria coastline and form the four magnificent gulfs of Porto, Sagone, Ajaccio and Valinco. At their heads the rivers have built up small alluvial plains which are unfortunately isolated from each other by the mountain ridges.

3 The depression which separates the Hercynian block from the Alpine fold area runs from the centre of the east coast near the town of Solenzaro to Corte, Ponte Leccia and to the north coast east of L'Ile Rousse. It is not drained by a single river but consists of a series of basins with intervening low watersheds that are utilized by streams for parts of their course. It is probably the only area in Corsica where soil forms to any great depth.

Climate and vegetation

Even this small Mediterranean island shows departures in detail from the main climatic type. The mild wet winters, and the hot arid summers are modified in Corsica by its insularity and mountainous character. (Tables, pp. 212–13.)

The encircling sea considerably increases the winter warmth of the coastlands and slightly reduces their summer heat so that at Ajaccio mean winter temperatures average 10°C and the hottest monthly mean, retarded by maritime influence until August, is under 24°C. The island has an average elevation of 550 m and 40 peaks above 2000 m so that all the year round there is on the mountains a large general temperature decrease due to altitude. In fact snow lies for six months on the highest peaks where it often lasts until May. As a result, there is a pronounced gradation of vegetation belts with ascent.

The usual characteristics of mountain climates are present here. Exposed peaks and mountain shoulders suffer climatic extremes diurnally and seasonally, and basins and valleys experience mountain and valley winds and temperature inversions. At the same time the rainfall is redistributed so that exposed slopes are reasonably well watered and sheltered ones lie in the rainshadow and receive below average falls. Only in very favoured localities is precipitation sufficient for crop growth throughout the year, although the snow cover maintains the groundwater of upper slopes and the flow of streams until the end of June.

Rainfall (600–1000 mm) tends to be slightly higher than average for these latitudes because the island lies in the direct path of depressions moving east along the Mediterranean during late autumn and winter. The front of the 'lows' brings hot humid sirocco airflow from the south and the rear is marked by cold, blustery maestrale winds from the northwest.

The summer drought is severe for three or four months and coupled with constant deforestation, excessive soil erosion and the poor pastoral economy keeps the coastlands up to about 550 m altitude, and over half the island as a whole, under a thick evergreen maquis, 'a strange mixture of hard and soft things, of sudden aromatic carpets of herb, of eruptions of hard grey boulders, of soft arbutus and cystus and then of spiny cactus. But mostly it feels soft, looking so moss-green, the hill-top a high ridge of green fur against the blue sky: the air smells sweet as so many odorous plants are crushed by the heel' (W. Sansom, *The South*). This expanse of maquis is broken in favoured tracts by ilex and olives and near Porto Vecchio by cork oaks that thrive on the siliceous schistose soils.

Above this lowland zone and reaching to about 1000 m are forests where chestnuts dominate on the more siliceous lands of the northeast and evergreen oak and deciduous beech on the granites. These gradually give way with height to pines which form nearly half the woodland of Corsica, the chief species being the magnificent Corsican pine (*P. laricio*). At about 1800 m a narrow sub-Alpine zone of stunted beech, fir and alder grades into the Alpine pastures which cover one-seventh of the islands. Some of these forests are private but the majority are communal and do not interfere with transhumance to the alpine meadows.

Land use and agriculture

Since the late nineteenth century the arable proportion of Corsica has decreased from over 30 per cent to less than

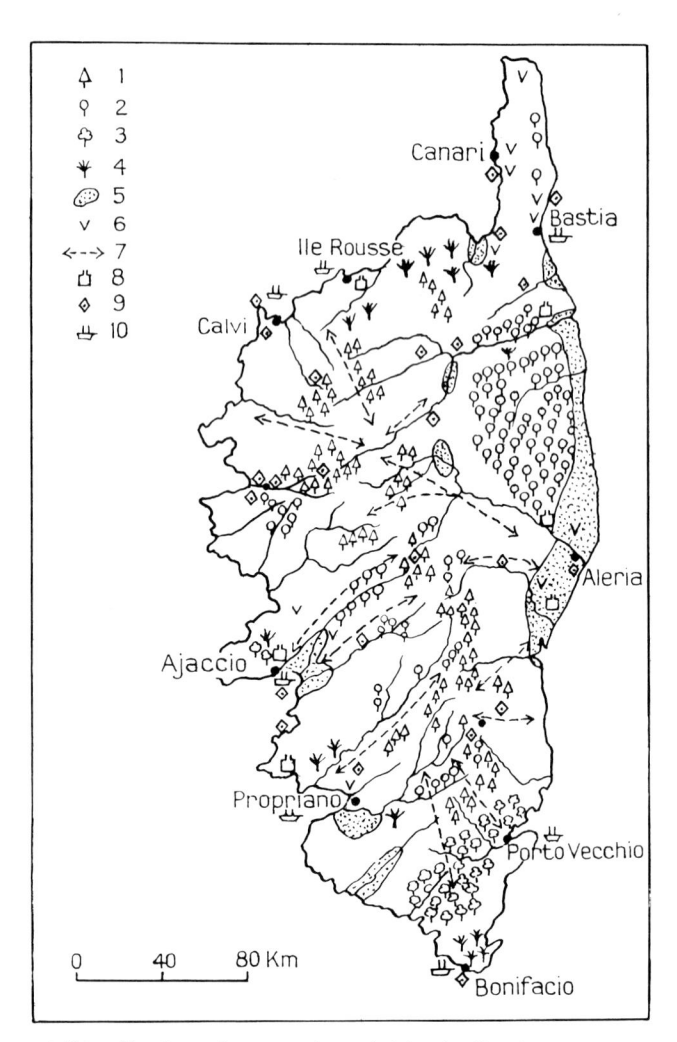

85 *Distribution of economic activities in Corsica*
   1 *main forests, largely pines*
   2 *chestnut woods*
   3 *cork-oak groves*
   4 *olive groves*
   5 *intensive cultivation and agriculture in clearings*
   6 *vineyards*
   7 *main transhumance routes*
   8 *chief milk collecting and processing factories*
   9 *industrial concerns, including hydro-electric*
   10 *ports*

3 per cent and the rural population has declined by one-half. Today cultivated fields cover a mere 22 000 ha or 2·5 per cent of the island, a proportion less even than that of Sardinia (2·8 per cent). The tillage is concerned mainly with wheat, olive and vine but some citrus, almond and peach orchards do occur and some gardens of early vegetables (fig. 85). Cereals occupy about half the arable land, with at least one year in three left in fallow. The remainder is under olive and other tree crops with vegetables grown as a polyculture between the trees. Nearly all the cultivated fields are in small rings around the villages and towns and the majority belong to holdings of less than 5 ha. Most of the olive groves are near Bonifacio and in the Balagne area

of the northwest which, because of its cultivated air, has been called 'the garden of Corsica'. Vineyards are best developed on the slopes of the Cap Corse peninsula.

A curious kind of shifting agriculture is practised in the maquis where crops will be grown on cleared patches for a year or two, and the land then allowed to revert to its original cover. Prolonged fallow (*achère morte*), which is only another form of shifting agriculture, is a common practice here wherever crops are grown. Sheep and goat herding in a transhumant pastoralism is also general, the pastures being held in common (see Chapter 7). Ewes' milk for Roquefort cheese is the main commercial enterprise, about 1800 tons of *pâté*, or 15 per cent of the French output, being produced annually. Other smaller agricultural surpluses include silk filament, hides, wool, hair, oil, wine, cork and chestnuts, the last forming an important item of the Corsican diet. The subsistence nature and the poor technical quality of the farming generally may be judged from the fact that 8000 of the 10 000 holdings are under 5 ha and only 1 farmer in 10 has a tractor.

The recent economic problems of Corsica became acute in the late nineteenth and early twentieth centuries when cheap grain from the New World rendered its cereal fields redundant. All told approximately 150 000 ha of land used at some time for cereals were gradually allowed to revert to maquis. There were also other adverse factors such as soil erosion and chronic unemployment which encouraged emigration, and ever since the early eighteenth century Corsican emigrants to France have found employment there especially as civil servants or administrators. The low standard of living and of employment on the island have since 1957 resulted in the formulation of five-year plans. The schemes for improvement include the rehabilitation of agriculture and the expansion of tourism.

The chief hope for agricultural expansion lay in the northeast coastal strip from Ghisonaccia northward to Bastia where the alluvial soils are reasonably fertile if well drained or irrigated. Here about 40 000 ha are potentially productive, including 30 000 ha which could be irrigated. Already 7000 ha have been cleared of maquis and provided with water and over 200 farms either newly created or completely reorganized. Many of these were allotted to repatriates from Algeria, most of whom instead of growing citrus, especially clementines, as was hoped, have cultivated vines. Of the water-supply reservoirs, that on the Golo river also provides hydro-electric power.

The schemes for expanding tourism involved the building of high-class hotels in picturesque coastal sites, particularly at L'Ile Rousse, Porticcio and Propiano (south of Ajaccio), and Porto Vecchio. Since 1961, with the inauguration of new car ferries and better air services, the number of visitors has doubled to over 300 000 and will probably double again within a few years particularly if the winter sports facilities are adequately developed. This influx will provide a market for at least some of the surpluses produced by the new agricultural schemes, which will specialize in fruit and cattle. It will also lessen the dependence (nearly half the labour force) on agricultural employment. Fishing can also be expanded. In 1966, the Corsican industry involved 284 boats and 675 fishermen and the catch of 775 tons was notable for the unimportance of sardines and the fine high-value landings of crustaceans (75 tons). It comprised 9 per cent of the French Mediterranean fish harvest by weight and 22 per cent by value.

Distribution of population (figs. 86 and 87)

There are considerable differences between the number of people recorded in the official statistics of Corsica and those actually resident in the island. The official returns based on the electoral lists are unreliable and many emigrants return home on polling day. However, it is certain that since 1880 the population has declined by at least 100 000, or 35 per cent. The *estimated* population was 275 000 in 1881; about 160 000 in 1954; and 716 000 in 1962, when the official returns recorded 275 000 residents. As many as 130 000 Corsicans are estimated to be working abroad, mainly in France as civil servants. This exodus is undertaken largely by younger people from rural districts,

86 *Distribution of towns and nucleated villages in Corsica*

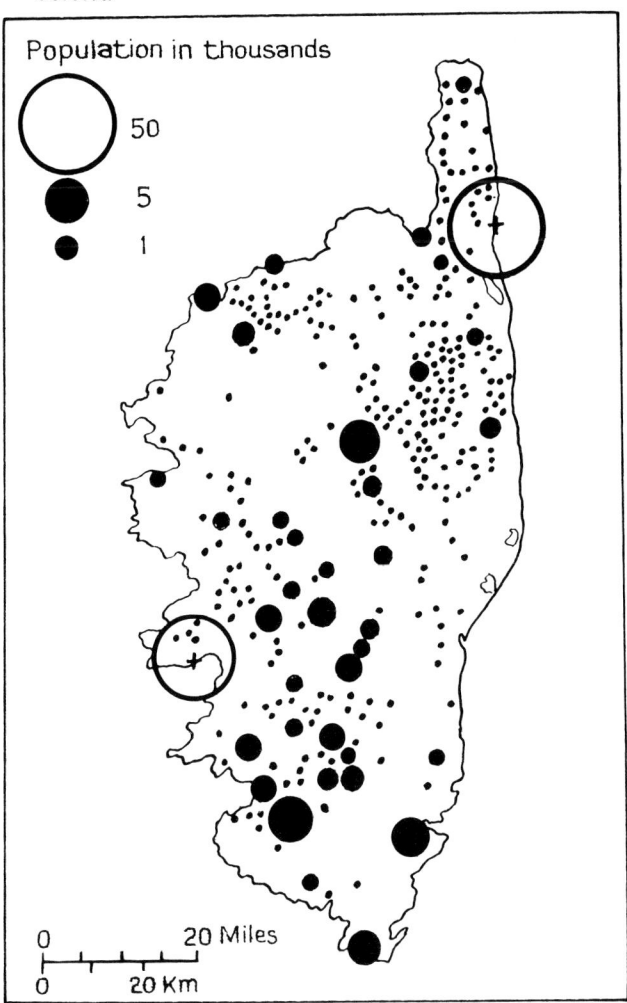

xxxiii *Corte, Corsica*

leaving in residence an increasing proportion of elderly agriculturalists and retired persons. The emigrants, however, usually retain their love and ownership of their native farmsteads and return to them maybe once a year for a holiday, or on polling day, and usually permanently on retirement. Since 1962, the population decline has been halted by the arrival of 12 000 repatriates from Algeria.

The population density of all Corsica averages only 20 per sq km, and over large areas drops to under 10 per sq km, a reminder that this is the most mountainous of all the larger Mediterranean islands. The lofty highlands have always been thinly peopled and have also separated the two areas of densest population, the Banda di dentro on the east and the Banda di fuori on the west, that are also differentiated by separate dialects.

Originally all land in Corsica was communally owned, but the mountainous terrain imposed its own isolation by separating the various cultivated areas from each other and from the pastures above the tree line. The population is today concentrated in the less marshy lowlands with their fringing slopes and in villages in mountain basins especially in the structural depression between the two main mountain chains. Four separate population patterns may be distinguished:

1 The southwest and southern coastlands and uplands with densities of 20 per sq km, distributed in the upper valleys of numerous rivers, except for a few fishing settlements on sheltered inlets along the ria coast (e.g. Bonifacio 4000). At the head of the largest inlet is Ajaccio (42 000), which grew up as the market and distributing centre for the fertile Campo di Oro. Most of its growth has occurred since 1811 when it became the capital of Corsica. Its port facilities have been improved and the harbour which is deep enough to shelter large ships is protected on the south by a peninsula and the citadel. The town is the seat of a bishopric and a prefecture, and has industries concerned with tourism, tobacco, macaroni, shipbuilding and sardine and coral fishing. It is linked by rail to Bastia, the Corsican port nearest to France.

2 The northwest coastlands, a small area near Calvi and L'Ile Rousse, with a population density of about 40 per sq km, where maritime and tourist activities add to a relatively productive agriculture in the Balagne. Here too at Calenzana a start has been made with food-processing industries.

3 The central mountain mass, a thinly peopled rugged terrain with much snow in winter on its higher tracts. Only in the mountain-girt basins in the line of the structural depression extending from the Ostriconi to the Solenzara rivers do sizeable villages occur.

4 The northeast uplands and coastlands, with population densities of 20 to 50 per sq km over large areas. Here the extensive coastal lowland was malarial until 1952 so that it was only cultivated at seasonal intervals by villagers from the mountain basins lying at 600 to 1000 m. Each community had its own coastal settlement. One of these upland basins, Castagniccia, rich as its name implies in chestnut forests, has 114 villages out of the Corsican total of 204. It was to these upland villages that the population fled for safety and sustenance during the struggles for

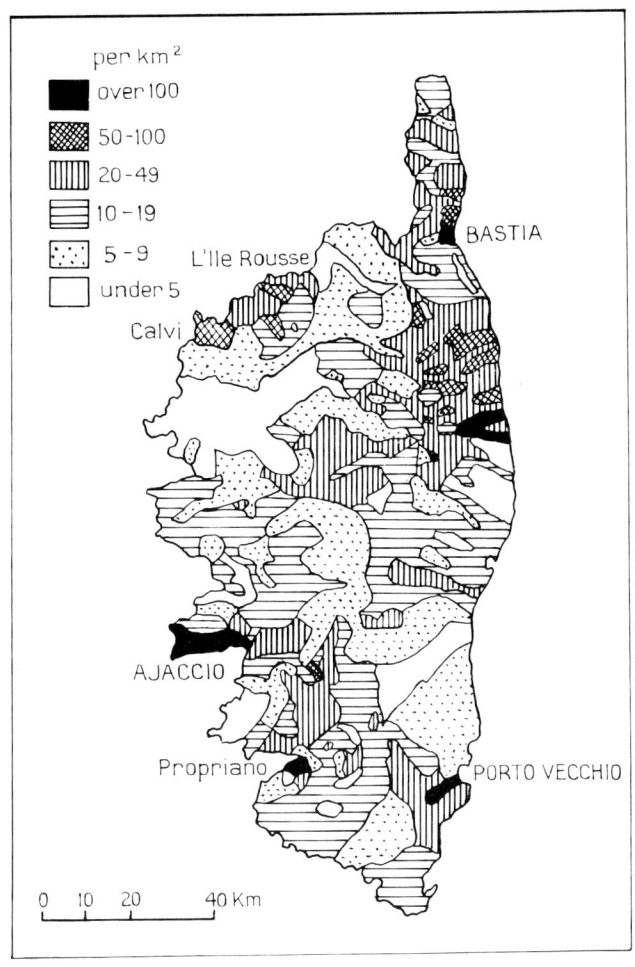

87 Corsica: density of population, average 1950 and 1960

independence. Traditional houses in this district show cow byres in the ground floor and chestnut lofts in the upper storey.

On this coast is Bastia (52 000), the largest town in Corsica and its nearest port to France. Like most of Corsica's oldest towns it has a hilltop fortified quarter, (*haute ville*) and a newer low-level quarter (*basseville*). The former, built and walled mainly by the Genoese between 1380 and 1520, has narrow, curving streets; the latter has a gracious rectangular plan with tree-shaded squares and wide boulevards, and dates mainly from post-Napoleonic times. The ancient harbour beneath the citadel has been extended by modern breakwaters and in 1966 Bastia handled 327 000 tons of cargo of which only 20 per cent were exports. It is the main entry for boats from Nice and Leghorn and has a small industrial sector which deals with most of the agricultural surpluses of north-eastern Corsica.

THE MALTESE ISLANDS
Achievement of political independence

Malta has many remains of the Neolithic and Bronze Age and later was colonized in part or whole by the Phoenicians mainly from Carthage. In 218 BC the Maltese surrendered willingly to the Romans who made Malta and Gozo a municipium with selfgovernment and a considerable measure of independence. The prosperity under this liberal regime is attested by several beautiful buildings such as the villa at Città Vecchiá. In AD 60 the Apostle Paul was wrecked off Malta and brought Christianity to the island. Publius, the chief islander, became its first Christian bishop and the present basilica in Città Vecchiá (Notabile) stands on the site of his house.

In AD 870 the Saracens, after slaughtering a large garrison, took over the island and established a fortress at Medina. They failed to popularize Islam but bequeathed to the island their language, a few customs and most of the placenames. The later incursions of other languages such as Italian, French and Greek have made Maltese a separate and rather difficult language.

In 1091 the Norman Roger Guiscard, then master of Sicily, extended his crusade to Malta where the Christians welcomed him as a liberator. He made the island a commune on the lines of an almost independent city state. Upon the extinction of the Norman line, Malta became a fief of Sicily and so eventually passed to Aragón. During the Sicilian interlude in 1223, Malta received about 3000 persons deported from the city of Célano in the Apennine Abruzzi which was razed to the ground by the Emperor Frederick II as retaliation for the revolt of its local overlord. In 1530 Charles V gave permission to the Knights of St John of Jerusalem, expelled by the Turks from Rhodes, to set up a base on Malta where they built a fortress on the south shore of the Grand Harbour. They brought with them about 4000 Greeks from the isle of Rhodes and thereafter grew rich from harrying the Turks. In 1565 Suliman I unsuccessfully tried to take the island and as a result of his siege Valletta was built. Over the centuries the Knights of St John grew less tolerant and less competent as rulers but their control ended almost by chance when Napoleon on his way to Egypt annexed the island and left a French commander in charge. French was then made the official language.

After the defeat of Napoleon, the Maltese accepted the sovereignty of the British Crown with certain provisos including religious freedom. The British gradually introduced a legislature, a judicial system and democratic institutions. Malta benefited commercially from the opening of the Suez canal in 1869.

During the twentieth century, in the 1914–18 war Malta provided many soldiers, sailors and workers for the allied cause, and in the Second World War withstood a three-year siege and aerial bombardment with such fortitude that it was awarded the George Cross. In 1947 the island acquired almost complete selfgovernment. In 1968 the British decided to run down gradually their large naval, military and aircraft bases and for the first time in their history the Maltese had hopes of eventually taking over their own

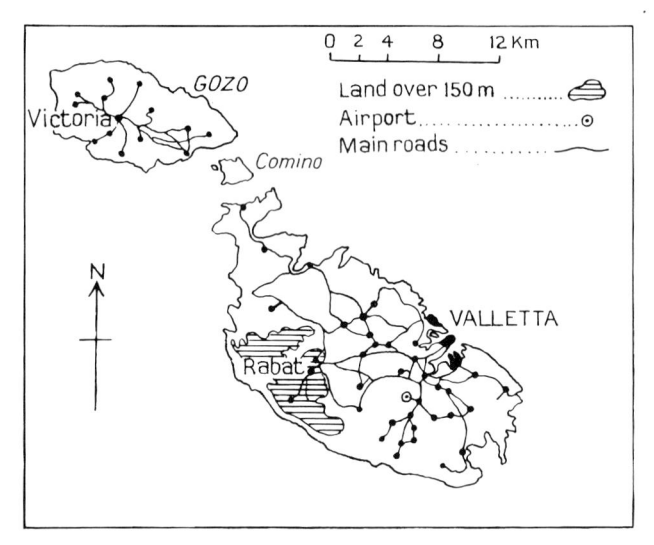

88 The Maltese Islands

insular inheritance. The bases are now financed mainly by NATO.

Physical geography (fig. 88)

The Maltese islands lie 96 km south of Sicily and 320 km north of Libya near the eastern approach to the Sicilian strait. The islands extend over a distance of 47 km and include Malta (247 sq km) and the islets of Gozo (65 sq km) and Comino (3 sq km).

Malta is built of low faulted blocks of porous Tertiary limestones that are tilted to the southeast. The geological structure consists largely of a fairly resistant bed of Corallian limestone overlying a thick bed of soft limestone. A great west–east fault crosses the northern part and forms a major escarpment that divides the island topographically into a southern two-thirds which is about 120 m higher than the northern one-third. The western part of the major southern fault block, is capped by Corallian limestone and rises to 240 m above sea level and presents spectacular cliffs to the sea. But in its eastern part where this limestone bed has been removed by erosion, the soft underlying limestones have weathered into a lower, rolling landscape that ends at the east coast in sheltered bays and inlets including the deep, spacious harbour at Valletta. The Corallian limestones of the higher western plateau include a thin bed of blue clay that throws out springs which are vital to the island's water supply as everywhere on the limestone drainage is underground. Valletta gets its drinking water from this area as well as from deep wells and recently in a big way from the desalination of sea water (see Chapter 5).

The average rainfall of Malta is about 430 mm but varies from 690 mm to as low as 300 mm and droughts lasting several years are not unknown. The rainfall comes mainly in the winter months. The average January temperatures (11°C) allow growth to proceed all the year but the summers are hot (August, 29°C) and evaporation excessive, especially as some air movement is usually present. The most notorious winds are the gregale from the northwest which is commonest at the equinoxes and often reaches gale force; and the sirocco from the southeast in summer which is a cT airmass modified by passage over the warm Mediterranean, so that it is muggy as well as hot.

Soils generally are thin and stony except in the eastern half of the island. Small patches of terra rossa do occur and are carefully preserved, as are all soils, and terraced and walled against erosion. Soil has even been imported from Sicily. Such natural vegetation as exists is xerophytic and a low-growing debased form of garrigue. The carob is the commonest tree. Despite these formidable disadvantages Malta is unique among Mediterranean islands in having broken out of the poverty or mediocrity imposed by either an animal husbandry or by cultivation based simply on wheat, vine and olive. The presence of the British naval and air bases made its influence felt at every level of island life and not least on the unpromising agricultural background.

Of the total labour force of 88 000, about 8 per cent are engaged in agriculture and fishing and 16 000 ha are under cultivation. The farm holdings are small, 4 out of 5 being less than 2 ha and the majority being under 1 ha. Individual holdings often consist of separate strips and most fields are terraced when on slopes and enclosed by drystone walls on flat areas. The hoe is the universal tool for every operation but some rotary cultivators are in successful use as well as threshing machines which make it possible for one-quarter of the farmers to find time for subsidiary employment. Half of the arable land is under a rotation of wheat and barley followed by a fodder crop of clover, or barley harvested green, or sulla, a tall leguminous crop unique to Malta. A further 39 per cent of the cropland is devoted to vegetables, especially tomatoes, onions, beans, peppers, courgettes and potatoes to supply formerly the needs of the military bases and today the increasing numbers of tourists. Considerable quantities of potatoes and onions coming onto the market early in the season are exported to Britain, and potatoes form the chief individual farm crop. Over 4 per cent of this market garden land is irrigated by either the pot method or primitive water wheels. On slopes too steep for other crops, vines for wine and table grapes are grown, and near Valletta small orchards of oranges, figs, apricots, nectarines and peaches have been established, along with ground crops of vegetables. The orchards age quickly in this climate and while new ones are established cotton crops can be harvested. The Maltese have a saying that crops grow on Malta by the grace of 'sun, sweat and sirocco', and undoubtedly the endowments of rain and soil are meagre. The cut-flower industry has been greatly stimulated by regular air services to Britain.

The animal husbandry is well managed. The breeds have all been upgraded by imports of pedigree cattle, sheep, goats and pigs from Britain. Goats are the most

XXXIV *Aerial view of Malta, showing terraced slopes and recent coastal expansion of settlement*

numerous of the domestic animals, numbering 26 000. It was here that the bacillus carried by the milk of goats responsible for Malta fever was isolated. In 1966 cattle numbered 7000, sheep 10 000 and pigs 15 000. In addition, 337 000 poultry were reared and over 40 000 rabbits, but despite the intensive land use the food produced only feeds the population for three months. Fishing provides a significant additional amount of high-class food, the catch being 26 000 tons and the personnel about 900. However there are still considerable imports of fish, notably of canned tunny and salmon from Japan.

Imports outweigh exports in the proportion of 5 to 1. The balance in the past was made up by the presence of the British bases which were the mainstay of the Maltese economy for 100 years. The naval dockyard was established at Grand Harbour and Marsamxett, and an aircraft station at Luga, and these, with an army base in addition, once provided employment for 62 000 persons as well as supporting the activities of the rest of the island. Valletta expanded with the influx of service personnel to include in its sphere the villages of Sliema (24 000), Hamrun, Paula (20 000), Cospicua and Zabbar. It has now itself a population of 19 000, but there are about 190 000 people in its neighbourhood. Since the strategy underlying the need for a British naval base at Malta is outmoded, the dockyard has been converted to civilian use for shipbuilding and repair and Valletta has been made an entrepôt with increased loading and storage facilities at the harbour. The recent closure of the Suez canal and the discovery of oil in Libya have coincided with the need for the giant oil tanker of 200 000 tons, container shipping and bulk-cargo carriers. Quick turnover in ports is another necessity in order to make these giant transport units economic. Furthermore a successful container operation involves full cargo in both directions, and therefore ports built and designed to serve these special carriers must have a populous industrial hinterland. Whether Malta could be developed as one of these terminals to serve the eastern Mediterranean remains to be seen. Valletta has an excellent deep harbour and room for expansion and development, and could become a vast free entrepôt. At present about 2562 vessels use the port every year.

Already more than 20 new factories have been built concerned with miscellaneous engineering. Consumer goods like plastics, buttons, pipes, cigarettes, furniture and hosiery are being made and light engineering connected with automobile assembly, air conditioners and insulating materials established. The fresh flower and seed industry has also been expanded.

Malta has 293 000 inhabitants and Gozo about 26 000, so their density of population is remarkable, being about 1150 for the former and 400 per sq km for Gozo. The long-standing dependency on service departments is rapidly declining, while that on manufacturing is rising. At the end of 1965, a transition year, the labour force was divided between the Malta government (17%), armed services (12%), private service departments (11%), agriculture and fishing (8%), construction and quarrying (12%), manufacturing (21%), distributive trades (14%) and transport and communications (5%). During the last twenty years, over 100 000 Maltese have emigrated, mainly to Australia (58% total), the United Kingdom (23%) and Canada (12%). On the islands most of the dwellings are small but are supplied with electricity from thermal stations and a pure drinking-water supply is now also ensured. Motor buses serve all the settlements. Tourism grows rapidly due partly to the very warm sea and abundance of sunshine, and to a strategic position on Mediterranean sea routes and European–African air routes. The policy of inducing foreigners to invest in properties and holiday homes in Malta has also had a beneficial effect on the island's economy.

III
The Balkan Peninsula

20 General Balkan Patterns

The rise of the nation states

The Balkan peninsula was at the first point of entry into western Europe of the great Slavonic invasions. Moreover along the lowlands leading from the Russian steppe to the Danube valley it was wide open to such migrations. On the east and southeast the broad embayments of the Black Sea and Aegean linked by the Bosporus invite entry from the Danube delta to the rias of southern Greece. On the northwest the valley of the Danube provides a passageway leading from central Europe, while the Postojna and Pear Tree (Hrušica) passes formed one of the most important links between the Mediterranean world and the lower Danube valley. The eastern shores of the Adriatic and the land behind it as far as the Danube was an area of great importance especially during the sixth and seventh centuries when the barbarians broke through the Roman defences. The ethnology of the Balkan Peninsula is therefore more involved than Iberia's or Italy's. At present it includes the South or Yugo–Slavs (the Bosnians, Croats, Serbs, Slovenes, Montenegrins and others) alongside the Greeks, Albanians and Turks as major groupings together with Jewish and gypsy minorities.

Two factors hindered their fusion into one nation. Firstly, the fragmented physical structure of the Balkan peninsula tended to isolate the various migratory groups so that by later historic times they had developed different languages and acquired different cultural heritages for which they desired a territorial base.

Secondly, the Romans who conquered and held the Balkans from the Danube to the Aegean, from earliest times had never pursued a policy of integration. The various conquered peoples had been allowed to keep their identities and to maintain their tribal organizations. In some cases they had their own regional autonomy. The Greeks had been confined to Macedonia since 146 BC and the four provinces of Pannonia, Moesia, Illyricum and Thrace were organized and governed in the Roman imperial manner by the planting of towns and the building of a military road network. The two frontier stations of Carnuntum between Vienna and Bratislava on the Danube and Aquincum upstream from Budapest show the splendour, quality and style of their occupation. On the islands and shores of the channels of the Adriatic coast many municipalities and military towns sprang up such as Pola, Albona, Zadar and above all Split where Diocletian built himself a vast palace with a water-supply system adequate to the city's needs to this day. The road network they planned and built remained a basis for the road network of the twentieth century and the termini they selected, Trieste, Durrës, Salonika and Istanbul are still important route junctions.

Parted from the western empire by Diocletian (AD 284–305), who was a native of Dalmatia, the eastern or Byzantine empire lasted over one thousand years but throughout that incredible length of time it lay exposed to pressures from the Slav migrations.

The original inhabitants of the Balkan peninsula, the Thracians and Illyrians had been submerged in the great wave of Slav migrations that reached their climax between the third and seventh centuries. Tumuli in various parts of the country, fortified towns, stone tablets, and inscriptions bear mute witness to these people but echoes of their tongues linger in certain speech usages of the Romanian and Bulgarian languages.

Most of the Slav movements were beyond the effective control of the emperor. In the sixth century a group of Slavs penetrated to the Morea; in the seventh century Slavonic tribes had moved to the northwest of the Balkan peninsula and settled in Croatia, Serbia, Bosnia, Hercegovina, Montenegro and northern Albania, thrusting out the older Roman colonies on the Adriatic. The emperor made peace with them and confirmed them in the possession of their lands. In the same century the Bulgars crossed the Danube and arrived in the district corresponding broadly to present-day Bulgaria. They were fierce, wild horsemen governed despotically by their khans, the leaders, and their boyars, the noblemen. After subjugating Moesia, they moved swiftly south to Constantinople and Salonika. The emperor was forced to cede Moesia to them, and having settled there, after a time they were gradually absorbed into the older, more civilized Slavonic culture. They gave their name and system of government to Bulgaria but the language, and the customs and local institutions until recently, were those of the people they supposedly conquered.

Their territory at first occupied both banks of the Danube and probably extended into eastern Hungary and even into Transylvania but with the arrival of the Magyars in the ninth century the lands north of the Danube had to be abandoned and much of the south bank evacuated likewise and left as neutral territory.

The Bulgarians therefore thrust south again and west as far as the Morava valley but although under constant threat from the north, the Balkan Mountains (Stara Planina) and the Maritsa valley remained always the heartland of the Bulgarian country. Even when Christianity arrived, in order to maintain the independence of the Bulgarian church the Bulgarians chose to follow the Orthodox rite and the Bulgarian patriarch established his see successively at Preslav, Sofia, Voden, Prespa, and finally at Ochrid (Ohrid). Under the Tsar Simeon (893–927) Bulgarian territory reached its greatest extent. It spread from the Adriatic to the Black Sea and northwest to the Sava and southwest to the Drin rivers. Preslav, the capital, then rivalled Constantinople in magnificence and the country attained 'a rank among the civilized nations of the earth'. As such it posed a threat to the authority of the Byzantine emperor who defeated the Bulgarians in battle in AD 972. By 1185, however, Bulgaria had again extended its frontiers to include the cities of Belgrade, Niš and Skopje and the lands of Albania, Epirus, Macedonia and Thrace. Trnovo (in present-day Bulgaria) was made the

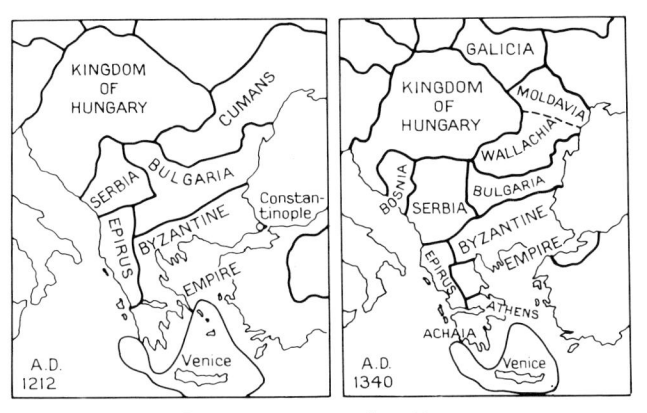

89 The Balkan States in 1212 and 1340

capital and, to judge by the art, architecture, roads and markets, the civilization and material standards were as high as anywhere in Europe at this time (fig. 89).

The Serbs in the northwest were a loosely-knit organization of tribes who for the first five centuries of their settlement in the Balkan peninsula were fully occupied in establishing their claims to their tribal lands, and in a struggle for union and centralization. At the same time a constant struggle went on between Venice and the Roman catholic church on the one hand and Byzantium and the Orthodox church on the other to gain supremacy over the Serbs. By the ninth century Serbia was united and beginning its long conflict with Bulgaria over the frontiers dividing them. Unfortunately after 1280, with the decline of the Byzantine empire, all the Balkan peoples lapsed into tribal warfare and so became an easy prey to Turkish dominion.

In 1340 the Turks ravaged the Maritsa valley (fig. 89) and in 1389 fought the decisive battle of Kosovo Polje in which a Christian league between Serbia and Bulgaria tried to halt the Turkish advance. The battle resulted in the defeat and subjection of the Serbs and Bulgarians and decided the fate of the peninsula for the next three centuries. The flower of the Serbian manhood was slaughtered and Bulgarian independence extinguished, while the peninsula was absorbed into the Ottoman empire. The first wave of Turkish conquest was devastating with towns, villages and all ecclesiastical buildings sacked and razed. Many inhabitants fled to the mountains of Montenegro and Albania or across the Danube for refuge and Turkish communities settled on the best of the land. The government was autocratic and oppressive. Men were heavily taxed, the tribute being exacted from the produce of their agriculture but as peaceful conditions returned so did the life of the peasant improve. Roads were built and commerce developed and while the Ottoman empire was secure peace at least ensued.

At its greatest extent the Ottoman empire reached to the Danube and to the outskirts of Vienna, but although that city was beseiged it was never taken and after 1683, the date of the final attempt to capture it, the Ottoman empire began to crumble (fig. 90). Its gradual withdrawal south was followed by anarchy and the Balkan territory was ravaged by marauding and guerrilla bands of soldiery.

The decline of the Ottoman empire left a power vacuum which presented problems for the four great powers with vested interests in the area. The Russians were anxious to have access to the eastern Mediterranean because as long as the Turks controlled the Bosporus Russian egress from the Black Sea was limited. Austria saw

90 The Balkan States in 1520 and 1800

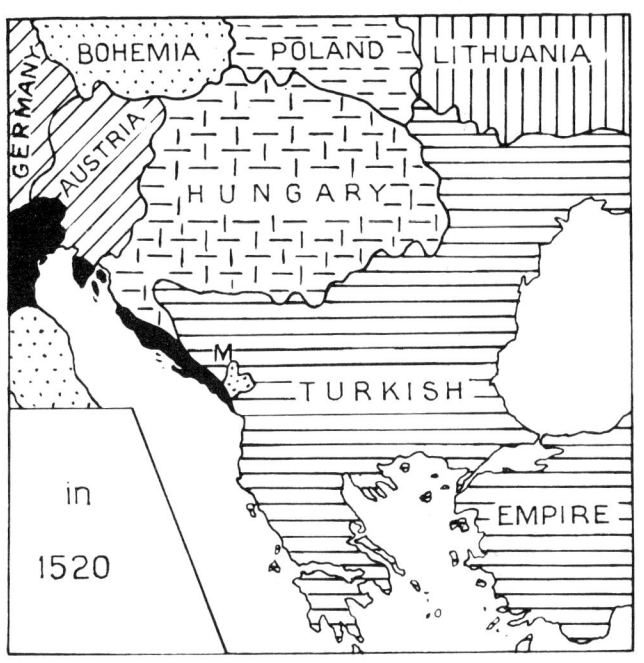

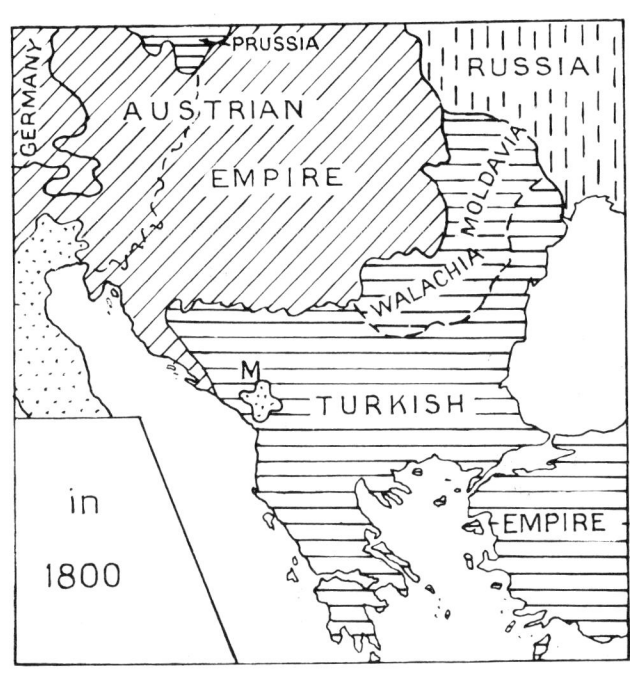

in the retreat of the Ottoman empire an opportunity also to extend its influence south and to gain access to the Mediterranean preferably by acquiring territory with a frontage on the Adriatic. France and Great Britain also were interested in maintaining peaceful commercial relations in the eastern Mediterranean.

For the greater part of the eighteenth century independent nationalities began to develop in the Balkans and in the nineteenth century, the 'springtime of nations', those peoples who had retained their language, religion and customs acquired a more acute national consciousness and gained their independence (fig. 90). Greece became independent in 1829, Serbia in 1833, Bulgaria and Montenegro in 1878, though not without prolonged rebellion and bitter wars against the Turks.

The British expedition to the Crimea had thwarted Russian ambitions temporarily in the Bosporus. Later when the Bulgarians rebelled against Turkish rule in 1876 they were assisted by the Russians who were able to dictate the peace of San Stefano (1878) and so gained for Bulgaria a much greater territory than it could reasonably claim. The country stretched from the Black Sea to Lake Ochrid and from the Danube to the Aegean. The Russians hoped they themselves had thus acquired access to the Mediterranean.

The Berlin Congress in the same year however adjusted the situation so that Bosnia and Hercegovina in the northwest of the peninsula were given to the protection of Austria–Hungary. The independence of Montenegro was ratified and the Bulgarian southern frontier moved back from the Aegean and the area south of it restored to Turkey. The Berlin Congress settled nothing. Austrian and Russian ambitions were still unsatisfied while Greece aimed at liberating Thessaly and Macedonia.

In 1908 Austria claimed complete sovereignty over Bosnia and Hercegovina. In 1912 the countries of Greece, Serbia, Montenegro and Bulgaria formed the Balkan League to drive the Turks out of the Balkan peninsula and left them finally with a rather larger area than the present Turkey-in-Europe, which is all that remains of the vast European Ottoman empire. But, this accomplished, the members of the League began to quarrel among themselves. Serbia, an inland country, wished for an access and seaport either to the Aegean or the Adriatic. The lower Maritsa valley, the obvious way to the Aegean, was claimed by Greece, and the Drin valley, the obvious route for Serbia to the Adriatic, ran through the territory that is now Albania.

The 1913 conference in London which settled the dispute confirmed Greece in the possession of the lower Maritsa Valley and made Albania into an independent country. A second Balkan war broke out and at its conclusion Greece acquired much territory at the expense of Turkey, Bulgaria had access to the Aegean, and Serbia gained territory at the expense of Macedonia. Bosnia and Hercegovina remained Austrian and Montenegro and Albania were unaltered.

Discontent again erupted in the Balkans during the 1914–18 war. Towards its end at the Corfu conference the Serbs, Croats and Slovenes decided to revolt against Austrian rule and to unite the southern Slavs into one state. This was the beginning of the federal state of Yugoslavia in which the republics of Bosnia–Hercegovina, Serbia, Montenegro, Macedonia, Croatia, Slovenia, Vojvodina and Kosovo–Metohija finally became a peoples' democracy. The boundary between Austria and Yugoslavia follows the crest line of the Karawanken Alps, that with Hungary the line of the Drava and Danube rivers, but elsewhere as far as possible ethnic boundaries were drawn following plebiscites. The peace treaty at the end of the Second World War (1946) gave the Istria peninsula and the port of Fiume, now Rijeka, to Yugoslavia, and Bulgaria ceded her territory along the Aegean to Greece and became the small, compact and ethnically homogeneous country facing the Black Sea. Each Balkan state now faces one of the three seas surrounding the peninsula while Yugoslavia has free trading rights in the port of Salonika giving the interior country of Serbia an outlet finally to the sea.

The interesting post-1946 history of these countries is discussed in the individual chapters dealing with them. Suffice it to say that none is a member of COMECON as Albania withdrew from membership in 1963 but in 1964 Yugoslavia became an observer of that body and so is allowed to participate in certain spheres of its activity. Greece in 1961 became an associate of the European Economic Community or Common Market.

THE GEOGRAPHICAL BACKGROUND
Physical structure and relief

The word Balkan is Turkish and means 'mountain'. Today confined to a single range running from east to west, south of the lower Danube, in the distant past it referred to the great mass of mountainous country that extends across the peninsula from the Adriatic to the Black Sea. North of the mountains lived the barbarous peoples of Pannonia, south of them the maritime nations of Macedonia, Thrace, Hellas and Thessaly.

Although this peninsula has a very different physical background it nevertheless contains the same main structural elements as Iberia. Two separate areas of alpine folds here flank peneplaned Hercynian blocks which lie at the core of the peninsula (fig. 91). This Hercynian massif, much fractured and compressed, consists of three separate entities. The most northerly or Pannonian block has foundered and underlies the lowland drained by the Danube and its tributaries, the Drava and Sava. It formed in Tertiary times the site of a great body of water that has since drained via the magnificent gorge cut through the fold mountains at the Iron Gate. Old shorelines can be traced at various places throughout the basins and the

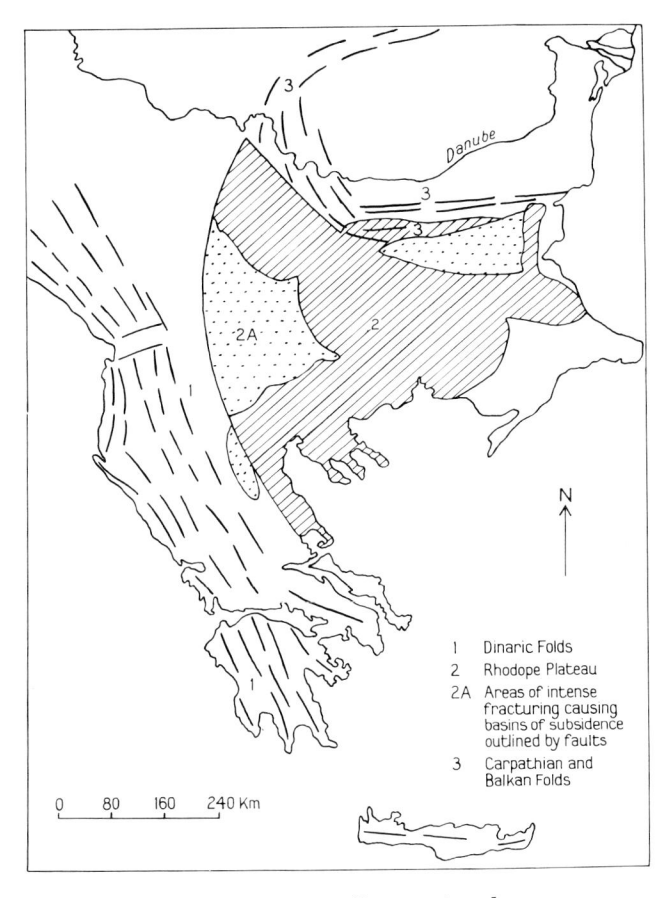

91 *General structure of the Balkan peninsula*

basement of crystalline rocks pierces the newer sediments in the high country between the Sava and the Drava rivers.

South of the Pannonian block is the upstanding Rhodope plateau that forms the heart of the country. Triangular in shape with its apices in Belgrade, Salonika and Alexandroúpolis, and its base in the rugged upland between Macedonia and Thrace, it is the main reason why two-thirds of the Balkan peninsula lies at over 500 m and half at over 1000 m. Formed of metamorphosed schists with large outcrops of granite, gabbro and serpentine it was greatly disturbed by the Alpine foldings. Evidence for this abounds in the great east–west faults as exemplified in the west–east courses of the west Morava river and of the Maritsa–Tundzha trough; in the hundreds of fault basins along the various rivers, and, to use positive landforms, in the famous horst mountains of Olympus, Ossa and Pelion. Vulcanism sometimes accompanied the fracturing and disturbances which continue actively today as in the Skopje earthquake of 1966. Furthermore the wide northern area which lies between the Dinaric Alps and the curving sweep of the Stara Planina (Balkan Mountains) is furrowed by a great tectonic trench that is drained to the north by the Morava and to the south by the Vardar. In fact the whole plateau south of Niš may best be described as a fault landscape in which are numerous small basins (*graben*) floored with fertile Tertiary and more recent deposits.

The Aegean Sea covers a third portion of the vast Hercynian block which here has foundered save for islands such as the Cyclades and Northern Sporades. However, it reappears in the large tableland of Asia Minor.

Alpine fold system. Fringing the Aegean foundered block on the south are the unsubmerged remnants of the Alpine folds, as in Crete and Rhodes, that continue eastward into Asia Minor as the Taurus mountains (fig. 91). The main Hercynian or median mass of the Balkans has on its western or Adriatic flank the Dinaric Alps and their lofty prolongations into the Peloponnesus. Of 1500 m average height and attaining 2500 m locally, they falter in their barrier function in Albania only, where they permit the development of surface drainage and lowlands to fringe the Adriatic.

Reduced to a peneplain and thickly covered with sediments in mid-Tertiary times, this Alpine fold area was later elevated by a general epeirogenic movement that however left the folds alongside the Adriatic submerged, and these today form the world's finest example of a Dalmatian coastline. The unsubmerged crests of the longitudinal folds are the long islands fringing the coast, and the synclinal downfolds occupied by the sea form straits or canali with their characteristic right-angled entrances. Inland the mountains create a most effective barrier accentuated by the karstic surfaces developed on the limestones of which they are largely built.

Alpine fold mountains intrude again on the northeast where the Carpathian system swings south and east in a great curve to the Black Sea. Pierced at the Iron Gate by the Danube, they are the only mountains today to which the Balkans have given their name but are also called the Stara Planina. With an average height of 1350 m they rarely reach summits of 2000 m and have numerous cols at less than that altitude, including the notable Shipka pass. Their folds are eaten into on the Black Sea coast where they present high cliff walls.

Although the same structural elements as occur in the Iberian peninsula are present here even to the sub-alpine triangular fault trough, their arrangement in the Balkans is such that no real barrier exists between the peninsular countries and central Europe. In fact two important international routeways traverse the peninsula following the Morava–Vardar corridor to Salonika and the Isker–Maritsa valleys to Istanbul. Furthermore the northern land link is extremely wide stretching through 1200 km from coast to coast. Continental influences penetrate therefore more intimately than in Iberia and peninsular Italy, and this is especially so from the point of view of climate, vegetation and ethnic composition.

Added to this is the fact that the peninsula disintegrates in the south into a series of many small peninsulas, isthmuses, fragmented coastal areas and islands so that a northern continental area made up of units largely isolated from each other adjoins a southern mosaic of small plains and mountains deeply penetrated by the sea. The Balkan peninsula presents two faces to the world, the one central European and continental, the other Mediterranean and maritime.

Climate. The difference between the continental and maritime aspects of the Balkan peninsula is most strongly revealed in its climate. But the maritime influence comes mainly from the west and south. In winter when the interior is cold and the Mediterranean relatively warm, a sequence of cyclonic depressions passes southeastward down the Adriatic. Most of them cross southern Greece and may proceed up the Aegean through the Dardanelles. The airmasses in front of these depressions may come from over the Mediterranean as southerly airflow and be warm and wet. However the rear of the 'lows' may bring cold air from over the Balkan interior or even farther north and violent chilly weather ensues with bora winds. Equally in winter cold air moves outward from the plateau and the Danube plains southward and the bora of the Adriatic has its equivalent northerly wind in the Aegean known as the vardarac, which often brings cold, dry and sunny weather (see Chapter 4).

In spring and summer the Balkan interior and especially the Danube plains heat up quickly and the relatively low pressure encourages cyclonic depressions to pass from the head of the Adriatic due eastward to the Danube lowlands and bring spring and early summer rains. But then very few depressions pass southeastward down the Adriatic which in summer experiences a strong decrease in rainfall.

There is in the hot season another important factor at work. The very low pressures over the Persian Gulf area then attract airflow from the north and this movement, often as strong northeast etesian winds, is characteristic of the eastern half only of the Mediterranean. The Aegean is especially the realm of the etesians.

When the results of the complex interactions of the major pressure systems and airflow are combined with those of regional relief and local landforms, the following vague climatic regions may be distinguished.

1 *Central European type of climate of the interior.* This covers a large area in the north and even penetrates into northern Greece. It is characterized by winters with the coldest month with means below freezing point and frequent snow cover, and by a maximum of rainfall in the warm season.

Its regional variations include:

- *a* the Danube plains where the Eurasian influence is strongest and can bring very cold (cP) spells in winter and hot tropical spells occasionally in summer
- *b* the interior mountains and plateaux where altitude increases precipitation and the length of winter cold and snow. Here local wind direction depends partly or largely on the orientation of the relief

2 *The Dinaric Alps or Western Highlands.* Here the precipitation, cloud cover and frequency of raindays are quite exceptional for the Balkans and for the Mediterranean. Over large areas the average rainfall exceeds 1200 mm annually and many districts have from 2000 to 3000 mm. The greater part of the precipitation falls in the cooler months of the year, particularly in northwestern Greece.

3 *The Mediterranean Coastlands.* The coastal margins of the Adriatic south of Istria all the way to the southern peninsulas and islands of Greece, and the coastal lowlands of the Aegean have a Mediterranean climate with a summer drought and cool-season maximum of rainfall. Yet these coastal strips show appreciable climatic differences. The Adriatic coast, although getting drier southwards, remains decidedly wet in autumn and winter and has a relatively short but strong summer drought and relatively small seasonal range of temperature. The southern Aegean coastal lowlands are also truly Mediterranean but the winters here tend to be cooler and the annual precipitation less. The plains of Thessaly have a climate transitional to a north Aegean Mediterranean type in coastal Macedonia and Thrace. Here the temperatures and precipitation show distinct interior or continental features. Mean January temperatures remain well above freezing point and the olive grows in sheltered places but generally winters are quite chilly with frosts common and occasionally quite severe. The summers have appreciable rainfall (20 to 30 mm average in driest months) but not enough to break the drought. Farther north on the coastal strip of the Black Sea in Bulgaria the continentality increases still further, partly due to cold winds from the east, and the annual precipitation decreases. But there is still a drought in summer just severe enough to qualify for inclusion with a Mediterranean type. It is noticeable that the olive will fruit on all the coastal strip of the Aegean but is not grown on the Black Sea littoral.

Vegetation and land use

The complexity of the relief and climate has produced a similar entanglement of vegetation. In this peninsula Asiatic and central European types mingle with Mediterranean flora. At the same time there exists a considerable number of endemic plants indicating that most of the Rhodope plateau escaped glaciation during the Quaternary period. The peninsula therefore has one of the richest floras in Europe (see Chapter 6). That it is so relatively poor today in tall forests reflects a strong biotic influence which will be described later.

The present vegetation associations closely reflect the climate. All the coastal strip with a Mediterranean summer drought supports, or used to support, a typical association of evergreen oak and pine (Aleppo, stone, etc.). Inland on the wetter west coastal flanks where altitude increases the winter coolness, and on the drier east coast in Thessaly,

Macedonia and Thrace where frosts are frequent in winter, a mixed forest mainly of evergreen and deciduous oaks prevails.

Everywhere away from the coastlands, the interior plains, plateau and ranges are, or used to be, clothed in deciduous oakwoods with beechwoods on the higher and wetter summits.

The plant associations show the usual gradations or zoning with altitude but their limits are much higher in the south and on south-facing than on north-facing slopes. Thus in Yugoslavia the alpine herbaceous communities above the treeline often begin at 1800 m in the north, 1900 m in the centre and about 2000 m in the south. Some details of these distributions and of the general flora have been given in Chapter 6, and further details will be given under the accounts of the individual countries. Here, however, we must reiterate the tremendous effect of man and his animals on the Balkan landscape. As in Spain, most of the plant cover is manmade and tall trees, and small edible animals, are mainly conspicuous by their absence. The forest clearance began in a big way in prehistoric times and was pursued actively by Phoenicians, Greeks and Romans. The Slav invasions of the sixth and seventh centuries brought a pastoral people. lovers of open spaces, with enormous flocks and herds. Ever since, these native pastoralists have been aided in the destruction of tall forests by exploitive invaders, such as the Venetians who needed timber for their flotillas. Vast areas of forest have been degraded into bushgrowth and inevitably soil erosion has caused natural regeneration to become slower and slower until the biotic influence has virtually prohibited it. Modern reaction to this misuse of protective vegetation cover is seen in recent attempts at re-afforestation and in the creation of forest reserves. Perhaps the greatest lesson that the classical civilizations taught mankind is that the natural environment is rapidly destroyed. The barren and bush-clad expanses of the Balkans are a sign of cultures ignorant of the elementary principles of land conservation.

Today the visitor may well be amazed at the vast monochromatic expanses of low bushgrowth, which, for example, covers more than half of Greece. In the Balkans, it can be classified into six categories:

1 Maquis, or macchie, consists largely of evergreen shrubs and represents the former domain of Mediterranean evergreen forests. In Greece it lies mainly below 800 m. Its taller growth contains the oleander, tulip tree, kermes oak, myrtle, olive and strawberry tree.

2 Phrygana is a lower, sparser growth, particularly of aromatic shrubs such as thyme, lavender, rosemary, cistus and broom, that replaces maquis on the drier limestone outcrops and on the heavily overgrazed areas. At a height of 700 to about 800 m it gives way to a denser deciduous–evergreen bushgrowth. In the western Mediterranean phrygana would be classified as garrigue.

3 Pseudo-maquis or pseudo-macchie dominates inland and upward of the maquis proper and often replaces maquis in lowland Macedonia and Thrace. It consists of a mixture of hardier evergreen and deciduous species, including box, cherry laurel (*Prunus laurocerasus*), jasmine, juniper and various oaks. It lacks myrtle and the olive but is often quite as luxuriant as the maquis.

4 Shiblyak is a Serbian word denoting deciduous brushwood. This shrubgrowth occurs widely in the hilly country of the interior Balkans, including Thessaly, Macedonia and Thrace. Among its dominant species are oaks, Christ's thorn, berberry and sumac, many of which will grow to a height of 2 or 3 m.

5 A more natural form of shiblyak occurs widely on the deforested karstic highlands particularly in Yugoslavia where a low bushgrowth is dominated mainly by deciduous oaks and manna or flowering ash.

6 On the high mountains above 1500 m sub-Alpine stunted bushgrowth of beech and various conifers occurs especially on crystalline or non-calcareous rocks. It has been greatly extended by deforestation and often becomes dense bushes similar to shiblyak but more naturally stunted.

All these forms of bushgrowth are useful for pastoral purposes and, as in the Apennine *bosco* and Iberian *matorral*, are often used as rough herbage for livestock. What has recently been said of Kokkinopolis, a badland in northwestern Greece, has a wide application. 'All the thickets in this area are intensively browsed by the goat herds that pass daily across Kokkinopolis on their way to and from pasturing on the nearby hillsides. The animals crop the shrubs including the thorniest kermes oak bushes, into low pyramidal mounds from which only an occasional leading shoot succeeds in growing above browsing level. Even the trees are not exempt from the depredations of the goats which mount any conveniently low side-branches, install themselves into the crown of the tree and strip all leaves within reach' (D. Harris & C. Vita-Finzi, 1968).

The land use by countries, given in the table in Chapter 7, shows clearly the difficulty of deciding what is forest and what is pasture. The arable or permanent crops are more easily delimited. In 1966, 32 per cent of Yugoslavia, 29 per cent of Greece and 18 per cent of Albania were under sown crops, These statistics, however, do not fully emphasize the fact that large tracts of Yugoslavia lie in the Danube plains, and indeed that this country belongs mainly to the Danube drainage. The general land use of the whole Balkan peninsula can in Europe be only rationally compared with that of Iberia which is slightly larger (59 000 to 53 000 sq km). Forty-two per cent of Iberia is cropland against 33 per cent in the Balkans, including Bulgaria. For livestock the following table demonstrates the relative pastoral qualities and predilections of the Balkan and Iberian peninsulas, the statistics being in thousands for 1966, with the human population added for purposes of comparison.

	Bovines	Ovines	Caprines	Equines	Humans
Iberia	4 606	24 578	2 897	1 909	41 206
Balkans	8 547	29 728	5 719	2 929	38 520

The extra 13 million beasts in the Balkans make a noticeable difference to the animation of the countryside.

Human patterns

The strong regional differences in the physical landscape and ease of access for invading peoples by land from the Danube plains and by an island-dotted sea have favoured a complex mosaic of different races, religions, languages and cultures. The Turkish domination although prolonged was never able to fuse these into one nation. The large majority of the 39 million people are Slav with Asiatic minorities such as Turks and Tartars (Tâtars). Minority groups of other racial make-up occur, for example, in Vojvodina with its enclaves of Czechs. However, these minority problems have greatly lessened in the twentieth century and the multiplicity of languages remains a more serious problem for intercourse, particularly in Yugoslavia. The general pattern of racial and linguistic distributions is discussed in Chapter 9. The recent cultural and political developments emphasize the differences in regional consciousness. Yugoslavia, a federation of different-speaking provinces, follows its own brand of Soviet socialism; Albania prefers a Chinese version; while Greece leans politically toward the western democracies. Clarity is best served by dealing with the present economy of each country separately.

21 Yugoslavia

The present Yugoslavia originated late in 1918 when the Serbs, Croats and Slovenes joined together as one state, which territorially comprised Serbia, Montenegro, Croatia (a former province of Hungary), Slovenia and Dalmatia (former provinces of Austria) and Bosnia–Hercegovina. The boundaries were for the most part fixed at the 1919 Paris Peace Conference and the only notable changes since have been in the northwest where in 1947 the Istrian peninsula and most of the Julian Alps were provisionally added to Yugoslavia. In 1954 this territory, together with a small southern suburb of Trieste, was permanently recognised as Yugoslavian (fig. 66). Yugoslavia was welded more closely into a nation by the struggle of the partisans against German and Italian aggression during the Second World War. In 1945 the monarchy was abolished and a republic set up under Marshal Tito who in 1963 was made president for life.

The Socialist Federal Republic covers an area of 255 804 sq km and has 20 million inhabitants mainly of Slavonic racial and linguistic affinities. It unites the six republics of Bosnia–Hercegovina, Croatia, Macedonia, Montenegro, Serbia, and Slovenia. The Republic of Serbia contains within its frontiers the two small autonomous areas of Vojvodina (beyond the Danube in the northeast) and Kosovo–Metohija around the headstreams of the river Drin where there are considerable minorities of Hungarians and Albanians respectively.

Structure and relief

The major physical features of Yugoslavia represent the main structural elements of southern Europe. Here two lines of Alpine fold mountains, the Dinaric Alps and southern Carpathians come close together before the Carpathians swing eastward as the Balkan Mountains. To the south of this conjunction lies the Rhodope plateau, the Hercynian core of the Balkans; to the north of it the Hercynian massif has subsided and is now covered by the broad Pannonian plain. Within its frontiers therefore Yugoslavia has two kinds of Alpine chains, an older to the east and a younger on the west, and two aspects of the Hercynian blocks, the subsident and the fractured uplifted types.

Three-quarters of the country is mountainous. On the west, occupying one-third of Yugoslavia, the Dinaric Alps extend from the Alps proper through Slovenia and Montenegro and interpose a high barrier between the Adriatic and the interior Balkans. The highest point is Triglav (2863 m) in the Julian Alps but another 140 peaks exceed 2000 m. On the northeast the Carpathians thrust across the Iron Gate and continue beyond the Danube as the Balkan Mountains or Stara Planina. However only a small part of this range is in Yugoslavia and the boundary along the crestline reaches 2000 m at one spot only and usually lies below 1200 m.

Both Alpine chains display bare limestone topography but the High Karst overlooking the Adriatic develops karstic landforms on a scale unequalled elsewhere in Europe. With a width of 80 km and length of 600 km, it rises sharply to over 2000 m in folded ranges that have been peneplaned into rather bare stony plateaux (*planina*) that alternate with elongated basins (*polja* or *polya*). The northwest to southeast relief and structural trends stand athwart the prevailing rainbringing winds and give exceptional falls, locally of over 3000 and 4000 mm. Being built of unusually porous, dazzling white limestone, it is so devoid of surface drainage that the only river to cross the mountain range is the Neretva which does so because it has cut down to the water table. A network of subterranean water channels and dripping caverns with stalactites and stalagmites lies underground while the waterless stony surface yields only a thin soil that supports a meagre pasture for herds of sheep and goats. The caves at Postojna are world famous. A thicker soil has accumulated in the polja or longitudinal depressions (polje means field in Serbo–Croat) and these are defined by a cover of grey-green shrubs. The polja tend to fill with water when the water table rises in the late winter and early spring and this limits the uses to which they can be put. An example is lake Cerknica which becomes a lake in spring and autumn, is grazed by cattle in summer and is covered by a sheet of ice in winter. Lake Skhoder (Scutari) is a polje permanently filled with water lying at the level of the water table. Solution hollows in the limestone form doline (singular dolina) and these are usually floored with a bright red soil, terra rossa, which is much more fertile. Both types of depression usually display sporadic patches of cultivation in an otherwise poor landscape. Similar areas of karst appear again in the Alpine mountains beyond the Morava valley.

The Rhodope massif that makes up the core of the country is traversed by major fractures running northwest–southeast some of which are followed by great rift trenches and drained by rivers like the Morava and the Vardar. The plateau is further cut up by fractures and fault-lines at right angles to the main trend lines so that the country presents a landscape of mountain basins and blocks, or graben and horsts, lying at different levels and linked by gorge sections.

Here, and in adjacent territories, earth movements are still active and earthquakes, probably caused by uplift or sinking on fracture lines, are relatively common. Earthquakes largely destroyed Skopje in 1963 and seriously damaged Banja Luka in November 1969.

The Pannonian lowland in the north of Yugoslavia is a remarkable gathering ground of rivers. The Danube flows through Yugoslavia for a distance of 590 km from near Mohács to the Iron Gate where it forms the frontier with Romania. In its passage through the country it is joined by the Tisa, or Tisza, from the northern Carpathians, the Morava from the Balkans in the south, and the Drava and Sava from the Slovene Alps. Thus the Danube drainage links up the main relief elements of the country

xxxv *View of the Dinaric Alps and Yugoslavian karst*

and the master river was the obvious site for the capital city, Belgrade.

Climate

Less than one-third of Yugoslavia is below 200 m and over half the country is above 300 m, with the result that the influence of relief on climate is strong. The chief source of winter warmth and of precipitation is the Adriatic which is largely excluded from influencing the Balkan interior by the great barrier of the Dinaric Alps except near the Postojna gap where opposite Rijeka the chain narrows and lowers. This wide mountain saddle encourages Adriatic 'lows' to move eastward especially in spring and summer. But it also allows cold airflow in winter from the great continental anticyclone to rush into the Mediterranean. These cold easterly winds in the Danube plains of Vojvodina bring devastating chill and are called the kosava or scythe; along the Adriatic coast they are known as the *bura* (Italian, bora) which may blow violently for days on end and bring communications to a halt. These interesting weather situations and local winds are described in Chapter 4. The climate may be judged from the table of statistics which have been selected to represent all parts of Yugoslavia.

Mediterranean summer-drought climate is confined to a narrow strip 25 to 50 km wide commencing some distance south of Rijeka which has a climate transitional between that of the north Italian plain and the true Mediterranean peninsulas. Mean January temperatures along this coast are from 6° to 8°C but they drop rapidly inland with altitude and northeastward where at least one month averages below zero. Spring arrives early on the Adriatic coast where flowers are everywhere in bloom by March whereas in the more continental interior plant life begins to revive about a month later.

Summer temperatures are high throughout the country, the July means being about 22° and up to 25°C in the south. The Adriatic coast is hotter than would be expected for a maritime location in summer, partly because the shallow sea warms up to 23°C and partly because the dazzling white limestone slopes devoid of surface water heat up rapidly under the summer sun. This coast has an exceptionally high number of cloudless days.

The large expanses of Yugoslavia lying above 1000 m have a distinct mountain influence with increased diurnal and seasonal temperature ranges. Temperature inversions prevail during calm weather in all the thousands of mountain hollows and with the multitudinous variations of site

and aspect cause a like multiplicity of local climates. But two generalizations hold true for all:
1 winter temperatures increase in length and severity towards the northeast
2 precipitation beyond the main coastal ranges decreases rapidly towards the east

Annual rainfall amounts on the coast are quite exceptional. Rijeka, in the full path of cyclonic depressions moving eastward and southeastward, has over 1500 mm. Southward a Mediterranean regime takes over but the annual precipitation exceeds 800 mm. On the limestone mountains the totals increase to over 4600 mm and in parts near Kotor to over 5000 mm. Eastward of the coastal slopes, the falls decline to under 700 mm on the lower parts and to under 500 mm in some of the sheltered basins. The remarkable fact is that the frequency of raindays scarcely changes except in the severest rainshadows, and in the sunny Mediterranean littoral. Sarajevo and Belgrade have

MEAN MONTHLY TEMPERATURES (centigrade)

	Altitude	J	F	M	A	M	J	Jy	A	S	O	N	D	Absolute Max	Absolute Min	Annual days with frost
Rijeka	3	5·6	6·1	8·9	12·8	16·7	20·0	23·3	22·8	18·9	14·4	9·4	6·7			
Šibenik	37	6·4	7·4	10·0	13·7	18·2	22·3	24·9	24·1	20·5	15·9	11·0	7·8			
Split	128	7·0	8·0	10·5	13·0	18·7	22·7	25·4	24·8	21·2	16·8	12·0	8·8			5
Crkvice	1100	0·0	0·6	3·3	6·7	11·1	15·0	18·3	17·8	15·0	10·6	5·6	1·7			
Ljubljana	320	−1·1	2·8	5·6	8·9	14·4	17·8	20·0	18·9	15·6	10·6	5·0	0·0			
Bihać	227	0·8	1·3	6·8	10·9	15·7	18·9	21·2	19·9	16·1	11·1	6·7	3·0			68
Bjelašnica	2067	−7·8	−7·8	−5·0	−2·2	2·8	6·7	9·4	9·4	6·7	2·2	−2·2	−5·6			200
Sarajevo	637	−0·7	0·2	5·7	9·5	13·5	16·5	18·8	18·5	15·0	10·1	5·6	1·0	11·5	−16·0	85
Niš	214	−1·0	1·1	5·4	11·4	16·7	19·5	22·5	21·5	17·7	13·0	5·6	1·2	38·6	−17·6	
Skopje	245	−1·4	1·2	7·3	11·8	16·7	20·4	23·2	22·3	19·1	13·9	6·1	1·1	36·0	−16·6	
Bitola	617	−0·9	2·7	6·7	11·1	16·4	19·7	22·1	21·9	18·3	13·3	6·4	2·5	36·2	−17·6	
Zagreb	163	−0·1	2·1	6·7	11·5	16·3	19·3	21·6	20·7	16·9	11·7	5·7	1·6	33·4	−13·2	
Banjaluka	163	−1·9	1·2	7·6	12·7	18·9	22·5	24·8	23·8	18·6	13·7	6·5	1·9	33·3	−14·5	
Belgrade	138	−0·7	1·0	6·6	11·4	16·6	19·7	21·8	21·2	17·1	12·4	6·2	1·9	41·7	−25·6	69

MEAN MONTHLY PRECIPITATION (millimetres)

	J	F	M	A	M	J	Jy	A	S	O	N	D	Mean annual Total	Raindays	Days with snowfall
Rijeka	94	97	125	122	119	132	71	107	178	234	175	140	1593	143	
Šibenik	63	55	57	61	73	60	37	27	72	101	111	85	802	93	
Split	75	62	76	85	68	55	30	42	74	112	106	92	877	102	0·6
Crkvice	502	474	510	422	263	164	70	70	264	550	664	673	4626	143	41·7
Ljubljana	74	69	99	99	109	145	140	145	140	168	114	104	1407		
Bihać	82	66	81	107	108	120	86	94	126	149	147	107	1294	117	22·8
Bjelašnica	182	192	195	194	165	162	105	103	133	182	180	193	2012	179	106·7
Sarajevo	55	51	60	76	83	101	65	60	77	89	85	75	890	151	36·2
Niš	33	45	37	56	49	70	43	45	43	64	61	35	587	118	
Skopje	35	30	19	43	56	57	36	36	30	51	38	48	487	68	13·3
Bitola	49	72	45	67	68	60	39	36	36	75	71	62	688	113	
Zagreb	46	47	59	72	79	100	81	81	86	98	79	61	900	140	29·4
Banjaluka	49	58	74	91	126	120	83	76	85	119	60	61	1019	146	35·5
Belgrade	33	33	37	60	71	74	61	52	44	56	49	44	623	156	26·1

as many raindays as the lofty Dinaric Alps and one-fifth or one-sixth their rainfall.

A line joining Tolmin with Titograd divides Yugoslavia into a western region with maximum rainfall in the cool season, mainly in late autumn, and an eastern region with maximum in the warm season, either late spring or early summer. However, even in the interior there is a secondary increase in late autumn. As the rainfall declines in amount it decreases in reliability and efficiency so that inland districts also have to guard against drought and soil erosion.

Snowfall and frost are rare on the Mediterranean littoral but become common on the interior mountains, as the statistics for Bjelašnica show. The deep snow cover of the Julian Alps is being developed for skiing.

Land use: agriculture and forestry

Given the mountainous character of Yugoslavia it is not surprising that only 57 per cent of the state is classed as agricultural land but this includes pastures (25 per cent total) and much that is marginally productive. Forests cover a further 34 per cent and waste and built-up areas the remaining 7 per cent. It is perhaps more surprising that agriculture, forestry and fishing actively engage 4·7 million persons or just over half the total labour force and yet produce only 27 per cent of the gross national product.

In 1945 when agrarian reform was first initiated, Yugoslavia was a country of peasants and two-thirds of all land holdings averaged 1 hectare and supported 1 million families. No less than ½ million families were landless. In that year the large estates were abolished and about 1·6 million ha were redistributed, of which over 50 per cent went to individual farmers in holdings of up to 45 ha of arable and forest; about 25 per cent went to socialized farms and nearly another 25 per cent went to the forestry fund, largely for reafforestation. This land reform, however, proved insufficiently radical since there was severe overpopulation in all the rural districts, the farm techniques were backward and the peasants had no purchasing power and could make only a subsistence livelihood on their holdings. They needed as well as land, fertilizer, seed, stock, equipment and to make an overall increase in productivity. The familiar pattern of underemployment, unemployment and high total man-hours per unit of farm product was prevalent everywhere. An attempt to found and maintain peasant labour cooperatives had to be wound up in 1951.

Two years later the second main agricultural reform plan limited the area of private holdings to a maximum of 10 ha, and 275 000 ha were bought for redistribution. The chief aims were modernization, collectivization and general cooperation on farms alongside an industrial programme to drain off surplus workers from the land. But here, again, as Tito pointed out, 'The psychology of men has been formed during thousands of years. It cannot be changed in a year or two. It is impossible: men are human beings.' By 1953 besides 850 state farms for plant and stockbreeding, only about 8 per cent of the farmland with 1937 cooperatives had been collectivized, and this was mainly in the northeast in the Pannonian lowlands. As today, nearly three-quarters of the farmers preferred to remain independent, with on average 4·6 ha of land each. Traditional in their methods, with little understanding of soil maintenance, their yields remained low. On each holding about 3·5 ha were cultivated and the remainder was fallow. Their holdings were fragmented, their tools few and most of their farm animals draft.

One of the problems of Yugoslav agriculture is to combine pastoral and arable farming practices, as traditionally the cultivation of crops has proceeded independently of animal husbandry. The result is that the slopes are overgrazed and in early spring fodder is very scarce. In the High Karst late in the year leaves and twigs replace grass for feed. To curtail grazing and protect the slopes from the resultant soil erosion, the state compulsorily acquired much mountain pasture and planted the slopes with timber trees and 150 000 ha with fruit trees. In the past the fragmentation and small size of the arable farms forced farmers to extend their crops on to marginal land or on to slopes too steep to hold their frail soil cover and this has also led to soil erosion and consequent flooding on lowlands. The basin floors and valley bottoms in any case were flooded each spring after the winter rains had raised the water table so that although these areas were most fertile their season of growth was curtailed and they were merely grazed while the fragile soils of the hillslopes were loosened by ploughing for crops.

Over large areas further reorganization of agriculture is essential but the establishment of mixed farming needs a great deal of capital expenditure on the consolidation of land holdings, farm buildings, electrification, water supplies, transport and machinery. It can only proceed slowly. The greatest progress has been made in the Pannonian lowlands where socialist farms occupy one-third of all farmland. One of the finest is the Belje State farm on the Danube (fig. 92). Today Yugoslavian agriculture still consists of a large private sector and a smaller socialized sector. In 1966 the private sector owned 86 per cent of the arable land and 90 per cent of the livestock; it produced 70 per cent of the total agricultural product but only 56 per cent of the marketed production.

The socialized sector consists of two types, first, socialized farms, and second general agricultural cooperatives. The first includes specialized farms and all state holdings, both of which have been since 1960 steadily decreasing in number and growing in size. In 1966, there were 278 such farms, averaging over 4000 ha and well over 500 employees. The second, or general agriculture cooperatives, rose to importance as the peasant labour cooperatives declined. They too have decreased in number and risen greatly in size. In 1966, the average general agricultural cooperative covered more than 500 ha and employed 62 workers. These

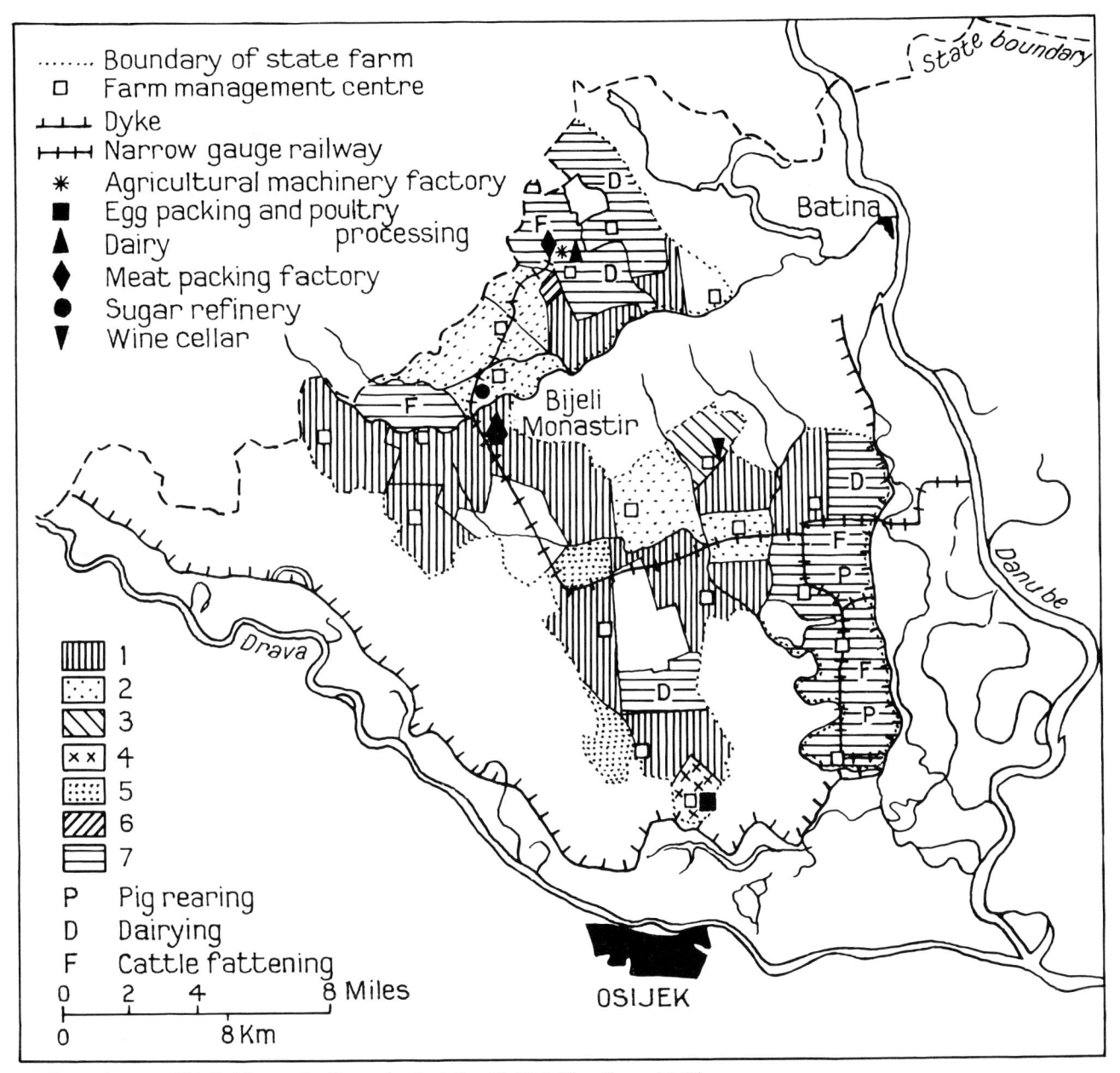

92 *State farm at Bijeli Manastir, Yugoslavia* (after F. E. I. Hamilton, 1968)

1 *wheat and maize*
2 *sugar-beet*
3 *vineyards*
4 *poultry*
5 *industrial crops, hemp, flax*
6 *seeds and nursery plants*
7 *silage maize, clover, or meadow*
The floodplains within the dykes are nearly entirely woodland or marsh.

socialized agricultural organizations provide services and buy and sell for private farmers if they wish as well as for the socialized sector.

At present in Yugoslavia just over 4 million ha of farmland are state-owned, nearly half of which is in the karst and under reclamation or afforestation or improvement. The remainder is in state farms (1·2 million ha) and general cooperatives, *agro-combinati* (1 million ha) which to date produce 30 per cent of all farm output, own half the machinery and use one-third of the fertilizer added to the soil. Some are run by agricultural organizations, and some are tractor and machine stations that also manage marketing schemes. They also provide advice, seed, fertilizer and livestock. This socialist sector is steadily being increased by land purchase, new reclamation, leases and gifts.

Already ½ million agricultural workers have been absorbed into industry since 1953 plus the natural increase and in all the republics the percentage of agricultural workers shows a marked decline.

	Serbia	Croatia	Slovenia	Bosnia–Hercegovina	Macedonia	Montenegro	Kosmet
1948	74	66	49	77	72	76	83
1961	59	45	31	50	51	47	64

This has resulted in a migration of population into the towns in search of higher living standards, and also in daily commuting from the agricultural villages to the nearby industrial towns since dual employment is quite common.

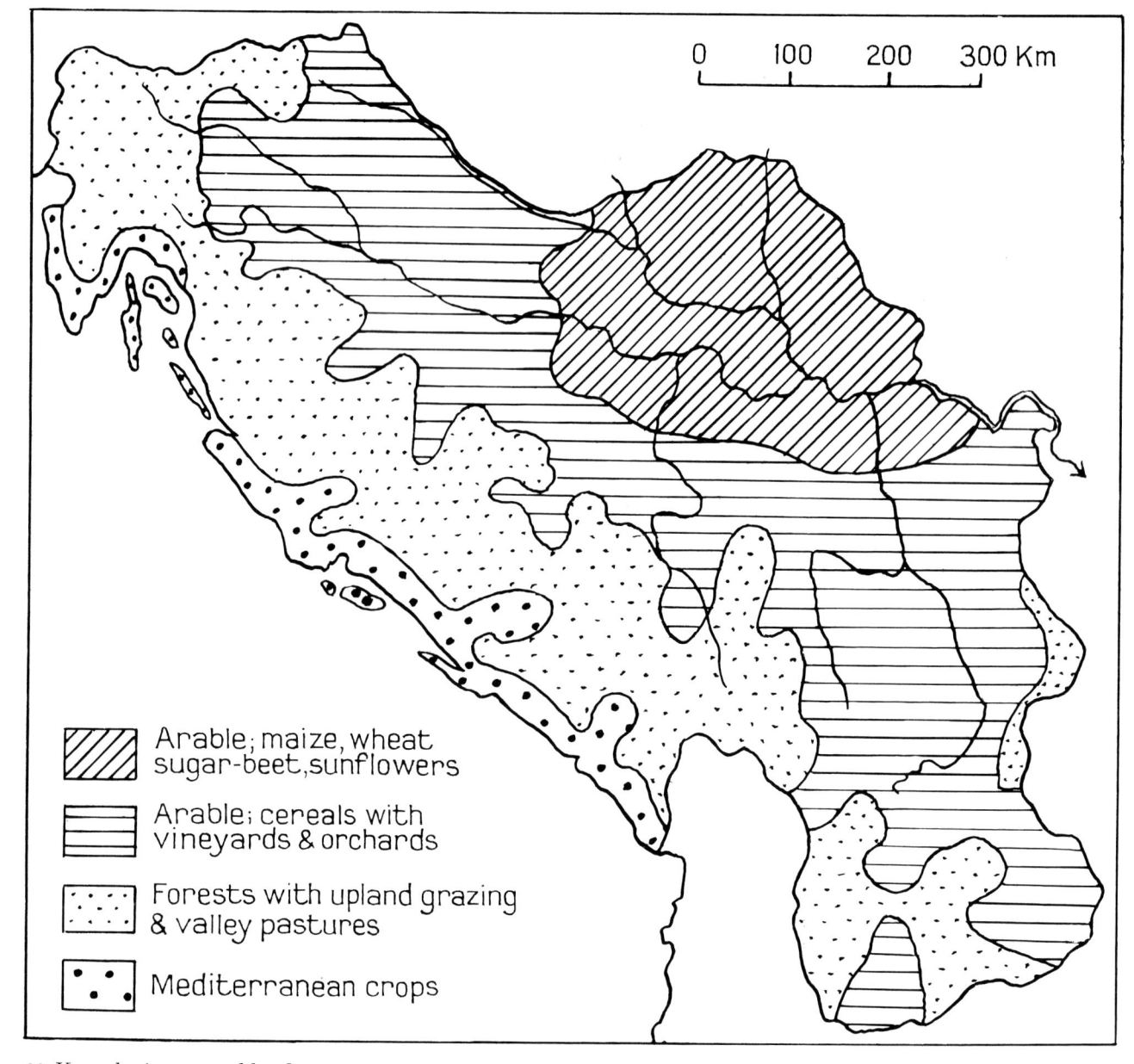

93 Yugoslavia: general land use

Meantime the private sector is encouraged to increase yield and efficiency by a special system of taxation. The farmer pays direct taxes only on the output norm of his farm and not on any surplus he achieves over and above that norm. He also pays direct tax on draft animals, hired labour and on farm implements to encourage him to join a cooperative. He gains rebates on land improvement, planting with fruit trees or vines, leasing to cooperatives, and on industrial crops, fodder or vegetables.

Gradually a change has come about and mixed and protein farming, specialized dairy and fruit farming and livestock fattening may all be found. The general result is that over several years the agricultural production has risen at 8 per cent per annum. The least progress has been made in the pastoral sector in the karst where improvement of stock, transport and pastures needs to be implemented so that products other than wool and hides can be marketed. Population in these areas is declining. From the accounts given above it seems clear that in Yugoslavia 'socialism wears a human face' since the sovietization of agriculture has not been ruthlessly pursued but instead trends have been established and allowed to grow at their own pace.

Agriculture remains predominantly private and its socialization is much more advanced in the plains and the more productive agricultural areas. Around the industrial towns specialist dairying and market gardening may be discerned, all organised in collectives which deal with harvesting, packing and despatch. Beyond the immediate hinterland of the towns large-scale specialized production of cereals or industrial crops or less often orchards organized as cooperatives, is gradually becoming the pattern wherever conditions make large-scale farming of a single product possible.

The crop use of the agricultural land in 1965 is given in the table and in figure 93. In 1966 the grain output for the whole country amounted to 14 million tons and of this 8 million tons consisted of maize (see Chapter 8). Since it tolerates podsols and high altitudes, maize growing is general throughout Yugoslavia for fodder as well as

YUGOSLAVIA: DISTRIBUTION OF THE TOTAL AGRICULTURAL AREA IN 1965

		Thousand ha	Per cent
I	Total agricultural area	14 800	100
	of which arable land	7 610	51·4
	orchards	435	2·9
	vineyards	260	1·8
	meadows	1 950	13·2
	rough grazings	4 500	30·4
	others		0·3
II	Distribution of arable land areas:		
	Total	7 610	100
	Cereals	5 230	68·7
	of which wheat	1 680	22·1
	rye	146	1·9
	barley	405	5·3
	oats	321	4·2
	maize	2 550	33·5
	Industrial crops	414	5·4
	of which hemp	47	0·6
	sugar-beet	80	1·1
	tobacco	61	0·8
	sunflower	159	2·1
	Market garden crops	588	7·8
	of which potatoes	320	4·2
	Fodder crops	814	10·7
	of which lucerne	347	4·6
	clover	216	2·8
	Fallow land & uncultivated crop land	564	7·4

Source: Statistical Bulletin, No 485/1967; OECD *Agricultural Development in Southern Europe*, 1969

subsistence. Increasing cold upwards and drought southwards makes it give way to wheat, and on the poorer soils and wetter colder uplands to barley, oats and rye. The Danubian plains form a vast corn belt, interspersed with sunflowers and alfalfa. Rice is only grown under irrigation and is restricted to riverine tracts in Pannonia and Macedonia.

The industrial crops introduced since 1945 are an expression of the idea that maximum use should be made of all resources and that every region is capable of complex development. These crops are sugar-beet, sunflowers, tobacco, hops, hemp and flax. Sugar-beet, which is grown in rotation in Pannonia with wheat and maize and, in some districts transported by water, is refined at numerous factories including those at Sremska Mitrovica, Novi Bečej, Novi Sad, Sombor and Požarevac. The residue is used for cattle feed. Sunflowers and poppies are also crushed for oil and animal feed. Fine Balkan tobacco is grown in the Morava valley and south–central Macedonia. Textile crops include hemp grown in Bačka, cotton in Macedonia and flax in Posavina, the valley plain of the Sava. Also produced are hops, paprika, lavender for oil and pyrethrum for insecticides, the last-named being a speciality of the island of Hvar.

Vegetables are increasingly grown but the brassicas, pulses and roots come mainly from the cooler, wetter northwest where they are intercropped with maize fodder and used in conjunction with dairying and pig farming.

The fruit, largely at present of plums, comes mainly from Bosnia and Serbia where the harvest is used for brandy-making, preserves and conserves. The plum orchards are however being gradually modified to include other temperate stone and pip fruits and to allow intercropping in spring and early summer with vegetables (*primeurs*). South of Split on the coast temperate fruits give way to subtropical citrus, figs, olives and vines. Packing and marketing are still crude and poor transport limits the prosperity of these horticultural enterprises.

Viticulture is widespread on south-facing slopes too steep for easy cultivation, in Pannonia, around Subotica and on the slopes of the Banat Hills and the Fruška Gora, and in the Morava and Timok valleys. This northern area produces 40 per cent of the grape harvest.

Animal husbandry

The development of beef and dairy farming is hampered in Yugoslavia by a crucial lack of fodder. Two-thirds of the fodder crops are grown in the Danube lowlands, Vojvodina and Slavonia, but succulent natural grasses are lacking. Three-quarters of the hay crop is produced in the Dinaric Alps and the Rhodope massif but here arable fodder crops are absent or scarce. Neither area can produce therefore outstanding beef or dairy breeds. Good strains need to be imported and improved animal dietary devised. In 1965, over 8 million tons of fodder were produced which was a notable achievement and since that year, meat and milk production have increased steadily but milk and butter-fat yields are low, and the lean meat quality of the beef stock not yet as high as elsewhere in the Balkan Peninsula notably Bulgaria nor as in countries like Hungary, Poland and Austria.

The best beef cattle and milch kine of the national Yugoslav beef and dairy herd of 5·7 million are bred mostly in the northern districts from Slovenia to the Stara Planina where the best fodder supplies are supplemented by the richest mountain pastures. Fortunately here also are the best transport facilities and the chief home markets in Zagreb and Belgrade. Processing factories have been established to deal with milk and cream products and with meat packing and animal byproducts.

South of the Sava–Danube line, the cattle industry

diminishes. A hardy local breed called the Busa survives the mountainous terrain, the cold winters, the dusty summers and the lean diet but its yield of milk and meat is mediocre. A few buffaloes are kept in the south.

Pigs are bred north of the Sava–Danube line along with cattle in the fodder country, and 5·2 million a year are kept, but to the south of the line poultry take the place of swine as a supplementary source of income and diet for the peasant proprietors. Slovenia has started the broiler chicken industry using broken maize meal and skim milk for feed. Sheep are on the decline particularly where the state takes over the impoverished karst lands and brings them under reclamation schemes. Today they number about 10 million. Transhumance still takes place in some of the remoter and poorer areas but the animals produce poor quality fleeces that command low prices. Investment in transport in rural areas tends to be low and refrigeration is weakly developed. The bulk of the food industries are in northern Yugoslavia, the food-processing factories generally being established on cooperative farms to save transport costs. Tobacco is therefore cured at Skopje and Titograd, tomatoes are canned at Negotin (Serbia) and Opuzen (Macedonia) and sugar refined at Bitola and Peć.

Land reclamation schemes (fig. 94)

Federal investment is largely reserved for long-term projects and agronomists tend to balance short-term increased productivity against long-term needs. Land surveys indicate that 3 million ha are liable to periodic flooding, 4 million ha to soil erosion, 5 million ha in need

94 *Land reclamation and improvement schemes in Yugoslavia*

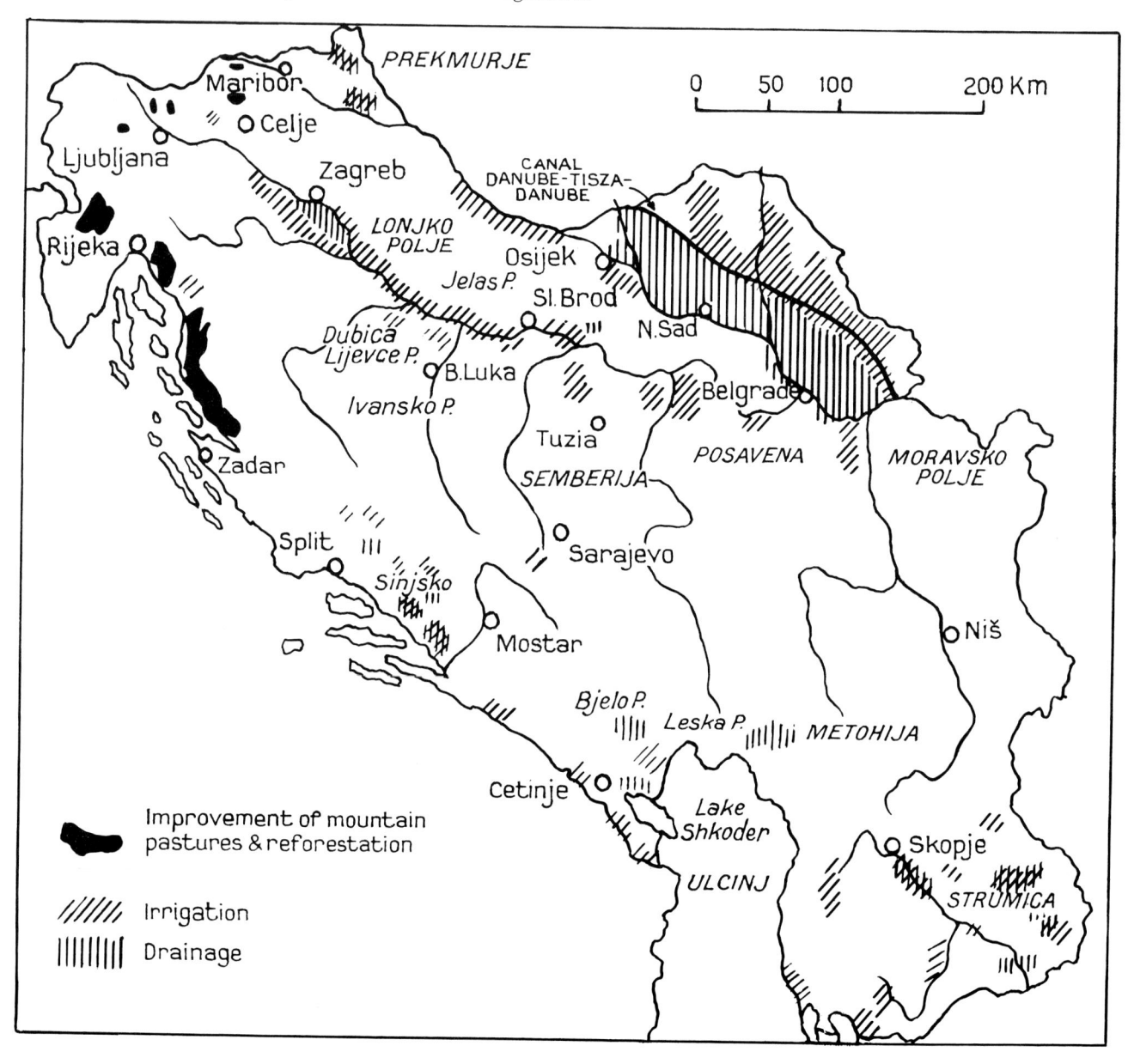

of constant fertilizer, and another 1·5 million ha exhausted. Elsewhere prolonged summer drought and autumn and summer floods make protective measures necessary. So far 22 land improvement schemes have been devised of which those outlined below are typical. Several of these will be finished within the next few years.

1 The building of the Danube–Tisa–Danube canal and 400 km of drainage channels, has allowed the reclamation of nearly 1 million ha in Bačka and the Banat, and the surplus water irrigates ½ million ha in the far northeast where the annual rainfall is under 400 mm.
2 The reclamation of 240 000 ha in Macedonia by flood control and river regulation. Of this about one-third will be irrigated.
3 The drainage of the salt marshes of the Pelagonia, Struga and Skopje polja and irrigating 250 000 ha in 14 polja, for example, Kočani, Ovče (Štip), Tikveš and Strumica.
4 Reclamation of floodplains by regulating the Drava, Sava and Morava rivers.
5 Developing the polja of the south by draining and protecting them from flood.

Land-use regions

The country has been further planned for agricultural development under five land-use regions.

The Pannonian lowlands

The portion of these lowlands which lies in Yugoslavia extends from the frontier with Romania and Austria to an undulating platform in the south which probably marks an old shoreline of the lake that once covered the whole area. Outlined on the east and south by the Morava river and its tributary the west Morava this shoreline extends as far west as the Kolubara river. The whole drainage basin crossed by the water systems of the Drava and Sava has developed rich chernozem soils on the loess-covered tracts similar to those of the Ukraine and forms the most productive region of Yugoslavia. With 65 per cent of its surface devoted to cereals and only 12 per cent to meadow and pasture it is the country's granary or cornbelt. Maize is the chief grain crop with market-garden and fodder crops in rotation and with pigs as the chief farm animals. Industrial crops are grown in specialized localities, as for example tobacco in the Tisa valley in the Banat, sugar-beet and vines in the Morava lowlands and flax in the lower Drina and Kolubara. North of the Sava river as temperatures and length of growing season decrease maize gives way to wheat, both being highly mechanized cultivations. The rich agriculture yields surpluses for food-processing industries such as flour milling, sugar refining, meat packing and brewing as well as for rope and linen manufacture from hemp and flax harvests. This is the most important area of state farms in Yugoslavia.

The foothill zone surrounding the Danube basin

This area, closely linked to the Pannonian basin and being the gathering ground of rivers draining to the Danube, still has bread-grains occupying half the arable land, but grass for stock raising and dairying occupies one-third of the farmland while fodder crops are grown on a further 7 per cent of land under the plough. Orchards (plums), vineyards and vegetables account for the remainder. On the alpine shoulders and upland valleys a Swiss-type dairy industry has evolved especially in the northwest that is organized into cooperatives. Manufactures based on dairy produce have developed in Maribor and Ljubljana with additional subsidiary industries such as candles, soap, margarine and meat packing. The more sheltered and south-facing valleys have orchards, walnut and chestnut groves, softfruit plantations and vineyards and food-processing activities based on these crops are also sited in Maribor.

The Central Highland area

Here because of the mountainous and dissected terrain, the population clusters in small scattered villages in the various polja which form sporadic patches of cultivation in a blighted landscape. These polja can only be tilled when the water table has subsided; then, during the summer grazing season, poor crops of wheat and maize or even simply pasture are the chief uses to which they can be put. Otherwise over the major part of the region sheep rearing on natural pastures is the chief farming activity. The processing of agricultural surpluses is concerned with wool, mutton, tallow and hides at towns like Mostar.

The Morava–Vardar corridor

This is a series of tectonic basins linked by the Morava flowing north to the Danube and the Vardar flowing south to the Aegean. The basins originally held lakes and are floored with thick, fertile soils derived from lacustrine deposits and river silts. These enclosed basins and plains at different levels in the mountains recall the district of Umbria in Italy and in the same way each basin shows a varied agricultural development and has given rise to a market town. Such are Niš, Leskovac and Skopje on the Morava–Vardar route, and Priština and Raška on the almost parallel route in the Ibar drainage corridor west of the Kopaonik mountains. Usually the mountain shoulders and slopes are under pasturage, often of bushes, for cattle and sheep, the wool of the latter being of good quality and prized for the famous carpet manufactures of Pirot. The lower and flatter tracts are under cereals, vegetables and fruit. The plum harvest is noted for its liqueur *slivovitz* (*šljivovica*). The crops become increasingly subtropical with distance south, wheat and maize yielding to rice, brassicas to tomatoes and peppers, and plums to melons and vines. Although cereals account for one-third of the arable, certain areas are given over to specialised production, the Bregalnica valley near Štip being noted for its opium poppies, the plains near Kočani for rice and the Strumica valley for cotton. Near the Greek frontier mulberry orchards have given rise to a well-known silk industry at Gevgelija. Food-processing industries, opium manufacture as well as agricultural research are centred at Skopje.

The Mediterranean coastal strip

The limestone folds along the Adriatic coastline suffered a recent marine transgression and the Dalmatian coast for 632 km is the prototype of its kind, with a total shoreline of 9600 km. The unsubmerged crests of the folds form 725 elongated islands and islets where the drowned valleys form the interposing *canali* or water straits. The Gulf of Kotor is an exception being the only fiord in the Mediterranean. This whole Dalmatian strip is also distinguished from the rest of Yugoslavia by its purely mediterranean climate. Since the drainage is underground and silt-free the coastal plain is very narrow. Never wider than 60 km it sometimes dwindles to less than a kilometre. Two or three areas alone provide sufficient space for cultivation and settlement; namely the Neretva valley, the Riviera of the Seven Castles, and the Dubrovnik district. These narrow strips lying at the foot of barren karstic mountains are under olives, vines, figs, tobacco, vegetables and oil seeds. Even so cereal crops occupy one-seventh of the ploughland and nearly one-tenth lies fallow each year. In the Neretva valley irrigation water is available but the water issuing by springs from deep underground sources has to be allowed to warm up before being led on to the fields. North of the Krka river cultivation is rendered impossible by the withering effects of the cold bora.

The towns along this coast are superb natural anchorages and, sheltered from the bora, were mostly bases used by Venice in the days of her maritime greatness. The word argosy is thought to be derived from the name of the town Ragusa which was known in Tudor England as Arragosa. They are beautiful little walled mediaeval towns with renaissance cathedrals, palaces and villas, of which Dubrovnik is a perfect specimen. They have specialist food-processing industries today such as winemaking (riesling), maraschino manufacture from cherries at Split, Šibenik and Zadar, and honey production and the making of pyrethrum powder or other aromatic substances from the shrubs of the lower mountain slopes.

Along this magnificent coast, tourism as an alternative to agriculture is being fostered and already is the second largest currency earner. Nearly 5 million foreign tourists a year are catered for in such centres as Split and Dubrovnik which can exploit the climate and the scenery, and in the Dinaric Alps and in Slovenia at Celje for winter sports, hunting and caving. Spa resources are being developed at certain centres where mineral springs exist, especially in Serbia and in the upper Sava valley. The government has in addition set aside four National Parks to attract foreign visitors throughout the year at Mavrovo, Galicia, Durmitor and Plitvice and in 1967 visas were abolished.

The home tourist industry has not been neglected and holidays with pay, one free holiday journey a year and trade union rest houses have been established, and the cultural inheritance of Yugoslavia with its Roman, Greek, Venetian, Moslem and Byzantine historical remains made available to foreigners and natives alike.

Forestry

Yugoslavia ranks third in Europe after Scandinavia and Austria for its forest reserves. Just over one-third of the country is forested and although 3 million ha are classed as degraded woodland or as coppice or shelter belts, large stands of tall forest remain, as in the upper Drina river basin. About three-quarters of the growth is deciduous hardwood, mainly oak and beech, and the coniferous remainder is concentrated as one might expect in the north on the slopes of the Slovene Alps. Since 70 per cent of the forests are under the care of the State, forestry has become one of its conservation measures and much of the tree growth on steep or exposed slopes has been, as far as possible preserved. Replanting in fact has been carried out over $\frac{1}{2}$ million ha and a further 1 million ha are in process of being replanted. The plantations are mainly in conifers or poplars since these are quick-growing and easier to process for pulp, but methods and plans to process the deciduous hardwoods are being developed, and the timber industry ranks third in value of Yugoslavian industries. Forest covers over half the surface of five specific areas: the mountain slopes of Slovenia, and of the Carpathians; central Slavonia; highland Bosnia; and central Bosnia–Hercegovina in a belt stretching from the Adriatic about the Neretva valley inland in the Drina basin to the Spreča river. In 1966, no less than 18 million cubic metres of wood was cut (see table, Chapter 7) and of these 8 million went for fuel and 10 million to feed 250 saw mills. Secondary woodfinishing manufactures are integrated with these saw mills and produce wood pulp, veneers and plywood as at Bihać, Blažuj–Sarajevo, Foca and Zavidoviči. Furniture making tends to be separated from the saw-milling centres and to migrate closer to urban markets.

The sulphate–cellulose paper industry is concentrated in Bosnia at Prijedor, Drvar (sulphite), Banja Luka and Maglaj, and in Montenegro at Ivangrad. A large newsprint factory importing Bosnian fir and spruce and cellulose has been established at Videm–Krško in Slovenia where hydroelectric supplies are plentiful.

Fishing

As in the Greek islands so here along the Dalmatian coast the chief resource of the islanders is fishing but the great need of the industry is capital expenditure on larger fishing craft, refrigeration and canning units, and a more flexible transport system to the major population centres. At present its main outlet is the tourist trade. As with 'mediterranean' waters elsewhere the chief catch is sardine in summer, tunny in winter and mackerel at other seasons. Smaller quantities of shell fish, especially oysters, are also harvested. Since catching methods for the three main types of fish vary, fishing becomes a part-time occupation for most of the employees. At present the catch is landed at tiny island ports like Vis and Kornat, or at Šibenik and other ports on the mainland. Several of these have canneries and over 20 000 tons of canned fish are produced annually. Fish farms have been established in all large inland water bodies as at lakes Shkoder (Shkodra) and Ohrid and also at points adjoining State farms along the Danube. The irrigated fields also produce fish as a second source of food and in order to maintain fertility. As the table in Chapter 3 shows, the total catch in 1966 was nearly 46 000 tons and of this nearly 40 per cent came from freshwater sources conveniently near the main markets.

Power resources: minerals and manufactures

The avowed aim of the Yugoslav federal peoples' republic has always been the full utilization of the natural and human resources in order to eradicate poverty and economic inequality. To establish industry was to strike at the twin problems of unemployment and at the extravagant use of manual labour in agriculture, and 40 per cent of the national income has been invested in industrial development. Despite the former almost total lack of skilled factory labour and management, 18 per cent of the population is today engaged in industry. The old craft and artisan activities concerned with food delicacies and textiles and wood carving are everywhere being replaced by sophisticated metallurgical and engineering industries.

The prime concern is with power resources. Water-power potential is considerable in two main areas; the Slovene Alps which being glaciated provide optimum conditions for the establishment of hydro-electric schemes, and the Dinaric Alps where the rivers are swift, the rainfall heavy and the runoff rapid. Every major river has been analysed hydrologically and the more important brought under control. In the karst areas the underground sources of water have been channelled to feed a few large rivers like the Cetina, Neretva, Trebišnjica and Morava. Four-fifths of the water power, actual and potential, is concentrated in the rivers Drina, Neretva, Cetina, Upper Sava, Upper Drava, Morava, Bosna and Vardar. The Drina has the largest share with one-sixth the total national potential. Great engineering skill has been necessary to control the rivers and the underground water supplies of the Dinaric karst, where 60 power stations have been built. Limestone areas are notoriously difficult to harness for hydro-electricity since the rock fractures easily and is porous. Yugoslavia has led the world in demonstrating that it is possible to overcome these difficulties. A high-tension grid system is being developed for the whole country to permit regional exchanges of electricity. During the winter when rainfall is heavy on the Dinaric Alps current is transmitted to north and central Yugoslavia, and in summer when drought afflicts the west coast, current is fed into the Dinaric region. The Romania–Yugoslavia Iron Gate project on the Danube was opened in 1972 (see Chapter 10). The hydro-electric output increased from 9800 million kWh in 1966 to 11 800 million kWh in 1968. Despite the great progress made, Yugoslavia has as yet tapped only 17 per cent of her potential hydro-electric power which, however, forms 58 per cent of the present national electricity output.

Coal in Yugoslavia is widespread, nowhere being more than 40 km from a source of fuel but nearly 99 per cent of it is poor-quality brown coal or lignite (fig. 95). Bituminous coal in small quantities is mined in three fields, at Raša, north of the Gulf of Kvarner; Rtanj–Timok, north of Dimitrovgrad; and Ibar, near Priština. Production is estimated at just under 1 million tons a year. The brown coal is dispersed in 130 small fields and this offsets transport problems to a certain extent but the seams in many of the fields are thin, distorted and difficult to mine because the overburden is so often fragmented. Many mines are troubled by gas and water. Mining is expensive and the coal industry has had to be rationalized. Small uneconomic pits have been closed and only large ones in which mechanization is possible are exploited. Today 33 brown-coal fields are in production, 13 of them possessing four-fifths of the total reserves. The most important producing fields are at Zenica–Sarajevo and Banovici, in east–central Bosnia, with 70 per cent of the national reserves; Despotovac-Senjski Rudnik, in east–central Serbia with 15 per cent of the reserves; and around Aleksinac and Zagorje in central Slovenia. The product is continuously being experimented with to develop processes suitable for industry. The greatest fuel resources are, however, in lignite and this is used almost exclusively for generating electricity. There are five large fields, Velenje, Kreka, Kolubara, Kostolac and finally Kosovo which is probably the largest lignite field in Europe. Of the lignite reserves, 38 per cent occur in Bosnia, 27 per cent in Serbia and 20 per cent in Slovenia. In 1968, the total production of solid fuel, largely lignite, was 27 million tons.

For oil the Zletovo and Aleksinac petroleum field with reserves estimated at 42 million tons, is the oldest field but has declined in importance since 1950 with the opening of the Croatia–Slavonia field farther downstream and the Banat field beyond the Tisa valley. 2·5 million tons of crude petroleum and 384 million cubic metres of gas were produced in 1968. An oil pipeline has been laid to Bakar, a new deepwater petroleum port near Rijeka, from the inland refineries established at Brod, Pančevo and Sisak.

However, as elsewhere, electricity is seen as the lynchpin of economic progress as it allows the dispersal of industry and overcomes the rudimentary transport system.

Mineral ores

A wide range of commercial ores is available in Yugoslavia both ferrous and non-ferrous. The annual production

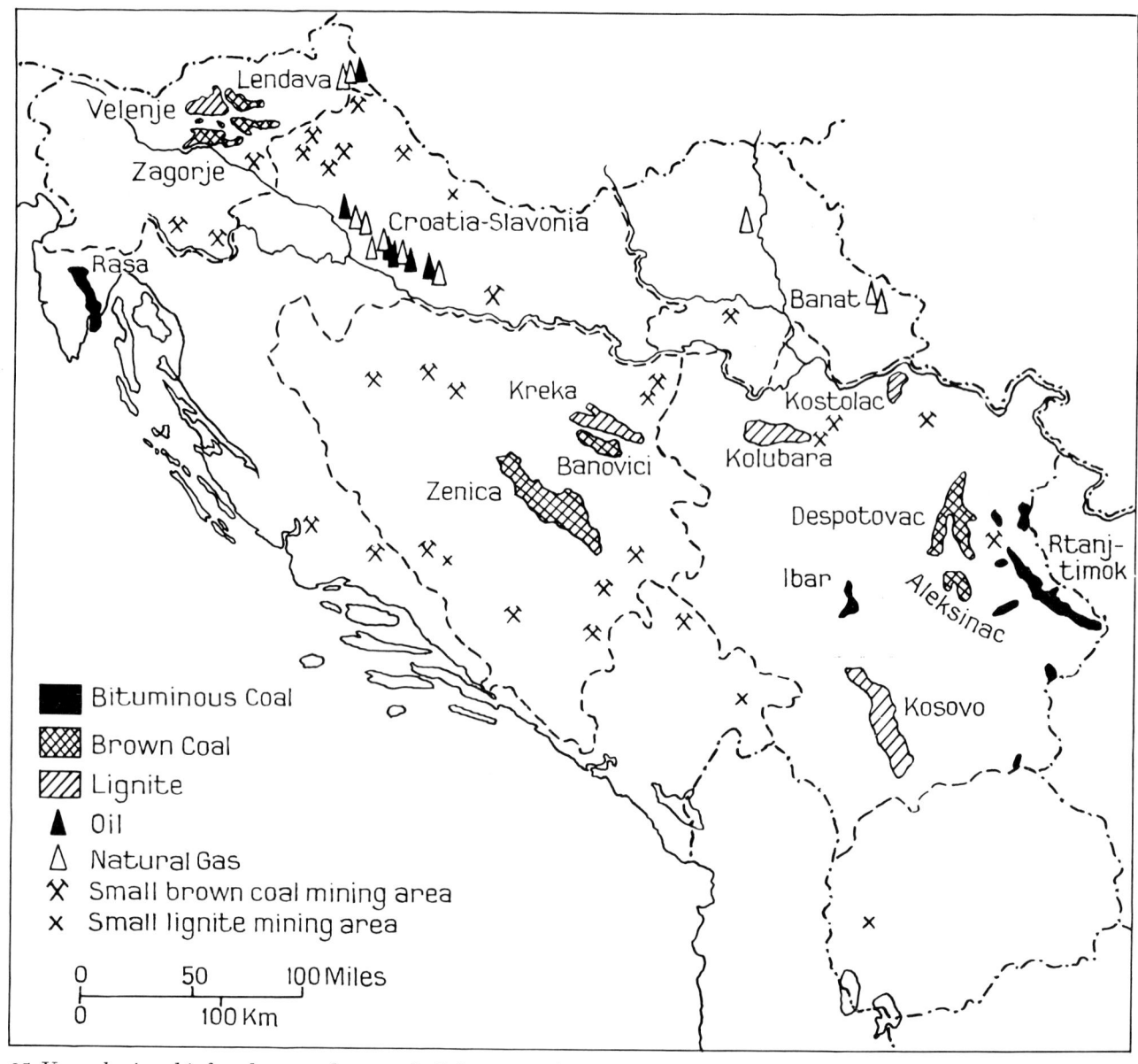

95 *Yugoslavia: chief coal, natural gas and oil deposits* (after F. E. I. Hamilton, 1968)

of iron ore is 2·5 million tons and it is planned to increase this eventually to 9 million tons. At present the largest workings and half the production are at Vares in eastern Bosnia where the ores are haematite with 37 per cent iron content, and are mined opencast. The remainder comes mainly from the Sanski or Sana basin (near and upstream of Prijedor) where the mining conditions are more difficult and production runs at about ½ million tons a year but the reserves are considerable. Elsewhere iron ore is widespread but is not as yet greatly exploited owing to mining difficulties, the isolated position of the richer areas and the phosphoric quality of the ores.

The country is self-sufficient in a great many industrial alloys, being the first European producer of antimony, chrome, copper, lead and molybdenum and a leading producer of mercury, bauxite and zinc. These deposits tend to occur where volcanic activity has taken place at the junction of the major structural units. Of the more important, copper ore is mined at Bor and Majdanpek in the mountains of northeast Serbia just south of the Iron Gate on the Danube. The ore is low grade but the reserves are large and can be worked opencast, especially at Bor, which is one of the biggest single copper mines in Europe. Total production of copper ore in Yugoslavia in recent years has been about 6 million tons a year, with a copper content of 60 000 to 70 000 tons. Lead and zinc ores are likewise widespread but the chief source at present is the Trepča mine in the Kopaonik mountains a few kilometres northeast of the town of Kosovka Mitrovica. The national output is 2·4 million tons of ore annually. Chrome is mined at Skopje, manganese at Cevljavonici and cinnabar, or ore of mercury, at Idrija. There has been, in response to world scarcity and demand, a great increase in the mining of all metals except antimony but the reserves of many are estimated to have a limited life. Bauxite with a high alumina content occurs mainly as massive beds in karstic hollows in the Dinaric Alps between Istria and the Albanian border. It is mined at Drniš, Rovinj, Mostar and Nikšić, and amounts to 1·9 million tons ore a year.

Manufactures

Metallurgy engages 92 500 workers, and over 1 million

tons of pig iron and nearly 2 million tons of crude steel are produced annually. Over half the steel capacity is in central Bosnia with Zenika as the major steel-producing town but new integrated plants, albeit small, are being built at Sisak, Ilyas, Nikšić and Skopje, and a larger integrated steel plant, capable of handling 500 000 tons of ore, has been set up at Smederovo on the banks of the Danube downstream from Belgrade. Non-ferrous ores provide one-fifth of the national income and their refining forms an important occupation at various centres throughout the federal republic, for example at Sevojno which refines copper mined at Bor, at Šabac with lead–zinc refineries, at Tetovo which manufactures chrome and at Strnišće, Šibenik, Titograd and Mostar which make aluminium. Titovo Užice has a copper rolling mill.

Engineering which employs the greatest number of workers (310 000) is found mainly in a belt from Trieste along the Sava valley to Belgrade and from thence up the Morava valley to Leskovac. It is concerned chiefly with building equipment, agricultural machinery, electrical engineering and ball bearings. Centres outside this main belt tend to specialize in particular products, for example, earthmoving equipment at Kruševac, railway rolling-stock at Brod, and cranes and marine engineering at Rijeka.

The rapid expansion in the electricity and mining industries and the basic industries established on them, stimulated a great demand for building materials and manufactures, such as cement, glass, asphalt, bricks, kaolin and sand. Federal funds are only available for the energy, cement, metallurgical and chemical industries and so it follows that these show the best regional dispersal and development, but the chemical manufactures tend to be sited where there are abundant power, water and raw materials. This industry includes the production of heavy acids, explosives and electro– and petro–chemicals. Fertilizers are made at Pančevo using byproducts from the natural gas industry, and at Prahovo from copper byproducts, while petro–chemicals are made at Zagreb using the byproducts from the oil refinery at Sisak.

Industry and manufacturing started in Yugoslavia with poor marketing facilities, a small domestic market, a shortage of skilled labour and low technical and management skills. By and large despite regional planning, power-intensive industries are based near the power supplies, and engineering where a pool of skilled labour exists, that is in the north. Elsewhere industries tend to be located near sources of raw material or in established urban centres. The main concentrations of industry lie north of the Sava–Danube line in northeast and in northwest Yugoslavia: thus the metallurgical, chemical and textile industries are established in a group of Slovene towns, Ljubljana, Kranj, Maribor and Celje and in the Croatian towns of Karlovac, Varaždin and Zagreb. The northeastern concentration accounts for 11 per cent of all Yugoslavian industry in Belgrade, Osijek, Novi Sad and Subotica. South of the main industrial belt industry is concentrated around Sarajevo in central Bosnia which has mining, metallurgical and timber-processing industries; at Niš, Leskovac and Kragujevac in central Serbia which has metallurgical, tobacco and textile manufactures; and at Split and Šibenik in Dalmatia where cement, chemical, aluminium and shipbuilding industries are present. Taking advantage of Yugoslav investment law whereby foreigners may hold up to 49 per cent of the capital in a local enterprise, Fiat and a Yugoslav firm, Crvena Zastava (Red Flag), assemble motor vehicles at Kragujevac.

New factory towns have been established in undeveloped central districts of the country where labour-intensive concerns could be effective, such as light engineering, leather and textile products. These newly industrialized towns include Titov Veles, Valjevo, Titograd, Vranje, Mostar, Banja Luka and Bihać. It is in the south and core areas of Yugoslavia that art and craft trades still survive. Consumer industries, which are the easiest to establish, are found mainly north of the Sava–Danube line where the largest home market also exists.

INDUSTRIAL EMPLOYMENT IN YUGOSLAVIA, 1966–8

	Value of (millions of new dinars)		Number employed (annual average)
	Social Product	Installations (1967)	
Metal industry	6 164	10 154	269 609
Textiles	3 981	6 121	230 210
Food processing	3 323	6 746	125 023
Electrical energy	2 638	19 999	36,737
Chemicals	2 631	6 257	70 960
Timber	2 069	3 544	136 614
Electrical industry	1 831	2 262	73 882
Building materials	1 630	3 021	73 379
Non-ferrous metallurgy	1 589	3 931	43 527
Printing	1 415	1 135	43 147
Coal and coke	1 337	4 281	65 567
Ferrous metallurgy	1 083	5 018	41 127
Mining (non-metals)	920	1 784	43 936
Petroleum	845	2 622	9 525
Leather	840	799	48 242
Tobacco	668	655	17 079
Paper	636	3 315	26 043
Shipbuilding	557	908	19 735
Rubber	422	546	15 409
Motion pictures	75	60	1 441
Other manufacturing industries	187	192	8 185
Total	34 841	83 350	1 399 377

These manufactures are chiefly in textiles, leather, clothing and rubber goods, and are represented in numerous towns including Zagreb, Maribor, Osijek and Belgrade. Clothing is made at Varaždin and shoes at Borovo–Vukovar. The actual and relative importance of the main branches of manufacturing in 1966–8 may be judged approximately from the table.

Population and political patterns

Yugoslavia is basically a federation of the South Slav people. Initially it was the Kingdom of the Serbs, Croats and Slovenes and these three nationalities form the majority of the population in the three states of Serbia, Croatia and Slovenia. Serbs also form a large percentage of the ethnic make-up of Bosnia, Montenegro and Macedonia. They comprise 42 per cent of the total Yugoslav population, with the Croats forming 23 per cent and Slovenes 9 per cent. The remaining peoples include at least 15 identifiable minority groups, the chief being Albanian, Magyar, Turk, Slovak, Gypsy, Romanian, Bulgar, German and Jew. The small Vlach minority is described in Chapter 8 under pastoral farming. The chief language is Serbo–Croat which is written in the Latin alphabet in Croatia and in the Cyrillic in Serbia. There are, however, many minority languages and numerous dialects. The chief religious adherences are Orthodox, Roman Catholic and Moslem.

The federation consists of six self-governing republics, and two internal autonomous regions (fig. 96).

	Area (thousand sq km)	Population (thousands)	Capital
Bosnia and Hercegovina	51	3 667	Sarajevo
Crna Gora (Montenegro)	14	520	Titograd
Croatia	57	4 314	Zagreb
Macedonia	26	1 530	Skopje
Serbia*	88	8 048	Belgrade
Slovenia	20	1 662	Ljubljana
Yugoslavia	256	19 741	

* With two autonomous regions

The autonomous areas within Serbia are the province of Vojvodina (22 000 sq km: 1 715 000 population) and the region of Kosovo–Metohija (11 000 sq km: 810 000 population). The most northerly of the republics is *Slovenia* which occupies the northwest Alpine region and the upper drainage of the Sava and Drava rivers. The people are Slovene, speaking a Slovene language that has several dialects all written in the Latin script. It includes territory at the Prekmurje that was formerly Hungarian and part of the former free territory of Trieste. To the south lies *Croatia* that includes the greater part of the Dalmatian coast and Istria formerly Austrian territory, the northern Dinaric Alps, and a large area of the Danube lowlands that

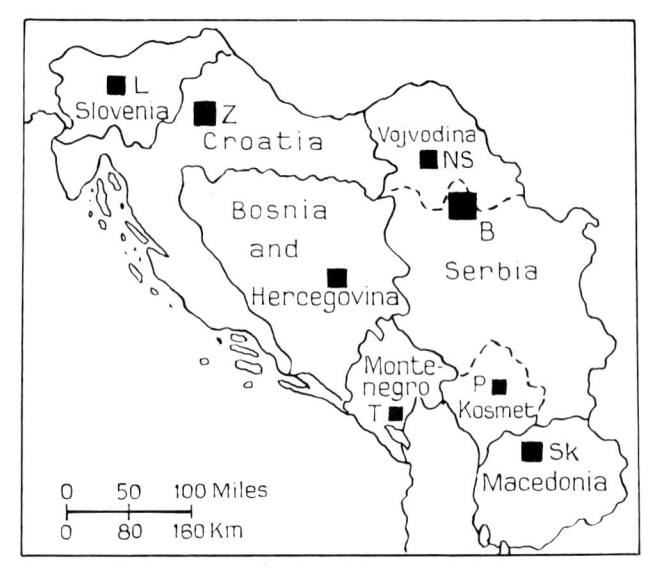

96 *Yugoslavia: the republics and the autonomous areas, with their capitals*

formerly belonged to Hungary. It is an awkward shape geographically and very diverse in character but exists by virtue of its almost entirely Croatian and Roman Catholic population.

The largest of the republics is *Serbia* which occupies the heart of the country, and with its autonomous provinces extends from the Danube southward, straddling the Morava–Vardar river route and occupying the horst and graben country of the Rhodope massif. Serbian in race and Orthodox in religion the population includes two main minority groups, namely 60 000 Bulgars in the Pirot district, and 30 000 Romanians in the Banat.

Prior to 1918 the autonomous province of *Vojvodina*, the lower Tisa basin, was Hungarian territory and so there is a significant Hungarian, Roman Catholic minority of about ½ million in addition to a residual group of Germans, partly Protestant, estimated at 50 000 today. The latter is the only minority group to have diminished rapidly in numbers having decreased dramatically from 575 000 in 1940.

Kosovo–Metohija (Kosmet) in the extreme south of Serbia, adjoins Albania and a majority of the population is Albanian and Moslem and most of the remainder Serbian and Orthodox.

Bosnia–Hercegovina lies between the Dalmatian coast and Serbia and occupies the greater part of the southern Dinaric Alps but with a relatively small outlet on the Adriatic. It is the only republic that is not established on a nationality basis although nearly half its population is Serb, and one-quarter are Croats. Most of the remainder are of Turkish descent. The Serbs are mainly Orthodox by religion, the Croats Roman Catholic, and the Turks Moslem.

Montenegro occupies the high mountain core of the

Dinaric alps north of Albania with a frontage on the Adriatic. Although small, it was the chief centre of resistance to Turkish and later invasions. Its population is Serbian and Orthodox.

Macedonia, farthest south and entirely inland, is a part of the territory that was liberated from Turkish rule in 1913 and partitioned between Greece, Bulgaria and Serbia. Later Bulgaria ceded the area around Strumica to Serbia but the division between Greece and Serbia remains unaltered. The population is Macedonian with Albanian and Turkish minorities and speaks a language that has affinities with both Serbo–Croat and Bulgarian and is written in a Cyrillic script.

Population and settlements

The average national density of population is about 77 per sq km and falls to under 40 in large areas of the Dinaric Alps (fig 97). The only large areas with a population density exceeding 100 per sq km are in the northeast Danube plains from Belgrade to the northern boundary at Subotica and in the northwest Sava plains near Zagreb.

Rural settlement

As just over half the labour force is engaged in agriculture, the major part of the population lives in rural settlements and these assume numerous forms according to the varied physical backgrounds and the many cultural influences to which they have been exposed. Basically they may be classified into nucleated or dispersed types and broadly speaking the dispersed type is found in mountain districts with pastoral occupations and the nucleated where arable farming predominates in the plains, vales and basins.

It is recognized that the *Stari Vlah* form the basal type of loosely knit settlement in the mountain tracts. They consist of small numbers of houses grouped spaciously without any obvious pattern occasionally having sufficient proximity to become large sprawling villages along a road, in a valley or on a hillside. Sometimes they appear to be the relic of a former large estate or zadruga. In either case they are surrounded by unenclosed arable land. Nucleation of these villages seems to occur where the arable land increases in extent and fertility, and there is evidence of pressures to do so as electricity, piped water and other services are made available. In the karst the scattered houses are irregularly disposed around a polje, sited on exposed limestone to avoid flood and precious cropland. Occasionally small more tightly-packed hamlets of a few dwellings will occupy hill summits or the lip of a gorge forming a dispersed general arrangement of settlements. These small summit nucleations form the *ibar* type.

In the Danube lowlands and in the wide Sava and Morava valleys, nucleated settlements are of three main types. First, the *timok*, a cluster of closely built houses irregularly laid out with unplanned twisting lanes. Originally pastoral villages they are now agricultural and are probably the result of in-filling. Second, the *ciflik* villages, which owe their origin to the Turkish occupation. The houses are built around a small mud-walled courtyard in one corner of which is the house and watchtower of the former Turkish owner. These are characteristic of the upper Morava, the lower Vardar, the Leskovac neighbourhood and of Kosovo–Metohija. The third lowland nucleated type, is the planned gridiron village, built by estate owners mainly in the eighteenth century on land recaptured from the Turks. These are common only on the northern plains, and are today often agro–towns amid wide agricultural fields.

In most of southern Yugoslavia, especially towards the Adriatic, a Mediterranean type of closely-knit hilltop village predominates. Particularly in highland Macedonia and Metohija the stone built dwellings with their orchards and walled enclosures tend to cluster around a square or market place and are sited on the tops of spurs and promontories where surprise attack was less likely and incidentally frost drainage and shadow effects were least. Many villagers here practise a kind of inverse transhumance, coming to the lowland fields for summer and living then in huts near their crops and returning to their main dwelling on the higher ground in winter.

Over the greater part of Yugoslavia the small hamlets and scattered dwellings support larger market centres of varying sizes, few of them with markedly urban facilities or varied industrial activity. However, in the rich plains of the Sava and Danube, the large defensive agro–towns or compact nucleated villages built against the Turks and other marauders, actually later nourished a dispersal of isolated farms (*tanya*) and hamlets in the wide plains of Bačka and Banat.

Urban settlement (fig. 98)

The Yugoslavian society is not highly urbanized in a

97 *Yugoslavia: density and general distribution of population* (estimated for 1968)

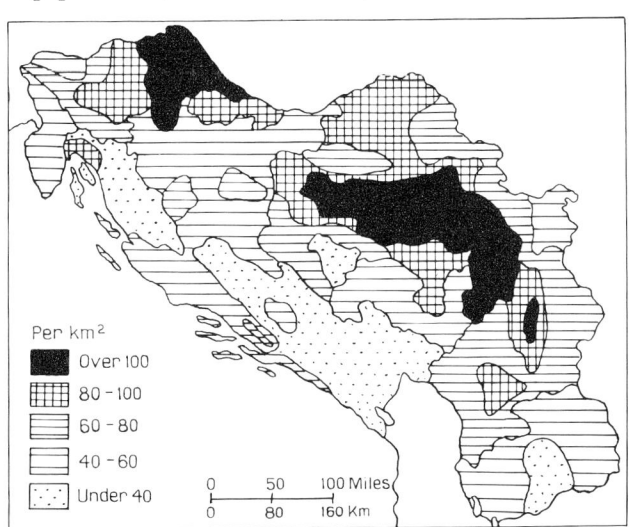

XXXVI *Townscape in Belgrade, Yugoslavia*

west European sense. At the 1961 census, of the 222 settlements with more than 5000 inhabitants about half could be classed as predominantly agricultural or rural in form and function. There were then only 13 towns definitely over 50 000 in population and, apart from the southern state capitals, nearly all were situated on the northeastern and northwestern plains and less so in the Morava–Vardar basins.

98 *Yugoslavia: distribution of towns with over 50 000 inhabitants*

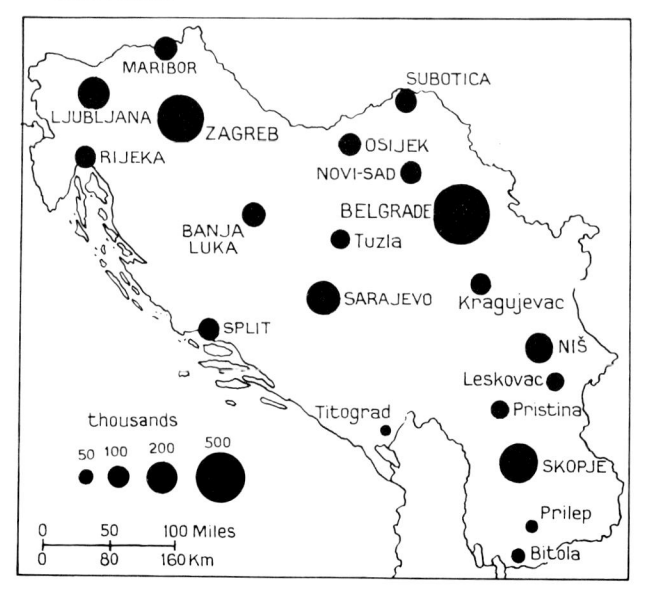

The chief cities are Belgrade (Beograd), Zagreb, Skopje, Sarajevo and Ljubljana but at least 8 other towns exceed 100 000 and 4 or 5 others are over 90 000.

Belgrade, the national capital, has nearly 700 000 inhabitants or 3·5 per cent of the total population of Yugoslavia but probably well over 1 million people live in its conurbation, which has grown rapidly in recent years with large suburbs of high-rise apartment blocks, and large satellites such as Pančevo. The city grew up on a spur of the Rudnik mountains at the junction of the Sava and Danube rivers. It is the largest city in a most productive and industrial hinterland for which it is the chief commercial and market centre. Moreover it has developed riverport facilities on the Danube and Sava and is the undisputed centre of the national road, rail and air networks. It has rail links with three ports on the Adriatic, Rijeka, Split and Šibenik, and with Salonika and Athens on the Aegean as well as with Sofia, Budapest and Bucharest. Besides functions associated with all capital cities it has a fine university and is also engaged in food processing, ceramics, and leather and textile industries. Coal, lead ores and building stone are all close at hand.

Zagreb (503 000, with 900 000 in its conurbation) is the second largest city in Yugoslavia. It stands on the northern valleyside spurs and floodplain of the Sava in a rich valley with abundant water power. Besides being an important financial and military centre it has a miscellany of industries including linen, carpets, foot- and headwear, pharmaceuticals, food processing, chemicals and varnishes. Its urban layout consists today of an agglomeration of two old spur-top settlements with a fairly old lower city that has been conjoined with a railway station complex and a new industrial quarter near the Sava. The township was never occupied by the Turks and has remained essentially Croatian.

Skopje (270 000), capital of the Macedonian Federal Republic, lies in the Vardar valley on the Orient Express route between the Danube plains and Salonika and is probably the chief route centre of the Balkans. It has railway workshops as well as food-processing industries concerned mainly with flour, sugar and brewing. Local silver mining has encouraged extensive art and craft manufactures. It is a noted centre for research in agriculture and tropical medicine. Since the disastrous earthquake of mid-1963 most of the city has been rebuilt as high-rise structures in an international style.

Sarajevo (227 000), capital of Bosnia, occupies a beautiful site in a basin at 550 m commanding the valley of the Bosna river, a tributary of the Danube which is followed by a main road and railway that continue southward over a relatively low pass (967 m) to the Neretva valley and so the Adriatic. Its chief industries are pottery, silk, flour milling, sugar refining, ball bearings, brewing, tobacco curing and embroidery and carpet weaving. It has in parts a decided eastern aspect with numerous mosques of the Bosnian Moslems.

Ljubljana (206 000), capital of Slovenia, is an important regional centre situated in the basin of the Ljubljanica river, a tributary of the Sava. Legend says that it was founded by Jason but its earliest certain beginning was as a Roman legionary fort. It remains a notable route centre with numerous industrial and cultural establishments and many attractive baroque buildings. Its main manufactures are concerned with textiles, pharmaceuticals, ceramics, paper, leather, timber, tobacco, building materials and food processing.

Brief mention only can be made here of the other Yugoslav cities with over 100 000 inhabitants. Maribor (153 000), northeast of Ljubljana, is a regional centre in a mountain basin formed by the Plij river, a Drava tributary. It has much the same industries as at Ljubljana, aided likewise by local hydro-electric developments.

In northeastern Yugoslavia Novi Sad (162 000), capital of the Vojvodina Autonomous Region, is a prosperous agricultural centre with textile, ceramic and food-processing concerns as well as port activities on the Danube. Farther north, Subotica (122 000) also undertakes food processing and engineering and is the main centre of the Hungarian minority. Osijek (119 000) combines agricultural industries with port activities on the navigable Sava river.

Farther south in the Morava corridor, Niš (145 000) is sited athwart the great route to the Vardar valley and the Aegean. Here the main railway and modern highway system bifurcate, sending branches to Sofia and to Skopje.

Not surprisingly it was a strong fortress under the Romans and Turks. Today it is mainly concerned with railway repairs, locomotive making and iron foundring, and has a local thermal power station.

Among the Bosnian mountains, Banja Luka on the Vrbas river, and on the Adriatic coast Split and Rijeka, each has about 130 000 inhabitants.

Rijeka (formerly Fiume), the most northerly Yugoslav port, is the natural outlet for the middle Danube lands to the Mediterranean, and favourable freight rates on Yugoslav railways ensure for it a considerable transit trade from Czechoslovakia and Hungary. It handles about 2 million tons of cargo annually. It is also the sea gateway for the rich agricultural and industrial areas of Vojvodina and Slavonia, and deals with over half of Yugoslavia's total seaborne trade. Rijeka shares with Pula and Split the Yugoslav shipbuilding industry that employs all told 20 000 workers and produces ships of up to 250 000 tons mainly for foreign merchants.

Split (Italian Spalato) has the finest harbour on the Adriatic with a deep safe anchorage in a broad bay. Added to this it occupies a central position reinforced by rail and road communications across the mountains to the Morava–Vardar corridor. Its manufactures include cement, aluminium and chemicals, using local raw materials and hydroelectricity from the Jadar, Krka and Cetina rivers. It handles about one-quarter of the national seaborne trade. The town focuses upon the waterfront palace of the Emperor Diocletian and is fortunate in having a sizeable coastal plain for modern expansion, a facility lacking at Rijeka.

Communications and trade

Transport facilities rival power resources as a key to economic development and unfortunately Yugoslavia in this respect has the lowest per capita railway length in Europe outside Albania. All but 6 per cent of the railways are single track with trains crossing at stations, and about one-fifth of the total length is narrow gauge. Moreover the permanent way and rolling stock generally are such that carrying capacity and speed are both low. Since 1960 conversion of narrow gauge to standard gauge has been active and several new railway links built and some stretches electrified.

The Dinaric area, a large triangular core with its wide base along the crest of the Dinaric mountains and its apex at Belgrade, is poorly served by railways. On the three sides of this triangle, railway systems show infinitely better development.

The Pannonian lowlands in the north handle approximately two-thirds of all railway freight and the line from Belgrade to Ljubljana has been double-tracked and electrified. The Morava–Vardar corridor and the Dalmatian coast handle each about one-sixth of the traffic. Railways across the core area linking these three concentrations are difficult and expensive, requiring frequent tunnelling, embanking and bridging. The line from Sarajevo to Ploče took eight years to construct and in the distance of 195 kilometres needed 71 viaducts and 106 tunnels. However, three important links have already been built to serve this broken mountain country, in which the grain runs against the direction the railway lines need to take. They are from Belgrade to Salonika by a more westerly route through the mountains via Kragujevac and Kosovska Mitrovica; from Zagreb to Split and Šibenik; and from Belgrade to Dubrovnik linking up Višegrad and Titovo Užice. Planned for the future is a longitudinal railway line from Rijeka to Prahovo, and a line to link Split more directly with the Pannonian lowlands. At present most traffic travels from Belgrade to Rijeka on the Adriatic since freight charges are low on these lines. On the other hand motor roads with good trucking and passenger service are probably the more suitable transport provision for the highland districts. The Brotherhood and Unity Highway already runs from Austria to Greece via Jesenice, Ljubljana, Zagreb, Belgrade, Skopje and Gevgelija. An Adriatic Highway follows the Mediterranean coast, but more inter-republic roads and good motor roads in mountainous districts are sorely needed.

Water transport is operative on the Danube (585 km) and Yugoslavia has a fleet of over 1000 river craft. The completion of the Danube–Tisa–Danube canal to take 1000-ton barges has increased the traffic of Belgrade which handles about 2·5 million tons of riverborne cargo annually. The completion late in 1972 of the dam and power station at Djerdap at the Iron Gate has made the journey easy for barges of 5000 tons whereas formerly the 4 million tons of cargo proceeding upstream needed assistance in the Sip canal cutting. The Sava is navigable to Sisak, the Tisa throughout its course (160 km) in Yugoslavia, and the Drava to above Osijek. Thus many river ports, such as Osijek, Sisak, Bosanski Brod, Novi Sad, Smederovo and Prahovo, handle over ½ million tons of traffic annually, particularly fertilizers, mineral ores, petroleum, sand and building materials and coal. The main waterway hazards are spring floods and ice floes in midwinter.

Foreign trade

The greater part of Yugoslavia's foreign trade is carried by rail and road. In the early 1960s about 35 per cent of the exports went to West Europe and 41 per cent to East Europe. These consisted by value mainly of manufactured goods of all kinds (60 per cent total), foodstuffs (20 per cent total), timber and non-ferrous ores and concentrates. The imports came mainly from East Europe (40 per cent) and West Europe (29 per cent) and were formed largely of machinery and other manufactured goods (over 50 per cent) and foodstuffs, particularly bread grains in years of poor local harvests. The table (left) shows the trend in recent years towards more trade with the capitalist world, although the type of goods remain the same. The external trade movements, present and future, are not entirely uninfluenced by federal politics, especially as new port facilities are needed on the Adriatic, where the main ports are in Croatia. Koper, near Trieste, is being developed by Slovenia but has a poor harbour and, as yet, no railway. Ploce near the mouth of the Neretva river is being developed by Bosnia–Hercegovina as a bulk-cargo port. It actually lies in Croatia but is very close to the Hercegovina boundary, and not far from the main railway up the Neretva valley, which has recently been rebuilt on the standard gauge. In the far south, Bar in Montenegro is being developed as an Adriatic port for Serbia. It is already joined to Titograd by a standard-gauge railway and financial loans from the World Bank have been made available for a line from Titograd up the Morača, Lim and middle Drina valleys to Belgrade and the general Yugoslav network.

This new port and railway development, due partly or largely to inter-republic aspirations, has reduced the role of Rijeka in Yugoslavian foreign trade from just over half in 1951 to about one-quarter. The problems here, as with

RECENT TRENDS IN YUGOSLAVIA'S FOREIGN TRADE AND FOREIGN EXCHANGE INCOME

Foreign Trade	Exports 1968	Exports 1969	Imports 1968	Imports 1969
Italy	2 207	2 537	3 357	3 612
West Germany	1 519	1 807	3 998	4 380
Other EEC countries	690	915	1 410	1 440
United Kingdom	759	952	1 092	1 334
Other Efta countries	1 183	1 205	2 006	2 428
Comecon	5 398	5 013	6 072	5 758
Other European countries	630	560	514	680
Afro-Asian countries	1 946	1 923	1 972	2 490
United States	1 118	1 087	1 126	1 033
Other countries	293	464	906	937
Total visible trade	15 743	16 463	22 453	24 092

Value in millions of new dinars

Figures cover the period January–November for each year
Source: *Indeks* (Belgrade)

Foreign Exchange Income	1968	1969
Savings of Yugoslav workers abroad	122·3	205·9
Emigrants remittances	40·6	45·3
Tourism	187·7	241·5
Other sources	33·3	44·6
Total	383·9	537·3

Value in millions of dollars

Source: *Ekonomska Politika*

those concerned with improving the internal transport system, involve more than economic considerations. Modern Yugoslavia originated, we repeat, as a kingdom in 1918, and today is a Socialist or People's Federal Republic consisting of six republics and two autonomous districts formed in November 1945 and working mainly under a constitution of early 1946. For centuries previously in the Balkans the effects of Turkish rule or neglect, the lack of political cohesion and stability, and the relative absence of factory industrialization had, with the aid of geographical difficulties of terrain and of differing cultural roots, created a greatly underdeveloped area, mistrusted by foreign investors. Since Yugoslavia federated under Marshal Tito, the internal progress has been phenomenal and the external attitude to it has steadily changed. In 1968 the motor firm of Crvena Zastava attracted the investment and further inflow of foreign capital (Fiat's) and within twelve months signs of increasing private foreign investment in Yugoslavia became truly realistic. Foreign government or international credit had already been forthcoming for economic installations and for highway and rail modernization. In the spending of this and other funds there can be seen the influence of federal policy and of the needs of the individual republics.

The federal government policy is to develop industries which will improve the income level and living standard in areas where rural overpopulation and unemployment are most serious. Needless to say many of these districts are in the mountains, especially in the south. But the federal government has also to pay attention to strategic and political factors. First, vital industries are best placed away from troublesome frontiers; and second and more important, attention must be paid to the needs of each individual republic and autonomous region in order to dispel or forestall possible political discontent. It is the latter influence which has encouraged, for example, the development of several ports on the Adriatic, such as Bar, and the widespread distribution of iron and steel factories. True, the Adriatic coast is long but the cost in the south of linking inland territories to it by transmontane railway is enormous. Similarly the widespread distribution of raw materials for iron and steel making in Yugoslavia may encourage a widespread distribution of such metallurgical concerns but it appears to be governmental, social and political policy rather than truly economic advantages that have caused steelworks to be established, for example, at Nickšić in Montenegro and at Skopje in Macedonia. At the latter the ore is very low grade and the coke supply is derived from the Kosovo lignite field, some distance away. However, everyone praises the resultant decrease of part- or under-employment in agricultural occupations.

The improvement of communications and greater use of motor transport (the number of motor vehicles rose from 55 000 in 1956 to 530 000 in 1968) will be advantageous to federal unity and to the progress of each republic. At the same time tourism brings increasing wealth. On the Adriatic coast and in the Julian Alps it has long been a prime source of revenue and is becoming embarrassingly popular, the problem being how far new facilities such as international-style skyscrapers should be allowed to alter the existing landscape. Yet even in the north on the flatter Sava–Drava–Danube plains and in the east along the great Morava–Vardar corridor the main roads are thronged with traffic at least in the warmer months. Here new hotels, as at Maribor and Leskovac, offer highly competitive prices and enjoy a well-deserved popularity. They remind us that Yugoslavia links not only the Mediterranean and Danubian countries but also western Europe and the lower Danube and Asia. Given peace and political cohesion, it is internationally by far the best placed of all the Mediterranean peninsulas. In early times for several millenia the valley plains of the Danube system formed the greatest overland route into western Europe from the southeast. Today traffic flows most freely south of the COMECON countries and the Sava–Drava–Danube routeway has resumed its ancient predominance. The result is reflected statistically in the table of foreign exchange income which shows that tourism will soon yield profits in excess of those derived from the remittances and earnings of Yugoslavs abroad. The latter number at least $\frac{1}{2}$ million and, of course, are few compared with the emigrants of for example Italy and Spain. But it is today important to notice that in the national balance sheet of Mediterranean countries they earn large amounts by selling their services abroad as well as by serving at home visitors from foreign countries.

22 Albania

Rise of the modern state

Apart from one short break, from 1385 to 1912, the territory occupied by modern Albania was under Turkish control. The present state took its origin from a League for the Defence of the Rights of the Albanian Nation formed by a group of Albanian leaders in 1878. The Turks, however, reacted to the League by more severe oppression including restrictions on the use of the Albanian language and in these were stoutly supported by the Greek Orthodox church. Gradually the Albanian revolts became more frequent and in 1912 the Turks agreed to make a separate autonomous province of the four vilayets of Shkodër, Kosovo, Ionanina and Monastir (Bitola) and to recognize the use of an official Albanian language. But when the Turks were thoroughly defeated in the Balkans (see pp. 261–2) the Albanians declared themselves independent of the Ottoman Empire.

In 1913, at the London conference following the first Balkan War, Albania was recognized as independent. Disaster then struck the country. In the First World War it was devastated, in part by a retreating Serbian army; then the Austians in the north, the Italians on the coast, and the French in the south occupied it. In 1921, the great powers reaffirmed the 1913 boundary which was now demarcated largely along crestlines around the mainly Albanian-inhabited territories.

When the Second World War broke out the Italians took over the government of Albania and when they withdrew in 1943 were replaced by the Germans. Albanian partisans, with great aid from Yugoslav partisans, largely caused the withdrawal of the Germans in 1944. Two years later Albania became a People's Republic and very close ties were established with Yugoslavia, although difficulties later began to arise between the two over Kosmet (Kosovo–Metohija) in Yugoslavia which had an Albanian majority.

In 1948, Stalin expelled Yugoslavia from the Cominform chiefly because it was too independent and Albania then chose to join the Cominform. However, in 1961 the Albanians broke off diplomatic relations with the U.S.S.R. and aligned themselves more closely than before with the Chinese People's Republic, which has since provided them with financial, technical and material aid. Albania has not attended meetings of COMECON and of the Warsaw Pact since 1962 but is presumably still a titular member of those bodies.

Structure and relief

From the Drin valley near the northern Albanian frontier to the Vijosë valley in southern Albania the Dinaric Alps change their northwest–southeast direction and run instead from north to south for about 170 km before returning to their original trend in Greece. In this stretch the mountains overlook a triangular coastal lowland that is for the most part gently undulating with ridges ending at the coast in low headlands divided from each other by swampy depressions. Thus this coastline is the only discordant stretch on the eastern shore of the Adriatic.

The mountains encircling this lowland average about 1000 m in height but many summits exceed 2000 m and the highest peak reaches 2694 m. Beyond this rugged relief in the southeast is a series of fault basins (graben) still occupied in parts by Lakes Ohrid and Prespa. This lake region is shared with Yugoslavia.

Not only is the coastline of Albania different from that farther north, and the coastal lowland more extensive than elsewhere on the eastern Adriatic shore, but the structure, whether of Tertiary strata on the lowland or of older Jurassic–Cretaceous sedimentaries on the mountains, is much poorer in calcareous rocks. The relief in the east, and especially in the drainage of the Black Drin, is made extremely rugged by the intrusion of igneous rocks such as serpentine and gabbro and by intensive glaciation in the past. The drainage pattern is also very different from that in the Yugoslavian Dinaric Alps. On the Dalmatian coastline only the Neretva river breaches the whole system, whereas in Albania it is common for streams like the Drin, Shkumbi, Devoll and Vijose to flow from the interior across the main highland range. It is this drainage enclave on the Adriatic side of the Balkans that forms most of the independent country of Albania, 'land of the mountain eagle' with an area of 28 748 sq km and a population of 2 million.

Climate

Albania is nowhere more than about 150 km from the Adriatic and its climate on the lowlands is typically Mediterranean, with mild winters and a marked summer drought. Even near Lake Ohrid the mean of the coldest month remains about zero (see statistics for Florina in northern Greece). The rainfall is everywhere abundant as the arc-shaped lowlands surrounded by their high mountain amphitheatre are exposed to bora airflow down the Adriatic as well as to hot southerly sirocco winds which here can bring heavy rains. Most of the lowland has over 1000 mm of rainfall annually and this increases rapidly to 1700 mm on the foothills and to well over 2000 or 3000 mm on the higher mountains. Probably only in the mountain-girt lake basins and valleys of the east near Korçë does the mean annual precipitation not exceed 800 mm. The heaviest rains on the coastal-facing slopes are in autumn and winter whereas in the east a slight resurgence of rainfall occurs in late spring and early summer.

Land use and agriculture (fig. 99)

Prior to the declaration of the People's Republic in 1946 Albania reflected the general condition and way of Mediterranean life before the age of steam. Ninety per cent of the population was engaged in agriculture, one-third of them growing food crops, mainly cereals, and the remainder occupied with transhumant pastoralism. The total area of cultivable land, largely along the Adriatic littoral and in a few inland valleys, amounted to 2·8 million ha but of this

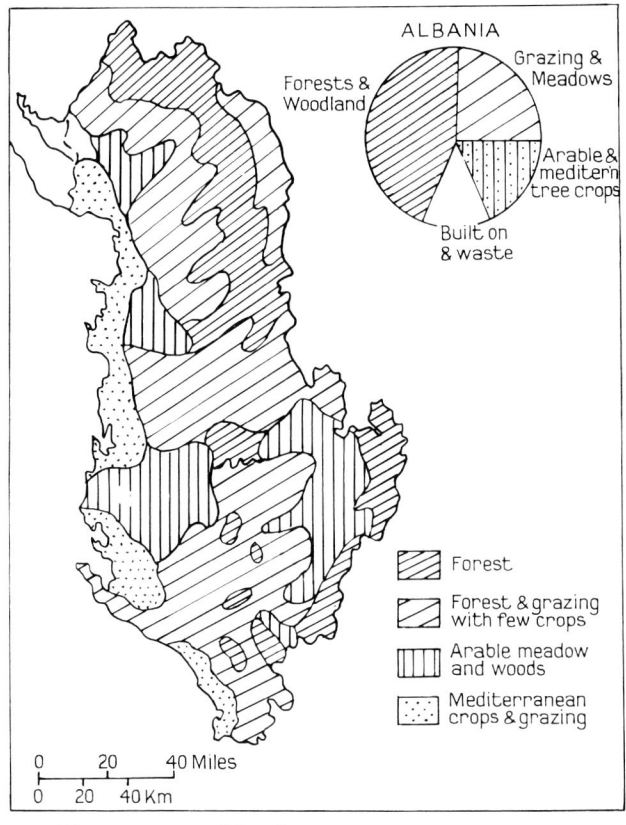

99 Albania: general land use

a mere 6 per cent was under field crops, another 6 per cent under tree crops and the large remainder under unimproved pasture. The rest of the country being wild, rugged and mountainous was under forest, waste and swamp or lake.

Cultivation was extensive in character, backward and declining in yield, through lack of new seed strains, fertilizer and soil maintenance. The work was normally undertaken by the women with ox-drawn ploughs or hand hoes while the men cared for the beasts. The farming was entirely subsistent with animal products providing necessary cash. The farms were small and poor, ranging from 3 to 7 hectares and the unenclosed fields ran the risk of being trampled and cropped by stray beasts. The animals were grazed on open mountain pastures that cover one-third of the country and consists of maquis, shrubs and shiblyak. No winter feed was grown specifically for the animals which depended for autumn and winter feed on maize stubble and on leaves and branches.

After 1946, under various five-year plans agricultural reform led to most of the farmland being organized (collectivization) into 24 state farms responsible for 69 400 ha, 1484 cooperatives with 325 600 ha, and 250 'local agricultural enterprises' with 10 800 ha. Private farms still number 43 500 but work only 49 000 ha, a very small proportion of the cropland. Most of these are isolated plots in mountainous areas not yet organized into groups for mutual help and improvement. By 1964 about 7400 tractors were in use.

The national land use in recent years has been divided as follows: agricultural crops, 18 per cent; pastures 25 per cent; forest 44 per cent; and built-on and waste 13 per cent. The nature of the crops grown has not altered but their relative importance has. Wheat and maize (about 121 000 ha each) are by far the chief cereals, and are followed in area sown by tobacco and cotton (23 000 ha each). Sugar-beet occupies about 6000 ha and rice nearly 4000 ha. The same crops are grown in mountain basins as in coastlands, the chief difference being that as the low-lying tracts are drained and reclaimed the yields are higher and the cultivation more advanced than in isolated districts inland where the methods tend to be more biblical in character. Rye and oats are largely restricted to the highlands.

The only types and areas of specialized production are olives on the coastal plains; vines on the sides of the Saranda and Semen valleys too steep for the plough; citrus fruit around Elbazan and Vlonë (Vlorë); tobacco around Shkodër (Scutari); and rice on the drained coastal swamps. Elsewhere each farm grows its own needs.

The remarkably high percentage of Albania under pastures (25 per cent) and forest (44 per cent) reflects the great importance of animal husbandry. In 1968, there were 3 641 000 farm animals or nearly 2 per inhabitant. These included over 120 000 equines, 428 000 bovines, 1·7 million sheep and 1·2 million goats. The horses and cattle were largely for draught, and the sheep and goats for milk, cheese, meat, wool, mohair and hides. The maize stubble and sugar-beet residue supply additional fodder for lowland cattle but most of the sheep and goats are confined to mountain pastures. The remarkably few pigs (150 000), as also the infrequent specialization in the vine, reflect the beliefs of the predominantly Moslem society who neither drink wine nor eat pork.

Most of the forests are secondary bushgrowth (*shiblyak*) used for pastoral purposes and for fuel much as is the *bosco* of the Italian Apennines. Two-thirds of the annual timber cut of 1·5 million cubic metres is consumed for fuel. Tall forests survive only on the higher, more inaccessible mountains, especially in the north. The great need for reafforestation to prevent soil erosion and as a profitable crop has led to some recent planting.

Power resources, mining and manufacturing (fig. 100)

Since 1941 a few small thermal electricity stations have been set up and have been supplemented by 6 hydro-electric plants, 3 being in the south and 3 near Tiranë. In 1967, of the total electricity output (550 million kWh) about 70 per cent was from water power. The potential of hydro-electricity is large and its development would easily supply all power needs for a long time to come, a fortunate condition for Albania which is strikingly lacking in solid fuels. About 340 000 tons of lignite are produced annually

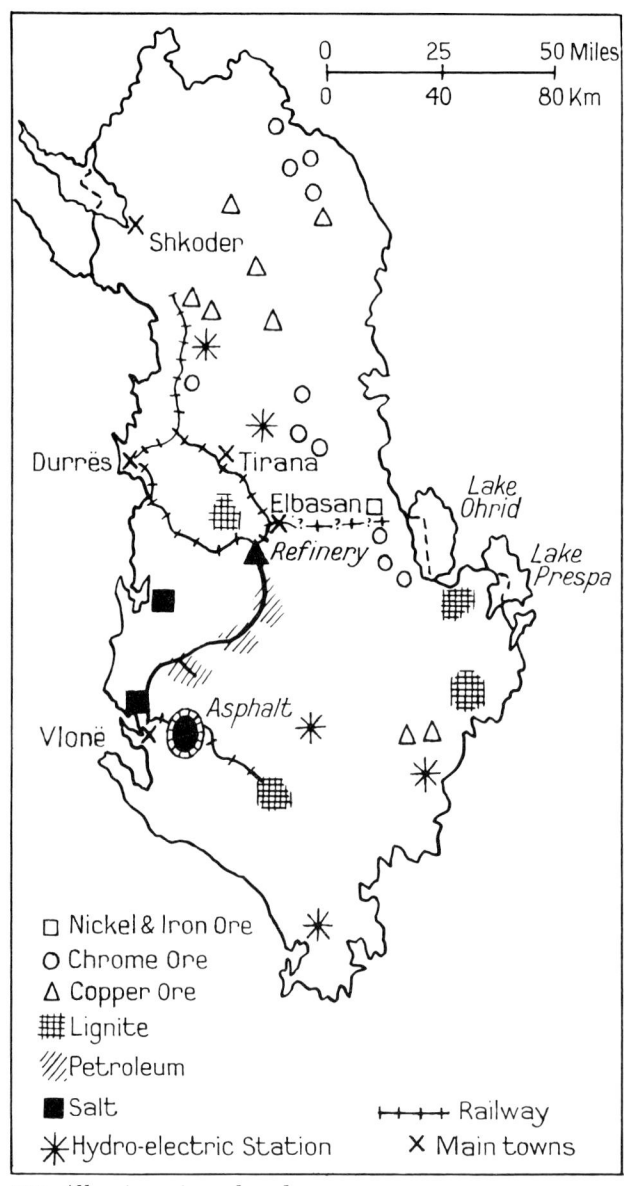

100 *Albania: mineral and power resources*

recent years. In addition, rock salt deposits near Vlonë and Durrës are large enough to be exploited commercially.

For metals, the main products are chrome ore (300 000 tons), copper ore (145 000 tons) and nickel iron ore (380 000 tons). The chromite occurs widely in association with igneous intrusions in the eastern mountains and is most worked in the high ranges overlooking Lake Ohrid and rising steeply east of the Drin valley in northern Albania. Low-grade copper ores are mined between the Drin and Mat valleys in northern Albania and a new copper refinery has been built at Kukës and an electric wire and cable factory at Shkodër. Some copper ore is also mined near Erseka in the southeast. Most of Albania's iron ore deposits are of low grade, except those near Pogradec close to Lake Ohrid. This area, formerly, worked by the Italians for export to Italy, today produces nickel iron ores which under a recent five-year plan are destined for a metallurgical plant at Elbasan.

Manufacturing industries

As yet the chief industries that have been established are concerned with processing agricultural surpluses such as textile fibres, sugar, cereals, grapes and tobacco; but, with aid from China, new engineering, chemical and cement works have been built. A fertilizer factory has been established at Lezhe; cement plants at Krujë, Shkodër and Vlonë; a copper smeltery at Kukës; a chemical plant for caustic soda at Vlonë; a paper factory at Lushnjë; and textile concerns at Tiranë and Berat. The best that can be said of the distribution of these new plants, as distinct from the widespread distribution of handicrafts of all kinds, is that they are concentrated on the coastal lowland between Elbasan and Durrës in the centre, and to a lesser extent around Shkodër in the north and Vlonë in the south. However, for some decades yet many of the consumer goods will be supplied by local craftsmen who have been organized into cooperatives and given output quotas to aim at. The tremendous increase in the output of cement, bricks and tiles exemplifies what may be expected of the larger industrial schemes.

In the absence of exact statistics it is difficult to judge the real progress of industrialization, as the people who took up factory jobs would often have been formerly engaged in handicrafts at least part-time. In 1965 of the total labour force of 840 000 persons about 490 000 or 58 per cent were primarily engaged in agriculture and fishing. Ten years earlier the proportion was 70 per cent. Whether the difference expresses the growth of industrial employment is another matter as transport, building and services have expanded rapidly as well as manufacturing. Probably today manufacturing and mining employ more than 15 per cent of the labour force. However, more important is their impact on the national economy. In 1955, they provided 28 per cent of the gross national product against an estimated 48 per cent in 1969.

from deposits near Tiranë and in the south in the Vijosë valley. For liquid fuels the country is better supplied as petroleum deposits have been found along the junction of the coastal plain and foothills, especially between Elbasan and Vlonë (Vlorë). A pipeline runs from Vlonë to a refinery at Çerrik just south of Elbasan. The annual output of crude petroleum has increased recently to 764 000 tons in 1964 and to nearly 1 million tons in 1968.

Albania possesses considerable other mineral wealth and, as with water power, their commercial working on any scale began after the First World War under Italian concessions. The asphalt bitumen deposits at Selenicë were worked in classical times and today are sent by narrow-gauge railway for processing at Vlonë. The Albanian output of crude naphtha has exceeded ¾ million tons in

Population and settlements

The total population of Albania in 1966 was about 1·9 million or the equivalent to 67 persons per square kilometre (fig. 101). Most of these are Albanians who claim to be descended from the original Balkan inhabitants who were later pushed into their mountain fastnesses by waves of steppe and other invaders. However, today, except in the most isolated mountain districts of the north, they have some Turkish, Italian, Slav and Greek elements in their ethnic make-up. The Albanians lacked an alphabet and when attempts were made to devise one in the late nineteenth century, the people of the north, mainly Roman Catholics, favoured Roman characters and those of the south, strongly influenced by the Greek Orthodox church, preferred Greek (Cyrillic) characters. Eventually in the early twentieth century agreement was reached to use a Latin script.

However, as well as numerous dialects, the Albanian language still retained two branches which although mutually intelligible differ appreciably in syntax and vocabulary. Today *Gheg* is spoken by most people north of the Skhumbi river and *Tosk* by most people south of it, the number of speakers being about equal. The difference may be seen in the place names, for example Vlonë (Gheg) and Vlorë (Tosk). The chief minority groups are Greek in the south, although most of these do speak Tosk; a few Slav-speaking Bulgars near Lake Ohrid; a few thousand transhumant Vlachs in the far south; and small bands of gypsies who follow an itinerant way of life. Moslems are by far the dominant religious group (65 per cent total) followed by Greek Orthodox (25 per cent total) and Roman Catholic. The last-named survives especially in the mountainous northeast.

The social organization and customs differ appreciably between north and south. In the north, and particularly in the lofty mountains, inbred tribes tend to control individual valleys. This has led to endless conflicts governed by the unwritten law of Lek which among other rules gave the execution of redress for wrongs to the kin of the wronged. The inevitable blood feuds today persist only in the mountainous north. Yet in parts the men feel naked without an ammunition belt and a musket and they look more like hunters than herders. It remains to be seen how far and how rapidly the new social code, social welfare and education will eradicate the worst of the ancient customs of the north.

The need for defence and certain unwritten laws prohibiting light-over-light or windows overlooking other people's private yards make the rural houses crudely strong and almost without windows, and turn small villages into loosely grouped dwellings. Isolated houses are rare. In the north and centre the domestic buildings sometimes become *kullë* or towers up to four storeys high similar to the pele towers on the Scottish–English border. However, in central Albania these were often built partly as watchtowers by powerful beys (landowners) during the

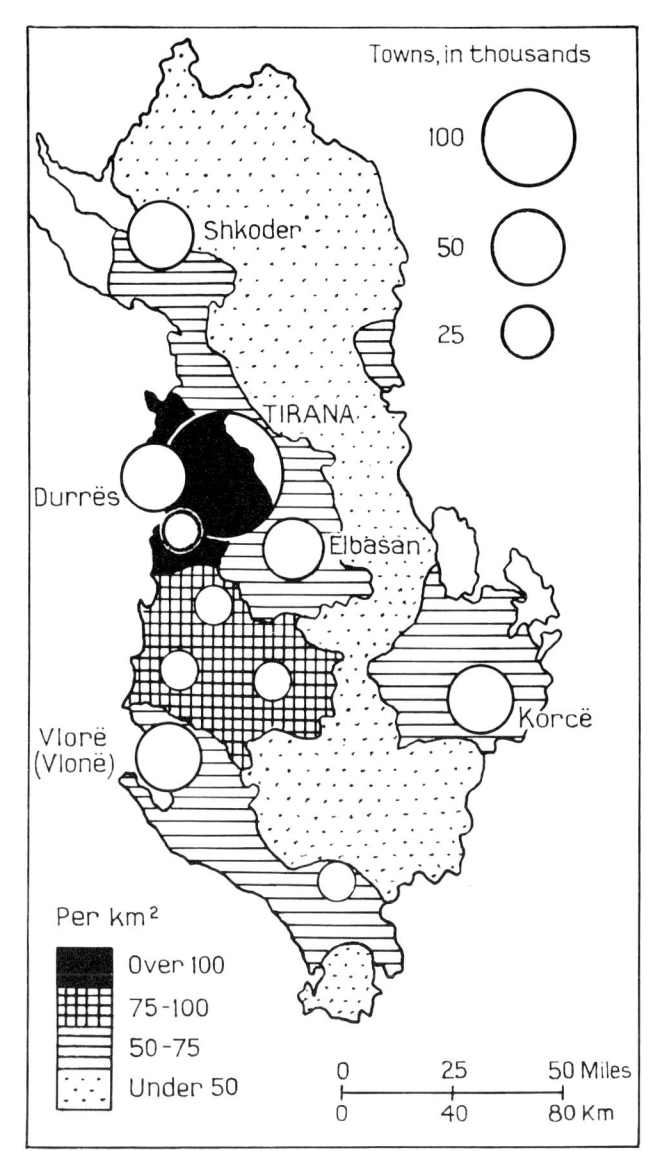

101 *Albania: density of population and chief towns* (estimated for 1969)
Towns with 40 000 or more inhabitants are named

long Turkish domination.

Town life has never developed strongly in Albania although the ports would be expected to stimulate urbanization. Today for local administration the State is divided into 26 regions, all too small to encourage the growth of large regional centres. The communications system emphasizes the nature of the towns. Nearly all transport is done by road and horsetrack. The road network has always been a legacy of invaders and conquerors. The Romans left a fine road from Durrës to Lake Ohrid and so to Constantinople. The Austro–Hungarians during the First World War left a motor-road network, as also did the Italians during the Second World War. Whereas the earlier legacies were soon largely destroyed, the last has been recently repaired and extended. Although for the most part not asphalted, about 3500 km are today suitable for motor traffic and only the steeper and mountain districts remain inaccessible to wheeled vehicles. The standard-gauge railway system was begun by the Italians in 1940 and carried on later with the aid of the Yugoslavs and Russians. It consists of a circular stretch from Durrës to Tiranë and Elbasan with a few short branch lines, including one northward to Milot on the Mat river, and another under construction to Polgradec on Lake Ohrid. There is also a short new line from Vlonë (Vlorë) inland to Milot and to Memaliaj on a lignite field in the Vijosë valley. With a total length of 151 km these railways carry 1½ million tons of freight annually or less than one-tenth of that carried by road.

Thus the urbanization owes little to highways and railways. Up to the 1930s when Italian influence began to be felt in the larger towns, most were unpaved and lacking in modern service facilities. Since 1950, considerable urban growth and civic progress have occurred but many towns remain handicraft centres with local markets and few or no urban services and amenities. It is hard to decide when some of the settlements cease to be purely agricultural and rural since domestic handicrafts are widespread. In 1961 about 31 per cent of the population was classed as urban in the official census.

All the larger towns are on the coastal lowland except Korçë which lies in a mountain basin south of Lake Ohrid (fig. 102). The only city is Tiranë (160 000), the national capital, that stands on the inner edge of the central coastal enclave amid the best arable soils in Albania. Its administrative core is typically west European but soon merges into almost rural suburbs rich in gardens and olive groves. It is the chief Albanian centre for light engineering, textiles, brewing and food processing and produces about half the national manufacturing output by value.

Five other towns have a population of between 40 000 and 50 000. Probably the second largest Albanian town is Durrës (Durazzo) which faces a shallow Adriatic bay that is the best harbour in the country. It acts as the port for the capital and the chief industrialized region. Shkodër (Scutari), which is almost as large, lies on a narrow plain east of the lake of that name. It is accessible only by shallow-draught barges but has been stimulated recently by new cement factories. Vlorë or Vlonë (Italian Valona) lies a short distance inland from a bay on which it has a small port. It is connected by narrow-gauge railways to lignite and asphalt fields inland and its industries include chemicals and cement. Korçë, the centre of the Lake Ohrid–Prespa region in southeast Albania, has new industries in metal mining and petroleum. The sixth town in Albania by size of population is probably Elbasan (about 40 000), with a markedly Turkish and Moslem aspect and recent additions of ferrous metallurgy and light engineering.

The trade of Albania emphasizes the as yet relatively small industrial function of the towns and the present political orientation of the republic towards China. In recent years the greater part of the exports by value have been copper, other mineral ores, tobacco and cigarettes, wine and wool. Forty per cent goes to China, 19 per cent to Czechoslovakia, 10 per cent each to East Germany and Poland, and the rest mainly to other west European countries. The imports are predominantly machinery and equipment, raw materials, chemicals and some foodstuffs. Sixty-three per cent comes from the Chinese People's Republic, 10 per cent from Czechoslovakia, 7 per cent from Poland and most of the remainder from nearby European countries. The close connexion with China is unique in Europe and Albania is as it were the champion and mouthpiece of that great country in the western hemisphere. The continued closing of the Suez Canal considerably lengthens the sea link between the two. It is to be hoped that Albania derives large financial aid from its ally as it sorely needs abundant capital investment.

There has recently been a strong governmental effort to alter or amend old customs and habits such as child marriage and the growing of long hair and beards. The last mosques and churches remaining open were closed down in 1967; the vendetta was outlawed, and disapproval expressed against the traditional costume of baggy trousers (*citjanes*) for women and the fez, embroidered waistcoats and calf-high boots for men. Ironically most of these traits were then becoming popular with the *avant garde* in several western countries. In 1969, on the twenty-fifth anniversary of the establishment of the Albania People's Republic, personal income tax was abolished throughout the state. At the same time tourism seems to be growing more freely on the fine Mediterranean beaches.

23 Greece

The modern state

The early history of Greece and the political setting in which the present state achieved independence have already been discussed. The modern state became autonomous in 1827 after the Battle of Navarino and its independence was recognized by the Sultan of Turkey in 1829. Greece then had about ¾ million inhabitants or about 20 persons per sq km. Subsequent boundary changes, included the annexation of the Ionian islands in 1864, of Thessaly and Arta in 1881, and of Macedonia, Epirus, Crete and the eastern Aegean islands in 1912–13.

During the First World War the Greeks tried hard to remain neutral but were eventually in 1917 forced by the Entente to fight against the central European powers. As a result of this, Greece in 1919 acquired Thrace, the Dodecanese (temporarily as it happened) and were allowed to send troops to occupy Smyrna (Izmir). In 1921–2 there followed a disastrous war with Turkey in which the Greeks were decimated and Smyrna was virtually destroyed. At the treaty of Lausanne in 1923 Greece and Turkey agreed to an exchange of Orthodox for Moslem populations, and nearly 1½ million Christians were for the most part compelled to migrate to Greece where a great many became squatters in the suburbs of Athens strongly dependent on the International Red Cross. As Greece and Bulgaria had in 1919 on a voluntary basis exchanged populations in this way, the ethnic competition of Greece was now much more homogeneous but the economy had suffered grievously during the wars with Turkey and had few openings to provide employment for 1½ million extra citizens.

Since 1923 the country politically has alternated between a republic and a monarchy and socially between communism and right-wing democracy. There have been intermittent internal troubles which were heightened during the Second World War when the country was occupied by Germans and Italians from 1941 to 1944. Partisan resistance to the invaders was most heroic but there were already signs of a serious civil war which, when the Germans evacuated the country, developed into a full-scale attempt at a communist takeover from 1944 to 1949. The destruction and disruption of life were especially serious in northern Greece where considerable areas lost most of their resident population. With vast U.S. aid, eventually peace prevailed and a large-scale reconstruction of the economy was begun. Nevertheless between 1940 and 1951 the national population of over 7 million increased by only 286 000. The postwar recovery was delayed by several disastrous earthquakes in 1953 but it was making itself clearly seen in the life and landscape by 1960.

In the meanwhile Greece's relations with its neighbours were improving. Italy finally ceded the Dodecanese islands in 1947; soon afterwards Greece joined the Council of Europe and NATO. In 1954, agreements of friendship were signed with Turkey and Yugoslavia, and in 1960, after Cyprus gained independence, friendly relations were resumed with Britain. In December 1967, the Greek king fled to Rome and the country, after a military rule, was declared a republic in July 1973.

Politics still loom large in the talk during the noonday siesta but the present controllers have cancelled the debts owed by the small landowners to the government and made agricultural and other credit available and Greece as a whole has never been more prosperous. The great American assistance is today augmented by loans from the World Bank and other international financial concerns and, as the following account hopes to show, given peace and sufficient aid, the modern Greeks may yet succeed in restoring to viability a landscape already devastated in classical times.

'Poverty,' said Herodotus, 'is the inheritance of Hellas', and modern Greece has been described as 'a poor country with all its bones showing'. It consists of mainland Greece (107 000 sq km: 7 million people) and insular Greece (25 000 sq km: 1½ million people). Even the mainland is fragmented, being interdigitated by the waters of the Aegean into numerous peninsulas, coves and inlets that are expressed inland as ribs and basins, some lake-filled and others partially or completely drained. However, the whole is under the suzerainty of the Greek government, although the tip of one of the longer mountainous peninsulas at Mount Athos manages its own affairs under the administration of a religious community.

Structure and relief

Eighty per cent of mainland Greece is mountainous as might be expected of the southernmost extension of the Balkan peninsula. The structural elements of Bulgaria, Yugoslavia and Albania continue here with little difference except close contiguity. On the west the Epirus mountains continue the line of the Alpine or Dinaric fold system and form the mountainous backbone of Greece. Consisting like their northern counterpart of a granite core, they throw off their limestone flanks on the east in separate and distinct ranges that run southeastward to form peninsulas and island fragments with separate lowlands between, so giving rise to the poetic expression that Greece is made up of three elements, mountain, plain and sea.

In Macedonia and Thrace the Rhodope plateau also continues south in a series of horsts separated by northwest–southeast fractures that form longitudinal depressions and rift valleys (for example, the Vardar and Struma valleys) and are segmented further by east–west faults creating isolated enclosed mountain basins. This faulting was accompanied by vulcanism, as seen in the isolated volcanic craters of the islands of Patmos, Nisyros and Melos as well as in lava flows (Methana), hot springs (Loutraki and Thermia) and earthquakes, to say nothing of the oracle at Delphi. The Aegean sea covers the foundered Hercynian and Alpine ranges which can be traced through the island arcs and clusters so that the sea coast here is discordant and forms a remarkable series of bays and headlands. These bays are in process of being silted up and the coastal plains are, for the most part, deltaic, as for example the combined deltas of the Axios (Vardar) and the Aliakmon which are gradually being extended into the Gulf of Salonika. The Pass of Thermopylae, once a narrow strip of beach that could be defended by a handful of men, now extends in coastal deposition for a width of five kilometres.

The Greek mainland may be divided into four structural–relief regions:
1 the Peloponnesus, built mainly of limestones and Tertiary sandstones, flysch and conglomerates
2 central and southeastern Greece, a mosaic of limestones, Tertiary conglomerates and crystalline outcrops
3 Ionian or western Greece, an alignment of limestones and Tertiary flysch and sandstones
4 northern Macedonia and Thrace, a conjunction of crystalline blocks, limestone masses, Tertiary strata and alluvial lowlands

The Peloponnesus is a mountainous peninsula linked to Greece proper by an isthmus only 6 km wide separating the gulfs of Corinth and Aegina that has been pierced by the Corinth canal. The core of the island is the barren limestone plateau of Arcadia that sends long ridges south and east out to sea to form a remarkable ria coastline with long narrow peninsulas deeply interpenetrated by the sea. The western coast is smooth, unbroken and concordant and upon it the longshore drift deposits great beaches of sand. The island covers 21 650 sq km and averages about 1500 m, its greatest height being Mount Taygetos, 2407 m. Four types of lowland may be distinguished in the Peloponnesus.

First, a coastal plain follows the west and north shore in Elis (Ilia) and Achaia, where with ampler precipitation the bulk of the Greek currant crop is grown.

Second, enclosed or partially enclosed mountain basins that rely for their fertility on springs thrown out at the foot of the surrounding limestone hills. Such are Laconia and Messenia and in all of them the centre is dry and dusty or else ill-drained and malarial.

Third, partially enclosed basins that form part of the course of a river such as that of Megalopolis drained by the Alfiós river.

Fourth, *katovothras* or plains which drain by underground means such as the high enclosed plain of Tripolis.

Central and southeastern Greece, the most difficult, broken and discontinuous terrain on the mainland, basically consists of four mountain chains that lead off from the Pindus range towards the southeast and of the lowlands between them. The northeastern chain is part of the crystalline Hercynian core and runs parallel to the coast from Salonika to the Gulf of Volos in a series of isolated horsts that include Mounts Olympus (2911 m), Ossa (1978 m) and Pelion (1651 m) and continues out to sea in the islands of Skiathos, Skopelos and Skyros of the northern Sporades. Enclosed by and within the mountains are old lake basins, Halmyros, Trikkala and Larissa, the last two drained to the Aegean by the river Pinios through the Vale of Tempe that lies between Ossa and Olympus.

Three other ranges run parallel with the first, the Othris, Oeta, and the chain linking Parnassus (2457 m), Helicon and Hymettos. Subsidence and solution of the limestone have broken these ranges into separate mountain blocks and likewise caused the formation of enclosed mountain basins and high plains such as along the Spercheios basin, and the lowland plains of Phocis and Boeotia drained respectively by the Cephisus (Kifissos) and the Asophos (Asopos).

Ionian or western Greece. The limestone ranges on the western side of the Epirus watershed lie in parallel folds and the longitudinal valleys between them have been developed in softer Tertiary rocks, partly sandstones and flysch. The drainage is however mainly underground, rainwater and streams disappearing down swallow holes, called *katovothras*, except where polja have developed. These are the most fertile areas in the region being generally floored by clay or terra rossa. The largest of them, at Yannina, is linked to the sea by the defile cut by the river Arakhthos. Settlements are also at the sites of springs or water issues.

Northern Macedonia and Thrace consists for the most part of the lower courses of the principal Greek rivers and is the only extensive lowland in Greece. In the east the Mesta (Nestos), Struma and Vardar (Axios) flow in a general southeast direction along fracture lines. Faults developed at right angles to this general trend have created the rift furrow occupied by lakes Koroneia and Bessikion which separates Chalcidice from central Greece. Old lake floors within scarp-enclosed mountain basins and indicated by old shorelines, some drained, some, where the exit is restricted, still swampy and malarial, are common features of the landscape. The largest are generally the most fertile, for example the plains of Serrai and Drama on the lower Struma, but beyond the Vermion hills the drier basins tend to be barren. The largest alluvial tract at the combined deltas of the Aliakmon and the Vardar is also the most swampy. Like all Greek lowland rivers except in the west, these are slow-flowing and in the drier months thread their way between banks of shingle and pebbles and clumps of oleander.

The structure and relief of *Insular Greece* are described separately at the end of this chapter.

Climate

The table of climatic statistics shows that Greece has a Mediterranean summer-drought climate everywhere below about 1000 m. No part of the country is more than 90 km from the sea so that the maritime influence is strong. However, the relief is very mountainous and especially in the northeast is often under a northerly airstream coming from the Danube lands and interior Balkans.

During winter when cyclonic depressions often take a track eastward just south of or across Greece, the weather is mild especially on the islands and along the west coast. On

GREECE: CLIMATIC STATISTICS, MEAN MONTHLY TEMPERATURES (centigrade)

	Altitude if over 100 m	J	F	M	A	M	J	Jy	A	S	O	N	D	Absolute Max	Absolute Min	Annual days with frost
Florina	620	0·8	1·7	5·6	9·2	15·3	18·9	23·0	22·3	17·9	12·6	7·2	1·8	40·8	−23	
Kozani	667	2·2	3·1	6·8	9·9	16·7	20·3	23·7	23·2	18·8	13·2	7·5	3·9	40·8	−19	55
Yannina	466	5·1	6·3	9·4	12·9	17·1	20·7	24·1	24·6	24·7	15·3	10·1	6·5	40·8	−9·9	31
Tripolis	661	5·1	6·0	8·5	12·4	16·4	20·7	24·2	24·2	20·4	15·1	10·4	7·1	48·0	−17·0	42
Kerkyra (Corfu)		10·4	10·8	12·5	15·3	19·2	23·1	25·8	26·0	23·1	19·0	15·2	12·1	38·8	−5	2
Arta		8·5	9·4	11·9	15·3	19·4	23·2	26·5	26·9	23·2	18·2	13·2	10·0	44·2	−8·9	15
Zakinthos		11·4	11·5	13·1	15·4	20·0	23·9	26·8	27·0	24·0	19·9	16·0	13·1	39·2	−1·0	0·2
Patrai (Patras)		10·1	10·7	12·7	16·0	20·0	23·9	26·9	27·0	23·9	19·1	14·8	11·8	41·7	−5·0	3
Thessaloniki (Salonika)		5·6	6·7	10·4	14·6	19·3	23·6	26·3	25·9	22·3	17·2	11·4	7·6	41·6	−9·5	20
Larissa		5·4	7·1	10·5	14·8	19·4	24·2	27·2	26·8	22·4	16·8	11·3	7·3	45·0	−13	36
Lamia		7·6	8·9	11·5	15·3	19·8	24·6	27·4	27·0	23·0	18·2	12·7	9·7	44·3	−8·2	14
Athens	107	9·1	9·5	11·5	14·9	19·3	23·7	26·9	26·7	23·3	19·0	14·3	11·0	43·0	−6·7	3
Skiros		11·4	11·6	13·1	16·2	19·9	24·1	26·6	26·5	23·5	19·9	16·1	13·1	40·0	−2·0	0·8
Thira (Santorini)	229	10·5	10·7	12·1	14·7	18·1	21·9	24·4	24·5	21·9	18·8	15·3	12·4	39·9	−4·8	0·9
Iraklion (Candia)		12·2	12·3	14·0	16·8	20·0	23·8	26·1	26·3	23·7	20·8	17·3	14·0	45·7	0·1	0·1
Anoyia	776	7·1	6·8	9·2	13·2	16·3	21·0	22·5	22·4	19·4	16·2	12·8	9·0	37·3	−5·0	8

MEAN MONTHLY PRECIPITATION (millimetres)

	J	F	M	A	M	J	Jy	A	S	O	N	D	Mean annual Total	Raindays	Days with snowfall
Yannina	122	126	105	97	106	80	27	14	59	133	156	172	1195	137	4·8
Tripolis	126	89	69	54	50	37	16	15	26	78	120	129	809	108	9·0
Kerkyra (Corfu)	159	139	93	80	48	26	7	19	63	176	161	201	1172	97	0·7
Arta	144	124	103	76	66	28	10	11	42	147	153	174	1080	101	0·8
Zakinthos	182	134	88	55	30	8	2	11	35	130	207	234	1115	100	0·8
Patrai (Patras)	98	78	67	52	34	17	4	5	28	94	113	120	707	103	0·7
Thessaloniki (Salonika)	35	31	38	42	53	35	25	24	33	59	64	47	486	93	4·8
Larissa	49	43	37	37	53	34	23	19	26	65	72	61	518	98	3·8
Lamia	67	65	53	35	43	37	16	16	24	76	76	76	584	86	3·3
Athens	53	40	30	20	21	16	4	8	16	40	66	69	384	82	3·5
Skiros	102	66	48	27	17	5	1	3	9	15	75	93	487	65	4·4
Thira (Santorini)	77	49	32	21	13	2	—	—	7	24	55	78	357	56	1·8
Iraklion (Candia)	86	72	46	27	23	2	1	7	18	39	100	92	510	97	3·0
Anoyia	215	173	119	54	63	6	2	14	17	71	157	232	1124	130	4·0

lowlands frosts are rare in the west and south but occur on about 20 to 30 days annually on the coastal plains of northern Macedonia and Thrace where the valleys may conduct cold air tongues from the interior down to the seaboard. Inland most of the mountains above 600 m have more than 40 days with frost annually and all the higher parts are snowclad in the coldest months. Snow occurs commonly at Delphi but never lies long, while at Athens schoolchildren may be rushed joyfully to play in the temporary snow on the nearby hills.

Spring is continental in its swift arrival and short duration and by May summer has come with mean monthly temperatures of 19°C or more. Within a few weeks the land is burned dry and dust becomes more frequent. By

July and August mean temperatures on the lowlands rise to 26° or 27°C and occasionally the heat exceeds 40°C. On the coastlands the sea breezes now increase in strength and frequency but the prevailing airflow everywhere is the etesian from the northeast, steady and almost cloudless.

Autumn is long delayed and definitely hot with average Septembers of 23°C and an abundance of sunshine. The islands are especially warm in October and November.

The precipitation shows variations similar to those of temperature with regard to altitude and exposure to westerly and continental airflow. The dominant characteristic is the summer drought which is fierce and prolonged in the south and severe and long in the north. The west and south, including the islands, have a marked winter maximum with a sudden onset of rains in late autumn. The drier northern interior and northeast coastlands have an autumn maximum with a secondary increase in late spring.

In addition to mountain climates above 1000 or 1500 m, it is possible to distinguish three climatic variants of which the most typically Mediterranean occurs on the east in Attica and in the islands of the Aegean. Here the monthly temperatures are relatively high and clear blue skies and a brilliant atmosphere are characteristic. Rainfall is below 600 mm and in parts less than 400 mm and all the hotter months bring a long pause to agricultural activities unless there are alternative water supplies.

The west coast has the mild winters and hot dry summers of the Mediterranean climate but a greater abundance of rainfall in the cool months since the mountain ranges interpose their height across the direction of the rain-bearing westerly winds. Over 1000 mm annually fall normally in the north and over 700 mm in the drier south. A richer and more varied vegetative response makes these

102 *Greece: general land use*

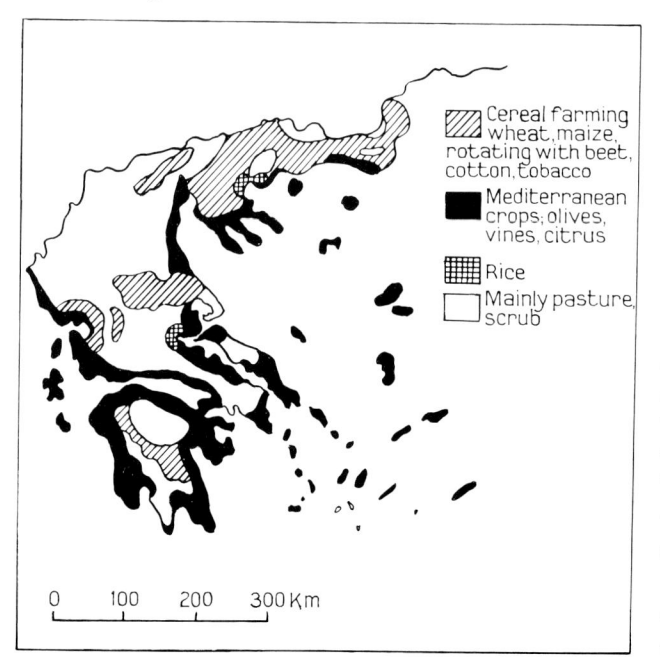

areas approximate more closely to the French and Italian Riviera. To the east and northeast of these two variants lies a more continental type in the rainshadow of the Epirus mountains and of the Albanian highlands. Here in most localities the temperatures swing between cool winters and scorching summers and in all cases the rainfall is less than 550 mm. The olive is rarely grown except on the coastal strip and these areas are usually under dry farming, producing cereals, thinly cultivated in strips and alternating with fodder or fallow so as to use two seasons' rainfall for one crop of grain.

Further details of the weather patterns, climate and local winds are given in Chapters 4 and 20. The characteristic vegetation is also discussed in Chapters 6 and 20 where some details on soils are given.

Land use (fig. 102)

Lowland covers only one-fifth of Greece and less than one-third of the national territory is suitable for cultivation but these disabilities are by no means the complete reason for the natural poverty of the Greek lands. During the course of the human occupation of the country the physical environment has been profoundly altered to the point where it is doubtful if it can ever be fully restored to its former productivity and fertility. From the earliest times the forests of the more accessible slopes have been cut for timber and fuelwood and natural regeneration has been retarded or prevented entirely by soil erosion and by the grazing of goats. This continues today and in rural districts loads of firewood are taken daily, often on the backs of donkeys or mules, to feed bread ovens in nearby villages. At present only one-fifth of Greece is classified as forest and the bulk of this is loosely spaced pinewoods. The only true tall forests surviving are on the most inaccessible and rainiest mountains. The annual cut of timber (2·7 million cubic metres) is the lowest of any south European country except Albania and nearly 90 per cent of this is hardwood for fuel (fig. 27).

Throughout the millenia, rapid soil erosion followed, as it still does, the clearance of the vegetative cover on slopes exposed to summer drought and to heavy autumn and winter rains. The rivers swollen by more rapid runoff laden with debris, eroded their banks more effectively and caused havoc by flood and by spreads of gravel and boulders. The soils upon which this layer of granular material was spread were less capable of holding moisture and therefore less fertile. Irrigation became more necessary and recourse had to be made to the supplies of groundwater by sinking wells. This led to the lowering of the water table and the further impoverishment of the landscape. The process continues at the present time. At the foot of the slopes the lowlands tend to become ill-drained or swampy.

Because limestone is present over so great an area of the country permanent surface streams are largely absent and the long summer drought and heavy winter rainstorms

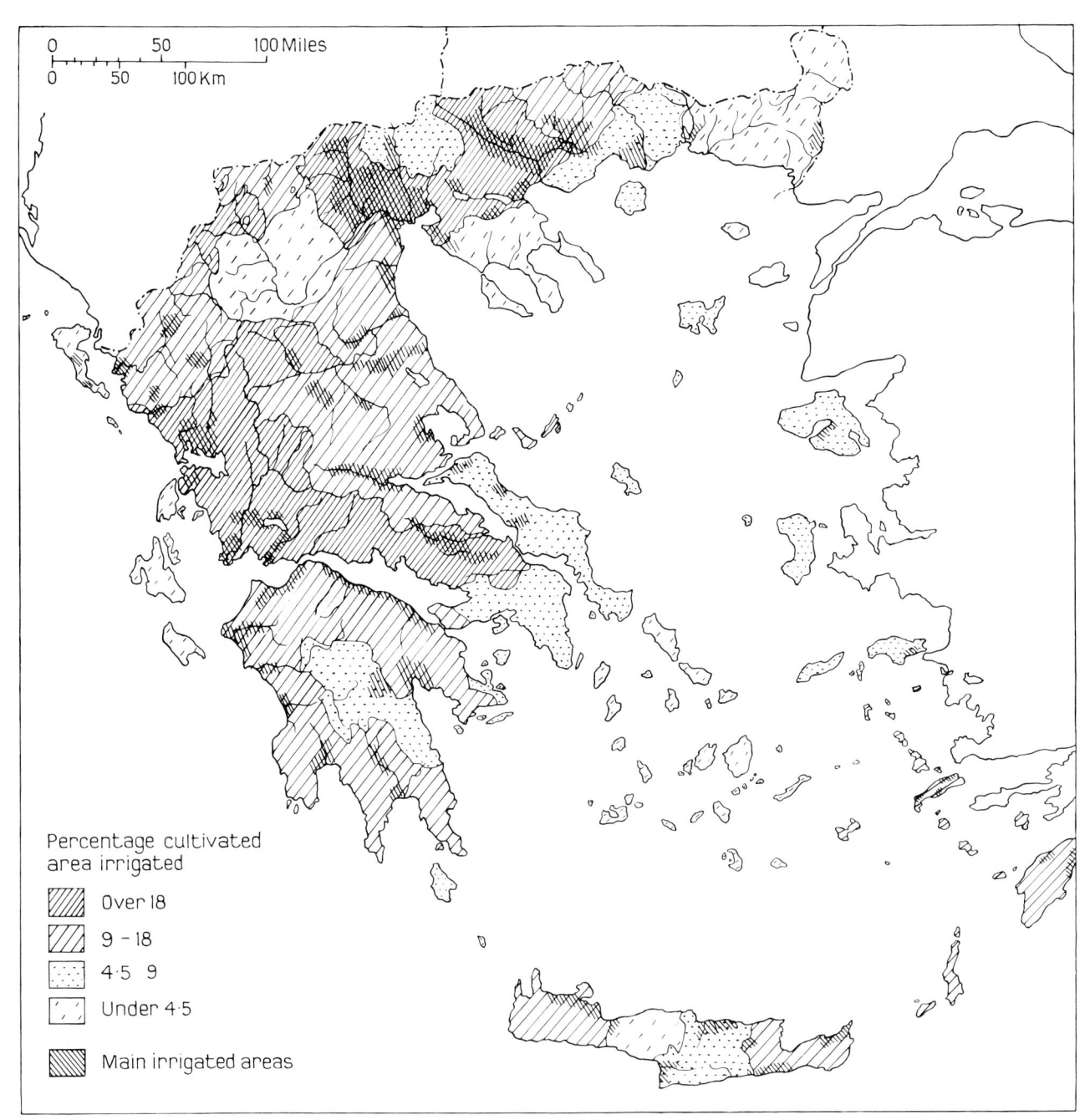

103 *Percentage cultivated area irrigated* (modified from *Economic and Social Atlas of Greece*, ed. B. Kayser and K. Thompson, 1964)

cause the regimes of those that are permanent to be irregular and unreliable. Elsewhere the rivers plunge down rocky channels and spread their debris far and wide in flood or disappear down swallow holes or end in deltaic flats.

Over large areas soils are either thin and poor or ill-drained and only 29 per cent of the country, or 3·9 million ha, is suitable for cultivation and a further 37 per cent is classed as pastoral. Since grass is only present in the spring this latter also includes maquis types of vegetation, and probably explains why the grassy Elysian fields represented peace and prosperity to the heroes of ancient Greece. They would still do so to farmers today. On the other hand the wide areas covered in aromatic shrubs also explain why the 'honey of Hymettos' finds a ready market in other countries of Europe.

Despite scarcity of suitable land and of water and soil poverty about 50 per cent of the population is directly dependent upon agriculture for a livelihood and agriculture contributes 24 per cent of the gross national product. Ultimately it is a question of water supplies. Limestone does not readily lend itself to water conservation or the building of dams, though remarkable progress has been made in similar areas of Yugoslavia. At present, Greek irrigated agriculture is largely dependent upon wells or springs and small enterprises, but large reclamation schemes are being planned (fig. 103).

Eighty km northeast of Athens, 20 000 ha of lake floor were reclaimed by the drainage of Lake Copias, a marshy lagoon tract in the plain of Phocis watered by the Cephisus river. Under the control of a British firm of engineers, the waters from rivers feeding the lake have been trapped in a

GREECE: STRUCTURE OF FARMS BY REGIONS

	Total area (hectares)	Number of farms*	Percentage of area leased to tenant farmers	Fragmentation Average number of parcels	Average area of parcels (hectares)	Average area of farms (hectares)
Thrace	250 693·8	67 832	10·5	7·6	0·5	3·7
Macedonia	940 526·4	294 122	15·7	6·7	0·5	3·2
Thessaly	465 716·9	111 528	7·4	6·9	0·6	4·2
Epirus	127 547·9	62 336	9·5	5·5	0·4	2·0
Central Greece	626 942·2	185 831	10·5	7·3	0·5	3·4
Peloponnesus	681 918·5	197 585	9·5	6·7	0·5	3·5
Ionian Islands	76 406·8	40 385	12·3	6·1	0·3	1·9
Aegean Islands	201 576·9	92 096	24·0	7·3	0·3	2·2
Crete	301 946·2	104 457	8·7	10·5	0·3	2·9
Total	3 673 275·6	1 156 172	11·8	7·1	0·4	3·2

* including holdings without any agricultural land

Source: 1961 Agricultural Census

GREECE: DISTRIBUTION OF FARMS BY SIZE IN 1961*

Size category	Farms		Total area	
Hectares	Number	Per cent	Hectares	Per cent
0·1– 0·9	261 772	23·0	131 988·2	3·6
1·0– 4·9	658 432	57·8	1 658 339·2	45·1
5·0– 9·9	172 745	15·1	1 143 208·9	31·1
10·0–19·9	38 912	3·4	498 130·8	13·6
20·0–49·9	6 863	0·6	185 291·8	5·1
50·0 & over	655	0·1	56 316·7	1·5
Total	1 139 379	100·0	3 673 275·6	100·0

* Source: 1961 Agricultural Census

circular canal around the perimeter of the basin. These are then led into a system of irrigating channels and away by a series of drainage tunnels and ditches. A second large scheme has resulted in 40 000 ha of malarial swampy lowlands being brought into use on both sides of Salonika. A large drainage plan was put into operation involving the lower courses of the Struma, Vardar and Aliakmon which were regulated to provide irrigation water. The whole area was then planned for development and settlement. Gradually in this way all the small plains of Greece are being protected against both drought and flood and productivity is increased.

The extension of land suitable for cultivation is matched by an improvement in farming techniques and a redistribution of farm holdings. United Nations Relief and Works Agency (UNRWA) has provided the Greek government with a great deal of agricultural machinery including 50 000 tractors. The familiar difficulties of miniscule and scattered plots especially in the mountains are everywhere apparent as well as conservative and ignorant attitudes to farming. Eighty per cent of farms are 5 ha or less and in fact average ¾ ha and instead of crop rotation and soil management much land is left fallow each year which eventually results in soil degeneration and erosion.

Greece exemplifies the problems of minifundia and excessive fragmentation of holdings better than any other country in the Mediterranean and probably in the world. In 1961, of 1 140 000 farms 23 per cent were under 1 ha and occupied 4 per cent of the total farmland. A further 58 per cent were 1 to 5 ha in size and covered 45 per cent of the farmed area (see tables above). The difficulties created by the relative smallness of the farms are augmented by their further subdivision into holdings, usually at least 6 or 7 per farm of under ½ ha each. In Crete the subdivisions average 10 per farm.

This excessive fragmentation is an important part of Greek life. First, the rules of inheritance demand that all sons get an equal share of their father's estate. In practice if the heirs cannot agree on a broad allocation of the land parcels, some parcels may be split up between them so creating very small plots. Second, the system of dowries for daughters on marriage is widespread in Greece and often leads to the fragmentation of already fragmented farms. Third, recent agrarian reforms and the abolition of large estates were followed by land redistribution in small plots of about equal value. Fourth, in 1952 an upper limit of 30 ha arable land was placed on farms and soon caused a decrease by 25 per cent in the number of farms exceeding 20 ha.

Since 1953 in an attempt to offset excessive fragmentation the government has met all costs of consolidation which could be undertaken if over half the owners of over half the land wanted it. From 1959 onwards these voluntary schemes have been supplemented by compulsory

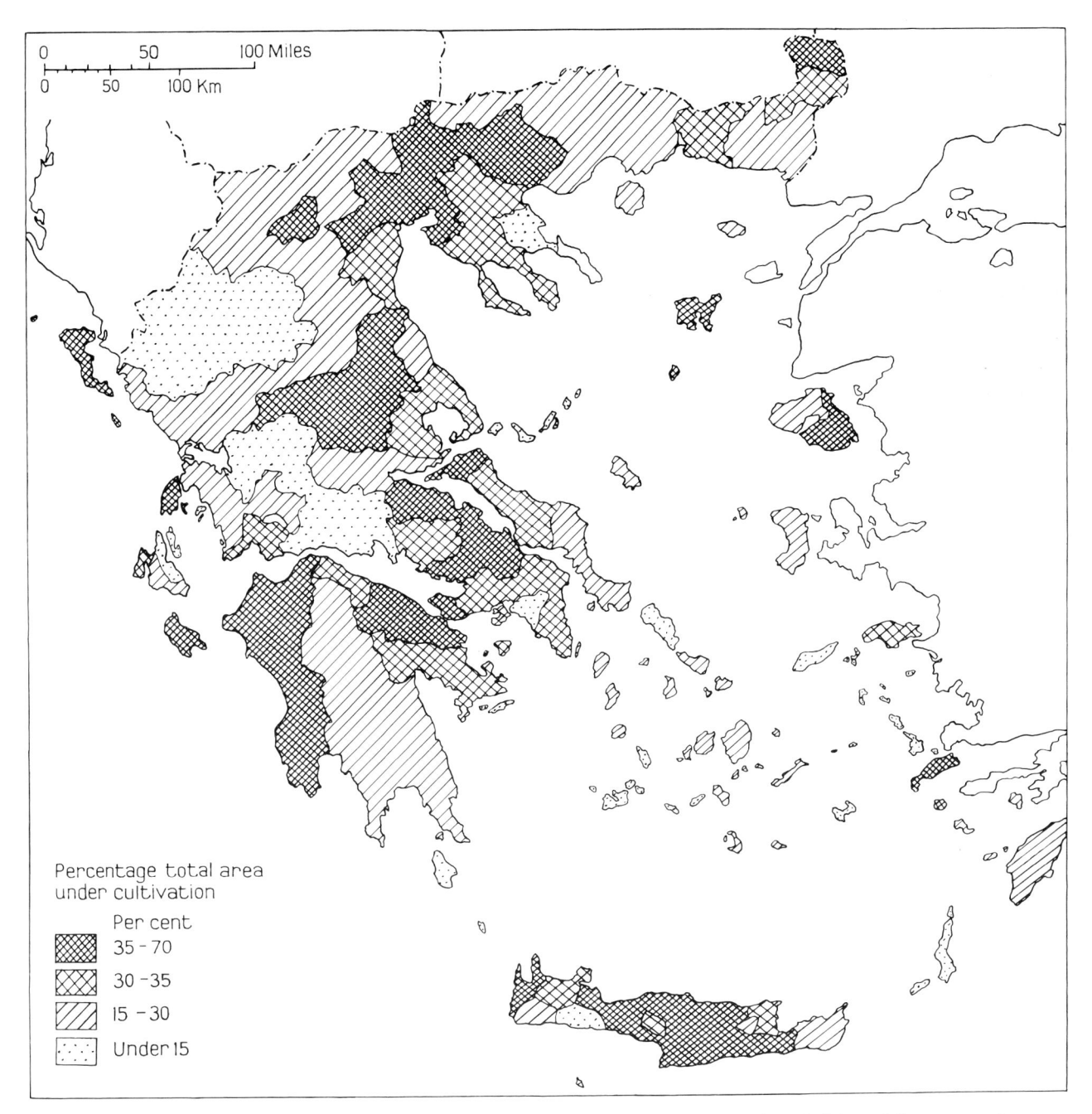

104 *Greece: percentage total area under cultivation* (modified from *Economic and Social Atlas of Greece*, 1964)

schemes, including large public irrigation and land reclamation projects. Between 1959 and 1967 over 205 000 ha in 227 villages were consolidated, about half being on a voluntary basis. However, this involved only 15 per cent of the farmland that needed consolidation and, of course, it is constantly being diminished by the custom of inheritance and dowries.

Arable farming (fig. 104)

Of the total land area of Greece 29 per cent is under field crops and the major part of this lies in the plains and the south. Cereals occupy about 44 per cent of the cropland, by far the chief being wheat which in 1966 occupied over 1 million ha and yielded 2 million tons of grain (see table, page 298). The second cereal is barley which is in demand for brewing at Athens and elsewhere, and the third, maize which is growing in importance with the spread of sprinkler irrigation. Of other crops the most widespread are the olive and the vine. The olive occupies 970 000 ha or about 28 per cent of the cropland but often is underplanted with cereals. It abounds in southern Greece where also are the 45 000 ha given over to table olives (fig. 31). The annual oil production is about 180 000 to 200 000 tons a year and makes Greece the third world and European producer after Spain and Italy. The vine (230 000 ha; 1 439 000 tons of grapes) rivals the olive for the hillslopes in the warmer south and often dominates the warmer hillslopes in the north where the winters are colder. The production includes the specialized currant which is discussed, with other fruits, in more detail in Chapter 8. Its marketing centres on Patras and Corinth.

Much of the best land is given over to special crops yielding quick cash returns or foreign currency. The considerable production of citrus fruits (over ½ million

GREECE: BREAKDOWN OF TOTAL AGRICULTURAL AREA AND ARABLE LAND (thousands of hectares)

	1961	1963	1965	1967
I Total agricultural area	8995	8893		9091
Area under cultivation	3708	3683	3685	
of which: arable land	2809	2755	2746	2996
market gardens and horticulture	117	116	113	
vines (wine and dessert grapes)	247	239	234	232
Forestry	535	573	592	
Permanent grassland and pasture	5287	5210		5239
II Total arable land	2809	2755	2746	2996
Cereals *total*	1766	1621	1774	1664
of which: wheat	1773	1078	1258	1052
rye	25	20	16	11
oats	149	126	120	111
barley	189	175	203	351
maize	191	185	144	139
Pulses	161	172	124	165
Fodder seeds	100	90	72	–
Fodder crops	327	369	379	–
Industrial crops *total*	365	442	310	–
of which: tobacco	105	147	132	125
cotton	216	233	136	141
sugar-beet	2·5	9·5	17	17

tons, of which one-fifth are lemons), of tobacco (130 000 tons), which forms the chief single export commodity and of cotton are all aimed partly at earning foreign exchange. Market-garden crops, particularly melons, tomatoes, aubergines, peppers and potatoes, temperate fruits, especially peaches, apricots and apples, and sugar-beet are among the other important crops. All when grown as a summer or autumn crop need much irrigation water. Rice is grown successfully on 17 000 hectares of easily inundated plain. Figs are important locally and in the drier areas carobs. The regional controls are availability of water for irrigation and depth of soil and flatness rather than climate. The Athens market is a great incentive to market gardening and horticulture in the south. Tobacco cultivation and curing and cotton growing are especially popular in Macedonia and Thrace where the many tedious hand pickings seem to suit well the abundance of labour. The Peloponnessus strongly dominates in citrus, olives and most other fruits, while the flat plains of Larissa are mostly under wheat often cultivated on a large scale with ample machinery. However, with cereals without irrigation water there is always a long lag before the next crop can be grown. In the meantime sheep and goats graze the stubble.

Animal husbandry

About 37 per cent of Greece is classified as pasture but probably more than one-third of this is only maquis and phrygana. Animal husbandry is not very advanced and the land has a low carrying capacity. Yet the total number of larger farm animals, nearly 14 million in all, far exceeds that of people and animal products yield 26 per cent of the total value of the agricultural output. As described in Chapter 8, the donkey, horse and mule are the main beasts of burden, the donkey being the Grecian jeep. The mountain pastures, bushgrowth and arable stubbles and fallows are given over to 8 million sheep and nearly 4 million goats, the latter being the largest herd of any nation in Europe. The forests and low bushgrowth of the mountains are used as summer pastures and the herders occasionally fire the shrubs to promote tender shoots. Grazing may become so scarce in winter that branches of trees are cut and stored for fodder. Probably half of all the sheep and goats in Greece graze in the forests at some time of the year. Transhumance, daily and seasonal, is usual but is increasingly being made more difficult as lowland areas come under irrigated crops that are usually for cash or industrial purposes and not for fodder. However, irrigation is likely to favour the growing of more alfalfa, maize and sugar-beet. The maize will benefit mainly pigs of which there are at the moment over $\frac{1}{2}$ million, some of which feed partly on acorns. The alfalfa, sugar-beet residue and other green fodders will benefit the cattle husbandry. Greece had over 1·1 million cattle in 1966 and of these 400 000 were used solely for draught, the others being cows or cow heifers mainly for milk, together with 38 000 Indian buffaloes. The dairy industry is growing in spite of the absence in country districts of piped water, means of sterilization and good communications. The quality of the herds is increasing rapidly. Between 1960 and 1966 the number of unimproved breeds of native cattle decreased by 300 000 and that of native breeds crossed with brown Alpine, Jersey or Frisian and pure imported breeds increased by 320 000.

Butter production is very small, being little used in a land of olive-oil experts. Milk production is about 1 million tons a year and cheese output about 100 000 tons. But the major proportion of this is from goats and sheep. Cheese made from ewe's milk is widely consumed in Greece either as fetta, a soft cheese soaked in brine, or as kefelatyri, an exceptionally hard variety. The Greeks do their best to produce tender beef steaks for the hordes of American and British tourists but their tastiest meats are veal and kid. The meat supplies to the prime Athens market are well organized in fully refrigerated container lorries. The specialities in the agrarian economy of the various islands are discussed later. However, we will attempt here, with the aid of the table and figure 104, a general summary of the main agricultural regions of Greece.

1 The three island groups where the olive, grape and citrus dominate and wheat plays a very minor role. Irrigation is scarce, the population density high, the farms relatively small and one farmer in every four a tenant, a proportion double that on the mainland.
2 Peloponnesus, where farming is based on the citrus, olive, grape and horticultural crops. The holdings are medium to large for Greece and farmers are relatively prosperous.
3 Epirus, where pasture predominates and yields are low. Crops, apart from oranges, are relatively unimportant. Minifundia prevails (average holding is 2·2 ha or less) and there is a considerable rural emigration to Athens, Larissa and Patras.
4 Central Greece, a chequerboard of mountain and plain, with diversified agriculture and medium to large holdings.
5 Thessaly, where cereals dominate and tobacco and cotton are also important. Irrigation is expanding and the agriculture on the flatter parts is relatively prosperous.
6 Macedonia and Thrace, where 70 per cent of the tobacco and 50 per cent of the nation's wheat and cotton are grown. This is Greece's chief agricultural region, including livestock. Irrigation is increasing rapidly and farms are relatively large for Greece and fragmentation least acute.

Agricultural emigration is fairly low or absent. The livestock industry is greatly aided by irrigated fodder crops.

GREECE: REGIONAL DISTRIBUTION OF THE PRINCIPAL CROPS

	Percentage share of total volume of production in 1964						
	Wheat	Tobacco	Cotton	Peaches	Oranges	Olives	Olive oil	Grapes
Peloponnesus	8·4	4·5	4·0	3·4	42·9	16·6	43·7	30·7
Central Greece	12·2	15·9	28·7	0·9	4·5	52·6	15·3	16·5
Thessaly	24·2	8·3	14·3	3·4	0·5	8·3	1·5	7·4
Epirus	1·0	1·5	4·9	0·9	31·7	2·8	1·9	0·6
Macedonia	40·4	60·6	45·3	88·8	–	8·3	2·2	12·8
Thrace	11·8	7·6	1·4	0·9	–	2·8	0·1	2·6
Aegean islands	0·9	1·5	1·4	0·9	2·6	5·5	18·7	5·0
Crete	0·8	–	–	0·9	11·3	2·8	12·8	20·1
Ionian islands	0·3	–	–	–	1·3	–	3·7	4·4

Power resources, minerals and manufactures

With a population of 8·6 million, Greece needs alternative employment to agriculture but the development of manufactures is hindered by the lack of power resources. The only solid fuel is mediocre to poor quality lignite mined in Euboea (5 million tons in 1965) which goes mainly to supply the largest thermal-electric station in Greece at Aliverion. This station supplies about half the current used in the Athens–Piraeus conurbation.

In 1954, four small hydro plants were opened at Aliverion, Ladon, Louros and Agras with a total installed capacity of 205 000 kW. In 1960, two further stations, near Ptolemais and Tairopos, were completed and work was started on a large station (400 000 kW) on the Acheloos river at Kremasta. Until this plant came into operation the percentage of the Greek electricity generated by water power remained at about 17 per cent. In 1967–8 of the total annual production of electricity (7000 million kWh) about 21 per cent was from hydro stations.

In 1953, oil was struck near Kleisura and a refinery has been built at Aspropirgos.

Mineral ores

Greece has a wide variety of mineral ores but the deposits generally occur in such small amounts that mining is relatively expensive. Bauxite (1¼ million tons) is mined near Delphi and is refined nearby at a new refinery, with an annual capacity of 200 000 tons of alumina and 72 000 tons of aluminium, at Dhistomon on the Gulf of Corinth. Appreciable quantities of magnesite (314 000 tons) are mined in Euboea and near Salonika, and small amounts of iron ore and iron pyrites at Stratoniki, of copper, zinc, lead and silver ores near Laurion, and of chromite near Eretria. It is, however, certain that apart from the bauxite, and perhaps also the magnesite, none of these deposits is likely to encourage large-scale industries. All told they employ about 22 000 persons.

Manufactures

Many manufactures in Greece, especially in rural districts, are still in the handicraft stage, being done in the homes or small workshops usually as a part-time and supplementary occupation. Large-scale modern factories are largely restricted to the vicinity of Athens and Salonika, the only other towns with notable manufactures being Patros and Volos. Over four-fifths of the factory products by value come from the Athens conurbation, which has a wide variety of manufacturing concerns as well as a large output of handicraft goods from small workshops.

The chief industries are concerned with the processing of agricultural raw materials, especially tobacco, cotton, wool and flour, to which are added a few larger units for shipbuilding, petroleum refining, chemicals, steel, aluminium and cement. Concerns based on national agricultural products are widespread, especially textiles, although Athens predominates. In 1966, Greek factories had 7272 cotton looms of which one-third were automatic. In small towns such as Lamina and Levadia textiles and clothing of cotton and wool are the main products and depend heavily on local sales to tourists. But most towns also engage in ceramics and quite a number in iron and copper ware.

Most of the large new factories have been built by American, or international, or Greek–American finances. They include a blast furnace at Piraeus which produced 210 000 tons of crude steel in 1966, mainly from scrap; an alumina refinery and aluminium factory at Dhishtomon near Patras; and an oil refinery and an ammonia plant in the Salonika area. Cement production has also increased rapidly and now exceeds $3\frac{1}{2}$ million tons a year. Shipbuilding and repairing are major occupations in the Piraeus. Even in ancient Greece, as agriculture declined the Greeks took up trading in manufactured goods throughout the shores of the Mediterranean and Black Sea. The tradition has never died; every Greek yearns to own a ship and today Aristotle Onassis and Stavros Niarchos are fleet-owning magnates of incredible wealth and power. The Greek mercantile marine is an important prop of the national economy. Greek-owned merchant vessels number about 1160, including many giant tankers, and represent over 7 million tons of shipping. This fleet deals with about one-tenth of the world's seaborne trade and also makes Greece the fifth passenger-carrying country after Japan, Great Britain, the United States and France. Greek coastal shipping plays a vital part in the trade of the Mediterranean, particularly between Athens and the Aegean islands (fig. 107).

It is difficult to give a true picture of the relative importance of manufacturing in Greek life and the Greek economy. In recent years, out of a total labour force of about 3·7 million, about 1·9 millions (53 per cent) were engaged in agriculture, and nearly $\frac{1}{2}$ million (14 per cent) solely in manufacturing. Modern financial investment will certainly increase the industrial sector. The relative value of the gross national produce at factor cost in 1974 has been estimated as follows:

	Percentage total value	
	1961	1974
Agriculture	30	20
Mining	1	2
Manufacture	18	24
Construction	6	7
Electricity, gas and water	2	3
Total industry	27	36
Total services	43	44

The emphasis given so far in this account tends to hide the significance of fishing and of tourism, always very difficult aspects to express statistically. The Greeks are great fishermen and have normally caught two-thirds of the country's requirements, the catch being mainly tunny, octopi, squid and red mullet, as well as crustaceans and sponges. However, in the last decade Greek ships have gone increasingly into the Atlantic for cod and in 1965 the total fish catch had already risen to 124 000 tons (see Chapter 3). Fishing in coastal ports and coastal boat services can be no more separated from tourism than can handicrafts, foodstuffs and catering. Today well over 1 million tourists visit Greece each year and tourism rivals shipping as a main prop of the national economy. The number of visitors has doubled since 1962 and facilities and amenities have increased likewise; the antiquities, sunshine and friendliness of the seafaring nation remain unexcelled.

Population and settlements

The total population of Greece has increased from 7·6 million in 1951 to over 8·6 million in 1966. In the latter year the average density of population was 65 per sq km and the distribution between the mainland and islands was roughly as follows:

	Area	Population	Density/sq km
Mainland	106 778	7 150 000	67
Islands	25 166	1 460 000	58
Greece	131 944	8 610 000	65

The density is well above average in Attica around Athens and Piraeus, in the northeast from Salonika to Kavala, along the west coast especially of the Peloponnesus and on the adjacent islands, and in central Crete and the Aegean islands of Chios and Samos.

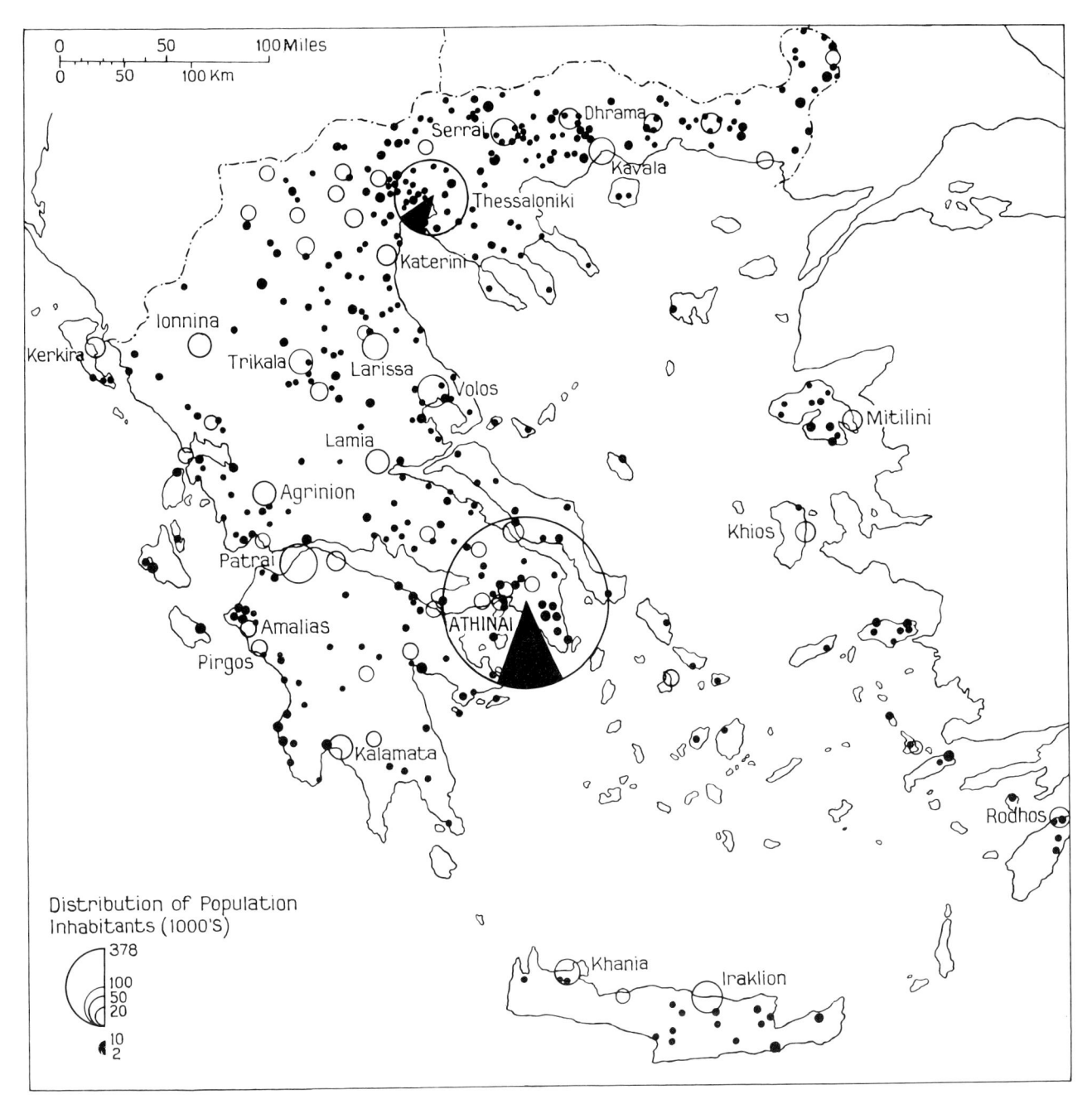

105 *Greece: distribution of settlements with over 2000 inhabitants* (modified from *Economic and Social Atlas of Greece*, 1964)
The place-names are in modern Greek

Urban population. The outstanding feature of the distribution is the great relative size of Athens, a unique condition among Balkan countries as the following statistics demonstrate:

		Percentage Population	
		Urban	Total National
Greece	Athens	52	20
Yugoslavia	Belgrade	13	3
Bulgaria	Sofia	29	8.5

Athens is also quite abnormally large compared with the second and third most populous towns, Salonika and Patras respectively (fig. 105). However, these three cities together have two-thirds of the total population of Greece.

Athens, the capital, has an estimated population of 2·8 million in its conurbation and is the administrative, industrial, commercial and cultural centre of the country with various industries, such as leather, industrial ceramics, textiles and carpets. The ancient city grew up on and around the Acropolis but the modern town has spread between the hills of the Acropolis and Lykabettos and reached the slopes of Hymettos and Parnis. Most of the administrative buildings lie between Syndagma and Omonia squares. Big modern apartment and office blocks have been built and arterial roads cut, but an incomparable opportunity for imaginative town planning has been missed and although luxury flats are in evidence there is also in parts a picturesque confusion of twisting streets. Some loosely spaced outer suburbs are still not fully serviced with piped water and it is rather ironic to be able to

XXXVII *Athens and the Acropolis*

contemplate on sale in the Plaka at the foot of the Acropolis sanitary buckets and old-fashioned water pumps. A great modern expansion of fine dwellings has spread along the coast southeastward to Ellinikon and Voula in the shelter of Hymettos. There is a short stretch of underground railway beneath the central city.

Piraeus, the port of Athens, on the other hand, was laid out on a rectilinear plan by Themistocles before 450 B.C. Today it is joined to Athens by road and rail as well as by built-up areas, and also acts as an entrepôt for the whole of the Aegean with 30 steamship lines serving 80 ports, linking the islands to the mainland and each other (fig. 107). It is a busy Mediterranean market handling fruit products and live animal cargoes, such as chickens and sheep. It has attracted a considerable variety of modern industries mainly these concerned with metallurgy and shipbuilding, food processing, especially flour and olive oil, chemicals, including fertilizers, textiles, engineering, printing and paper and cement. It fronts onto a fine capacious harbour which includes a basin for small boats, with a well laid out waterfront.

Salonika (Thessaloniki), with 400 000 people in its conurbation, is the second city in Greece. Finely situated on the Gulf of Thermaikos, its site assures the city of prosperity since it stands on the old *Via Egnatia* linking Istanbul to the Adriatic coast and is the terminus of the great valley route along the Morava–Vardar from central Europe. As a Roman capital it once ranked second only to Istanbul in importance. During the Crusades it became a base of operations against the Saracens. Today its industries are concerned with petroleum refining, chemicals, leather, textiles, brewing, flour milling and tobacco. By virtue of its magnificent Byzantine relics it is an important centre for tourism. It has been an entrepôt since 1925 and a free port for Yugoslavia since 1929. Salonika could have a prosperous future as the outlet for the Balkans and modernization of its port facilities to handle bulk cargoes, container traffic and the building of large storage depots could double the amount of cargo handled. A large new Esso refinery has been built here to deal with $4\frac{1}{2}$ million tons of crude petroleum annually.

Patras, with 102 000 people in its built-up area, is the third Greek port after Piraeus and Salonika (fig. 105). It stands on a gulf on the west coast at the entrance to the Gulf of Corinth and the Corinth canal and has regular shipping services, among others, to Corfu and to Brindisi. It handles much coastal traffic in Mediterranean produce and is the chief trading centre for Greek currants. Here the ancient acropolis is crowned by a ruined Byzantine castle rebuilt in turn by Franks, Venetians and Turks.

The other three chief towns of mainland Greece are Volos (about 65 000), probably the fourth Greek port and

XXXVIII *Currants spread out to dry at Amalias, Peloponnesus, Greece*

a main centre for tourism (Pelion is nearby) and seaside holidays, and a ferry station with regular shipping services to the northern Sporades; Larissa (60 000), a provincial capital of a wide lowland, one of the chief regional route centres in Greece; and Kavala (60 000), an ancient port built on an amphitheatre of hills facing the north Aegean and the island of Thasos, near the site of the battle of Philippi (42 B C), and itself the landing place of St Paul. These have small industries including tobacco, foodstuffs, textiles and construction materials.

Rural population. One-third of the population of Greece is rural and lives in small villages. Ninety-seven per cent of these villages have less than 2000 inhabitants, more than half of them are hamlets of less than 200 and all are isolated from each other by distance, difficult terrain and lack of communications. Of these settlements, 11 516 are of 10 dwellings only, harbouring approximately 50 people. Nevertheless the Greek rural population is, apart from the three largest cities, as large as the urban. The typical 'country town' or large village tends to be huddled on the hillside in a good defensive position above cultivable lowlands and usually is surrounded by olive groves, vineyards and terraced fields. All have their Orthodox churches, their markets and streetside stalls and cafés. The streets are crowded, lively and noisy and seem full of small shopkeepers, salesmen, hucksters and kiosk holders.

The shopkeepers sit at the doors of their shops which often open into caverns (basements) below the pavement level and are entered down a flight of steps. Not infrequently a proportion of the goods are displayed on the exterior.

Greece is, of course, a mountainous country and its lowlands are for the most part segmented by the disposition of the relief, and poorly supplied with water. They are fragmented further by the system of land holding. Village growth was not fostered by this discontinuity of the agricultural land parcels and, in addition, village settlement came late because so much of the land is pastoral or merely of marginal agricultural value. Nomadic and semi-nomadic pursuits are therefore part of the rural background and sedentary agricultural occupations, in the absence of water, electricity supplies and communications, are naturally difficult to establish. The villages that do develop are tiny with no administrative and a minimal market function.

A remarkable feature of Greek life is the mobility of the population. Not only is there a large annual emigration abroad to better living conditions but there are also large internal movements of population. Three aspects of this internal migration may be observed.

1 *Rural mobility.* This is of an extreme kind in Greece. In addition to the normal pattern of transhumance to the summit pastures in spring and to the lowlands in winter,

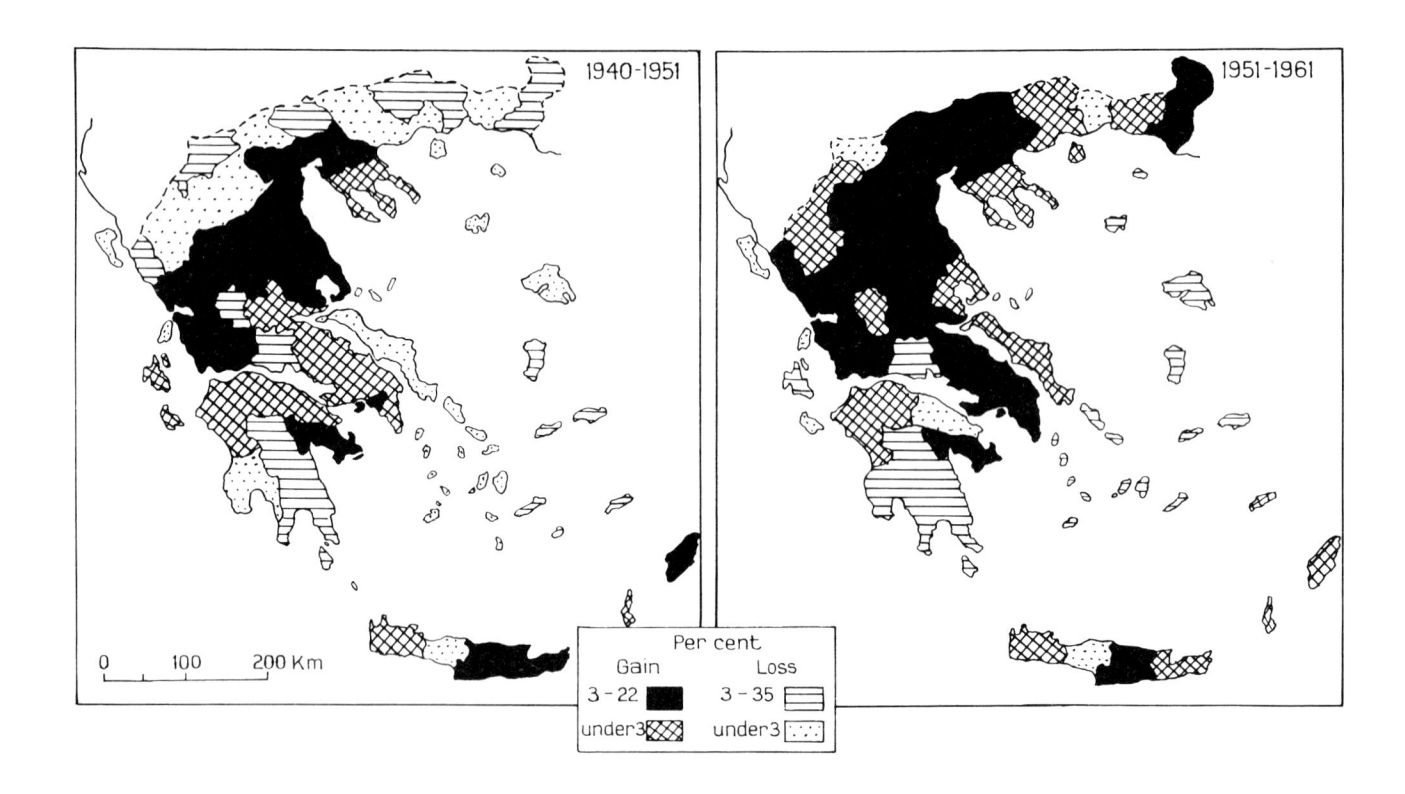

106 *Greece: variations in population density 1940–51 and 1951–61* (simplified from *Economic and Social Atlas of Greece*, 1964)

there are widespread daily migrations of a whole community to work in a distant tobacco or cotton field, as well as seasonal migrations of workers to the olive and orange groves of Epirus and to the vineyards of Peloponnessus at harvest time.

2 *Movement to better agricultural land.* Ever since the drainage and irrigation schemes began in 1930 the Greek people have been drawn down from the mountains onto the plains now rendered fertile and free from malarial and intestinal diseases. In 1920, when no less than 77 per cent of the Greek population was rural, half lived in the mountains; in 1951, the rural population represented 62 per cent of the population and 27 per cent lived below 100 m; in 1961, about 59 per cent of the population was living below 100 m. Everywhere in Greece there is among the rural communities a tendency to leave the impoverished mountain pastures and the marginal agricultural lands and move to employment in the plains and valleys (fig. 106). The main reception areas have been in the plains of Macedonia, the Soufi plain on the Turkish border and the lower valley of the Thiamis opposite Corfu. Whole communities will migrate in search of better districts and small groups will 'hive off' from a village community and settle as squatters in villages deserted in their turn by a rural exodus to the towns.

3 *Rural migration into towns.* The flight from the land gravitated invariably towards Athens whither the ferry links also brought migrants from the Aegean islands. Even before the Second World War and the civil war that succeeded it in Greece, the drift of people into Athens caused its population to increase by 3 per cent per annum and by 1940 Athens and Piraeus had merged and covered 420 sq km. The political instability which has been evident in the last decades in Greece has magnified this population growth. The population of this metropolitan city is now out of all proportion to the size of Greece and the number of its people.

Two results are apparent in the landscape. First, the annual increase in the population of Athens outstrips its capacity to house its inhabitants. The city population therefore increases in density, as the dwellings spread towards more difficult ground and more expensive lay-outs Also the outer suburban tract of shanty dwellings does not tend to decrease in width.

Secondly, in the rural landscape, abandoned villages become more common with the years. Between 1951 and 1961 about 160 villages were totally evacuated. To these must be added the villages partially abandoned, with some or many empty houses that fall into decay (fig. 106). Surrounding all these villages are the parcels of land held by former occupants left to revert to wilderness so that agrarian reform and improvement is hampered. The people who have fled the countryside are lost to an agricultural industry which is still insufficiently mechanized.

The external emigration by Greeks averaged just over 100 000 persons a year from 1963 to 1966 and of these 66 per cent went to Germany, 15 per cent to Australia and most of the remainder to North America.

Communications and commerce

Communications in Greece are hindered by the mountainous nature of the country. Of the approximately 14 000 km of road, one-quarter is unsurfaced track and although there are 8500 buses in Greece providing public transport, in many highland districts the horse, mule and donkey are the only carriers. In 1965, the motor fleet consisted of 65 000 lorries and 105 000 private motor vehicles of which about half were privately owned and most of the rest were taxis plying in the large towns. The great national highway of Greece runs from Corinth via the neighbourhood of Athens, Volos, Larissa and Salonika and so north into Yugoslavia for Belgrade. The railways extend for 2584 km and half of this network is state-owned. One runs the length of the country from Kalamata in the Peloponnesus via Athens and Salonika to Alexandroupolis on the eastern frontier.

With a coastline of 15 000 km and hundreds of islands the Greeks naturally rely largely on sea-borne traffic both for domestic trade and as an external lifeline (fig. 107). Nine-tenths of all Greek foreign trade goes by sea. Around the coasts of the mainland and the islands are spaced 158 small ports but only 28 are economically viable and these handle four-fifths of the cargoes. Ninety per cent of the total traffic moves through the ports of Piraeus and Salonika.

Improvement in all means of communications is one of the keys to the increase in industrial production and development in Greece. This involves the building of a larger merchant navy since sea transport must always be the basic means of communication in Greece and also provides an important source of revenue. The Greek merchant fleet in June 1967 consisted of 1230 cargo boats of $5\frac{1}{4}$ million tons gross registered tonnage, 121 passenger ships of 418 000 tons, 206 tankers of 2 077 000 tons and 229 other boats totalling 108 000 tons. Most of these cargo boats and tankers carry for other nations. Invisible earnings for shipping services and the profits from tourism approximately cover the large Greek trading deficit, exports being one-third the value of imports. The chief exports are foodstuffs and live animals (32 per cent total value), tobacco (30 per cent total) and crude materials (20 per cent). The chief buyers are West Germany (20 per cent), the United States (10 per cent) and France and the United Kingdom (6 per cent each). The main imports are machinery and transport equipment (37 per cent), other manufactured goods (18 per cent) and foodstuffs (13 per cent). Again West Germany (17 per cent) is the main supplier followed by the United States, Italy and the United Kingdom with about 10 per cent each. Greece was the first country to become an associate member of the Common Market and these ties are clearly reflected in its trading relations.

INSULAR GREECE

The Greek islands cover 25 166 sq km or one-fifth the total area of Greece and consist of a scattered group off the west coast in the Ionian Sea and various groups in the Aegean Sea between mainland Greece and Asia Minor. The only large islands are Crete and Euboea.

Structure and relief

The various island groups represent the unsubmerged portions of either an Alpine fold range or of a Hercynian median mass of the type seen on the mainland as horsts in Mounts Ossa, Pelion and Olympus and on a vast scale in the Rhodope plateau. All that remains of this former great Hercynian massif in the Aegean is in parts of Euboea and in the 220 small islands of the Cyclades and northern Sporades, an approximate area of 3000 sq km. These groups contain the islands Andros, Tinos, Mykonos, Milos, Thira, Skiros, Skiathos, Skopelos and so on. All have basement rocks of schists, gneiss and granite weathered in places into china clay and with small local concentrations of metals, such as iron and copper on Serifos and emery on Naxos. Over some parts, as in eastern Naxos and Paros, the igneous and metamorphic base is capped with hard limestones, often marble, that yield more fertile soils and a varied relief as well as small karstic tracts with subterranean drainage. Elsewhere the shattering of the Hercynian massif was accompanied by vulcanism and several of the islands display volcanic features such as lava flows and cones. It is still a region of instability, subject to earthquakes although far less so than the Ionian group. The fine circular harbour of Milos is a partially drowned crater and sulphur is mined on the island. The breached crater rim of Thira (Santorini) surrounds the still active vent of the tiny Kaimeni islands (burnt islands) which appeared in 1573 and 1712. Eruptions occurred as late as 1926 and 1928 and a great earthquake in 1956.

With the above group should probably also be included Samothrace, Lemnos, Lesbos and Chios, which appear to be the fractured and disjointed remnants of various mountain ranges that once stretched southeastward from the Greek mainland to Asia Minor. They are for the most part covered with limestone or marble but Tertiary conglomerates predominate on Lemnos and southeast Chios, and there are large areas of igneous intrusions and volcanic lavas on Lesbos (Mytilene), Samothrace and Lemnos.

In the Peloponnesus and south of it the Alpine or Dinaride ranges swing in a vast broken curve through Kythera, Crete, Karpathos and Rhodes so enclosing the deep trough of the Crete sea to the north. These consist largely of limestones and Tertiary sands and clays although western Crete has a crystalline mass.

The seven Ionian Islands off western Greece, the largest of which are Corfu (Kerkyra), Cephalonia (Kefallinia), Zante (Zakynthos), Levcas (Leukadia) and Ithaca (Ithaki), are still more clearly detached portions of a large Dinaric range. These are entirely of Jurassic–Cretaceous limestone and Tertiary sands and clays, the latter predominating only on Corfu (Kerkyra). The lowlands, as in Crete and low-lying coastal areas generally, are floored mainly with Pliocene deposits.

To describe the relief and landforms of these islands is quite beyond the scope of this book. They are for the most part high, steep or precipitous and often cliff-bound. If it is assumed that they are very small the following summit heights will give some idea of their hilly or mountainous nature. Few fail to reach 800 m, many exceed 1200 m; several 1400 m and Crete rises to 2456 m.

Climate and vegetation

The Ionian islands, as the statistics for Kerkyra and Zakinthos show, have ample rainfall except in summer, whereas the Aegean islands, except on the highest parts, normally have less than 550 mm and have long periods of severe drought. Southern Crete has in parts less than 300 mm of rainfall a year. This, however, is the main difference as both groups are sunny and experience steady etesian winds in the hotter months. Of the individual islands many have peculiar local winds and strange lee-wave eddies (see Chapter 4), as well as föhn effects, and canalization of prevailing airflow. The islands have been depleted of their trees except in the Ionian group where the higher rainfall permits natural regeneration. Corfu (Kerkyra) is the most closely wooded island in Greece with dark firs to the summits of the mountains. In Crete all that remains of a former forest cover are a few cypress groves in the west, scattered copses of Aleppo pines in the east, and some stands of ilex on the eastern slopes of the Lassithi mountains. Spanish chestnuts still cover steeper slopes along the south coast but elsewhere and on the greater part of all the islands low-growing phrygana has taken over.

Crete

With an area of 8331 sq km, Crete is the fifth largest island in the Mediterranean and the largest and most southerly of the Aegean islands. Its maximum dimensions are only 250 km from west to east, and 57 km from north to south dwindling to 12 km in the east. It lies 100 km from the Greek mainland and 180 km by sea from Asia Minor and from Athens. It is also halfway between ancient Troy and the Nile mouth and halfway between Sicily and Cyprus.

The whole island is traversed from east to west by a mountain chain that has been fractured into four regions of high relief, the White Mountains or Lefka Ori (2452 m), the Psiloriti massif (2456 m in Mount Ida), the Lassithi mountains (2148 m) and the Sitia mountains (1237 m). These upland masses separate areas of lowland floored mainly with Pliocene deposits. The island is so tilted that the mountains lie closest to the south coast which is high and rises steeply from the sea, and most of the lowlands are on the north. An upland belt of 800 m links the Psiloriti mountains to the Lassithi range, south of which is the plain of Messara and Monofatsion drained partly by the Yeropotamos which gives at its mouth the only commodious harbour on the south coast. North of the mountain backbone small plains where cultivation is possible occur at Retimo (Rethimnon), Canea, Iraklion and in a few localities on the Hierapetra peninsula in the east. In addition the mountains hold many upland basins, such as Omalos in the White Mountains, Nidea in the Psiloriti group, and the Lassithi plateau, which all provide summer grazing. The porous limestone overlying the impermeable crystalline base provides the other characteristic landforms of Crete, namely those typical of karst. These include the Samaria gorge with walls 300 m high and caves and much subterranean drainage in eastern Crete. On the Lassithi plateau groundwater is pumped to the surface by windmills.

The majority of the $\frac{1}{2}$ million people live on the northern half of the island concentrating on water supplies. On the south coast there is little room between the mountains and the sea for habitations and cultivation except in the plain of Messara which with irrigation water from the Yeropotamos is the most fertile tract in Crete. The chief crops are wheat and barley, olives, grapes and citrus fruits. Sheep and goats are kept on the upland basins and mountain pastures. Crete has a total cultivated area of just over 130 000 hectares on which it produces about 19 per cent of the olives, 16 per cent of the citrus fruits and 7 per cent of the grapes grown in Greece.

The rise of western civilization in Crete

For the geographer and historian Crete seems to hold the key to an understanding of the interaction between man and his environment because this island was the stepping-stone or threshold by which civilization reached Mediterranean Europe. Probably it was from Crete that the Neolithic culture advanced to the Franchthi cave on the Gulf of Argolis in southern mainland Greece between 6500 and 6000 B C. The succeeding copper and bronze-using cultures formed a truly magnificent Aegean civilization which spread to the whole Mediterranean and from thence over much of Europe. It is today recognized as having two phases, the Minoan or Cretan and the Mycenean or mainland phase. Both were distinct and it was only towards the end of the Minoan that its influence can be detected in the later mainland branch. The Minoan civilization lasted from 3000 to 400 B C, from the Neolithic, which arrived in Crete from Egypt and the Middle East, through nine periods of Minoan culture each marked by some definite advance in the potter's art.

The evidence for a high standard of civilization is to be found in the archaeological remains, some of which are of peripheral interest to the geographer as such:
1 the ruins of palaces, villas, houses, farms, tombs, religious enclosures and fortifications; the palace of Minos at Knossos consisted of several storeys containing over 1000 rooms, and there were similar palaces in the other cities (e.g. Mallia and Gournia). An extensive city surrounded the palace at Knossos

2 the domestic and ritualistic furniture
3 the articles of domestic ware, such as pottery
4 the tools such as the potter's wheel and the plough, and instruments and weapons made of copper and bronze
5 the public works such as drainage systems and roads and bridges to take wheeled vehicles
6 writing and specialized skills and crafts, such as weaving
7 the decorative arts displayed in their public buildings and in the objects of their personal use

These material things were combined with a high degree of government and organization under a central authority.

It is only possible to suggest features in the Cretan landscape that were especially favourable to Minoan culture since the environment that witnessed its rise has been radically changed by human misuse or misunderstanding of the natural resources. Possibly the most important apart from its favourable situation with regard to receiving new techniques and discoveries from Egypt and the Middle East, was the varied geological structure which provided a variety of cultivable landscapes, besides yielding abundant timber and mineral resources such as metals, building stone and china clay.

The beneficent climate was certainly an important factor in that outdoor activities could be pursued throughout the year with the minimum of clothing. Shelter for man and beast needed only to be minimal and, in any case, during the long period of neolithic farming, was provided by the dry limestone caves of which there are over one thousand. Zeus himself was reputed to have been born in the cave Dikteon Antron in the Lassithi mountains and his aegis or shield was made of a goatskin. Each small lowland plain fostered a city as it does to this day, making the miniature city states characteristic of Greek civilization. In the words of Homer it was a 'rich island, populous beyond compute with ninety cities'.

The most important of these were Knossos on the Iraklion plain, and Phaestos, the harbour city, on the Messara lowland. The towns grew up generally on barren rocky slopes invisible from the sea where the houses had the benefit of cool evening breezes, were close to the herds, above the malarial lowlands, and relatively secure from piracy. Even today the population is urban with agricultural functions except in a few towns with administrative, commercial and market activities. The largest are Iraklion (Candia) with 65 000 inhabitants and Khania (Canea) with 40 000. These ports on the northern coast, as well as Retimo (Rethimnon), also engage in sea fishing for coral and sponges and sea food in season but this industry is merging more and more into larger capitalized units carried out in factory ships based on Athens.

In fact the ancient Minoans, like many of the present-day islanders, soon outstripped the capacity of Crete and the islands to support their exploitive economy. Deforestation led to soil depletion, overgrazing by goats reduced the natural vegetation, the surface runoff was increased in amount and rapidity, and the lowlands became subject to floods, and ill-drained and unhealthy. Slight climatic change in the precipitation regime or amount may well have aided the erosion but in any event the natural environment had been gravely impoverished long before Knossos burned and the palace of Minos was destroyed. Fortunately by then civilization had spread to the Greek mainland and into more recent history.

Other islands

The remaining Greek islands cover an area of 16 885 sq km and had a population in 1966 estimated at 955 000. Only central Crete, Chios and the Ionian Islands had densities of over 70 persons per sq km or well above the national average. Figure 106 shows that in recent decades the smaller and more distant islands have lost many people through emigration either to Athens and to a lesser extent to Thessaloniki or to North America and Australia.

The Ionian islands (2237 sq km) have a cultivated area of about 31 000 ha and a total population of about 200 000. Their high population densities which on Corfu average over 160 per sq km and on Zakinthos or Zante over 90 per sq km, are attributable largely to ample cool-season rainfall and a highly specialized agriculture. Zakinthos and Cephalonia specialize in currants, Corfu (Kerkyra) and Leukadia in olives and the others (Ithaca, etc.), mainly in citrus. Argostolion (10 000 inhabitants) has a large sheltered harbour and has developed as a port for the harbourless Adriatic mainland of northwest Greece. The town of Corfu (35 000) is the chief port of the Ionian islands and has regular sea connections with Brindisi, Patras and Piraeus.

In the Aegean, Euboea is usually grouped statistically with central Greece from which it is separated by a channel that narrows to 60 m. It has an area of 3580 sq km and a population of about 180 000. Khalkis (27 000) is the main town and port. The lowlands are rich but severely restricted and grow mainly olive, vine, fig, wheat and maize. However, large parts of the island afford grazing for sheep and goats. In the south Karystos has been famous for green marble since Roman times.

The Northern Sporades have about 50 inhabitants per sq km whereas the larger Aegean islands nearer Turkey, except Samothrace, have densities of 65 or over. Here as elsewhere on the more fertile lowlands, especially where irrigation water is available, tobacco and cotton supplement the normal Mediterranean crops of cereals and tree fruits. Normally the cereals are consumed locally but much of the harvests of grape, citrus, tobacco, olive and cotton are exported to the mainland. A few small food industries have grown up based on local products such as mastic gum from Chios for use in making Turkish delight, or the production of sultana raisins, Greek wines and liqueurs in Samos, Mytilene and Chios. Handicraft industries of a wide variety are carried on, often in the sunshine. Textiles, leather work,

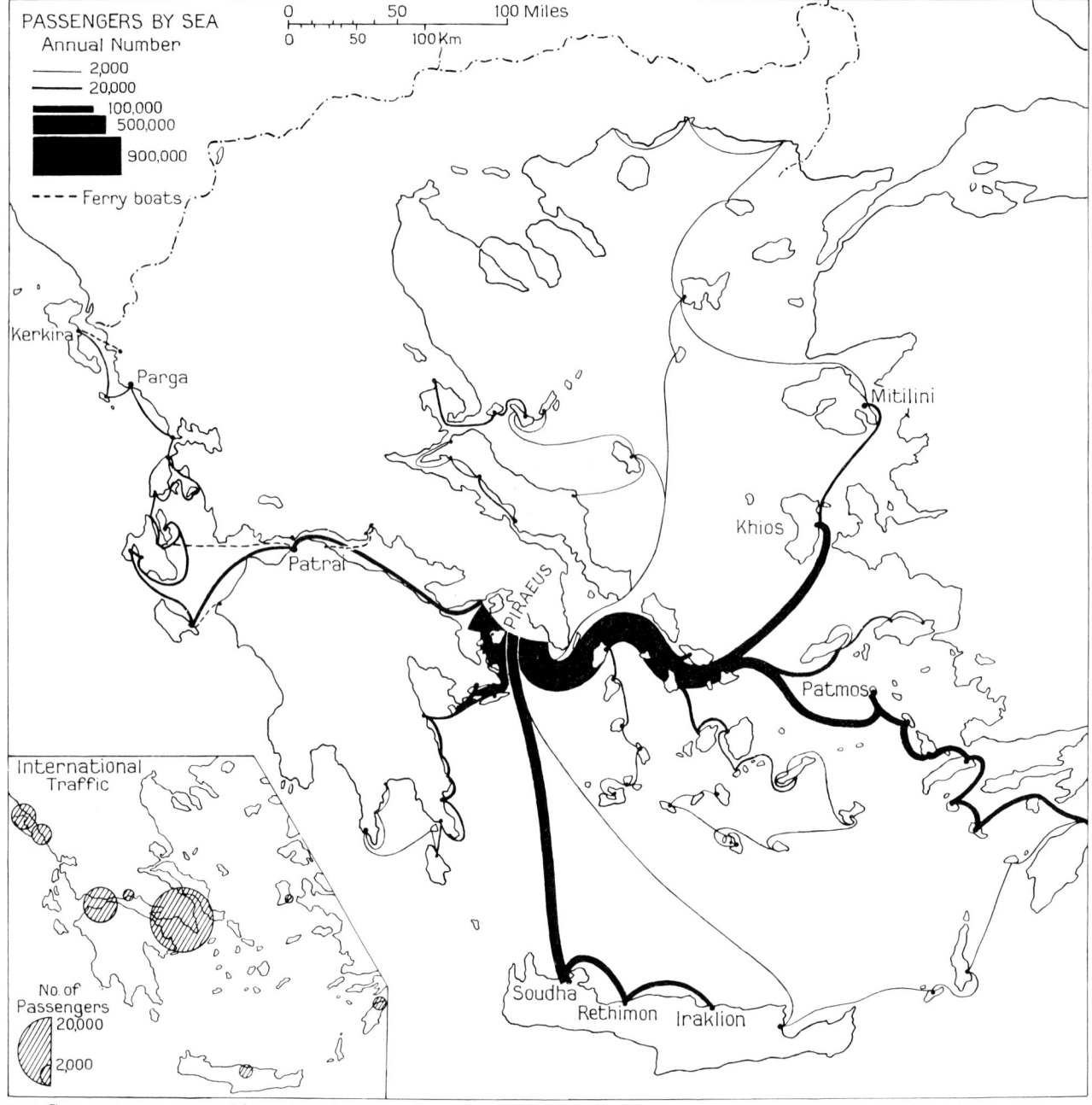

107 *Greece: international (inset) and national passenger traffic by sea* (after *Economic and Social Atlas of Greece*, 1964)

metal work and pottery depend heavily on the tourist trade which fortunately is increasing rapidly with increasing amenities. Yet the towns and ports remain small. Chios with a naval school, ferry services and shipping lines and the advantage of being the centre of one of the most populous islands, has about 27 000 inhabitants.

The Cyclades have developed town life on a small scale but they are thinly peopled. The chief and normally the only town on each island is the port, which handles the ferry services and perhaps small local exports such as pumice from Santorini, barytes from Melos, and emery from Naxos. Many of the Cyclades are entirely pastoral. Syros or Ermoupolis (21 000) the chief town, is the nome, or capital, of the island group and was a trading entrepôt for the mainland and insular Greece before the rise to importance of Aegina, and later of Athens and Piraeus.

The Dodecanese are distant from the homeland but are well served by sea links (fig. 107). Rhodes, the largest island (540 sq km), is one of the least fertile but the 'city'

(30 000 inhabitants) is nicely planned and includes the fortress of the Knights of Rhodes. It hopes to expand tourism further in all seasons and has excellent hotels and an efficient road system. It is developing ship building and repairing industries and may become a container entrepôt for the eastern Mediterranean.

There is perhaps no need to mention other maritime aspects of the insular life. Sea bathing and suntanning are today among the chief occupations. All islands have their fishers and small boat owners and in the clear silt-free water sponge gathering and more amateurish aqualunging flourish. The Greek islanders, as well as the coastal mainlanders, remain sailors at heart. To them Greece, in the words of a medieval manuscript, is

> Clustered grape and honeyed garden
> Marriage bed and banquet room
> Wind and ship and breeze and haven
> Moon and lamp and coming home.

Appendix 1

TIME SEQUENCE OF GEOLOGICAL PERIODS AND MAJOR EARTH MOVEMENTS

Eras	Periods		Approx. age million years	Earth movements
				much little
Quaternary	Recent			
	Pleistocene (Ice Age)		2	ALPINE
Cainozoic, or Tertiary	Pliocene		12	ALPINE
	Miocene		25	ALPINE
	Oligocene		40	ALPINE
	Eocene		60	ALPINE
	Paleocene		70	ALPINE
Mesozoic, or Secondary	Cretaceous	Upper		ALPINE
		Lower	135	
	Jurassic	Upper		
		Middle		
		Lias	180	
	Triassic	Keuper		
		Muschelkalk		
		Bunter	225	
Palæozoic	Permian		270	Hercynian
	Carboniferous		350	
	Devonian		400	
	Silurian		440	Caledonian
	Ordovician		500	
	Cambrian		600	
Archæan	Pre-Cambrian			Charnian

Notice that the chronological columns are not drawn to a time-scale. Thus the Hercynian earth movements were as prolonged as those of the Alps.

Appendix II: Conversion Tables

SI UNITS TO IMPERIAL UNITS

LENGTH AND HEIGHT 1 km = 1000 m   1 m = 100 cm = 1000 mm
    1 kilometre = 1094 yds = 3281 ft = approx $\frac{5}{8}$ mile or 0·62 mile
    1 metre = 1·094 yds = 3·281 ft = 39·37 ins = approx $\frac{10}{3}$ feet
    1 centimetre = 0·3937 inch = approx 0·4 inch
    1 millimetre = 0·03937 inch = approx 0·04 inch

AREA 1 sq km (km²) = 100 ha = 1 million sq. m
    1 sq kilometre = 0·386 sq mile = 247 acres
    1 hectare = 2·47 acres

WEIGHT 1 metric ton = 10 quintals = 1000 kilograms
    1 metric ton = 2205 lb = 19·7 cwt
    1 kilogram = 2·205 lb
    1 gramme (g) = 0·035 oz
        Rough rule: to turn metric tons into British tons deduct $1\frac{1}{2}$ per cent

YIELDS PER AREA
    1 metric ton per hectare = 0·398 British ton per acre
    1 quintal per hectare = 0·0398 British ton per acre

NUMBERS per sq km to per sq mile
    1 unit per sq kilometre = 2·59 units per sq mile

DEGREES CENTIGRADE TO FAHRENHEIT

°C	−20	−10	0	10	20	30	40	50
°F	−4	14	32	50	68	86	104	122

$t°C = \frac{9}{5}t + 32°F$

VOLUME 1 cub metre = 1000 litres = 35·315 cu ft = 220 U.K. gallons
    1 litre = 0·22 U.K. gallon = 0·353 cub ft

Appendix III: Bibliography

ABBREVIATIONS

A.A.A.G.	*Annals of the Association of American Geographers* (Albany, New York)
A.C.G.I.	*Atti del Congresso dei Geografi Italiani*
A. de G.	*Annales de Géographie* (Paris)
B.A.G.F.	*Bulletin de l'Association Géographes Français* (Paris)
B.S.G.F.	*Bulletin de la Societé Géologique de la France* (Paris)
B.S.G.I.	*Bollettino della Societa Geografica Italiana* (Rome)
C.R.	*Comptes Rendus* International Geographical Congress
Econ. Geog.	*Economic Geography* (Worcester, Mass.)
Erdk.	*Erdkunde* (Bonn)
Est. Geog.	*Estudios Geográficos* (Madrid)
Ét. Rhod.	*Études Rhodaniennes* (Lyon)
Geog.	*Geography* (Sheffield)
G.J.	*Geographical Journal* (London)
G.M.	*Geographical Magazine* (London)
G.R.	*Geographical Review* (New York)
I.B.G.	*Transactions, Institute of British Geographers* (Philip, London)
INEA	Istituto Nazionale di Economia Agraria
La Géog.	*La Géographie* (Paris)
Méd.	*Méditerranée* (Gap, Hautes Alpes, France)
Met. R.	*Meteorologische Rundschau* (Berlin)
Pet. Mitt.	*Petermanns Geographische Mitteilungun* (Leipzig)
Proc. Geol. Ass.	*Proceedings of the Geological Association* (London)
R.G.A.	*Revue de Géographie Alpine* (Grenoble)
R.G.I.	*Rivista Geografica Italiana* (Florence)
R.G.L.	*Revue de Géographie* (Lyon)
R.G.Pyr.	*Revue Géographique des Pyrénées et du Sud-Ouest* (Toulouse)
R. Met. Aer.	*Rivista di Meteorologia Aeronautica* (Rome)
S.G.M.	*Scottish Geographical Magazine* (Edinburgh)
UTET	Union Tipografico Editrice Torinese

Chapters 1 and 2: Structure, Relief and Landforms

BAILEY, E. B. *Tectonic Essays mainly Alpine*, Clarendon Press, Oxford, 1935.

BIROT, P. *La Méditerranée et Le Moyen-Orient*, Presses Universitaires de France, Paris, 1964.

COE, K. (ed.) *Some Aspects of the Variscan Fold Belt*, Manchester University Press, Manchester, 1962.

COLLET, L. W. *The Structure of the Alps*, Arnold, London, 1935.

FURON, R. *La Paléogéographie*, Payot, Paris, 1959, pp. 351–82.

JUDSON, S. 'Erosion and deposition of Italian stream valleys during historic time,' *Science*, 1963, pp. 898–9; 1968, p. 1444.

KOBER, L. *Bau und Enstehung der Alpen*, Franz Deuticke, Vienna, 1955.

MACHATSCHEK, F. *Das Relief der Erde*, Vol. 1, Borntraeger, Berlin, 1955.

OXBURGH, R. E. 'An outline of the geology of the central Eastern Alps', *Proc. Geol. Ass.*, 1968, pp. 1–127.

OXFORD UNIVERSITY EXPLORATION CLUB 'Methana, Greece', *Bulletin* No. 16, 1968, pp. 1–14.

RAMSAY, J. G. 'Stratigraphy, structure and metamorphism in the Western Alps', *Proc. Geol. Ass.*, 1963., pp. 357–91.

WILLS, L. J. *A Paleogeographical Atlas*, Blackie, London, 1952.

VITA-FINZI, C. *The Mediterranean Valleys: Geological Changes in Historical Times*, Cambridge University Press, Cambridge, 1969.

Chapter 3: The Submerged Landscape: Marine Life and Fisheries

BARBAZA, Y. 'La pêche méditerranéene', *A. de G.*, 1961, p. 60–70.

BARNES, H. (ed.) *Oceanography and Marine Biology. Annual Review*, Vol. 5, Allen & Unwin, London, 1967.

BARTZ, F. *Die Grossen Fischereiräume der Welt*, Vol. 1, F. Steiner, Wiesbaden, 1964.

BESANÇON, J. *Géographie de la Pêche*, Gallimard, Paris, 1965.

BIROT, P. *La Méditerranée* . . . , 1964, Vol. I, pp. 25–44.

BOURCART, J. 'Essai de carte sous-marine de l'Ouest de la Corse', *Rev. de Géog. physique et Géol. dynamique*, Paris, 1957, pp. 31–6.

DOUMENGE, F. 'Problèmes de la pêche en Méditerranée occidentale', *B.A.G.F.*, 1958, pp. 7–23.

FAO *Yearbook of Fishery Statistics*, Rome.

FAO *Fisheries in the Food Economy*, Basic Study No. 19, Rome, 1968.

GAMULIN-BRIDA, H. 'The benthic fauna of the Adriatic Sea', in Barnes, H. (ed.), pp. 535–68. With fine maps.

PERES, J. M. 'The Mediterranean benthos', in Barnes, H. (ed.), pp. 449–533.

ROUCHE, J. *La Méditerranée*, Flammarion, Paris, 1946.

SION, J. 'Le rôle des articulations littorales en Méditerranée', *A. de G.*, 1934, pp. 372–9.

SVERDRUP, H. U. *et al. The Oceans* . . . , Prentice-Hall, New York, 1942, pp. 642–9.

ULLYOT, P. & ILGAZ, O. 'The Hydrography of the Bosporus', *G.R.*, 1946, pp. 44–66.

ZENKOVITCH, V. R. *Processes of Coastal Development*, Oliver and Boyd, Edinburgh, 1967.

Chapter 4: Climatic Characteristics

AIR MINISTRY, BRITISH METEOROLOGICAL OFFICE *Weather in the Mediterranean: General Meteorology*, Vol. 1, HMSO, 1962; Vols. 2–4, 1957.

BÉNÉVENT, A. 'Bora et mistral', *A. de G.*, 1930, pp. 286–98.

BÉRENGER, M. *Essai d'étude météorologique du bassin méditerranéen*, Paris, 1955 (Memoire de la Météorologie Nationale No. 40).

BIEL, E. R. *Climatology of the Mediterranean Area*, Chicago University Press, Chicago, 1944.

BIROT, P. *La Méditerranée...*, Presses Universitaires de France, Paris, 1964, pp. 45–65.

BILLAUT, M. et al. 'Problèmes climatiques sur la bordure Norde du monde méditerranéen,' *A. de G.*, 1956, pp. 15–39.

CONRAD, V. 'The Climate of the Mediterranean region', *Bulletin American Meteorological Society*, 1943, pp. 127–45.

HARE, F. K. *The Restless Atmosphere*, Hutchinson, London, 1966.

HUTTARY, J. 'Die verteilung der niederschlage... im Mittelmeergebiet', *Met. R.*, 1950, pp. 111–19.

KOPPEN, W. & GEIGER, R. *Handbuch der Klimatologie*, Tome III, E. Alt, *Mittel und Sud Europa*, Borntraeger, Berlin, 1932.

TREWARTHA, G. T. *The Earth's Problem Climates*, Methuen, London, 1961.

REICHEL, E. 'Die niederschlagshaufigkeit im Mittelmeergebiet', *Met. R.*, 1949, pp. 129–42.

SCHNEIDER-CARIUS, E. 'Die Etesian', *Met. R.*, 1948, pp. 464–70.

THRAN, P. & BROEKHUIZEN, S. *Agro-Climatic Atlas of Europe*, Elsevier, Wageningen, Netherlands, 1965.

Chapter 5: Rivers and Water Supply
Rivers and river transport

ALMAGIÀ, R. *L'Italia*, Tome 1, U.T.E.T., Turin, 1959.

BECKINSALE, R. P. 'River regimes', in *Land, Air and Ocean*, Duckworth, London, 1966, pp. 312–23; and *Introduction to Physical Hydrology* (ed. R. J. Chorley), Methuen, London, 1971, pp. 176–92.

Comptes Rendus XVI Congrès Internat. Géogr. Lisbon, Tome 2, 1944.

FAUCHER, D. *L'Homme et Le Rhône*, Gallimard, Paris, 1968.

HUGENTOBLER, E. *Le Rhône navigable du Leman à la Méditerranée*, Franco-Suisse, Ambilly-Annemasse, 1949.

PARDÉ, M. *Le Régime du Rhône*, University of Lyon, Lyon, 2 vols., 1925; 'Le régime du Tibre', *R.G.A.*, 1933, pp. 289–335; 'Les régimes fluviaux de la peninsule Ibérique', *R.G.L.*, 1964, pp. 129–82.

SERVIZIO IDROGRAFICO *Dati caratteristici dei Corsi d'Acqua italiani*, Rome.

UN *Annual Bulletin of Transport Statistics for Europe*, New York.

Irrigation and land reclamation

ANTONIETTE, A., D'ALANNO, A. & VANZETTI, C. *Carta delle Irrigazioni d'Italia*, INEA, Rome, 1965 (an indispensable account with fine maps).

BECKINSALE, R. P. 'Human responses to river regimes', in *Introduction to Geographical Hydrology* (ed. R. J. Chorley), Methuen, London, 1971, pp. 142–64.

CARÈRRE, P. & DUGRAND, R. *La Région méditerranéenne*, Presses Universitaires de France, Paris, 1967.

FAO *Production Yearbook*, Rome (annual).

FAO *Mediterranean Development Project*, Rome, 1960.

HOUSTON, J. M. *The Western Mediterranean World*, Longmans, London, 1964, pp. 104–57.

NAYLON, J. 'Irrigation and internal colonization in Spain, *G.J.*, 1967, pp. 178–91.

MOHRMANN, J. C. H. & KESSLER, J. *Water Deficiencies in European Agriculture*. International Inst. for Land Reclamation and Improvement, Publ. 5, Wageningen, Netherlands, 1959.

Water supply

BOLTZ, C. L. *Desalination*, British Aqua Chem. Ltd, London, 1967.

MILLER, D. G. *Desalination for Water Supply*, Water Research Assoc., Medmenham, Bucks, 1962.

POPKIN, R. *Desalinization: Water for the World's Future*, Praeger, New York and London, 1969.

WASHINGTON, D.C. *First International Symposium on Water Desalination*, 1965.

Chapter 6: Vegetation and Soils

BIROT, P. *La Méditerranée et le Moyen-Orient*, Tome I, Presses Universitaires de France, Paris, 1964, pp. 66–129.

BORDAS, J. & MATHIEU, G. 'Les Sols de la région du Bas-Rhône', *Monographie du Ministère de l'Agric.*, Paris, 1943.

DUCHAUFOUR, PH. *Précis de Pédologie*, Masson, Paris, 1965 (with full bibliography).

FAO *Soil Map of Europe*, Rome, 1965–6 (with explanatory text).

GAUSSEN, H. 'Les sols et le climat Méditerranéen de France', *Rev. la Chêne*, 1931, pp. 71–97.

KUBIENA, W. L. *The Soils of Europe*, Murby, London, 1953.

RIKLI, M. *Das Pflanzenkleid der Mittelmeerländer*, 3 vols., Hans Huber, Bern, 1943–8.

SCHIMPER, A. F. W. & FABER, F. C. *Pflanzen-Geographie*, 2 vols., Fischer, Jena, 1935.

TAMÉS, C. *Los grupos principales de suelos de la España peninsular*, Ministr. de Agricultura, Madrid, 1957.

TERÁN, M. DE (ed.) *Geografía de España y Portugal*, Tome 2, *La Vegetacion*, by P. Font Quer, Montaner y Simon, Barcelona, 1954, pp. 145–271.

UNESCO *Vegetation Map of the Mediterranean Region*, FA, Rome, 1963; *Bioclimatic Map of the Mediterranean Region*, FAO, Rome, 1969 (each with an explanatory booklet).

VILAR, E. H. DEL & ROBINSON, G. W. *The Soils of the Lusitano–Iberian Peninsula*, Murby, London, 1938.

Chapters 7 and 8: Land Use

General and forestry

BIROT, P. *La Méditerranée . . .*, Vol. I, Presses Universitaires de France, Paris, 1964, pp. 133–6.

FAO *Agricultural Yearbook*, Rome (annual).

FAO *Horticulture in the Mediterranean Area*, Rome, 1968.

FAO *World Forest Product Statistics*, 1954–63, Rome, 1965.

FAO *Yearbook of Forest Products*, Rome (annual).

HOLLISTER, R. *A Technical Evaluation of the First Stage of the Mediterranean Regional Project*, OECD, Paris, 1967.

HOUSTON, J. M. *The Western Mediterranean World*, Longmans, London, 1964.

PRENTICE, A. 'Re-afforestation in Greece', *S.G.M.*, 1956, pp. 25–31.

UN *Statistical Yearbook*, New York (annual).

Livestock and pastures

CARRIER, E. H. *Water and Grass: A Study of the Pastoral Economy of Southern Europe*, Christophers, London, 1932.

DAVIES, E. 'The patterns of transhumance in Europe', *Geog.*, 1941, pp. 155–68.

MATLEY, I. M. 'Transhumance in Bosnia and Herzegovina', *G.R.*, 1968, pp. 231–61.

MULLER, E. 'Die herdenwanderungen im Mittelmeergebiet', *Pet. Mitt.*, 1938, pp. 364–70.

OEEC *Pasture and Fodder Development in Mediterranean Countries*, Paris, 1951.

PRENTICE, A. 'Livestock and forage production in Central Macedonia', *S.G.M.*, 1957–8, pp. 146–57.

Agricultural crops and landholding

ANDREWS, A. C. 'Acclimatization of citrus fruits in the Mediterranean region', *Agric. Hist.*, 1961, pp. 35–46.

FISCHER, T. 'Der Oelbaum', *Pet. Mitt.*, heft 147, 1904.

FRANKLIN, S. H. *The European Peasantry, The Final Phase*, Methuen, London, 1969.

GALTIER, M. G. 'Le vignoble espagnol d'aujourd'hui', *B.A.G.F.*, 1950, pp. 96–103.

HOFFMAN, G. W. 'Problems of agricultural change in south-eastern Europe', *G.R.*, 1965, pp. 428–31.

OECD *Agricultural Development in Southern Europe*, Paris, 1969.

WARRINER, D. *The Economics of Peasant Farming*, Oxford University Press, London, 1964.

Chapter 9: Settlers and Settlement

Races and languages

AUTY, R. 'Community and divergence in the history of the Slavonic languages', *The Slavonic and East European Rev.*, 1963, pp. 257–73.

BOYD, W. C. *Genetics and the Races of Man*, Blackwell, Oxford, 1950.

COON, C. S. *The Races of Europe*, Macmillan, London, 1939.

GEIPEL, J. *The Europeans, an ethnohistorical survey*, Longmans, Harlow, 1969; Pegasus, New York, 1970.

HENCKEN, H. 'Indo-European Languages and Archaeology', *Amer. Anthrop. Assoc. Mem.*, No. 84, New York, 1955.

HÉRAUD, G. *Peuples et langues de l'Europe*, Denoël, Paris, 1968.

MOURANT, A. E. *The Distribution of Human Blood Groups*, Blackwell, Oxford, 1954.

Settlement and historical geography

BLOCH, R. *The Etruscans*, Thames and Hudson, London, 1958.

BLOCH, R. *Origins of Rome*, Thames and Hudson, London, 1960.

BRADFORD, J. *Ancient Landscapes*, Bell, London, 1957.

BRAUDEL, F. *La Méditerranée et le monde méditerranéen a l'époque de Philippe II*, 2 vols. Colin, Paris, 1966–7.

'Cambridge Ancient History', vols. 5–12, with four volumes of illustrations, Cambridge University Press, Cambridge, 1927–39.

'Cambridge Economic History of Europe', vols. 1–3, 1963– .

'Cambridge Mediaeval History', 8 vols. 1911–67; summarized in Previté-Orton, C. W. *The Shorter Mediaeval History*, 2 vols., Cambridge University Press, Cambridge, 1952–3.

'Cambridge New Modern History', 12 vols., Cambridge University Press, Cambridge, 1957–68.

CHILDE, V. G. *The Prehistory of European Society*, Penguin, Harmondsworth, 1962.

CLARK, J. G. D. *Prehistoric Europe, The Economic Bases*, Methuen, London, 1952.

CARY, M. *The Geographic Background to Greek and Roman History*, Clarendon Press, Oxford, 1949.

DANIEL, G. E. *The Megalith Builders of Western Europe*, Hutchinson, London, 1958.

DVORNIK, F. *The Slavs in European History and Civilization*, Rutgers University Press, New Brunswick, N.J., 1962.

EAST, W. G. *An Historical Geography of Europe*, Methuen, London, 1966.

HARDEN, D. *The Phoenicians*, Thames and Hudson, London, 1962.

LAVEDAN, P. *Histoire de l'Urbanisme. Antiquité–Moyen Age*, Henri Laurens, Paris, 1926.

POWELL, T. G. E. *The Celts*, Thames and Hudson, London, 1958.

SMITH, C. T. *An Historical Geography of Western Europe before 1800*, Longmans, London, 1967.

WARD-PERKINS, J. 'Etruscan towns . . .', *G.J.*, 1962, pp. 389–405.

WHITE, M. E. 'Greek colonisation', *Jour. Econ. Hist.* 1961, pp. 443–54.

WOODHEAD, A. G. *The Greeks in the West*, Thames and Hudson, London, 1962.

Chapter 10: International and economic patterns

ANNUAL REPORTS, ECSC, EEC, EFTA, OECD.

CALMAN, J. (ed.) *Western Europe, a Handbook*, Anthony Blond, London, 1967.

CARONE, G. *Il Turismo nell'economia internazionale*, A. Guiffrè, Milan, 1959.

Europa Year Book, Vol. I, Europa Publications, London (annual).

NEUNDORFER, L. *Atlas of Social and Economic Regions of Europe*, A. Lutzeyer, Baden-Baden, 1964.

OECD *Agriculture and Economic Growth*, Paris, 1965.

OECD *Agricultural Development in Southern Europe*, Paris, 1969.

OECD *Le Tourisme dans les pays de l'OCDE*, Paris (annual).

SINCLAIR, D. J. 'Steel in Europe', *G.M.*, 1969, pp. 610–11.

UN *Statistical Yearbook*, New York (annual).

YATES, P. L. *Food, Land and Manpower in Western Europe*, Macmillan, London, 1960.

Chapter 11: Spain; Basal developments and patterns

Anuario Estadístico de España, Instit, Nacional, Madrid (annual).

Atlas Nacional de España, Instit, Geog. y Catastral, Madrid, 1965– .

BENNETT, H. H. 'Soil erosion in Spain', *G.R.*, 1960, pp. 59–72.

BIROT, P. *La Méditerranée . . .*, Vol. I, Presses Universitaires de France, Paris, 1964, 189–354.

BIROT, P. 'Progrès récents dans la connaissance de la géomorphologie et de la structure de l'Espagne, septentrionale', *A. de G.*, 1960, pp. 548–50.

CHILCOTE, R. H. 'The Spanish iron and steel industry', *Geog.*, 1967, pp. 60–4.

DAUMAS, M. 'Où en est le remembrement rural en Espagne?', *R. G. Pyr.*, 1971, pp. 213–27.

DEFFONTAINES, P. 'Transformations récents du delta de l'Ebre pour l'irrigation et la riziculture', *C. R. Internat. Géog. Congrès*, Lisbon, 1949.

GALTIER, M. G. 'Le vignoble espagnol d'aujourd'hui', *B.A.G.F.*, 1950., pp. 96–103.

Geographical Handbook British Admiralty: Naval Intelligence Division. *Spain and Portugal: The Iberian Peninsula*, Vol. I, 1941; *Spain*, Vol. 3, 1944.

GUTKIND, E. A. *Urban Development in South Europe: Spain and Portugal*, Collier-Macmillan, London and New York, 1967.

HERNÁNDEZ-PACHECO, E. *Sintesis Fisiográfica y Geologica de España*, Museo Nacional de Ciencas Nat., Ser. geológica, Madrid, 1934.

HOUSTON, J. M. 'Irrigation as a solution to agrarian problems in modern Spain', *G.J.*, 1950, pp. 55–63.

HOUSTON, J. M. *The Western Mediterranean World*, Longmans, London, 1964, pp. 161–335.

JÜRGENS, O. *Spanische Städte*, Friederichsen, Hamburg, 1927, with atlas.

LAUTENSACH, H. 'Die niederschlagshöhen auf der Iberischen Halbinsel', *Pet. Mitt.*, 1951, pp. 145–60.

LAUTENSACH, H. & MAYER, E. 'Iberische Meseta and Iberische masse'. *Zeitschr für Geomorphol.*, 1961, pp. 161–80.

NAYLON, J. 'Land consolidation in Spain', *A.A.A.G.*, 1959, pp. 361–73.

NAYLON, J. 'Progress in land consolidation in Spain', *A.A.A.G.*, 1961, pp. 335–8.

NAYLON, J. 'Irrigation and internal colonisation in Spain', *G.J.*, 1967, pp. 178–91.

NAYLON, J. 'Tourism· Spain's most important industry', *Geog.*, 1967, pp. 23–40.

REUS, C. F. *Geografía Económica de España*, Miguel Arimany, Barcelona, 1952.

SORRE, M. *Géographie Universelle*. Tome VII, *Méditerranée: Espagne–Portugal*, Colin, Paris, 1934.

TAMÉS, C. *Los grupos principales de suelos de la España peninsular*, Ministr. de Agricultura, Madrid, 1957.

TERÁN, M. de (ed.) *Geografía de España y Portugal*, 5 vols. some with several parts, Montaner y Simón, Barcelona, 1952–

Chapter 12: Spain; Major geographical regions

General

BIROT, P. *La Méditerranée . . .*, 1964.

HOUSTON, J. M. *The Western Mediterranean World*, 1964. pp. 251–335.

TERÁN, M. DE (ed.) *Geografía de España y Portugal*, Tome IV. *España: Geografía Regional*, Montaner y Simón, Barcelona, 1954– (consists of several volumes).

North, Meseta, Catalonia

COURTENAY, P. P. 'Madrid: The circumstances of its growth', *Geog.*, 1959, pp. 22–34.

CUILLE, G. 'L'industrie textile de la région de Sabadell–Tarrassa (Catalogne)', *Méd.*, 1960, pp. 3–41.

DOBBY, E. H. G. 'Catalonia . . .', *G.R.*, 1938, pp. 224–49.

DOBBY, E. H. G. 'Galicia', *G.R.*, 1936, pp. 555–80.

DOBBY, E. H. G. 'The Ebro delta', *G.J.*, 1936, pp. 455–69.

DRESSER, B. 'The Sierra de Gredos', *S.G.M.*, 1958–9, pp. 175–9.

GLADFELTER, B. G. *Meseta and Campiña Landforms in Central Spain*, University of Chicago Press, Chicago, 1971.

HUETZ DE LEMPS, A. *Vignobles et Vins du Nord-Ouest de l'Espagne*, 2 vols., Institut de Géographie, Bordeaux, 1967.

KLEIN, J. *The Mesta*, Harvard University Press, Cambridge, Mass., 1920.

NAYLON, J. 'The Badajoz plan . . .', *Erdk.*, 1966, pp. 44–59.

NONN, H. 'Évolution géomorphologique et types de relief en Galice . . .', *Rev. Géog. Physique et Géol. Dynamique*, 1969, pp. 31–50.
REGALES, M. F. & LEDO, A. P. 'El proceso de urbanización en el Pais Vasco y Navarra: Polución y contaminación del ambiente', *Geographica*, Inst. de Geog. Aplicada, Madrid, 1971, pp. 125–41.
ROBERT, D. 'La région de Santander', *A. de G.*, 1936, pp. 1–18.
SOLÉ, I. SABARÍS *Los Pirineos: El medio y el hombre*, Martin, Barcelona, 1956.
SOLÉ, I. SABARÍS *Geografía de Catalunya*, Vol. 4, *Enciclopèdia Catalana Aedos*, Barcelona, 1958.
VILAR, P. 'Le port de Barcelone', *A. de G.*, 1934, pp. 489–509.

Andorra
CORTS PEYRET, J. *Geografía e Historia de Andorra* Editorial Labor, Barcelona, 1945.
LLOBET, S. *El medio y la vida en Andorra*, Consejo Superior de Investigaciones Cientificas, Barcelona, 1947.

South-East, South, Balearic Islands
COMÍN, A. C. *España del Sur*, Editorial Tecnos, Madrid, 1965.
GILBERT, E. W. 'The human geography of Mallorca', *S.G.M.*, 1934, pp. 129–47.
GILBERT, E. W. 'Influences of the British occupation on the human geography of Menorca', *S.G.M.*, 1936, pp. 375–90.
HERMET, G. *Le Problème Méridional de l'Espagne*, Colin, Paris, 1965.
HOUSTON, J. M. 'Urban geography of Valencia . . .', *I.B.G.*, 1949, pp. 19–35.
MARTINEZ, M. C. *Fundamentos del Desarrollo Economico de Andalucia*, Consejo Superior de Investigaciones Cientificas, Madrid, 1963.
MAUREL, J. B. *Geografía Urbana de Granada*, Inst. de Juan Sebastian Elcano, Zaragoza, 1962.
NAYLON, J. 'The Campo de Gibraltar development plan', *I.B.G.: Area*, No. I, 1959, pp. 21–2.
NIEMEIER, G. *Siedlungsgeographische Untersuchungen in Niederandalusien*, de Gruyter, Hamburg, 1935.
PATTERSON, R. M. 'The Balearic islands', *Journ. Geog.*, 1946, pp. 153–6.
SERMET, J. *L'Espagne du Sud*, Arthaud, Paris, 1953.

Gibraltar
Annual Report on Gibraltar, HMSO, London.
BECKINSALE, R. P. 'Gibraltar'; and 'Strait of Gibraltar', *Encyclopedia Americana*, New York, 1969.
Gibraltar Directory and Guide Book, Gibraltar (annual).
JESSEN, O. *Die Strasse von Gibraltar*, Dietrich Reimer, Berlin, 1927.

Chapter 13: Portugal
Anuário Estatístico, Lisbon (annual).
BIROT, P. *Le Portugal*, Colin, Paris, 1963.
CHILCOTE, R. H. 'Portugal's new iron and steel industry', *Geog.*, 1963, pp. 188–90.
C.R. Internat. Géog. Congrès, Lisbon, 1949. Has many articles on Portugal and a series of geographical guidebooks.
DOBBY, E. H. G. 'Economic geography of the port wine region', *Econ. Geog.*, 1936, pp. 311–23.
GEOGRAPHICAL HANDBOOK British Admiralty: Naval Intelligence Division, *Spain and Portugal*, Vol. 2, *Portugal*, 1942.
GIRÃO, A. DE A. *Geografía de Portugal*, Portucalense Editôra, Oporto, 1952.
HAYES, R. D. 'A peasant economy in north-west Portugal', *G.J.*, 1956, pp. 54–70.
HOUSTON, J. M. *The Western Mediterranean World*, 1964, pp. 336–68.
LAUTENSACH, H. 'Portugal . . . I. Das Land als Ganzes', *Pet.Mitt.* Heft 213, 1932; II, 'Die portugiesischen Landschaften', *Pet.Mitt.* Heft 230, 1937.
PEREIRA, A. G. 'Les vignobles du nord du Portugal', *R. G. Pyr.*, 1932, pp. 202–33.
RIBEIRO, A. G. 'Le site et la croissance de Lisbonne', *B.A.G.F.* 1938, pp. 99–103.
RIBEIRO, O. *Portugal*, Tome V. *Geografía de España y Portugal*, ed. M. de Terán, Montaner y Simón, Barcelona, 1955.
STANISLAWSKI, D. *The Individuality of Portugal*, University of Texas Press, Austin, and Nelson, Edinburgh, 1959.
STANISLAWSKI, D. *The Algarve*, University of Texas Press, Austin and London, 1965.
STANISLAWSKI, D. *Landscapes of Bacchus: The Vine in Portugal*, University of Texas Press, Austin and London, 1970.

Chapters 14 and 15: Southern France
General
AMBASSADE DE FRANCE, London *Energy in France*. A/60/6/8; *The French Tourist Industry*. A/56/2/8.
Annuaire Statistique, Instit. Nat. de la Statistique, Paris.
Atlas de France, Comm. Nat. de Géog., Paris, 1961– .
DEMANGEON, A. *France, économique et humaine. Géog. Univ.*, Tome VI., pt. I, Colin, Paris, 1946.
FAUCHER, D. (ed.) *La France Géographie Tourisme*, Vol. I, Larousse, Paris, 1951.
HOUSTON, J. M. *The Western Mediterranean World*, Longmans, London, 1964.
MARTONNE, E. DE *France Physique. Géog. Univ.*, Tome VI, Colin, Paris, 1947.
MONKHOUSE, F. J. *A Regional Geography of Western Europe*, Longmans, London, 1968.
PINCHEMEL, PH. *France*, Bell, London, 1969.

THOMPSON, I. B. *Modern France*, Butterworth, London, 1970.

Mountains: Alps and Jura

AGER, D. V. & EVAMY, B. D. 'The geology of the southern French Jura', *Proc. Geol. Ass.*, 1963, pp. 325–55; 483–96.

ALLEFRESDE, M. 'Les fabrications fromagères en Haute-Savoie', *R.G.A.*, 1952, pp. 625–41.

AMBASSADE DE FRANCE, London. *The City of Grenoble.* A/54/12/7.

BACCONNET, D. 'L'industrialisation d'une grande vallée alpestre... le Grésivaudan', *R.G.A.*, 1956, pp. 99–166.

BLANC, A. et al. *Les régions de l'Est*, Presses Universitaires de France, Paris, 1970.

BLANCHARD, R. *Les Alpes Occidentales*, 12 vols. Arthaud, Grenoble, 1944–56.

VEYRET, R. G. & GERMAIN, F. *Grenoble, capitale Alpine*, Arthaud, Grenoble, 1967.

VEYRET, P. & G. *Au Coeur de l'Europe: Les Alpes*, Flammarion, Paris, 1969.

French Pyrenees

ARQUÉ, P. *Géographie des Pyrénées françaises*, Presses Universitaires de France, 1943.

BOESCH, H. 'Die Natur im Baskenland', *Geog. Helvetica*, 1955, pp. 136–44.

CAZES, G. *Le Tourisme à Luchon et dans le Luchonnais*, Institut de Géog., Toulouse, 1964.

CHEVALIER, M. *La Vie Humaine dans les Pyrénées ariègeoises*, Génin, Paris, 1956.

GAUSSEN, H. *Géographie botanique et agricole des Pyrénées Orientales*, Paul Lechevalier, Paris, 1934.

Revue Géographique des Pyrénées et du Sud-Ouest is indispensable.

TAILLEFER, F. 'Glaciaire pyrénéen...', *R.G. Pyr.*, 1957, pp. 221–44.

Coastlands, Lowlands and Plains

AGNEW, S. 'Rural settlement in Bas Languedoc', *Geog.*, 1946, pp. 67–77.

AGNEW, S. 'The vine in Bas Languedoc', *G.R.*, 1946, pp. 67–79.

AMBASSADE DE FRANCE, London. *Tourist Development in Languedoc and Roussillon.* B/33/2/7.

BADOUIN, R. et al. *L'économie du Languedoc–Roussillon*, Gauthier-Villar, Paris, 1968.

BENIAMINO, O. 'Grasse...', *R.G.A.*, 1957, pp. 763–74.

BETHEMONT, J. 'Le riz et la mise en valeur de la Camargue', *R.G.L.*, 1962, pp. 153–206.

BORDAS, J. 'Les Sols de la région du Bas-Rhône', *Monographie du Ministère de l'Agric.*, Paris, 1943.

CARRÈRE, P. & DUGRAND, R. *La Région Méditerranéenne*, Presses Universitaires de France, Paris, 1967.

CAVERIVIÈRE, CH. 'L'Exploitation des étangs de Sigean', *R. G. Pyr.*, 1950, pp. 61–80.

CHABOT, G. *La Bourgogne*, Colin, Paris, 1945.

CORBEL, J. 'La violence des vents dans le couloir Rhodanien', *R.G.L.* 1962, pp. 273–86.

DUGRAND, R. 'L'Aménagement du Bas-Rhône', *A. de G.*, 1953, pp. 368–73.

DUGRAND, R. *Villes et Campagnes en Bas-Languedoc*, Presses Universitaires de France, Paris, 1963.

GALTIER, G. *Le Vignoble de Languedoc méditerranéen et du Roussillon*, 2 vols., Causse-Graille, Montpellier, 1960.

GEORGE, P. *La Région du Bas-Rhône*, J.-B. Baillère, Paris, 1935.

GEORGE, P. *Études géographiques sur le Bas Languedoc. La Région Montpelliéraine*, H.-G. Peyre, Paris, 1938.

GIBB, R. W. 'La Camargue', *Geog.* 1942, pp. 63–6.

HOYLE, B. C. 'Changes in the Durance valley', *Geog.*, 1960, pp. 110–13.

JONES, I. E. 'The development of the Rhône', *Geog.*, 1969, pp. 446–51.

JOURNAUX, A. *Les Plaines de la Saône,... Mâconnais, Côte d'Or....*, Impr. Caron, Caen, 1956.

KAYSER, B. *Campagnes et Villes de la Côte d'Azur*, Rocher, Monaco, 1958.

LABASSE, J. & LAFERRÈRE, M. *La Région Lyonnaise*, Presses Universitaires de France Paris, 1966.

LIVET, R. *Habitat Rurale et Structures Agraires en Basse Provence*, University of Aix–Marseille, Ophrys, Gap, 1962.

Lyon et sa Région..., with portfolio of maps by Comité pour l'Aménagement et expansion économique de la Région Lyonnaise, Lyon, 1955.

NICOD, J. 'Grandeur et decadence de l'oléiculture provençale', *R.G.A.*, 1956, pp. 247–95.

PRÉVOT, V. 'La culture du riz de Camargue', *L'Inform. Géog.*, 1953, pp. 13–20.

PIERY, M. (ed.) *Le Climat de Lyon et de la Région Lyonnaise*, Cartier, Lyon, 1946.

RONCAYOLO, M. 'Évolution de la banlieue marseillaise dans la basse vallée de l'Huveaune', *A. de. G.*, 1952, pp. 342–56.

RUSSELL, R. J. 'Geomorphology of the Rhône delta', *A.A.A.G.*, 1942, pp. 149–254.

SERVICE MARITIME DES BOUCHES DU RHÔNE *Port de Marseille*, 1958.

SUCHEL, J.-B. 'L'Hydraulique agricole... entre Vienne et Bollène...', *R.G.L.*, 1957, pp. 201–26.

Monaco

HANDLEY-TAYLOR, G. *Bibliography of Monaco*, St James Press, Chicago, 1968.

KOLLER, F. *Histoire de la Principauté de Monaco*, Feniks, Brussels, 1963.

LA GORCE, P. M. DE *Monaco*, Éditions Rencontre, Lausanne, 1963.

SHERMAN, C. L. *The five little countries of Europe . . .*, Doubleday, New York, 1969.

Chapter 16: Switzerland and Liechtenstein

Annuaire Statistique de la Suisse, Basle.
Atlas der Schweiz, ed. E. Imhof, Eidgenössischen Landestopographie, Waber–Bern, 1965–
BOESCH, H. & HOFER, P. *Villes Suisses à vol d'oiseau*, Kümmerly & Frey, Berne, 1963.
BRUNNER, P. *Les Chemins de Fer aux prises avec la Nature Alpestre*, Grenoble Univ., Allier: Grenoble, 1935.
CAROL, H. V. & SENN, U. 'Jura, Mittelland und Alpen: Ihr anteil an flache und bevolkerung der Schweiz', *Geog. Helvetica*, 1950, pp. 129–36.
COKER, J. A. 'Tourism and the peasant in the Grisons', *S.G.M.*, 1950, pp. 107–16.
DICKINSON, R. E. *The West European City*, Routledge & Kegan Paul, London, 1961.
EGLI, E. *Die Schweiz, eine Landeskunde*, Haupt, Bern, 1966.
FRÜH, J. *Geographie der Schweiz*, 3 vols. Fehr'sche Buchhandlung, Sankt Gallen, 1930–8 (also in French).
FULLER, G. J. 'The Trient valley . . .' *Geog.*, 1955, pp. 28–39.
GABERT, P. & GUICHONNET, P. *Les Alpes et Les États Alpins*, Presses Universitaires de France, Paris, 1965.
GARNETT, A. 'Insolation and Relief', *I.B.G.* 1937.
GEORGE, P. & TRICART, J. *L'Europe Centrale*, Orbis, Paris, 1954, pp. 449–95.
GUTERSOHN, H. *Geographie der Schweiz*, Vol. I, *Jura*, 1958; Vol. 2, 2 parts, *Alpen*, 1961–4; Vol. 3, 2 parts, *Mittelland*, 1968–9, Kummerly & Frey, Bern.
GUTKIND, E. A. *International History of City Development: Alpine and Scandinavian Countries*, Collier-Macmillan, London & New York, 1966.
MARTONNE, E. DE *Les Alpes*, Colin, Paris, 1926.
MAYER, K. B. *The Population of Switzerland*, Columbia University Press, New York, 1952.
MAYER, K. B. 'Recent demographic developments in Switzerland', *Social Research*, 1957, pp. 331–53.
MOUNTJOY, A. B. 'Milk pipelines and mountain economies', *Geog.*, 1959, pp. 124–6.
MUTTON, A. *Central Europe*, Longman, London, 1968, pp. 57–139.
RATHGENS, C. 'Die wirtschaftlandschaften der Schweiz', *Geog. Rundschau*, 1951, pp. 170–5.
THOMAS, W. S. G. 'Remaniement Parcellaire in Switzerland', *Geog.*, 1967, pp. 307–10.
VEYRET, P. & G. *Au cœur de l'Europe: Les Alpes*, Flammarion, Paris, 1967.

Liechtenstein

Tatigkeits-und Rechenschaftsberichte. . . . Vaduz (annual).

KRANZ, W. (ed.) *The Principality of Liechtenstein*, Gov. Information Office, Vaduz, 1969.
MALIN, G. *Kunstführer Fürstentum Liechtenstein*, Kümmerly & Frey, Bern, 1968.
RATON, P. *Le Liechtenstein, Histoire et Institutions*, Droz, Geneva, 1967.
SHERMAN, C. L. *The five little countries of Europe. . .*, Doubleday, New York, 1969.

Chapters 17 and 18: Italian Mainland

General

Annuario Statistico Italiano, Rome (annual).
ALMAGIÀ, R. *L'Italia*, 2 vols., U.T.E.T., Turin, 1959.
ANTONIETTI, A., D'ALANNO, A. & VANZETTI, C. *Carta delle Irrigazioni d'Italia*, I.N.E.A., Rome, 1965 (text and fine maps).
ANTONIETTI, A. & VANZETTI, C. *Carta della Utilizzazione del Suolo d'Italia*, Feltrinelli, Milan, 1961.
BARBERIS, C. *Le migrazioni rurali in Italia*, Feltrinelli, Milan, 1960.
BIROT, P. & GABERT, P. *La Méditerranée . . .* , Presses Universitaires de France, Paris. 1964, pp. 357–539.
DAINELLI, G. (ed.) *Atlante Fisico–Economico d'Italia*, Consociazione Turistica Italiana, Milan, 1940.
Geographical Handbook British Admiralty · Naval Intelligence Division, *Italy*, 4 vols., 1944–5.
GRINDROD, M. *The Rebuilding of Italy*, 1945–55, Royal Inst. Internat. Affairs, London and New York, 1955.
GRINDROD, M. *Italy*, Benn, London, 1968.
GUTKIND, E. A. *Urban Development: Italy and Greece*, Free Press, London, 1969.
HILDEBRAND, G. *Growth and Structure in the Economy of Modern Italy*, Harvard University Press, Cambridge, Mass., 1965.
HOUSTON, J. M. *The Western Mediterranean World*, Longman, London, 1964, pp. 370–544.
LUTZ, V. *Italy, a Study in Economic Development*, Oxford University Press, London, 1962.
MEDICI, G. *Land Property and Land Tenure in Italy*, I.N.E.A., Edizioni Agricole, Bologna, 1952.
MILONE, F. *L'Italia nell'economia delle sue regioni*, Einaudi, Turin, 1958.
MORI, A. & CORI, B. 'L'area di attrazione delle maggiori città italiane', *R.G.I.*, 1969, pp. 3–14.
PINNA, M. 'Contributo alla classificazione del clima d'Italia', *R.G.I.*, 1970, pp. 129–52.
TIRONE, L. 'La Vigne dans l'Exploitation Agricole en Italie', *Méd.* 1970, pp. 339–62.
TOURING CLUB ITALIANO *Conosci Italia*. Vol. 1, *L'Italia Fisica*, Milan, 1957; Vol. 2, *La Flora*, Milan, 1958; Vol. 7, *Il Paesaggio*, Milan, 1963.
WALKER, D. S. *Italy*, Methuen, London, 1967.

Regional: North

BONATO, C. *L'Economia agraria della Lombardia*, I.N.E.A., Milan, 1952.

COLEMAN, A. 'The town of Lecco. Urban structure...', *Geog.*, 1958, pp. 243–52.
DALMASSO, E. *Milan: Capitale Economique de l'Italie*, Ophrys, Paris, 1971.
DEMATTEIS, G. *Le Località centrali nella geografia urbana de Torino*, University of Turin Press, Turin, 1966.
EDWARDS, L. F. 'Trieste, international city', *Geog. Mag.*, 1947, pp. 464–72.
GRIBAUDI, D. *Piedmonte e Val d'Aosta*, U.T.E.T., Turin, 1960.
MAINARDI, R. *The Urban Network of Northern Italy*, Centro di Documentazione, Milan, 1970.
MERLO, C. *Liguria*, U.T.E.T., Turin, 1962.
MIGLIORINI, E. *Veneto*, U.T.E.T., Turin, 1962.
MIHELIĆ, D. *The Political Element in the port geography of Trieste*, University of Chicago Press, Chicago, 1969.
ORTOLANI, M. *La Pianura ferrarese*, Mem. di Geog. econ., XV, Naples, 1956.
PECORA, A. 'Contributi allo studio geografico della citta di Milano', *B.S.G.I.*, 1953, pp. 419–24.
PELLEGRINI, G. C. *Studi o Ricerche sulla Regione Turistica: i lidi ferraresi*, Vita e Pensiero, Milan, 1968.
PRACCHI, R. *Lombardia*, U.T.E.T., Turin, 1960.
ROBERTSON, C. J. 'Agricultural regions of the North Italian Plain', *G.R.*, 1938, pp. 573–96.
RODGERS, A. L. 'The port of Genoa', *A.A.A.G.*, 1958, pp. 319–51.
RODGERS, A. L. *The Industrial Geography of the Port of Genova*, University of Chicago Press, Chicago, 1960.
TOSCHI, U. *Emilia–Romagna*, U.T.E.T., Turin, 1961.
TIZZONI, A. 'Distribuzione topografica della grandi industri di Milano', *Atti Congr. Geog. Ital.*, 1957.
WOODCOCK, A. C. 'The methane gas industry', *Geog.*, 1956, pp. 265–7.
WOODWARD, G. 'The new Italian Autostrade network', *Geog.*, 1963, pp. 68–70.
VEYRET, P. & G. *Au cœur de l'Europe: Les Alpes*, Flammarion, Paris, 1967.

Peninsula
AUDIN, A. 'La naissance de Roma', *R.G.A.*, 1956, pp. 21–31.
BALDACCI, O. *Puglia*, U.T.E.T., Turin, 1962.
BEVILACQUA, E. *Marche*, U.T.E.T., Turin, 1961.
CARLYLE, M. *The Re-Awakening of Southern Italy*, Oxford University Press, London, 1962.
CASTAGNOLI, F. et al. *Storia di Roma*, Vol. 22 of *Topografica e Urbanistica*, Ist. Studi Romani, Rome, 1960.
DESPLANQUES, H. *Campagnes Ombriennes*, Colin, Paris, 1969.
DESPLANQUES, H. 'La reforme agraire italienne', *A. de G.* 1957, pp. 310–27.
DICKINSON, R. E. *The Population Problem of Southern Italy*, Syracuse University Press, Syracuse, U.S.A., 1955.
INEA *L'Economia agraria della Campania*, Rome, 1948.
MCNEE, R. B. 'Rural development in the Italian south', *A.A.A.G.*, 1955, pp. 127–51.
MEYRIAT, J. (ed.) *La Calabre*, Colin, Paris, 1960.
MILONE, F. *Mem. Illustrativa della Carta di Utilizzazione del Suolo della Calabria*, sheets 19–20, Cons. Naz. Ricerche, Naples, 1956.
MOUNTJOY, A. B. 'Planning and development in Apulia', *Geog.*, 1966, pp. 369–72.
RANIERI, L. *Basilicata*, U.T.E.T., Turin, 1961.
RUOCCO, D. *I Campi Flegrei. Studio di geografia agraria*, Mem. Geog. econ. univ., XI, Naples, 1954.
SERONDE, A. M. 'Rome. Étude d'évolution urbaine', *B.A.G.F.*, 1954, pp. 121–7.
SERONDE, A. M. 'Les régions sous-développées de l'Italie', *A. de G.*, 1956, pp. 187–98.
SIMONCELLI, R. *Il Molise*, Fac. Econ. Univ. Roma, Rome, 1969.
TOSCHI, U. 'The Vatican city, from the standpoint of political geography', *G.R.*, 1931, pp. 529–38.
UNGER, L. 'Rural settlement in the Campania', *G.R.*, 1953, pp. 506–24.
WIRTH, E. 'Die Murgia dei Trulli, Apulien', *Die Erde*, 1962, pp. 249–78.

Chapter 19: Islands of the Western Mediterranean
General
HOUSTON, J. M. *The Western Mediterranean World*, Longman, London, 1964, pp. 534–44; 611–43.
KALLA-BISHOF, P. M. *Mediterranean Island Railways*, David & Charles, Newton Abbot, 1970.
MACIVER, D. R. *Greek Cities in Italy and Sicily*, Clarendon Press, Oxford, 1931.

Sicily
MILONE, F. *Memoria illustrativa della carta della utilizzazione del suolo della Sicilia*, C.N.R., Rome, 1959.
MILONE, F. *Sicilia: la natura e l'uomo*, Boringhieri, Turin, 1960.
PHILIPPSON, A. 'Die Landschaften Siziliens', *Erdk.*, 1934, pp. 321–42.
PRESTIANNI, N. 'Latifondo e riforme agraria', *L'Italia Agricola*, Rome, 1947, pp. 611–25.
ROCHEFORT, R. 'Le pétrole en Sicile', *A. de G.*, 1960, pp. 22–33.
ROCHEFORT, R. *Le Travail en Sicile; étude de géographie sociale*, Presses Universitaires de France, Paris, 1961.
SERONDE, A. M. 'Les régions sous developpées de l'Italie', *A. de G.*, 1956, pp. 197–8.

Smaller Italian Islands
See article by R. P. Beckinsale in Geographical Handbook. British Admiralty: Naval Intelligence Division, *Italy*, vol. 4, 1945, pp. 655–739.

MIKUS, W. 'Aspetti e problemi della geografia della popolazione nelle isole minori dell'Italia meridionale', *R.G.I.*, 1969, pp. 15–54.

Sardinia

BALDACCI, O. *La Casa Rurale in Sardegna*, Centro di Studi per la Geografia Ethnologica, Florence, 1952.

DOZIER, C. L. 'Establishing a framework for development in Sardinia: The Campidano', *G.R.*, 1957, pp. 490–506.

LE LANNOU, M. 'Sardaigne', *R.G.L.*, 1951, pp. 113–29.

LILLIU, G. 'The nuraghi of Sardinia', *Antiquity*, 1959, pp. 32–8.

MORI, A. *Le Saline della Sardegna*, Mem. Geog. Econ. Univ., Naples, 1950.

MORI, A. & SPANO, B. *I Porti della Sardegna*, Mem. Geog. Econ. Univ., Naples, 1952.

PAMPALONI, E. *L'Economia agraria della Sardegna*, I.N.E.A., Rome, 1947.

PELLETIER, J. *Le Relief de la Sardaigne*, Mem. et Doc. de l'Inst. d'Études rhod., Lyon, 1960.

PINNA, M. *Il Clima della Sardegna*, Ist. Geog. Univ., Pisa, 1954.

PINNA, M. & CORDA, L. *La Distribuzione della popolazione . . . della Sardegna*, Ist. Geog. Univ., Pisa, 1956–7.

Corsica

FAUCHER, D. (ed.) *La France Géographie Tourisme*, I. Larousse, Paris, 1951, pp. 261–7.

Geographical Handbook. British Admiralty. Naval Intelligence Division, *Corsica*, 1942.

LEFEBVRE, P. 'La population de la Corse', *R.G.A.*, 1957, pp. 557–75.

METRO, A. *Les Suberais de la Corse*, FAO, Rome, 1958.

PERRY, P. J. 'Economy, landscape and society in La Castagniccia (Corsica) . . .', *I.B.G.*, 1967, pp. 209–22.

RENUCCI, J. 'La Corse et le tourisme', *R.G.L.*, 1962, pp. 207–24.

RONDEAU, A. 'Problèmes de morphologie régionale en Corse', *B.A.G.F.*, 1956, pp. 49–61.

THOMPSON, I. B. *Modern France*, Butterworth, London, 1970.

The Maltese Islands

Malta Year Book and *Statistical Abstracts of the Maltese Islands*, Valletta, Malta (annual).

BLOUET, B. *The Story of Malta*, Faber, London, 1967.

BLOUET, B. W. 'Rural settlement in Malta', *Geog.*, 1971, pp. 112–18.

BOWEN-JONES, H. et al. *Malta*, Durham University Dept. Geog., Durham, 1961.

FLEMING, J. B. 'Notes on rural Malta', *S.G.M.*, 1946, pp. 56–60.

GRANDVILLE, O. DE LA *Malte*, Librairie Droz, Geneva, 1968.

LANG, D. M. *Soils of Malta and Gozo*, HMSO, London, 1960.

ROBINSON, G. W. S. 'The distribution of population in the Maltese Islands', *Geog.*, 1948, pp. 69–78.

Chapter 20: The Balkans

ANCEL, J. *Peuples et Nations des Balkans*, Colin, Paris, 1926.

BEAVER, S. H. 'Railways in the Balkan Peninsula', *G.J.*, 1941, pp. 273–94.

BLANC, A. *Géographie des Balkans*, Presses Universitaires de France, Paris, 1965.

CHATAIGNEAU, Y. & SION, J. *Les Pays Balkaniques. Géog. Univ.* Tome 7, part 2, Colin, Paris, 1934.

CVIJIĆ, J. *La Péninsule Balkanique: géographie humaine*, Colin, Paris, 1918.

GEORGE, P. *Géographie de l'Europe central slave et danubienne*, Presses Universitaires de France, Paris, 1964.

JELAVICH, C. & B. (eds.) *The Balkans in Transition*, University of California Press, Berkeley, 1963.

KIROV, K. T. 'Les limites des influences climatiques dans la Péninsule Balkanique', *C.R. Congrès Géog. et Ethnog. Slaves*, Sofia, 1936, pp. 119–25.

SCHECHTMAN, J. B. *European Population Transfers, 1939–1945*, Oxford University Press, London, 1946.

SCHECHTMAN, J. B. *Post-war Population Transfers in Europe, 1945–1955*, Oxford University Press, London, 1962.

TURRILL, W. B. *The Plant Life in the Balkan Peninsula*, Clarendon Press, Oxford, 1929.

WOLFF, R. L. *The Balkans in our Time*, Harvard University Press, Cambridge, Mass., 1956.

Chapter 21: Yugoslavia

Annual Yearbook *Statištički Godišnjal* SFRJ, Belgrade (Beograd).

Atlas *Geografski Atlas Jugoslavije*, ed. Mardešič, P. and Dugački, Z., Znanje, Zagreb, 1961.

AVSENEK, I. *The Iron and Steel Industry in Yugoslavia 1939–1953*, Mid-European Studies Center, New York, 1953.

ASENEK, I. *Yugoslav Metallurgical Industries*, Mid-European Studies Center, New York, 1955.

AVSENEK, I. *Yugoslavia's Postwar Industrialisation*, Library of Congress, Washington, 1956.

BILANDŽIC, D. *Management of Yugoslav Economy, 1945–1966*, Yugoslav Trade Unions, Beograd.

BLANC, A. *La Croatie Occidentale*, Institut d'Études Slaves, Paris, 1957.

BLAŠKOVIĆ, V. *Ekonomska geografija Jugoslavije*, Birozavod, Zagreb, 1952.

CAESAR, A. A. 'Yugoslavia: Geography and post-war planning', *I.B.G.*, 1962, pp. 33–43.

ETEROVICH, F. H. & SPALATIN, C. (eds.) *Croatia: Land, People, Culture*, 2 vols., University of Toronto Press, Toronto, 1964.

FISHER, J. C. 'Urban analysis: Zagreb', *A.A.A.G.*, 1963, pp. 255–84.

FISHER, J. C. *Yugoslavia: A Multinational State...*, Chandler, San Francisco, 1966.

GAMS, I. 'Some morphological characteristics of the Dinaric karst', *G.J.*, 1969, pp. 563–72.

Geographical Handbook. British Admiralty: Naval Intelligence Division, *Yugoslavia*, 3 vols., 1944–5.

HALPERN, J. M. *A Serbian Village*, Oxford University Press, London, 1958.

HAMILTON, F. E. I. 'Yugoslavia's hydro-electric power industry', *Geog.*, 1963, pp. 70–3.

HAMILTON, F. E. I. 'Changing pattern of Yugoslavia's manufacturing industry', *Tijdschr. econ. soc. Geogr.*, 1963, pp. 96–106.

HAMILTON, F. E. I. 'Location factors in the Yugoslav iron and steel industry', *Econ. Geog.*, 1964, pp. 46–64.

HAMILTON, F. E. I. *Yugoslavia: Patterns of Economic Activity*, Bell, London, 1968.

HOFFMAN, G. W. 'Yugoslavia in transition...', *Econ. Geog.*, 1956, pp. 294–315.

HOFFMAN, G. W. 'Changes in the agricultural geography of Yugoslavia', in *Geographical Essays on Eastern Europe*, ed. N. J. G. Pounds, Indiana University Press, Bloomington, 1961.

HOFFMAN, G. W. 'The Problem of the Underdeveloped Regions in southeast Europe: A Comparative Analysis of Romania, Yugoslavia and Greece', *A.A.A.G.*, 1967, pp. 637–66.

HOFFMAN, G. W. (ed.) *Eastern Europe. Essays in Geographical Problems*, Methuen, London, 1971.

KARGER, A. 'Die Entwicklung der Siedlungen im Westlichen Slawonien', *Kölner Geographische Arbeiten*, heft 15, Wiesbaden, 1963.

KMETIĆ, M. *Self-Management in the Enterprise*, Medunarodna Štampa, Interpress, Beograd, 1967.

LAKIĆEVIĆ, O. (ed.) 'Facts about Yugoslavia', REVIEW, Belgrade, 1970.

LEFEVRE, M. A. 'La Zadruga...', *A. de G.*, 1930, pp. 316–20.

MELLEN, M. & WINSTON, V. H. *The Coal Resources of Yugoslavia*, Praeger, New York, 1956.

MILOJEVIĆ, B. Ž. 'Types of villages and village houses in Yugoslavia', *Prof. Geog.*, 1953, pp. 13–17.

MILOJEVIĆ, B. Ž. *Geography of Yugoslavia, a Selective Bibliography*, Library of Congress, Washington, 1955.

MILOJEVIĆ, B. Ž. *La Yougoslavie. Aperçu Géographique*, Federal Institute for Statistics, Belgrade, 1956.

ORTOLANI, M. 'Le isole dalmate', *R.G.I.*, 1948, pp. 186–203; 255–76.

OSBORNE, R. H. *East Central Europe*, Chatto & Windus, London, 1967, pp. 184–223.

POUNDS, N. J. G. *Eastern Europe*, Longman, London, 1969, pp. 616–734.

ROGLIĆ, J. 'The Yugoslav littoral,' in Houston, J. M. *The Western Mediterranean World*, Longman, London, 1964, pp. 546–79.

SAVORY, H. J. 'Settlement in the Glamočko Polje', *G.J.*, 1958, pp. 41–55.

TOMASEICH, J. *Peasants, Politics and Economic Change in Yugoslavia*, Oxford University Press, London, 1955.

WILKINSON, R. H. 'Yugoslav Macedonia in transition', *G.J.*, 1952, pp. 389–405.

WILKINSON, R. H. 'Jugoslav Kosmet: The evolution of a frontier province', *I.B.G.*, 1955, pp. 171–93.

WILKINSON, R. H. *Maps and Politics: a Review of the Ethnographic Cartography of Macedonia*, Liverpool University Press, Liverpool, 1957.

Chapter 22: Albania

Annual Statistical Yearbook *Vjetari Statistikor*, Tiranë.

BLANC, A. 'Naissance et évolution des paysages agraires en Albanie', *Geografiska Annaler*, 1961, pp. 8–16.

BLANC, A. 'L'Évolution contemporaine de la vie pastorale en Albanie méridionale', *R.G.A.*, 1963, pp. 429–61.

COON, S. *The Mountains of Giants*, Harvard University Press, Cambridge, Mass., 1950.

Geographical Handbook. British Admiralty: Naval Intelligence Division, *Albania*, 1945.

HAMM, H. *Albania: China's Beachhead in Europe*, Weidenfeld & Nicolson, London, 1963.

HASLUCK, M. *The Unwritten Law in Albania*, Cambridge University Press, Cambridge, 1954.

NICOL, D. M. *The Despotate of Epirus*, Blackwell, Oxford, 1957.

POUNDS, N. J. G. *Eastern Europe*, Longman, London, 1969, pp. 814–58.

Reale Soc. Geog. Ital. *L'Albania*, Zanichelli, Bologna, 1943.

SKENDI, S. *The Albanian National Awakening*, Princeton University Press, Princeton, 1967.

Chapter 23: Greece

Annual and Monthly Statistics *Bulletin de Statistiques*, Athens.

Atlas *Economic and Social Atlas of Greece*, ed. Kayser, B., Thompson, K. *et al.*, National Statistical Service, Athens, 1964. A fine large atlas with a text and captions in Greek, English and French.

COMMON, R. 'The sample use of land use in Greece 1958', *Geog.*, 1962, pp. 299–301.

Geographical Handbook. British Admiralty: Naval Intelligence Division, *Greece*, 3 vols. 1944–5; *The Dodecanese*, 1943.

GUTKIND, E. A. *Urban Development: Italy and Greece*, Free Press, London, 1969.

HAMMOND, N. G. L. *Epirus*, Oxford University Press, London, 1967.

HARRIS, D. R. & VITA-FINZI, C. 'Kokkinopolis. A Greek badland', *G.J.*, 1968, pp. 537–46.

HOFFMAN, G. W. 'Thessaloniki; the impact of a changing hinterland', *East European Quarterly*, 1968, pp. 1–27.

KASPERSON, R. E. *The Dodecanese*, Chicago University Press, Chicago, 1966.

KAYSER, B. *Géographie Humaine de la Grèce*, Presses Universitaires de France, Paris, 1964.

MASON, F. C. *Greece*, Overseas Econ. Surv., HMSO, London, 1956.

OGILVIE, A. G. 'Population density in Greece', *G.J.*, 1943, pp. 251–60.

PHILLIPSON, A. *Das Klima Griechenlands*, Dümmlers, Bonn, 1948.

PHILLIPSON, A. *Die griechischen Landschaften: ein Landeskunde*, 4 vols., Klostermann, Frankfurt-am-Main, 1951–9.

PRENTICE, A. 'Reafforestation in Greece', *S.G.M.*, 1956, pp. 25–31.

PRENTICE, A. 'Livestock and forage production in central Macedonia,' *S.G.M.*, 1957, pp. 146–57.

TOURRATON, J. 'Le centrale hydroélectrique d'Agra (Macedoine) et le programme de l'électricité en Grèce', *A. de G.*, 1955, pp. 293–5.

WAGSTAFF, J. M. 'Rural migration in Greece', *Geog.*, 1968, pp. 175–9.

Index

Index

Aar–Saint-Gotthard massif, 5, 203, 204, 205
Abies, 54, 55
Abruzzi, 32, 72, 212, 220, 233, 234
Adige, river, 45, 219, 225
Adriatic Sea, 6, 14, 15
Aegean Islands, 296, 299, 305–18, 320–1
    Sea, 7, 14, 15, 291, 305
Aegina, 46, 292
agrarian reform, 122–3, 182, 197, 214, 270–3, 317
agrarian systems, *see* collectivization; cooperative; polyculture; state farms; tercio
agriculture, 63–82, 313; *see* individual countries
airflow, *see* wind
Aisne river, 165
Ajaccio, 42, 212, 213
Albacete, 118, 119
Albania, 286–90, 320
    agriculture, 75, 286–7
    climate, 286, 293
    fisheries, 19
    forestry, 65, 67
    historical development, 286
    industries, 196, 288
    irrigation, 44
    land use, 63–82, 286–7 *passim*
    livestock, 69–72
    minerals, 7, 288
    population, 98, 289
    power resources, 287–8
    structure: relief, 286
    trade, 290
    tree crops, 78, 79, 81
    urbanization, 290
Albufera, La, 18, 143
Albula pass, 204
Alcobaça, 153
Alentejo, 147, 150, 152, 154, 163–4
Alfa, 52, 121
Algarve, 147, 149, 164, 315
Algeciras, 141, 142
Alhambra, 95, 141
Alicante, 128, 145
Almadén, 127
Almeria, 128, 145
alpage, 166–7, 169, 171, 196
Alpine foreland, 5, 168–9
    orogeny, 2–7
Alps
    Albanian, 286
    Austro–Italian, 5
    Balkan, 263–4
    Dinaric: Dinarides, 4, 5, 263, 286, 291
    Eastern, 5–6, 193
    Franco–Italian, 3–4, 225–6
    glaciation in, 8, 9–10, 309
    Grisonides, 5
    Helvetides, 4, 5, 6
    Italian, 225
    Maritime, 3, 169
    passes across, 109
    Pennides, 4
    Pre-Alps, 5, 168, 192
    structure: relief, 2–11, 311, 316
    Tirolides, 5–6
    trans-Alpine routes, 107–9, 203–5
Alto Adige, 45, 219, 225
anchovy, 20, 21

Andalusia, 37, 126, 140–1, 145, 315
Andorra, 98, 134–5, 315
animal husbandry, 69–75
    Albania, 69–72, 287
    French Alps, 69–74, 167, 169, 171
    French Pyrenees, 176
    Greece, 69–72, 298–9
    Italy, 69–72, 218–19
    Jura, 166–7
    Portugal, 69–72, 154–5
    Spain, 69–73, 123–4
    Switzerland, 69–72, 197–8
    Yugoslavia, 69–74, 273–4
Apennines, 8, 36, 209, 211–12, 231–6
    central, 232–4
    northern, 193, 232
    southern 233–4
Apulia, 214, 217, 218, 234, 238, 318
agriculture, 18–19, 20, 63–82; *see* individual country
Aragon, 115
Aranjuez, 136
Arenas Gordas, 12
Argille Scagliose, 212
Arles, 74, 92, 183, 187
Arno river, 17, 37, 236
Aspromonte, 233, 234
Asti, 216
Asturias, 132
Athens, 26, 33, 293, 301–2
Athos, Mount, 14, 291
Atlantic, 15, 20, 35, 118
Augusta, 224
Austria, 109
Aveiro, 154, 156, 159, 162
Avignon, 179, 188

Badajoz, 44, 118, 119, 130, 135
Baetic: Betic, Cordillera, 8, 117, 120, 140–1
Badalona, 127, 143
Balearic basin, 14
    islands, 145–6, 315
Balkan Peninsula, 260–6, 319
    animal husbandry, 265–6
    climate, 264, 269
    drainage, 37–8
    human patterns, 266
    land use, 63–82, 256–6
    national development, 260–62
    structure: relief, 262–4
    vegetation, 52, 54, 55, 264–5, 319
Banat lowlands, 45
Barcelona, 30, 143, 128–9, 144, 315
Bari, 219, 240
Bas Languedoc, 73, 178–82; *see* Languedoc
Basilicata: Lucania, 215, 233, 318
Basle: Basel, 41, 195, 201, 202, 203
Basque, 130, 133
    regions of, 133–4
Bastia, 255
bauxite, 173, 182, 220, 299
Beira, 154, 163
Beja, 163–4
Belgrade: Beograd, 280, 282
Belledonne massif, 3
Bel Paese, *see* cheese
Bergamo, 221, 223, 229

Bermeo, 20
Berne: Bern, 42, 195, 201, 202, 204
Besançon, 167, 168
Bilbao, 128, 129, 133
Black Sea, 13, 14–15, 16
    fisheries, 19, 22
Blanes, 143
Bologna, 228, 230
Bonifacio, strait, 248, 251
bora, see wind
bosco, 53–4, 265
Bosnia–Hercegovina, 262, 280
Bosporus, 13, 15, 16, 311
Braceros, 122
Braga, 156, 161
Brenner pass, 107, 193, 225
Bresse, 182
Brindisi, 240
Bugey, 187
Bulgar, 83, 280, 289
Bulgaria, 44, 63
Burgos, 29, 130

Cadibona pass, 211
Cadiz, 118, 119, 128, 140
Cagliari, 212, 213, 248, 249, 250
Calabria, 233, 234, 318
Calanche, 212
Caledonian orogeny, 2, 309
Caltanissetta, 243, 244–5
Camargue, 45, 183–4, 186, 316
Campagna, Roman, 45, 46, 236–7
Campania, 46, 237, 318
Campidano, 45, 248, 249
Campo, 163
Canal Lodosa, 126, 139
Cantabrian mountains, 20, 73, 132–3
Capri, isle, 226, 238
Carbonia, 249
carob: locust, 51, 153
Carpathian mountains, 6, 8
Cartagena, 128, 129, 145
Carthaginians, see settlement
Cassa del Mezzogiorno, 98, 215, 224, 234, 238–40, 249
Castelo Branco, 154, 156, 159
Castile, 32–3, 72–3
Catalonia, 127, 139–40, 143, 314
Catania, 245
Causses, 73
Cavour canal, 45
Celts, 88; see settlement
Central Massif, 3, 4, 165, 180, 182
Central Sierras, 136, 138
centuriation, 91
Cephalonia, 305
Cévennes, 73
Chalcidice, 14, 292
Chamrousse, 35
cheese, 70–71
    Bel Paese, 71
    Emmentaler, 71, 197
    fetta, 299
    gruyère, 171, 197
    Gorgonzola, 71, 228
    Mozzarella, 70, 219
    Parmesan, 71, 228
    Pecorini, 71, 218
    Roquefort, 171, 253
Chiana, Val di, 236
ciflik, 281
cités–ouvrières, 189
citrus fruit, 30, 79–80, 124–5, 313
    in Greece, 297–8
    in Italy, 79–80, 217–18
    in Portugal, 79, 153
    in Spain, 79–80, 124–5, 143, 145
Ciudad Real, 118, 119
climate, general, 23–26, 311–12, 319
    annual moisture budget, 42
    air masses, 23–6
    Cyclonic depressions, 23–6
    jet stream, 24
    of Albania, 286, 293
    of Balkans, 264
    of Greece, 292–4
    of Italy, 212–14, 317
    of Portugal, 148–9
    of Sicily, 243
    of Southern France, 166, 169–70
    of Spain, 30, 117–19
    of Switzerland, 194–6
    of Yugoslavia, 268–70
    potential evaporation, 42
    precipitation, 30–3
    precipitation deficit, 42–4
    rainfall regime, 31
climatic change, 8–10
climatic regions, 35–6, 118, 214, 270
Cluse, 165
CNR, 41, 103–4, 182
coal: lignite, 7, 133, 135, 157, 187–8, 198, 220, 249, 277, 285, 287–8, 320
coast
    benches, 11–12, 132
    delta, 17, 186–7
    lagoon, 18, 145, 157, 162, 178, 316
Coimbra, 33, 147, 149, 156, 159
Col de la Perche, 174, 175
    de Perthus, 175
    du Pourtalet, 175
    de Puymorens, 175
collectivization, 270, 287
colmate naturali, 227
COMECON, 79, 99, 100, 106, 111–12, 286
communal farming, 253–4
communications, 108–9; see individual country
Como, lake: town, 9, 19, 210
cooperatives, 68, 122, 151, 154, 158, 168, 171, 179–80, 184, 214, 249, 270, 287
Cordoba: Cordova, 62, 140
Corfu, 305–6
Corinth, canal, 292
cork: cork oak, 48–9, 149–50, 319
Corsica, 250–5, 319
    agriculture, 252–3
    climate, 251–2
    fisheries, 253
    historical development, 250–1
    industries, 255
    land use, 252–3
    population: towns, 253–5
    transhumance, 252
    structure: relief, 251

Corsica—*continued*
   tourism, 253
   vegetation, 252
Corte, 254
Cortijo, 140
Corunna, 132
Costa Brava, 128
Côte: Burgundian, 185, 316
Côte d'Azur, 169, 173
cotton cultivation, 74, 77, 125, 140
Crau, 45, 171, 184, 186–7
Cremona, 227
Crete, 296, 299, 306–7
Croatia, 280, 319
cuniculi, 89
currant, 78–9, 297, 303
Cyclades, 14, 305, 308

dairying, *see* animal husbandry
Dalmatia, 276
Danube, 19, 37, 41, 267–8, 284
Dardanelles, 13, 15, 16
date palm, 125
Delphi, 291, 299
depressions, cyclonic, 24–6
desalination, 35, 46, 107, 250, 256
Dijon, 184, 187
Digne, 171
Dinaric Alps: Dinarides, 8, 26, 32, 286, 291
Dodecanese Islands, 308, 320, 321
Dolomites, 10, 226
Dombes, Pays de, 183, 185
Donzère–Mondragon H.E. site, 41, 102, 172, 183
Dora Riparia, 193, 230
Doubs river, 165
Douro: Duero, river: basin, 37, 135, 151, 157, 160, 162–3
dry farming
   secano, 69, 124, 126, 135
Dubrovnik: Ragusa, 74, 276
Durance river: basin, 40, 45, 172, 179
Durrës: Durazzo, 289, 290

earthquakes, 160, 263, 291, 305
Ebro, basin, 37, 44, 39, 116, 138–9
   irrigation, 116, 126, 138–9
Egadi islands, 245
Elba, 220, 237
Elbasan, 390
Elche, 145
El Ferrol, 132
El Grao, 144
Elvas, 29, 153, 163
emigration, 253, 304
employment, 64
   agricultural, 64
   industrial, 106–7
   see individual country
energy resources, 11, 99–106, 158, 244, 277, 318
   coal, *see* coal
   gas, natural, 7, 158, 219–20, 318
   geothermal, 11, 100
   hydro-electricity, *see* water power
   nuclear, 100, 187
   petroleum, 7, 127, 158, 244, 277, 288, 299, 318
engineering, *see* individual country

Ente Maremma, 214
Epirus, 296, 299
Escorial, 136, 138
esparto, 52, 121
Esterel, 169, 171
Estrêla, Serra de, *see* under Serra
Estremadura, 163
Étang de, Berre, 41, 172
   Thau, 18
   Vaccerès, 186
etesian wind, 24, 26, 28
Etna, 11, 243
Etruscans, *see* settlement
Euboea: Evvoia, 305, 307
eucalypts, 67
Euratom, 111
European Coal and Steel Community (ECSC), 111, 314
European Economic Community (EEC), 98–9, 107, 111–12, 314
European Free Trade Association (EFTA), 111–12, 203, 314
Euskari, 133
evergreen oak, *see* oak
Evora, 149, 150, 156, 159, 163–4
Extremadura, 72, 73, 123, 124, 135

FAO, 45, 311, 312, 313
Faro, 157, 159
Ferrara, 221
ferretto, 59
FIAT, 221, 229
fig, 79, 153
Figueira da Foz, 157
fir, *see Abies*
fisheries, 17–22, 311
   Albania, 19
   Atlantic, 19, 20
   Black Sea, 19
   coastal lagoons, 18, 156–7, 162
   freshwater, 18–19, 156
   in Corsica, 253
   in Greece, 19, 22, 300
   in Italy, 19, 20
   in Languedoc, 19, 181, 186
   in Mediterranean, 17–22
   in Portugal, 19, 20, 156
   in Sicily, 245
   in southern France, 19
   in Spain, 19, 20, 21, 127, 140
   in Switzerland, 19
   in Yugoslavia, 19, 20, 21, 22, 276–7
   oysters, 18, 20, 157
   sardines, 21, 148, 156–7
   tunny, 18, 156–7
Fiume: Rijeka, 209, 262, 269, 283, 284
Florence: Firenze, 37, 97, 207, 208, 223, 224, 236
flysch, 3, 4
Foggia, 223
föhn: foehn, *see* wind
fontanili, 45, 227, 228
forestry, 65–7, 313, 321
   in Greece, 65–7
   in Portugal, 149–52
   in Yugoslavia, 276
Fos, 189, 190
France, Southern, 165–90, 315–16
   agriculture, 63–82 *passim*, 166–71
   animal husbandry, 69–74, 171

climate, 166, 169, 170
fisheries, 19, 181, 186
forestry, 67
industries, 106, 181, 187
irrigation, 44, 45, 180, 186
olive, 81, 178
population, 98, 173
tourism, 109–10, 172–3
urbanization, 189
vine, 78, 178, 185, 186
France, Southern, regions, 165–90
   Alps, 168–73
   Jura, 165–8
   Languedoc, 178–82
   Pyrenees, 174–7
   Rhône–Saône valley, 182–8
   Rousillon, 178–82
frane, landslip, 56, 212
Fréjus pass, 109
French Alps, 168–73, 316
   agriculture, 166–71
   animal husbandry, 69–74, 171
   climate, 169–70
   industries, 172–3
   land use, 166, 171
   structure: relief, 168–9
   tourism, 172–3
   urbanization, 173
   vegetation, 170–1
   water power, 172

Galicia, 20, 131–2, 314
Gallura, 246, 248
Garda, Lake, 9, 19, 193, 210
Gargano, 233, 234
Garrigue, 51–2, 171, 176, 179, 180
gas, natural: methane, 7, 158, 219–20, 318
Geneva, 201, 203
Gennargentu, 248
Genoa, 96, 97, 224, 232, 250, 318
Genoese, 96–7
Gibraltar, 98, 118, 119, 141–3, 315
   Strait of, 7, 8, 11, 16, 17–18, 141, 315
Gijón, 128, 129, 132–3
glaciation, 3, 9–10, 169, 176, 194, 211, 251, 316
glacier, 10, 169, 176, 196
Gorgonzola, *see* cheese
Gozo, island, 256, 257
graben, 2, 267
Granada, 130, 141, 315
Granollers, 127
Gran Sasso, 33, 72, 211, 234
Grasse, 171
Great St Bernard Pass, 193, 203, 205, 223
greco: grecale: gregale, 28, 256
Greece, 19, 21, 44, 45, 46, 59, 63–82 *passim*, 98, 99, 106, 109–10, 291–308, 320–1
   agriculture, 63–82, 297–9
   animal husbandry, 69–72, 298–9
   climate, 292–4
   communications, 305
   commerce, 305
   fisheries, 19, 21, 22, 300
   forestry, 65–7, 321
   industries, 106, 300
   irrigation, 44, 45, 295–6

   land holdings, 296–7
   land reclamation, 46, 295
   land use, 63–82, 294–9
   minerals, 299–300
   population, 98, 300–4
   power resources, 101, 299
   rural population, 303–4
   soils, 59
   structure: relief, 291–2, 305–6
   tourism, 109–10, 300, 302–3, 307
   tree crops, 77–82
   urbanization, 301–3
Greek Islands, 305–8, 320–1
Grenoble, 173, 316
Grésivaudan, 168, 171, 173
Guadalajara, 127, 138
Guadalquivir, 37, 44, 140, 141
Guadiana, 37, 44, 135
Guadix, 141
Guimaraes, 159
Guipúzcoa, 133

Henry the Navigator, 147, 164
Hercynian massifs, 8, 169, 175–6, 247–8, 251, 262–3, 309
Hérault, 178, 179
horst, 2, 267
Huelva, 67, 140
huerta, 125, 126, 138, 143, 144, 145, 244
hydro-electricity, *see* water power
hydrology, *see* river regime; water supply
Hymettos, 295

Iberia, 37, 121, 124, 125, 266; *see* Portugal; Spain
Iberian mountains, 138
Ibiza, 145
Iglesiente, 249
intercropping, *see* polyculture
Ionian basin, 14, 34
   Islands, 296, 299, 305–6, 307
Iraklion, 29, 293, 307
Iron Gates, 262, 267, 277
iron industry, *see* steel; ore
irrigation, 42–5, 312
   Albania, 44
   Camargue, 186
   Greece, 44, 45, 295–6
   Italy, 44, 45, 46, 228–9
   Malta, 44
   Portugal, 44, 155–6
   Southern France, 44, 45, 180, 186
   Spain, 42, 44, 125–6, 138–9, 140–4, 145
   Switzerland, 44
   Yugoslavia, 44, 45
Irún, 133, 175
Isère river, 168, 172, 173, 177, 183, 185
Italy, 206–50, 317–19
   agriculture, 63–82 *passim*, 214–19
   animal husbandry, 69–72, 218–19
   climate, 212–14
   communications, 223
   economic problems, 98–9
   farm holdings, 214–15
   fisheries, 19, 20, 21
   forestry, 65–7
   industries, 106, 219–22

Italy—*continued*
   irrigation, 44, 45, 46, 228–9
   land reclamation, 46, 214, 249
   land use, 63–82 *passim*, 214–19
   national development 206–9
   population, 98, 223–4
   ports, 224
   port traffic, 224
   power resources, 100, 101, 219–20
   rivers, 37
   structure: relief, 209–12
   textiles, 229–30
   tourism, 109–10, 222–3, 231
   tree crops, 77–80, 216–18
   urbanization, 230–1
   viticulture, 78–9, 216–17
Italian Regions
   Alps, 209, 213, 214, 225
   Apennines, 211–12, 213, 214, 219, 231–46
   Islands, 241–50
   North Italian Plain, 209–11, 213, 214, 219, 225–31
   Peninsular coastal lowlands, 236–8

Jaca, 134
Jaen, 81, 140
Jerez, 140, 142
jet stream, 24
Jungfrau, 5, 10, 11
Jura, 5, 165–8, 193, 316
   agriculture, 166–7
   animal husbandry, 166–7
   climate, 166
   communications, 168
   industries, 167–8
   land use, 166–7
   population, 168
   structure: relief, 165
   water-power, 167

Karpathos, 14, 305
karst, 10–11, 165, 168, 176, 183, 211, 267, 277, 305, 320
katavothras, 292
Knossos, 306, 307
Korcë: Koritsa, 290
Kosovo–Metohija, 280, 281, 286, 320
Kotor, 269, 276
Kullë, 289

Lagos, 157
La Mancha, 122, 135; *see* Meseta
land consolidation, 122–3, 182, 197, 270
land holdings: farm size, 122–3, 155–6, 196–7, 214–15, 296–7
land parcellation, 123, 182, 317
land reclamation: drainage, 45–6, 155, 214, 249, 274–5, 295–6, 312
land splintering, 162, 215, 249, 270, 296
land tenure, 122, 155, 162, 213, 214–15, 296, 317
land use, 63–82, 313; *see* individual country
landforms, 8–12, 165; *see also* glaciation; periglaciation
Langres, 183
languages, 83–5
   Euskari, 133
   Gheg, 289
   Tosk, 289

Languedoc, 178–82, 316
   agriculture, 178–81
   animal husbandry, 179 180–1
   climate, 178
   fisheries, 18, 181
   market gardening, 180
   oleiculture, 180
   population, 181–2
   structure, 178
   tourism, 182
   urbanization, 181–2
   viticulture, 179–80
Lannemezan, 176
La Nurra, 248, 249, 250
La Planasse, 14, 18
Larissa, 293, 303
Las Marismas, 12, 46, 126, 140
latifundia: latifondi, 122, 155–6, 206, 243–4, 318; *see also* land holdings
Lausanne, 203
Lauterbrunnen valley, 9
Lavéra, 189, 190
Lecco, 230, 318
lee-depressions, atmospheric, 23–4
Leghorn: Livorno, 223, 224, 236, 255
Leiria, 150, 156, 159
Leixoes, 160, 161
Lemnos, 14, 305
lemon, *see* citrus
León, 114, 122, 130
Lerida, 139
Les Alpilles, 169
L'Estaque, 188
Lesbos: Lesoos, 305
Levante region, 20
levante wind, 1, 28
Levantine basin, 14, 15
leveche, 27
Lezirias, 156, 163
libeccio, *see* wind
Liechtenstein, 98, 205, 317
lignite, *see* coal
Liguria, 38, 214, 232, 318
Linares, 135
Lipari Islands, 245–6, 319
Lisbon, 149, 159–60, 315
Little St Bernard Pass, 107, 109, 193, 203, 205, 223
livestock, 69–75, 313; *see also* animal husbandry
Ljubljana, 109
Llanos de Urgel, 139
Llivia, 134
Llobregat river, 17, 139, 140, 143
loess, 12
Logroño, 139
Lombardy, plain, 8, 35–6, 212, 213, 214, 216, 318, 219, 221–4 *passim*, 225–31, 317–18
Lorca, 145
Lucca, 207, 218, 221
Lucerne, lake, town, 200, 201, 203
Lusitania, 147
Lyon: Lyons, 41, 186, 187, 316

Macedonia, 280, 281, 292, 296, 299
Madrid, 29, 30, 33, 136–8, 314
Maggiore, lake, 4, 9, 19, 193, 210
maize, 74, 75–6, 153–4, 216, 271, 272, 273, 275, 287, 297, 298

Málaga, 6, 128, 145
Majorca: Mallorca, 145–6, 315
Maltese Islands, 255–8, 319
　agriculture, 75–8, 256
　animal husbandry, 69–70, 256–7
　climate, 256
　Comino, 256
　fisheries, 19
　Gozo, 256, 257
　historical development, 255–6
　industries, 257
　irrigation, 44
　population, 98
　structure, 256
　tourism, 258
　trade, 257
Manresa, 140
Manzanares river, 37, 136
maquis: macchia: matorral, 50–1, 120–1, 154, 171, 176, 218, 234, 238, 243, 248, 252, 265, 295
Mar Menor, 18, 145
marcite pastures, 68
Maremma, 45, 214, 237
Maribor, 283, 285
Maritime Alps, 169, 170, 171, 172–3
Marmara, Sea, 13, 15
Marseille: Marseilles, 29, 188–90, 316
Massif Central, 3, 4, 165, 180, 182
Matterhorn, 10, 11, 192
Maures massif, 169, 171
median mass concept, 6–7
Medina del Campo, 135
Mediterranean Sea, 13–22, 311
　airmasses over, 23–6, 213
　currents, 15–18
　fisheries, 17–21, 311
　salinity, 15–17
　submarine relief, 13–15
　surface circulation, 16–17
　temperature, 15–17
　tides, 16
　water budget, 15–16
　waves, 16
mercantile marine, 128–9, 224, 300
Mercantour massif, 3, 168
mercury, metal, 127, 220
Merida, 44, 114
merino sheep, 72–3
Meseta, Iberian, 72, 73, 116–17, 135–8, 314
Messina, 16, 245
Mesta, 72, 314
metayage: share cropping, 122, 155
Methana, 14
Methuen treaty, 147
mezzadria, 215
Mezzogiorno, 98–9, 215, 234, 238–40, 244; *see* Cassa
Milan: Milano, 228, 229, 230, 318
minerals, 7; *see* individual country
Minho valley, 37, 151, 154, 159, 162
minifundia, 122–3, 155–6, 162, 196, 296–7
Minorca: Menorca, 145–6, 315
Minos, 306
mistral: maestrale, 27, 28, 29, 170, 248
molasse, 3, 4, 6
Monaco, 98, 174, 316–17
Mondego valley, 154, 156
Monferrato hills, 193, 225

Mont: Monte, Blanc, 3, 5, 8, 168, 169
　Canigou, 175–6
　Corno, 8, 211
　Pelvoux, 3, 168
　Rosa, 4, 192
montado, 163
Mont Cenis pass, 107, 109, 193
monte bajo, 120
Montélimar, 27, 183, 185
Montenegro: Crna Gora, 262, 271, 280–1, 285
Mont Louis, 35
Montpellier, 179, 181, 182, 316
Moors: Arabs, *see* settlement, 94–5, 114–15, 140, 142, 145, 147, 207, 208, 241–2, 246, 250, 255
Morava–Vardar corridor, 275–6
Mostar, 74, 275, 279
Motril, 145
mountain building, 2–7
　Alpine, 2–7
　Armorican, 2
　Caledonian, 2, 8
　Hercynian, 2, 4–5, 6, 263
　Variscan, 2
mountain climate, 36, 117, 135, 148, 166, 169–70, 176, 194–6, 213, 252, 264, 268–70, 286, 294
mountain vegetation, 54–5, 119, 134, 138, 166, 172, 176, 196, 243, 248, 252, 256, 265
Mulhacén, 8, 141
Murcia, 43, 144, 145

nagelfluh, 3
Naples, 224, 237–8
nappes, 3, 168
Narbonne, 89, 181, 182
NATO, 111, 116, 160
natural gas: methane, 158, 219–20, 224
Navarra: Navarre, 114, 315
Naxos, 305
Nice, 170, 173, 255
Nîmes, 181, 182
Niš: Nish, 269, 275, 283
noria, 42
North Italian Plain, 193, 209–11, 225–31, 317–18
Novi Sad, 283
nuclear power, 100, 187
nuraghi, 246, 319

oak: *Quercus*, 48–9
　cork (*suber*), 48, 49, 120, 149–50, 248, 319
　evergreen: holm (*ilex*), 48–9, 120, 151
　hairy (*pubescens*), 52
　kermes (*coccifera*), 49
　pedunculate (*pedunculata*), 52, 53
　Portuguese (*lusitanica*), 52
　sessile (*sessiliflora*), 52, 53
　Turkey (*cerris*), 52
　Valona (*aegylops*), 49
Odeillo, 35
OECD, 65, 99, 314
oil, *see* petroleum
Olbia, 248, 250
oleander, 51
olive
　oleiculture, 30, 48, 76, 80–2, 124, 140, 152–3, 313
　in Albania, 287

olive—*continued*
   in Corsica, 252
   in Greece, 80–2, 297, 298, 299
   in Italy, 80–2, 217–18
   in Portugal, 80–2, 152–3
   in Sardinia, 249
   in Sicily, 244
   in Southern France, 81, 178, 180, 316
   in Spain, 80–2, 124, 140
   in Yugoslavia, 273
Oporto, 149, 160–1
Orange, 185
oranges, *see* citrus fruit
Orense, 132
orogeny, *see* mountain building
Ossa, 292, 305
Ottoman, *see* settlement
Otranto, Strait, 15
Oviedo, 130, 133

Padua: Padova, 221, 223
Pajares pass, 132
Palermo, 29, 245
palm, date, 145
   dwarf, 52
Palma, 118, 119, 146
Pamplona, 33, 134
Pannonian lowland, 267–8, 275
Parmesan, *see* cheese
Parnassus, 292
pass: col, *see* under own name
pastoralism, *see* animal husbandry; transhumance
pasture, 313; *see* land use; animal husbandry
Patras, 293, 307, 302
Pavia, 207
Pays de Dombes, 183, 185
Peloponnesus: Morea, 32, 292, 296, 299, 305
periglaciation, 9
Perpignan, 134, 170
Perugia, 223
petroleum, 7, 127, 158, 244, 277, 288, 299, 318
Phlegrean Fields, 209, 318
Phoenicians, *see* settlement
phrygana, 52, 265
Piacenza, 223, 227
pianura alta, 226
   bassa, 228
Pic d'Aneto, 8
pine: *Pinus*, 49–50
   Aleppo (*halepensis*), 49–50, 120
   Corsican (*laricio*), 252
   domestic: stone (*pinea*), 49, 120
   maritime (*pinaster*), 49–50, 120
   Scots (*silvestris*), 53
   Serbian (*nigra*), 54
Pinios river, 38
pipeline, 108–9, 189, 199, 277, 288, 317
Piraeus, 302
Pisa, 96, 223, 250
plums, 77–8, 153, 163, 275
Po basin, 7, 18, 31, 36, 209–10; *see* North Italian Plain
   river, 37, 40–1, 45, 209–10
polyculture, 121, 143, 182, 214, 217, 252
Pombal, Marquis de, 147–8, 164
Pompeii, 11, 238
Pontevedra, 118, 119, 124

Pontine marshes, 46, 214–15
Port de Bouc, 41, 189
Port Mahón, 146
Porto Empedocle, 245
Porto Foxi, 224, 250
Porto Marghera, 221, 224, 230, 231
Porto Torres, 46, 250
Portugal, 147–64, 315
   agriculture, 63–82 *passim*, 153–6
   animal husbandry, 69–72, 154–5
   climate, 30, 33, 148–9
   employment, 156
   energy resources, 100, 106, 157–8
   fisheries, 19, 20, 21, 156–7
   forestry, 65–7, 149–51
   historical development, 147–8
   industries, 106, 158–9
   irrigation, 44, 155–6
   land holdings, 155–6
   land use, 63–82 *passim*, 121, 149–56
   minerals, 158
   population, 98, 159–62
   ports, 160–1
   structure: relief, 148
   tourism, 109–10
   tree crops, 78, 79, 80–2, 153
   urbanization, 159–62
   vegetation, 149
   viticulture, 78, 151–2
Portugal, Regions of, 162
   Alentejo, 163–4
   Alto Douro, 162
   Central Serra, 163
   Coastal Douro, 162
   Minho, 162,
   Tras-os-Montes, 162–3
   southern Portugal, 163
port wine, 151–2, 161, 315
Posidonia meadows, 18
Potenza, 233
Pre-Alps, 5, 168, 192, 193
precipitation, 30–3
   annual rainfall, 30
   raindays, 31–2
   rainfall intensity, 31–2
   seasonal regimes, 31
   snowfall, 32–3
primeurs, 30, 185–6, 273
Provence, 35, 94, 171, 316
Puigcerda, 134, 175
Pyrenees, 8, 10, 108, 134, 174–7, 315, 316
   agriculture, 176
   animal husbandry, 176
   climate, 170, 176
   drainage, 176
   glaciation, 134, 176
   hydro-electricity, 134, 177
   industries, 177
   land use, 134, 176
   population, 177
   routes across, 134, 174–5
   structure: relief, 134, 174–6
   tourism, 134, 175
   vegetation, 176

Quaternary Ice Age, 8–9
Quercus spp., *see* oak

races: racial traits, 83, 313
rabelo, 151
rainfall, *see* precipitation
raisin, 79, 219, 232
Ravenna, 18, 96, 211, 219, 220, 221, 223
rax, 3
Reconquista, 96, 114, 147
regadio, 126
Reggio di Calabria, 223, 240
Reinosa, 133
remembrement, 122–3, 182, 197, 270
rendzina, 61–2
Rethimnon: Retimo, 307
retsina, 50
Rhodes, 14, 308
Rhodope plateau, 14, 263, 264, 267, 291
Rhône, 3, 37, 39–41, 183–8
Rhône–Saône valley, 179, 182–8, 316
   agriculture, 179, 184–5
   animal husbandry, 179, 180–1
   climate, 170, 178, 184
   corridor, 185–6
   delta, 183
   hydro-electricity, 101–4, 187
   industries, 187–9
   land use, 179, 184–5
   river regimes, 39–40
   structure: relief, 182–4
   river transport, 41
   urbanization, 181, 187–90
Ribatejo, 154, 159, 163
rice, 44–5, 76–7, 124, 143, 154, 186, 216, 228, 298, 316
Riccione, 34
Rijeka: Fiume, 31, 283, 284–5
Rimini, 211, 223, 230
Rio Tinto, 127
river regimes, 38–41, 312
   Ebro, 39
   glacial, 38–9
   Italy, 40–1
   Mediterranean, 38
   nival, 38–9
   nivo–pluvial, 39–40
   pluvial, 39
   Po, 40–1
   Rhône–Saône, 39–40
rivers, Southern Europe, 5–46
river terraces, 12, 69, 163, 311
river transport, 41, 128, 148, 284, 312
Riviera di Levante, 223, 232
   di Ponente, 223, 232
Rome, 33, 223, 234, 236, 237, 318
Roncesvalles pass, 134, 175
Roussillon, 178–82, 316

Sabadell, 127, 143
Sagres, 164
Sagunto, 127, 144
Saint Bernard passes, 193, 205, 223
Saint Gotthard, massif, pass, 107, 193, 203, 204, 295
Salonika: Thessaloniki, 293, 301, 302, 321
salt pans, 35, 181, 186, 245, 249
   pastures, 186
Samos, 307
San Marino, 98, 233, 234–6
San Sebastian, 133

Santander, 133, 315
Santiago, 114, 132
Saône basin: river, *see* Rhône–Saône
Santarém, 151, 153, 155, 156
Santorin: Santorini, 14, 293, 305, 308
Sarajevo, 74, 280, 283
sardines, 18, 21, 148, 156–7, 159; *see* fisheries
Sardinia, 45, 246–50, 319
   agriculture, 249
   animal husbandry, 249
   climate, 212, 213, 249
   historical development, 246
   industries, 250
   irrigation, 45
   land use, 248
   mining, 249–50
   population: towns, 250
   structure: relief, 247–8
   transhumance, 249
   transport, 250
Sassari, 212, 213, 223, 248, 250
Savona, 224, 232
schistes lustrés, 168
sea level changes, 11–12
secano, 69, 124, 126, 135
Sele plain, 46
Semmering pass, 107, 193
Serbia, 260–1, 268, 280
sericulture, 187, 221, 229
Serra da Estrêla, 148, 149, 154, 158, 162, 163
Serra de Monchique, 148, 164
   de Sintra, 151, 163
Serra d'Ossa, 148
Séte, 181, 190
settlement
   historical, 85–97, 313–14
   Bulgars, 260–1
   Carthaginians, 144, 241, 246
   Celts, 88, 313
   city states, 95–7, 188, 207
   Etruscans, 89, 250, 313
   Genoese, 96–7
   Goths, 142, 206
   Greeks, 86–8, 114, 241, 245, 313–14
   Moors: Arabs, 94–5, 114–15, 140, 142, 145, 147, 207, 208,
     241–2, 246, 250, 255
   Normans, 242, 255
   pre-historic, 85–6, 114, 142, 241, 296, 250, 306
   Phoenicians, 88–99, 114, 145, 179, 241, 246, 255, 313
   Romans, 89–93, 114, 140, 144, 145, 181, 206–7, 241, 245, 260
   Serbs, 261
   Slavs, 260, 313
   Turks: Ottoman, 97, 261, 286
   Vandals, 241, 246, 250
   Visigoths, 114–15, 147
   Vlachs, 74, 84
   Völkwanderung, 93–4
Setubal, 151, 153, 157, 159, 161
Seville, 29, 41, 116, 140
sheep, 69, 71, 72–4; *see* animal husbandry; transhumance
sherry, *see* Jerez
shiblyak, 265, 287
Sicily, 45, 241–5, 318
   agriculture, 243–4
   climate, 243
   fisheries, 245
   historical development, 241–2

Sicily—*continued*
  industries, 244–59
  land use, 243–4
  minerals, 7, 244–5
  population, 245
  structure: relief, 242–3
  urbanization, 245
Siena, 207, 236
Sierra Morena, 120, 148, 158
  Nevada, 8, 33, 37, 141
Sierra de Cazorla, 120
  Gata, 148
  Gredos, 138
  Guadalupe, 148
  Guadarrama, 119, 138
  Monchique, 148, 164
Sila, La, 214, 234
silk, 187, 221, 229, 276; *see* textiles
Simplon pass, 107, 193, 204
sirocco: scirocco, *see* wind
Skopje, 280, 283, 285
sljivovica, 77–8, 275
Slovenia, 280
snow, 9, 32–3, 36, 195–6, 234
soil erosion, 215, 253, 294
soil moisture deficiency, 30
soils, 55–66, 312
  black earth: tirs, 62, 141
  brown forest, 57, 58
  calcimorphic, 61–2
  chernozem, 61
  halomorphic, 61
  hydromorphic, 61
  podzolic, 57–8
  rendzinas, 56, 57
  terra rossa, 56, 57, 58–61
solar furnace, 35
Somport pass, 134, 175
Spain, 63–82 *passim*, 98, 99, 114–46, 314–15
  agriculture, 75–7, 124–6
  animal husbandry, 69–73, 123–4
  climate, 30, 117–19
  communications, 128–9
  drainage, 117
  economic problems, 99
  energy resources, 100, 104–6, 126–7
  fisheries, 19–20, 121, 132, 140
  forestry, 65–7, 119–20
  historical development, 114–16
  industries, 106, 127, 143
  irrigation, 42, 44, 125–6, 143
  land holdings, 122–3
  land reclamation, 46
  land use, 63–82 *passim*, 121–6
  minerals, 127
  population, 98, 129–30
  ports: port traffic, 128–9
  Reconquista, 96, 114–15, 147
  soft fruit, 70–8
  structure: relief, 116–17
  tourism, 109–11, 128
  transhumance, 72–3
  tree crops, 70–82
  urbanization, 129–30
  vegetation, 119–21
  viticulture, 78–9, 124, 135, 140, 143
Spain, Regions, 131–46, 314

  Andalusian Trough, 140
  Andalusia: Baetic (Betic) Cordillera, 140–1
  Balearic Islands, 145–6
  Basque Mountains, 133–4
  Cantabrians, 132–3
  Catalonian Mountains, 139–40
  Central Sierras, 138
  Ebro Trough, 138–9
  Galicia, 131–2
  Iberian Highlands, 138
  Mediterranean Coastlands, 143–5
  Meseta, 135–8
  Pyrenees, 134
Sperkhios river, 14
Spezia, La, 232, 236
Split: Spalato, 18, 283
Sporades, 14, 305, 307–8
Stara Vlah, 281
state farms, 287
steel: iron production, 106–7, 127, 144, 158, 177, 199, 220, 279, 300, 319
steppe, 52
Stromboli, 243
submarine relief, 13–15
Subotica, 281, 283
sulphur, 220–1, 244–5
sunshine, hours of, 33–5
Suria, 140
sweet chestnut, 52, 53, 149, 150, 243, 244, 254
Switzerland, 191–205, 317
  agriculture, 63–82 *passim*, 196–8
  animal husbandry, 69–72, 197–8
  climate, 194–6
  communications, 199, 203–5
  economic problems, 99
  employment, 197–8, 200, 201
  fisheries, 19
  forestry, 65, 67
  glaciation, 9–10, 194
  historical growth, 191–2
  hydro-electricity, 198–9
  industries, 106, 199–201
  irrigation, 44
  land holdings, 196–7
  land use, 63–82 *passim*, 196
  livestock, 69–72, 197–8
  metallurgy, 199–201
  population, 98, 201–3
  power resources, 100, 101, 198–9
  relief: landforms, 8–12, 192–4
  structure, 192–4; *see* Alps
  textiles, 200–1
  tourism, 109–11, 200–1
  trans-Alpine routes, 107–9, 193, 203–5
  transhumance, 72, 196, 197
  urbanization, 201–4
  vegetation, 54–5, 196
  viticulture, 78
  watches: clocks, 199–201
Syracuse: Siracusa, 34, 223, 245
Syros: Siros, 46

Tagus: Tajo, 37, 44, 148, 149, 154, 155, 157, 160, 163
tanya, 281
Taranto, 46, 224, 239–40
Tarragona, 143
temperature, 28–30; *see* climate

tercio system, 140
Terni, 220, 223
terraces, coastal, 11–12, 132
    riverine, 12, 69, 163
terra rossa, 56, 57, 58–61
textiles, 106, 158–9, 187, 200–1, 220, 221, 229–30, 276, 282, 288, 300
Thermopylae, 14, 292
Thessaly, 264, 296, 299
Thrace, 292, 296, 299
Tiber, 37, 206, 236, 237
timok, 281
Tiranë: Tirana, 288, 289, 290
tirs, 62
Tirso plain, 46, 248, 249
Titograd, 280
tobacco, 74, 77, 135, 216, 273, 275, 279, 283, 287, 298, 300
Toledo, 91, 135–6
tomillares, 121
Tosk, 289
Toulon, 173
tourism, 32, 41, 109–11, 314–19 *passim*
    Austria, 109
    Bulgaria, 109, 110
    Corsica, 253
    France, southern, 109–10, 172–3, 182
    French Alps, 172–3
    French Pyrenees, 175
    Greece, 109–10, 300, 302–3, 307
    Italy, 109–10, 222–3, 238
    Languedoc, 182
    Liechtenstein, 205
    Malta, 258
    Portugal, 109–10
    Spain, 109–10, 128, 134, 135, 136, 143, 146
    Switzerland, 109–10, 200–1
    Yugoslavia, 109–10, 276
towns: cities, *see* individual entries
tramontana, 27, 28, 213
transhumance, 36, 72–4, 176, 313
    Corsica, 252, 253
    French Alps, 72, 73, 74, 171
    Greece, 298–9, 303
    Italy, 72, 73
    Pyrenees, 72
    Sardinia, 249
    Spain, 72, 73, 135
    Switzerland, 72, 197
    Yugoslavia, 73–4
Trapani, 245
Tras-os-Montes, 154, 159, 162–3
tree crops, 77–82, 143, 153, 171, 216–18, 244
    citrus, 79–80; *see* citrus
    cork, 149–50
    olive, 80–2; *see* oleiculture
    soft fruit, 77–8
    vine, 78–9, 151–2; *see* viticulture
Trentino–Alto Adige, 219, 225
Trieste, 26, 27, 223, 224, 318
tunny: tuna, 17, 20, 21, 148
Turin: Torino, 33, 221, 223, 229, 230, 318
Tuscany, 237
Tyrrhenian Sea, 6, 7, 14

urbanization, 138, 159, 173, 314, 317, 318
    see individual entries for cities and countries

Val d'Aosta, 208, 220, 223, 318
    de Chamonix, 9
Valence, 27, 40, 184
Valencia, 143–4
Valladolid, 118, 119, 130, 135
Valletta, 257
Vardar, river, 38, 257, 267, 275–6, 283, 291, 292
vardarac, 27–8
Vascongadas, 133
Vatican City State, 98, 237, 318
Vaucluse, 45
vega, 138, 141, 145
vegetation, 47–55, 119, 120–1, 171, 264–5, 312, 319
Venezia, 208, 223
Venice, 96, 207, 202, 211, 224, 230–1
Vercelli, 228
Verona, 223
Vesuvius, 11, 238
Viana do Castelo, 156, 159, 162
Vigo, 128, 132
Vila Nova de Gaia, 161
vine, *see* viticulture
vinho verde, 151
Viscaya, 133
Visigoths, *see* settlement
viticulture, 30, 78–9, 124, 313, 315
    in Greece, 78, 79, 297, 298
    in Italy, 79, 216–17, 317
    in Portugal, 78, 151–2, 315
    in southern France, 78, 185, 186, 316
    in Spain, 78–9, 124, 135, 140, 143, 314
    in Switzerland, 78
    in Yugoslavia, 78, 273
Vitoria, 133
Vlachs, 74, 84
Vlonë: Valona, 287, 288, 289
Vojvodina, 45, 267, 280, 283
Vouga basin, 154
Vulcano, 243, 246
vulcanism, 7, 11, 14, 212, 226, 238, 243

water power: hydro-electricity, 99–106
    in Albania, 100, 287–8
    Danube, 101, 262, 267, 277
    France, 100, 101–4, 187
    French Alps, 172
    Greece, 101, 299, 321
    Italy, 100, 101, 219–20
    Jura, 167
    Portugal, 100, 106, 157–8
    Pyrenees, 177
    Spain, 100, 104–6, 126–7
    Switzerland, 100, 101, 198–9
    Yugoslavia, 100, 101, 277, 320
water supply, 46, 312
    see desalination
wheat, 75–6, 153–4
wind: airflow, 26–8
    bise, 196
    bora, 27–8, 29, 268
    etesian, 24, 26, 28, 29
    föhn, 26, 29, 196
    greco: gregale, 28, 256
    levante: levanter, 28
    leveche, 27

wind: airflow—*continued*
  libeccio, 28
  marin, 27
  meltemi, 26
  mistral: maestrale, 27, 28, 29, 170, 248
  mountain breeze, 28, 196
  nortada, 148
  orographic influence, 26
  sea breeze, 28
  sirocco: scirocco, 26–7, 213, 243, 248
  tramontana, 27, 28, 213
  vardarac, 27–8
  vendavale, 28
wolfram, 158
World Health Organization, 203

Yannina, 293
Yeropotamos, 306
Yugoslavia, 267–85, 319–20
  agriculture, 63–82 *passim*, 270–6
  animal husbandry, 69–74, 273–4
  climate, 268–70
  communications, 283–4
  economic problems, 99, 270–5, 277, 284–5
  employment, 64, 106, 279
  fisheries, 19, 20–1, 276–7
  forestry, 65, 67, 276
  historical growth, 260–2, 267
  industries, 106, 278–80
  irrigation, 44, 45, 274–5
  land reclamation, 46, 274–5
  land use, 63–82 *passim*, 270–3, 275–6
  livestock, 69–74
  minerals, 277–8
  population, 98, 280–3
  power resources, 100, 101, 277
  structure: relief, 267–8; *see* Dinaric Alps
  tourism, 109–10, 276
  trade, 284–5
  transhumance, 72–4
  tree crops, 77–81
  urbanization, 281–3

Zagreb, 280, 282–3, 320
Zakinthos, 293, 305–6
Zante, 79, 305
Zaragoza: Saragossa, 29, 42, 130, 138–9
Zézere river, 157, 163